Fundamentals of
Composites Manufacturing

Fundamentals of Composites Manufacturing

Materials, Methods, and Applications

Second Edition

A. Brent Strong

Society of Manufacturing Engineers
Dearborn, Michigan

9

Library of Congress Catalog Card Number: 2007935302

International Standard Book Number (ISBN): 0-87263-854-5

ISBN 13: 978-087263854-9

Additional copies may be obtained by contacting:
Society of Manufacturing Engineers
Customer Service
One SME Drive, P.O. Box 930
Dearborn, Michigan 48121
1-800-733-4763
www.sme.org/store

Online video
Visit www.sme.org/focm

SME staff who participated in producing this book:
Rosemary Csizmadia, Senior Production Editor
Steve Bollinger, Manager, Book & Video Publications
Frank Bania, Cover Design
Frances Kania, Administrative Coordinator

Printed in the United States of America

Acknowledgments

Special thanks to Rik Heselhurst, Ph.D. at the School of Aerospace, Civil and Mechanical Engineering, University College, The University of New South Wales who was the principal author for the chapter on design and who contributed significantly to the chapter on repair. Also thanks to Scott Beckwith who made special contributions to the chapter on resin transfer molding (RTM) and to many figures and tables throughout the text.

Many thanks also to the American Composites Manufacturers Association (ACMA). In particular, thanks to Andy Rusnak, the editor of *Composites Manufacturing* magazine. I have been a contributing editor to this magazine and have had many published articles. With ACMA's permission, substantial portions of these articles have been used in this text. In that regard, I also give thanks to the co-authors that assisted in writing many of the ACMA articles. Those co-authors include: Scott Beckwith, Perry Carter, Tom Erekson, John Green, Charles Harrell, Val Hawks, Mike Hoke, David Jensen, Barry Lunt, William McCarvill, Mike Miles, Frédérique Mutel, John Richards, Christopher Rotz, Deryl Snyder, and Troy Takash.

Similarly, I am the author of a text entitled *Plastics: Materials and Processing*, which is currently in its third edition. As some of the topics in that plastics text and in this present composites text overlap, I have borrowed some concepts and text from that text. My thanks go to Prentice Hall, the publisher of this text, for permission to use these materials.

Special thanks to David Sorensen of the Utah Manufacturing Extension Partnership who has developed and taught me much of the information on the methods of planning. I also thank Bill Angstadt of Angstadt Consulting for insights into the new planning methods that must be used in today's world. Further thanks to Dixon Abell of Poly Processing Company for his insightful discussions. And thanks to Whitney Lewis of the Manufacturing Leadership Forum who helped greatly in the production of the figures.

I want to especially thank Rosemary Csizmadia of the Society of Manufacturing Engineers who has edited this book. She has gone far beyond just checking for grammar and agreement errors. She has also given insightful help in content and organization.

Contents

Preface

THE PURPOSE OF THIS BOOK

People entering the composites field need a way to learn the basics. Whether they are learning about composites in a college course or in the workplace, they need a simple text that gives enough detail to help with understanding. They also need a reference book that presents material in an easily understandable format. The first edition of this book has filled that need but is now over a decade old. This second edition describes advances that have been developed in the recent past, expands upon the explanations of the first edition to give further understanding, and adds key information that will serve as a reference for refreshing and expanding your knowledge of composites.

THE IMPORTANCE OF MATERIALS IN COMPOSITES

The field of composites is about materials and the way they are made into products. (In the aerospace world, this is called materials and processing or M&P.) Therefore, the first focus of this text is on materials. This is reflected in several major chapters devoted to understanding the basic materials that are put together to create composite structures. Too often the materials side is viewed by composite engineers as merely a way to get data that will allow design calculations to be made. This text tries to simply and clearly present the details of why composite materials behave the way they do. Thus composites designers will gain an understanding of the causes of material performance and be able to choose from a wider set of acceptable materials than might otherwise be possible.

THE IMPORTANCE OF COMPOSITES MANUFACTURING

Improving manufacturing technology is the greatest challenge today in the field of composites. When composites are chosen for an application principally because of their properties, it is natural that the manufacturing methods would be chosen to optimize those properties, even to the point where good manufacturing methods might be adversely affected. This practice has been evident in the aerospace industry where many composite parts have been made by processes that require high labor and costly techniques. Up to recent years, even large airplanes have been primarily hand-made.

In some applications, the practice of property optimization will continue in spite of the problems that might arise in manufacturing. However, those in charge of manufacturing have the challenge of improving the manufacturing process so that the quality of the performance of the part can be maintained. Competitive pressures may also play a part to encourage reduction of costs through manufacturing improvements. These manufacturing improvements must be done while still maintaining or improving part properties. Applications involving most advanced composites fit into this category.

For other composite parts a different situation occurs with respect to manufacturing. In these the choice of composites is based on *both* material performance and manufacturing efficiencies. Parts made by these criteria are generally engineering composite products. In these parts the need to improve manufacturing is critically important because of the inevitable pressures of the marketplace for reduced costs and improved throughputs.

Regardless of the situation, good initial choices about the type of manufacturing process used and then subsequent improvements to it are critically important. Too often the initial choice of manufacturing method is made based on previous experience or on available equipment. A rigorous method for evaluating the choice of manufacturing process is rarely used. Then, after the method is chosen, there is great reluctance to change it significantly. This is especially true in situations where the part and the manufacturing process have been approved by a governmental agency or a major assembler like an airplane company. These products are locked into technologies that are difficult to modify. However, as experience with long-term performance of composite parts grows and the data bases of composite properties and designs increases, the reluctance to change manufacturing processes will decrease. Experience diminishes risk.

The composites industry will continue struggling to cope with the requirements for superb mechanical properties and the need for economical manufacturing methods. The manufacturing side is changing rapidly and a premium is being placed on the innovative individual.

Manufacturing technologies have been borrowed from many other industries. From plastics came the concepts used in resin curing, composites molding, extrusion, and finishing. From metals came ideas in casting, forming, and mold making. Some of the reinforcement products are, of course, from the textile industry but other concepts include textile fiber handling, cloth pattern cutting, and lay-up. A broad spectrum of engineering disciplines has contributed—from bridge building to laser cutting. Insight and innovation are key elements to progress and success in the field of composites manufacturing.

FEATURES OF THE BOOK

For classroom use or individual reading, the book contains many features to make learning easier.

1. Each chapter begins with an overview of the key points to be addressed. The order of the chapter follows the order of the overview so that you can see what is in the chapter and easily find it.
2. Each chapter contains a case study discussing a specific application of one or more principles taught in the chapter.
3. A summary is given in each chapter so that the key learning points can be reviewed.
4. For assistance to college professors, each chapter has a laboratory experiment dealing with the concepts of that chapter.
5. A set of questions is given in each chapter so that individuals can assess learning and professors can use them in preparing questions for exams.
6. A bibliography details additional reading for each chapter.
7. A glossary of important terms is found at the back of book to assist in understanding or remembering terms given in the book. Each of the terms defined in the glossary is marked in bold print the first time it is used in the text.

8. Improving upon the first edition, figures have been enhanced and added to expand the breadth and depth of coverage.
9.  Providing a valuable additional resource, the *Composites Manufacturing* video series produced by the Society of Manufacturing Engineers (SME 2005) complements this text. The series offers excellent visual representations of the materials and manufacturing methods addressed in the text. A selection of SME video clips, excerpted from the video programs referenced in the bibliographies within the chapters, is available online to complement study. Visit www.sme.org/focm.

1

Introduction to Composites

CHAPTER OVERVIEW

This chapter examines the following:

- The concept of composites
- Roles of the matrix and reinforcement in composites
- A history of composites
- Composite types—engineering and advanced
- Markets
- Composites industry structure

THE CONCEPT OF COMPOSITES

In the broad sense, the term **composite materials** refers to all solid materials composed of more than one component wherein those components are in separate phases. This definition includes a wide assortment of materials, such as: fiber reinforced plastics, regular and steel reinforced concrete, particle filled plastics, rubber reinforced plastics, wood laminates, ceramic mixtures, and even some alloys. The breadth of the materials encompassed within this definition precludes their examination at anything except a cursory level within the pages of a single book. Therefore, this book will focus on one major branch of composite materials—those that are fiber reinforced—a much narrower definition. So, when the terms "composites" or "composite materials" are used, the definition envisioned is: *Composite materials are those solid materials composed of a* ***binder*** *or* ***matrix*** *that surrounds and holds in place* ***reinforcements****.*

The binders or matrices (these terms are considered equivalent) of most importance in the marketplace are polymeric, and are, therefore, the ones of most interest in this book. Some attention also will be given to both metal and ceramic matrix composites. Generally, for all of these matrix materials, the reinforcements considered are fibers. Some discussion will, however, be devoted to particle filled composites and nano fillers in their various forms.

Some writers have suggested an alternate definition of composites: Mixtures of two or more solid materials that are mechanically separable, at least in theory, and possessing *complementary* properties. This definition emphasizes the improvements in properties possible when composites are made. The more narrow definition of composites suggested previously fits within this alternate definition and, as will become clear, the complementary nature of matrix and reinforcement is the reason composites are so important commercially. However, not all properties and characteristics are advantageous when composites are made. For each application the advantages and disadvantages should be weighed.

Some of the advantages and disadvantages of composites are listed in Table 1-1. As the properties and characteristics of composites are explored throughout this book, these advantages and disadvantages will become more obvious. Some important properties of composites and metals are compared in Figure 1-1. Note the low weight, low thermal expansion, high stiffness, high strength, and high fatigue resistance of composites versus steel and aluminum. These graphs lump all composites into one group, and all types of steel and aluminum into two others, but the data correctly reflect the general trends.

Of great value is that the separate characteristics of the matrix and reinforcements contribute synergistically to the overall properties of the composite. Moreover, because so many different matrix and reinforcement materials can be chosen, a wide range of properties is possible. Within a particular choice of matrix and reinforcement, the orientations of the reinforcements, manufacturing method, processing conditions, and combinations made with other materials all give additional variety in the properties available. (The reader who is not familiar with composites might benefit from viewing the series of videos from the Society of Manufacturing Engineers, which complement this textbook. The particular video entitled "Composite Materials" would be especially useful for an introduction to composites [Society of Manufacturing Engineers 2005].)

ROLES OF THE MATRIX AND REINFORCEMENT IN COMPOSITES

The **matrix** is the continuous phase of the composite. Its principal role is to give shape

Table 1-1. Advantages and disadvantages of composites.

Advantages	Disadvantages
• Lightweight	• Cost of materials
• High specific stiffness	• Lack of well-proven design rules
• High specific strength	• Metal and composite designs are seldom directly interchangeable
• Tailored properties (anisotropic)	• Long development time
• Easily moldable to complex (net) shapes	• Manufacturing difficulties (manual, slow, environmentally problematic, poor reliability)
• Part consolidation leading to lower overall system cost	• Fasteners
• Easily bondable	• Low ductility (joints inefficient, stress risers more critical than in metals)
• Good fatigue resistance	• Solvent/moisture attack
• Good damping	• Temperature limits
• Crash worthiness	• Damage susceptibility
• Internal energy storage and release	• Hidden damage
• Low thermal expansion	• EMI shielding sometimes required
• Low electrical conductivity	
• Stealth (low radar visibility)	
• Thermal transport (carbon fiber only)	

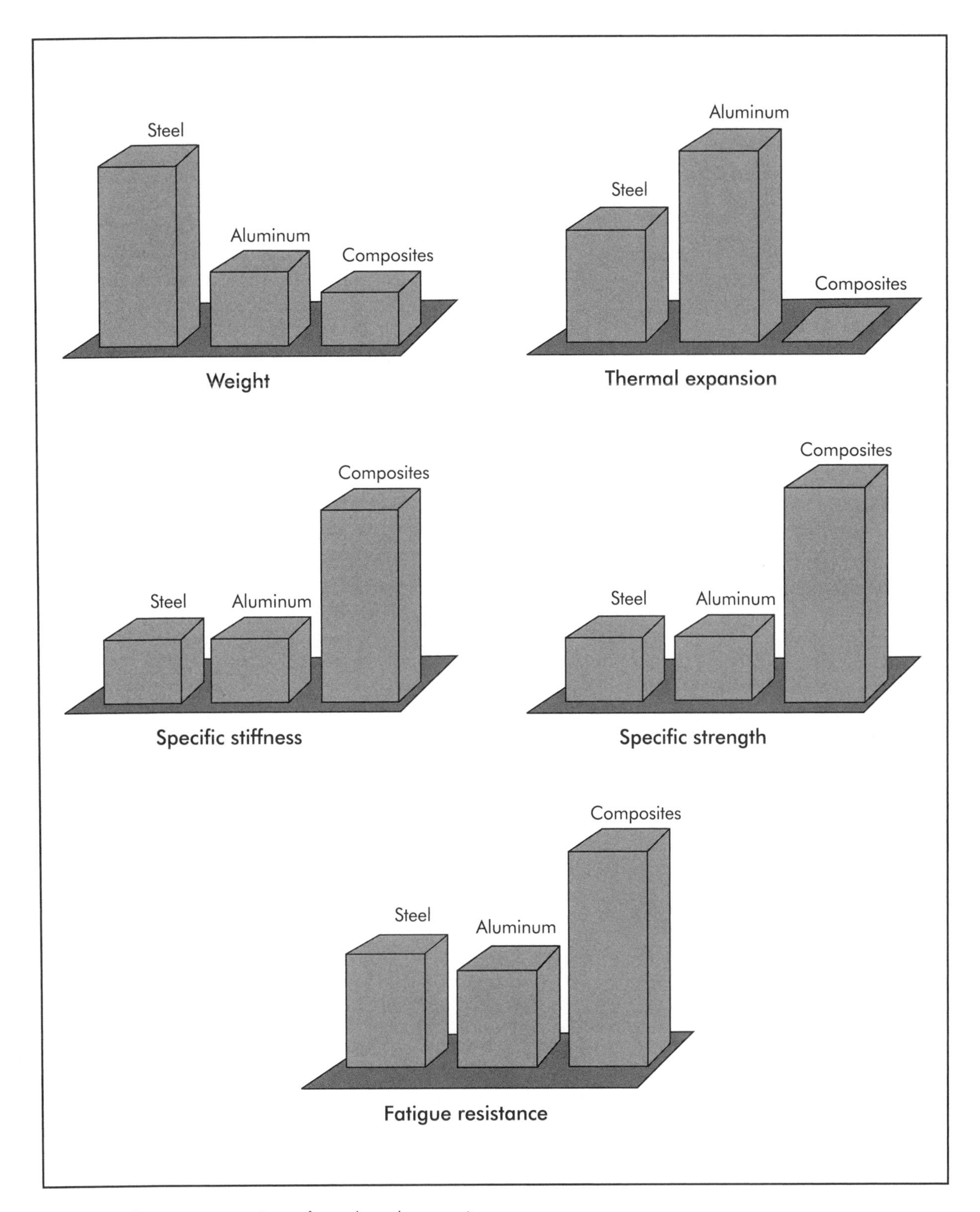

Figure 1-1. Property comparison of metals and composites.

to the structure. Therefore, matrix materials that can be easily shaped and then hold that shape are especially useful. The most common materials with this characteristic are polymers. Therefore, well over 90% of modern composites have polymeric materials (sometimes referred to as plastics or resins) as their matrix.

As the continuous phase, the matrix surrounds and covers the reinforcements. Hence, the matrix is the component of the composite exposed directly to the environment. Another role of the matrix is, therefore, to protect the reinforcements from the environment. The degree of protection desired is one of the key considerations in choosing the type of polymeric matrix for the composite. For example, polymeric matrices give good protection against moderately hostile conditions but may be inadequate when high temperatures or some aggressive solvents are present. These extreme conditions may require a ceramic or metal matrix composite.

The matrix is the component of the composite that first encounters whatever forces might be imposed. Generally, the matrix is not as strong as the fibers and is not expected to withstand these imposed forces. However, the matrix must transfer the imposed loads onto the fibers. The effectiveness of load transfer is one of the most important keys to the proper performance of the composite. Almost all of the common commercial polymeric, metallic, and ceramic matrix materials adequately transfer the loads onto the fibers.

The principal role of the *reinforcement* is to provide strength, stiffness, and other mechanical properties to the composite. Generally, the mechanical properties are highest in the direction of orientation of the fibers. For example, if all the fibers in a composite are oriented in the long direction of the part (like strands in a rope), the composite is strongest when pulled in the long direction. This characteristic of composites allows the part designer to specify certain percentages of the fibers to be in certain exact orientations for a particular application. If the forces on the part would come from all directions, then the designer would specify randomization or multi-directionality of the fibers. The composite part manufacturer has the obligation to control fiber orientation so the directional and percentage specifications of fiber can be satisfied. This manufacturing task has led to several manufacturing methods that are unique for composites. These and other more traditional manufacturing methods, which can be modified to accommodate composites, will be examined later in this book.

All of the properties of composites arise, to some extent, from the interaction or presence of both the matrix and the reinforcement. However, as just discussed, some properties are dominated by either the matrix or the reinforcement. Even when one component of the composite dominates, both components generally must work in concert to obtain optimal performance. For example, if the matrix does not bond well to the fibers, or the fibers are not strong, little improvement in strength is obtained over just the strength of the matrix. Some properties such as toughness, electrical properties, and damping arise from a strong combination or interaction of the matrix and the reinforcement. These combination properties will be discussed with a particular emphasis on the nature of the interactions required for enhanced performance and the expected results from such interactions. The roles of matrix and reinforcement are summarized in Table 1-2.

HISTORY OF COMPOSITES

Early in history it was found that combinations of materials could produce properties superior to those of the separate components. For example, mud bricks reinforced with straw were used by ancient Israelites

Table 1-2. Roles of the matrix and reinforcement in a composite.

Matrix	Reinforcements
• Gives shape to the composite part • Protects the reinforcements from the environment • Transfers loads to the reinforcements • Contributes to properties that depend upon both the matrix and the reinforcements, such as toughness	• Give strength, stiffness, and other mechanical properties to the composite • Dominate other properties such as the coefficient of thermal expansion, conductivity, and thermal transport

in Egypt. Mongols made composite bows by bonding together (using glue made from animal hoofs and bones) five pieces of wood to form the core of a bow (center grip, two arms, two tips) to which cattle tendons were bonded on the tension side and strips of cattle horns on the compression side. This assembly was steamed and bent into the proper shape, wrapped with silk thread, and then cooled slowly (for up to a year) to create some of the most powerful bows ever made. Bridges and walls are constructed today of steel reinforced concrete. Nature uses the same principle in celery and, less obviously, in trees where the pith surrounds a fibrous (reinforcing) cellulose material. This is what gives the celery, and wood, strength.

The history of modern composites (of the type discussed in this book) probably began in 1937 when salesmen from the Owens-Corning Fiberglas® Company began to sell fiberglass to interested parties around the United States and those customers found that the fiberglass could serve as a reinforcement. Fiberglass had been made, almost by accident in 1930, when an engineer became intrigued by a fiber that was formed during the process of applying lettering to a glass milk bottle. The initial product made of this finely drawn molten glass was insulation (glass wool), but structural products soon followed.

The fiberglass salesmen soon realized that the aircraft industry was, in particular, a likely customer for this new type of material. Many of the small and vigorous aircraft companies seemed to be creating new aircraft designs and innovative concepts in manufacturing almost daily, and many of these new innovations required new materials.

One company, Douglas Aircraft, bought one of the first rolls of fiberglass because its engineers believed the material would help solve a production problem. They were trying to resolve a bottleneck in the making of metal molds for its sheet forming process. Each changed aircraft design needed new molds, and metal molds were expensive and had long lead times. Douglas engineers tried using cast plastic molds, but they could not withstand the forces of the metal-forming process. The engineers reasoned that maybe if the plastic molds were reinforced with fiberglass they would be strong enough to allow at least a few parts to be made so the new designs could be quickly verified. If the parts proved to be acceptable, then metal dies could be made for full production runs. Dies were made using the new fiberglass material and phenolic resin (the only resin available at the time). What a success!

Other applications in tooling for aircraft soon followed. Many of the tools (molds, jigs and fixtures) for forming and holding aircraft sections and assemblies needed to be strong, thin, and highly shaped, often with compound curves. Metals did not easily meet these criteria, so fiberglass reinforced phenolic production tooling became the preferred material for many of these aircraft manufacturing applications.

Not long afterward, unsaturated polyester resins became available (patented in 1936). They eventually (although not immediately) became the preferred resin because of their relative ease of curing when compared to phenolics. Peroxide curing systems, needed for the polyesters, were already available with benzoyl peroxide being patented in 1927 and many other peroxides following not too long afterward. Higher performance resins also became available about this time with the invention of epoxies in 1938. The materials and the applications seemed to be converging at the same time.

The fast pace of composites development accelerated even more during World War II. Not only were even more aircraft being developed and, therefore, composites more widely used in tooling, but the use of composites for structural and semi-structural parts of the airplanes themselves was being explored and adopted. For instance, in the frantic days of the war, among the last parts on an aircraft to be designed were the ducts. Since all the other systems within the aircraft were already fixed, the ducts were required to go around them. This often resulted in ducts that twisted and turned around the other components, usually in the most difficult-to-access locations. Metal ducts just could not be easily made in these convoluted shapes. Composites seemed to be the answer. The composites were hand laid-up on plaster mandrels (internal molds), which were made in the required shape. Then, after the resin cured, the plaster mandrels were broken out of the composite parts. Literally thousands of such ducts were made in numerous manufacturing plants clustered around the aircraft manufacturing/assembly facilities.

Other early WWII applications included engine nacelles (covers), radomes (domes to protect aircraft radar antennas), a structural wing box for the PT-19 airplane, and non-aircraft applications such as cotton-phenolic ship bearings, asbestos-phenolic switch gears, cotton-asbestos-phenolic brake linings, cotton-acetate bayonet scabbards, and thousands of other applications.

At about this time (1942), the United States government became concerned that the supplies of metals for aircraft might not be available. A major effort was initiated to develop the design rules and manufacturing methods for composites as possible replacements for aircraft metals. Several critical parts (including wing sets for the AT-6 trainer) were made to prove out the design concepts and manufacturing methods, which included the development of filament winding and spray-up, sandwich structures, fire-resistant composites, and prepreg materials.

When the war effort came to a sudden halt, many companies who had been active in making war materials were faced with an acute problem. They needed to quickly identify new markets and new products that utilized the expertise they had developed. Some of this development work was supported by the materials suppliers. Some war-oriented applications were converted directly to commercial application, such as fiberglass reinforced polyester boats.

Almost everyone agreed that the pent-up demand for automobiles was a logical application for composites. By 1947 a completely composite automobile body had been made and tested. This car was reasonably successful and led to the development of the Corvette® in 1953, which was made using fiberglass preforms impregnated with resin and molded in matched metal dies. Eventually the dominant molding method for automobile parts was compression molding of sheet molding compound (SMC) or bulk molding compound (BMC)—both to be explained in a later chapter.

Some of the products made during the post-war era now represent the major markets for composite materials. In addition to aircraft, these include boats, automobiles, tub and shower assemblies, non-corrosive pipes, appliance parts, storage containers, and furniture.

The push for aerospace dominance that began in the 1950s (during the Cold War) and picked up speed in the 1960s and 1970s gave a new impetus to further composite development. Hercules, Inc. acquired filament winding technology from W. M. Kellogg Company and began making small rocket motors. In 1961, a patent was issued for producing the first carbon (graphite) fiber, which then was used in many of the rocket motors and in aircraft. Other important fibers were also developed during this period, including boron fibers in 1965 and aramid fibers (DuPont's Kevlar®) in 1971. In 1978, the crowning jewel of this period was the development of the first fully filament wound aircraft fuselage, the Beech Starship.

Recent material and process improvements and the development of higher performing fibers and resins have led to tremendous advances in aerospace, armor (structural and personal), sports equipment, medical devices, and many other high-performance applications. Composites seem to have found their place in the world. They are finding increased use, especially in products where performance is critical. In 1989, the National Academy of Engineering issued its list of the top 10 engineering achievements of its lifetime (the 25 years prior to 1989). It ranked the development of advanced composites as number six on that list and cited especially advanced composite materials such as graphite-epoxy materials used to make lighter and stronger aircraft, bicycles, golf clubs, and other products. The academy said composites add lightness and strength to a wide range of everyday objects.

Some composite product applications are very new, including stealth aircraft, space structures, the wrapping of concrete structures with composites for improvement of earthquake performance, wholly composite bridges, and other construction edifices. It is interesting that construction is still a major market for composites, just as it was in 1500 B.C. when the Egyptians and Israelites used straw to reinforce mud bricks.

COMPOSITE TYPES—ADVANCED AND ENGINEERING

Polymeric composites can be divided into two groups, advanced composites and engineering composites. These differ principally in the type and length of the fiber reinforcement and in the performance characteristics of the resins used. No sharp division boundary exists, however, and so one type of composite transitions gradually into the other.

The **advanced composites** are generally characterized by long, high-performance reinforcements and by resins with superior thermal and mechanical properties. Advanced composites are typically used for aerospace applications (like rocket motor cases and airplane parts), but are also used for high-performance sporting goods (golf clubs, tennis rackets, bows and arrows), as well as an increasing number of other parts that require the superb properties (primarily high strength-to-weight and high stiffness-to-weight ratios) obtained from these materials.

The stealth bomber is probably the most sophisticated of all products made with composites. In this aircraft, composites are used for their strength, stiffness, and other mechanical properties. In fact, some of the stealth properties of the bomber depend upon the unique nature of composite materials. The value of stealth technology, which includes many other factors, such as shape, can be illustrated by comparing the radar cross-sections (RCS) of various airborne objects as shown in Table 1-3. The RCS is the area seen on radar. The smaller the area, the more difficult the object is to see. When comparing the various RCSs, note that the B-52 has approximately the same RCS as a train boxcar (108 ft^2 [10 m^2]), while the B-2 has about the same wingspan as the B-52 but has the RCS of a small bird (0.1 ft^2 [.01 m^2]).

Wrapping concrete highway columns for protection against corrosion and to improve performance of the columns during earthquakes is a relatively new application. Highway overpass columns are wrapped with carbon fiber and epoxy composite material and then cured to give a hard, strong overwrap. This overwrap on the column decreases the tendency of the concrete to buckle and collapse in an earthquake.

Table 1-3. Radar cross-sections (RCSs) of typical airborne objects.

Airborne Object	RCS, ft^2 (m^2)
Jumbo jet	1,076.4 (100)
B-17 Flying Fortress	861 (80)
B-47 bomber	431 (40)
B-52 bomber	108 (10)
B-1B bomber	11 (1)
Large fighter	54-65 (5-6)
Small fighters	22-32 (2-3)
Small single-engine plane	11 (1)
Man	11 (1)
Small bird	0.1 (.01)
Insect	0.0001 (.00001)
F-117A Stealth Fighter	1.1 (.1)
B-2 Stealth Bomber	0.1 (.01)

The second major group of composite materials is **engineering composites**. These materials are characterized by fibers that are both shorter in length and lower in mechanical properties. Generally, they are either made with low-cost thermoset resins or fiberglass (called fiber reinforced plastics or FRP), or thermoplastic resins with short fibers (called fiber reinforced thermoplastics).

Products made from **thermoset** resins with fiberglass (FRP) include train cars, boat hulls, canoes, tubs, shower stalls, spas, fuel storage tanks, and other parts made from conventional thermoset resins. These structures illustrate the capability of engineering composite materials to give good strength and good durability at moderately low prices. The processes for manufacturing FRP products are generally less precise in maintaining exact fiber orientations and resin/fiber content than the advanced fiber processes.

The fiber reinforced **thermoplastics** are normally molded by the same processes as their non-reinforced thermoplastic analogues. Such conventional processes include injection molding, extrusion, and so on. The products made from fiber reinforced thermoplastics are generally the same as their non-reinforced counterparts, but have the reinforcement added to gain strength, stiffness, or some other mechanical property. For example, an injection molded nylon gear might be cracking because of high torque requirements. This problem might be solved by using a fiberglass reinforced nylon, which could result in three times more strength.

A summary of the differences and similarities in materials technologies of engineering and advanced composites is given in Table 1-4. A comparison of the processing and fabrication methodologies of advanced and engineering composites is given in Table 1-5. Table 1-6 provides a comparison

Table 1-4. Comparisons of materials technologies.

Property	FRP Composites	Advanced Composites
Reinforcement fiber	Predominantly E-glass products C-glass for corrosion Corrosion versions of E- and C-glass Limited "natural" fiber usage Often referred to as: FRP = fiber reinforced plastics GRP = glass reinforced plastics GFRP = glass fiber reinforced plastics	High-strength S-glass (R- and T-glass versions) Carbon and graphite Aramid (like Kevlar™) Boron Quartz (fused silica) Polyethylene (like Spectra™) Advanced polymeric fibers Ceramic fibers
Typical fiber form(s)	Continuous fiber roving and chopped fibers dominate Woven roving products Tightweave glass fabrics (numerous) Glass veil materials Commercial mat and stitched/bonded materials	Continuous fiber roving and tow materials dominate Prepreg materials (unidirectional tapes and fabrics) Higher fiber volumes (55–65%) Complex structural textile products
Resin matrix	Polyester and vinyl ester leading candidates Wide variety of polyester formulations Phenolics for fire applications Limited epoxy use in certain areas Entirely thermosets—except for some glass fiber thermoplastics	Epoxies generally baseline Bismaleimides (BMI) and cyanate esters for higher temperatures Polyimides Numerous high-performance thermoplastics Toughened resin system interest and use for damage tolerance
Application temperature and conditions	Ambient temperature range (25–140° F [4–60° C] typical) Exposure to UV and weather conditions Exposure to fuels, waste materials, water, sea water, and processing chemicals	Broader temperature exposure range Space environments Aerospace service conditions Exotic fuels, solvents, fluids Extreme conditions: 180–700° F (82–371° C) environmental conditions, much more possible Cryogenic conditions
Core material usage	Balsa and foam core materials Core materials consistent with near-ambient use conditions Core use technology built upon marine and boat market	Extensive use of honeycomb core materials Aluminum Nomex™ (aramid) Carbon fiber Extensive design and manufacturing technology base

Table 1-5. Comparisons of composites processing and fabrication methodologies.

Property	FRP Composites	Advanced Composites
Production quantities	Production quantities often quite large: 100s to 1000s or more Pipe—usually miles/kilometers Tank systems—usually on the order of several hundred	Production quantities often much smaller: typically less than a few hundred Complex parts often less than a dozen
Manufacturing processes	Generally simple and requires only moderate labor skills Filament winding, pultrusion, contact molding, "chop and spray," resin infusion (resin transfer molding [RTM], vacuum-assisted resin transfer molding [VARTM], Seemann Composites resin infusion molding process [SCRIMP®])	Processes more complex and tightly controlled with paperwork Filament winding, fiber placement, hand lay-up and vacuum bagging, tape lay-up, resin infusion (RTM, VARTM, resin film infusion [RFI], some SCRIMP™), thermoforming Considerably more quality control and in-process inspections Autoclave processing of many parts
Tooling materials and methods	Steel and aluminum molds and tooling Fiberglass tooling common Tool surfaces often gel coated Simple tooling fabricated for prototyping as well as some production Overall tool cost significantly lower	Invar, steel, and aluminum tooling Tight dimensional controls Extensive tooling design efforts Highly polished tool surfaces Composite tooling using carbon-fiber laminates Higher temperature tooling requirements
Cure methods	Often ambient cure Oven cure on some parts Steam, heated platens, heated oil, and electric heaters used in some tooling applications	Controlled oven or autoclave curing typically Precise temperature-time-pressure-vacuum cycles for many parts Heated tooling during most process steps

of the market applications and philosophical differences between engineering and advanced composites. These tables will serve as references for understanding the nature of composites throughout the remainder of this text.

MARKETS

Composite materials are part of our lives every day. We ride in cars and light rail systems with composite panels; we fly in planes with composite parts; our homes have composite showers and numerous composite parts used in hidden places, such as in washers and dryers; and we enjoy recreation with composite tennis racquets, golf clubs, and boats.

The breadth of the composites market can be seen in the data compiled by the American Composites Manufacturing Association

Table 1-6. Market applications and philosophical differences of composites.

Property	FRP Composites	Advanced Composites
Application temperature	Generally ambient (25–140° F [4–60° C]) conditions Some extremes as a result of fluids and chemicals being processed	Typically high performance over broad temperature ranges: –65 to 165° F [–54 to 74° C] for many defense-related products Cryogenic and space applications incur lower temperatures High-temperature operating conditions in 300–700° F [149–371° C] range for aerospace parts
Economic considerations	Low cost—materials less expensive FRP products are lower-cost items	More performance than cost driven Materials considerably higher cost
Safety factor (S_F)	Typically use S_F = 2.0–5.0 Built to endure more robust handling Safety factors dependent upon certification standards	Typically use S_F = 1.25–3.0 Operates at extremes of performance capability Safety factors dependent upon rigid customer or industry standards
Design factors	Generally use vendor data or data derived from limited testing S-basis or B-basis design approach using nominal properties Designers often talk in terms of "fiber and resin weight" percent when preparing manufacturing specifications	Generally an extensive data base is developed from numerous tests A-basis often used, B-basis used with backup testing to verify performance Designers use "fiber volume" and "resin weight" percent when preparing specifications Designs specify numerous environmental factors (damage, temperature, pressure, etc.)
Testing and certification	Materials testing often limited to quality control upon material receipt Testing often uses material supplier data and past experience Certification to industry standards: American Society for Testing and Materials (ASTM), American Petroleum Institute (API), Underwriters Laboratories (UL), Society of Automotive Engineers (SAE), American National Standards Institute (ANSI), American Society of Mechanical Engineers (ASME), British Standards (BS)	Materials and structural component testing often extensive Significant testing at various environmental or load limits Less reliance on vendor-supplied data Certification rigid, following customer and governing agency requirements: Department of Defense (DOD), National Aeronautics and Space Administration (NASA), Federal Aviation Authority (FAA), Department of Transportation (DOT), National Institute of Standards and Technology (NIST)

(ACMA), which is presented in Figure 1-2. This data, which is for the United States, is similar to data worldwide. The largest market for composites is transportation (32%), thus suggesting the large amount of composites in automobiles and other land vehicles, which compose this category. Construction (20%), corrosion-resistant (12%), marine (10%), electrical/electronic equipment (10%), consumer (7%), and appliance/business equipment (5%) are also large markets. The aircraft/aerospace market represents only 1%, which is surprisingly small in light of its importance in the origin of composites. Of course, the aerospace products are fewer in number but much higher in profit (per item) than other products.

The growth of the carbon fiber market is possibly a key to understanding the composites market in the future. Shipments of carbon fiber have grown geometrically since its introduction in the early 1970s. As the volume of carbon fiber has increased, its price has steadily declined. Both trends, volume growth and price decline, are expected to continue at least until the price of carbon fibers reaches a minimum based on raw material cost. Carbon fiber usage is expected to expand into markets traditionally held by fiberglass, such as automotive. Should this

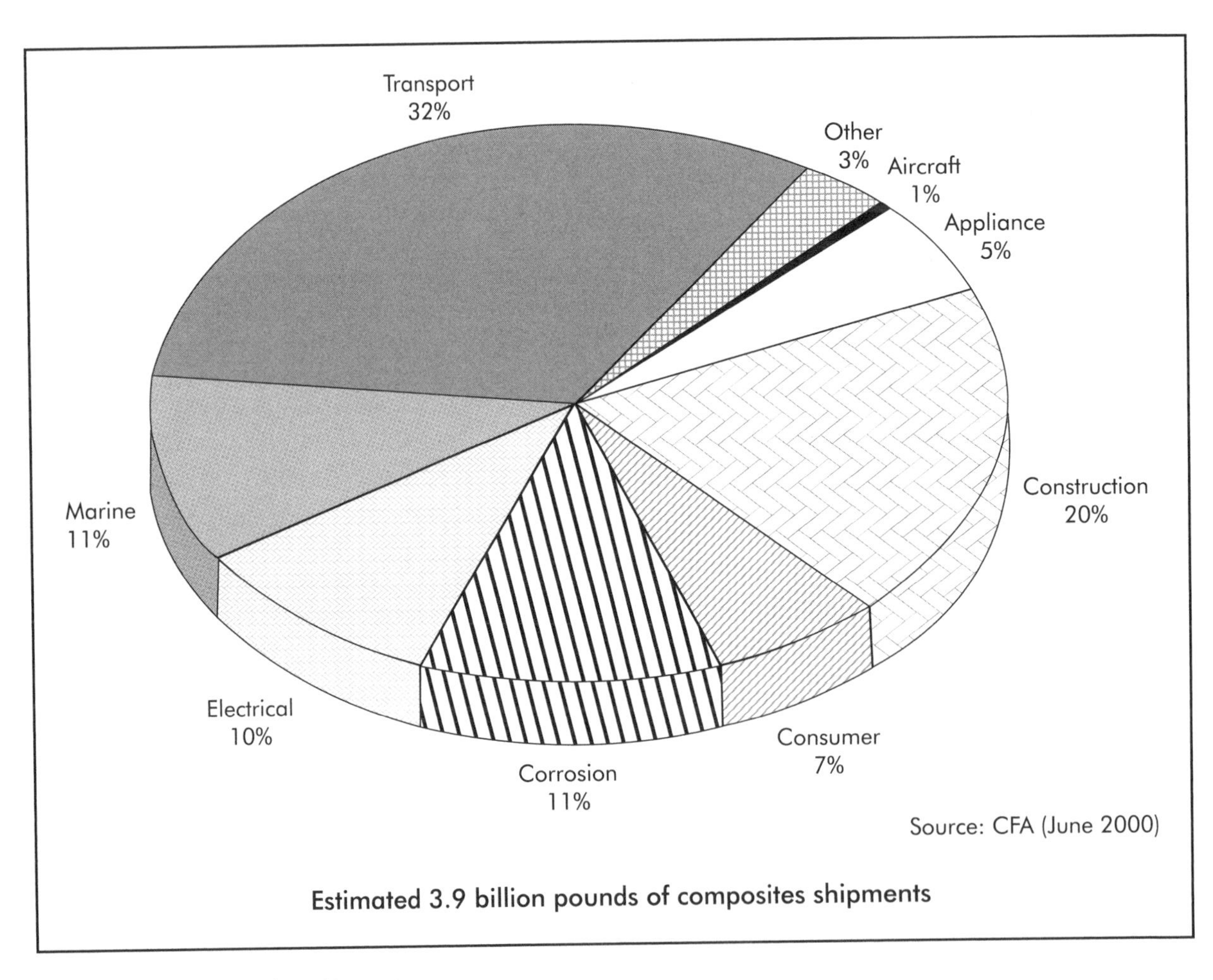

Figure 1-2. U.S. composites shipments.

happen, some major restructuring could occur within the raw material manufacturers and/or within the pricing structure for fiberglass.

Trends in the major composites markets are not necessarily the same. Although, of course, all markets would like to have, simultaneously, lower costs, higher production capabilities, and higher performance, in reality they tend to emphasize one or the other of these factors. For example, the aerospace industry desires products that are high in performance and, as a result, they are also high in cost and low in production rate. In contrast, the automotive market seeks composite parts that are high in production rate and low in cost. These automotive products will generally be lower in performance than the aerospace products. The tub/shower market also optimizes on cost.

COMPOSITES INDUSTRY STRUCTURE

A diagram of the structure of the composite industry is given in Figure 1-3. The manufacturers of fibers and resin supply the raw materials to the remainder of the industry. These manufacturers are generally large companies utilizing complex plants to make the fibers and resins. Clearly, the

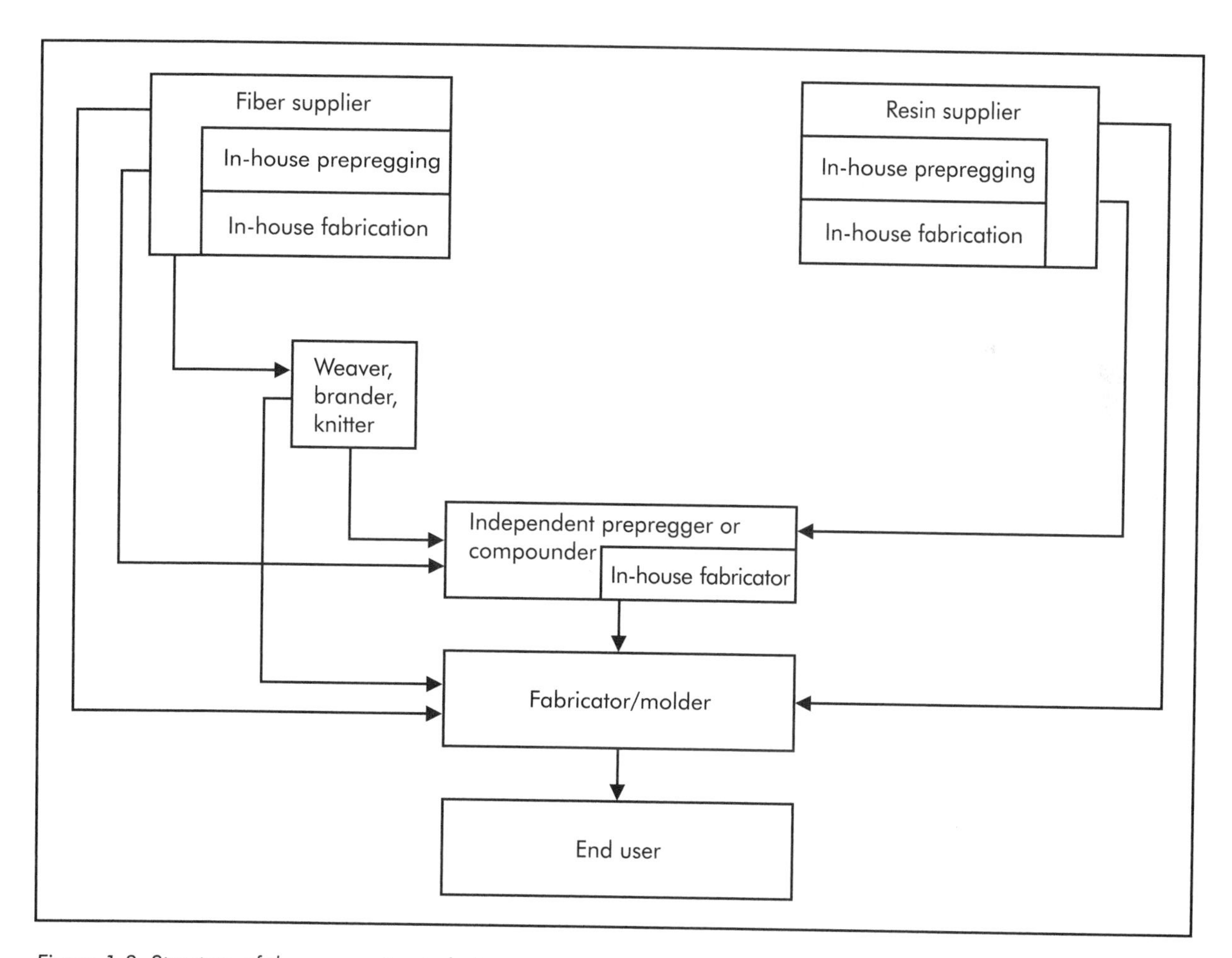

Figure 1-3. Structure of the composites industry.

capital investment would be high for a company seeking to begin the manufacture of fibers or resins. Therefore, the number of resin and fiber manufacturers is small.

After the raw materials are made, they must be combined together to make the composite material. Then the composite material is molded to create the composite part. This combining of raw materials can be done just prior to or during the molding process. Or, it can be done as a separate step before molding, often by someone other than the molder. When done just prior to or as part of the molding process, the raw materials go directly from the manufacturer to the fabricator/molder. Not shown are distribution companies that might provide service for the manufacturers by warehousing the raw materials. These companies allow the fabricator/molder to buy in smaller lots and offer improved delivery service over the raw material suppliers.

Several different types of intermediate companies might be involved in combining the resin and the fibers prior to molding. Two general types of combinations are made.

For advanced composites, the combination of resin with reinforcement is called a **prepreg**, which is an abbreviation of "preimpregnated fibers." Prepregs are made with precise resin and fiber contents and orientations, and careful wetting of the fibers. The prepregs, in a sheet form, can be made using cloth or some other woven, braided, or knitted fabric, which is impregnated with the resin. Or, alternately, they can be made using unidirectional fibers held in a parallel array and then coated with the resin. When using the most common resins, the prepregs must be kept refrigerated so they do not cure prematurely. The fabricator/molder lays the prepreg sheets in the mold, compresses them to remove the air, and then heats the prepreg material to mold and cure. The details of this consolidating and molding process are discussed later in this book in the chapter on open molding (long fibers). The fabricator/molders who use prepregs justify the additional cost by pointing to their carefully controlled resin-fiber content, the elimination of the problems associated with wet resin, and their overall ease of fabrication. Nevertheless, the higher cost generally limits the use of prepregs to only high-value applications such as aerospace and sporting goods products.

The second type of combination system for composites is compounding or manufacture of molding compound. This system is associated with engineering composites or fiberglass reinforced plastic (FRP). Just as with the prepreggers, the compounders are often able to make more precise molding compounds than could be conveniently made by the fabricator/molder. The molding compounds can be either in sheet form or in a bulk material, often sold as a log. Molding compounds usually contain chopped fiberglass (rather the continuous fibers of prepregs). Hence, performance of molding compounds is lower than with prepregs. Molding compounds are, therefore, associated with engineering composites whereas prepregs are associated with advanced composites.

Each class of company can be involved in more than one step in the overall composites business. For instance, the raw material suppliers, especially the resin manufacturers, are often involved in compounding. This is especially true with fiber reinforced thermoplastics. Some raw material suppliers are also involved in making sheet and bulk molding compounds and prepregs.

A few of the raw material suppliers mold parts directly. Some independent prepreggers and compounders also mold composite parts. In most of these cases, however, a separate division of the raw material supplier or the independent prepregger/compounder does the molding, thus preserving the idea that the company is not competing directly against its customers.

CASE STUDY 1-1

Flying Car

Developments in the field of composites after WWII were critical to establishing the materials' position in the world today. One of those developments was especially innovative but had only a brief period of success. However, the concept, while no longer commercial, is still being pursued and may someday be a major factor in modern technology. That development was the auto/plane.

The development of the auto/plane was led by Convair Aircraft Company in the years following WWII. Convair reasoned that the many returning wartime pilots would like to continue with their flying, but also would like to combine that flying with family vacations. Hence, Convair made an automobile with an all-composite body (for weight savings) that would allow a special wing assembly to be attached. The car/plane is shown in Figure 1-4.

The concept was to build the car/plane so the wings could be attached to the car whenever the car was to be used as the cabin of the airplane. Otherwise, the car would be used as a standard automobile. The wings would be available for rent at various airports, thus permitting the driver to rent a wing assembly at one airport, fly to the vacation site, turn in the wing assembly, and drive away.

Prototypes were made and successfully demonstrated. The concept, although never fully commercialized, is still alive. Modern versions tend to have the wings permanently attached to the car but capable of being retracted or folded into a storage compartment inside the car.

What a boon these car/planes would be today in Los Angeles and other cities with massive freeway problems and long commutes. On the other hand, maybe the problems

Figure 1-4. Flying car. (Courtesy Brandt Goldsworthy)

would simply be moved from the surface to the air above the city, making the skies more dangerous than the roads!

SUMMARY

When reinforcements are added to binding matrix materials, the resulting combination materials are called composites. The properties of composites vary widely depending on the nature of the reinforcement (length, concentration, and type) and the type of matrix material (polymer/plastic, ceramic, or metal). The most common composites are made with fiber reinforcements and polymer matrices. In general, the materials combine synergistically; that is, the properties are better in the combined state than they are separately. In fact, the combination is often better than just the simple addition of the properties of each element.

The principal roles of the reinforcements are to give mechanical strength and stiffness to the composite. The reinforcements also dominate a few other properties such as thermal expansion. The principal roles of the matrix are to give shape to the part, protect the reinforcements from the environment, and transfer loads to the reinforcements. The matrix also contributes to properties such as toughness.

At low fiber concentrations and when the fibers are short, the properties of the composite are dominated by the matrix. When the matrix is a thermoplastic, these composites (fiber reinforced thermoplastics) can be viewed as specialty grades of traditional thermoplastics with improved mechanical properties. When the fibers are somewhat longer and encapsulated by a low-cost thermoset matrix, the composite material is called fiberglass reinforced plastic (FRP). Fiber reinforced thermoplastics and FRPs are engineering composites.

If the fibers are longer and higher in concentration, and especially if they have high mechanical properties, they dominate the properties of the composite structure. These composites are called advanced composites. The matrices for advanced composites have higher thermal and protective capabilities than those used in engineering composites. Usually the advanced composite matrices are thermosets, but some advanced thermoplastic matrices are also available.

The applications of advanced composites are chiefly in aerospace and high-value products where performance is a premium. Applications for engineering composites are more cost sensitive and in much higher volume. Typical engineering composite applications would include automotive parts, construction products, appliance parts, tubs and showers, and the many other products that have been collectively identified as fiberglass reinforced plastics.

The composites market is multi-tiered in structure. The heart of the business is the molders and fabricators of the composite parts. These molders/fabricators can either combine the reinforcements and the matrix materials at the time of molding or use intermediate materials in which the reinforcements and matrix have already been combined by another company. These intermediate materials are called prepregs (fibers and resins only) or molding compounds (fibers, resins, and filler). In some instances the raw material manufacturers supply their materials directly to the molder and in others they supply them to a maker of the intermediates. Some raw material suppliers also make the intermediate materials as part of their product line. The raw material suppliers and the intermediate material manufacturers also can serve as molders and fabricators.

The composite market is growing. It has been strongly influenced in the past by governmental policy, chiefly in aerospace, but now seems to be dominated by consumer and industrial products. The growth of the composites business is strong and may become even stronger as the price of some raw materials (especially carbon fibers) drops with improved manufacturing technologies and higher volumes.

LABORATORY EXPERIMENT

Property Differences Between Reinforced and Non-reinforced Materials

Objective: Discover the differences in the properties of composites and non-reinforced products by examining their properties as obtained from a typical database, either written or electronic.

Procedure: Using either a written database (such as that found in *Modern Plastics Encyclopedia,* The McGraw-Hill Companies, Inc., 212-512-2000) or an electronic database (such as Prospector available through IDES, 800-788-4668) fill out Table 1-7.

Compare the various properties of typical thermoplastic and thermoset resins at two levels of reinforcement, neat = 0% and 30%, which are the typical reinforcement levels.

QUESTIONS

1. What is the purpose of the resin in parts with long fibers at high fiber concentrations?
2. Identify three advantages of composites over metals for structural applications.
3. Indicate two major forces that led to the development of composites during WWII.
4. Distinguish between a prepreg and a molding compound.
5. Distinguish between engineering composites and advanced composites.
6. What are two factors that contribute to the stealth of an airplane?

BIBLIOGRAPHY

Society of Manufacturing Engineers (SME). 2005. "Composite Materials" DVD from the Composites Manufacturing video series. Dearborn, MI: Society of Manufacturing Engineers, www.sme.org/cmvs.

Strong, A. Brent. 2006. *Plastics: Materials and Processing*, 3rd Edition, Upper Saddle River, NJ: Prentice-Hall, Inc.

Table 1-7. Property differences between reinforced and non-reinfroced materials.

Property	Resin Type	Neat Resin	Reinforced (30%)
	Nylon		
Tensile strength			
Tensile modulus			
Flexural strength			
Flexural modulus			
Izod impact			
Heat distortion temperature			
	Polycarbonate		
Tensile strength			
Tensile modulus			
Flexural strength			
Flexural modulus			
Izod impact			
Heat distortion temperature			
	Unsaturated Polyester		
Tensile strength			
Tensile modulus			
Flexural strength			
Flexural modulus			
Izod impact			
Heat distortion temperature			
	Epoxy		
Tensile strength			
Tensile modulus			
Flexural strength			
Flexural modulus			
Izod impact			
Heat distortion temperature			

2

Matrices and their Properties

CHAPTER OVERVIEW

This chapter examines the following concepts:

- Polymers, plastics and resins defined
- Polymerization, naming, characteristics, and molecular weight
- Solidification, melting, and other thermal considerations
- Thermoplastics and thermosets
- Aromatic and aliphatic materials
- Matrix-dominated properties
- Wet-out of fibers
- Additives: fillers, pigments, viscosity control agents, surface agents

POLYMERS, PLASTICS AND RESINS DEFINED

Matrix materials can be polymeric, metallic, or ceramic. This chapter will discuss the general nature and properties of polymeric matrices. (The following few chapters will discuss specific types of polymeric matrices. Then, one chapter will be devoted to ceramic and metallic matrix composites, thus completing the chapters on matrices.)

The terms *plastics* and *resins* are often used to refer to polymer matrices. Some people in the composites industry use all these terms—polymer, resin, and plastic—interchangeably, while others make distinctions. This text will define the terms and explain the differences seen by some, and then attempt to adhere to those differences even though a more relaxed usage is often accepted.

The term **polymer** means "many units" and refers to the way in which individual molecular units are joined together into a chain-like structure where each unit is like a link in the chain. A single molecular unit that can be made into a polymer is called a **monomer**. The monomers are made of atoms that have a specific and consistent grouping. When the monomers are linked together into a polymer chain, the single long chain is called a **polymer molecule**. Figure 2-1 illustrates the relationship between a group of monomers and a polymer molecule. Although the concept of a single polymer molecule is simple and a convenient method to represent polymers, in actuality, single polymers are rarely if ever found isolated in nature. The polymers are found as mixtures of many similar chains clustered in a group or pile. These mixtures of polymers are somewhat analogous to a plate of spaghetti.

Some single polymer chains can be composed of several thousand monomer units, while other polymer molecules may have only a few monomer units. When only a few monomer units are joined together, the name applied to these short polymers is **oligomers**. The normal situation is, however, that polymers have long chains and, generally, even those polymers with only a few units will later be joined together to form long chains when they are molded into a finished product. Therefore, even when oligomers

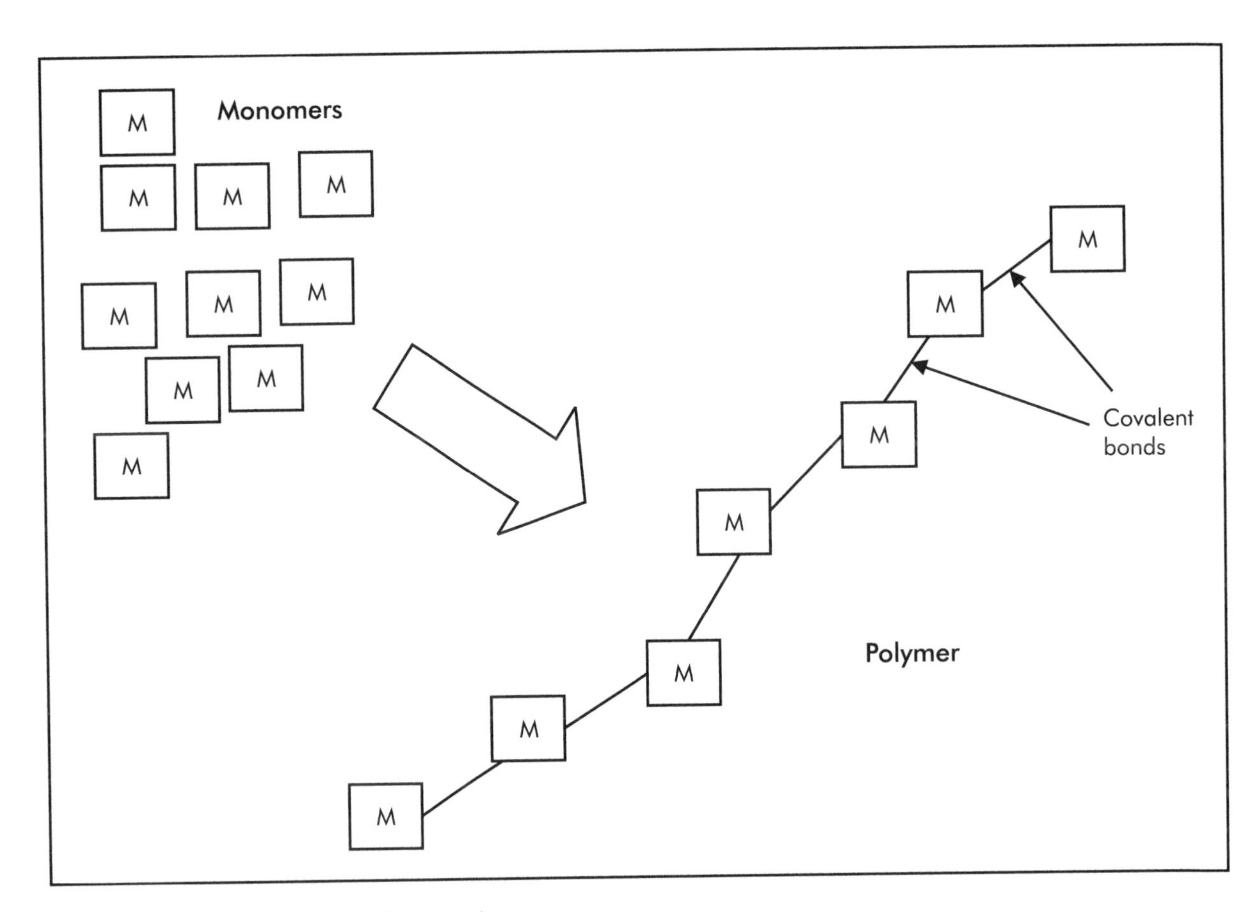

Figure 2-1. Monomers uniting to form a polymer.

are present, they are sometimes called **pre-polymers** in anticipation of their eventual linkage into much longer chains.

The term **resin** refers to polymeric materials that are not yet in their final form and shape. Most of the common resins are liquids, although they also can be solid materials that will be subsequently heated to melting and then formed or shaped. The term "resin" originated from the Greek word meaning tree sap. This reflects that, anciently, tree sap was used to hold other materials together and was also shaped into amulets and other decorative pieces.

The term **plastics** refers to polymeric materials after they have been finally shaped. (The word "plastics" comes from the Greek *plastikos*, meaning to form or mold.) Hence, a resin is changed into a plastic through some forming or molding process. Whereas the terms "polymer" and "resin" can refer to both naturally occurring and synthetic materials, plastic is used only to refer to synthetics. Composite materials made from fiberglass and low-cost resins are often called fiberglass reinforced plastics (FRP). In this sense, plastics might refer to any of the common matrix materials used with fiberglass. Because of this association of the term "plastic" with low-cost composites, the matrix in high-performance composites is not referred to as a plastic. The matrix for these high-performance composites is called a "resin," sometimes even after it has been shaped. (See the "Composite Materials," video program, Society of Manufacturing Engineers, 2005 for a good review of the nature of matrix materials.)

POLYMERIZATION, NAMING, CHARACTERISTICS, AND MOLECULAR WEIGHT

The process of linking the small units (monomers) into long chains (polymers) is called **polymerization**. Polymerization can proceed by several different mechanisms. The two most common are **addition polymerization** (also called free-radical polymerization and chain-growth polymerization) and **condensation polymerization** (also called step-wise polymerization). Simplified representations of addition and condensation polymerization are presented in Figures 2-2 and 2-3.

Addition Polymerization

The addition polymerization reaction begins with the activation of some molecule, usually a peroxide, which can form a highly reactive species such as a **free-radical**. Free-radicals are molecular species that have an unpaired electron. The activation step forming the free-radical usually occurs simply by heating the peroxide, although it could also occur through the action of certain chemicals. This activation is shown in Step 1 of Figure 2-2.

The unpaired electron in the free-radical has a strong driving force to pair with another electron and form a bond. Normally, the free-radical will pull an electron from a weak bond in a nearby molecule (usually a monomer molecule) and pair that electron with its own unpaired electron to create a new bond between the free-radical and the nearby molecule. The nearby molecule is, of course, one of the monomers that has been chosen because it has a bond from which an electron can be easily extracted and it is in high concentration and thus readily available. The monomer will then bond with the free-radical. After the bond is formed between the free-radical and the monomer, the monomer has an unpaired electron, which was the second electron in the weak bond. It is said, therefore, that the monomer has been activated, as shown in Step 2 of Figure 2-2. Only certain types of monomers have the kind of weak bonds that can be easily attacked by a free-radical and are thus appropriate as starting materials for the polymerization reaction. The most common of the appropriate monomers are those that contain a carbon-carbon double bond.

The unpaired electron on the newly bonded monomer is like the electron in the original free-radical. That is, it is highly reactive and seeks to bond with another electron from yet another weak bond. The activated monomer can react with any easily removed electron in its vicinity. Usually, the most abundant source of easily removed electrons is another monomer which, as has been pointed out, usually contains a carbon-carbon double bond. Therefore, the activated monomer reacts with a new monomer. The addition of a new monomer has the effect of lengthening the chain by one monomer unit (Step 3, Figure 2-2).

When the new monomer is added, a new free electron is created, thus creating another active site on the chain. This same process of an activated chain reacting with a new monomer can continue many times and the polymer can grow to be very long, perhaps thousands of monomer units (Step 4, Figure 2-2). This chain growth is usually promoted by performing the polymerization reaction with high monomer concentrations and high heat and/or with the use of a catalyst that facilitates chain growth.

Eventually, a terminating agent either is introduced into the mixture or simply occurs through contamination or reactions with

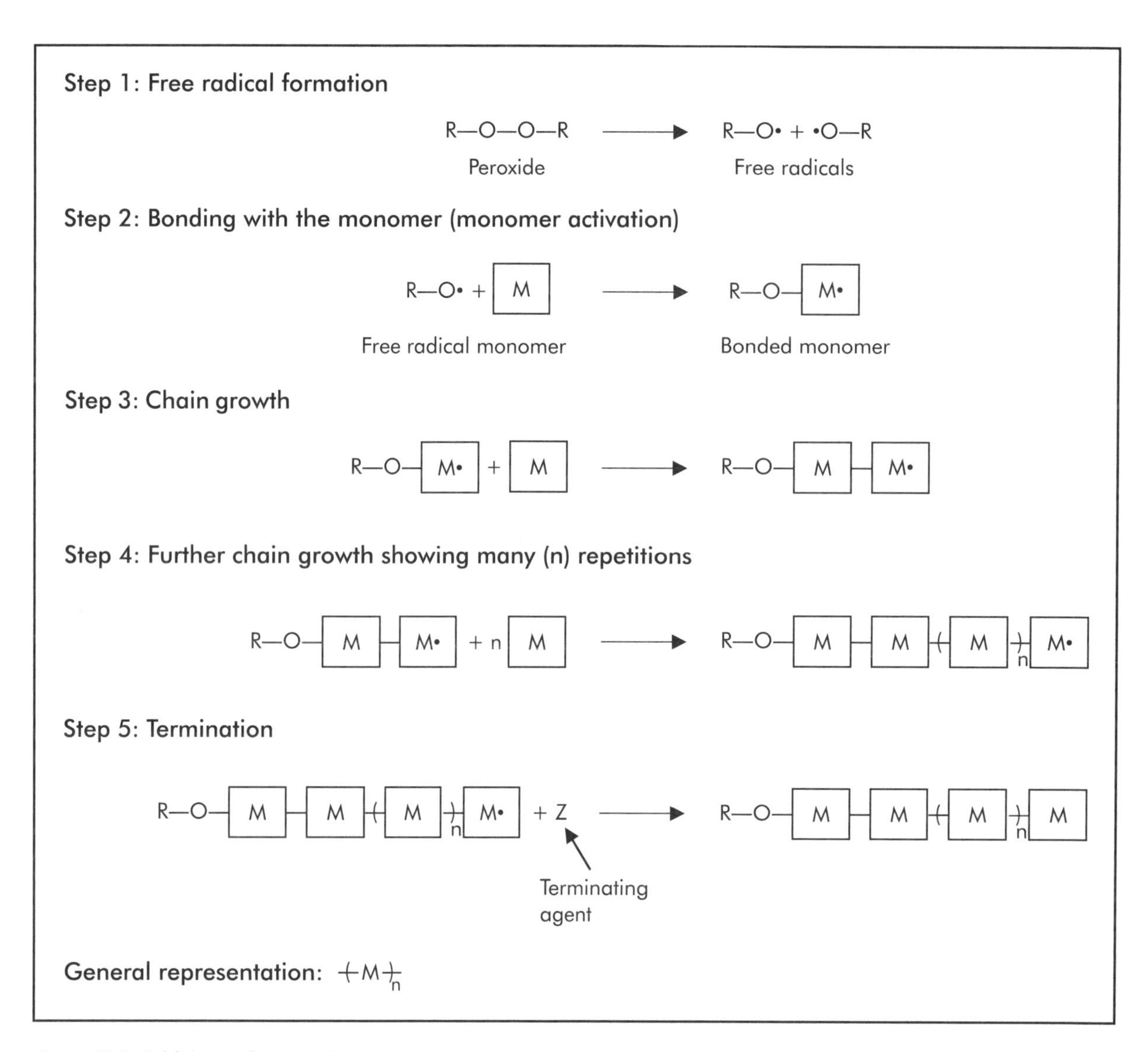

Figure 2-2. Addition polymerization.

minor constituents of the mixture. When the terminating agent is encountered by the activated chain, the activation is quenched and further reaction stops (Step 5, Figure 2-2). Care must be taken during the polymerization reaction to ensure that premature termination does not occur because of combination with unreacted peroxide, which is a powerful terminating agent. Therefore, the initial concentration of the peroxide is usually just high enough to initiate the chain reaction on several monomers, which each grow into separate chains, but not so high that the peroxide will compete with the monomers in the chain growth steps. Eventually, when most of the monomer is used up, termination from some natural or introduced termination agent occurs.

The general representation of the polymer, as shown at the bottom of Figure 2-2,

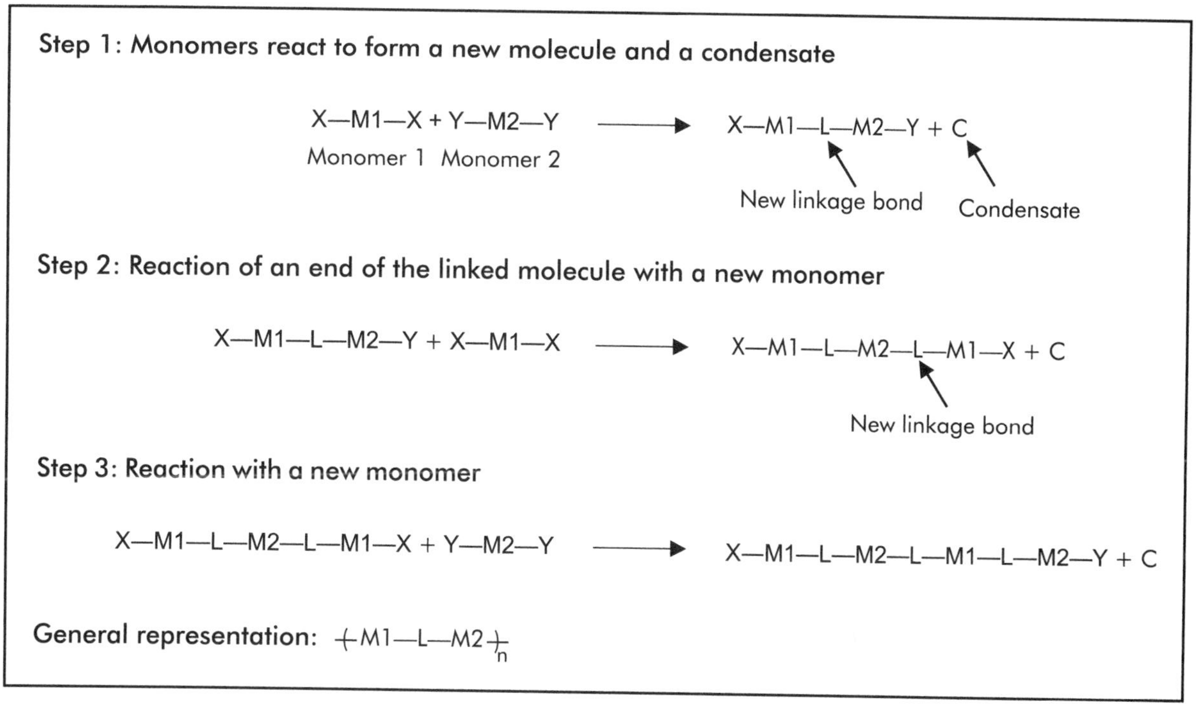

Step 1: Monomers react to form a new molecule and a condensate

X—M1—X + Y—M2—Y ⟶ X—M1—L—M2—Y + C

Monomer 1 Monomer 2

New linkage bond Condensate

Step 2: Reaction of an end of the linked molecule with a new monomer

X—M1—L—M2—Y + X—M1—X ⟶ X—M1—L—M2—L—M1—X + C

New linkage bond

Step 3: Reaction with a new monomer

X—M1—L—M2—L—M1—X + Y—M2—Y ⟶ X—M1—L—M2—L—M1—L—M2—Y + C

General representation: $\left[\text{M1—L—M2}\right]_n$

Figure 2-3. Condensation polymerization.

does not show either the original link with the peroxide on one end, or the link with the terminator on the other end. These end groups are usually of minor importance in the properties of the molecule and so are left out of the general formula. What is shown, however, is the nature of the basic monomer that has combined to form the polymer. The parentheses and horizontal extensions through the parentheses are meant to indicate that the monomer is bonded to other, identical monomers on either side. The total number of units bonded together is represented by n, a number that is usually an average number because polymers tend to have a statistical distribution of the actual number of monomers in each molecule.

Condensation Polymerization

The reactions that occur in condensation polymerization are quite different from those occurring during addition polymerization although, of course, both reaction sequences result in long polymer chains. In condensation polymerization, two different monomer types, M1 and M2, must be present. Each must have certain characteristics, which are present in all monomers that react by condensation polymerization. The first characteristic is that each monomer must have two active end groups, as shown in Figure 2-3 (the Xs and Ys). Usually, one monomer will have both ends that are one reactive group and the other polymer will have both ends of the second reactive group. The second characteristic of condensation monomers is that the active end group, X, of one monomer must react with the active end group, Y, of another to form a linkage element, L, which joins the two monomers (M1 and M2) into a linked molecule. A condensation by-product or condensate, C, is also formed (Step 1, Figure 2-3).

Note that the linked molecule resulting from this reaction (Step 1) has two active ends—an X and a Y. Therefore, the linked molecule can react with additional monomers of either type M1 or M2.

The reaction of the linked molecule with another monomer (of type M1 in the case shown in Step 2, Figure 2-3) again results in a combined state product in which a new linkage element forms and the new monomer is added to the previous linked molecule. Another condensate by-product is formed. The resulting product is referred to as a longer linked molecule as shown in Step 3, Figure 2-3.

The only new feature in the reaction sequence to be noted in Step 3 of Figure 2-3 is that the longer linked chain has both ends the same, X. Therefore, it can only react with type 2 monomers. This should not be a problem as both monomer 1 and monomer 2 are usually present in high concentrations. The reaction that occurs with the longer linked chain and the new monomer is analogous to the others. That is, a new linkage element is formed that joins a new monomer to the previous molecule and a condensate is created.

The sequence of steps shown in Figure 2-3 can continue as long as sufficient monomers are present to react with the ends of the linked chains. In practice, as the polymer gets longer and the concentration of monomer gets smaller, the probability of the linked chain encountering a new monomer and making an effective encounter is greatly diminished. This is especially true if the temperature is relatively cool since the movement of the linked chain can be slow and encounters between the active ends and monomers would be rare. Therefore, the polymerization reaction is usually conducted at elevated temperatures and in high monomer concentrations to improve reaction efficiencies. The mixture can be cooled to stop the reaction when the desired chain length is reached. It is also possible to quench the reaction by the addition of end-termination molecules.

Polymer Naming Systems

Names for polymers are largely dependent upon the type of the polymerization reaction used to make the polymer. The most common method for naming polymers made by addition polymerization is to simply combine "poly" with the name of the monomer, as with polyethylene and polypropylene, which are made from ethylene and propylene monomers, respectively. Polymers created by the condensation polymerization reaction are usually named by combining "poly" with the name of the linkage element formed when the monomers join together. Examples of this type are polyesters and polyamides, which contain the ester and amide linkages, respectively. A few polymers, especially those not formed by these two principal methods, are named for some unique characteristic of the molecules. Examples of this type are epoxy and silicone, because these molecules contain the epoxy group and the silicone group, respectively.

Polymer Characteristics

Since the addition and condensation reactions proceed by such different mechanisms, the characteristics of the polymers made by the two methods are quite different. The chain-reaction steps typical of addition polymerization often result in the formation of relatively few but very long-chain polymers. This is because of the low concentration of initiating free-radical used (hence only a few chains are initiated) and the highly reactive nature and rapidity of the chain-growth steps (causing the chains to grow very long). In contrast, condensation polymerization usually results in fewer and shorter molecules. Many chains begin in condensation polymerizations because any two monomers can react to begin new chains, thus creating a large number of chains, each competing for new monomers to add. Achieving very long chains was originally a problem with condensation polymerization reactions. However, the major producers seem to have solved this problem today.

The length of the chain is usually characterized by a quantity called **molecular weight**, which is a measure of the number of monomer units in the chain. Although somewhat difficult to quantify in terms of the actual number of monomer units present in any one particular chain, the molecular weight can be treated and understood statistically. The effects of molecular weight on the properties of polymers are evident and important. For instance, with all other factors being constant, higher molecular weight usually results in higher strength, stiffness, toughness, hardness, abrasion resistance, melt viscosity, and melting point, just to name a few of the most important properties. (A rough rule of thumb is that mechanical properties are improved by increasing molecular weight. However, in composites, the presence of the reinforcement may be even more dominant in increasing mechanical properties than molecular weight.) Most of these properties will be discussed in the chapters on mechanical properties. However, two of the properties—thermal behavior (especially melting) and toughness—are so fundamental and highly dependent upon the nature of the matrix (rather than the reinforcement) that they are considered in this chapter.

Solidification, Melting, and Other Thermal Considerations

The major difference between any liquid (including liquid polymer resins but also traditional liquids such as water) and its associated solid is the ability of the molecules in the liquid to move about with relative freedom while the molecules in the solid are more fixed in position. Liquids can be poured and will change shape to match the container, whereas solids resist changes in shape and do not need a container.

The most common method of changing from a liquid to a solid is by cooling the liquid until it freezes. This cooling removes heat energy from the liquid, which slows the motions of the molecules. Eventually the molecular motion (especially translational motion) becomes so slow that the forces of attraction *between* the molecules overcome many of the molecular motions (especially translation) and the molecules lock into set positions. Molecules attract each other but these attractions do not have the same strength as the primary bonds that hold the atoms of the molecules together as a molecule. Therefore, the attractions that cause the molecules to become a solid are usually called either **attractions** or **secondary bonds** to distinguish them from the primary bonds linking the atoms in the chains themselves. After becoming locked in place, the molecules are restricted in their lateral or translational movements, but their atoms can still vibrate and rotate to a limited extent. The temperature at which the molecules lock in place and the material becomes a solid is called the **freezing point**.

In strict definition, because a mass of polymer molecules varies so greatly in length, there is no *single*, sharply defined temperature that can be called the melting point or freezing point. This broadness of the melting/freezing point is especially true if the molecules are highly entangled and randomly arranged. Such polymers are said to be **amorphous**. When the molecules are arranged in orderly arrays, which are called crystalline regions (to be covered in more detail later), the melting/freezing point is sharper. The amorphous "melting point" is the temperature at which the material can be treated like a liquid in molding processes. For simplicity, the term **melting point** will be used for both the amorphous and crystalline types of polymers. It refers to the temperature at which the material becomes liquid-like.

If the frozen solid material were to be heated again, the molecules would again gain energy, first in greater vibrations and rotations, and then with increased lateral and translational movements. With continued heating the molecules would break loose from their intermolecular attractions and become free to move about. This is the melting point. Normally, the temperature at which the molecules lock in and break loose is the same. Thus, the freezing point and melting point are the same temperature. Remember, the melting/freezing points are actually regions of temperatures and the numbers reported are merely representative of the regions.

The melting point of a polymeric material is strongly dependent upon its molecular weight. As the molecular weight increases, the melting point also increases. This dependence can be understood by realizing that as the molecular weight increases and the polymer chains get longer, they will entangle or intertwine to a greater extent with neighboring polymer molecules, thus restricting the translational movements of all the molecules. Hence, more heat will be required to break loose these longer molecules since they must overcome both the natural intermolecular attractions and the greater entanglement. This higher amount of energy required to reach this disentangled and free movement state is seen as a higher melting point. Therefore, the solid region of a high-molecular-weight polymer extends to a higher temperature than does the solid region of a low-molecular-weight polymer. Simply put, as the molecular weight increases, so does the melting point. This relationship can be seen in Figure 2-4.

The **decomposition point** is another important thermal transition in polymers. At the decomposition temperature, sufficient heat has been put into the polymer so that the motions of the atoms (vibrations and translations) are great enough to break the primary bonds between the monomer units. When this happens, the polymer breaks apart and degrades. Rarely, however, does the polymer revert to monomers under these conditions. Because this breaking apart occurs at such high temperatures, the polymer is more likely to oxidize and degrade. The oxidation results in some off-gases and the formation of char, a charcoal-like material that is a non-meltable solid residue. A similar reaction occurs when coal or wood is heated excessively without catching fire to form charcoal. When polymers degrade, they no longer possess many of the desirable properties of the original polymer. Hence, care should be taken to not allow polymers to reach their decomposition temperature. The decomposition temperature is not highly dependent on molecular weight because the strengths of the bonds broken upon decomposition are the same whether the molecule is large or small. The decomposition temperatures for small and large molecules are shown in Figure 2-4.

Other thermal transitions and properties can be important in polymer matrices. The most important of these other properties is the **glass transition temperature**, often abbreviated as T_g. This thermal transition occurs in the solid phase and marks a change from a rigid solid to one that is more pliable. The motions of the atoms in a rigid, solid material below the T_g are generally simple vibrations in rigidly fixed positions. Above T_g but still below the melting point, the vibrations and rotations are more expansive and the atoms may move laterally for short distances. However, they are still locked into relatively fixed positions because of continued entanglements. Hence, between the T_g and the melting point, the material becomes somewhat pliable, although still a solid. In some materials the change in physical state can be quite apparent. The state above the glass transition is referred to in these materials as the "leathery state."

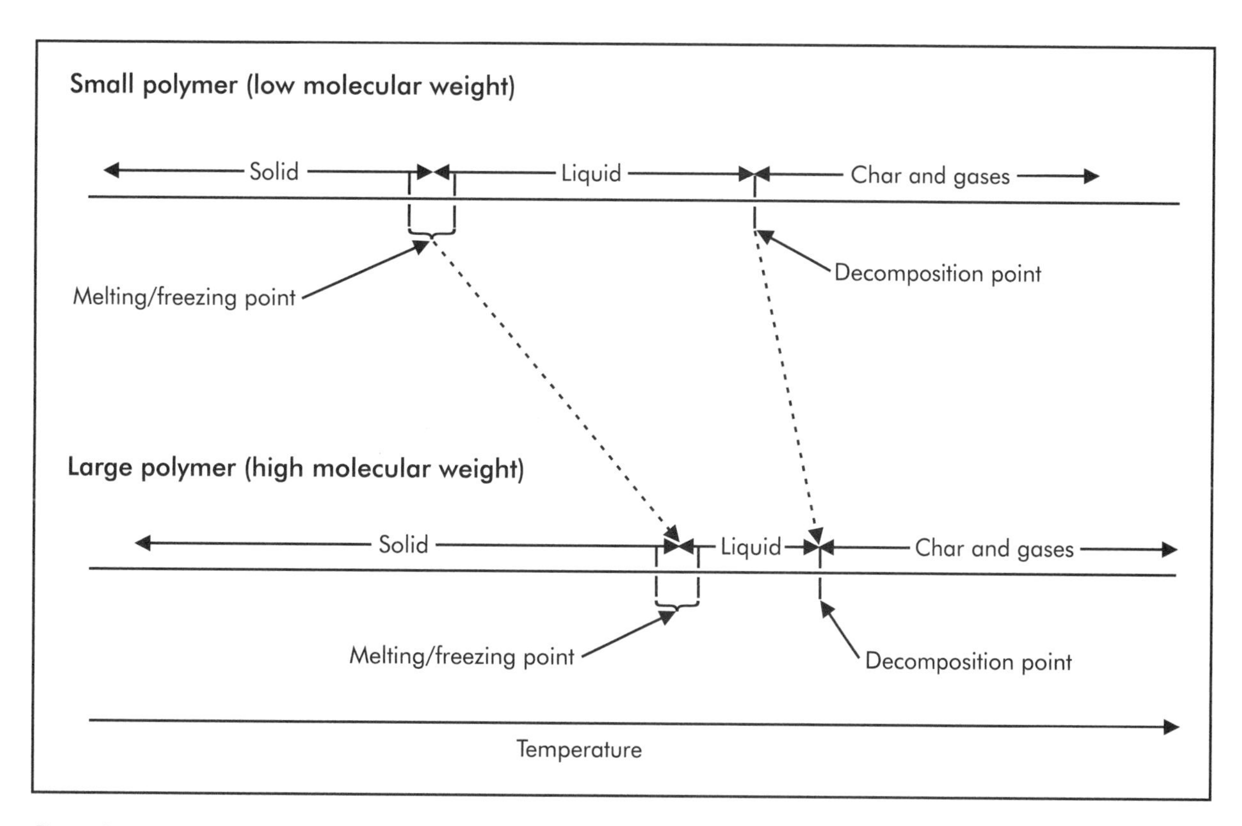

Figure 2-4. Temperature transitions for small and large polymers.

Crystalline materials do not exhibit a strong T_g transition because they are locked into position by the crystalline structure, which is not broken apart until the melting point. However, even highly crystalline polymers typically have some amorphous regions, and these regions exhibit a T_g transition.

A property closely related to the glass transition temperature is the **heat distortion temperature** (HDT), which is defined by the deflection temperature under load test (ASTM D648). (Note: ASTM is the acronym for the American Society for Testing and Materials—a society that supervises the establishment and dissemination of testing standards for materials.) While not a thermal transition in the same sense as the melting point or the glass transition temperature, the HDT is a convenient measure of the maximum use temperature for a polymer. This property is determined using a test apparatus that heats a sample of the composite material (5 × 0.50 × 0.25 in. [12.7 × 1.3 × 0.64 cm]) in a heated bath of mineral oil or some other liquid that is thermally stable at the temperatures to be used and will transfer heat readily to the sample. A weight (designated by the test procedure and dependent upon the type of plastic material being tested) is then placed on the sample. The entire apparatus is then heated, usually with stirring to ensure good heat uniformity in the liquid. The temperature at which the sample deflects (bends) a specified distance under these conditions is the HDT.

THERMOPLASTICS AND THERMOSETS

As has been evident from the discussion of the thermal properties of polymers and the preliminary suggestions about the mechanical properties (to be discussed in more detail later in this textbook), the interactions between molecules, such as entanglements and intermolecular attractions, have major effects on key polymer properties. Another type of interaction between molecules has such a strong effect on the properties of polymers that all polymers can be classified according to whether or not this interaction is present. The interaction is the formation of actual bonds between the molecules. These bonds between molecules are called **crosslinks**. Polymers in which crosslinks *are not present* are called thermoplastics and polymers in which the crosslinks *are present* are called thermosets. Following is a discussion of some properties of thermoplastics and thermosets and their differences, especially their performance under different thermal conditions. The crosslinks are bonds *between* the polymer molecules. They have approximately the same strength as the bonds between the atoms of the polymer molecule itself.

Thermoplastics are resins that are solids at room temperature. They are melted or softened by heating, placed into a mold or other shaping device when in the molten stage, and then cooled to give the desired shape. Several molding processes, such as extrusion, injection molding, blow molding, and thermoforming, are based on this type of resin behavior. Some of the common thermoplastics include: polyethylene, polypropylene, nylon, polycarbonate, polyethylene terephthalate (PET), polyvinyl chloride (PVC), acrylic, and acetal. Even after molding, thermoplastics can be melted by reheating, so the thermal environment of thermoplastic resins is a consideration in its use. If the temperature is too high, the product could undesirably soften, distort, and lose other key properties. Hence, thermoplastics become *plastic* (moldable) with heat.

Thermosets are resins that are usually liquids (or easily melted solids) at room temperature. They are placed into a mold and then solidified into the desired shape by a process other than freezing. The molded material is set in its shape by a heating process in which bonds are formed between the molecules. These bonds or crosslinks change the basic nature of the material. After the polymer material has been heated and the crosslink bonds formed, which is called **curing**, the thermoset material can no longer be melted. Hence, thermosets will not melt upon heating in contrast to thermoplastics, which are capable of being remelted. Therefore, thermosets become fixed (set) with heat.

A typical example of a thermoset reaction is the baking (curing) of a cake, where the thermoset polymer is the gluten in wheat flour. The flour is dissolved in water with other additives to make a batter. The liquid batter is then placed into a cake pan (mold) and baked (cured). After curing, the cake cannot be remelted. If it is heated again, the cake burns or chars (degrades).

Several molding processes, such as hand lay-up, spray-up, compression molding, filament winding, resin transfer molding, and pultrusion, are based on thermoset behavior. Some of the most common thermosets include: unsaturated polyester, vinyl ester, epoxy, polyimide, phenolic, and cyano-acrylate. Many of these thermoset resins are especially important in composites.

The vastly differing behaviors of thermoplastics and thermosets arise from a basic difference in the chemical structure of the thermoset and thermoplastic resins. That significant difference arises because thermoset resins have sites along the polymer chain that can be activated to become reactive. These sites react in such a way that chemical bonds are formed between adjacent polymer molecules. Because these sites exist on all the thermoset molecules, the potential exists for the entire group of polymer molecules to become joined together, which is often, although not always, the result. The process of crosslink formation is called *curing*.

When crosslinks form, they restrict the movement of polymer molecules and the atoms within the molecules to an even greater degree than either intermolecular attractions or entanglement. However, below the glass transition temperature of the polymer, the molecules are locked in and the movements of the atoms within them are so restricted that the additional restrictions from the crosslinks are sometimes not apparent. As the temperature is raised, the crosslinks restrict the movement of the atoms such that when the temperature is reached that would have been the T_g in a non-crosslinked polymer (assuming the polymers are identical except for the crosslinks), the restrictions are so great that the normal flexibility achieved at T_g cannot be obtained. In other words, T_g is increased by the presence of crosslinks. As more crosslinks form, the molecules become ever more restricted and the heat energy required for movement continues to increase. Therefore, T_g will continue to increase as the amount of crosslinking increases. The examination of T_g has become a convenient method for following the extent of crosslinking in some polymers.

The behavior differences between thermoplastics and thermosets also can be understood in terms of molecular weight. When a crosslink forms, thus linking two polymer molecules, the molecular weight of the combined molecule becomes approximately twice as much as it was when the two were separate. The molecular weight continues to increase as more molecules are linked together. It is evident that with only a moderate number of crosslinks, many polymer chains can be joined together, resulting in a high molecular weight. The effect of this higher molecular weight is to increase the melting point, as shown in Figure 2-4 for low- and high-molecular-weight polymers. However, in the case of crosslinking, the molecular weight can become extremely high, increasing the melting point so much that it will actually occur above the decomposition point, as depicted in Figure 2-5. Therefore, the molecular weight increase is so large that a thermoset material will decompose before it melts. In other words, the thermoset is not meltable and will decompose if heated to a high temperature.

Another method of comparing thermoplastics and thermosets is to observe the viscosity profiles of the two materials as they are molded. Profiles of this type are shown in Figure 2-6 for a typical thermoplastic and a typical thermoset. Note that the axes of the graph are viscosity versus time or temperature. This double label on the x-coordinate merely reflects that the time and temperature are usually increased together, so the independent variable could be either one.

The initial viscosity of the thermoplastic is above the liquid-solid line, indicating that at room temperature the thermoplastic is a solid (high viscosity). Heating of the thermoplastic causes motion in the molecules and eventually results in melting, shown as the point where the thermoplastic curve drops below the liquid-solid line. The viscosity of the thermoplastic continues to drop as the temperature is increased as indicated in the thermoplastic curve. At any point along the curve, the temperature could be reversed and the material would move upward along the same curve. That is, the thermoplastic can be repeatedly heated and cooled, melted and solidified, without changing the basic nature of the material. This is suggested by the double-direction arrow associated with the thermoplastic curve.

The initial viscosity of the thermoset is usually just below the liquid-solid line. However, it may be an easily melted solid (just above the line). For both thermoplastics and

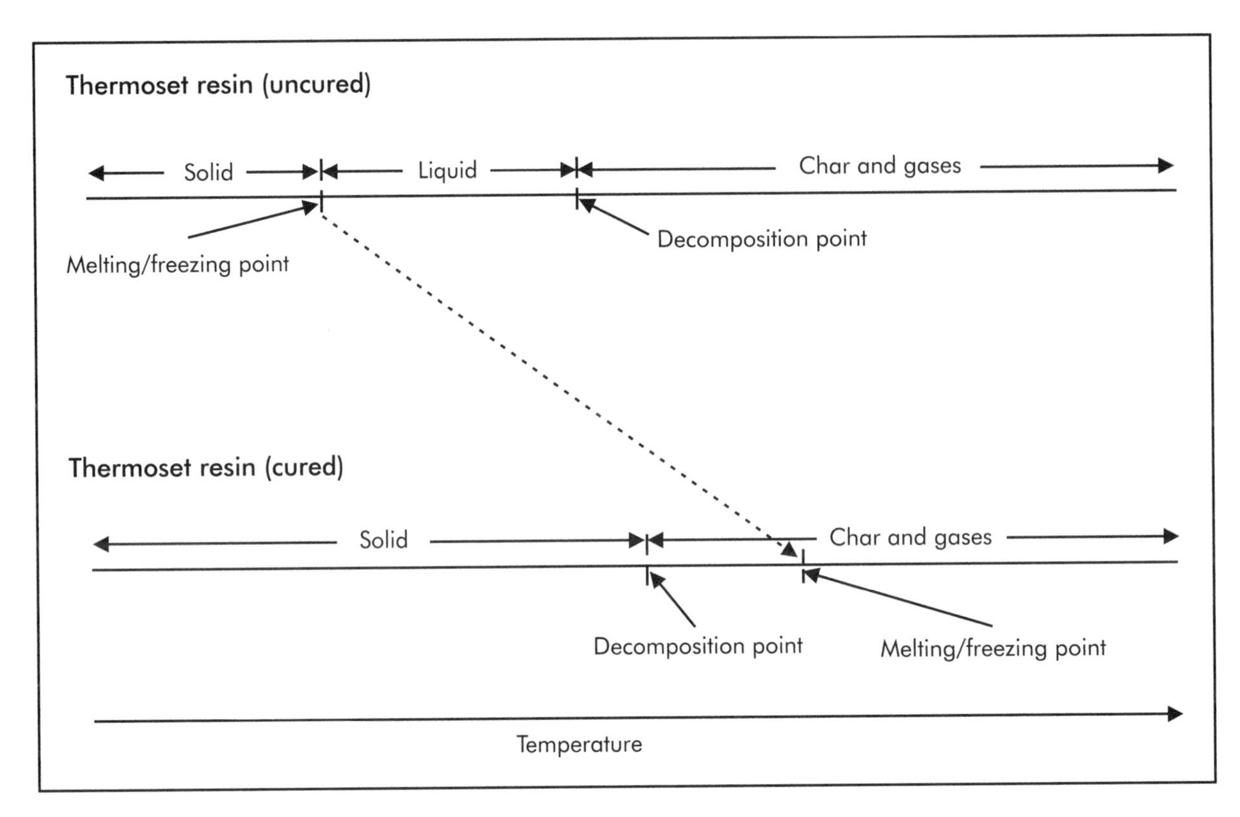

Figure 2-5. Changes in the melting point when curing a thermoset resin.

thermosets, the initial effect of increasing temperature is to lower viscosity. But the sharper drop in the thermoset line is because of the lower initial molecular weight of thermosets, which means that thermal effects are more pronounced. In contrast to the behavior of thermoplastics, the viscosity of the thermoset resin reaches a minimum and then rapidly increases to a position substantially above the liquid-solid line. The behavior of the thermoset line is actually a combination of two separate phenomena that add together to give the thermoset effect. These two effects are represented as two additional lines (given only for explanation purposes) in Figure 2-6.

As the temperature is initially raised, the molecules begin to move and melting/thinning occurs. This period is identified in Figure 2-6 as Region A and the curve representing the thinning due to temperature slopes sharply downward. This curve dictates the nature of the first part of the thermoset curve. With continued heating, the molecules start to react and crosslinks begin to form. These take some time to develop, thus giving the initial drop in viscosity from purely thermal thinning. The crosslinking line shows how the viscosity increases as the crosslinks develop (see Region B). Eventually the crosslinking curve crosses the liquid-solid line. At this point, the material is a cured solid. The summation of the temperature thinning curve and the crosslinking curve results in the initial drop and then the sharp rise in viscosity of the thermoset curve. The thermoset material is irreversibly changed from a liquid to a solid as indicated by the single-direction arrow associated with the thermoset curve. This is in contrast to

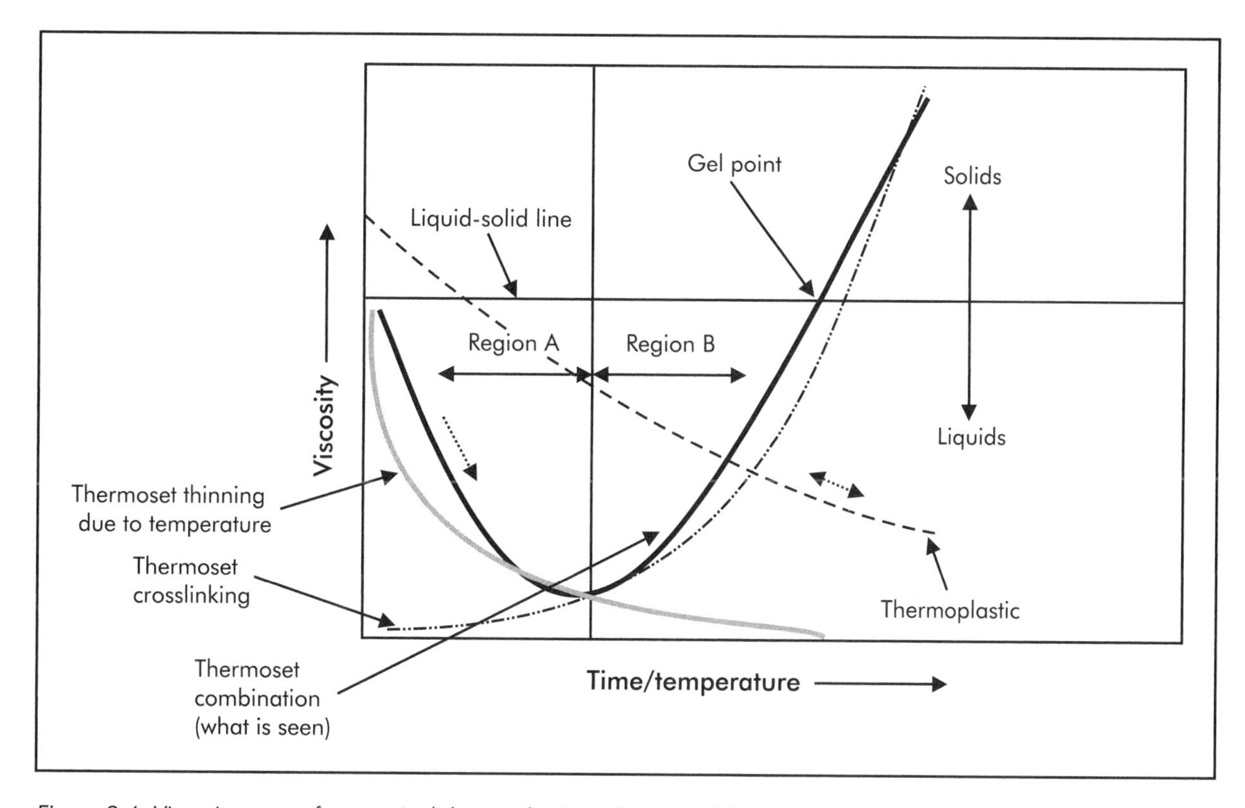

Figure 2-6. Viscosity curves for a typical thermoplastic and a typical thermoset.

the double-ended arrow associated with the thermoplastic line.

The degree of crosslinking is sometimes called the **crosslink density**. With most polymers, the theoretical number of crosslinks can be calculated. This then can be related to some thermal or mechanical property of the polymer that is easily measured, such as HDT or T_g. This allows other samples from production to be measured as a quality control standard to see if they have achieved the desired crosslink density.

AROMATIC AND ALIPHATIC MATERIALS

Before considering other matrix-dominated properties, a basic difference in polymers that affects many of the properties should be explained. That difference is the extent to which a polymer contains the aromatic group. The **aromatic** group and various representations of that group in polymers are pictured in Figure 2-7. The name "aromatic" is used because most of the molecules containing this group have a strong odor. Years ago they were called aromatic because they were easily identified by their smell. However, not all chemicals with strong smells are aromatic in the structural sense.

The aromatic group is formed from six carbon atoms and attached hydrogen atoms, one per carbon. When not attached to another part of a molecule, the aromatic group is the benzene molecule. When attached to another part of the molecule, the hydrogen at the attachment point is removed. Three representations are given

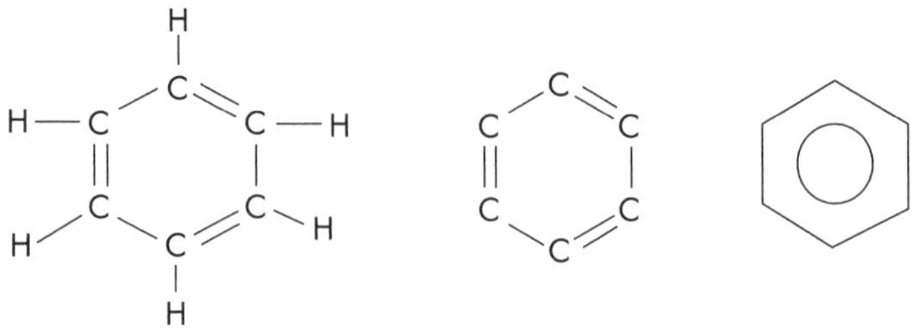

a. Aromatic group (benzene)

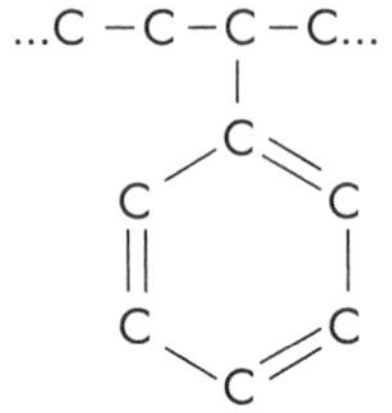

b. Polystyrene (pendant aromatic)

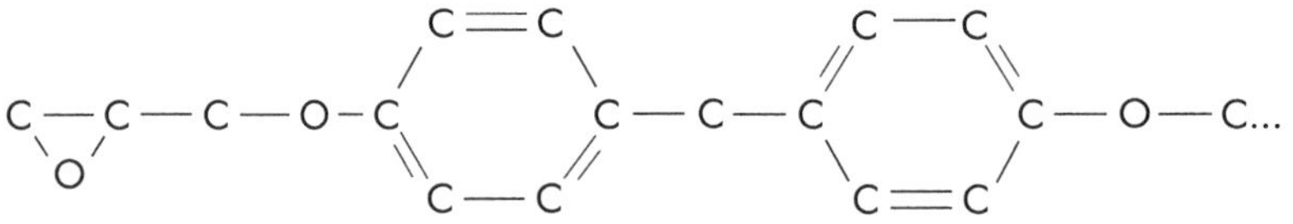

c. Epoxy (aromatic backbone)

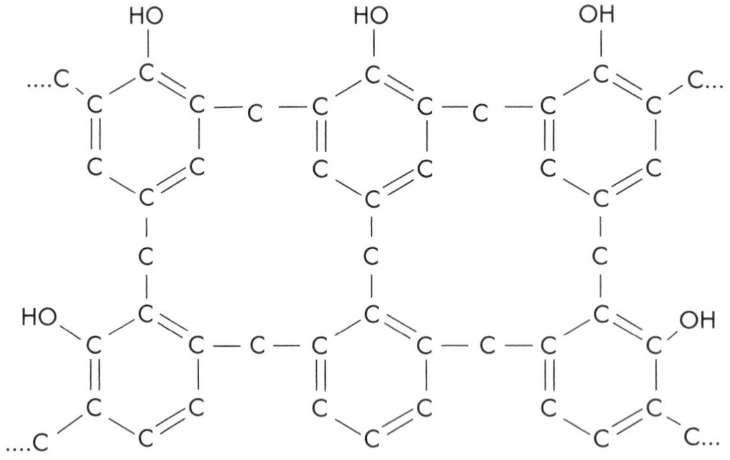

d. Phenolic (aromatic network)

Figure 2-7. Aromatic molecules.

in Figure 2-7a and all three are equivalent. The later two representations are merely shorthand versions of the first.

Because of its physical nature, the aromatic group imparts special properties. It is a rather large group of atoms, very stable and resistant to most of the chemical processes that a polymer might encounter in use. Because of its size and shape (flat and stiff), the aromatic group imparts stiffness and strength to the molecules in which it resides. More aromatic groups usually mean more stiffness and strength.

The aromatic group can be attached to the polymer in several ways. One attachment method, shown in Figure 2-7b, is dangling from the polymer backbone. This is called the **pendant** attachment. Another method of incorporating the group into a polymer is shown in Figure 2-7c where the aromatic group is part of the backbone. A third method is when the aromatic molecules form a network. The stiffness and strength of the polymer are increased in all methods of attachment. However, the attachment as part of the backbone has a greater effect than the pendant, and the network of aromatic rings has more effect than either the backbone or the pendant.

Some polymers have no aromatic groups at all. These polymers are said to be **aliphatic**. "Aliphatic" simply means the absence of aromatic. As the number of aromatic groups increases in the polymer, it is said to have greater aromatic content. At some point, the aromatic content will become important in determining the properties of the polymer. At that point, the polymer is usually referred to as simply aromatic in nature.

Other polymer properties affected by the aromatic content are discussed in the following sections.

Matrix-dominated Properties

Most composite properties are dependent upon the combination of the matrix and reinforcement together. However, in some properties, the reinforcement will largely determine the property with little contribution from the matrix, while in others, the matrix will dominate. Therefore, it makes sense to discuss the matrix-dominated properties in this chapter on matrices and to discuss the reinforcement-dominated properties in the chapter where reinforcements are discussed in detail. Those situations when the property is strongly affected by both materials or when the interaction of the materials is critical to the composite's performance will be discussed in chapters on composite properties and fiber-matrix interfaces.

Even though composite properties are the primary focus of this book, an understanding of the properties of the non-reinforced matrix can be useful in predicting matrix-dominated composite properties. It is also useful in choosing which matrix material should be chosen for a particular application. Therefore, the properties for a wide variety of representative matrix materials (non-reinforced) are given in Table 2-1. The details of each of the matrix materials presented will be discussed in the next few chapters of this book.

The discussion in the previous chapter of the role of the matrix as a component of a composite indicated that the matrix must protect the reinforcement from the environment. Some of the environmental conditions that might affect composite properties include the following: heat, solvents, staining agents, gases, fire, electricity, and light. Another key property that has a strong relationship to the matrix is toughness, although this property is different from the others because it is also strongly dependent upon the reinforcement. Each of the environmental conditions or threats is listed in Table 2-2 and will be addressed in the following sections. If the matrix is the dominant factor in each of these properties,

Table 2–1. Properties of various matrix materials (non–reinforced).

Property	Polyesters	Epoxies	Polyimides	Phenolics	Thermoplastics	Ceramics	Metals
Maximum use temperature, ° F (° C)	175–285 (79–141)	200–350 (93–177)	400–600 (204–316)	300–400 (149–204)	340–450 (171–232)	1,500–4,000 (815–2,204)	1,000–2,000 (538–1,093)
Flexural strength, ksi (MPa)	8–23 (55–159)	45–55 (310–379)	15–25 (103–172)	7–14 (48–96)	14–18 (96–124)	6–10 (41–69)	—
Flexural modulus, Msi (GPa)	.5–.6 (0.3–0.4)	2.0–2.4 (1.4–1.7)	.5–.8 (0.3–0.6)	1.0–1.2 (0.7–0.8)	.4–.5 (0.3–0.4)	—	—
Density (g/cc)	1.1–1.2	1.2–1.25	1.3–1.4	1.3–1.4	1.1–1.2	1.8–2.5	2.0–6.0
Tensile strength, ksi (MPa)	3–13 (21–89)	5–15 (34–103)	6–13 (41–89)	5–9 (34–62)	5–20 (34–138)	4–20 (28–138)	7–70 (48–482)
Tensile modulus, Msi (GPa)	.3–.6 (0.2–0.4)	.6–.7 (0.4–0.5)	.5–.6 (0.3–0.4)	.8–1.7 (0.6–1.2)	.3–.7 (0.2–0.5)	.5–17 (0.3–12)	7–70 (5–50)
Izod impact, ft–lb/in. (kJ/m)	.2–.4 (3.2–6.4)	.2–.3 (3.2–4.8)	1.0–1.5 (16.0–24.0)	.2–.6 (3.0–9.6)	.4–1.3 (6.4–21.0)	<1 (<16.0)	1.5–2.2 (24.0–35.8)
Moisture absorption, %	0.15–0.60	0.1–0.7	1.1–1.2	—	0.01–0.3	—	—
Elongation, %	1.4–4.0	—	1.1	0.5–0.8	—	—	—
Coefficient of thermal expansion ($\times 10^{-5}$)	—	6–7	3–5	3–5	3–5	0.5–0.6	0.6–1.2

then the specific nature of the matrix itself is likely to be a primary determining factor in how the material responds to the various environmental conditions. A general discussion about the natures of the polymers and how they might affect properties are useful and will be provided in this chapter. The specific details will be discussed when each particular polymer is introduced in later chapters.

Thermal Properties

The most important matrix-dominated properties are clearly those associated with thermal conditions. The matrix might melt (if a thermoplastic) or char (if a thermoset) when the temperature of the composite material is raised excessively.

As previously discussed, when interactions between molecules are full bonds or

Table 2-2. Environmental agents that can strongly affect the matrix.

Environmental Agent	Possible Effects on the Matrix
Heat	Thermoplastics are melted with the melting point being dependent upon the specific nature of the polymer, including the molecular weight and the amount of secondary bonding. Thermosets are softened but not melted. Excessively high temperatures can result in degradation (decomposition) of thermosets and thermoplastics.
Solvents	The degree of interaction of the polymer with the solvent is dependent on the chemical reactivity of the polymer and how similar it is to the solvent. In general, highly polar molecules react strongly with highly polar solvents.
Staining agents	Polar staining agents will have stronger interactions with polar matrix materials and the resulting stains will be more strongly embedded.
Gases	Oxidation can readily degrade some polymers and the effect increases with the reactivity of the bonds along the backbone. This effect is enhanced if the permeation of the gas into the polymer is high. Permeation is increased when the chemical nature of the gas is similar to the chemical nature of the polymer. Permeation is also increased when the atoms in the polymer are further apart. Thus, crosslinking reduces permeation as does crystallinity.
Fire	Fire-retardant properties depend upon many polymer factors. The most common ways to improve fire-retardant properties are to increase aromatic content, add fillers, and add halogens.
Electricity	Almost all polymeric materials have high electrical resistance, high dielectric strength, and low conductivity. Changes in polymer type result in minor changes in these properties, but all are significantly different from those of metallic materials.
Light	Ultraviolet (UV) radiation can degrade polymers. Aromatics are somewhat more sensitive to UV light than aliphatics.

crosslinks, for example, in thermosets, the melting point is raised above the decomposition temperature and no melting occurs. Within the realm of thermoplastics, several types of interactions less permanent than crosslinks can also occur. In general, any interaction or association between molecules will increase the melting point of the polymer. One such interaction is entanglement or intertwining, which is typical of polymers and increases as the molecular weight of the polymer increases. Other intermolecular interactions, such as secondary bonding, the most common type being hydrogen bonding, will also raise the melting point. Secondary bonds are broken by heating and are not, therefore, of the same strength as crosslinks.

Some polymers pack closely together in the solid state and form structures that are analogous to crystals in ceramic and metal materials. These regions of close packing are therefore called **crystalline regions**. The crystalline regions are held together by secondary bonds, or crystalline bonds, which increase the melting point in these regions. Most polymers having crystalline regions are thermoplastics; this is because the crosslinks in thermosets tend to interfere with the packing of the polymers. Molecular motion in a crystalline region is highly restricted. Therefore, the type of molecular motion needed for a glass transition is missing within the crystalline region. The crystalline part of a group of molecules would not have a T_g, but other regions in the same molecular mass where no crystallinity exists could have a T_g. Such mixed polymers are called **semi-crystalline**. The regions of non-crystallinity are called **amorphous regions**. Figure 2-8 shows representations of the crystalline and amorphous regions.

The presence of aromatic content increases the melting point of most polymers. This is simply understood because more thermal energy is required to get the large, stiff, aromatic group to move freely as it must in a liquid. Hence, the melting point is higher. Decomposition points are lowered slightly

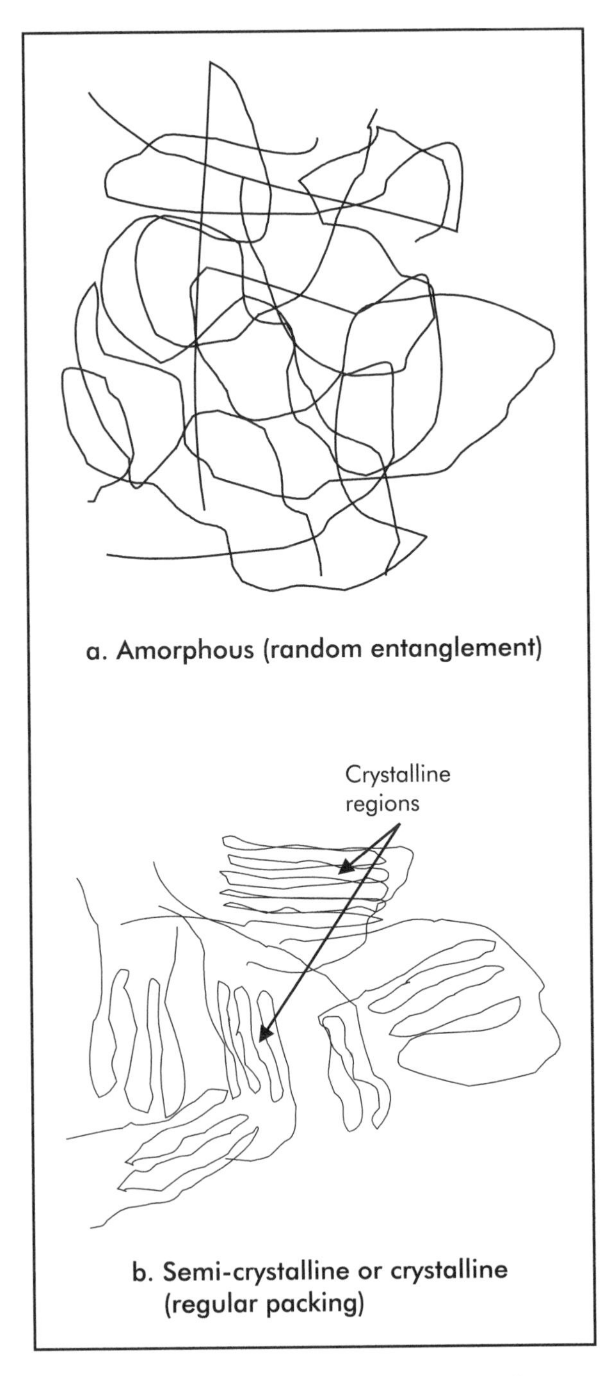

Figure 2-8. Representations of crystallinity in polymers.

by aromatic content because the bond where the aromatic group is attached to the polymer is slightly less stable than most of the other polymeric bonds.

Resistance to Solvents and Water

Resistance to chemicals depends upon the chemical nature of the matrix. In general, the more the chemical nature of the polymer resembles the environmental chemical that is interacting with it, the more the effect of the environmental chemical. The chemical natures of materials are often complex and, therefore, difficult to describe in brief. However, one aspect is so important that it can sometimes be the only aspect of the chemical nature that needs to be considered. That aspect is **chemical polarity**.

Chemical polarity, or simply **polarity**, is a measure of how evenly the electrons are distributed throughout the molecule. In some molecules the electrons are naturally drawn to particular atoms within the molecule. Molecules with an oxygen or nitrogen atom will often be polar because electrons are often drawn to the site of these atoms. When some site within the molecule is rich in electrons, some other site must be scarce since the total number of electrons in the molecule is a constant. The electron-rich sites are negative and the electron-poor sites are positive. When this polarity exists, two polar molecules will have strong attractions because the negative site on one molecule will be attracted to the positive site on the other molecule. Therefore, a polar solvent molecule can react strongly with a polar polymer molecule because the positive site on one molecule will be attracted to the negative site on the other. The stronger the attraction, the more the polymer's properties are changed by the solvent. If the attraction is slight, the interaction might only be seen over long times or at elevated temperatures. Raising the temperature swells the polymer and allows the solvent molecules to penetrate further, thus increasing the interaction between polymer and solvent. Stronger interactions could result in swelling, even at room temperatures, and the loss of some key properties. Even stronger interactions could result in the polymer actually dissolving, in which case it would cease being a matrix for the composite. The polymer could also react with the solvent to form bonds, thus changing the basic nature of the polymer; this usually results in significant deterioration of key properties of the composite.

Water is a strongly polar molecule. Therefore, polar matrix materials, such as unsaturated polyesters, are more susceptible to water attack than non-polar molecules, such as polypropylene.

The rule of similarities between polymer and solvent also applies to aromatic polymers. Therefore, highly aromatic polymers are readily attacked by highly aromatic solvents. Other factors can also come into effect, but the rule of similarities is good as a first indication of whether a material will be attacked by a particular solvent.

Some thermoplastics, especially those associated with high-performance applications, are especially solvent resistant. This resistance arises because these materials have compact polymeric packing; that is, the polymeric molecules are tightly packed with many intermolecular interactions, but not crosslinks as these molecules are still thermoplastics. Further, the molecules are often highly aromatic and have great stability. Therefore, there is little capability for solvents to attack, except in the case of the most aggressive aromatic solvents and then only under hot conditions.

The use of composites for their chemical resistance is widely applied and a rapidly growing market. Some industries, such as pulp and paper manufacturing, oil and gas transport and refining, chemical processing, and waste-water treatment are major users of composite pipes, fittings, and tanks. These industries rely on the corrosion resistance of composites as well as their strength and relatively low cost. The choice of matrix resin is largely dependent on the type of chemical environment expected in the application. Most of the chemical-resistant resins are categorized as "moderate" as to the degree of chemical attack they can withstand. Many resins can adequately meet the required standards. About one-third of the total tank and pipe market requires premium resins where chemical attack is strong; thus special resins with specific resistances are required. Specifications for use in these applications often require that special tests be passed by the intended composite material.

Staining agents have interactions with polymers that are much like solvents. Polar staining agents are more likely to interact with polar matrices. When they interact, the permanence of the stain is stronger.

Permeability and Resistance to Gases

The most important interactions with gases are those that cause chemical changes in the polymer from the chemical attack by the gas. In this regard, the effect of gases is much like the effect of solvents.

The most important reactive gas is oxygen. When oxygen reacts with the polymer, degradation usually occurs and the polymer experiences a significant reduction in useful properties. Some specialty chemicals, such as Irganox®, are highly effective in preventing or retarding polymer oxidation. These types of **antioxidants** are usually added at small concentrations (typically 1–2%).

Another interaction between matrix materials and gases is **permeability**, that is, the ability of a gas to pass into or through a material. This tendency is increased when the gas and the polymer have similar chemical natures. The increase in permeability is probably from a swelling that occurs from solvent-like effects between the gas and the polymer. Permeability is also increased as the space between the atoms in the polymer increases or as the size of the permeating gas decreases. This occurs simply because the gas molecules can work their way through the polymer more easily when the spaces are bigger. Therefore, crosslinking, which pulls the molecules tighter together, will decrease permeability. Crystallinity also decreases permeability.

Fire Resistance

The resistance of a polymer to fire depends on several factors. Some types of polymers burn more readily than others do. Most aliphatic polymers burn readily while aromatic polymers burn with greater difficulty. Highly aromatic molecules tend to burn slowly, if at all, and form chars when they burn. Those chars resist further burning. Therefore, if the non-flammability of a material is a desired property, that property can be enhanced by adding aromatic content. If the polymer desired is aliphatic, and even for some aromatics, some other method of imparting the flame-retardant property is needed for many applications. The most common method of enhancing fire retardation is to include in the composite mixture compounds containing the **halogen** atoms—fluorine, chlorine, bromine, and iodine. The halogens unite with hydrogen during the burning process to create a dense gas that smothers the flame. The problem with suppressing a flame by using halogens is that the gas created is hazardous and optically dark and dense. Nevertheless, halogen materials can be beneficial in reducing flame spread and are used extensively.

Halogens can be added to resins in several ways. These are understood by considering the various methods of adding halogens to a polyester resin mixture. One method is by bonding halogen atoms to one or more of the components of the polyester resin. Any of the components of the polyester mixture—diacid, glycol, or solvent—can be the source of halogenated atoms; all will work in reducing flammability. (This is discussed more extensively in the chapter on polyesters.) Halogens also can be added as fillers, such as magnesium chloride or calcium bromide. Some materials, especially those containing phosphorous, enhance the effectiveness of halogen-containing materials in suppressing flammability; thus they are frequently added to resin mixes along with the halogens.

Another method to reduce flammability is to use some non-halogenated fillers that have other beneficial properties that fight fires. The most common of these fillers is alumina trihydrate (ATH), also called alumina, aluminum oxide, and hydrated alumina. ATH is a molecule in which three water molecules are loosely bound. When heated, these water molecules are released and can help to suppress a flame by cooling and smothering it. ATH is relatively inexpensive and has the further advantage that it does not contribute to the toxicity of the smoke given off.

Electrical Properties

The electrical properties of a polymer are most often important in the electronics industry where composites are used as circuit boards or other components or sealants. Almost all polymers are nonconductive and have high electrical resistance and dielectric strength. Some minor electrical property changes result from the chemical nature of the polymers. However, all are significantly better in electrical resistance than metals and have been used without significant modifications regardless of the type of polymer.

Some electrical properties, such as arc resistance, are improved by the addition of certain fillers. For instance, ATH, which is used as a flame-retardant filler, will substantially raise the resistance of a composite material to surface arcs. Other powdered minerals can be added to increase dielectric properties.

For some applications, the nonconductive nature of composites is a problem. One important example of this is on the surface of airplanes where conductivity to prevent damage from lightning strikes is necessary. Composite airplane skins can be made conductive by adding metallic powders as fillers, including metal-coated reinforcements, or including a wire mesh within the composite laminate. All of these methods are used commercially.

UV Resistance and Optical Properties

The most important interaction of light with the polymer matrix is from the ultraviolet (UV) component of light. The UV range of frequencies is just higher than visible light and is a major component of sunlight. UV frequencies are near the natural frequencies of the electrons in the chemical bonds along the polymer backbone. Therefore, the interactions between this UV light and the electrons are strong, often resulting in excitation of the electrons and a resultant breaking of the bond. Hence, UV light can degrade polymers.

The nature of the atoms has some effect on the tendency of the electrons to become excited by the UV light and degrade. Generally, aromatic polymers are more easily degraded by UV light than are aliphatic polymers. Some applications may require a particular resin to be used because of cost reasons or for its particular physical properties. If the polymer has poor UV resistance and the application will subject the resin to UV light, some chemicals can be added to the resin mixture to enhance UV stability. These chemicals are called **UV inhibitors**. A typical UV inhibitor is Tinuvin®.

Other properties that pertain to light are relatively unimportant in composites. For example, whereas many plastics are clear or translucent, most composites are not because of the interaction of the reinforcements with the incident light or because of the frequent inclusion of fillers in the composite mix. Usually, the use requirements of the polymer are that the material be opaque and that the reinforcements not be seen. Opacity is usually accomplished with filler or a finely woven reinforcement material. However, it also can be achieved by adding pigments to the polymer matrix.

Toughness

Toughness is a key property for many composite applications. It is related to strength and therefore strongly dependent upon the nature of the reinforcement. However, toughness is also strongly related to the ability of the matrix to absorb energy. Therefore, toughness is dependent on the combination of both matrix and reinforcement properties.

Toughness is a measure of the ability of the material to absorb energy. The most important type of toughness is **impact toughness** or the ability to absorb an impact, such as from a falling or striking object. The ability to absorb energy, without breaking the matrix material or the fiber reinforcement, depends upon the ability of the matrix to quickly diffuse the energy of the impact, and the ability of the matrix and fibers to withstand the accumulation of energy at a particular point. This latter property is related to the strength of the composite which, as has been already pointed out, is one of the factors that affects toughness. But the other property, the ability to quickly diffuse energy, is of most importance here because it is highly dependent upon the matrix.

The matrix can diffuse energy by converting the impact energy to some other energy form, preferably heat and motion. This is done when the matrix stretches, rotates, vibrates, or otherwise moves in response to the impact. Hence, the more efficient the movement and the spreading of the movement widely among the atoms of the molecules, the more likely the material is to diffuse the energy, and thus prevent a localized accumulation, which may be sufficient to break the matrix or the fibers at some specific location.

The best situation is to have both the matrix and reinforcements spread the energy. As will be discussed in the chapter on reinforcement types, aramid fibers are the choice when toughness is a premium. However, for some applications, other fibers have to be used because of such requirements as stiffness or cost. Therefore, changes in the toughness of the matrix are the key to improving the toughness of the composite.

A key matrix property that affects toughness is elongation, that is, the ability of the material to stretch when a force is applied. Polymers can be improved in their elongation in several ways.

One method is simply to use a polymer that has a flexible backbone. Polymers of this type are usually aliphatic with few secondary bonds to neighboring molecules and few pendant groups. In traditional resins, the elastomers or rubber materials are of this type and they obviously have high elongation or stretch. One problem is that strength and stiffness are often decreased significantly and so other critical polymer properties, such as the ability to pass loads onto the reinforcements, might be compromised.

Another way to improve elongation is to mix some rubber or elastomeric material into the polymer resin. These materials are called **tougheners** or **toughening agents**. Elongation is usually improved but the resulting material is often two phases and becomes highly susceptible to attack by chemicals and UV light.

Yet another method to improve elongation is to decrease the amount of interpolymer interactions so that elongations can proceed more freely. In thermosets, this usually means reducing the number of crosslinks. While this may improve toughness, it also decreases the thermal stability of the material, its solvent and gas permeation resistance, and some key mechanical properties. Therefore, some limitations exist in this method. Of course, thermoplastics have no crosslinks so they naturally have higher elongation than do thermosets. Therefore, the majority of thermoplastic composites find use when toughness is an important property.

The most sophisticated and probably the best method to increase toughness in a polymer is to modify the nature of the polymer so it has a naturally high elongation without compromising (much) the other properties. Some thermoset materials have incorporated aliphatic segments within an otherwise aromatic molecule to meet toughness standards. This type of molecule seems to have the balance of properties required for many applications. It is also possible to mix highly elastomeric substances (rubbers) into the polymer matrix to achieve improved toughness.

WET-OUT OF FIBERS

The matrix cannot perform its duties in the composite unless the fibers have been well wetted. Dry fibers do not receive the transfer of loads that must occur for the fibers to do their job. Therefore, a critical function of the matrix is simply to wet and bond to the fibers and form a continuous phase that interconnects all parts of the composite.

The **viscosity** of the resin is the single most important factor in **wet-out**. If the viscosity is too high, this means the resin is too thick; it cannot penetrate the fiber bundles, which are like groups of filaments in a string or yarn. Therefore, control of resin viscosity is an important parameter to achieving success in composite preparation.

One problem immediately encountered is that as the molecular weight of the polymer increases, so too does the viscosity. This is especially important in thermoplastics because they do not change their molecular weight as they are processed. (Remember that thermoplastics do not undergo any chemical reaction when they are molded; rather, they are melted, put into a mold, and then cooled to solidify.) As a result, wet-out of the fibers with thermoplastics can be difficult. This is a major limitation in the use of thermoplastics and one that is addressed in detail in the chapter on thermoplastic composites manufacturing processes.

Some thermoplastic (and even some thermoset) resins are applied from a solvent, which gives good wet-out, but also requires that the solvent be extracted either before or during cure. Not only does this require additional processing steps, but the health and environmental problems associated with the presence and recovery of the solvent often make this method unpractical.

Another method of wetting the fibers with thermoplastics is to apply it as a solid and then achieve fiber wetting during the molding stage where the temperature of the material is higher and the thermoplastic is a liquid. Several methods of introducing the thermoplastic have been tried, including dusting the fibers with thermoplastic powder, pressing a thermoplastic sheet onto the fibers as they are arranged in their layered structure, and intermingling thermoplastic fibers with the reinforcement fibers. All of these methods have advantages and disadvantages and all are currently used commercially. As the molecular weight and aromatic content of the thermoplastic resin increases, the viscosity also increases. At high molecular weights or aromatic contents, the viscosity of some thermoplastics is so high that wet-out is achieved only with great difficulty. However, the excellent properties that can be achieved with high-performance thermoplastic resins gives incentives to companies to resolve the wet-out problem.

Because thermosets have a low molecular weight before curing, they are ideal for use with composites because the fibers are wetted and then the material is cured. This fact alone may have led to the overwhelming use of thermosets with long fiber composites. (The short fiber composites can be wetted more easily and therefore thermoplastics are often used.) Even with the ease of fiber wetting using thermosets, care must always be taken to achieve good wet-out. Fortunately, the color or sheen of the reinforcement fibers often changes when they are wetted, thus assisting the molder in knowing when wetting has occurred.

ADDITIVES: FILLERS, PIGMENTS, VISCOSITY CONTROL AGENTS, SURFACE AGENTS

Additives are the materials added to the basic composite materials (resin system and reinforcements) to improve some specific property. Most additives are present in only small concentrations; although one, fillers, can be a major component of the mixture. Nevertheless, even when present in relatively high concentrations, additives are not considered to be essential components of the mix, at least not to the same extent as the resin and fibers. Note that in this sense, the term "resin system" includes the resin, solvent, and cure system, which are all essential to the basic formation of a thermoset material.

Some additives have already been discussed in this chapter as they related to specific matrix-dominated properties. Additives include flame-retardant materials, such as halogen-containing resins or special fillers, electrical property enhancers, UV inhibitors, and tougheners. Many other additives are used with matrix systems. Some are general and find wide use, while others are used only with one particular resin. The general types are considered in this section and those for a specific resin type are considered when that resin is discussed later.

Fillers

Fillers are solid materials ground to fine powders and added to the resin mix to reduce overall cost and, occasionally, to impart some other beneficial property. The most common fillers are common minerals, such as limestone and talc, which have been purified and ground. Fillers usually have no particular length-to-width ratio and are not, therefore, fibers, but are, instead, particles. Fillers usually cost much less than the resin and therefore any combination of the resin and filler will be less costly in terms of total weight or total volume than the pure resin itself. Cost reduction is usually the most important reason to use fillers in composite materials.

Fillers are usually mixed into the liquid or molten resin. This can be done by the resin manufacturer or by a special compounder, such as a manufacturer of molding compounds, but is most often done in thermoset formulations by the molder. The most obvious effect of the filler is to thicken the resin mixture. For some applications this thicker mixture is desirable, such as when the material needs to have a particular viscosity for some molding operation. However, the higher viscosity usually reduces the ease with which fibers are wetted. Therefore, care should be taken in adding fillers to ensure that the resulting mixture will still wet-out the fibers. One method to ensure that the filler/resin mixture will still wet the fibers is to control the viscosity of the mixture so it stays within acceptable limits.

The effect on mechanical properties of adding reasonably low concentrations of fillers to the resin mixture is generally not large, especially in the presence of fiber reinforcements; although some increased stiffness or reduced strength and reduced elongation is common. Fillers can also impart some special benefits such as fire retardation, color, or dimensional control (called low profile and discussed in the section on molding compounds in the polyester chapter).

Pigments and Dyes

The color of a composite is affected by the way light is absorbed or diffracted by either the polymer, fibers, or additives in the material. Additives that cause specific light to be absorbed are called **colorants**. The colorants can be organic, based on carbon-containing molecules, in which case they are also known as **dyes**. Alternately, the materials can be ground, inorganic powders called **pigments**. Dyes and pigments are available in a wide range of colors and shades. Generally, dyes are more subtle in color than pigments but pigments are more intense (vibrant). Pigments are also more stable at high temperatures.

Viscosity Control Agents

Viscosity control is important for the proper molding of resin mixtures to make composite parts. For instance, if the viscosity is too low, the material might run down the side of the mold or drip off the fibers. Fillers might be added to thicken the mixture; but sometimes they cannot be used or, alternately, even with fillers the material is still too thin. In these cases, a **thixotrope** can be used to significantly increase the viscosity of the mixture with the addition of only a small amount. A common material of this type is fumed silica, which can give a significant increase in viscosity either by itself or with the addition of special thixotrope enhancers. This is accomplished because there is some secondary bonding induced between the polymer chains. These secondary bonds will decompose with heat and are therefore not important in the final product, but during processing they add considerable viscosity to the resin mixture.

An equally troubling problem is when a certain amount of filler is needed and the viscosity of the mixture is too thick. In this case, additives can be used to thin the material. Such thinning is especially important for fiber wet-out and spraying. Using viscosity reducers also allows higher filler levels to be used for purposes such as cost or flammability control.

Surface Agents

Some resins have a tendency to foam, thus complicating their application especially with spraying. Hence, additives are available that reduce the tendency of materials to foam. These additives usually work by changing the charge or the energy on the surface of the resin or filler particles in the mixture. Silicones are materials that can serve as **anti-foaming agents** of this type.

Another problem associated with the surface energy and charge is the tendency of some surfaces to not be wetted by the resin mixture. Often, this problem can be solved by adding a **surfactant** to the resin mixture. The surfactant changes the charge or energy of the surface to be coated and facilitates wetting by the resin. Surfaces that might have this problem include molds and fibers.

CASE STUDY

Toughened Resins

Some resins have been modified to become tougher. The results of this type of modification can be seen by comparing the extent of damage that occurs with a particular type of impact. A comparison of three materials clearly illustrates the effect of the resin in reducing the amount of impact. The results of a standard impact on three composites are presented in Table 2-3. As can be seen, the standard epoxy has more extensive damage than either the toughened epoxy or the thermoplastic. The toughened epoxy has been well formulated as can be seen from its superior performance against the normally excellent thermoplastic.

Table 2-3. Comparison of three matrix materials after a standard impact.

Resin Type	Standard Epoxy	Toughened Epoxy	Thermoplastic
Damage area, in.^{2} (cm^2)	3.5 (22)	1.5 (10)	2 (13)
Weight gain with water, oz. (g)	1.3 (37)	1.3 (37)	0.2 (5.7)

The chemical nature of the materials is assessed with the results of the solvent interaction test also shown in Table 2-3. The weight gain of water under a standard soak test is a measure of the amount of solvent interaction with polar solvents. As can be seen, the standard and toughened epoxies have the same amount of moisture pickup. This is an indication that their chemical nature is still about the same. The thermoplastic is obviously different, having much lower water absorption.

SUMMARY

The matrix can be called a resin and may even be called a plastic. Strictly speaking, a resin is a polymeric material that has not yet been molded to its final, useful form. A plastic is the material after it has its final shape. However, all the terms, resin, plastic, polymer, and matrix, are used interchangeably by many people.

Matrix materials dominate some properties of composites. The most important of those properties is associated with thermal transitions. The melting point or freezing point is dependent on the nature of the matrix and on the amount of interactions between molecules. As the intermolecular interactions increase, the melting point increases. The most common interaction is entanglement or intertwining, which is higher when the molecular weight is greater, when secondary bonding or intermolecular attractions are present, and when there is primary bonding or crosslinking. When crosslinking occurs, the molecular weight is dramatically increased. This will usually raise the melting point so high that it is above the decomposition temperature. Hence, crosslinked materials cannot be melted—they decompose first.

Matrix materials are often classified by whether they have crosslinks or not. Those with crosslinks are called thermosets and those without them are called thermoplastics. The thermosets cannot be melted after the crosslinks are formed. They are molded by mixing the resin, usually a liquid at room temperature, with various other additives and the fibers. Then the mixture is placed into a mold where it is heated so that the crosslinks form. The process of forming the crosslinks is called curing. Thermoplastics are usually solids at room temperature. They are melted, mixed with additives and reinforcements, and then placed into a mold and cooled to solidify. Thermoplastics can be melted again because no crosslinks are formed.

Other matrix-dominated properties are mostly associated with agents in the environment of the composite. The matrix dominates the resistance of the composite to solvents, stains, gases, fire, and light (UV).

Toughness is a property dependent upon both the matrix and the reinforcement. When a matrix can effectively diffuse the energy from an impact, toughness is improved. This ability to diffuse energy is closely associated with elongation of the polymer. Therefore, changes in the polymer to improve its elongation will improve toughness. These changes may include adding rubber material, modifying the polymer so it has segments that have high elongation, and reducing the intermolecular interactions, including the elimination of crosslinks (that is, using a thermoplastic).

The matrix cannot do its job if it does not wet the fibers completely. Therefore, viscosity control of the resin is important. If the matrix viscosity is too high for good fiber wetting, the heat can be raised, a solvent added, or some other modification made that will give good fiber wetting.

Several additives can improve the performance or cost of the resin mixture. Some of the more common additives are fillers, pigments, viscosity control agents, and surface treatment materials.

LABORATORY EXPERIMENT 2-1

Determining the Heat Distortion Temperature

Objective: Determine the heat distortion temperature (HDT) of a composite, which is a simple and effective method of assessing the maximum use temperature for a composite.

Procedure:

1. Make several composite samples by laying up sheets of fiberglass cloth with a room-temperature-cure unsaturated polyester, a room-temperature-cure epoxy, and two samples with a thermal-cure epoxy. Cure the thermal-cure epoxy at two different temperatures. Make sure the percent weight of resin and fiber are the same in all the samples.
2. Cut the samples into the specified dimensions (see American Society for Testing and Materials standard, ASTM D648).
3. Place a sample on the test apparatus and heat to the point of deflection as specified in the test. Note that the HDT apparatus can be made quite inexpensively. It can be contained in a battery jar or some other heat-resistant glass laboratory container. The supports, pressure rod, and deflection gage are readily obtained from laboratory supply houses.
4. The materials should all have significantly different HDTs. The higher cured epoxy should have a higher HDT than the same epoxy cured at a low temperature because of the higher number of crosslinks formed at the higher temperature. The HDT of the unsaturated polyester will typically be lower than the epoxy because of the nature of polyesters versus epoxies. However, a high amount of crosslinking could alter the order somewhat.

LABORATORY EXPERIMENT 2-2

Fiber Wet-out

Objective: Determine the differences in using two different types of resins—a thermoset and a thermoplastic—to make composites.

Procedure:

1. Obtain several sheets, about 12 × 12 in. (30.5 × 30.5 cm) of a standard fiberglass mat. Make several piles comprised of two layers each of the mat.
2. Obtain some unsaturated polyester resin of a standard viscosity. Pour the resin onto one of the piles of mat and work it into the fibers so they are fully wetted. Note the time and effort required to wet the fibers.
3. Obtain some thermoplastic resin, preferably polyethylene. This will be in pellets. Heat the pellets to melting and then pour the melted resin onto a pile of fiberglass mat. Work the resin into the fibers so they are fully wetted. Note the time and effort required.
4. Discuss the problems and advantages of each of the resin wet-out experiments.

QUESTIONS

1. Explain the differences between the terms resin, plastic, and polymer.
2. Explain why a crosslinked material cannot be remelted at high temperature.
3. Discuss the concept of chemical similarity between solvent and polymer and how that similarity might affect the solvent sensitivity of the polymer.
4. Give two factors that determine the flammability of a polymer and how those might be useful or not in actual products.
5. Explain how toughness can be improved in composites.
6. What can be done to improve wet-out of fibers if the viscosity of the resin is too high?
7. Why would the addition polymerization reaction be expected to create fewer but longer polymer chains than the condensation reaction?
8. Discuss how crosslinking might affect the following properties: toughness, gas permeability, and strength.
9. What is aromatic content and what are its effects on flammability and strength?
10. What is molecular weight? Discuss why it must be determined as a statistical average.
11. Explain the room-temperature condition of uncured thermosets (liquids) and thermoplastics (solids) in terms of molecular weight. What happens to thermosets and thermoplastics when they are cooled below room temperature? What happens to thermosets and thermoplastics when they are heated above room temperature (under curing conditions)?
12. Some thermoset materials are solids at room temperature. What is the likely cause of this condition, focusing on the molecular properties of the polymers? What happens when these materials are heated above room temperature?
13. Some thermoplastic materials, such as Teflon®, can only be melted with great difficulty because they begin to degrade when heated. Explain why this is so.
14. Indicate what is meant by a polar material.
15. What is an example of another property, besides molecular weight, which can raise the melting point of polymers?

BIBLIOGRAPHY

ASM International. 1987. *Engineered Materials Handbook,* Vol. 1, "Composites." Materials Park, OH: ASM International.

Society of Manufacturing Engineers (SME). 2005. "Composite Materials" DVD from the Composites Manufacturing video series. Dearborn, MI: Society of Manufacturing Engineers, www.sme.org/cmvs.

Strong, A. Brent. 1995. *Fundamentals of Polymer Resins for Composite Manufacturing*. Arlington, VA: Composites Fabricators Association.

Strong, A. Brent. 2006. *Plastics: Materials and Processing*, 3rd Edition. Upper Saddle River, NJ: Prentice-Hall, Inc.

3

Unsaturated Polyesters

CHAPTER OVERVIEW

This chapter examines the following concepts:

- Overview of polyester resins and their uses
- Polymerization of unsaturated polyesters
- Effects of various diacids, glycols, and specialized monomers
- Crosslinking mechanisms
- Crosslinking agents
- Cure control additives
- Molding compounds
- Property optimization

OVERVIEW OF POLYESTER RESINS AND THEIR USES

Unsaturated polyesters (UP) or, alternately and equivalently, **thermoset polyesters** (TPs) are, by far, the most commonly used thermosetting resins for composites. They are used in applications such as boats, corrugated sheets, and golf carts. This widespread application is due to many factors, not the least of which is low cost. Unsaturated polyesters are commonly priced about 25% less than vinyl esters and about 33–50% less than epoxies.

In addition to their cost advantage, polyester thermosets have advantages in ease of cure, ease of molding, a wide range of property possibilities, and a wide experience base in developing technologies and design parameters. However, polyester thermosets have several drawbacks or disadvantages in properties and manufacturing such as relatively poor durability, brittleness, and air pollution difficulties. In general, however, the advantages seem to outweigh the disadvantages and polyesters continue to be the resins of first choice for many composite products.

Several companies make polyester resins and therefore have vested interests in their continued marketplace dominance. These companies are active in improving polyesters by creating new permutations of the basic building blocks used to make them and by inventing new building blocks from which they can be made. It is just this possibility of widely varying properties that originally led to the adoption of unsaturated polyesters as the resin on which the composites industry was built. Other resins may continue to take portions of the composite market because of some specific property or combination of properties that uniquely fit a particular application. But these markets are not detrimental to the total market volume of unsaturated polyesters because of the overall growth of the composites industry, most of which is based on polyester thermosets as the resins of choice.

Polyester thermosets have become so prevalent that references to broad categories of composites, such as fiberglass reinforced plastics (FRP) or more simply just fiberglass (which seems confusing since this name does not address the existence of a matrix), and sheet molding compounds (SMCs) automatically imply that the resin is a thermosetting polyester.

No list of uses of polyester thermosets can possibly be complete, but a short list can give some appreciation for the scope of its market. Essentially every brand of automobile and nearly every automobile model use polyester thermoset parts; some automobiles use thermosets for almost their entire body structure. Most boats, especially small pleasure craft, use polyester as the principal material for the hull and often for the deck and, for sailing ships, many of the masts and other fittings. Most clothes washers, driers, and many other major appliances use polyester thermosets for pump housings, motor shrouds, and various other components. Tub and shower units, spas, building panels, and many other items of construction make polyester thermosets the major resin of the construction and home-building market. Polyester thermosets are one of the major resin types for tanks and pipes used to contain non-corrosive chemicals in industries such as paper and pulp, oil and gas, chemical processing, and water transport and treatment. Virtually every major manufacturing industry in the world uses polyester thermosets in some part of their total operations. It can be safely said that polyester thermosets are practically everywhere.

Note that no mention has been made of polyesters used for soda bottles or for fibers. Although these materials are made from "polyesters," the type of polyester used for these applications is different from those used in typical composites. The differences are those of thermosets versus thermoplastics. The polyesters that dominate the composite market have a reactive unsaturated site and are thermosets, hence the abbreviations for unsaturated polyester (UP) or thermoset polyester (TP). The polyesters used for soda bottles and fabric are thermoplastics, usually polyethylene terephthalate (PET). They do not form crosslinks, so they are not thermosets. Because this book focuses on composites and thermoset resins, the term "polyester" will refer to the thermoset resin unless specifically stated otherwise.

Another major point of clarification is appropriate before discussing the details of polyesters. Two different chemical processes are important in first making and then using thermoset polyesters. The first of these processes is **polymerization**, that is, the actual making of the unsaturated polyester resin. The polymerization step is usually done in large chemical plants by resin producers. An overview of polymerization will be given because it will assist in understanding the types of polyester resins available and in discovering some of their inherent limitations. The output product from the polymerization reaction is unsaturated polyester resin and it is usually sold in bulk (tank cars, barrels, etc.).

The second process important in unsaturated polyesters is **crosslinking** or **curing**. This process is done at the time of molding of the composite parts; that is, when the unsaturated resin is mixed with the reinforcement and molded to make a part. An understanding of the molding and curing process is necessary because this step also determines many of the final properties of the part and the molding economics strongly depend on how curing takes place. To avoid confusion, it is important to keep these two processes—polymerization and curing—clearly separate in your mind.

POLYMERIZATION OF UNSATURATED POLYESTERS

Polyester resins are made (polymerized) from various chemicals (monomers) using the condensation polymerization reaction. Therefore, as explained in Chapter Two on matrix materials, at least two types of monomers must be present. Further, each of the monomers must have two active ends (called **difunctional** where the prefix "di" means two and "functional" means reactive group). And, the two monomer types must react with each other to form a linking entity that ties them together. A condensation by-product is also formed. As was discussed in the chapter on matrix materials, the name of polymers formed by the condensation method is often "poly" plus the name of the new linking entity created in the reaction. This is exactly the case with polyesters, where the links formed as a result of the polymerization reaction are ester groups.

Polymerization Mechanism

The polymerization reactions for making a typical polyester are depicted in Figure 3-1. As shown at the beginning of the reaction sequence, the two monomers that combine to make polyesters are **glycols** and **diacids**. The reactive groups on glycols are the OH (alcohol) groups. They are located on either side of a central group labeled as a generic "G." "Glycol" is the name given to molecules with two alcohol groups. The reactive groups in the diacids (meaning two acid groups) are composed of the atoms COOH and sometimes called carboxylic acids. These reactive acid groups are shown on either side of the generic diacids, "A." Occasionally, the diacid will be referred to as simply "acids," but it

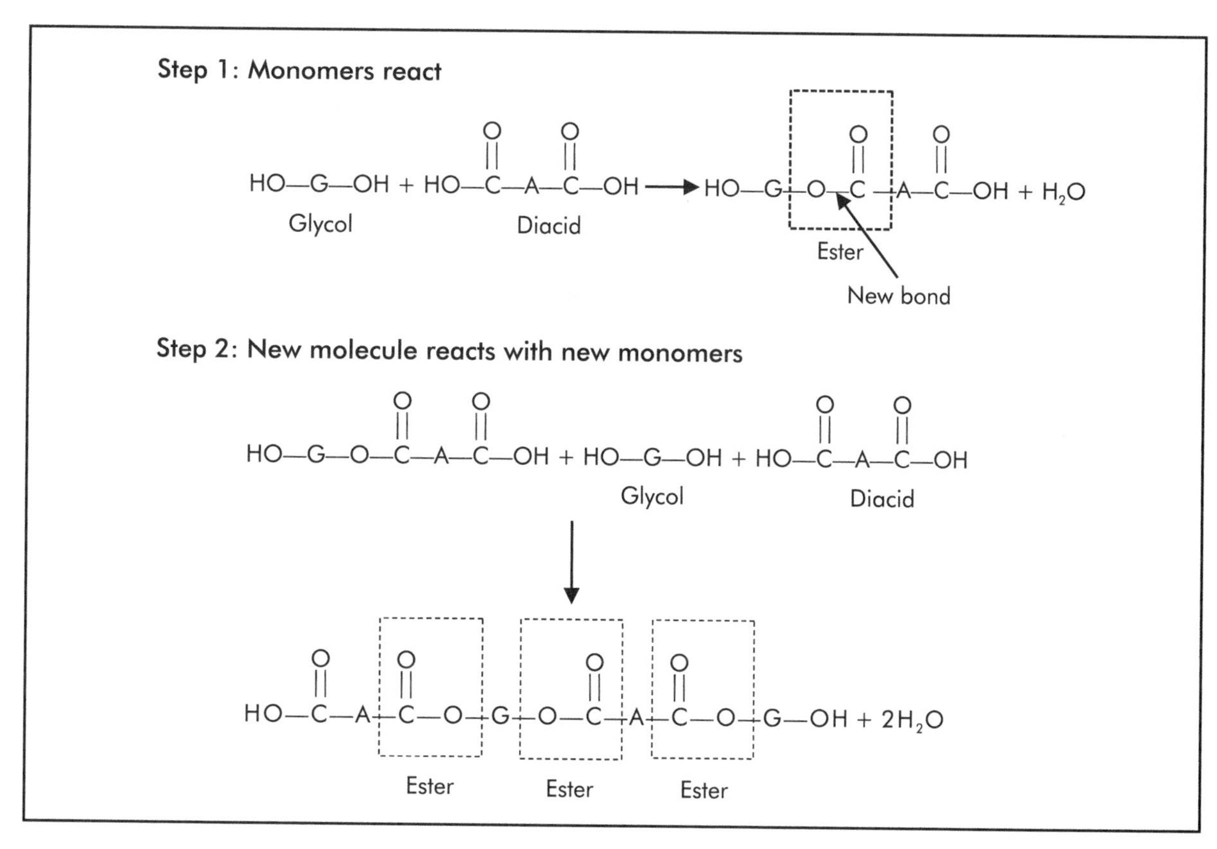

Figure 3-1. Polymerization of a typical polyester.

should be clearly understood that each acid must have two reactive groups; hence, "diacid" is the more appropriate term.

The term **alkyd** (pronounced *al'-kid*), derived from a contraction of alcohol and acid, is another name for unsaturated polyesters. Typically, however, the term "alkyd" refers to those polyester resins used for paints and quick-drying molding compounds.

When a glycol comes into contact with a diacid, the H on the glycol and the OH on the diacid join together and form water, H_2O, the condensate by-product. Simultaneously, the O on the glycol bonds with the C in the acid group. The arrangement of the atoms about this link is characteristic of the organic group called **esters** and so it is called an ester linkage. The initial linkage process is shown as Step 1 in Figure 3-1.

Note that Step 1 has resulted in a molecule that is a combination of the original glycol and the original diacid monomers. They are joined by an ester linkage. Also, note that after the new molecule has been formed, that new molecule still retains two active ends—an alcohol end (OH) and an acid end (COOH). Both ends can react with new monomers to further elongate the molecule.

Step 2 in Figure 3-1 shows the reactions of the active ends with new monomers. The alcohol end reacts with a new diacid monomer and the acid end reacts with a new glycol monomer. Each of the reactions in Step 2 results in the formation of a new ester linkage and the elimination of a water by-product. When Step 2 has been completed, the resulting molecule has three ester linkages joining together in an alternate arrangement, two acid groups, A, and two glycol groups, G.

The resulting molecule still has two active ends, thus permitting additional reactions with more new monomers. Each of the ensuing reactions will follow the same pattern; that is, alcohol ends react with diacid monomers and acid ends react with glycol monomers to form ester linkages and water by-products in each case. These reactions can continue in the same pattern until the concentrations of the diacid and glycol monomers become so low that continued growth is not possible. The polymerization reaction also could be stopped by simply lowering the temperature of the reaction, since these types of reactions are carried out at high temperatures to facilitate the effectiveness of the encounters of the monomers with the active ends of the growing polymers. Or, the reactions can be ended by adding a quenching agent that will react with the end of the chain but will not have an active site for further reactions.

At any time during the polymerization reaction, a new polymer chain can begin whenever a diacid encounters a glycol. This ease of beginning a chain often leads to many chains being formed, even in commercial polymerization operations. These chains can react with each other provided the alcohol end of one chain encounters the acid end of another. Chemists operating polymerization reactions have found conditions that favor the formation of long chains and can run the polymerization reactions to optimize the length of chain (molecular weight) as desired.

Polymer Properties and Control

The properties of polymers are dependent upon several factors. One is obviously the presence of the ester group. Ester groups are known in organic chemistry to have a set of typical properties, both chemical and physical. Many of these properties are present in the polymerized products as well as non-polymer ester compounds. For example, the ester linkage is quite polar and so some sensitivity to polar solvents, such as water, would be expected. Indeed, such sensitivity is seen in polyesters, especially at high temperatures. Polyester resins typically absorb significant water at about 200° F (93° C) and lose significant strength when that absorption occurs. Water absorption at low temperatures, as would be encountered with a motor boat, is usually not a problem.

The **acid number** is a measure of the amount of acid monomer that has not yet been reacted and, therefore, a measure of how far the polymerization reaction has proceeded. The amount of acid not reacted (acid number) can be determined using conventional chemistry techniques such as titration. Titration of the acid by a base can be done quickly and automatically, thus giving a rapid and reliable measurement of the state of the polymerization reaction. (See the American Society for Testing and Materials standard, ASTM D1639, for details of the test.)

High amounts of residual acid groups have been shown to be detrimental to several key properties, such as chemical resistance and stability of molecular weight. Residual acid content could be present because the reaction was not carried out to completion or because the amount of diacid monomer was greater than needed to react with the glycol. Generally, residual acid content is reduced by conducting the polymerization reaction in an excess of glycol and ensuring that the acid number is below a certain value.

Processing viscosity is another critical parameter monitored during the polymerization reaction along with the acid number. The processing viscosity is also a measure of the extent to which the polymerization reaction has proceeded; that is, how long the polymer chains have become. As the molecular weight of the polymer increases, the viscosity increases. The viscosity measurement during the polymerization reaction is taken when the polymer solution is hotter than the conditions used in defining a standard viscosity. Therefore, the name given is properly a "processing viscosity."

Under standard conditions, liquid unsaturated polyester resins range in viscosity from 50 centipoise (thin liquids) to 4,000 centipoise (thick liquids) and even higher for some of the low-melting solid resins. These viscosities are typically determined by **gel permeation chromatography** (GPC). Typical molecular weights for these materials are 800–10,000. Higher-performance resins tend to have higher molecular weights, but are more difficult to process.

EFFECTS OF VARIOUS DIACID, GLYCOL, AND SPECIALIZED MONOMERS

A factor that strongly affects the properties of the polyester is the choice of the specific types of glycol and diacid monomers. Figure 3-1 has represented these monomers in a generic form (with the letters A and G), not giving the actual formulas of the monomer molecules. This was done for simplicity in explaining the polymerization reaction.Of course, this does not represent any real system since specific monomers must be chosen when the polymerization is done.

One of the most important considerations in the choice of diacid or glycol is determining which of these monomers will contain the unsaturation required for crosslinking to occur. (The specific mechanism of this crosslinking will be discussed in a following section.) **Unsaturation** is a term applied to organic molecules that contain the carbon-carbon double bond. The name **unsaturated polyester** emphasizes the presence of these carbon-carbon double bonds in the polyester molecule. Normally, the diacid is the monomer that contains the carbon-carbon unsaturation. The term **olefinic unsaturation**, meaning carbon-carbon double bonds in aliphatic compounds, as is the case in most of the common diacids used in polyester thermosets, is another name that can be found in the literature.

Diacid Monomers

A list of the most commonly used *unsaturated diacids* is given in Figure 3-2. The

Unsaturated Diacid Monomers

Name	Structure	Comments
Fumaric acid	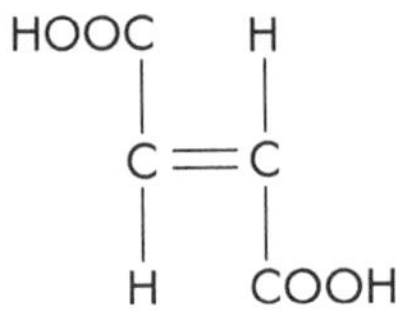 HOOC H | | C=C | | H COOH	Trans-isomer, highly reactive, crosslinkable
Maleic acid	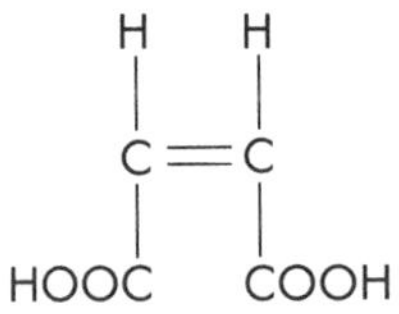 H H | | C=C | | HOOC COOH	Cis-isomer, converts to fumaric acid, crosslinkable
Maleic anhydride	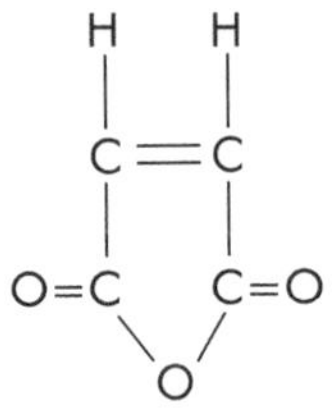 H H | | C=C | | O=C C=O \ / O	Readily converts to maleic acid and fumaric acid in presence of water, crosslinkable

Figure 3-2. Unsaturated diacid monomers typically used in making polyester thermosets.

carbon-carbon double bond (unsaturation) is in the center of each of the molecules. An acid group (COOH) and a hydrogen atom are attached to each of the unsaturated carbons. In the first molecule, **fumaric acid**, the acid groups are arranged so they are on opposite sides of the molecule. (Fumaric is pronounced *foo-mahr'-ic*). This arrangement is called the **trans-isomer** where the term "trans" means "across" and "isomer" means atomic arrangement. The second molecule, **maleic acid**, has the same atoms, but in this case the acid groups are on the same side of the carbon-carbon double bond. (Maleic is pronounced *mah-lay'-ic*.) This arrangement is called the **cis-isomer** where "cis" means "on the same side." The cis-isomer, which is not as stable as the trans-isomer, is converted into the trans-isomer under the conditions normally present when polymerization occurs. The third unsaturated molecule is **maleic anhydride**. This compound can be thought of as maleic acid with H_2O removed. (Anhydride means "without water.") Although not strictly a diacid in the form shown in Figure 3-2, maleic anhydride readily converts to maleic acid and then to fumaric acid under the conditions of polymerization. Because the cost to make maleic anhydride is substantially less than either fumaric or maleic acids, maleic anhydride is the most common of the unsaturated diacids used in commercial production of unsaturated polyesters.

As will be shown in a later section of this chapter, the presence of the carbon-carbon double bond (unsaturation site) is critical for the crosslinking of polyesters. Every carbon-carbon double bond is a potential site for crosslinking. Since the polymerization reaction results in having acids alternate with the glycols along the backbone of the polyester molecule, crosslinks could be formed at a tremendous number of locations on each molecule. Experience has shown that the best properties are obtained if some of these potential sites are not crosslinked. (Too many crosslinks can cause the entire polymer structure to be tightly tied together and impart brittleness to the final part.) Two methods are commonly used to accomplish this limited crosslinking.

1. Simply conduct the crosslinking reaction so that only some of the sites are reacted.
2. Replace some of the unsaturation sites with saturated sites; that is, use some diacid monomers in the polymerization reaction that do not contain the carbon-carbon double bond. This is most easily done by mixing the two types of diacids—those containing carbon-carbon double bonds and those that do not. Because of differences in the reactivities of the various types of diacid monomers, the order of addition and the conditions under which they are added is part of the technology of polymerization. This technology is used to optimize the properties of the polyester molecules.

Commercially, both methods of limiting crosslinking are used. The first method, limiting the number of sites that reacts, is controlled by the molder who can vary the concentrations and conditions of the crosslinking reactant to achieve the level of crosslinking desired. (More about this will be discussed later in this chapter.) The second method, polymerizing with both saturated and unsaturated diacids, is controlled by the manufacturer of the polyester resin. The various resin manufacturers have tried a multitude of different saturated diacids and achieved polymers with a wide range of properties. The effects of these various diacids will now be examined.

A list of the most common *saturated diacids* is given in Figure 3-3. (The typical mixture of diacids used to make a crosslinkable polyester would be one from the unsaturated diacids illustrated in Figure 3-2, and one or more from the saturated diacids illustrated in Figure 3-3.) Because of the widely varying properties that can be achieved from the choice of saturated diacids, the various polyesters are often named according to the type of saturated diacid used. For example, polyester resins are often referred to as "ortho resins" or "iso resins"—two of the most common saturated diacids.

The proper name of the monomer from which ortho resin is made is **orthophthalic acid** (pronounced *or-tho-thal'-ick*). It is a diacid where the two acid groups are attached to an aromatic ring on adjacent carbons around the ring. The properties of aromatic materials, discussed in the chapter on matrix materials, are present in this and the other aromatic diacids. Those properties include improved stiffness, improved strength, lower flammability, and lower resistance to UV light. Compared to the other common saturated diacids, ortho has low cost and good styrene compatibility. This latter property will become more meaningful as the crosslinking reaction is discussed. Unsaturated polyesters containing ortho are the most common of all the types commercially available. Ortho is, therefore, sometimes referred to as a general-purpose polyester thermoset.

A molecule related to orthophthalic acid is orthophthalic anhydride. It converts to the diacid under polymerization conditions and so its properties are the same as the ortho diacid.

Isophthalic acid (pronounced *eye-so-thal'-ick*) is the second most common of the

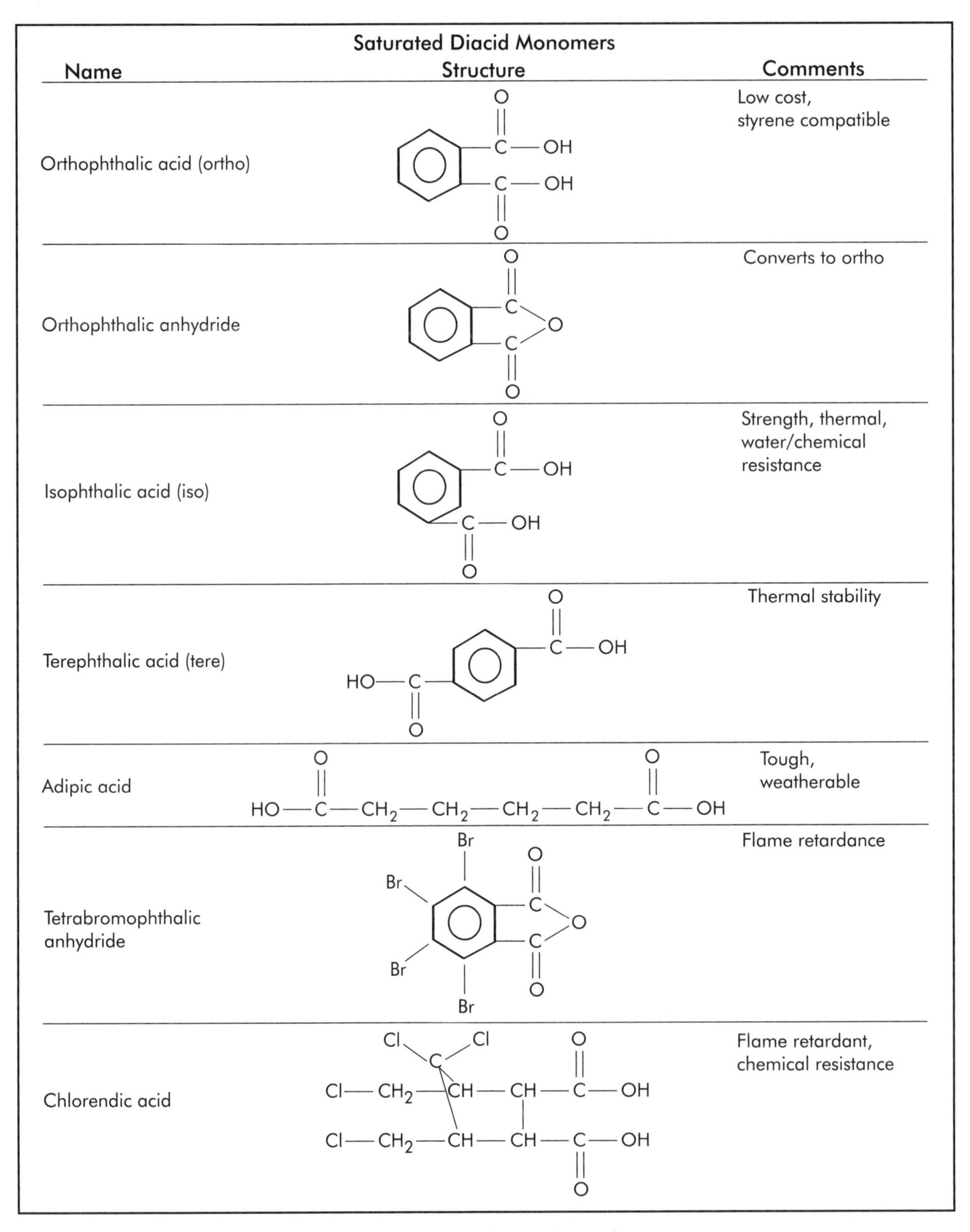

Figure 3-3. Saturated diacid monomers typically used in making polyester thermosets.

polyester thermosets (behind ortho). Like ortho, it has two acid groups attached to an aromatic ring. But in iso, the attachments are to carbons separated on the ring by one other carbon.

Iso resin has the properties of aromatics but has some important differences from ortho resins. It may be surprising that ortho and iso have different properties since they differ only in the points of attachment of the acid groups on the aromatic ring. However, when the diacid monomers are incorporated into the polyester polymers, the attachment points change the *shape* of the polymer, which affects the manner in which the polymers can pack together, intertwine, or otherwise interact. Since these intermolecular interactions affect many properties, small changes in the structures of the monomers can result in substantial changes in the polymer's properties. The major differences between iso and ortho are that iso has improved strength, toughness, and resistance to heat, water, and chemicals. The improvement of iso strength over ortho strength is as much as 34% in non-reinforced materials and 10% in composites.

Iso costs about 10% more than ortho because it is more costly to make. The most common application for iso polymer is as the outer coating for a composite part. This layer, called the **gel coat**, must be the first line of defense against the environment, and iso does a better job at that than does ortho. Iso resins are also used in applications where moderate chemical or water resistance is needed, but usually not at elevated temperatures.

Terephthalic acid (pronounced *tair-a-thal'-ick*) or tere is less common than either ortho or iso. Tere is the third possible arrangement of two acid groups on an aromatic ring with attachment on two carbons separated by two other carbons. Polymers containing the tere group have higher thermal resistance, usually measured by an increased heat distortion temperature (HDT).

Adipic acid (pronounced *ah-dip'-ick*) is very different from ortho, iso, and tere. It is not aromatic and, therefore, is less rigid (more flexible) and less strong. Adipic has much more elongation and is therefore tougher. It also has improved weathering (better resistance to UV light and oxygen) than the aromatic diacids. There are many other non-aromatic (that is, aliphatic) compounds that can be used for their toughness and weatherability properties besides adipic, but they are usually more expensive.

Tetrabromophthalic anhydride (pronounced *tet-rah-brom-oh-thal'-ick*) and **chlorendic acid** (pronounced *clor-en'-dick*) are two compounds that are diacids under the conditions of polymerization. Both compounds contain halogens and are therefore effective in improving flame resistance. (Halogens are F, Cl, Br, and I atoms.) As discussed in the chapter on matrix materials, the presence of halogens helps to smother the flame. Almost any halogen-containing compound is flame retardant, so the choices listed here are just two of many possible flame-retardant monomers. Chlorendic acid has the added advantage of imparting chemical resistance.

Glycol Monomers

The polymerization reaction for making polyesters is between diacids and glycols. Glycols are occasionally called "diols," a name that reflects the presence of two alcohol (OH) groups on the molecule. Several glycols can be chosen as monomers in making unsaturated polyesters. Some commercial glycol monomers are listed in Figure 3-4. The most commonly used glycol monomer in unsaturated polyesters is **ethylene glycol**, chiefly because of its low cost and generally good properties. If the number of carbons in the glycol chain is increased, additional flexibility and toughness are imparted to the polyester molecule. Therefore, the longer

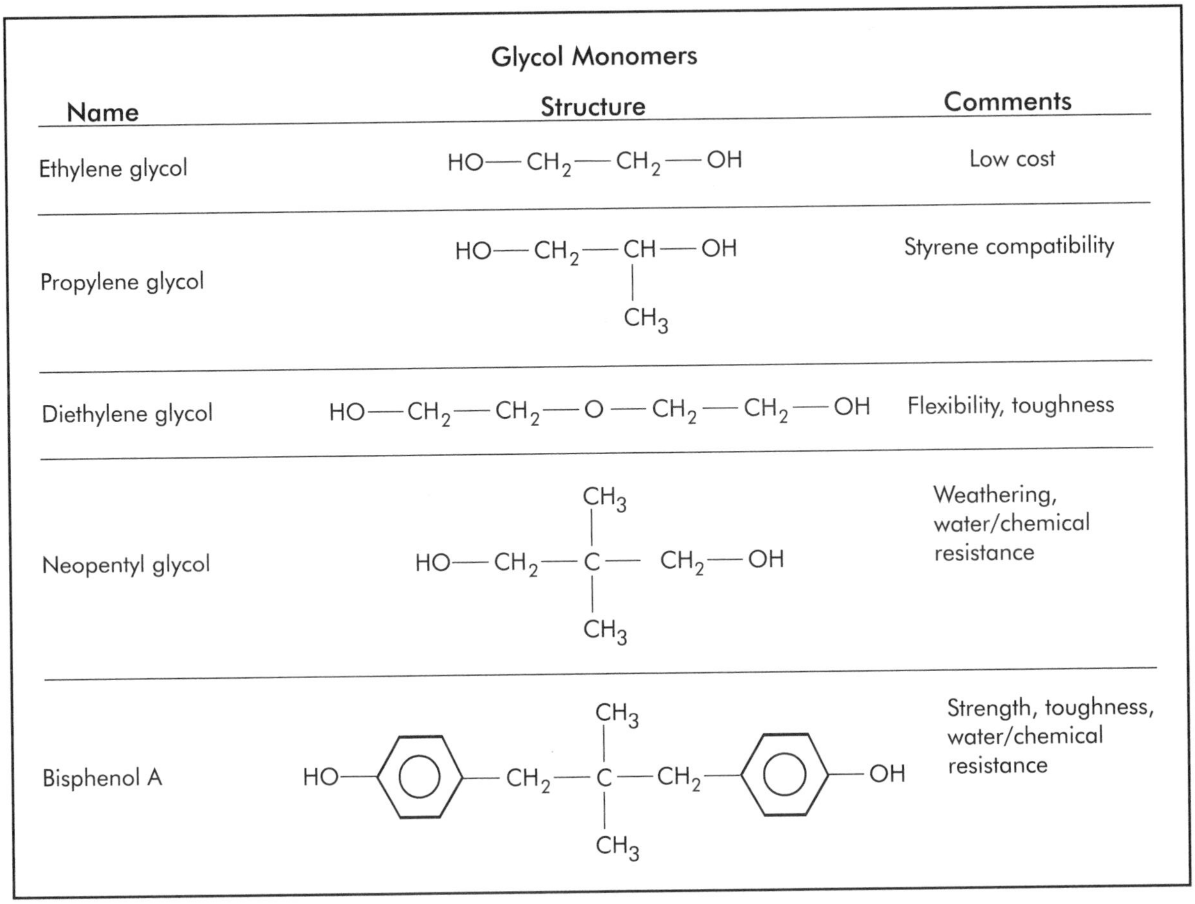

Glycol Monomers

Name	Structure	Comments
Ethylene glycol	$HO-CH_2-CH_2-OH$	Low cost
Propylene glycol	$HO-CH_2-CH(CH_3)-OH$	Styrene compatibility
Diethylene glycol	$HO-CH_2-CH_2-O-CH_2-CH_2-OH$	Flexibility, toughness
Neopentyl glycol	$HO-CH_2-C(CH_3)_2-CH_2-OH$	Weathering, water/chemical resistance
Bisphenol A	$HO-C_6H_4-CH_2-C(CH_3)_2-CH_2-C_6H_4-OH$	Strength, toughness, water/chemical resistance

Figure 3-4. Glycol monomers used in making polyester thermosets.

carbon atom content of **propylene glycol** contributes to its frequent use to impart greater toughness. Even longer and more flexible chains are seen with **diethylene glycol** and **neopentyl glycol**. Neopentyl also adds weathering and chemical/water resistance. These monomers are all aliphatic and illustrate that as aliphatic character is increased, the molecule becomes more flexible and tougher.

Highly aliphatic glycols with up to nine atoms in the chain are used commercially. However, as the glycol chain is lengthened, the resulting polymer becomes significantly softer. For example, as the glycol chain is increased from 2 to 6 carbons, the Barcol hardness (a measure of the hardness of the material) drops from about 50 to about 25.

A common glycol used to impart some aromatic content, with all of the normal aromatic characteristics, is **bisphenol A**, also shown in Figure 3-4. This particular aromatic compound also has some aliphatic content, the non-aromatic carbon groups on the ends, and so both strength and toughness can be improved over the other glycols already discussed. Cost is, however, somewhat higher than the simpler glycols.

Specialized Monomer Additives

Some monomers that are neither diacids nor glycols will react during the polyester polymerization process to give property improvements not possible with the standard diacids or glycols. One of the most interesting of these materials is **dicyclopentadiene** (DCPD). This chemical will react with acids and glycols by a unique ring-opening mechanism.

The mechanical properties of polyester resins made with DCPD as one of the monomers exhibit improvements in flexural modulus and flexural strength over conventional polyesters. Improvement in HDT is also typically achieved when DCPD is incorporated into the resin. Another advantage is the cost improvement over many of the specialty monomers used to achieve enhanced mechanical and physical properties. Some work has also indicated that the surface uniformity (lack of dimples or shrink marks) is also improved when DCPD is present.

DCPD can be polymerized by itself, that is, without the presence of glycols or diacids. Pure DCPD resin has many interesting properties that may lead to its widespread use in some of the markets now serviced by other thermoset resins.

When the diacid monomers (usually more than one), the glycol monomers (usually only one type), and any other specialized monomers are mixed together under the right conditions, polymerization occurs. The resulting polymer is a polyester and, because one of the monomers has a carbon-carbon double bond, it is unsaturated. It can, therefore, undergo crosslinking reactions and become thermoset. Two typical unsaturated polyesters, isophthalic polyester and bisphenol A fumaric acid polyester, are shown in Figure 3-5.

CROSSLINKING MECHANISMS

As previously indicated, polymerization is the first chemical process used in making polyester thermosets and it is conducted by large resin companies. The second chemical process is crosslinking or curing, which takes place at the time of molding. Crosslinking or curing has a major impact on the economics of the molding of parts and is critical to achieving good properties in the molded parts.

Some molders may believe that a simple recipe and single molding procedure can be followed to consistently result in good parts. Sadly, the curing reaction is a complicated process affected by many different variables, some of which cannot be fully controlled by any simple procedure. Inevitably molding conditions will change (humidity, weather, resin uniformity, conditions of ingredients as they are stored, suppliers, equipment conditions, etc.) and the circumstances under which the original procedure was developed may no longer exist. Sometimes poor parts are made without realizing it, at least until the marketplace responds. However, the principles of crosslinking can be understood and, with some training and careful monitoring of the reaction variables, the process can be kept within reasonable control parameters.

At the most basic level, crosslinking is easily understood. During the crosslinking reaction, bonds are formed between the polymer molecules, thus linking them together to form enormous interconnected networks. When this happens, the resulting material becomes solid with dramatic increases in strength, stiffness, hardness, and other desirable (usually) mechanical, physical, and chemical properties. The resulting solid material cannot be remelted, so any molding of the material must be done before or during the period when the crosslinks are forming. Also, reinforcing fibers must be introduced before curing so that they can be wetted by the liquid resin while it is still a low-viscosity fluid. The fibers also must be positioned properly before curing has finished.

The unsaturated polyesters are able to form crosslinks because they have some segments along the polymer chain that contain

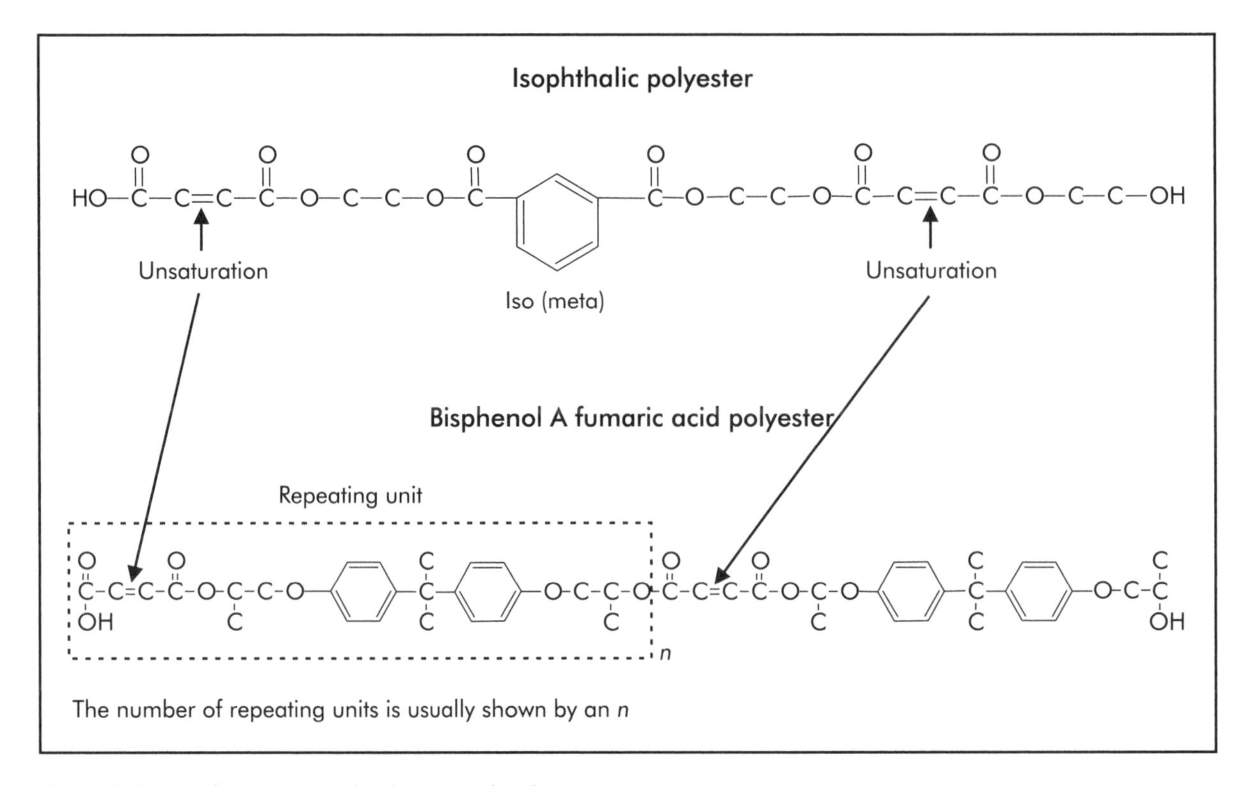

Figure 3-5. Specific unsaturated polyester molecules.

a carbon-carbon double bond. Crosslinks also can be formed by mechanisms that do not use carbon-carbon double bonds, but these do not apply to the polyester thermosets and are discussed in later chapters. The carbon-carbon double bonds were already present in some of the monomers that reacted to form the polyester polymer and were not affected by the polymerization reaction as explained previously. Since these double bonds are not affected by the polymerization reaction, a separate reaction is used after polymerization is completed to cause the carbon-carbon double bonds to react and crosslinking to occur. It is this capability—to crosslink at a time completely independent of the polymerization reaction and under relatively simple conditions—that makes polyester thermosets useful as matrix materials.

Crosslinking Steps

The creation of the crosslinks during curing involves a series of chemical steps. A simplified depiction of the crosslinking steps is given in Figure 3-6. A generic polyester polymer is shown in Figure 3-6a with the carbon-carbon double bonds specifically identified. For simplicity, other parts of the chemical chain, glycols and acids, are represented generically with the circles.

The curing or crosslinking reaction begins when the unsaturated polyester is mixed with a small amount (usually about 1.5 to 2%) of material that initiates the curing reaction. The most common **initiator** is an organic peroxide. When heated or chemically triggered, organic peroxides will split apart to form two segments, each of which has an unpaired electron. A chemical entity

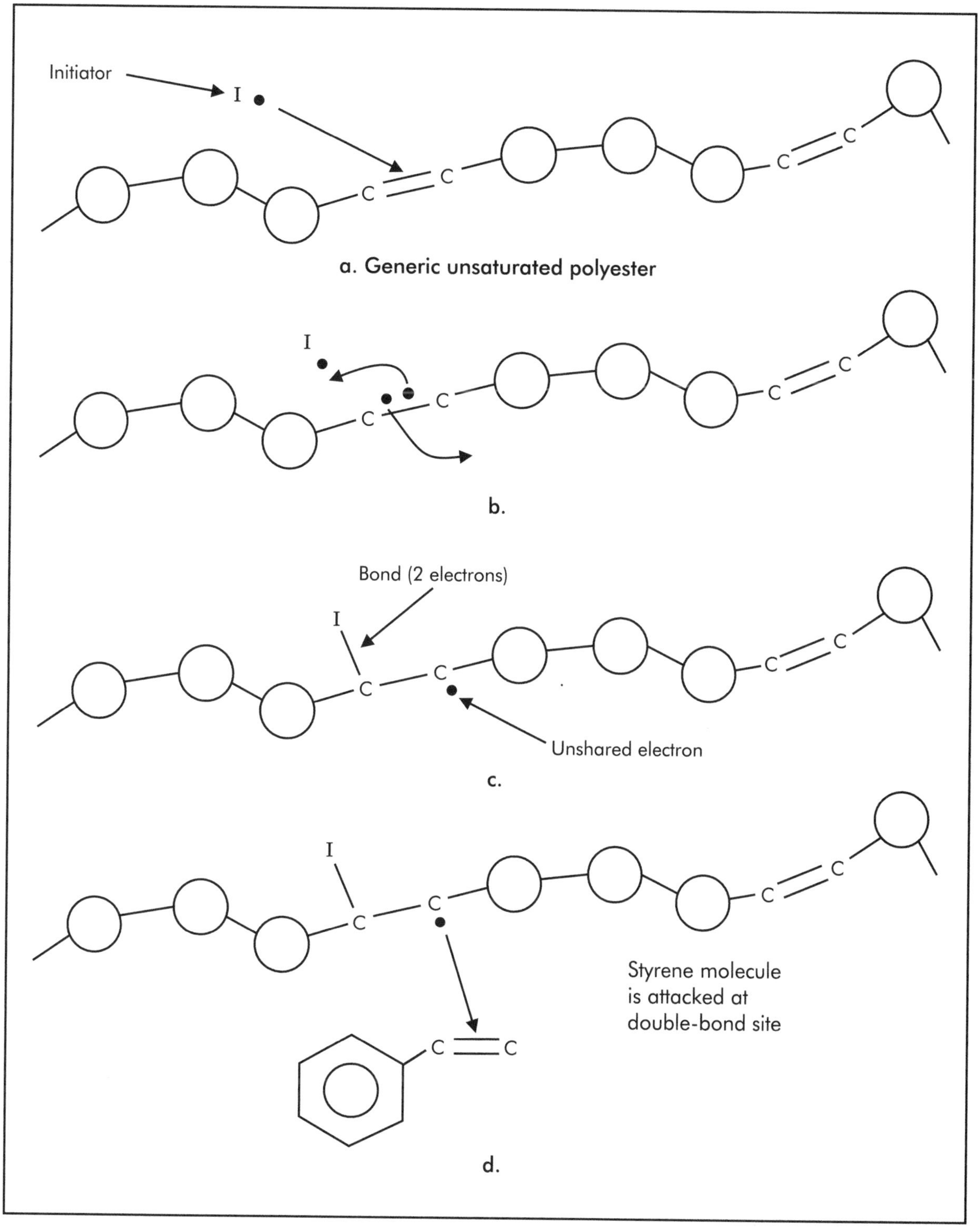

Initiator

I

C C C C

a. Generic unsaturated polyester

I

C C C C

b.

Bond (2 electrons)

I

C C C C

Unshared electron

c.

I

C C C C

Styrene molecule
is attacked at
double-bond site

C C

d.

Figure 3-6. Unsaturated polyester crosslinking reactions.

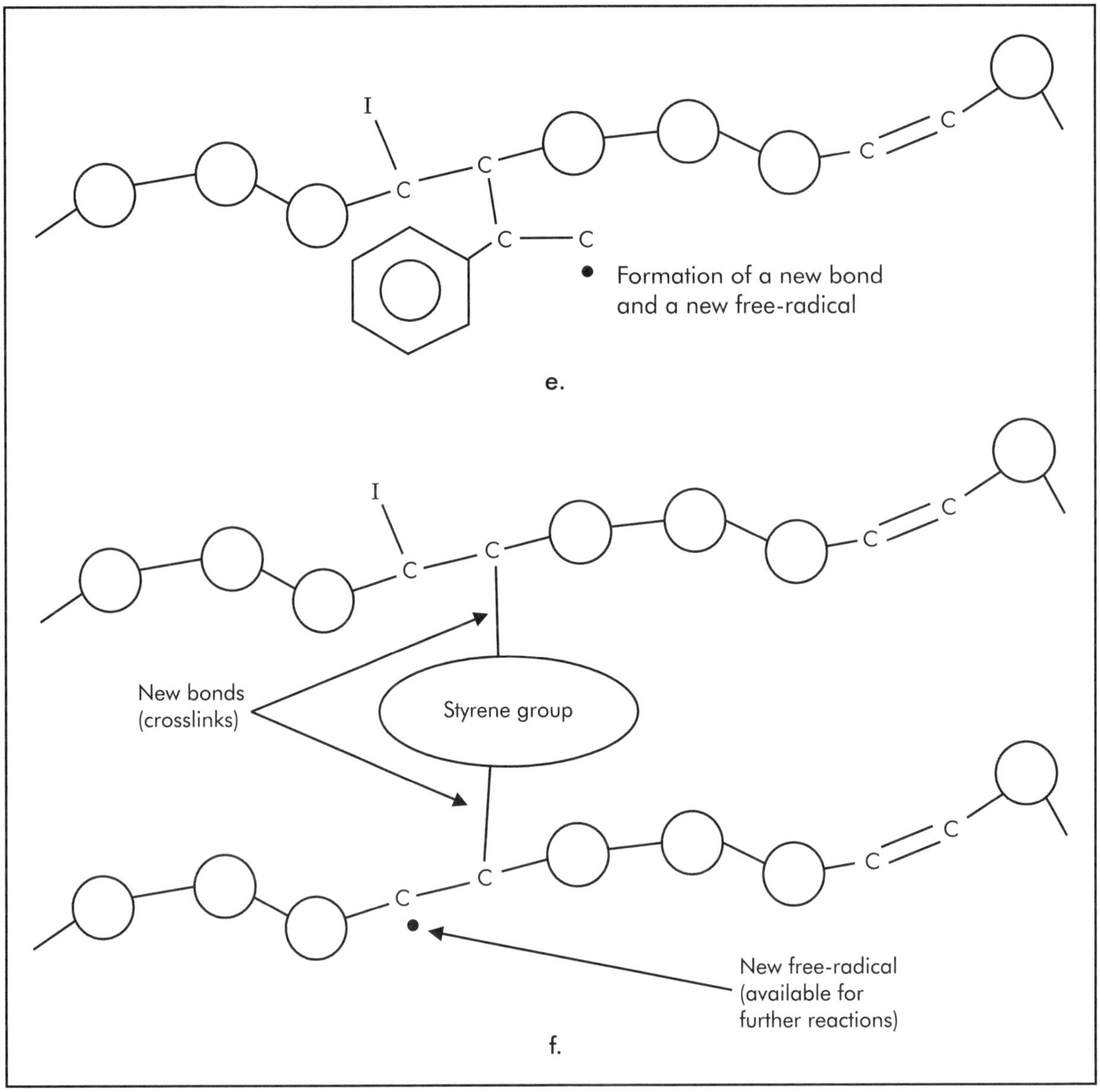

e.

f.

Figure 3-6. Continued.

containing an unpaired electron is called a **free-radical**. These segments containing unpaired electrons are reactive and seek loosely held electrons with which they can form a stable bond. The attack of this initiator on the carbon-carbon double bond, in which one of the bonds has loosely held electrons, is shown in Figure 3-6a.

As shown in Figure 3-6b, the unpaired electron on the initiator attracts one of the two electrons in double bond, thus forming a new bond between the initiator and one of the carbons in the carbon-carbon double bond. When one of the electrons has been taken from the carbon-carbon double bond, the double bond disappears, with only a single carbon-carbon bond remaining. The other electron, which was originally in the double bond, migrates to the second carbon where it becomes a new free-radical (unpaired electron). Figure 3-6c shows the bond and the new free-radical located on the carbon atom.

Note that initiators are consumed in the reactions they initiate by becoming bonded to the polymer. They are not, therefore, true **catalysts,** which improve the speed or efficiency of a reaction but are not consumed therein. However, common usage has long applied the term "catalysts" to these peroxide initiators, although that name in this case, by strict definition, is erroneous. These materials are referred to as initiators or peroxides throughout this text.

The new free-radical residing on the backbone carbon is also reactive and will aggressively seek to bond with some other electron that might be loosely held. Theoretically, the free-radical could react with a carbon-carbon double bond site on a neighboring unsaturated polymer. In practice, however, direct reactions between polymers are difficult because the polymers are relatively large molecules that do not move around easily to line up the unsaturation sites and allow bonding between them. Hence, this direct linkage between polymers occurs only rarely.

Because the free-radical can react with any material having an unsaturation site, small molecules containing these sites are added to the polymer mix as co-reactants and solvents to give additional movement to the polymers. These small molecules can move about freely between the polymer chains and will, therefore, react with the free-radical site on the polymer. A common free-radical attack on one of these small molecules is shown in Figure 3-6d. Because these small molecules are both solvents and reactive agents, they are called **reactive diluents**.

The most common reactive diluent is styrene—an aromatic ring attached to a carbon-carbon double bond. When a styrene molecule approaches the unpaired electron on the polymer, attack is made on the electrons in the double bond, a rearrangement occurs, a new bond is formed between the polymer and the styrene, and a new free-radical is formed on the styrene as shown in Figure 3-6e. The free-radical on the styrene can then react with any unsaturation site that it might encounter. The most probable reactant, simply because of mobility and concentration, is another styrene molecule. This could cause several styrene molecules to join in a small chain pendant from the backbone of a polymer. Eventually, the free-radical on a styrene will encounter an unsaturation site on another polymer. When this happens, the two polymer molecules become linked together through the styrene bridge.

The presence of several styrene molecules in the bridge is depicted in Figure 3-6f, where the styrene is not shown as a single molecule but as a group. Even when the crosslink between polymers is formed, the free-radical persists. It now resides on the second polymer as shown in Figure 3-6f. The free-radical might react with more styrene molecules and, eventually, with more polymers. This chain-reaction mechanism often continues until most or all of the polyester polymers are linked together.

The process is finally terminated when the free-radical reacts with something other than a carbon-carbon bond in styrene or another polymer. Some of these other reactants can include oxygen, contaminants, and other peroxide (free-radical) segments. Hence, if these other potential reactants are present in large concentrations, the crosslinking reactions could be terminated prematurely and the desired number of crosslinks might not be formed. This is one of the reasons it is so critical to control the concentration of initiator and any other material that might react with free-radicals.

Crosslinking Agents, Co-reactants, and Co-monomers

The small molecules, such as styrene, link the polymers together much like bridges linking two sides of a chasm. These **bridge molecules** are called **crosslinking agents**, **co-reactants**, or **co-monomers**. To not confuse the bridge molecules with the monomers used in creating the polyester polymer, this text will refer to them as bridge molecules or co-reactants. The characteristics of several of these bridging molecules will be discussed later in this chapter.

While this explanation has focused on just one polyester chain and the molecules it bonds with, in actual practice, several initiator (peroxide) molecules will react with several carbon-carbon double bonds to begin these crosslinking reactions on many different polymers throughout the mix. Therefore, only a few initiator molecules are needed to get the reactions started. Hence, the initiators are only needed in small quantities, usually 1.5 to 2% of the total resin weight.

Several different organic peroxides are used commercially as part of the cure system for polyester thermosets. The most common of these is methyl ethyl ketone peroxide (MEKP). It is relatively stable and has the capability of being chemically as well as thermally activated. Most room-temperature cures involve MEKP and some chemical accelerator (as will be described later in this chapter). Other common peroxides are dibenzoyl peroxide (BPO), 2,3-pentanedione peroxide (2,3 PDO), also known as acetyl acetone peroxide (AAP), and lauroyl peroxide (LPO). Each one differs from the others in the temperature of activation, the types of chemicals that affect the activation reaction, and the type of resins with which it is most commonly used. All peroxides are dangerous because of their tendency to decompose, sometimes with explosive force under certain conditions. Therefore, the peroxides should generally be stored in refrigerators and in small quantities.

During the curing process, heat is liberated as new bonds are formed. This liberated heat, called an **exotherm** (meaning heat is given off by the reaction), can be a problem under some circumstances, and should be considered in setting the conditions of the curing reaction. The first consideration is obviously the temperature at which the curing reaction is run. Because the exotherm adds heat to the system, some compensation should be given for the exotherm or the reaction may proceed too quickly. The dangers of reacting too quickly can include ignition and fire, degradation and loss of properties, and incomplete crosslinking. Another consideration is the dimensional thickness of the material that is reacted. Polymers are good heat insulators so heat is not dissipated well through them. Therefore, thick sections may retain exothermic heat and cause excessive heat build-up. If some areas are thick and others thin, differential heating could occur and the part could be warped. If an exothermic reaction begins to overheat, it is critical that the material be dispersed into a thinner configuration. Therefore, consideration should be given to this possibility when designing equipment for mixing and molding thermosets.

Co-reactants and Crosslinking Agents as Solvents and Diluents

As described previously, the co-reactant or crosslinking agent is important to the crosslinking reaction as a bridge between the polymers, thus facilitating their linkage. In addition to serving as a bridge, the crosslinking agents generally have a second major function in the polyester curing process. They are solvents or diluents for the resin which, in many cases, is so viscous that efficient curing reactions, wet-out of the fibers, and other processing steps would be difficult if not thinned by a solvent. The inclusion of the solvent greatly improves the mobility of the resin molecules, thus allowing them to align more readily to achieve effective linkages. It also lowers the overall viscosity of the resin so that fiber wet-out is easier.

A potential problem with the inclusion of a solvent in the curing mixture is that no solvent should be present in the final product as that could significantly erode the physical and mechanical properties of the composite. This problem has been solved by choosing those solvents that also serve as bridges; that is, they are also co-reactants. Because the common crosslinking agents are both solvents and co-reactants, most or all of the crosslinking agent will be used up in forming the crosslinks and no residual free solvent will be left in the polymer. This double nature has, therefore, led to the alternate term already introduced, "reactive diluent." A list of common crosslinking agents is given in Table 3-1.

Curing

The crosslinking agent can affect other characteristics of the curing reactions and of the finished product. However, these are generally less important than its functions of solvation and bridging (co-reaction). One effect is the change in the temperature at which curing takes place. The first chemicals listed in Table 3-1 are generally capable of crosslinking polyester thermosets at both room temperature and elevated temperatures. Those that contain the allyl group, which are listed later in the table, require elevated temperatures for curing. Even those crosslinking agents that allow room-temperature curing will, however, generally slow the reaction. This is because they add mass to the system and that mass must be heated to achieve the same reactivity level as would be present with just resin.

Some of the crosslinking agents are much less effective with peroxide-initiated curing reactions and tend to be used in other reaction methods such as with ionic catalysts. These alternate curing reactions will not be covered in this textbook as they are much less important than the peroxide-based cure systems.

Each of the crosslinking agents has a characteristic exotherm and a characteristic shrinkage; both are important for polyester thermoset applications. The polyester resin itself has shrinkage of about 3%. But when a crosslinking agent is present, the shrinkage increases by an amount characteristic of the particular co-reactant. The additional shrinkage from the crosslinking agent can range from a low of 7% to as much as 27%, adjusted for the amount of co-reactant present. Styrene, the most commonly used crosslinking agent, has shrinkage of about 17%.

Common Crosslinking Agents

The low cost of styrene and its generally beneficial properties are the basis for its widespread acceptance and use as the most common crosslinking agent. Styrene, which is aromatic, adds the typical aromatic properties of stiffness, strength, low water absorption, and thermal stability to the crosslinked polyester product. However, it also adds susceptibility to UV light, certain chemicals, and, under some conditions, brittleness. Careful research has indicated that as the amount of styrene increases relative to the amount of resin, most of the changes in properties attributable to styrene go through a maximum for that property. For example, strength initially increases with greater content but then decreases as the amount of styrene continues to increase relative to the amount of resin. This same type of behavior is exhibited by all of the crosslinking agents, although at different concentration levels and to different degrees. In general, therefore, the concentration level of the crosslinking agent should be optimized carefully for each particular grade of resin.

Table 3-1. Crosslinking agents (solvents/co-reactants) used with polyester thermosets.

Crosslinking Agent	Characteristics
Styrene	Low cost, high strength, and stiffness, good general properties and heat distortion temperature (HDT)
Methyl methacrylate (MMA)	Good weather resistance, low smoke, low odor
Butyl acrylate and butyl methacrylate	Good weather resistance
Alpha methyl styrene	Cooler cure, reduced exotherm, poor physical properties
Vinyl toluene	Low volatility, high flash point, low elongation, low shrinkage
Para-methyl styrene	Low volatility, high flash point
Diallyl phthalate (DAP)	Very low volatility, molding compounds (pastes)
Diallyl isophthalate (DAIP)	Very low volatility, molding compounds (pastes)
Octyl acrylamide	Solid co-reactant, mostly for molding compounds
Trimethylol propane triacrylate	UV and electron beam cures
Triallyl cyanurate (TAC)	High HDT, multi-site crosslinking
Triallyl isocyanurate	High HDT
Diallyl maleate	High HDT
Diallyl tetrabromophthalate	Fire retardant

A potentially serious problem with styrene is its effect as a hazardous material. The United States government has established exposure limits for styrene and the industry has adopted even more restrictive but voluntary standards. Some of the current strategies to reduce styrene emissions are discussed in Case Study 3-1. The emission problem with styrene and some of its other deficiencies has led to the adoption of several alternate crosslinking agents with polyester thermosets, as is shown in Table 3-1.

Methyl methacrylate (MMA) is probably the most common crosslinking agent other than styrene. MMA is well known as a superior material for weather and UV light resistance. MMA also has the advantage that it gives off little smoke when it burns, thus improving the smoke rating of some composite resins. A strategy to achieve improved overall flammability characteristics has been to use halogens to suppress the flame and MMA to obtain a lower smoke density, thus improving upon general-purpose resins in both flame and smoke ratings.

CURE CONTROL ADDITIVES

In addition to the major components of the polyester thermoset mixture that have already been discussed—polyester polymer, initiator, and crosslinking agent—many different materials are routinely mixed with thermoset resins to control the rate of the cure reaction. In some cases, the purpose of these additives is to slow the cure rate so that **shelf life** (storage time) will be longer. A molder may also want to extend the **pot life**, which is the time after the materials have all been mixed and before the cure has progressed so far that the material is hard. Extending the pot life is usually desired in cases where the process time for mixing the reinforcement or the time for placing the materials into the mold is long. On the other hand, the molder may want to shorten the time to cure or to cure at a lower temperature. All of these objectives can be achieved with proper choices of cure control additives. Additives that accomplish all of these functions are discussed in this section.

Inhibitors

Because polyesters crosslink by a free-radical chain-reaction mechanism, the creation of just small amounts of free-radicals in the mixture can be sufficient to initiate the crosslinking reaction. This is especially true because as the crosslinks are formed, they create heat (exotherm), which increases the reaction rate. Hence, to avoid premature crosslinking and extend the shelf life of the resin, materials are often added that deactivate (or react with) free-radicals, thus preventing their reaction with the polyester resin. These materials are called **inhibitors** because they inhibit crosslinking.

Several types of inhibitors exist. Two of the most common chemicals widely used as inhibitors are hydroquinone and catechol. Both belong to the general chemical category called free-radical scavengers.

Free-radicals can be created by several mechanisms other than by decomposition of organic peroxides. For example, some carbon-carbon double bonds, such as those in many polyester resins and styrene or other reactive diluents, are especially susceptible to decomposition by heat and they often decompose to form a free-radical. Therefore, whenever possible, resins should be stored in cool locations. If cool storage is not available, inhibitors are usually added. Resins made in the summer often have enhanced levels of inhibitor because they are more likely to become hot during transit or storage. Some resin manufacturers offer summer and winter resins because of the differing levels of inhibitor added. Free-radicals also can be created simply by the action of UV light on the resin. Therefore, resins are usually shipped in metal containers or dark glass bottles to prevent UV light from penetrating to the resin. Contaminants are also capable of creating free-radicals. Because of all the potential sources of free-radicals, which will shorten the shelf life of the resin, most resin mixtures contain a small amount of inhibitor when shipped from the resin manufacturer.

When formulating the resin mixture for curing, the level of inhibitor should be known so that the correct amount of peroxide initiator can be added. When high levels of inhibitor are present, higher levels of initiator should be added to overcome the inhibitor effect and carry out the crosslinking reaction properly.

Promoters

Most organic peroxides are activated to free-radicals by heating the material as it cures in the mold. Some molders desire to avoid that heating, such as when the mold is large and would require an enormous oven to heat it. Therefore, some chemical methods have been developed to react with the organic peroxides and cause them to split apart into free-radicals. These chemical materials are called **promoters** or **accelerators**.

The most common promoters are cobalt compounds, such as cobalt naphthenate or cobalt octoate. The cobalt compound reacts with peroxides, especially MEKP, to create free-radicals. In the process, the cobalt loses an electron (going from Co^{+2} to Co^{+3}) and changes from a purple or pink to a greenish brown color. However, the Co^{+3} is quite unstable under normal curing conditions and it reverts to the Co^{+2} form. It can then react with another MEKP molecules to make more free-radicals. Because of the repeating nature of the cobalt reactions with MEKP, little cobalt is needed, usually less than 0.5% based on the weight of the resin. Sometimes other metal ions, such as calcium, potassium, magnesium, and copper, will further improve the efficiency of the cobalt compound.

Other promoters are the tertiary amines, such as dimethyl analine (DMA) and diethyl analine (DEA). These materials are especially useful in promoting the decomposition of dibenzoyl peroxide (BPO) to form free-radicals. Some peroxides are more susceptible to decomposition from these amines than from cobalt compounds. On the other hand, some resins have been found to cure most efficiently with a combination of the cobalt and tertiary amine promotion systems.

One safety caution is extremely important in using promoters/accelerators. *Promoters should never be mixed directly with peroxides.* The mixture could explode. To ensure that this never happens, either the peroxide or the accelerator should be independently added into the resin/solvent mixture and thoroughly stirred before the other material is added. In practice, the promoter is usually added first, often by the resin manufacturer, since as soon as the peroxide is added, cross-linking may begin even when no promoter is present.

MOLDING COMPOUNDS

Unsaturated polyester molding compounds are made by mixing the resin or resin solution with fillers and/or reinforcements in a step that is separate from the molding operation. This step is often done by a manufacturer whose operation is solely dedicated to making these molding compounds. Such a manufacturer is often called a **compounder**. There are a few advantages to making a molding compound in a step separated from the actual molding operation.

- The compounder's expertise in the material and its properties are sometimes better than the molder.
- Specialized equipment can be used to mix the materials.
- Economies of scale may apply.
- Some time delay is required between compounding and molding, which can be better accommodated by the compounder than the molder in some cases.

Molding compounds are made so that they are ready for molding, without needing any additional mixing or preparation steps. To achieve this ready-to-use status, all the ingredients necessary for curing must be mixed into the molding compound when it is made. Hence, the compounder mixes the initiator and all other components of the curing system, such as the accelerator, with the resin, solvent, and other desired materials (such as fillers and reinforcements). Then the entire mixture is packaged for shipment and storage. The inclusion of the initiator in the mix limits the conditions of storage. If the molding compound gets hot enough to cause the peroxides to activate and form free-radicals, the molding compound will cure, perhaps prematurely. Therefore, storage of molding compounds is usually done at low temperatures. Even at low temperatures, curing will slowly proceed. Therefore, a time limit on storage is also applied to the molding compound.

The properties of the molded parts made with molding compound are dependent upon the resin type, the solvent, the reinforcement, the filler, and possibly even the minor constituents, just as they would be for any other composite product. For example, the long-term stability of polyester thermoset parts in hot water depends strongly on the selection of all the constituents in the molding compound. If any constituent is changed, the hot-water performance can be affected adversely. Even with the best of each type of constituent, polyester thermosets are likely to absorb significant moisture in hot water and lose mechanical and chemical properties. Thus performance in this type of environment requires careful attention to all constituents.

Polyester Thermoset Molding Compounds

Several types of polyester thermoset molding compounds are commonly used. These include alkyds, allylics (principally diallylphthalate [DAP] and diallyl isophthalate [DAIP]), bulk molding compound (BMC), and sheet molding compound (SMC). These differ in the types of unsaturated polyesters used, the types of active diluents, and the method of adding reinforcement, if any. Each is discussed briefly in this section.

Alkyds

Alkyds (pronounced *al'-kids*) are polyester molding compounds based on general-purpose resins. The alkyds are prepared by blending the resin with the initiator and then with cellulose pulp, mineral filler, lubricants, pigments, and possibly short fibers. The mixing is usually done in a heated roller mill that gives some curing to the polymer. When the proper degree of polymerization is reached, the semi-solid material is removed from the rollers, cooled, crushed, and ground into molding powder. This powder is usually compression molded, often requiring only low pressures. (Compression molding is discussed in detail later in this text.) Alkyd parts have high electrical resistance, low moisture sensitivity, and good dimensional stability. However, the alkyds have some tendency to absorb water, thus restricting their application to low-water environments.

When solvated, alkyds are used for paints. The paints have high durability and have been used for many years, but have found limited use because of their relatively slow drying rate.

Allylics

The **allylics** (pronounced *al-lil'-icks*) are a group of unsaturated polyesters based upon resins formed from a monomer containing the allyl group, a particular organic chemistry group containing the carbon-carbon double bond in a particular arrangement. The most common of these monomers are **diallylphthalate** (DAP) and **diallyl isophthalate** (DAIP). Both of these are listed in Table 3-1. When polymerized into a linear, uncrosslinked polymer and then combined with a filler and possibly a reinforcement, these materials are called allylic, DAP, or DAIP molding compounds. The reinforcements are usually short fibers or mineral fibers so that the paste-like consistency can be maintained.

Allylic pastes are used as body putty for repairing automobiles and for many other applications in which a high degree of shaping is required. Allylics are also used extensively for electrical parts, chiefly connectors, switches, bobbins, and insulators. These parts are made by compression or transfer molding (discussed in a later chapter). Allylics are also used for **potting**, the process of encasing an article or assembly in a resinous mass. This is done by placing the article to be potted into a container that serves as a mold, pouring the liquid or paste resin into the mold so that the desired portion of the part is covered with resin, and then curing the resin. After the casting is completed, the mold remains as part of the assembly. Potting is used extensively in the electrical industry to encapsulate wire ends.

Bulk Molding Compound

Bulk molding compound (BMC) or, alternately, dough molding compound (DMC), is made by combining the unsaturated polyester resin with an initiator, filler, and reinforcement fibers. The term **premix** also applies to these mixtures. The mixing of the components is done at room temperature to avoid premature curing and with as little agitation as possible so that long fibers (up to about 2 in. [5 cm]) can be used. The resultant materials are stored at low temperatures to prolong their shelf life. Typical concentrations of BMC components are: resin and styrene = 25%, filler = 45%, reinforcement = 30%, although these concentrations can vary widely depending upon the application.

BMC materials are sticky and have a dough-like consistency, which allows them to be metered, shot by shot, into an open mold. Therefore, the principal molding method is compression molding. This permits largely automated and rapid molding of parts with moderately high complexity. The automobile industry has adopted this method for manufacture of numerous parts, including body panels, grills, trim, air-conditioning ducts, electrical components, and various semi-structural members.

One problem inherent in the use of BMC is the limited amount of material movement that is possible within the mold. Generally, the charge of BMC is placed in the center of a mold and then the material moves to fill the mold as the closing mold presses against the material. Even with careful adjustment of the resin viscosity and using the proper amounts of resin, filler, and reinforcement, movement over long distances in the mold will tend to cause separation of the components. Therefore, parts made by BMC molding are limited to about 16 in. (40 cm) in their longest dimension.

Sheet Molding Compound

Larger parts can be made with **sheet molding compound (SMC)**. The composition of SMC is about the same as BMC, but the method of mixing the components is quite different. In making SMC, the components are not mixed together in a batch blending process. Rather, the pre-initiated resin/filler mixture is doctored onto a moving sheet of polyethylene film as shown in Figure 3-7. The layer of resin/filler mixture is about .25 in. (.6 cm) thick. The reinforcement fibers are chopped to the desired length (typically 1–3 in. [3–7 cm]) and sprinkled onto the resin/filler layer. A second polyethylene sheet that has also been doctored with a layer of resin/filler mixture is then placed on top of the chopped fibers. Thus a sandwich is formed with the reinforcements in the middle, surrounded by the resin/filler mixture, and enclosed by the polyethylene film. This sandwich is passed between rollers that push the fibers into the resin/filler mixture, wetting the fibers. The entire mixture is then rolled to an appropriate size for easy handling and removed for storage at a cool temperature. Fiber length can be longer in SMC because the mixing system is less likely to break the fibers. Also, because SMC has less movement in the mold, the longer fibers do not impede movement as can occur with BMC.

When SMC is to be used, the roll is taken to the molding station, usually a large compression molding machine, and unrolled and cut to the desired lengths. These lengths are typically about the same size in area as the part to be molded. The polyethylene sheets are removed from the SMC sandwich as the material is placed into the mold. Sufficient layers of SMC sheet are placed in the mold

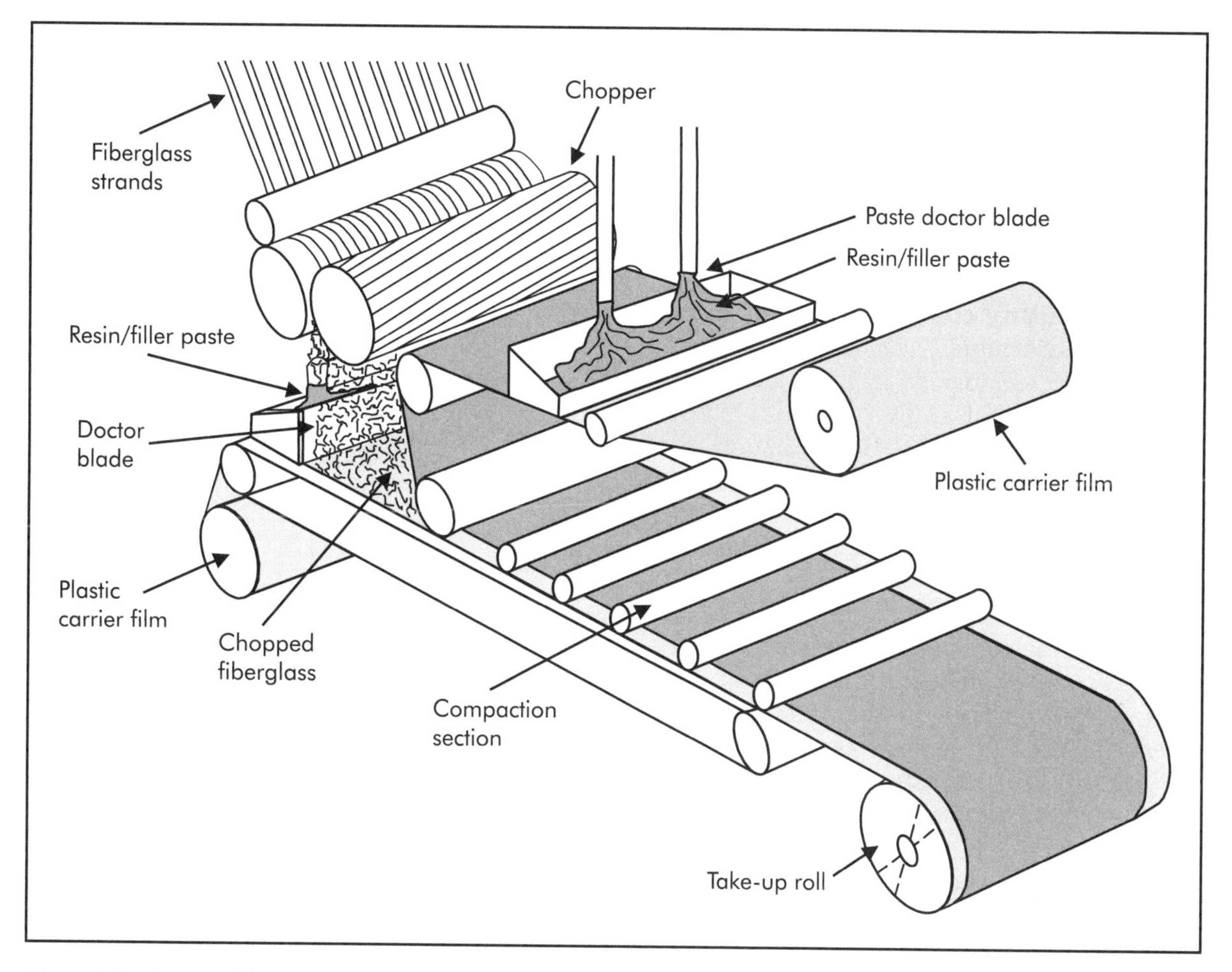

Figure 3-7. Sheet molding compound (SMC) machine.

to obtain the thickness desired. If additional thickness is needed in some locations but is not wanted in other places, smaller strips of SMC can be laid into the mold. When all of the material has been properly placed into the mold, the mold is closed, and the part is cured. Because the SMC is placed so that it covers the surface of the mold, little movement of material occurs within the mold. However, there must be enough movement of the resin to bind the various layers of SMC together. Large parts can be made by this process. A well-known example of a part made from SMC is the body of the Corvette® (General Motors).

Molding Systems

The use of BMC and SMC for exterior body panels in automobiles has required the development of a molding system that will result in a smooth, defect-free surface, often called a Class A surface. The inherent shrinkage of polyester materials as they crosslink (often 7–10%) normally results in fiber show-through and small sink marks and dimples that are unacceptable in the automotive panel market. A special additive system has been developed to solve this problem. It is called the **low-profile** or **low-shrink system**. The system is based upon the addition of about 2–5% low-profile additive consisting of thermoplastic resins and additional fillers containing divalent metal ions, such as Ca^{+2} and Mg^{+2}. The addition of these materials to polyester resins results in thickening of the resin/fiber/filler mixture. This process is called **ripening**, which results from the formation of secondary bonds between the polymer and the metal ions. Reactive chemistries, such as the reaction of urethanes, also have been used to provide ripening of polyesters in place of the metallic ion systems. The mechanisms for thickening and low-profile expansion have not been clearly defined and are, therefore, a matter of some ongoing research.

The thickening of BMC and SMC occurs to some extent even without the presence of the low-profile additives because of the normal curing that takes place with an initiated resin system. But the low-profile system gives control and specific parameters to the thickening and, of course, produces the expansion upon curing that compensates for normal polyester cure shrinkage. Thus the use of thermoset polyesters in automobiles has increased steadily for many years.

PROPERTY OPTIMIZATION

Viscosity

Some components are relatively minor in volume but can be important in optimizing the performance of the part. For example, viscosity control can be important to achieve uniform distribution of all the components in the mix. The resin might be so thin that on a vertical surface, the resin runs down, thereby leaving resin-poor areas near the top of the surface. This can be prevented by adding a material that thickens the resin. Such a material is called a **thixotrope**. One group of popular thixotropes works by increasing the hydrogen bonding in the mixture and, consequently, building viscosity.

Control of viscosity might be needed in just the opposite way—the mixture may be so viscous that fibers and fillers might not be fully wetted. It is possible to add a **thinner** to the mixture that will reduce the viscosity and allow wetting of the solid materials. Such a material can allow greater filler and fiber loading levels. A popular thinner consists of salts of organic acids in a solution that is easy to add to the resin before the fillers and fibers are mixed in.

Curing

The curing of a polyester resin is a complex process and is critically important to achieving good composite part properties. The cure can be affected by most of the components of the polyester system and by innumerable variables in the conditions under which curing takes place. Therefore, this section is intended to give an overall view of the effects of the most common variables in the system, components, and conditions, and to suggest ways in which these variables might be altered to achieve optimal part properties.

The characteristics of the cure of an actual part are not always easy to measure because it is difficult to put the appropriate monitoring devices (thermocouples, pressure transducers, etc.) in the right area without disrupting the cure or final properties. Therefore, some laboratory tests have been developed that give a reasonable approximation of cure characteristics. The most common laboratory tests are **cure time** and **exotherm peak** measurements. They are performed by placing a measured amount of the polyester resin mixture in a small cup that has been instrumented with a thermocouple. The cup is then placed into a thermal bath or on a hot plate that has been set to a particular temperature (usually about 180° F [82° C]). A stopwatch is started and the material is carefully stirred. After a few minutes the material will suddenly thicken. This is the onset of **gelation**—that point when the crosslinking reaction is dominating the viscosity and a permanent deformation of the resin is observed. This is the **gelation (or gel) point** and the time to this point is called the **gel time**. These relationships are shown in Figure 3-8.

The combination of the gelation and exotherm conditions can give a reasonable understanding of the conditions that exist during the cure of a real part. For instance, if the gel time is short, the part will generally cure in a faster time. A high-temperature exotherm indicates a hot mixture, and generally that the cure will proceed to completion provided the time to peak exotherm temperature is not too short, which would indicate that the peroxide initiator is breaking apart too quickly. Remember also that the gel/exotherm test is done using a set amount of material. Further, the real part to be molded is not only larger but often of a different thickness. In general, as the thickness increases, the peak exotherm temperature also increases. This is caused by the heat retention of the polymer material and the accompanying higher reactivity when heat is retained.

Increasing the initiator (MEKP) concentration along with a constant concentration of accelerator (cobalt naphthenate) will decrease the gel time since there are more free-radicals present and they initiate more crosslinking reactions. But this effect gradually levels off since the higher

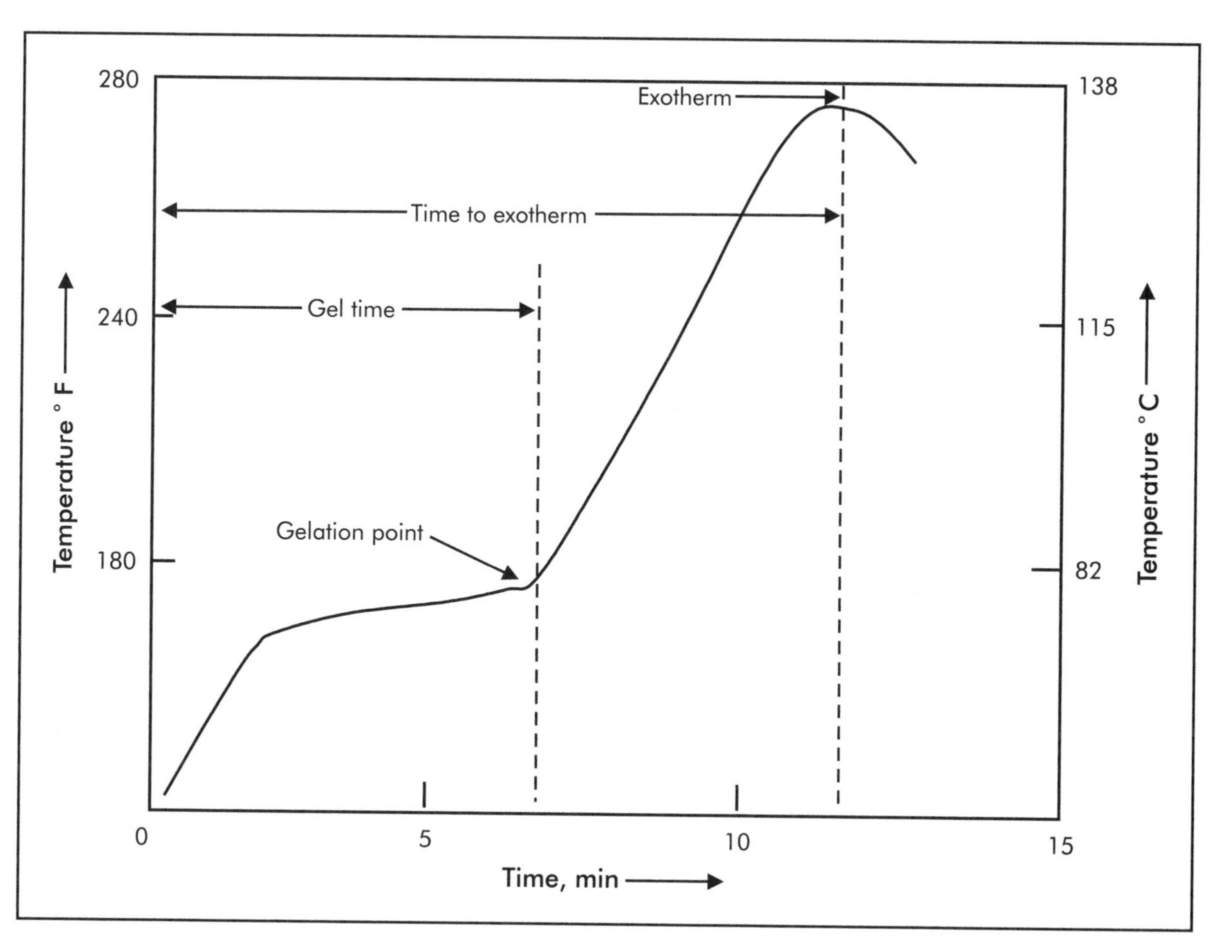

Figure 3-8. Gel time and exotherm measurements.

concentration of initiator will not all become active without also increasing the amount of accelerator. If the initiator concentration and the amount of accelerator (cobalt naphthenate) both increased, the decrease in gel time would be even greater. Remember, however, that the extent of crosslinking will eventually decrease with increasing initiator concentration. Not only will the cure reaction be much faster when the concentration of initiator is increased, but the peak exotherm temperature also will be higher. Hence, the reaction is both faster and hotter since the number of free-radicals available is greater.

Several of the most important factors in controlling the cure of an unsaturated polyester thermoset part are *summarized* in Figure 3-9. Each of the properties appears on a teeter-totter arranged according to whether it generally makes the curing reaction go slower or faster. The action of these factors will help to modify or interpret the gel/exotherm test results in light of the actual conditions that exist in the real part. Then the variables can be divided into three groups—those that make the cure reaction go slower, faster, or either way.

Following are the variables that make the curing reaction go slower.

- ***Inhibitors***. These are chemicals that absorb free-radicals. They are generally added to the resin mixture by the resin

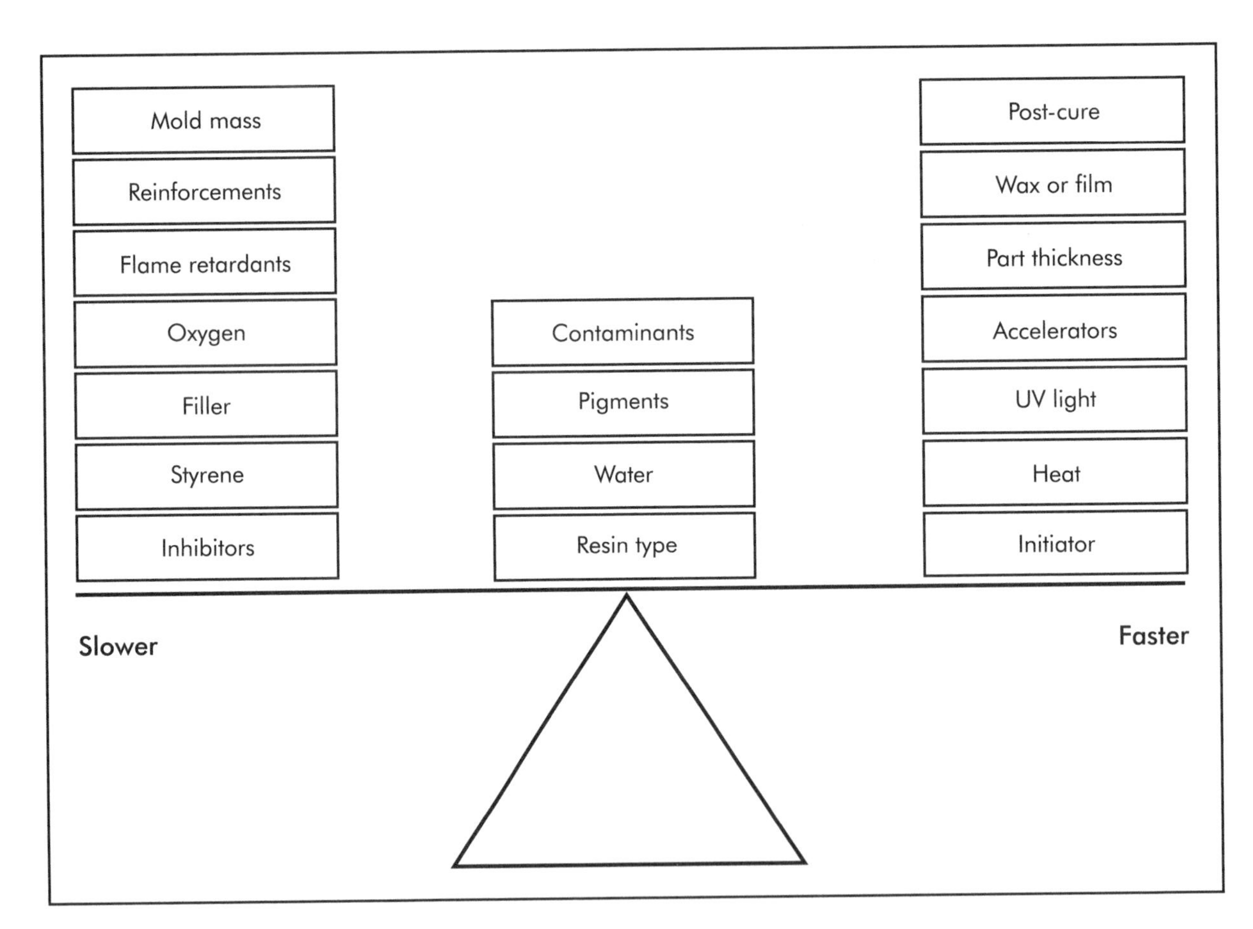

Figure 3-9. Additives to the polyester mix and their general initial effects on curing rate.

manufacturer to stop or interrupt any crosslinking chain reaction that might be started by the chance creation of free-radicals. Because free-radicals can be created through several different mechanisms, such as interactions between the molecules with heat and light, the possibility of having chance creations of free-radicals is quite high. Should these chance reactions occur while the resin is stored, the entire batch of resin could be prematurely crosslinked unless protected with inhibitors. When planned curing is to be done, sufficient peroxide initiator must be added first to react with any remaining inhibitor and then react with the resin. Some approximation of the amount of inhibitor remaining in the resin can be gained from gel/exotherm tests.

- ***Styrene or other crosslinking agents***. Two different effects are occurring simultaneously with styrene and the other crosslinking agents. The first is the dilution effect. That is, styrene is a *solvent* for the system. Therefore, increasing the amount of styrene will dilute the total system and add mass, which must be heated in the reaction. Therefore, additional styrene will slow the reaction. However, the second effect of styrene is as the *crosslinking agent*. In this role, the styrene makes the crosslinking occur more quickly and efficiently. This effect would generally speed up the reaction, although it is usually seen later in the overall curing cycle. Styrene is placed on the teeter-totter on the slower side (see Figure 3-8) because of its initial effect.
- ***Filler***. These materials are usually powdered inorganic materials such as calcium carbonate, calcium sulfate, and talc. They add mass to the system and absorb some of the heat generated by the crosslinking reactions, thus reducing the exotherm and slowing the overall reaction.
- ***Oxygen***. Normal oxygen gas is made up mostly of O_2 molecules but has a small amount of ozone (O_3) molecules present as well. These ozone molecules are like inhibitors in their ability to absorb free-radicals. Therefore, if oxygen is present or added to the system, the curing reaction is slowed. Sometimes oxygen is actually used to stop the reaction and in these instances it is called a **quench**. Since oxygen is almost always present, the amount of peroxide initiator used is normally adjusted to overcome its effects. Alternately, the oxygen can be excluded from the reaction area by using some surface-coating material like wax or a film.
- ***Flame retardants***. The action of these materials can be complicated, depending on the type of flame retardant used. Some flame retardants contain metals, which might serve as accelerators, while others might serve as free-radical absorbers and therefore inhibit. Therefore, the simple designation of them as favoring slowing the reaction is based primarily on their additional mass and assumes they have little direct chemical interaction in the crosslinking reactions. It is best, however, to check each flame-retardant type for specific reactivity.
- ***Reinforcements***. The fibers are usually inert to the crosslinking reactions. Therefore, their principal effect is added mass and the subsequent absorption of heat that occurs.
- ***Mold heat capacity***. The mold is composed of a massive amount of material that acts as a heat sink, perhaps the most important mass in the entire system. Because the mold is often at room

temperature, heat is absorbed out of the resin system and thus the curing reaction is retarded. In some cases, however, the mold is heated. When heated, the mold will accelerate the curing mechanism. Hence, the classification of the mold as slowing the reaction is simply a reflection of the normal, cold state that occurs during most molding.

Here are the variables that make the curing reaction go faster.

- ***Initiators***. These materials are generally peroxides. The peroxides are useful materials because they break apart easily with heat or chemicals to form free-radicals that initiate the crosslinking reaction. If a part is to be cured with heat, then a heat-activated peroxide is used. If the part is to be cured without heating, then a combination of a peroxide and an accelerator system is needed. Consideration should be given to final part characteristics, such as color, as these can be affected by the peroxide chosen.
- ***Heat***. Heat will increase the rate of almost any chemical reaction for several different, although interrelated, reasons. Molecules must collide to react, and when molecules are heated, they move more. Therefore, the chance they will collide is increased. Also, the viscosity of the mix is reduced by heating so the molecules move more freely and colliding likelihood is increased. Higher heat also increases the energy of movement of the molecules. This means that when the molecules collide, they will collide harder and, therefore, have a higher chance of creating an effective collision. (Not all collisions result in bonding.) The heating of the molecules increases the overall energy in the system, thus making it easier for the molecules to overcome the energy of activation, which is a threshold barrier for effective reactions to occur. Yet another effect of heating is to increase the break-up of the peroxide molecules, thus increasing the number of chains started and, therefore, the rate of the curing reaction. Heat also can cause direct formation of free-radicals at the carbon-carbon double bonds in the polyester molecules. It is this possibility that leads to the addition of inhibitors to prevent crosslinking from occurring prematurely.
- ***UV light (sunlight)***. This type of energy will create free-radicals at the carbon-carbon double bonds. To prevent this from happening, resins are usually stored in opaque containers and inhibitors are added. However, in some resin systems, such as those used in dental fillings, UV light is intentionally introduced into the system by shining a special light on the polymer. This UV light can cause crosslinking to occur without the use of peroxides or external heat.
- ***Accelerators (promoters)***. These materials are added to the system to improve the efficiency of the initiator. The most common accelerators are organic compounds, such as naphthanates, of cobalt or some other metal. Usually only a small amount is needed to greatly improve the efficiency of the peroxide. With accelerators, some peroxide systems do not need to be heated and room temperature cures can be done. Some accelerator systems are further improved by adding aniline compounds, such as DMA, which are sometimes called co-accelerators. With some peroxides, the co-accelerators are effective without the organic metal accelerators. Because of the sensitivities of peroxides to the accelerators, great care should be taken to control the accelerator system

concentrations and the order of addition of the peroxide and accelerators.

- ***Part thickness***. When parts are thick, the resin retains heat and therefore increases the overall temperature and reaction rate. This behavior can be a major problem when parts have significant differences in thickness. Not only will the reaction rate be higher in the thicker areas, but the amount of crosslinking that occurs could be higher in those thicker areas. This could cause embrittlement in the thick areas and internal stresses in the part because of the increase in part shrinkage that occurs with higher degrees of crosslinking. Furthermore, the higher heat in the thicker regions can cause greater thermal shrinkage and result in cracking or part deformations like sink marks and dimples. Hence, parts should be carefully designed to minimize thickness variations.
- ***Wax or films***. These materials are added to exclude oxygen and prevent its inhibition effect, which can occur when it reacts with free-radicals, thus quenching the reaction. Since the greatest inhibition effect is usually on the surface exposed to the oxygen, a wax coating or film can be added only to that surface to prevent a decrease in surface properties, such as hardness or gloss. Some polyester resin systems contain small amounts of wax. During the cure, the wax rises to the top and forms a protective layer that excludes the oxygen. This layer of wax can be a problem if the resultant part is to be bonded to some other object because bonding agents do not adhere well to most waxes.
- ***Post-cures***. Adding heat after the part is formed is called **post-curing**. This is done to generate additional movement of the molecules and provide greater efficiency of molecular collisions, thereby resulting in the formation of more crosslinks. The heating also can cause any residual peroxide to break apart and initiate some additional chains.

Following are the variables that can make the curing reaction go either faster or slower.

- ***Resin grade/type***. The resin is the most complicated component in the entire mixture and is, therefore, so variable in its effects that it cannot be classified on either side of the teeter-totter (see Figure 3-9). Some resins, especially those with a high degree of unsaturation, will cure quickly and result in high-temperature exotherms and high crosslink densities. Others, with less unsaturation or with unsaturation sites that are less reactive, proceed more slowly. Some resins are especially sensitive to certain initiators and therefore proceed rapidly when they are used but more slowly with others. The length of the chain of the resin before crosslinking can affect the reaction rate. Long chains can increase system viscosity and therefore slow the reaction rate. Resins might even be mixtures of two or more types of polymers, each with its own reactivity. Therefore, the manufacturer of the resin should be consulted about resin reactivity as it might occur in the particular system used.
- ***Water***. Water adds to the total mass in the system, thus slowing the reaction by absorbing heat. On the other hand, water can increase reactivity in some peroxide systems, although it can retard rates in others. Some resins are less reactive when water is present. Water can also affect the action of accelerator systems. The action of water is, therefore, difficult to predict. The best practice is to monitor and record humidity and

water content in the system and then attempt to correlate its effect on the overall system.

- ***Pigments***. Usually pigments have little effect on the system. However, since some contain metals, such as iron oxides, they can have an accelerating effect. In contrast to that, however, some metals will interfere with the basic crosslinking reactions. Pigments add mass, thus retarding the system, all else being equal.
- ***Contaminants***. Because the nature of the contaminant is not specified, the nature of the effect is not known. However, the fact that the effect is unknown invites variation in the process and, potentially, the production of poor quality parts. Some contaminants are significant parts of the environment in some manufacturing plants and, therefore, may have a more predictable and repeatable effect. It is good to be aware of the possibility for such contamination and act to control and limit its presence.

In general, if the reaction proceeds too slowly, the cure is likely to be lower than desired because contamination has had time to reduce the amount of crosslinking. Long cures will also increase the number of molds needed, thus increasing the cost to achieve the same production volumes as could be attained with faster cures.

If the cure proceeds too quickly because too much heat is generated, the part could be too brittle. Also, the crosslinking might proceed so quickly that all of the styrene is not used and the resulting part could have residual styrene content. Further, if the cure proceeds quickly because too much peroxide was used, incomplete crosslinking would likely result and that would lead to poor physical properties and poor solvent and water resistance. All of the other variables that may increase reaction rate could likewise cause problems with the part.

Cures that are too slow and too fast can be problems. The best situation is to understand, monitor, and control the effects of the various components in the mixture and the variables that control the rate. The cure reaction is at the heart of polyester molding and must be carefully controlled to make good quality parts over the long run.

Additives

Other additives might be added for long-term, cured-product improvement. Often specialty chemicals that protect against ultraviolet (UV) light degradation are added to the resin before mixing with other components. Antioxidant protective chemicals might similarly be added. Usually these materials are added in small percentages (less than 10%). Some fillers may have UV protective capabilities and could be added by themselves or in combination with other fillers. The most common UV-protecting filler is carbon black. Pigments (solid inorganic materials) and dyes (organic colorants) are frequently added to polyester mixtures. These are often in low concentrations (less than 10%) and generally have little effect on other properties. However, some colorants may affect the cure system (they can act as accelerators) and therefore caution and experimentation should be observed before changing to a new colorant.

CASE STUDY 3-1

Reduced Emissions Resins

Styrene is the most widely used co-reactant or reactive diluent for polyester systems. It works well and is relatively low cost. However, it also has a high vapor pressure and a potentially harmful vapor if breathed in excess. Therefore, the U.S. and other governments have established maximum air-quality limits for styrene. These limits have been gradually reduced so that complying with them is becoming more difficult for FRP molders. Hence, the resin companies and molders have embarked on several different paths to attempt to reduce the amount of styrene emissions.

Strategy 1: Modify existing polymer systems. The initial effort by almost all of the resin manufacturers was to modify their existing resin systems so that less styrene would be needed. This strategy was not only prudent commercially, it also gave quick results.

The resin manufacturers realized that if the viscosity of the base resin were lower, the amount of styrene needed to achieve a good viscosity for fiber wet-out could be reduced. The first method to reduce the viscosity of the base resin was, therefore, to shorten the molecular chains; that is, to reduce the molecular weight of the resin. Shorter chains have less intermolecular entanglement and, therefore, flow more easily. This means the viscosity is lower.

The amount of styrene needed in these lower-molecular-weight resin systems has been successfully reduced. That reduced level of styrene, which was dictated by viscosity control requirements, has also reduced the amount of styrene available for crosslinking. Therefore, the nature of the crosslinking has also changed. The lower amount of styrene available for crosslinking results in a "starved" situation during crosslinking. This starving means that the number of styrene molecules in each of the crosslinks is reduced. However, the shorter polymer chains allow the carbon-carbon double bonds to occur more frequently. This results in a higher number of crosslinks being formed overall. Therefore, the effects on brittleness from the reduction in styrene content and from the increase in crosslink density approximately cancel each other out. So, brittleness is about where it would be in the unmodified resin system.

Generally heat distortion temperature and solvent resistance will be increased by higher crosslink density. Other properties, such as blister resistance, ultraviolet light resistance, chalking, cracking, and surface smoothness, also may be affected. But the extent of these effects must be evaluated on a case-by-case basis with each resin and application. These effects are much more important in gel coats than in laminating.

Strategy 2: Develop new polymer systems. Polyesters are made from the reactions of diacids and glycols. Resin producers have found that by varying the types and quantities of diacids and glycols, the resin properties can be significantly changed. Some of the acids and glycols, such as neopentyl glycol, are known to be more soluble in styrene than the other common components. Therefore, they are used in higher concentrations in reduced-emission resins. At the same time, these alternate components might give additional benefits such as increased toughness or heat distortion temperature. Therefore, the new resins might have both reduced styrene content and improved physical properties, or at least no major reduction in properties as a result of the decrease in styrene.

Some resin manufacturers have found that new resin systems requiring lower styrene concentrations can be made by varying the sequence of the addition of the diacid and glycol components. These changes occur because mixtures of multiple types of diacids and glycols are often used in making polyester resins that have different reactivities. An example of such a change would occur under the following two scenarios. If two diacids are mixed together and then added to the glycols, the distribution of the acids could be random. However, if one of the diacids is added to the glycols and then the second is added later, the result could be a preferential positioning of the second diacid at the ends of the polymer. Hence, two very different polymers could be formed.

Another major area of focus is the blending of unsaturated polyesters with other polymer types. These blends can be done as monomers, often referred to as *copolymers*, or as blends after full polymerization has been done, referred to as *mixtures*. Some of the interesting blends with unsaturated polyesters include: polyurethanes, melamines, phenolics, epoxies, and vinyl esters. In general, these blended polymers are higher in cost than the basic polyester, but the properties of the blend can be substantially better, thus allowing for the use of less resin or for entering new markets.

Strategy 3: Use other monomers. Styrene is not the only monomer that acts as a reactive diluent for unsaturated polyesters and vinyl esters. Therefore, some alternate monomers, which have lower vapor pressures, are being investigated as styrene alternatives. One approach is to choose monomers that are similar to styrene in chemical reactivity but have a higher molecular weight and, therefore, a lower vapor pressure. For instance, vinyl toluene is similar to styrene. It has an additional methyl group on the benzene ring, but has a lower vapor pressure and, therefore, lower air-pollution potential. Another example is t-butyl styrene. The major drawback to using alternate monomers is their higher cost.

Another approach is to simply combine two or three styrene molecules together and then use these dimers and trimers as the reactive diluents. Short chains such as these are sometimes called **oligomers**. Performance of these materials is almost the same in the cured product, and wet-out of the fiberglass is still excellent since only enough is added to achieve the same viscosity used with the monomer. However, the cost of producing the dimers and trimers is significant.

Some have been looking at using completely new monomers that are not like styrene at all. The prime example is methyl methacrylate (MMA), sometimes known as *acrylic*. This material also has a high vapor pressure, so it too is considered a hazardous air pollutant (HAP). However, it may be more compatible with some polymer systems and, therefore, allow lower concentrations to be used to achieve acceptable viscosities. MMA also can give some improved properties such as lower smoke generation. Again, cost is a drawback.

Strategy 4: Form a surface layer to suppress the emission of styrene. Such layers, often formed by adding wax to the resin mixture, have been used for many years to exclude oxygen so that cures can be improved. Hence, this concept is really using an old technology for a new purpose. The problem with this method has always been the potential for poor bonding wherever the wax is present. This adhesion problem has been alleviated somewhat by creating adhesion-promoting films. These films contain carbon-carbon double bonds, thus allowing crosslinking to occur through the film. Adhesion between materials on both sides of the film is thus promoted.

CASE STUDY 3-2

New Curing Options

Some new molecular concepts have been shown to improve or, at least, present new options for curing polyesters. These innovative cure systems may not apply to all situations, but should at least be considered for the unique advantages that they offer.

A recent development by Landec Corporation, called the Intelimer® system, is a unique heat-activated polymer material that has a definite trigger point for commencement of the crosslinking reaction. This polymer is not a polyester and has nothing to do with the final polyester resin. However, Intelimer has the unique property of a sharp melting point that can be chosen to fit the needs of a particular system.

In the Intelimer system, one of the key components of the cure is encapsulated in the special polymer. The melting point of the polymer is chosen to be at the lower range of the temperatures used in curing the polyester. Systems with the following encapsulated components are available: organic cobalt (an accelerator), tertiary amine (co-accelerator), dibutyl tin dilaurate (accelerator), and tert-butyl perbenzoate (initiator). These materials, when encapsulated, are protected by the polymer and do not interact with the cure system. When the melting point of the encapsulating polymer is reached, the component is released and it immediately begins to interact with the other components of the cure system.

This type of system has several advantages. First, the shelf life of the resin is extended because some of the critical materials, like the initiator, can be encapsulated, thus making premature activation less likely. Another advantage is more rapid release of the critical component, which can result in more uniform curing and lower residual styrene content since all of the resin is at temperature when the curing begins. Physical properties of the final product are comparable, if not better. One improvement is the enhanced viscosity control in the thermally activated systems because only viscosity is dependent on the temperature until the trigger temperature is reached, thus improving the ability to wet the fibers in a repeatable manner.

SUMMARY

Polyester thermosets (PT) or unsaturated polyesters (UP) are the most common of all the resins used in composites. This market dominance is chiefly because of their low price, the ability to make many types of products with differing properties, and ease of molding with these resins.

The polymerization process begins with the reaction of a glycol with a diacid. When these monomers react, they form an ester group that links the two monomers together. Water is also formed as a by-product. The new molecule still has two reactive ends, which allow further reactions with additional monomers, thus creating additional ester links and longer chains. These reactions can continue as long as the monomer concentrations and the reaction conditions will support additional ester linkage formation.

Various diacids and glycols can be used as monomers in thermoset polyesters. These are summarized in Tables 3-2 and 3-3. Some of the monomers must contain a carbon-carbon double bond or unsaturation point for the crosslinking reaction to occur. Usually, the diacid is the monomer containing this feature and several different unsaturated diacids are used commercially. Experience has shown that the best polyester properties are achieved when some of the diacids are not unsaturated and not, therefore, potential crosslink sites. The most common of these saturated acids are ortho and iso. Several other diacids are also

Table 3-2. Common diacids used in polyesters and their attributes.

Diacids/ Anhydrides	Attributes
Maleic/fumaric	Unsaturation (crosslink site)
Orthophthalic	Low cost, styrene compatibility
Isophthalic	Strength, water/ chemical resistance
Terephthalic	Thermal stability
Adipic	Toughness, weatherability
Halogenated	Flammability resistance

Table 3-3. Attributes of glycols commonly used in polyester resins.

Glycols	Attributes
Ethylene	Low cost
Propylene	Styrene compatibility
Diethylene	Toughness
Neopentyl	Weathering, water/ chemical resistance
Bisphenol A	Water/chemical resistance, strength

Table 3-4. Purpose of some polymeric additives used in polyester resins.

Resin	Purpose
Dicyclopentadiene (DCPD)	Lower cost, improved stiffness
Styrene butadiene rubber (SBR)	Toughness
Thermoplastics	Surface quality, toughness

used commercially, each imparting its own particular advantages.

Various glycols can be used to modify and improve the properties of the polyesters. The most common properties added with the various glycols are toughness and resistance to environmental factors.

The curing or crosslinking reaction in unsaturated polyesters uses the same general mechanism as addition polymerization. An initiator, usually an organic peroxide, is activated by heat, time, or chemicals to form free-radicals, which can combine with carbon-carbon double bond unsaturation sites. The activated polymer can then combine with any other unsaturation site to create a new bond and another free-radical. Normally bridges of two or three styrene molecules link polymers together. Eventually all of the polymers in the mixture are linked together in a great mass.

Additional polymeric materials might be added for various purposes. The most common of these are summarized in Table 3-4.

Some additives to control the curing mechanism are important in polyester mixes. For example, inhibitors are used to extend the shelf life or pot life of polyesters. Accelerators are other important additives that make the curing reaction proceed faster or at lower temperatures.

Various molding compounds are sold as ready-to-use molding materials. Some are based on allylic materials, such as DAP, and others use general-purpose polyester resins, such as BMC and SMC. All the molding compounds have high filler contents. Some contain low-profile additives to improve the surface of the molded composite part. These low-profile systems are especially important in making automotive body panels.

The cure system is affected by many variables, including the following: mix ratios and types of resin, initiator, inhibitor, accelerator, and solvent/crosslinking agent; storage time after activation; thickness of the part; cure temperature; fillers; colorants; water; contaminants; and others.

Polyester resins are made from a series of building blocks of monomers and additives. Table 3-5 shows some generic recipes to illustrate how various components might be added to achieve particular properties.

LABORATORY EXPERIMENT 3-1

Effects of Concentration Changes in Curing Agents

Objective: Determine the effects of peroxide concentration and accelerator concentration on gel time and various physical properties (flexural strength, toughness, HDT, etc.).

Table 3-5. Generic recipes for unsaturated polyesters.

Purpose or Characteristic	Components				
	Unsaturated Acid	Saturated Acid	Glycol	Reactive Diluent	Specialty Monomer
Low cost	Maleic anhydride	Orthophthalic	Ethylene	Styrene	None
Flexible (low cost)	Maleic	Orthophthalic	Diethylene	Styrene	None
Gel coat	Maleic	Isophthalic	Propylene glycol and neopentyl	Styrene	None
High-temperature performance	Fumaric	Bisphenol-A	Ethylene	Styrene	None
Flame retardant	Maleic	Tetrabromophthalic anhydride	Ethylene	Styrene	None
High flexural properties, low cost	Maleic anhydride	Orthophthalic	Propylene	Styrene	DCPD
Weathering	Maleic	Adipic	Propylene	Methyl methacrylate	None

Procedure:

1. Obtain some general-purpose polyester resin and place it in six cups. Each cup should have the same weight of resin and the total volume should not be more than about 1 in. (2 cm) high in the cup.
2. Mix in three different concentrations of MEKP (initiator), two cups with each concentration. Suggested concentrations are 0.5%, 1.0%, and 2.0%.
3. After the peroxides are thoroughly mixed into the resins, add two different concentrations of cobalt naphthenate (accelerator) so that each of the MEKP concentrations has both accelerator concentrations and mix well with the other materials. Suggested concentrations are 0.004% and 0.008%.
4. Start a stopwatch and continuously stir each cup and note the time when the material becomes so thick that it takes a permanent set.
5. Put a thermocouple into each cup at the same time as the watch is started and measure the rise in temperature.
6. Plot the results on coordinates of gel time versus MEKP concentration (with lines for both accelerator levels), and temperature versus MEKP concentration (with lines for both accelerator concentrations).

QUESTIONS

1. What is the meaning of "unsaturated" with respect to polyesters and what is the component usually responsible for imparting unsaturation to the polymer?
2. What is a peroxide and how does it participate in polyester curing?
3. What is an inhibitor or stabilizer?
4. What is an accelerator or promoter?
5. Why is styrene especially useful as a diluent or solvent for polyesters?
6. What is an exotherm and how is the peak exotherm measured?
7. Why are exotherms greater when parts being cured are thicker?
8. What are the most important chemical differences between a thermoset polyester and a thermoplastic polyester, before and after curing?
9. Describe the difference between "summer" and "winter" polyester resins.
10. What is the difference between an initiator and a catalyst?
11. Give three methods to terminate a polyester curing reaction.
12. Describe the distinguishing characteristics of aromatic and aliphatic polyesters.
13. What does the name "polyester" refer to?
14. What is a low-profile additive used for?
15. What factor causes iso and ortho resins to behave differently?
16. Some companies heat SMC material just prior to putting it in the mold. Give a reason why this may be done.

BIBLIOGRAPHY

Landec Corporation. No date. "Intelimer® Catalysts Have a Pre-set 'Switch' Temperature." Brochure published by Landec Corporation, 3603 Haven Ave., Menlo Park, CA 94025, phone 800-817-7376.

Society of Manufacturing Engineers (SME). 2005. "Composite Materials" DVD from the Composites Manufacturing video series. Dearborn, MI: Society of Manufacturing Engineers, www.sme.org/cmvs.

Strong, A. Brent. 2006. *Plastics: Materials and Processing*, 3rd Edition. Upper Saddle River, NJ: Prentice-Hall, Inc.

Strong, A. Brent. 1995. *Fundamentals of Polymer Resins for Composite Manufacturing*. Arlington, VA: Composites Fabricators Association.

4

Epoxies

CHAPTER OVERVIEW

This chapter examines the following concepts:

- Overview of epoxy and its uses
- Structure of the polymer
- Crosslinking and processing parameters
- Physical and mechanical properties of cured epoxy composites
- Special composite applications (prepregs, tooling)

OVERVIEW OF EPOXY AND ITS USES

Epoxies (pronounced *ee-pox'-ees*) are the second most widely used family of thermosetting resins for composite applications (after polyester thermosets). In contrast to the polyester thermosets, about 50% of all epoxy resins are used for nonreinforced applications. The most important of these is for coatings (paint) on products such as ships, metal pipes and tanks, cars, appliances, industrial machines, and numerous devices that need an especially durable coating. Another major application for nonreinforced epoxy resins is as adhesives. In this application, epoxy resins are often used in conjunction with many other types of composite materials, not just those made with epoxy as the matrix.

The largest market for reinforced epoxy resins in laminate sheets, which is about 25% of the total of all epoxies produced, is for electrical circuit boards. Epoxies are especially useful for these boards because of their inherent low conductivity and high dielectric strength. Also, epoxies have a low tendency to emit gases even when subjected to an electrical discharge. Many other thermoset resins emit gases from unreacted crosslinking agent/solvent or from unreacted monomer. Another advantage of epoxies in this application is their thermal stability which, although not among the highest of thermoset materials, is higher than polyester thermosets and most other low-cost thermosets that might compete with them in the electrical circuit board market.

The relatively good thermal stability of epoxies, their excellent adhesive properties, and their good mechanical properties have led to their widespread use as the principal resin in most **high-performance composites** (chiefly those using carbon or graphite fibers as the reinforcement). The markets that most often use high-performance composites are aerospace, sporting goods, and medical devices. However, many other markets for advanced composite materials are emerging as the price of the carbon/graphite fibers decreases. This is discussed in more detail in the chapters on reinforcements and applications.

STRUCTURE OF THE POLYMER

The epoxy resins are characterized by the presence of the three-member ring **epoxy group**, sometimes called the **oxirane**

group, which is where crosslinking occurs in epoxies and which gives epoxy polymers many of their characteristic properties. This epoxy group can be attached directly to another organic group as illustrated in Figure 4-1a. (To be general in the representation, the organic group is simply represented with an "R." This is a standard representation in organic chemistry and is meant to indicate that many organic groups could be chosen. In some epoxy molecules, the R is a simple, small organic group, while in others the R is a complex group and could even be a polymer). If the epoxy ring is attached to another carbon, which is then attached to some organic group, the combination of the epoxy ring and the bridge carbon is called the **glycidyl group** as illustrated in Figure 4-1b.

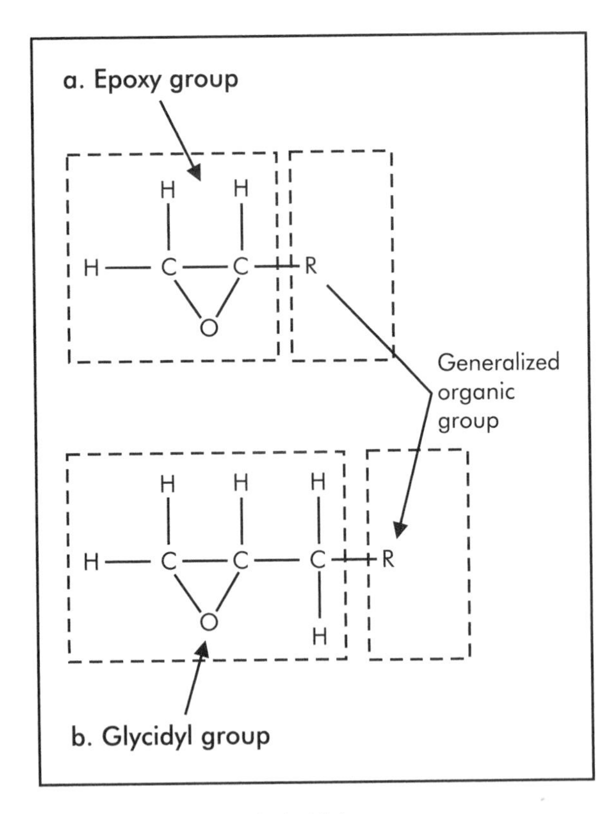

Figure 4-1. Epoxy and glycidyl groups.

The epoxy polymer can be thought of as being composed of two parts—the epoxy ring (either as just the three-member ring or with the bridge carbon) and the rest of the polymer, which is represented as the R group. The two parts of the molecule are considered separately to obtain a simplified understanding of epoxy resins. Then, to see a more complete picture, the nature of epoxies is considered when both the epoxy ring and the remainder of the polymer are jointly discussed.

Epoxy Ring

As indicated previously, the epoxy rings are the sites where crosslinking occurs. As such, these rings have been called "active sites" in previous chapters and are analogous in this sense to the carbon-carbon double bond sites in unsaturated polyesters. However, there is a major difference between the frequency of occurrence of the epoxy rings along the epoxy polymer backbone and the occurrence of the carbon-carbon double bonds in unsaturated polyesters. Whereas the carbon-carbon double bonds occur frequently along the polymer backbone and possibly as often as every 10-20 atoms in normal commercial polymers, the epoxy rings or glycidyl groups are only attached to the ends of the polymer or the ends of branch chains as illustrated in Figure 4-2.

Several methods are used to synthesize the epoxy polymer, that is, to add the epoxy ring to the ends of a polymer chain or to create a polymer containing the epoxy rings. These reactions are done by the manufacturer of the epoxy resin and are not part of the curing or molding operation. The molecular synthesis depicted in Figure 4-3 is the most common. The starting materials are bisphenol A and epichlorohydrin, a highly reactive monomer. The reaction bonds the epoxy rings, actual glycidyl groups, to both sides of the aromatic segment.

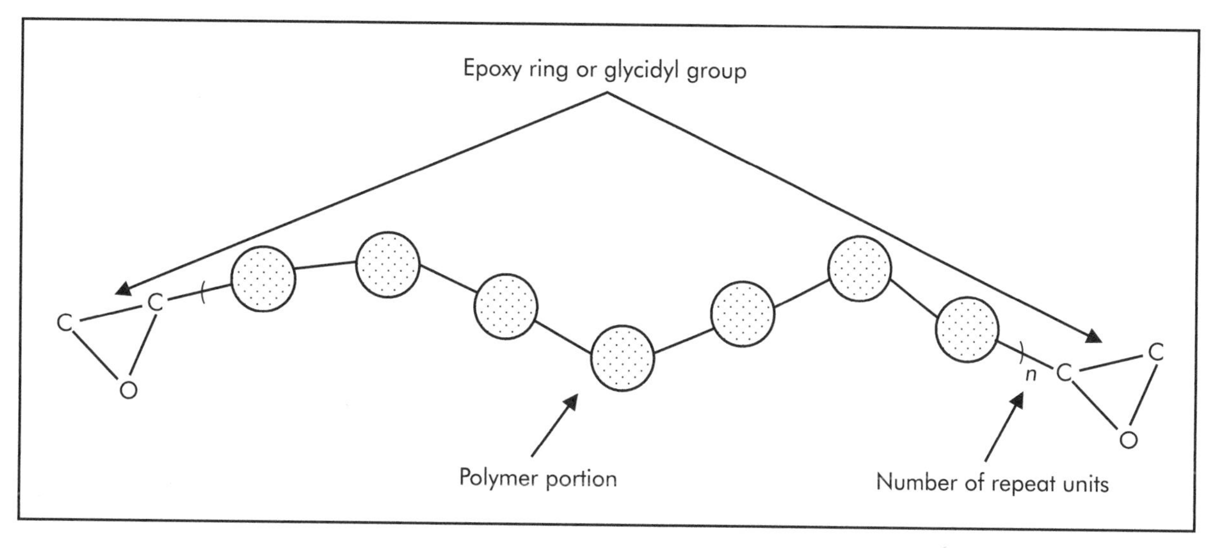

Figure 4-2. Generalized representation of an epoxy polymer.

The most common name system for epoxies indicates the various components as they exist in the final epoxy resin. The epoxy resin shown in Figure 4-3 is a good example. This molecule has two glycidyl groups on each end and so the name begins with "diglycidyl" where the "di" means "two." Each glycidyl group is attached to the polymer segment with an oxygen molecule in a linkage known in organic chemistry as "ether." Then the polymer segment is named and, in this case, it is the bisphenol A polymer. Therefore, the name of the epoxy is "**dig**lycidyl **e**ther of **bi**-**sp**henol **A**." These types of epoxy resins are often abbreviated by using various critical initials (as highlighted in the name), resulting in DGEBPA.

The reactions used to *synthesize* the epoxy resins, shown in Figure 4-3, are separate from the reactions used to *crosslink* the epoxy resins discussed in the following paragraphs. The synthesis reactions are done by chemical reactions in large plants and the output products are uncured epoxy resins. The crosslink reactions are done at the time the epoxy part is molded and the output is finished parts made from the cured epoxy resin.

Other than serving as the linkage point for the molecule used to cure the epoxy, the main contribution to the properties of the polymer from the epoxy ring is the addition of hydroxyl (OH) groups to the molecule, which are formed as part of the epoxy curing process by some (amine) curing agents. The hydroxyl groups add to some properties, such as adhesivity, of the final cured polymer.

The Non-epoxy Ring Portion

As previously indicated, epoxy molecules consist of two parts—the epoxy rings and remainder of the molecule. An understanding of the non-epoxy ring parts of the epoxy molecule can be gained by examining a few specific epoxy molecules. Referring again to Figure 4-3, in actual practice, the bisphenol A reactant would be a short-chain polymer having a few bisphenol A units between two OH groups on the ends (where n is typically 2-12). This situation would result in the same linkage of the epoxy rings as shown in Figure 4-3, but the segment between the rings would be longer. Because the number of repeat units in this segment can change, that is, n can be large or small, the properties of

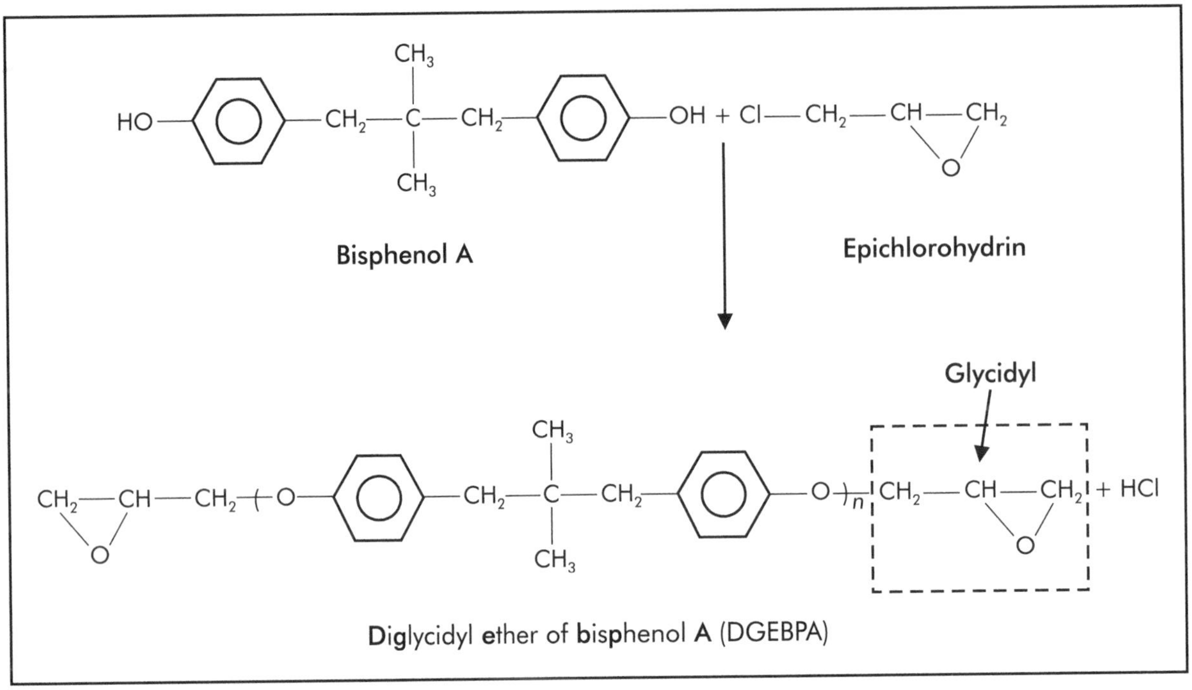

Figure 4-3. A common epoxy synthesis reaction.

the products based on these polymers change accordingly. This is another way of saying that the epoxy polymer's properties change as the molecular weight increases (as n gets larger). The properties of the uncrosslinked epoxy polymer (DGEBPA) as a function of n are given in Table 4-1. As expected, when the number of repeat units increases, thus increasing the molecular weight, the heat distortion temperature increases and the viscosity increases.

Several other common epoxies are represented to give an understanding of the breadth of variety capable with epoxy resins. The molecule depicted in Figure 4-4 has three aromatic rings in the center, each with a glycidyl group extending from the ring. This molecule has three ends to which epoxy rings can be attached. The amount of crosslinking is, therefore, higher than with the standard two-ended epoxy, DGEBPA.

Figure 4-5 illustrates a tetra-functional (four-ended) polymer. This material, called TGMDA in the United States and TGDDM in Great Britain, is popular when a part requiring high thermal tolerance is needed. The four crosslinking sites provide substantially more crosslinks than the standard DGEBPA,

Table 4-1. Properties of uncrosslinked DGEBPA epoxy resin as a function of the number of units in the polymer section, n

Number of Polymer units, n	Heat Distortion Temperature (HDT), °F (°C)	Viscosity (Physical State)
2	105 (40)	Semi-solid
4	160 (70)	Solid
9	265 (130)	Solid
12	300 (150)	Solid

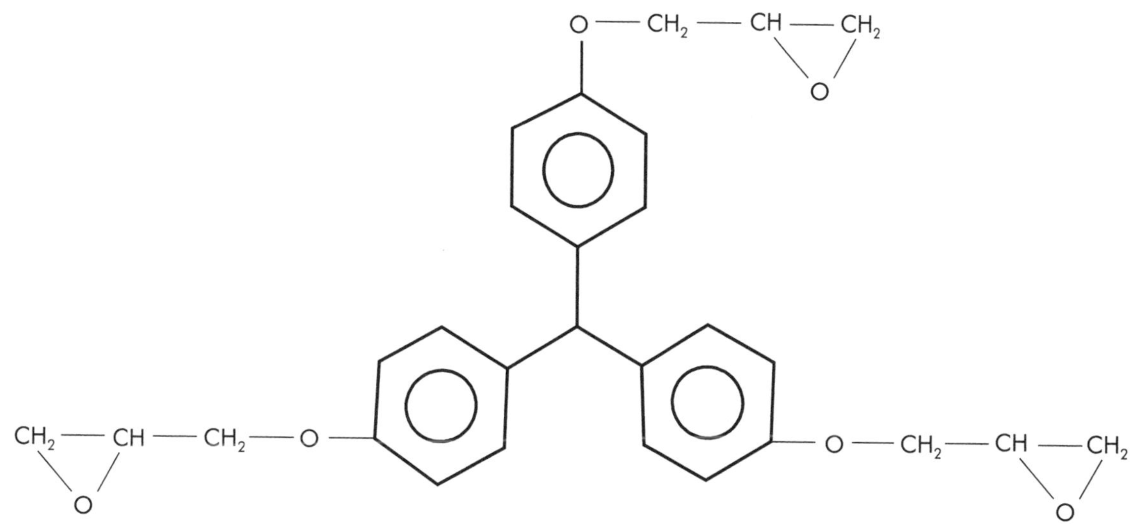

Figure 4-4. Tri-functional epoxy.

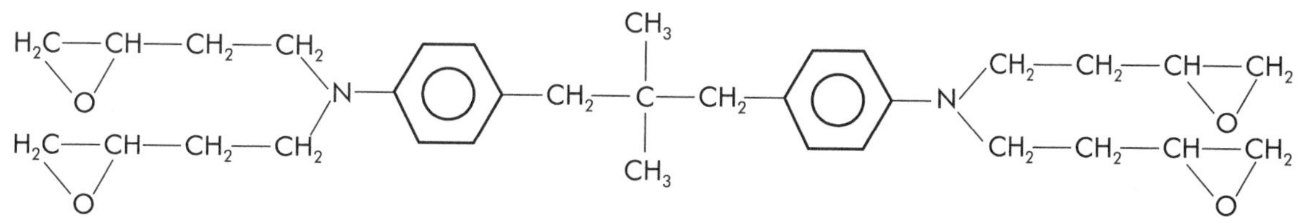

Tetraglycidylmethylenedianiline (TGMDA)
Tetraglycidyldiaminodiphenylmethane (TGDDM)

Figure 4-5. Tetra-functional epoxy.

which results in a tighter structure and requires more heat to cause movement.

The molecule depicted in Figure 4-6 is made from a polymer called a novolac that has several pendant chains on the backbone. Each of these pendant chains and the two ends of the backbone can be capped with an epoxy ring. Hence, the number of crosslinks in this epoxy can be quite high with resulting high thermal stability and excellent mechanical properties.

Some types of epoxies require properties other than those that can be obtained just by increasing the degree of crosslinking. One of these is a cycloaliphatic epoxy. This non-aromatic molecule is generally a short polymer, sometimes with no repeat units at all. Figure 4-7 illustrates a cycloaliphatic epoxy where the "S" means saturated; that is, there are no aromatic double bonds. It has good weathering properties, which are common to most aliphatics. It is also an excellent

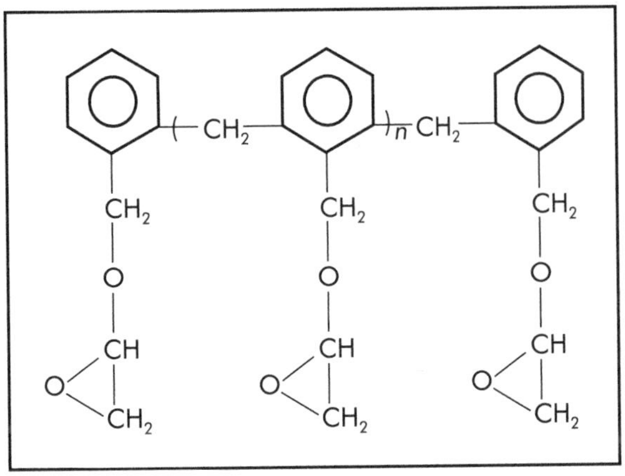

Figure 4-6. Novolac-based epoxy resin (epoxydized phenolic resin).

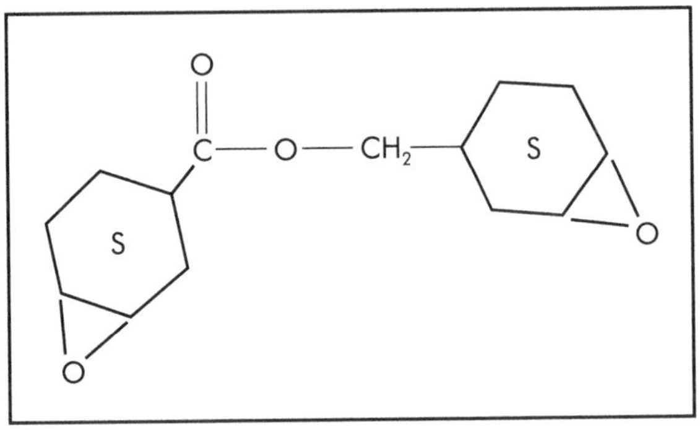

Figure 4-7. Cycloaliphatic epoxy.

solvent for other epoxies when a solvent is needed to improve wet-out of the fibers, or for some other reason.

The most common types of epoxy polymers, including those illustrated in the preceding figures, can be classed into four convenient groups: aliphatic (straight chain), aliphatic (cyclical), aromatic (straight chain), and aromatic (cyclical). These structures vary in thermal stability as measured by HDT (see Table 4-2). As expected, the higher the aromatic content, the higher the HDT. Also, cyclical polymers typically have higher thermal stability than linear polymers.

Table 4-2. Comparison of the heat distortion temperatures (HDTs) of various types of cured epoxy resins.

Epoxy Resin Type	HDT, ° F (° C)
Aliphatic (straight chain)	230–300 (110–150)
Aliphatic (cyclical)	255–270 (125–132)
Aromatic (straight chain)	265–290 (130–145)
Aromatic (cyclical)	290–350 (145–175)

Epoxy resins containing halogen atoms, chiefly Cl and Br, can be used for applications where low flame spread is required. A typical halogenated polymer is shown in Figure 4-8. Here the bromine atoms are attached to the aromatic rings. In a fire, these bromines will dissociate from the ring and combine with hydrogen atoms to form HBr, a heavy gas that smothers the fire. It should be noted that this gas is choking and irritating to the eyes and mucous membranes. Therefore, even though the flame spread is low, the smoke can be hazardous or toxic.

When an epoxy resin with high thermal properties is desired (that is, higher than can be obtained with the more traditional methods of increasing the amount of cross-linking), an imide-based polymer can be used as the basis of the epoxy polymer. The imide group is a five-member ring containing nitrogen, oxygen, and carbons as depicted in Figure 4-9. The imide group is stiff and stable, thus contributing to high stability in the epoxy molecule and increasing the thermal properties significantly. (Imide polymers that do not have the epoxy group attached will be discussed in the chapter on specialty resins.)

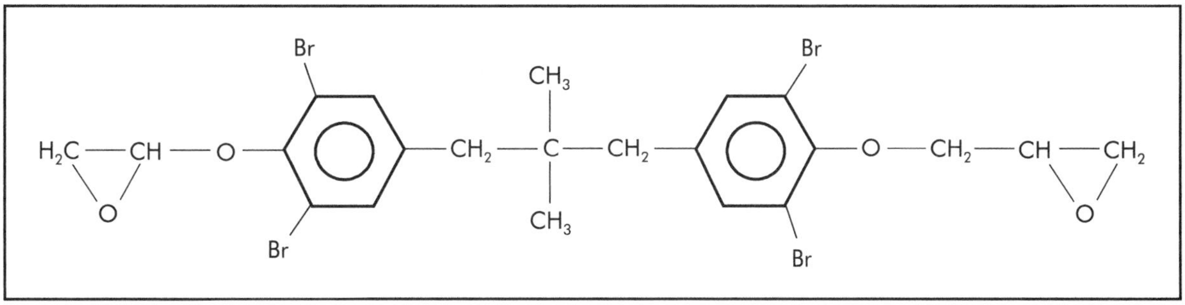

Figure 4-8. A flame-retardant epoxy.

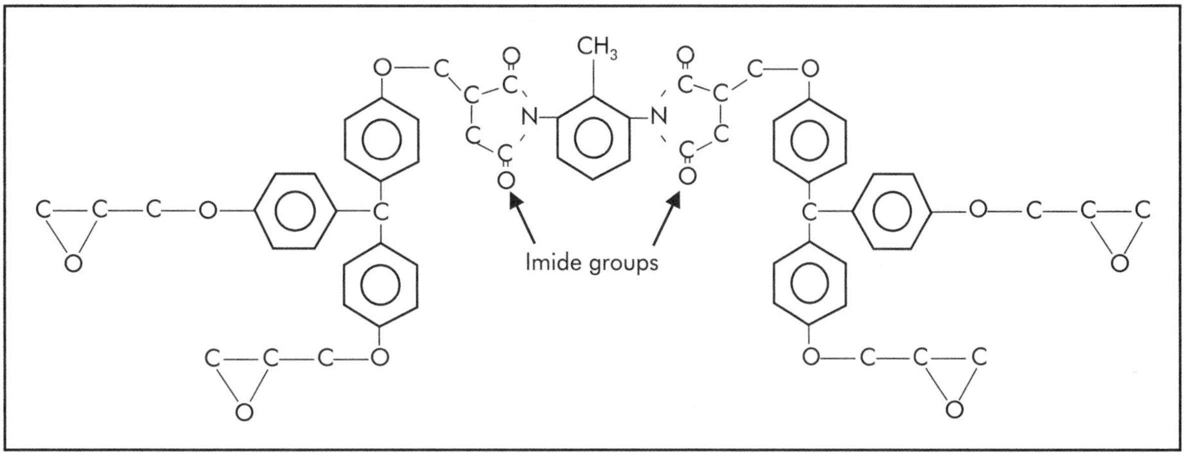

Figure 4-9. An imide-based epoxy.

The molecule shown in Figure 4-10 gives high toughness to epoxy functionality. This toughness is achieved by making the polymer chain more flexible. To do this, the aromatic character has been reduced in favor of higher amounts of linear aliphatic segments. Another method of adding flexibility to an epoxy is to mix together the epoxy resin with some rubber polymers. This mixture, while not ideal in terms of weathering and mechanical strength, has improved toughness over the epoxy resin itself. A modification to adding rubber polymers is to add thermoplastic polymers to the epoxy. This combination generally has better stability than merely the mix of rubber and epoxy and avoids some of the negatives of rubber tougheners. Still another method of adding toughness to an epoxy is to join a sheet of the epoxy with a sheet of rubber material. This is called the duplex method and, while not often used, does have some advantages in ease of fabrication and mixing.

CROSSLINKING AND PROCESSING PARAMETERS

To put the crosslinking of epoxies into a context of familiarity, the information presented in this section is summarized in Table 4-3, which shows comparisons of epoxy and unsaturated polyester crosslinking parameters. The headings of each subsection in this section of the text are the same as the major

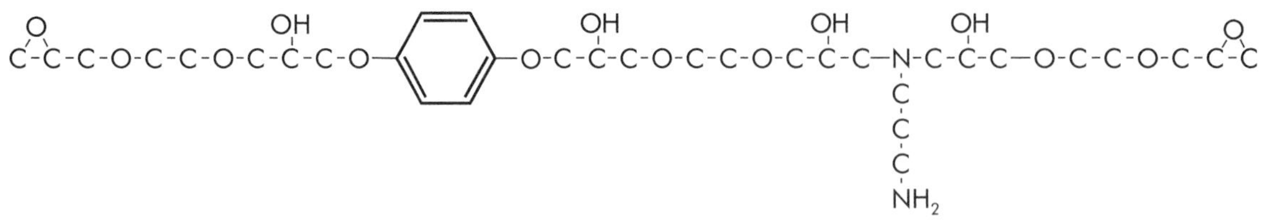

```
 O           OH                    OH            OH     OH                 O
C-C-C-C-O-C-C-O-C-C-C-O-<benzene>-O-C-C-C-O-C-C-O-C-C-C-N-C-C-C-O-C-C-O-C-C-C
                                                      C
                                                      C
                                                      C
                                                      NH2
```

Figure 4-10. A flexibilized epoxy.

entries in the table, thus allowing for easy comparisons between text and table. The table also serves as a convenient review of the crosslinking parameters of the unsaturated polyesters discussed in a previous chapter in this text. This item-by-item examination of crosslinking and processing parameters also will be useful for comparing and contrasting resins discussed later in the text.

Active Site on Polymer

The epoxy group is the site where crosslinking occurs in epoxy resins. This site is usually at the ends of the polymer chain or, in some instances, at the end of side chains. It is never in the middle of the chain because the epoxy ring only has one attachment point. If a second attachment is attempted, the ring will open. The placement of the epoxy ring is, of course, quite different from unsaturated polyesters where the active group, a carbon-carbon double bond, occurs along the backbone of the polymer in each of the polymer repeat units. Even when the epoxy group is present in each of the repeating groups of the epoxy polymer, as with the novolac epoxy, the ring is still at the end of a side chain. Because the epoxy group is on the end of the chain, there is much greater choice in the nature of the backbone for epoxies than for polyesters in which the crosslinking site is integral to each repeat unit.

Type of Crosslinking Reaction

The crosslinking reaction in epoxy resins is based upon the opening of the epoxy ring by a reactive group on the end of another molecule, called the **hardener**. A typical hardener is shown in Figure 4-11. It has amine groups (NH_2) on both of its ends. The presence of active groups like the amines on both ends allows the hardener to react with two epoxy groups on two different molecules, linking the two molecules together.

The ring-opening reaction is illustrated in Figure 4-12. When the end of the amine hardener comes into close proximity with the epoxy ring, the nitrogen on the hardener (which has a slight negative charge and is therefore called a **nucleophile**) seeks the slightly positive carbon in the epoxy ring. The nitrogen forms a bond with the carbon and thus breaks open the epoxy ring. The nitrogen loses a hydrogen molecule as it forms the bond with the carbon. The hydrogen, which is slightly positive, will then bond to the oxygen that was part of the epoxy ring. The new carbon-nitrogen bond is the critical bond for crosslinking.

The end carbon of the epoxy ring, which is the terminal carbon of the chain, is usually the more accessible of the two epoxy-ring carbons. It is therefore the atom that usually reacts with the nitrogen.

The epoxy ring opening reaction is quite different from the free-radical crosslinking reaction used to crosslink unsaturated polyesters. The epoxy reaction creates an OH group that is important to some of the properties of the epoxy resin, such as adhesivity. No such group is formed during the unsaturated polyester crosslinking reaction.

Table 4-3. Comparisons of epoxy and unsaturated polyester crosslinking parameters.

Processing Parameter	Epoxy	Unsaturated Polyester
Active site on polymer	Epoxy ring C—C (bridged by O)	Unsaturation C=C
Type of crosslinking reaction	Ring opening	Addition/free radical
By-products of the cure reaction	None	None
Reactive agent to begin cure	Hardeners (usually a bi-functional short polymer)	Initiators (usually a peroxide)
Amount of hardener or initiator	Usually 1:1 with polymer	1–2% of polymer
Toxicity of uncured reactants	Some are skin irritants and possible carcinogens	Few problems although styrene emissions can be a problem
Use of solvents/diluents	Less frequent	Usually styrene (active in the reaction)
Volatiles from the system	Low	Relatively high
Fiber wet-out/viscosity	Generally high viscosities and wet-out more difficult	Low viscosities and easy wet-out
Cure temperatures	Mostly elevated, some at room temperature	Mostly at room temperature or slightly elevated
Accelerators, promoters, and catalysts	Not common	Frequent
Cure rates	Generally moderate to long	Generally short to moderate
Pot life	Adjustable from minutes to hours	Generally short
Use of inhibitors	Rare (none)	Frequent
Use of fillers	Occasionally	Often in high concentrations
Degree of cure	Post-curing not uncommon	Rarely needs post-curing
Shrinkage	Low	High

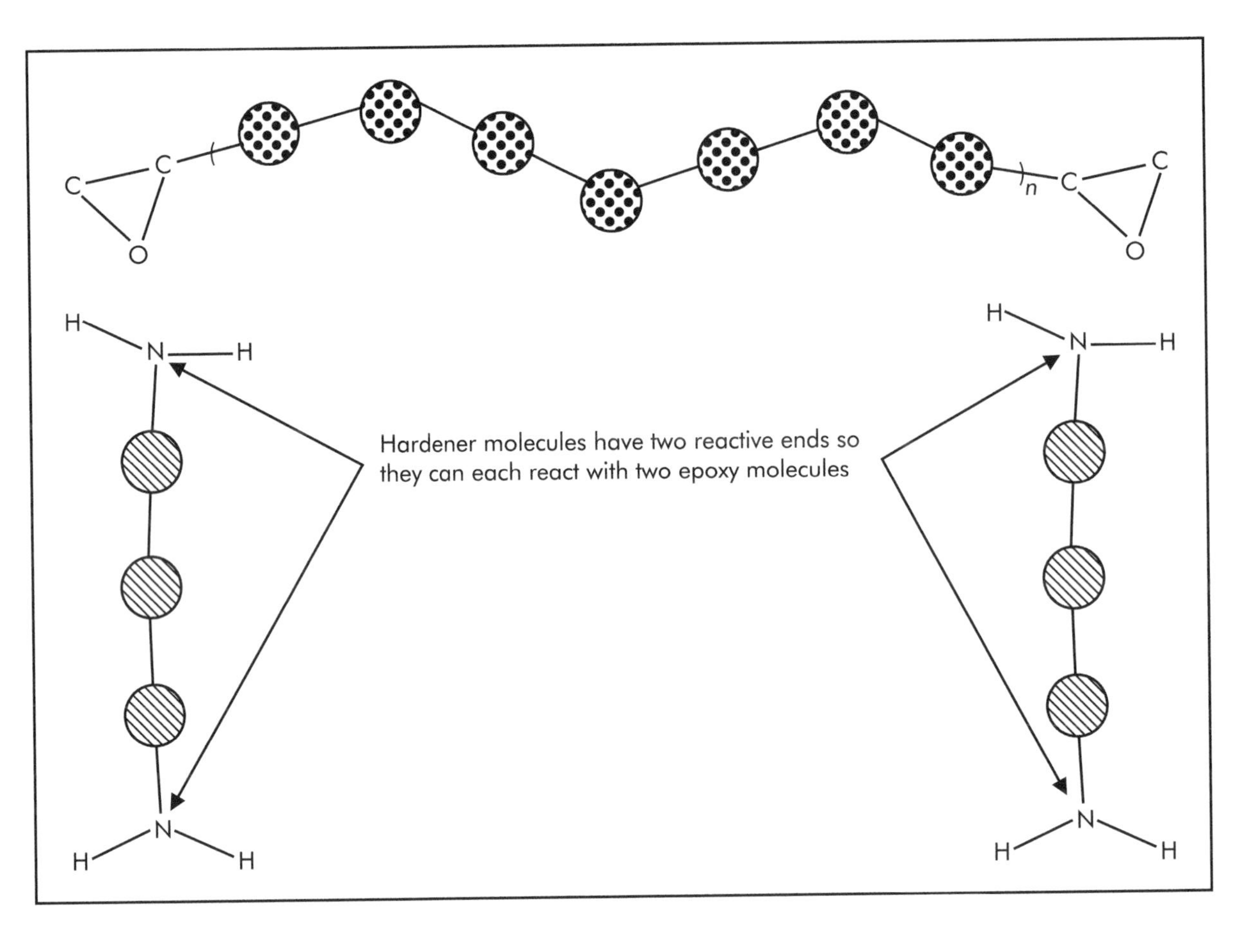

Figure 4-11. A typical epoxy molecule and amine hardeners.

The structure of the cured epoxy is illustrated in Figure 4-13.

By-products of the Cure Reaction

In one aspect the epoxy and thermoset polyester curing reactions are the same—neither curing reaction results in a by-product such as that created by condensation polymerizations. This greatly simplifies the curing reactions because no care needs to be taken for the elimination of by-products. Several other thermoset curing reactions do produce by-products, so this can be an important consideration.

The reactions used to *synthesize* the polymer should not be confused with the reactions to *cure* the polymer. Epoxies and thermoset polyester resins are synthesized by condensation reactions in which a by-product is formed. This by-product must be removed by the resin manufacturer to obtain the desired resin properties. However, little if any residue of the by-product remains in the resins shipped from the manufacturer. Therefore, by-product formation as part of polymer *synthesis* can be largely ignored by the molder.

Reactive Agent to Begin Cure

The comparison of reactive agents in Table 4-3 illustrates the greater variety of choices available for crosslinking epoxies versus unsaturated polyesters. With polyesters, only peroxides are commonly used

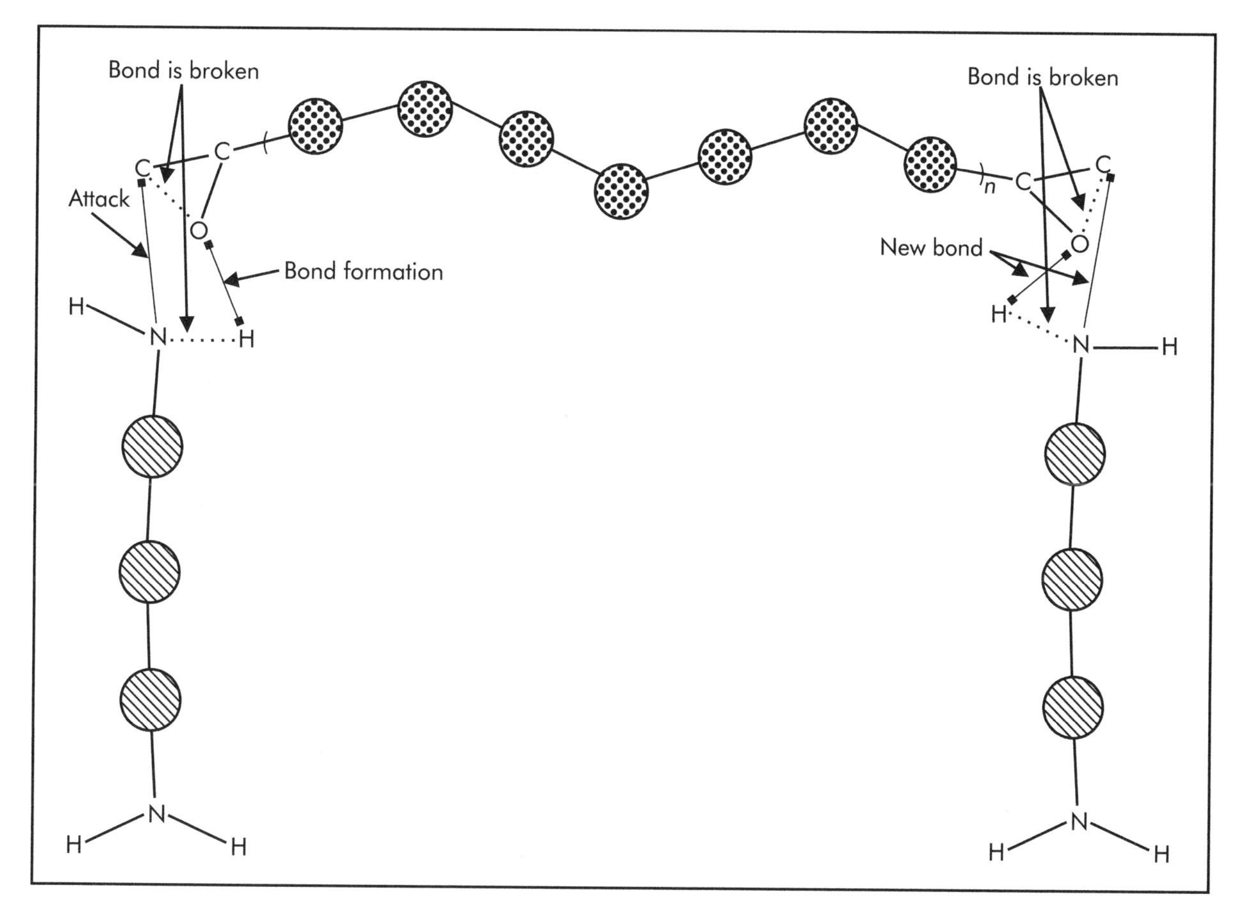

Figure 4-12. Epoxy curing reaction with amine hardeners.

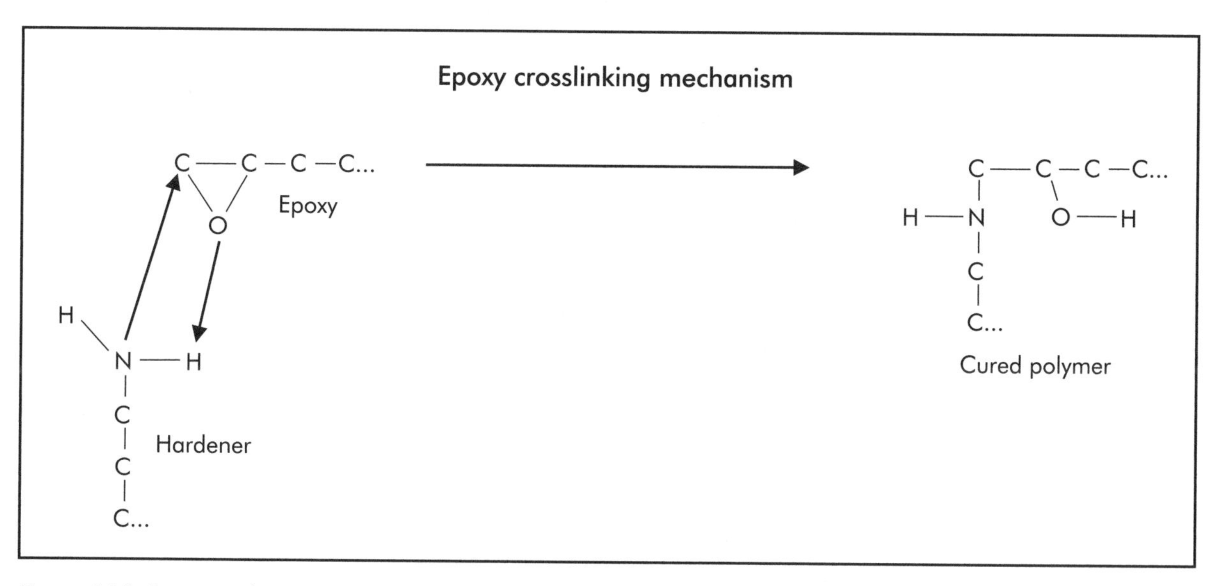

Figure 4-13. Reaction showing a cured epoxy.

as reactive agents, and then only a few of those are available. This means the major parameters that affect the crosslinking agent in polyester systems are those that determine the activation of the peroxide—the temperature at which the cure takes place and the concentration of reactants in the crosslinking reaction.

For epoxies, several types of reactive groups will open the epoxy ring and start the crosslinking sequence. The critical component of a reactive group (hardener) is that it contains a group that will open the epoxy ring. The most common reactive groups for this purpose are amines, amides, acids, phenols, analines, mercaptans (sulfides), and other molecules that with a simple reaction can form one of these reactive groups (such as an anhydride). Some catalytic ring opening materials are also available. There are, of course, many different molecules within each of these major types that could be used. Each type can affect the curing characteristics and final part properties, as will be discussed later in this chapter.

Molecules that have reactive groups used to cure epoxies are called **hardeners** or **curing agents**. The molecules containing the reactive groups can be small molecules or polymers, can have only two reactive groups or many, and can be slow or fast to react. In every case, however, the reaction is started merely by mixing the epoxy with the hardener, thus causing the active site on the polymer to come into proximity with the epoxy ring. To facilitate the movement of the hardener and the epoxy, the mixtures are often heated. Heating also may be required to activate the reaction of some hardeners. The epoxy and hardener are often referred to as Parts A and B. So, directions for curing epoxies are sometimes given as: combine Parts A and B and mix thoroughly and, if necessary, heat to complete the curing.

Many factors influence the choice of hardener, but the most important seem to be the following: compatibility with the epoxy resin, processing characteristics (time, temperature, etc.), and resultant properties. A list of the most common hardeners and their advantages is given in Table 4-4.

Most hardeners react with epoxy rings to create a hydroxyl (OH) group on one of the carbons that previously was part of the epoxy ring. Under certain conditions, these hydroxyl groups can react with other epoxy rings in a crosslinking reaction. Thus, under these conditions, once the epoxy ring is opened, the epoxy molecule can react with other epoxy molecules. These newly opened rings then have OH groups that can further react with other polymers. Hence, a chain-reaction system is established. This reaction is called **homopolymerization**. When these conditions occur, the hardener might be thought of as an initiator. However, in general, the hardener and homopolymerization reactions will both occur to cause crosslinking of the epoxy.

Amount of Hardener or Initiator

In simple epoxy systems the resin and hardener are mixed in roughly equal proportions, as with parts A and B of household epoxy adhesives. However, in more sophisticated systems and especially where properties need to be optimized, the number of active sites on the epoxy and on the hardener, as well as the molecular weight of both need to be carefully considered when deciding how much of each component to mix together. Just as some epoxy resins have more than two epoxy sites, some hardeners can have multiple active sites. The relative concentrations of epoxy groups in the resin and reactive sites on the hardener are important in determining the amount of crosslinking and, therefore, the physical and mechanical properties of the crosslinked polymer. The relative amounts of each material that will

Table 4-4. Advantages and disadvantages of important epoxy hardeners.

Hardeners	Advantages	Disadvantages
Aliphatic amines	Convenience, low cost, room-temperature cure, low viscosity	Skin irritant, critical mix ratios, blushes
Aromatic amines	Moderate heat resistance, chemical resistance	Solids at room temperature, long and elevated cures
Polyamides	Room-temperature cure, flexibility, toughness, low toxicity	High cost, high viscosity, low heat distortion temperature (HDT)
Amidoamines	Toughness	Poor HDT
Dicyandiamide	Good HDT, good electrical properties	Long, elevated cures
Anhydrides	Heat and chemical resistance	Long, elevated cures
Polysulfide	Moisture insensitive, quick set	Odor, poor HDT
Catalytic	Long pot life, high HDT	Long, elevated cures, poor moisture
Melamine	Hardness, flexibility	Elevated temperature cure
Urea	Adhesion, stability, color	Elevated temperature cure
Phenol	HDT, chemical resistance, hardness	Solid, weatherability

give exactly the right number of active sites for complete reaction at all sites is called the **stoichiometric ratio**.

The most convenient method for comparing the number of active groups in the polymer and the hardener is to consider the **epoxy molar mass**. This is the molecular weight of the epoxy (or of the hardener if considering the hardener molar mass) divided by the number of active sites per molecule. A small epoxy molar mass number means there are only a few atoms between the active groups. Therefore, when the epoxy mass number is small, the number of crosslinks formed per unit of weight will be high. Conversely, high epoxy molar mass numbers suggest that the number of crosslinks will be relatively few.

If the stoichiometric ratio is used, the maximum amount of crosslinking is achieved. Some epoxy resins and hardeners have a large number of active sites and, therefore, create a large number of crosslinks per unit length of polymer. The number of crosslinks per unit length is called the **crosslink density**. Generally, a high crosslink density is desired. However, brittleness can occur with increasing crosslink density so some reasonable limits are usually imposed. Higher crosslink density means the material is stronger, stiffer, less sensitive to solvent attack, and has a higher maximum use temperature, but it is often less tough. More about the properties of the epoxy polymers will be discussed later in this chapter.

The various cases of mixing a tetra-functional hardener (four active sites) and a bi-functional epoxy (two active sites) in less than, equal to, and more than the stoichiometric ratio are illustrated in Figure 4-14. When the concentration of the hardener's active sites is *much less* than the number of the epoxy's active sites, the crosslink density will be low and the polymer will be a simple adduct between the epoxy and the hardener. The physical properties of such a product are generally not as good as if it were highly crosslinked.

If the *same number* of epoxy rings and hardener active sites are mixed together (stoichiometric mixture), a crosslinked thermoset polymer is obtained with most physical properties at maximum. This gives a three-dimensional crosslinked matrix if either the epoxy resin or the hardener has more than two active sites per molecule. This is the case in Figure 4-14 where the hardener is tetra-functional.

When the concentration of curing agent reactive sites *exceeds* the number of epoxy rings, the material forms a thermoplastic

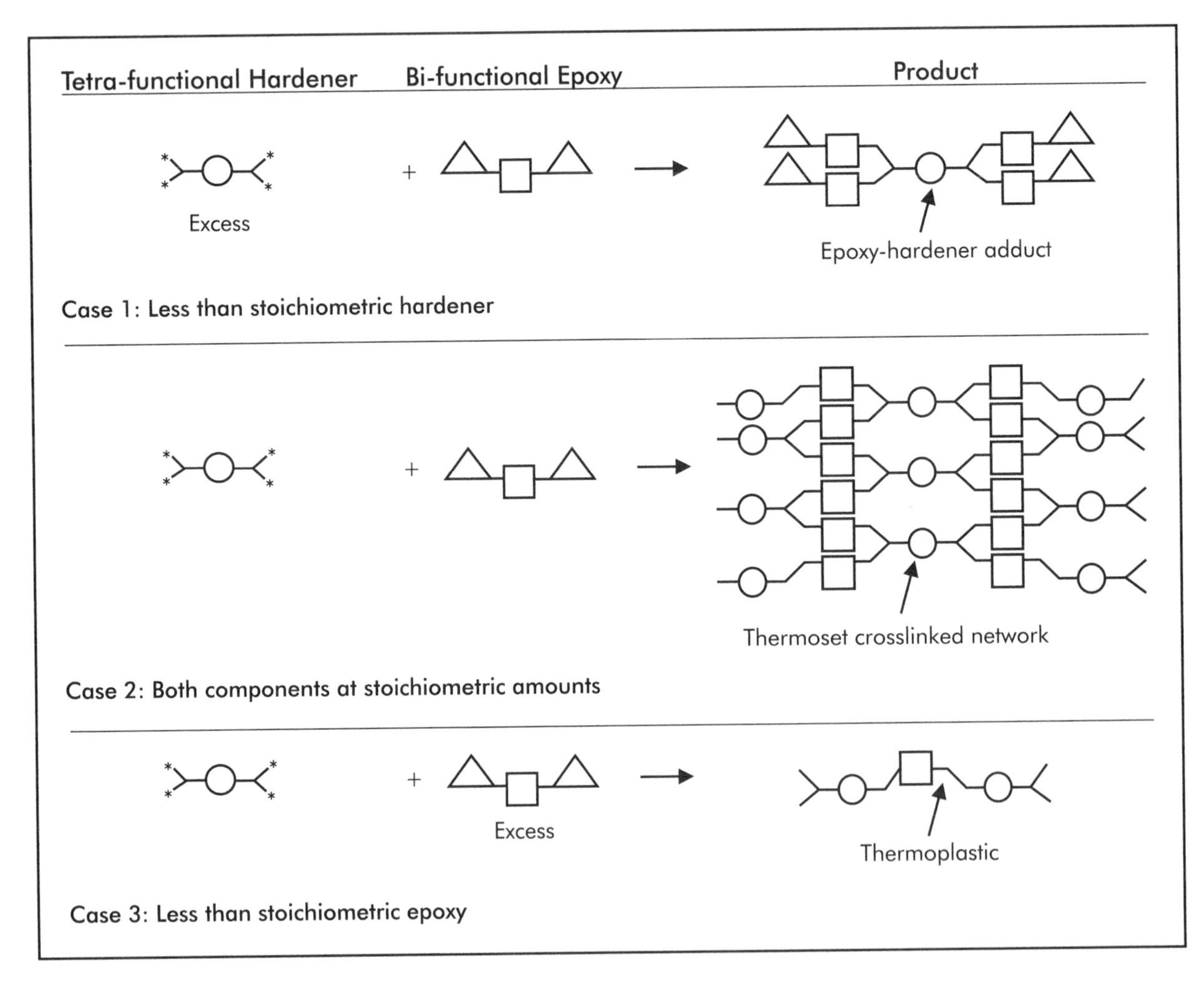

Figure 4-14. Schematic representing curing at various hardener/epoxy concentrations.

resin and the physical properties again reduce and linear polymers result. High excesses of hardener will result in nonpolymeric products that have poor properties.

When the epoxy polymer is capable of homopolymerization (as discussed previously in this chapter), the amount of hardener used is less than the stoichiometric amount. The appropriate amount of hardener is determined, in part, by the conditions under which the crosslinking reaction is run as homopolymerization usually requires higher cure temperatures than normal hardening reactions. In practice, most molders will use about 20% less hardener than resin on a molar mass basis to allow for homopolymerization and other methods of opening the epoxy ring, such as with contaminants.

Polyester thermosets offer a distinct contrast in the amount of initiator needed versus the amount of hardener used in an epoxy system. Remember that polyester thermosets use peroxide initiators, possibly with chemical accelerators. Only a small amount (1–2%) of initiator is required, versus the epoxy case where the hardener and resin are often 1:1 in concentration.

The two curing systems, epoxy and polyester, can be combined in some unique instances. Since almost any type of polymer can be used as the heart of an epoxy system, an innovative system uses an epoxy polymer that contains carbon-carbon double bonds. In this case, two crosslinking reactions can be carried out simultaneously or sequentially—one by the traditional epoxy method on the ends of the polymer, and one by the additional mechanism in the middle used by polyester thermosets. A solvent, such as styrene that will react in the addition reactions, is typically used because it is consumed in the curing reactions and solvent removal from the cured part is not necessary. This dual-cure system allows for both a high-molecular-weight polymer and high crosslink density.

Toxicity of Uncured Reactants

Some uncured epoxy resins and some hardeners are known to cause skin irritation. Therefore, direct contact should be avoided. Protection is given by the simple practice of wearing plastic or rubber gloves when handling the uncured resins and hardeners. After curing, no such problem exists.

Another problem is the potential toxicity of vapors or liquids if ingested. Researchers have suggested that some resins or hardeners are potential carcinogens. Fortunately, most of the epoxy resins and hardeners have low vapor pressures in contrast to polyesters where styrene vapors are often concentrated. Thus, the likelihood of breathing the vapors from epoxies is not high. Likewise, the chance of ingesting the material is low. Nevertheless, precautions, such as a respirator, may be appropriate for some of the most worrisome of the resins and hardeners. The material safety data sheet (MSDS) for the specific resin should be consulted before handling.

Use of Solvents/Diluents

Solvents have been used with epoxy resins to obtain thinning and easier fiber wet-out, but the environmental problems associated with them have forced their elimination in most cases. This has been possible because, in contrast to polyester thermosets, there is no need for a solvent to be a bridge molecule between epoxy chains as this linking role is filled by the hardener. Therefore, the high vapor pressures associated with solvents are not present in most epoxy resin curing systems.

However, if high-molecular-weight epoxy resins are used, possibly to reduce the brittleness that comes with high crosslink density, the pre-cured epoxy resin will likely be a solid. In these cases, solvents can be added to provide the capability to wet-out the reinforcement or mix easily with a filler. These solvents are removed from the finished parts.

Volatiles from the System

As already indicated, most epoxy resins and hardeners have low vapor pressures and are, therefore, inherently low in emitted volatiles. The normal absence of solvents in epoxy systems also leads to low volatility of the system. However, if solvents are used or if the resin or hardener is heated, the volatility will increase and caution should be taken to not become exposed to the vapors. This usually requires the use of a respirator.

Fiber Wet-out/Viscosity

Epoxy resins are generally more viscous than polyesters, especially because of the normal presence of styrene or some other solvent in the polyester system. Therefore, fiber wet-out is much more difficult with epoxies.

A method of reducing the viscosity of epoxy resins is to *combine different epoxies*. For instance, a low-viscosity resin based on bisphenol A can be combined with a higher-viscosity resin also based on bisphenol A. This allows the molder to obtain moderate-to-low viscosity and yet retain the acceptable physical properties typical of the higher-molecular-weight resin. If resins from different families are combined, they should be checked carefully for compatibility. Addition of a low-viscosity resin also helps prevent unwanted solidification (crystallization) of the polymer, which can occur upon prolonged storage when the freezing point of a resin is at or below room temperature. While this freezing does not produce changes in the chemical nature of the polymer, it is a bother in processing as the entire batch must be heated to melt the solid mass.

The problem of fiber wet-out with epoxies is sometimes solved by heating the resin before it is applied to the fibers. When heated, the viscosity drops, thus facilitating wet-out. However, the curing reaction also begins when the resin is heated, which may significantly shorten the pot life or working time. This complication can be alleviated by using a latent hardener and then choosing a heating temperature high enough to cause thinning but not so high that the hardener is activated. Conversely, the heating and wet-out can be done quickly so the crosslinking reaction simply does not have sufficient time to proceed very far.

An epoxy resin also can be diluted with solvent to reduce its viscosity. When this is done, the solvent is removed prior to or during the cure and adequate provisions must be made to control the solvent emissions. Reactive solvents, as used in polyesters, are not commonly used with epoxies. But some low-molecular-weight epoxies, which are, of course, reactive, have been used successfully as reactive diluents.

Cure Temperatures

Most epoxies are cured at elevated temperatures to facilitate the movement of the resins and hardeners, thus increasing the likelihood of their colliding and resulting in a crosslinking reaction. Since both epoxy and hardener are often polymers, although usually short ones, some heating may be required just to make them liquids at room temperature.

The temperature selected for curing the epoxy depends strongly on the nature of the resin and hardener and, somewhat, on the intended application. Some resins and, especially, some hardeners do not react readily at room temperature and therefore need higher temperatures to reduce the cure time to a reasonable level.

The performance of the polymer is often related to the cure temperature. A general rule is that the maximum use temperature should not exceed the glass transition of the polymer, which usually increases with increasing the cure temperature (and increasing crosslinking). Therefore, epoxies used at high temperatures need to be cured at even higher temperatures. The cure temperature for common epoxies used in composites is 250° F (121° C). For even higher performance or for resins more difficult to crosslink, the common high-temperature epoxies are cured at 350° F (177° C).

Some epoxy resins will cure at room temperature, initiated merely by the mixing of the resin and the hardener, which is often an aliphatic amine. These are called **room-temperature vulcanization (RTV)** resins. ("Vulcanization" is an old name for crosslinking still used occasionally. It was derived from the crosslinking reaction of natural rubber using sulfur as the crosslinking agent with extensive cooking. Vulcan was the Roman god of fire, hence the name.) Applications for RTV epoxies are grouting, household adhesives, electrical potting, and simple coatings.

Elevated temperature epoxies, sometimes called **high-temperature-vulcanization (HTV)** products, are used for composites, high-performance coatings, and high-performance adhesives. The high-temperature epoxies are usually made from multifunctional and highly aromatic epoxies and hardeners. They have high crosslink density.

Some hardeners react slowly with epoxies at room temperature but will react rapidly at high temperatures. These hardeners are often referred to as **latent** because their ability to react is "hidden." A typical example of a latent hardener is dicyandiamide (DICY). It is a solid curing agent, which, when ground and mixed into a liquid epoxy resin, provides stability for up to six months at ambient temperature. Cures in which DICY is the curing agent occur when the epoxy mixture is heated to at least 250° F (121° C). Note that the DICY is often suspended in the epoxy resin as a small-size solid, ground to particles. This means that processes in which the resin is forced through tightly packed fibers may filter out the hardener and result in incomplete cures. Therefore, DICY should be used only when the fibers are spread and easily wetted.

Accelerators, Promoters, and Catalysts

Because most epoxies are cured at elevated temperatures, the use of accelerators, promoters, and catalysts is less common than with polyesters. However, some epoxy-hardener reactions can be facilitated through the use of accelerators and/or catalysts to reduce the gel time.

Some *acids* act as accelerators by reacting with the epoxy ring to make it easier to open. The acid molecule donates a hydrogen molecule to the oxygen of the epoxy ring, which then makes the carbons in the epoxy ring slightly more positive. The terminal carbon is subsequently more easily attacked by the slightly negatively charged hardener. (Refer to Figure 4-13 for the reaction of a hardener and an epoxy.) This type of acid accelerator is most often used with amine and amine-derivative hardeners. Too much acid will, however, inhibit the reaction as the excess acid can react with the amine-derivative hardener to form an amine salt, which is less reactive than the original amine.

Another type of accelerator is *basic*. This type generally reacts with the hardener to make it a more effective attacking agent. For example, polysulfides are much more reactive with epoxies if a basic material, such as a tertiary amine, is present. This is because the tertiary amine takes a hydrogen molecule off the polysulfide and leaves a strongly negative sulfide behind.

The types of accelerators or promoters used in polyester thermoset systems, such as cobalt compounds, are not used in epoxies.

Cure Rates

Epoxy cures can be as short as a few minutes but are usually several hours long, even when heated. These curing reactions are generally slower than polyester cures. Even polyester cures that take several hours, as would be the case for a room-temperature cure of a large part, are faster than a comparably sized epoxy part, even if the epoxy were heated to cure and the polyester was cured at room temperature.

A typical epoxy cure would be a ramp up to the cure temperature over about one hour, holding at the cure temperature (often about 250° F [121° C] or 350° F [177° C]) for one to five hours, and then a gradual cool down for another hour. A typical curing profile for an epoxy is shown in Figure 4-15. The resin/fiber system is heated to about 130° F (54° C) where it is held at temperature for an hour. This allows the system to equilibrate and gives uniform curing. The system is then further heated to about 355° F (179° C) and held for about two hours to complete the cure. The cure rates and temperatures depend on both the resin and the hardener and are suggested by the resin manufacturer.

Pot Life

The **pot life** is the working time of the resin during fiber wet-out and other pre-cure operations. (This is a different time than ***shelf life***, which is the time that the resin can be stored without reacting.) The long curing times of epoxies and the need to heat the materials provides for long pot life in most epoxy systems. Occasionally, a system will be intentionally heated or use reactants that are especially active, which can result in some urgency to complete the molding of the material. Those epoxies that cure at room temperature will, of course, have a shorter pot life than high-temperature-cure epoxies. Usually this is not a problem as even in the RTV systems the pot life is sufficient for most manufacturing processes. Typical commercial composite processes that sometimes

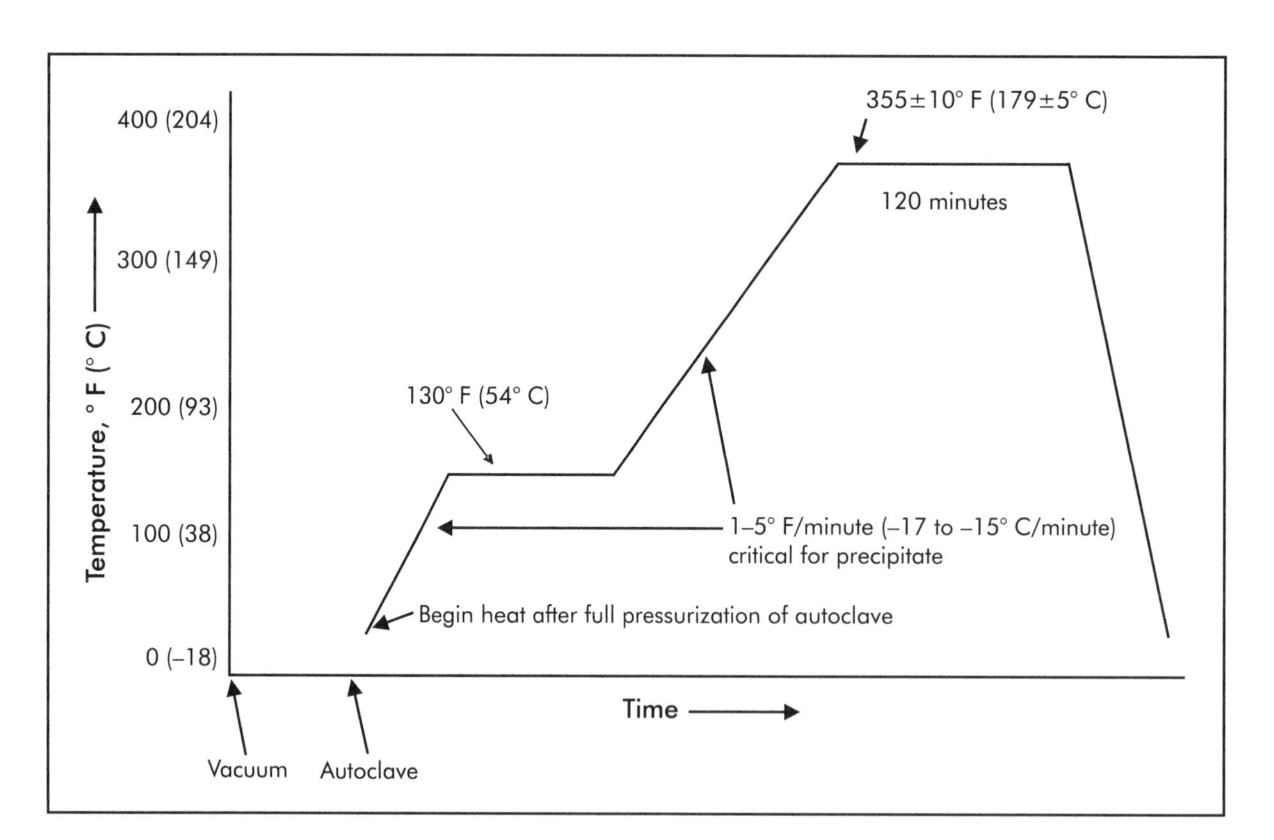

Figure 4-15. Typical curing profile for an epoxy.

use room-temperature-cure epoxies include filament winding, resin transfer molding, and patching repairs—all discussed later in this text.

The shelf life of epoxies can be a problem under some circumstances. Some epoxy products are made from materials consisting of resin and hardener that are pre-coated onto the reinforcements. These materials, called **prepregs**, are discussed later in this chapter. Prepregs and any other material that consist of mixed resin and hardener will gradually cure, even at low temperatures. Therefore, whenever resin and hardener are mixed, it should be anticipated that curing will occur. The best that can be done to improve the shelf life of such materials is to lower the temperature of the mixture during storage and thereby extend the shelf life. Even when kept refrigerated, the manufacturer usually sets a maximum shelf life after which the material is assumed to have become too hard (cured) to use. Normal industry practice is to discard the materials after they have reached this maximum shelf life.

Some epoxy resins have been developed that have extremely long shelf life. These resins advertise that they do not need to be stored at low temperatures. They have demonstrated up to one year of shelf life when kept at room temperature. These are valuable new resins because of this characteristic, which means that they will not have to be refrigerated and, even more, they do not have to be discarded when some maximum time limit has been exceeded. One mechanism to provide this long cure life is to use a hardener that is applied to the prepreg in a solution with resin and solvent. As the prepreg is dried by evaporating off the solvent, the hardener crystallizes and becomes separate from the resin, thus inhibiting the reaction between resin and hardener. When the prepreg is heated during molding, the *hardener crystals remelt* and mix with the resin, thus reinitiating the crosslinking reaction.

Use of Inhibitors

The long cure times, relatively stable pot life, lack of peroxide initiators in the curing system, and general need for heated curing make the use of inhibitors unnecessary in most epoxy systems. Inhibitors of the type used in polyester systems are unknown for epoxies.

Use of Fillers

In contrast to thermoset polyesters, epoxy composites rarely contain fillers. This arises from the generally high performance requirements placed on epoxy parts. This high performance is often highly weight sensitive, meaning that the epoxy composite is used where mechanical and physical properties are to be at maximum and weight a minimum. These performance factors usually outweigh the cost advantages that can be obtained from using fillers. With polyester thermosets, the cost of the product usually outweighs the performance requirement, so long as a minimum performance level is reached. In contrast to epoxies, weight is usually not a major consideration with polyesters.

Following are some of the most common advantages of using fillers with epoxies, (depending, to some extent, on the type of filler chosen): extended pot life, lowered exotherm, resistance to thermal shock (cracking that may come from extreme and rapid changes in the temperature of the composite part), reduced shrinkage, improved ability to machine the composite part, improved abrasion resistance, increased electrical conductivity in applications where a conductive material is needed, improved compressive strength, increased density, improved arc resistance, increased viscosity, and improved self-lubricating properties.

Degree of Cure

The slow cure rate for epoxy resins suggests that post-curing might be important as a method of getting more efficient use of the molds. **Post-curing** is done by removing the part from the mold after considerable curing has taken place so that the part is dimensionally stable, and then placing the part into an oven where it is given an extended additional curing time at an elevated temperature. The post-curing advances the cure; that is, it creates a higher degree of crosslinking in the part. This is done to improve the part's physical and mechanical properties.

Polyester parts also can be post-cured to advance the crosslinking, but this is comparatively rare. Polyester polymers do not require the same mobility to achieve crosslinking as do epoxy polymers. Therefore, polyester polymers can react adequately at room temperature provided there is sufficient solvent present to form the bridges between molecules. However, in cases where the polyester does not have the required mobility at room temperature, the cure is done at elevated temperatures, much like an epoxy.

Shrinkage

One advantage of epoxies over unsaturated polyesters is the low shrinkage that occurs during the molding process. The crosslinking reactions of polyesters tend to draw the molecules together tightly, whereas the crosslinking reactions of epoxies are far less tight. The greater tightening of the molecular mass with polyesters is due to the greater number of crosslinks and the relative frequency of the crosslink sites along the backbone, in comparison to epoxies where the crosslinks are only on the ends of the chains.

Lower crosslink density, even within a polymer type, will generally result in lower shrinkage. Low shrinkage improves molding capabilities, especially for parts in which dimensional tolerances are critical. This property is also important when the mold is made of a flexible material and shrinkage could change the shape of the mold itself.

PHYSICAL AND MECHANICAL PROPERTIES OF CURED EPOXY COMPOSITES

As was discussed in the chapter on matrix materials, some properties of composites are mainly dependent on the matrix and others on the reinforcement. Even within those considerations, the choice of matrix can make some differences, usually not major, in properties that are mainly dependent on the reinforcement. Since these changes are small, this section examines the trends in the properties of epoxy composites and compares those with the expected properties of polyester thermosets. The results of that comparison are given in Table 4-5, which can serve as a reference for comparisons of other matrix materials (not polyesters or epoxies) when they are discussed in later chapters.

Adhesion

The creation of the OH group during the crosslinking reaction (see Figure 4-13) and the possible presence of OH groups in the body of the polymer give epoxy resins the superior ability to adhere to many surfaces. The presence of the OH group creates high surface energy—a characteristic of materials that are good in adhesion. The reason for this good adhesion is that if a high-energy surface is bonded to some other surface, the net result is a lowering of the energy of the entire system—a condition favored by nature and described by the laws of thermodynamics. The higher the energy of the surface, the more likely it is to have a lowering of its energy when it is bonded to some other surface. Therefore, the high surface energy of epoxies results in a strong tendency to bond with other surfaces.

Table 4-5. Comparisons of unsaturated polyester and epoxy properties.

Property	Epoxy	Unsaturated Polyester
Adhesion	Excellent	Good
Shear strength (fiber-matrix bond)	Excellent	Good
Fatigue resistance	Excellent	Moderate
Strength/stiffness	Excellent	Good
Creep resistance	Moderate to good	Moderate
Toughness	Poor to good	Poor
Thermal stability	Good	Moderate
Electrical resistance	Excellent	Moderate
Water absorption resistance	Moderate	Poor to moderate
Solvent resistance	Good	Poor to moderate
UV resistance	Poor to moderate	Poor to moderate
Flammability resistance	Poor to moderate	Poor to moderate
Smoke	Moderately dense	Moderately dense
Cost	Moderate	Low

As previously mentioned, epoxy resins are used extensively for adhesives. These are used in many applications, including the bonding of composites to other composites as well as to other materials such as metals, ceramics, and plastics. Epoxies are, of course, also used as adhesives in systems that do not involve composites.

Possibly the most important aspect of adhesion in epoxy composites is the bond strength between the epoxy matrix and the reinforcement fibers. Just as with other surfaces, the epoxy material has a tendency to form strong bonds with the surface of the reinforcements. This strong bonding is seen in many of the properties of composites including strength, stiffness, creep resistance, and solvent resistance. The adhesion between the matrix and the fibers is a key aspect of these properties since if the matrix slips along the reinforcement and does not transfer the load, the reinforcement cannot do its job. Therefore, even though the reinforcement is the principal element for strength, stiffness, and other mechanical properties, differences between the adhesive capabilities of epoxies can make sufficient difference that the mechanical properties could be significant.

Shear Strength (Fiber-matrix Bond)

The composite property most closely identified with the bond strength between the resin and the fiber is the composite's shear strength. This property is measured by methods that attempt to separate the fibers from the resin that binds them. The methods for doing this will be discussed in detail in the chapter on resin-fiber bonding. The important consideration here is that epoxies have excellent shear strengths, especially in comparison with thermoset polyesters and most other matrix materials. This is one of the key reasons why epoxies are used in high-performance composites.

Fatigue Resistance

Fatigue resistance is a measure of how a material performs with repeated forced movements over long periods of time. Even when the movements imposed on the composite part are below the normal strength and stiffness of the materials, over time the part might fail. This failure is often the result of breakdown of the bonds between the resin and the fibers. As mentioned previously, the integrity of the fiber-matrix bond is critical to the successful performance of the composite.

Fatigue also depends on the ability of the matrix to withstand repeated flexing. This involves the ability of the matrix to elongate with the repeated movements of the composite part and the overall strength of the matrix. Although epoxies can be quite stiff, indicating poor elongation, some types have good elongation while still maintaining many of the most important strength and stiffness properties. Therefore, fatigue resistance in epoxies is usually excellent.

Strength/Stiffness

As with unsaturated polyesters, the strength and stiffness properties of composite products are chiefly dependent on the reinforcement, possibly more for these properties than any others. Even though the fibers are a critical factor, the fiber-matrix bond strength also affects strength and stiffness, as described previously.

Even considering the effects of the fiber and the fiber-matrix bond, the matrix strength and stiffness have some relevance to the strength and stiffness of the composite. If the matrix breaks or deflects under load, that will prevent it from transferring the load to the fibers. Epoxies are generally stronger and stiffer than most of the competitive resins, especially polyester thermosets. This strength and stiffness are attributable to the high aromatic character of most epoxies and the basic internal adherence of epoxy molecules to their neighbors, which is usually stronger and more stable than in polyesters.

Creep Resistance

Creep is the gradual movement of a material under long-term loads. This movement is usually a stretching or sliding of the material. It is strongly controlled by the presence of fiber reinforcements that have almost no creep under normal environmental conditions. However, because the matrix has the responsibility of holding the fibers in place, when the fibers are not continuous in the part, a slippage in the matrix can result in movement of the fibers or creep. Therefore, matrix stiffness under load is important in these cases and epoxies are known to have excellent resistance to this type of internal motion.

Normally, two important factors affect the amount of matrix creep. The first of these is the crosslink density. More crosslinks hold the matrix in place better. In this regard, then, polyesters are excellent because of their high crosslink density. The other factor is the stiffness of the polymer backbone itself. That stiffness is dependent on the aromatic nature of the resin and the overall structure. Epoxies are usually better than polyesters in this property. Overall, polyesters and epoxies are moderately good. But, if epoxies are strongly crosslinked, they can become slightly superior to polyesters.

Toughness

The normal procedure to improve mechanical and physical properties is to increase the molecular weight of the resin. This can be done in thermosets most easily by increasing the crosslink density until the crosslinked structure becomes so rigid that brittleness becomes a problem. Therefore, optimizing toughness in composites often requires a compromise with some other properties.

With epoxies, higher crosslinking can be achieved through the use of multifunctional reactants and more aggressive curing conditions, including higher temperatures and longer curing times. As with other thermosets, toughness eventually reaches a maximum value after which the rigidity of the highly crosslinked structure leads to brittleness.

In addition to the amount of crosslinking in the polymer, toughness of the epoxy part is dependent on the length of the polymer chain segment between the epoxy end groups. Long polymer segments are generally tougher than short polymer segments of the same chemical type. But this factor is counteracted by the presence of more ends with short polymer segments and, therefore, the ability, generally, to achieve higher crosslinking. Obviously a compromise must be reached on the length of polymer chain segment and crosslink density if toughness and the other properties are to be maximized simultaneously.

The toughness of the epoxy part is also dependent upon the amount of aromatic character in the epoxy and hardener molecules. Less aromaticity or higher aliphatic content generally means more flexibility in the molecules and, therefore, improved toughness but decreased overall strength, stiffness, and maximum use temperature in the final part.

A tactic for improving toughness without major sacrifices in other physical properties is to include aliphatic segments along the backbone of an otherwise highly aromatic epoxy polymer. These aliphatic segments act like internal springs, giving improved flexibility and, therefore, improved toughness. These types of materials are called **toughened epoxies**. An example of a toughened (or flexibilized) epoxy is given in Figure 4-10. The major problem with toughened epoxies is the higher cost associated with the more complex synthesis of the epoxy resin.

It is, of course, also possible to increase the toughness of an epoxy by using a highly flexible hardener. This gives flexibility to the crosslinks and, in the cured polymer these flexible crosslinks act the same as a flexible backbone.

Another method to increase the toughness of epoxies is to add rubber polymers. These can be copolymerized with the epoxy (mixing of monomers) or alloyed (mixed into the polymer melt). The problem with adding rubber polymers is that chemical resistance and strength are often negatively affected. This problem is especially acute if the rubber material is incompatible with the epoxy resin and forms a separate phase when the two polymers are mixed.

Closely associated with the adding of rubber tougheners is the addition of a thermoplastic polymer to an epoxy. Many thermoplastics can add toughness and their effectiveness is generally dependent on their compatibility with the epoxy resin. Some can be added during the polymerization stage to form a fully compatible polymer that has domains of thermoplastic surrounded by areas of thermosets. Many experts believe that the combination of an epoxy with a thermoplastic will be a major factor in the future of epoxy resins. This will combine the processing advantages of a thermoset, low viscosity for wet-out and ease of shaping, with the impact strength and damage tolerance of thermoplastics.

Thermal Stability

Two important factors strongly affect the thermal stability of epoxy composites.

One factor is the extent of crosslinking. The higher the density of the crosslinks, the higher the thermal stability or allowable use temperature.

The second important factor affecting the thermal stability is the basic nature of the polymer chain. For instance, the heat distortion temperatures (HDTs) of various epoxies given in Table 4-2 show that aromatics have higher thermal properties than aliphatics, and that cyclical aliphatics, which are stiff, but not as stiff as aromatics, have higher thermal properties than straight-chain aliphatics. Table 4-2 also shows that acid-cured, straight-chain aliphatic epoxies have higher HDTs than equivalent amine-cured polymers. Hence, the aromatic content of the polymer, the nature of the hardening agent, and the overall crosslink density can all affect the use temperature.

Still another factor in thermal stability, although not one that is related to most normal applications, is the temperature at which degradation occurs. Most composites have lost their stiffness and strength long before this temperature is reached. However, some applications see excursions to high temperatures and the degradation temperature can be important. Epoxies are only moderately good in these applications, but are better than thermoset polyesters. Other resins, discussed in the chapters on specialty and high-performance resins, are vastly superior to epoxies.

Electrical Resistance

Epoxies are the preferred resin for composites used in electrical applications. This is both a function of performance and of cost. Epoxies seem to have the right compromise in this regard. They have better electrical properties than the cheaper polyesters and are lower priced than the better performing, high-performance resins.

The electrical properties of epoxies are improved because of the lack of solvent or monomer residues, which can become volatile when heated or subjected to strong electrical fields and cause the electrical properties to decline precipitously. Even non-crosslinked epoxy and hardener polymers are large enough that they do not volatilize easily, thus removing this problem from epoxies in general. In contrast, polyesters may have considerable residual styrene content that can be vaporized if the polyester is heated.

Water Absorption Resistance

The high aromatic content of most epoxy resins helps them resist water absorption. However, countering that is the presence of the OH groups formed when epoxy resins are cured. These OH groups have an affinity for water and will, therefore, absorb water molecules. The net result of these two opposing forces is that water absorption resistance is only moderate in epoxies, with high-performance polymers being superior. Polyesters have a high number of doubly bonded oxygen molecules along their backbones. These oxygen molecules are highly polar. Therefore, they are highly susceptible to water absorption, especially at high temperatures where the natural swelling of the polymer further increases the ability of the water to penetrate. Epoxies are generally superior to polyesters in resisting water absorption.

Solvent Resistance

Epoxy resins are resistant to solvents that are dissimilar from their basic nature. Polar solvents, like water, will be slightly attracted to epoxies, whereas non-polar and non-aromatic solvents (that is, most organic solvents) have little effect on epoxies. The majority of solvents are reasonably resisted, leading to a characterization of the solvent resistance of epoxies as good.

UV Resistance

The high aromatic content common to most high-performance epoxies is not good for ultraviolet (UV) light resistance. The bond between the aromatic group and the molecule to which it is attached have about the same energy level as UV light. This results in excitation of the bond electrons by UV light and subsequent rupture of the bond. The same situation applies to aromatic polyesters.

Polyesters and epoxies can be made using aliphatic polymers. These are much more resistant to UV light and have significantly better weatherability. In common practice, the UV resistance of both materials is enhanced through the use of chemical additives that absorb UV light or absorb the products of UV light's attack on the resin, thus preserving the life of the resin.

Flammability Resistance

High aromatic content reduces the tendency of materials to burn. The aromatic materials tend to char rather than burn freely, which is seen as a reduction in both the ability of the material to be ignited and the rate at which the material burns. Epoxies and highly aromatic polyesters are similar in this property. In both types of resins, the addition of halogen atoms, usually added to the polymer during synthesis, will further reduce the flammability of the resin.

Smoke

Even though the burn rate is low for aromatics, the amount of smoke is often high. The smoke is quite dark (sooty) and can be emitted in large amounts even though burning is not vigorous (a smoldering effect). When halogens are used to give non-burning characteristics to the resins, the smoke can become acrid and potentially dangerous. Epoxies and polyesters are similar in smoke generation.

Cost

Without any question, cost is one of the major impediments to widespread adoption of epoxies as replacements for polyesters. Polyester resins are approximately half to one-third the cost of epoxies and, in applications where profits are low and volumes are high, the cost differential is critical. Therefore, the cost of epoxy resins, the greater difficulty in wetting the fibers, and the need to cure them with elevated temperatures are the most important factors that restrict their use to composite applications where cost is not the major factor in the purchase of the resin. These types of applications usually require high-performance materials.

SPECIAL COMPOSITE APPLICATIONS

Prepregs

A form of epoxy combined with reinforcement is called a **prepreg**. Used extensively in the aerospace and sports equipment industries, a prepreg is a sheet made of fibers or cloth onto which liquid epoxy resin mixed with hardener has been coated. In research and development, the coating process was accomplished on a machine that pulled the fabric or, if unidirectional fibers were used, pulled a group of carefully arrayed fibers through a resin bath resulting in the fibers being fully wetted by the resin. Excess resin was removed by squeezing the fiber/resin sheet. Alternately and more commonly in commercial practice, the dry fabric or arrayed fibers can be pressed against a sheet of release paper to which the resin/hardener mixture has previously been applied. This method is preferred in commercial practice over the resin bath because the amount of resin/hardener coated onto the paper can be carefully controlled by using a roller applicator system. The resultant prepreg material is tacky and so the release paper is used to keep the layers from sticking together. The sheet is rolled up and stored at low temperatures until use.

When the material is to be molded into a part, the roll is allowed to return to room temperature so it is tacky and pliable. It is then laid into the mold or otherwise shaped. The paper layer is removed as the prepreg material is laid into the mold. Several layers are often placed on top of each other until the desired thickness is reached. Because the design of composite parts often stipulates that the directions of the layers be different, the prepreg materials are often laid down at specific angles and may even be precut to fit specific shapes. The flat nature of the prepreg material and the use of a paper layer to separate the layers in the roll facilitate this cutting. The tacky nature of the material is advantageous in keeping it in place in the mold and in keeping the various layers in place on top of one another.

The laid-up material is then cured, often under pressure, squeezing the layers together. Autoclaves are conveniently used to cure these prepreg parts because heat and pressure can be applied simultaneously. Alternately, the materials can be enclosed in a vacuum bag and cured. These molding methods are discussed further in the chapter on molding of advanced materials.

Tooling

Many of the key performance characteristics of epoxies have led to their widespread adoption as a material for tooling (molds). In this application, epoxies are competing with metals. The ease of forming epoxies by molding gives them an inherent advantage over metals, especially for molds that must be made quickly and at low cost. However, the metals have improved surface hardness and durability over epoxies. Metals have the advantage for molds that will be used to make many parts or for molds where the dimensions are especially critical. Some molders will, however, point to the difference in thermal expansion properties between the metal molds and the composite part and will therefore prefer epoxy tooling because it has the same expansion characteristics as the part being cured. For these critical expansion situations, a metal with a thermal expansion coefficient the same as carbon-epoxy composites has been developed. The metal, Invar®, is very expensive and difficult to machine. However, the value of matching the thermal expansion of the mold with the part has given this metal a special niche where high volumes and precision parts are both important. For many other applications, epoxy molds have proven to be far less expensive and often durable or precise enough to meet part performance requirements.

Most epoxy molds are made by forming them over a positive model of the part to be molded. This model can be made by machining, casting, or can be the original part itself. The model is often made from epoxy filled with glass beads so it will machine easily and not be so heavy as to be cumbersome. After treating the model so the mold epoxy will release, the epoxy is laid onto the surface of the model and then cured. The thickness of the epoxy is determined by the desired stiffness of the surface of the mold. Curing is almost always done at the highest possible temperature to ensure that the epoxy will survive repeated moldings (even at lower temperatures). The higher-temperature cure gives a higher crosslink density in the epoxy and results in a more durable mold.

The epoxy mold is usually backed by stiffeners to give additional rigidity to the mold. These stiffeners can be metal, composite, or even wood and are arranged on the back of the epoxy mold so that critical dimensions are frozen in place by the stiffeners.

Some epoxy molds are stabilized by adding large amounts of filler, often metal particles. These molds have improved dimensional stability. They are often easy to machine, thus allowing the mold to be cut directly from a dimensional representation of the part as would be available from a computer-aided design (CAD) program.

For additional stability of the surface, epoxy molds can be metal coated. This coating is usually nickel, which is splattered onto the epoxy surface as a fine spray. After curing the metal, the surface becomes much more scratch resistant than even the most durable epoxy. The small amount of metal reduces the cost of the mold and allows the thermal expansion properties of the epoxy to be largely retained.

CASE STUDY 4-1

Cure and Shelf-life Monitoring

The long cure times for epoxy molding have suggested that molding cycles might be shortened if the optimal moment for stopping the cure can be determined explicitly. Reaching this optimum cure temperature/time would also allow careful control of the crosslink density, thus permitting the optimization of toughness and strength. Therefore, several techniques have been used for monitoring the state of the cure.

The first methods of monitoring the cure were not directly associated with the part. These methods used oven temperature or time as the method for determining when the cure was completed. While these have some value and are still dominant in the industry, they lack the precision of a method that directly senses the state of the part as it is being cured.

Initial experimental attempts used the viscosity of the resin to sense the cure state of the part. These were largely unsuccessful because the viscosity became high long before the part was fully cured, thus reducing the ability to distinguish, on the basis of viscosity, a nearly cured part from a fully cured part. The experiments led to the use of various electrical properties of the part as indicators of cure. It was found that many electrical properties, such as conductivity, resistivity, dielectric strength, permittivity, and others changed as the part became cured. Sadly, the cost of the equipment required to make these measurements is quite high, thus limiting the use of this technique.

An inexpensive and easy-to-use device (named CrossCheck™) has recently been developed that allows the cure state to be sensed. This device uses the resistance of the resin as the characteristic property to sense the changes in the cure state of the polymer. It detects changes in the ability of the polymer to transfer charge. Charge transfer capability is directly linked to changes in the material state such as crosslinking, viscosity, temperature, etc. The device uses multiple electronic techniques to reduce the inherent noise to a point where the signal-to-noise ratio is high enough to provide a stable and repeatable measurement. It is usually sold as a hand-held instrument.

The shelf-life period has traditionally defined whether or not a resin can be used. Many instances occur routinely where resin or prepreg material must be discarded because it is "out of date" and suspect on whether it is still acceptable for use. Another problem is the possibility of using material that is within the stated shelf-life limits but, because of some unknown storage condition, it has seriously degraded. The CrossCheck device and technique have been successfully used to monitor shelf life and out-life changes in several different types of thermosetting resins. The results of these studies have shown a gradual change in resistance followed by a period in which the resistance changes quickly. Comparisons with physical specimens have confirmed that the initial period was a time of stability and then the end period was when the polymer was getting harder (curing). Hence, the useable shelf life can be easily monitored.

Other resins that have been successfully monitored using CrossCheck include UV-cure epoxies and cyano-acrylates. The device/technique is especially convenient for measuring the rapid, room-temperature cures of these materials. In addition, the device has been used for measuring changes in the mixing of components in various polymeric formulations. In these studies, the resistivity of the base resin was measured and then monitored as it changed as various components (such as accelerators, pigments, and surfactants) were added.

Devices and techniques, such as Crosscheck, have wide applicability for monitoring shelf-life changes and normal cures. The changes, which were previously only detectable with costly equipment, are now easily detected with techniques and devices readily available at modest costs.

SUMMARY

Epoxy resins are the second most widely used matrix materials for composites. The excellent balance between good properties and moderate cost has led to this position. In fields where advanced composites are used and properties are at a premium, epoxies are the foremost resin.

The epoxy resins are characterized by the presence of the epoxy ring, a three-member ring with two carbon molecules and an oxygen molecule. This ring can be added to the chain end of a variety of resins to give a wide selection of properties. Usually, the epoxy polymers are highly aromatic so that the advantages of strength and stiffness can be used in the composite part.

The epoxy ring is the point at which the crosslinking reaction takes place. A molecule possessing active hydrogen is added to the epoxy resin and thoroughly mixed to begin the crosslinking reaction. These activating molecules are called hardeners. The crosslinking reaction involves the opening of the epoxy ring by the active hydrogen on the end of the hardener molecule. That hydrogen bonds to the epoxy oxygen. Simultaneously, a bond is formed from the atom that was previously attached to the hydrogen to the carbon that was previously attached to the oxygen.

Crosslinking occurs because both the epoxy and the hardener resins have at least two active groups. Hence, each hardener ties together at least two epoxy chains. The choice of hardeners is wide. Several of the most common are listed in Table 4-4 along with the advantages and disadvantages of each.

The processing of epoxy resins is somewhat more difficult than polyesters. The epoxies are usually cured at elevated temperatures. Wet-out of the fibers is more difficult with epoxies because the resins are usually high viscosity and solvents are rarely used. The cure rates are slow, thus leading to long cure cycles. However, shrinkage is low for epoxies.

The properties of epoxies are generally superior to polyester thermosets. Adhesion of epoxy resins to the fibers is good as is adhesion to other materials. Strength and stiffness are superior for epoxies, as is creep resistance, due in part to the excellent adhesion and to the high aromatic content of most epoxy resins. Toughness can be a problem with any composite material but a few means are available to toughen epoxies without sacrificing other key properties too much.

LABORATORY EXPERIMENT 4-1

Resin and Hardener Ratios

Objective: Determine the differences in properties from mixing resin and hardener in different proportions.

Procedure:

1. Pour equal 100 ml (3.4 oz.) portions of epoxy resin into four different metal or glass cups. (Pint canning bottles will work fine.)
2. Prepare various amounts of hardener (of a type suggested by the resin manufacturer) and place in separate cups. Suggested amounts are: 50 ml (1.7 oz), 100 ml (3.4 oz), 150 ml (5.1 oz), and 200 ml (6.8 oz).
3. Pour one of the hardener materials into one of the resin cups. Mix the materials thoroughly.
4. Pour the contents of the cup into a mold that forms a flat plate. (A convenient

mold is made from two pieces of plate glass covered with film and a simple metal frame.)

5. Insert a thermocouple into the edge of the mixture to monitor the temperature.
6. Heat the mixture to the cure temperature suggested by the resin manufacturer and hold for the suggested time.
7. After cool down, remove the cured plate from the mold.
8. Repeat steps 3 through 7 for each of the hardener/resin combinations.
9. Cut tensile specimens out of each plate. (It is recommended that at least three specimens be cut for each mixture.)
10. Test the tensile properties (strength, stiffness, and elongation) for each of the mixtures.
11. Report the properties and discuss the following: trends, the importance of the properties in composites even though no fibers are present in these samples, and the causes of differences.

QUESTIONS

1. Discuss three advantages of epoxy resins versus thermoset polyesters.
2. What is a hardener and how does it differ from an initiator used in polyester curing?
3. Describe the differences in crosslink density that would be expected in epoxies and in polyesters.
4. Describe the physical property changes in epoxies that come from increasing the aliphatic content of the resin. When would this typically be done?
5. Indicate three reasons why epoxies are the resin of choice in making composite parts for airplane structures such as wing and tail components.
6. Describe the naming system for epoxy resins.
7. Contrast the environmental differences between epoxy and polyester resins.
8. Discuss two methods that might be employed to make an epoxy resin tougher. What are the advantages and disadvantages of each of the methods?
9. Why is cure sensing more important in epoxies than in polyesters?

BIBLIOGRAPHY

"Dow Liquid Epoxy Resins," brochure. No date. Midland, MI: Dow Chemical Company, 2040 Willard Dow Center, Midland, MI, 48674, Form 296-00224-0199 WC+M.

"Epon® Epoxy Resins Turn Up FRP Performance," brochure. Houston, TX: Shell Chemical Company [now called Resolution], P.O. Box 2463, Houston, TX 77252, 800-832-3766, SC: 2445-96.

"Formulating with Dow Epoxy Resin," brochure. No date. Midland, MI: Dow Chemical Company, 2040 Willard Dow Center, Midland, MI, 48674, Form 296-346-1289.

Morgan, Peter. 2005. *Carbon Fibers and their Composites*. Boca Raton, FL: Taylor & Francis.

Society of Manufacturing Engineers (SME). 2005. "Composite Materials" DVD from the Composites Manufacturing video series. Dearborn, MI: Society of Manufacturing Engineers, www.sme.org/cmvs.

Strong, A. Brent. 2006. *Plastics: Materials and Processing*, 3rd Edition. Upper Saddle River, NJ: Prentice-Hall, Inc.

Strong, A. Brent. 1995. *Fundamentals of Polymer Resins for Composite Manufacturing*. Arlington, VA: Composites Fabricators Association.

5

Specialty and High-performance Thermosets

CHAPTER OVERVIEW

This chapter examines the following:

- Introduction
- Vinyl esters
- Phenolics
- Carbon matrix
- Polyimides and related polymers
- Cyanate esters
- Polyurethanes
- Silicones
- Dicyclopentadiene (DCPD)

INTRODUCTION

Although the composites market is dominated by unsaturated polyesters and epoxies, several specialty resins are used for applications in which the properties of the polyesters and epoxies are not quite appropriate. Even though specialty resins only represent about 20% of the total composites market, their contributions are important because of the criticality of some products that can only be made with these resins.

As the specialty resins are discussed, their unique properties will be examined. In general, these properties will be compared to polyesters and epoxies, which are convenient reference materials, although in some cases the properties will be compared to metals and other composite and non-composite resins. The molecular features of these resins that give rise to their special or outstanding properties also will be discussed.

In most cases, some particular property will be especially important in these materials and it will determine the major applications for the resin. One important property that has wide application is chemical resistance. Vinyl esters are popular because they have the combination of good chemical resistance and good price. Flammability is another key property that is especially important and the key resin in this case is phenolics. Phenolics also have good thermal stability as do carbon matrix composites, polyimides, and silicones. Excellent toughness is a characteristic of cyanate esters, silicones, and dicyclopentadiene (DCPD). All of these resins, including their manufacturing characteristics and curing mechanisms, will be discussed. In some cases, special manufacturing procedures, which might be used in molding products made from these resins, are identified.

VINYL ESTERS

Vinyl esters (pronounced *vine'-ell ess'-ters*) are a family of thermosetting resins that have many similarities to, and seem to fit between, both unsaturated polyesters and epoxies. Vinyl ester resins are slightly more expensive than unsaturated polyesters but are not as expensive as epoxies. The vinyl esters have superior toughness and corrosion resistance (to organic solvents and water) in comparison to polyesters, but are generally not as good in these properties as the epoxies. The vinyl esters cure much like unsaturated polyesters and are, therefore, easier to cure than the epoxies.

The combination of excellent processing characteristics, good properties, and reasonable cost suggest that vinyl esters be used in applications where cost is an important factor but where the chemical resistance and/or toughness of unsaturated polyesters are not adequate. Some of these applications include: pipes and tanks for the chemical industry, air-pollution control applications such as scrubbing towers, equipment housings where moderate impact performance is important, and marine applications where corrosion resistance is critical. Another important advantage of vinyl esters over traditional thermoset polyesters is the generally lower styrene content that can be used. This alleviates some of the problems of styrene emissions during polyester manufacturing.

Structure and Synthesis

The reasons for the intermediate nature of vinyl esters, seen as residing between epoxies and thermoset polyesters, are easily understood from their diagrammatic structure shown in Figure 5-1. The vinyl ester molecule is seen to be a combination of the epoxy backbone with vinyl ester groups, which contain a carbon-carbon double bond on each end replacing the epoxy rings. The name *vinyl ester* arises because the chemical group that connects the carbon-carbon groups to the main polymer chain is an ester. Occasionally, carbon-carbon double bonds are called "vinyl" groups and so the name *vinyl ester* was created to identify those molecules in which a carbon-carbon double bond was linked to the rest of the molecule by an ester linkage. Those carbon-carbon double bonds are sites that can be crosslinked using the same general methods used to crosslink unsaturated polyesters.

The synthesis of vinyl ester polymers is usually done by a large chemical company, not by the molder. These polymers are made by reacting an epoxy resin with acrylic acid, as shown in Figure 5-2. The acrylic acid opens the epoxy ring and forms new bonds with the oxygen and carbon at the epoxy ring site. Because acrylic acid has a carbon-carbon double bond, or unsaturation, in its back-

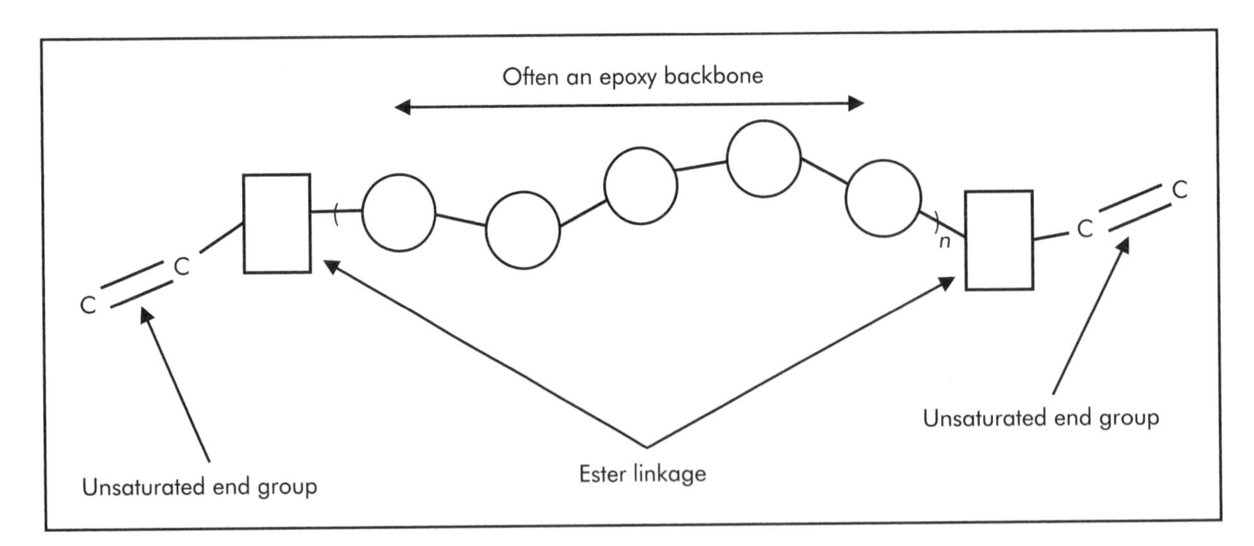

Figure 5-1. Representation of a typical vinyl ester polymer.

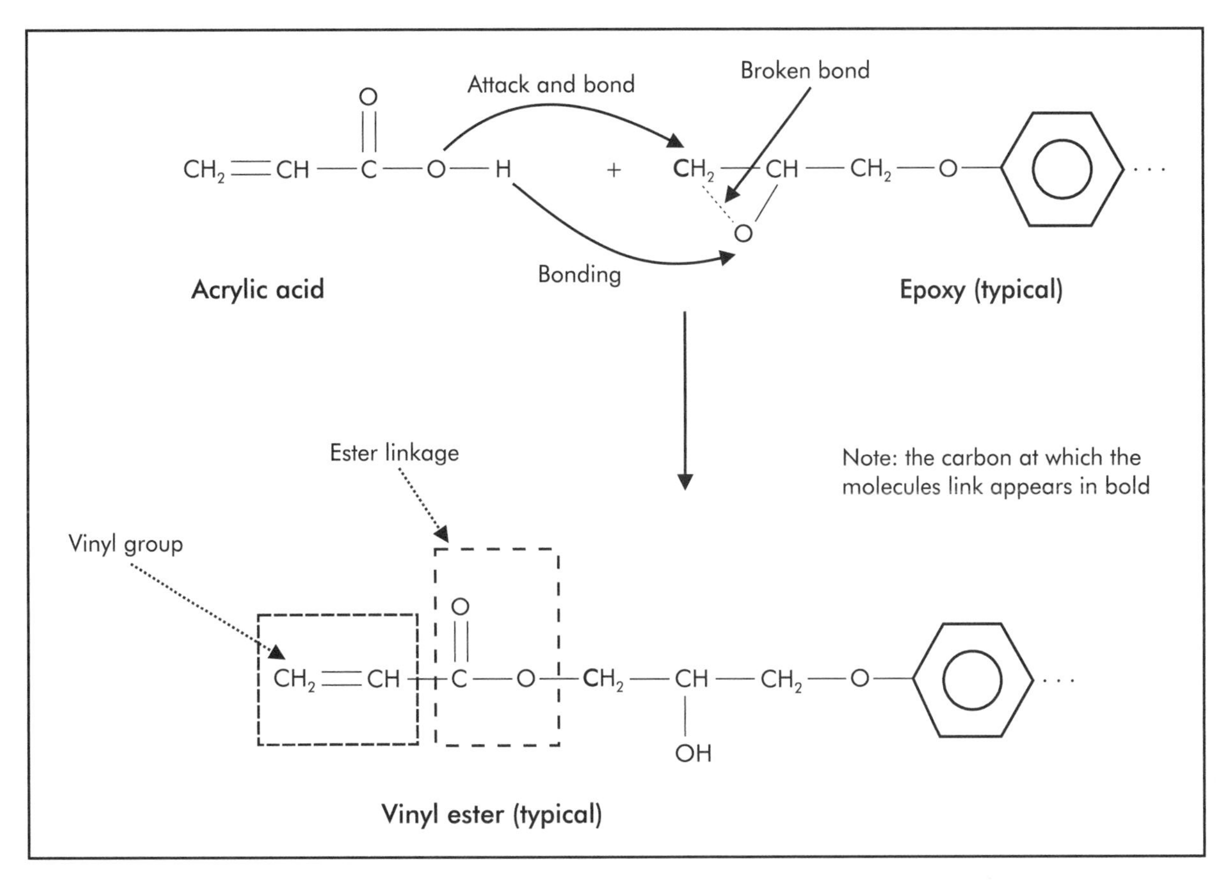

Figure 5-2. Synthesis of a vinyl ester from an epoxy.

bone, that unsaturation becomes attached to the end of the chain. The reaction of the acrylic acid occurs with all of the epoxy rings in the molecule, so multiple sites are available for subsequent crosslinking reactions. Thus, the vinyl ester group becomes the end group of each chain. The synthesis path of vinyl esters from epoxies and acrylic acid is sometimes emphasized by calling vinyl esters "acrylic-modified epoxies." The similarity of vinyl ester backbones and epoxy backbones can be seen in Figure 5-3 where the chemical structures of some common vinyl esters are shown. These should be compared with the similar epoxy molecules in Figures 4-3 and 4-6. In essence, the synthesis reaction has changed the polymer from an epoxy, which is crosslinked by using a hardener, into a vinyl ester that has carbon-carbon unsaturation at the ends and can be crosslinked like an unsaturated polyester with a peroxide initiator and styrene.

The beneficial nature of the often highly aromatic and strong backbone associated with epoxies is retained in the vinyl ester. But the properties of the crosslinked vinyl ester are somewhat different from those of epoxies because of two important reasons.

First, the ester linkages at each end of the molecule make the resin more susceptible to water and other solvent attack than the analogue epoxy, but not as susceptible as a polyester. The reason is simple. The ester is a point that is susceptible to attack by water,

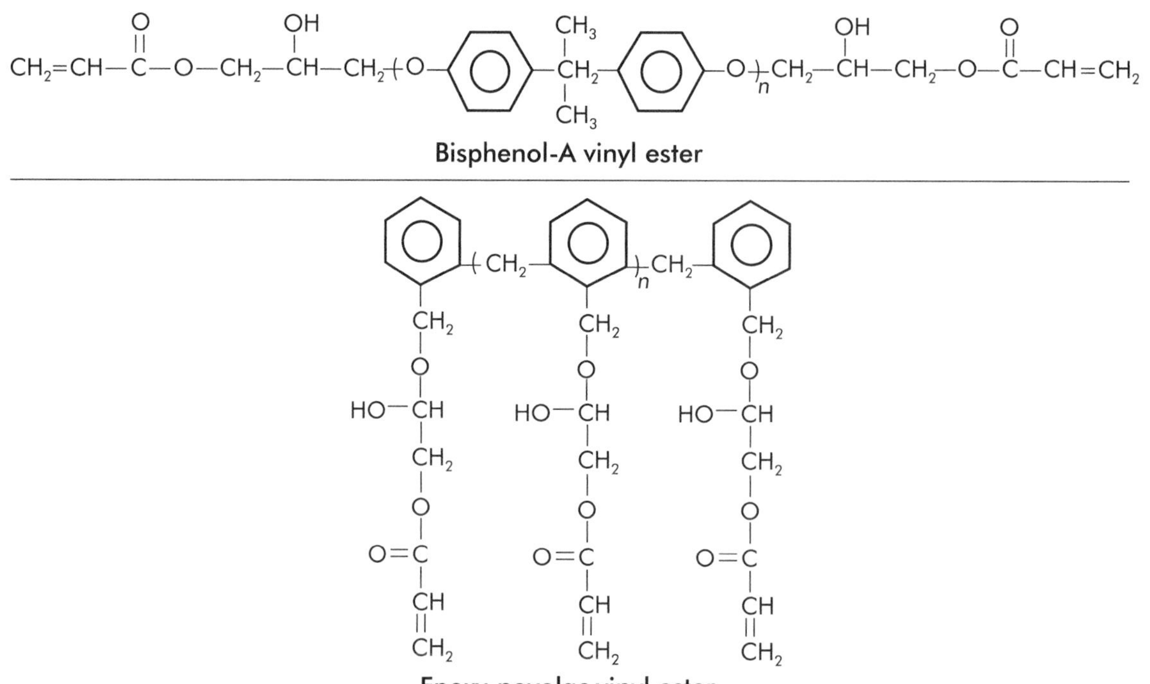

Bisphenol-A vinyl ester

Epoxy-novolac vinyl ester

Figure 5-3. Epoxy-based vinyl esters.

and solvents and the vinyl ester have more of these ester sites than the epoxy, but still fewer than the polyester. The esters occur in every repeating unit of the polyester but only near the ends of the chains of the vinyl esters.

The second important reason for the differences between epoxies and vinyl esters is that the crosslinking of vinyl esters is done in the same manner as in unsaturated polyesters with styrene as the bridge molecule(s) between polymer chains. Epoxies do not, of course, use styrene as their bridge molecule but, rather, use a hardener. Therefore, the properties associated with the polyester groups and the properties in the epoxy associated with the hardener would be changed when converting an epoxy to a vinyl ester. These two factors result in molecules that have some characteristics of the epoxy backbone but also some characteristics of the polyester thermoset. Hence, vinyl esters are intermediate between epoxies and polyesters as was stated earlier.

Crosslinking Reactions

The crosslinking reactions of vinyl esters, shown in Figure 5-4, begin with a peroxide initiator that splits into two parts, each containing a free radical. Those free radicals then react with the carbon-carbon double bonds on the ends of the vinyl ester polymers. This is shown as Step 1 in Figure 5-4. This reaction is, of course, typical of the chain-reaction nature of the free-radical reactions as was discussed in the chapter on unsaturated polyesters. Note that a new free radical is formed on the end of the polymer chain where the vinyl ester group is located.

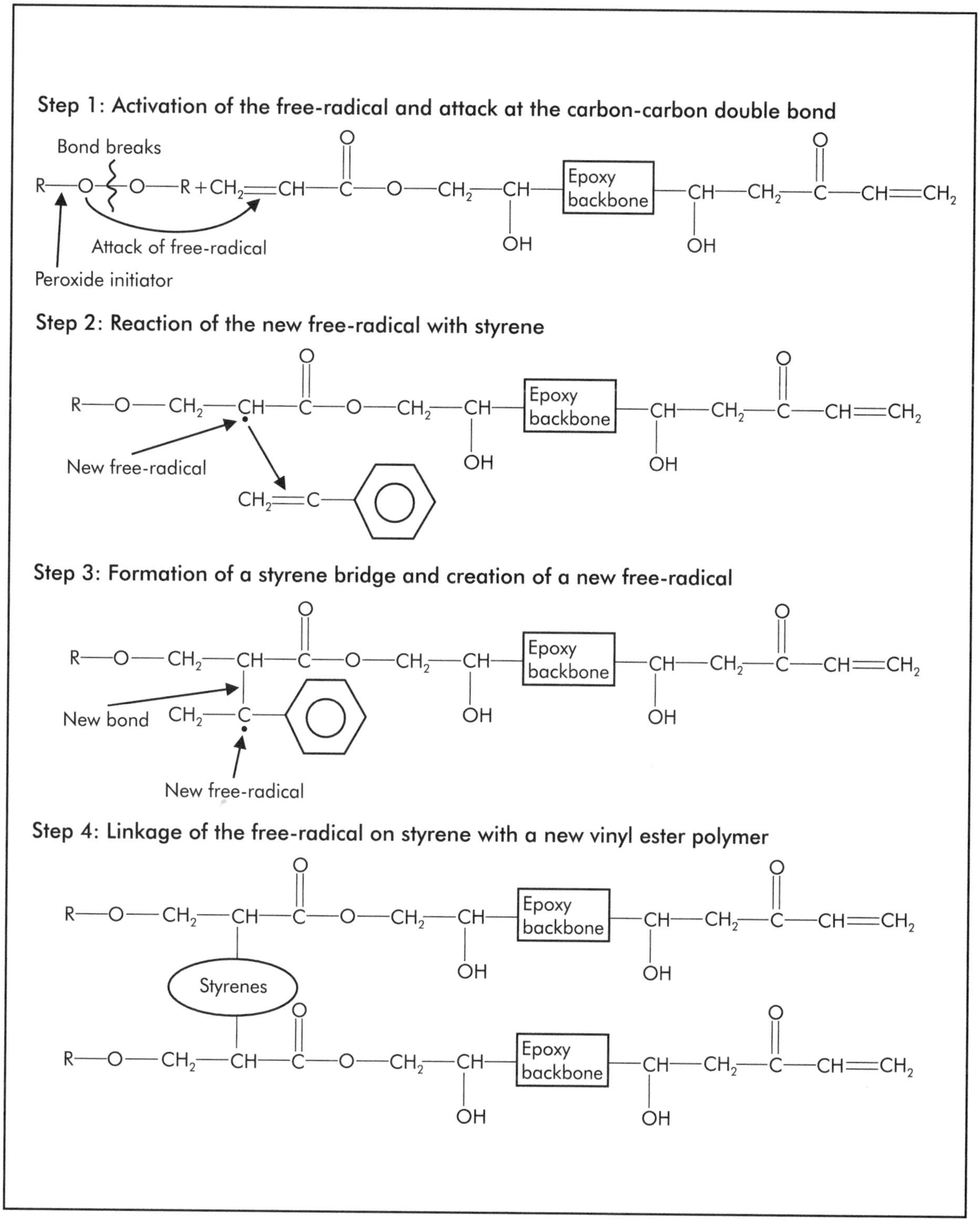

Step 1: Activation of the free-radical and attack at the carbon-carbon double bond

Bond breaks

Epoxy backbone

Attack of free-radical

OH OH

Peroxide initiator

Step 2: Reaction of the new free-radical with styrene

Epoxy backbone

New free-radical

OH OH

Step 3: Formation of a styrene bridge and creation of a new free-radical

Epoxy backbone

New bond

OH OH

New free-radical

Step 4: Linkage of the free-radical on styrene with a new vinyl ester polymer

Epoxy backbone

OH OH

Styrenes

Epoxy backbone

OH OH

Figure 5-4. Crosslinking reactions for a vinyl ester.

Step 2 in Figure 5-4 shows the attack of the free radical on the polymer end with a styrene molecule. Just as with unsaturated polyesters, the vinyl ester crosslinking reaction is carried out in a styrene solution. The attack is followed in Step 3 by a bonding between the styrene and the polymer.

Depending on the relative concentrations of polymer and styrene, and the conditions under which the reaction proceeds, one or several styrene molecules can form the crosslinking bridge as shown by the oval in Step 4. This step also shows the attachment of the first polymer to a second vinyl ester polymer. The non-reacted ends of the vinyl esters will likely join to different molecules, thus resulting in all of the molecules being joined into a large crosslinked network.

Because vinyl ester unsaturation is at the end of the polymer chain, the carbon-carbon double bonds are more reactive (accessible) than the carbon-carbon bonds in polyesters. As a result, vinyl ester resins cure rapidly and consistently to give fast gel times with nearly complete reaction of every double bond. This high reactivity of the vinyl ester groups allows some vinyl ester systems to be crosslinked without the need of a reactive solvent such as styrene. Furthermore, even when styrene is used, the lower number of double-bond reactive sites and the greater mobility (reactivity) of the end groups (versus mid-molecule polyester-type groups) allows a significant reduction in the amount of solvent (reactive diluent). Also, because there are fewer reactive groups in vinyl esters, the shrinkage is lower than that of most polyester resins.

The peroxide initiators, accelerators, and other additives typically used in polyester thermosets often can be used in crosslinking vinyl esters. However, the optimal concentrations and, in some cases, the optimal choice of ingredients can vary from those used in typical polyester systems. For instance, while methyl ethyl ketone peroxide (MEKP) is the most common peroxide for polyesters, the most common for vinyl esters is benzoyl peroxide (BPO). Also, instead of cobalt accelerators, the most commonly used accelerators in vinyl esters are the aromatic amines (analines).

Properties

Corrosion Resistance

The polyester sites on the polymer are the most vulnerable locations for chemical (corrosion) attack. Because vinyl esters have fewer polyester sites than traditional thermoset polyesters, they are inherently less susceptible to chemical attack. The typical high aromatic content of the vinyl ester backbone has some additional benefit in chemical resistance. The reduction in polyester groups reduces the polar character of the vinyl ester compared to the polyester, thus significantly reducing the amount of water the vinyl ester is likely to absorb. A plot of weight gain for a vinyl ester resin and a polyester resin are compared in Figure 5-5.

Corrosion resistance of vinyl ester parts sometimes can be further improved by some or all of the following methods:

- Use one or two layers of glass veil, a thin glass cloth material, and place it near the exposed side of the vinyl ester container.
- Post-cure the part to enhance the amount of crosslinking and therefore make the structure tighter.
- Use benzoyl peroxide and dimethyl analine as the curing system, which will lead to a high degree of crosslinking.
- Dilute the corrosive liquid that is attacking the polymer as much as possible.
- Remove corrosive fumes.
- Operate at the lowest possible temperature, thus reducing general chemical activity.

Unreacted carbon-carbon double bonds along the backbone of the polyester are

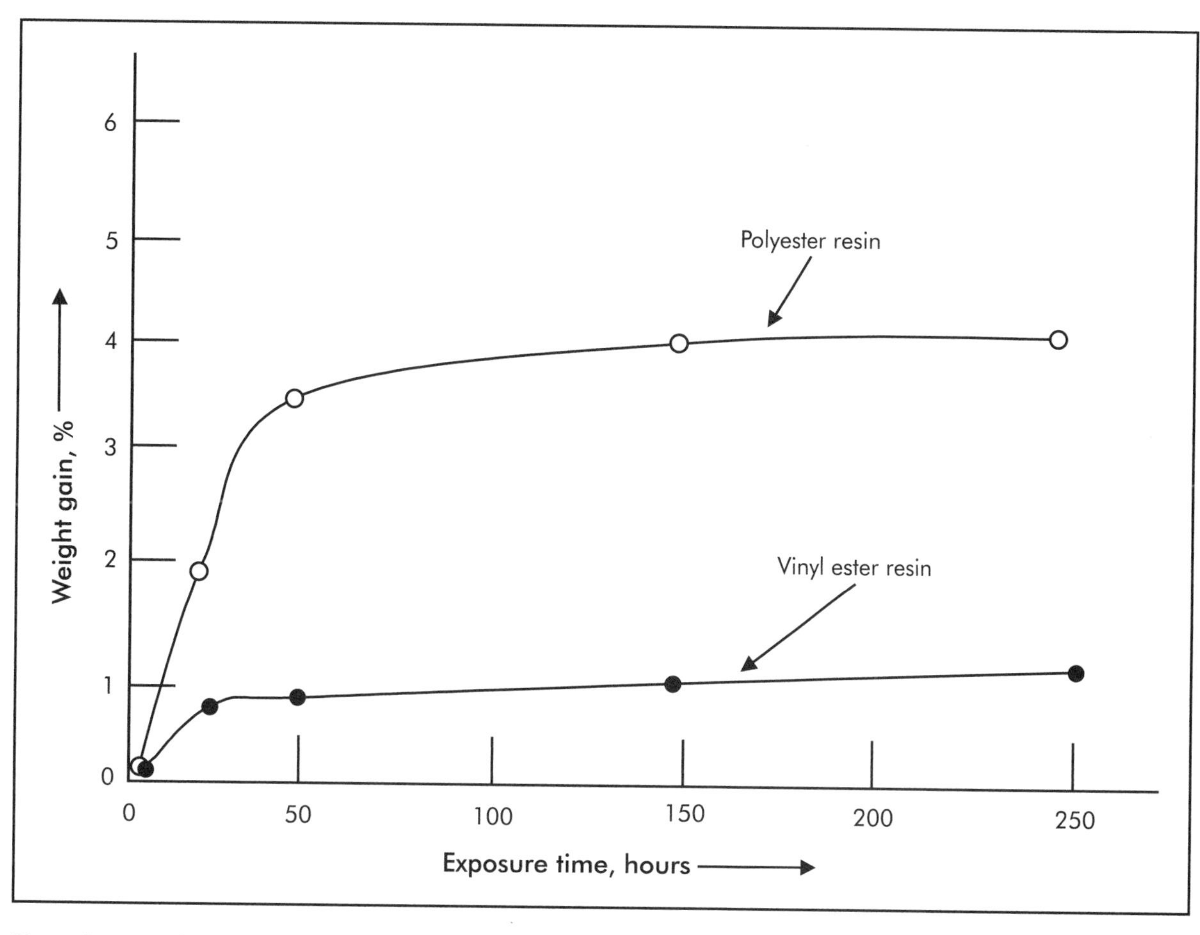

Figure 5-5. Weight gain of polyester vs. vinyl ester resins from water absorption.

highly susceptible sites for chemical attack, especially from agents such as oxygen and oxidizing acids. The effects of these agents are most likely to be seen in long-term thermal aging performance and resistance to aggressive environments. However, vinyl esters have carbon-carbon double bonds only on the ends of the chains. This reduces the number that are likely to be unreacted (compared with polyesters) and the severity of the problem should attack occur since the entire chain is not cleaved.

Adhesive Strength

The hydroxyl groups that are pendant from the polymer chains of the vinyl ester molecules (which are made when the epoxy is converted to a vinyl ester as shown in Figure 5-2) have a generally beneficial effect on the internal bonding of composite laminates. That is, they play a role in bonding the resin to the reinforcement, just as the hydroxyl groups do in epoxy resins. These hydroxyl (OH) groups react with the surface of the glass fibers, which results in excellent glass fiber wetting and good adhesion to the glass fibers. This is one of the factors responsible for the higher strengths obtained with laminates made of vinyl esters when compared with traditional polyesters. The hydroxyl groups also provide sites where adhesives can bond to vinyl esters, thus giving them greater adhesive strength, in general, than polyesters.

Flame Retardant

Vinyl esters are made flame retardant in the same way as with polyesters and epoxies. The most common methods are halogenation of the resins and adding halogenated or alumina trihydrate fillers.

Molding

Highly filled vinyl esters can be used as sheet molding compounds (SMCs) and bulk molding compounds (BMCs) and will compete with similar polyester molding compounds, especially for specialty resin applications. However, vinyl esters do not respond well to the divalent thickeners used in polyester SMC and BMC molding compounds. Other thickening agents have been developed and this change in materials does not generally affect molding properties.

Toughness

The presence of the crosslinkable sites (carbon-carbon double bonds) only on the ends of the vinyl ester polymer means that the crosslink density of vinyl esters will be substantially less than in polyesters. Therefore, vinyl esters will tend to be less brittle and tougher than polyesters.

The epoxy backbone in vinyl esters is important in applications where toughness is required. The improved toughness of vinyl esters over polyesters comes from the higher strength of the backbone and the higher elongation associated with terminal crosslinking. Because the vinyl esters are usually crosslinked in a solution with the reactive diluent, the molecular weight of the vinyl ester can be greater than in normal epoxies, which are often crosslinked without solvents. Hence, even compared with epoxies, the vinyl esters can have higher molecular weight, which leads to increased toughness.

PHENOLICS

Phenolics (pronounced *fen-ahl'-icks*) were the first thermoset materials synthesized (under the name of Bakelite® by Leo Bakeland in 1907). They are still among the most important thermosets, undoubtedly because their properties are unique and valuable. The cost of phenolic resins is typically 10–15% higher than polyester thermosets.

Phenolics have been used for many years as general, non-reinforced thermoset plastics in applications such as electrical switches, junction boxes, automotive molded parts, consumer appliance parts, handles for pots and pans, adhesives for plywood, and even billiard balls. Because phenolic resins have high crosslink densities, they have high shrinkage and are quite brittle. Almost all have fillers added to reduce shrinkage and improve toughness. In the vast majority of phenolic products, inexpensive fillers such as wood flour, sawdust, ground nut shells, and talc are used. Since the color of phenolics varies from brown to amber and the shade is difficult to control, most applications where the phenolic will be seen use a pigment to make the color uniform. Carbon black is the most common pigment, so most phenolic parts seen by customers are black. For instance, phenolics handles for cooking pans are black, but phenolic glue in plywood, which is really not seen by the consumer, does not contain a pigment.

As with the polyesters, vinyl esters, and epoxies, the properties of phenolics are best understood in light of the molecular structure of the polymer. They are, therefore, discussed in detail after the section on synthesis and crosslinking reactions. In brief, the major properties of phenolics are:

- excellent flammability performance, especially the combination of low flame spread, low smoke generation, and low smoke toxicity;
- low heat transfer;

- high thermal stability;
- high electrical resistance;
- excellent resistance to chlorinated solvents; and
- good adhesion.

Even though the properties and cost of phenolics are attractive and the market for non-reinforced phenolics is quite large, their use in fiber-reinforced composites is much less common than polyesters, epoxies, or vinyl esters. This lower usage results from the *difficulty in processing phenolics* when compared to the polyesters, vinyl esters, and epoxies. Nevertheless, certain key advantages of phenolics, especially low flammability, are so strong that their use as a matrix for both fiberglass and carbon fiber composites is increasing significantly.

The increased demand for phenolics has spurred work to simplify the molding of phenolic composites (discussed later). Phenolic composites can now be made by a variety of traditional composite processes with only minor allowances for the unique nature of their curing process.

Polymer Structure

Phenolics are formed from the condensation polymerization reaction between phenol, an aromatic molecule, and formaldehyde, a small organic compound often used as a solvent or preservative. Phenol has three active sites, all on the phenolic ring. Each site can react with a formaldehyde molecule to link phenolic rings together with a carbon bridge (called methylene because the carbon has two attached hydrogen molecules) as shown

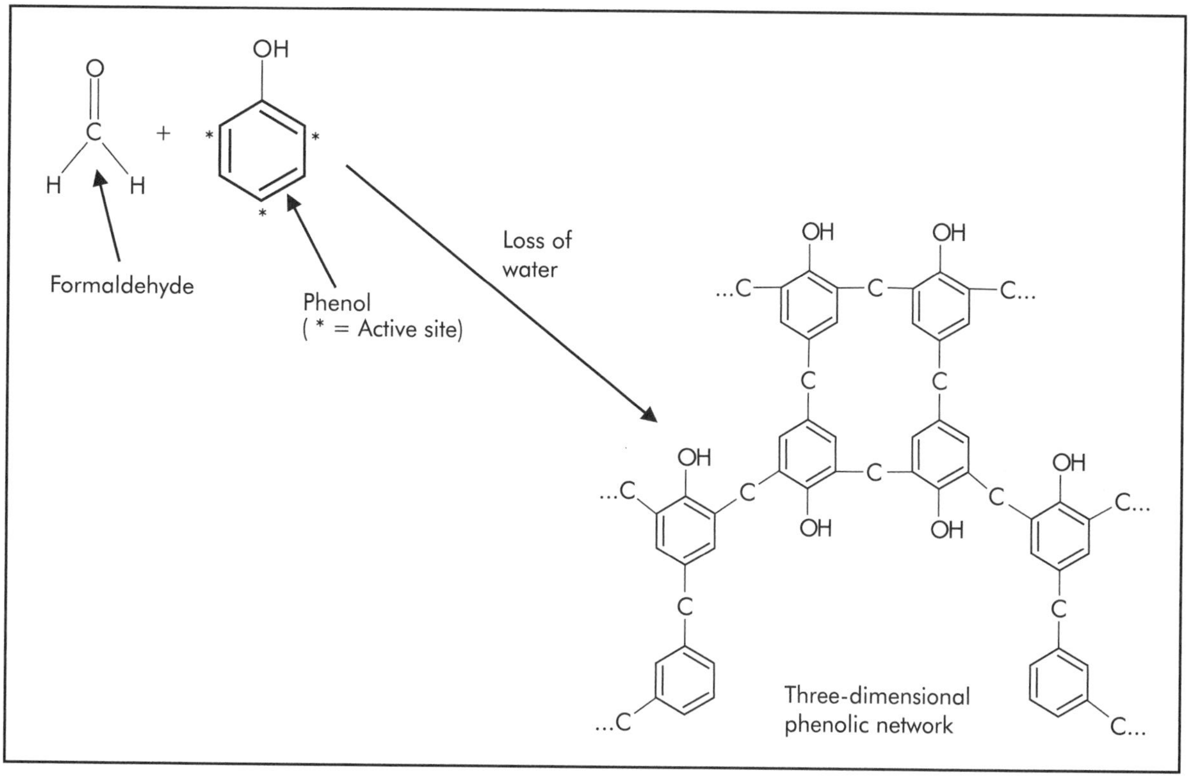

Figure 5-6. Phenolic polymerization and crosslinking.

in Figure 5-6. The multiple reaction sites on the phenolic ring allow each ring to be joined to two or three others, thus resulting in a tightly crosslinked, three-dimensional structure.

Crosslinking Reactions

The crosslinking reaction can be carried out under two different conditions that proceed by two slightly different crosslinking reaction sequences. In both reaction sequences, an intermediate is formed and the reaction can be *stopped at the intermediate stage*. It is convenient to stop at this stage because the intermediate materials are more convenient to handle and ship than are the monomers; but the intermediate stage is still highly moldable. This permits the intermediates to be made by a resin manufacturer who then ships them to the molders of the final parts.

When any crosslinking reaction is stopped at an intermediate stage, the process is called **B-staging** and the resin is said to have been **B-staged**. The name derives from the concept that the totally unreacted material, usually monomers, is the A-stage; the intermediate material is the B-stage; and the fully cured and crosslinked material is the C-stage.

One of the two reaction conditions for phenolic B-staging is with an excess of alkali (strongly basic) solution, which results when the concentration of the formaldehyde reactant is much greater than the concentration of the phenol reactant. This process can be carefully controlled to produce a viscous liquid intermediate product, called a **resole,** which is a minimally crosslinked, linear polymer of aromatic rings linked by carbon bridges. When desired, the resole can be crosslinked by simply reheating (to about 300° F [150° C]). Since no additional material needs to be added to the resole to accomplish this final crosslinking, resoles are called **one-step resins**. Although not strictly needed, an acid catalyst can reduce the gel time of many resole systems.

The uncured (B-staged) resole is a viscous, water-soluble liquid. It is widely used as a coating and impregnation material. Resoles can be used to coat fibers in composites and they give strength and body to paper when subsequently cured. They also can be stirred vigorously to form a foamed material, about the consistency of whipping cream, which then can be molded and cured into a rigid foam. These foams will readily absorb and hold water and are used widely by florists as embedding forms for making floral displays.

When used to coat reinforcement fibers, the resole/fiber mixture is usually catalyzed to reduce the cure time. When crosslinking is desired, the reaction proceeds with sufficient rapidity that even composite processes requiring a short reaction time, such as pultrusion, can be used. Filament winding, which requires a somewhat longer pot life, is best accomplished by using a latent catalyst system. These catalysts become activated when heated. The major difficulty with resoles used as composite resins is the formation of water as a condensate or by-product of the curing reaction. To obtain the good properties required of most composites, this water needs to be extracted. That can be done with some difficulty and an extra processing step is required.

The liquid nature of resole material permits the making of sheet molding compound (SMC) in an analogous fashion to that of unsaturated polyester resins. Most of the same storage restrictions common with polyester SMC also apply to phenolic SMC.

In the second type of B-staging reaction, the phenol is in excess compared to the formaldehyde reactant and an acid solution is used. These conditions are opposite of those used to form resoles. The resulting material, called a **novolac**, is an essentially non-crosslinked polymer in the form of a powder (after stripping out the water).

Novolacs will not crosslink with just the addition of heat, but require a curing agent. The most common of the curing agents for novolacs is hexamethylene tetramine or, simply, **hexa**. Hexa is heat activated, so molding of novolacs is usually done with pressure and heat to compress the powder and activate the crosslinking reaction. Because a second material (hexa) must be added to novolacs, they are called **two-step resins**.

Novolacs, as powders, are rarely used in making composites directly. However, the novolac-phenolic resins can be readily converted into epoxy and vinyl ester resins which are, of course, used as the resin in composites as previously discussed. When the novolac-phenolic resins are sold for conversion to other resins, they are often sold as solutions; the most common solvents are water, simple alcohols, such as ethanol, and simple aromatics, such as toluene. During crosslinking, these solvents are evaporated. The need to evaporate the solvent is one of the drawbacks in using phenolics for composite applications.

The most important difficulty in the processing of phenolics is the creation of the water by-product from the condensation crosslinking reaction, and this occurs with both resoles and novolacs. This condensate is formed throughout the polymer mixture and can result in small bubbles and voids in the part. To reduce the amount of condensate in the sample, a common practice when compression molding phenolics in a hot press is to open the mold slightly and quickly during the molding cycle, thus permitting the vaporized water to escape. This process is called **bumping the mold** or **breathing the mold**. To minimize the formation of condensate bubbles, phenolics are often initially cured at temperatures below the boiling point of water. Then, after a considerable amount of the crosslinking has been achieved, the part is post-cured at about 320° F (160° C) for 1–24 hours, depending on the thickness of the part. This advances the amount of crosslinking and allows the water by-product to diffuse out of the part.

The need for additional steps in processing phenolics has limited their use to mainly those composite applications where their unique properties are an overriding feature that dictates their use.

Properties

Phenolics have good chemical resistance. This capability is indicative of the highly crosslinked aromatic nature of the phenolic polymers. Phenolics have excellent resistance to alcohols, esters, ketones, ethers, chlorinated hydrocarbons, benzene, mineral oils, animal and vegetable fats, and oils. The resistance of phenolics to weak acids and bases is also good, but phenolics are not able to withstand strong acids or strong bases (alkaline solutions). Further, phenolics do not have good resistance to boiling water. As with other aromatic materials, the ultraviolet (UV) light resistance of phenolics is poor.

One of the limitations of phenolics is it has some generally lower mechanical properties, especially impact toughness, when compared to other high-volume thermosets. However, the manufacturing conditions used in making the phenolic, including the oven post-cure aging, make a major difference in properties. These conditions are less important in many of the other resins. Hence, direct property comparisons with other materials can be confusing because the data may not be truly comparable.

Flammability Performance

By far, the most important property for reinforced phenolics is *flammability performance*. Much of the impetus for the increased use of phenolics is the enactment of stricter specifications for flammability requirements in many applications, especially public transportation. For instance, in the

1980s, the Federal Aviation Administration tightened aircraft fire specifications to increase the safe evacuation time for passengers in airplane fires. Other transportation agencies have likewise increased their requirements. The flammability of plastics and composites is often measured by examining their *flame characteristics*, such as the flame spread index, and *smoke characteristics* as measured by smoke optical density tests and toxicity tests.

Typical requirements for transportation applications are for a flame spread index value (ASTM E 162), Is, of less then 35 and a smoke density, Ds, of less than 100 after 1.5 minutes and less than 200 after 4 minutes (per ASTM E 662). As shown in Table 5-1, phenolics easily meet these requirements. This table also lists off-gases, which are restricted in some transportation specifications, usually because of potential toxicity. As noted in the table, phenolic off-gases are well under the specified limits.

The results of tests on several typical resins are shown graphically in Figure 5-7. Here the superior performance of phenolics is obvious. The low smoke generation of a phenolic over time versus a polyester and an epoxy is shown in Figure 5-8.

The desirable combination of flame-retardant properties exhibited by phenolics is extremely hard to achieve with any other resin, especially when cost is also considered. These favorable flammability results can be understood by looking at the molecular structure of the phenolic resin shown in Figure 5-6. The first impression of this structure is that it is highly aromatic with many benzene rings, which are tightly interconnected into a three-dimensional structure. As discussed in previous chapters, high aromatic content is beneficial in many ways. One is a resistance to burning. Highly aromatic molecules are difficult to ignite and, when finally ignited, burn slowly. This is because they *do not readily volatilize*. Therefore, they

Table 5-1. Flammability specifications and test results for a typical phenolic resin.

Test	Typical Transportation Requirement	Typical Phenolic Result
Flame spread test (ASTM E 162), Is	<35	1.0
Smoke density (ASTM E 662)		
Ds (1.5 minutes)	<100	1.0
Ds (4.0 minutes)	<200	15
Ds (maximum)	—	51
Smoke chamber gas analysis (parts per million) of the National Bureau of Standards		
CO	<3,500	100
HF	<200	0
NO_2	<100	0
HCl	<500	0
HCN	<150	0
SO_2	<100	80

do not easily create flammable vapors that facilitate ignition and propagate the spread of the flame. Furthermore, highly aromatic compounds, especially those that are cross-linked, tend to burn incompletely in limited oxygen environments as is almost always the case in real fires. This limited burning results in the formation of a **char**, much like the char (charcoal) that forms when wood burns incompletely. This char is even more resistant to burning than the original phenolic. Hence, not only is the phenolic slow to burn, but when burned, it forms a product that further reduces burning. This low flammability is natural with phenolics and so the addition of halogen materials is not necessary as required with many other resins.

While the smoke from burning aromatics is quite dark, it is formed slowly and is chiefly composed of relatively harmless hydrocarbons. Furthermore, it does not contain the halogen compounds that exist in other flame-retardant polymers. These halogens form dense and choking smoke that is avoided with phenolics. Hence, the smoke density results of phenolics are also favorable.

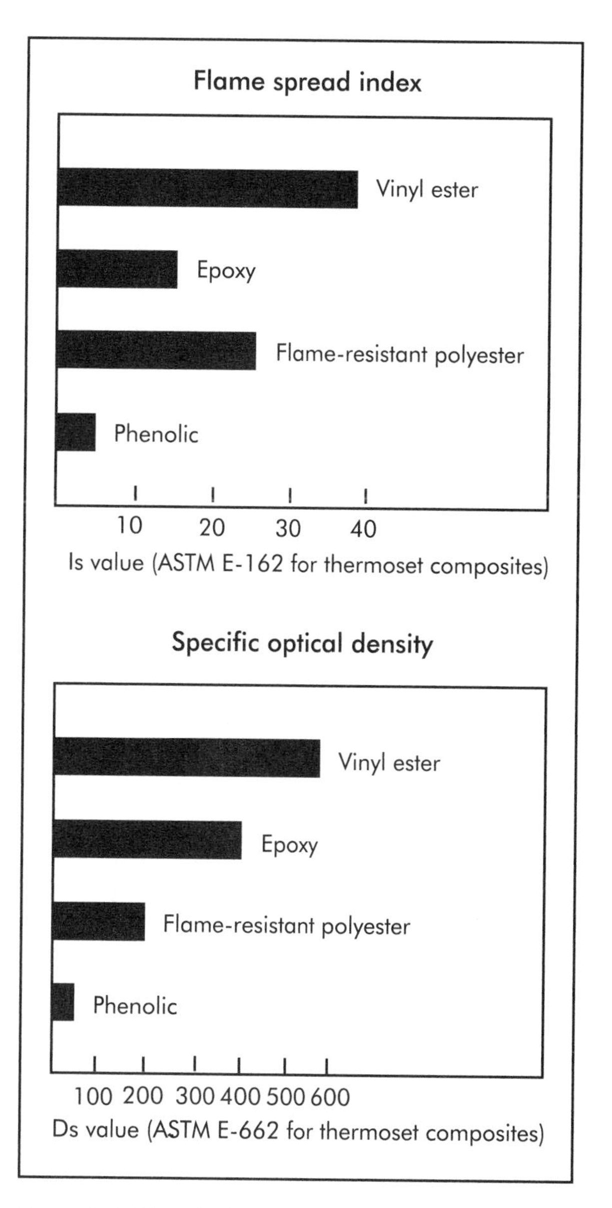

Figure 5-7. Phenolic flammability.

Applications

Because of their favorable flammability properties, phenolic composites are the material of choice for walls, ceilings, and floors of aircraft interiors, train interiors, and other public transportation panels.

An interesting and somewhat unique application for phenolic/carbon fiber composites is in the exit nozzles of rockets, which are depicted in Figure 5-9. In this application, the general advantage of weight savings with composites is augmented by the advantage of superior thermal insulation during firing of the rocket. When phenolic/carbon-fiber composites are exposed to high temperatures, as occur in the exhaust flow path of a rocket (as high as 6,000° F [3,300° C]), the phenolic resin degrades and turns into char, as explained previously. In addition to its excellent non-flammability, the char has superior thermal insulation characteristics and tends to contain the heat within the char zone. The thermal resistance is so effective that even though temperatures on the inside surface of the char reach 4,000° F (2,200° C), just 0.5 inches (13 mm) within the char, the temperature is

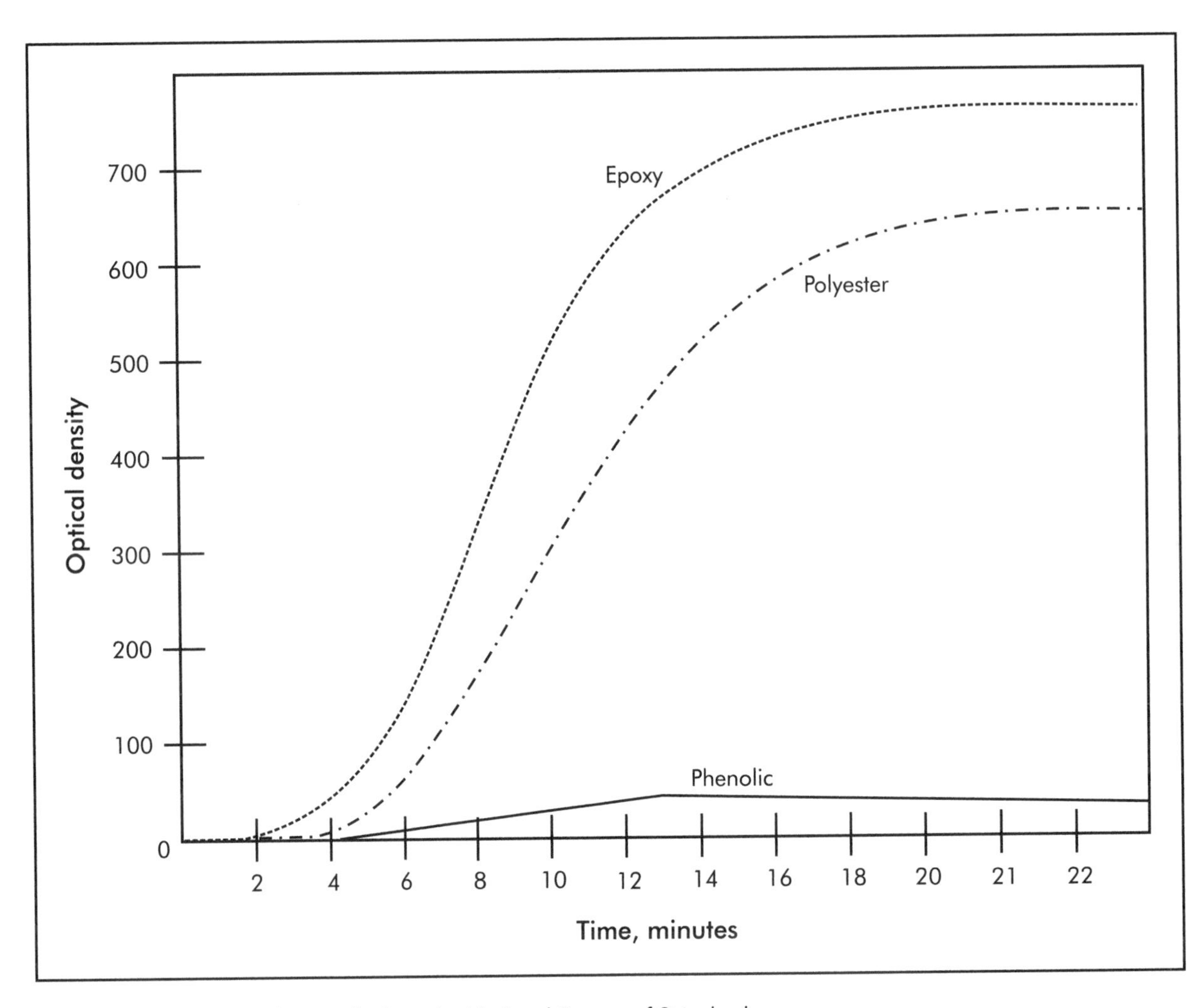

Figure 5-8. Smoke chamber results from the National Bureau of Standards.

only 500° F (260° C). On the outer skin of a normal nozzle, which is several inches thick, the temperature is elevated no more than 10° F (5° C). A non-reinforced char would be porous and weak and quickly wear away from the action of the hot and high-velocity exit gases coming from the rocket. But, when reinforced with a high-temperature-stable reinforcement like carbon fibers, the *char erodes slowly*, little by little in a process known as **ablation**. When used in this type of application, the phenolic composite is called an ablative material. As the char layer erodes away, a new layer of composite underneath the char is exposed, which then also chars, thus continuing the thermal resistance of the material until it is totally consumed by this ablative process. Rocket nozzles are designed so that the thickness of the phenolic/carbon fiber material is sufficient to last for the entire length of the burn (with a reasonable safety factor).

Phenolics are specified in non-aerospace transportation applications, such as public trains and subways, especially when some of the operation is in tunnels or underground. In 1979, BART, the rapid transit system of greater San Francisco, had a serious fire in a train while it was in a tunnel, resulting in a death and several injuries. Since then, phenolics have been used for interior surfaces—ceilings, side walls, and end walls—in all BART vehicles. Halogenated polyesters can meet the standards for exterior parts, but only with some difficulty. Specifications in Europe, which are even more stringent than in the United States, have led to the use of phenolic composites in most European mass-transit applications.

The overall mix of properties available from phenolics has led to many applications in addition to those directly related to low flammability. For instance, automotive manufacturers continue to use considerable amounts of phenolic for under-hood applications where low flammability, high heat performance under load, dimensional stability, and resistance to fuels and lubricants are the properties of most interest. These desirable properties are the result of the aromatic nature of phenolics and the tight crosslinking that prevents movement of the molecules (high heat stability and dimensional stability) and reduces chemical attack by solvents. The high crosslink density gives excellent creep resistance to phenolic composites. When other resins may require molded-in metal inserts for mounting the composite part to an assembly, phenolics can be mounted directly, thus reducing manufacturing costs. The recent development of some impact-resistant grades further increases the penetration of phenolic composites into the automotive market, usually displacing metals.

A non-transportation application that is growing rapidly is the use of phenolic composites as grating materials for decking on offshore oil rigs. These phenolic-fiberglass gratings can withstand brief direct flame contact without major structural damage. They have low thermal conductivity, good resistance to chemicals, and achieve the strength of steel at much lower weights.

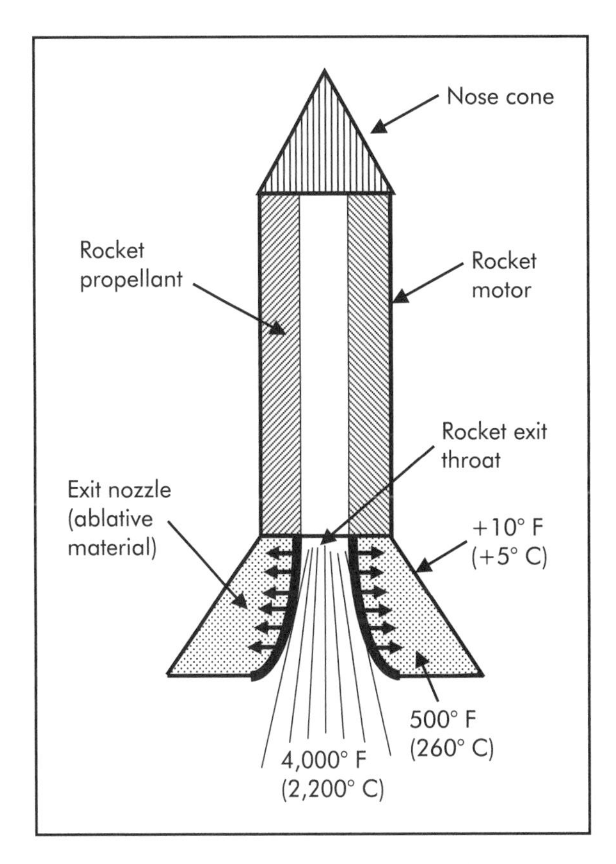

Figure 5-9. Rocket exit nozzle with phenolic as the ablative material.

One other application for phenolics that has already been alluded to in the discussions of epoxies and vinyl esters is the use of novolacs as reactants in making some important types of epoxy and vinyl ester polymers. The novolacs are a type of phenolic that is relatively low in cost. A novolac can have several chains pendant from its backbone. The structures are highly aromatic and, therefore, have good mechanical properties.

Some manufacturers have made products combining polyesters and phenolics. The polyester material forms the basic composite structure and then a phenolic is used to coat the surface. The phenolic is often a paste material that is smoothed onto the polyester surface and then painted to the desired color. This is a three-step process compared to the normal two-step process of gel coating and polyester lay-up, but the phenolic-polyester process does give the benefit of a non-flammable surface. Some phenolic pastes that can be pigmented are being developed to eliminate the painting step, thus simplifying the phenolic coating process.

The excellent thermal insulation and surprising strength of phenolic/carbon fiber composites has led to the use of these materials as the basic materials used in making carbon-carbon composites.

CARBON MATRIX

Carbon matrix materials are much different from all of the other matrix materials considered in this text. There is no "carbon" resin that can be purchased from a resin manufacturer and added to fibers to make a composite structure. Instead, some other resin is used and then, through a series of steps, that resin is *converted into a "carbon" matrix*. In theory, almost any resin that creates a high-carbon-containing char can be used to create the carbon matrix. The most common resin used in this conversion process is phenolic. Another material occasionally used as the basis for forming carbon matrix composites is **pitch**, the product left behind when the more volatile components of coal tar or crude oil have been removed through distillation. Pitch is also a precursor material for making carbon fibers, as will be discussed in the chapter on reinforcements.

Thermal Stability Characteristics

Before considering the details of carbon matrix materials, it is appropriate to discuss the general characteristics of thermally stable materials since thermal stability is the principle property imparted by using a carbon matrix. The characteristics of the carbon matrix materials will then be examined and related to the general characteristics for thermal stability. Other resins that show good thermal stability, such as the polyimides and silicones, will be discussed last.

Following are the main factors contributing to the high-temperature, thermal stability of polymers:

- high bond strength,
- high activation energy,
- stabilization by resonance or inductive effects,
- high heat capacity,
- absence of easily degraded atomic structures,
- intermolecular interactions, and
- backbone stiffness and steric hindrances.

Strong bonds in the polymer, especially along the backbone, are essential for thermal stability since it is *bond breakage that results in thermal degradation*. Therefore, if the bond strength is high, the energy to break the bond will be high, and the bond will be stable to higher temperatures. The **bond strength** is usually expressed in terms of the **bond energy**, which is the energy required to completely separate the atoms and break the bond. Among the highest bond energies are those between aromatic carbons and nitrogen atoms, aromatic carbons and oxygen atoms, aromatic carbons with other aromatic carbons, carbon and fluorine, and silicon with oxygen. Clearly, the presence of aromatic molecules usually indicates high bond strengths. But there are also other bonds that have been found to have high bond energies such as silicone with oxygen and aromatic carbons with nitrogen. These are important in polyimides and silicones, which are discussed later in this chapter.

The requirement for high **activation energy** relates to the tendency (or lack thereof) for bonds to be broken through some *chemical reaction*. For instance, some types of bonds are especially susceptible to oxidation and these would impart thermal weaknesses and have low activation energies. Resistance to oxidation is the most important type of chemical resistance in thermally stable polymers because of the natural tendency for oxidation to occur at high temperatures in the presence of air. However, some bonds such as double bonds in aromatic rings alternate around the ring (single, double, single, etc.) and this causes the electrons of the double bonds to be spread over several atoms. This property is called **electron conjugation** and great stability is achieved in the bonds that participate in the conjugation. The bonds around the aromatic ring resist most ordinary chemical attacks and, therefore, have high activation energies. This results in **stabilization** by resonance or inductive effects.

Heat capacity is a measure of the tendency of a material to absorb heat. When heat capacity is high, the atoms rotate and vibrate so that the energy is dissipated internally without causing internal damage. Thus, polymers with high heat capacity have the ability to absorb energy by internal mechanisms. The energy is not, therefore, directed to the bond, causing it to rupture and degradation to take place.

Some atomic structures are especially prone to volatilization under thermal conditions. When these atoms volatilize, they break their bond with the polymer and often combine with a neighboring atom to produce a stable gaseous molecule. A typical example is the thermal degradation of polyvinyl chloride (PVC) where the chlorine atoms break from the carbon backbone and combine with a neighboring hydrogen to form HCl. In aromatic molecules there are few of these easily volatilized atomic groups. Thus, aromatics are noted for their non-volatilizing nature.

The thermal stability of a polymer is also achieved when there are interactions or bonds between the molecules such as crosslinks. Stability is achieved because the crosslinks must be broken for *extensive degradation* to occur. While some of the bonds might be broken when heated extensively, the high number of total bonds means that the breakage of only a few bonds has little effect on the overall properties of the polymer. Thus, significant loss of properties requires more energy (heat) than would be the case without the crosslinks or other intermolecular attractions.

When a polymer backbone is rigid, mechanical forces are present in the molecule and hold the atoms in place. Therefore, high heat absorption is required before the polymer softens sufficiently to lose mechanical properties. This means the stiff polymer is more likely to have sufficiently high mechanical properties at high temperature. One method of developing stiffness along the backbone is to employ structures that are planar. This is the case with graphite materials and polyimides. The stiffness of the backbone of a polymer also can be caused by interferences between the atomic units along the backbone. In essence, the atoms get in each other's way. These types of interferences are caused by the shape of the atomic units and are, therefore called **steric** (shape) **interactions**.

The properties that lead to thermal stability are not completely independent. Thermally stable polymers usually possess several simultaneously. However, each property represents a different aspect of thermal stability and, therefore, it is instructive to consider each one independently.

Nominal Performance Criteria and Testing

A polymer is designated as thermally stable when it meets certain nominal (ideal) performance criteria. Such materials are those that retain their physical properties under the following thermal and time conditions: 500° F (260° C) for 1,000 hours, 1,000° F (538° C) for 1 hour, or 1,500° F (816° C) for 5 minutes. Thermally stable polymers also would be characterized by high melting or degradation points, resistance to oxidation at elevated temperatures, resistance to non-oxidative thermal breakdown (thermolytic processes), and stability when exposed to radiation and chemical agents.

The ability of various composite materials to retain their properties when exposed to the effects of time and temperature is presented in Figure 5-10. Carbon-carbon composites (carbon matrix materials reinforced by carbon fibers) are shown to retain properties at both high temperatures and high exposure times. Other materials, such as ablative composites (for example, carbon-reinforced phenolics), have good retention at high temperatures but only at relatively short durations, because, of course, they are ablating. Figure 5-10 also shows epoxy composites have good retention of properties at temperatures below 500° F (260° C). Polyimides, which also appear, have higher temperature capability than the epoxies and will be discussed later in this chapter. Metal matrix composites are also plotted in Figure 5-10 and will be discussed in a subsequent chapter.

The methods used to determine thermal stability are listed in Table 5-2 and include the following: weight loss, such as that measured by thermal gravimetric analysis; detection of volatile off-gases; thermal transition measurements, such as those measured by differential scanning calorimetry; and loss of properties as indicated by thermal mechanical analysis. Some basic properties of polymers determined by these tests include the glass transition temperature (T_g), the melting point (T_m), the heat distortion temperature (HDT), and decomposition temperature (T_d), as well as the more direct performance-related properties.

The characteristics of thermal stability in carbon matrix materials will be discussed

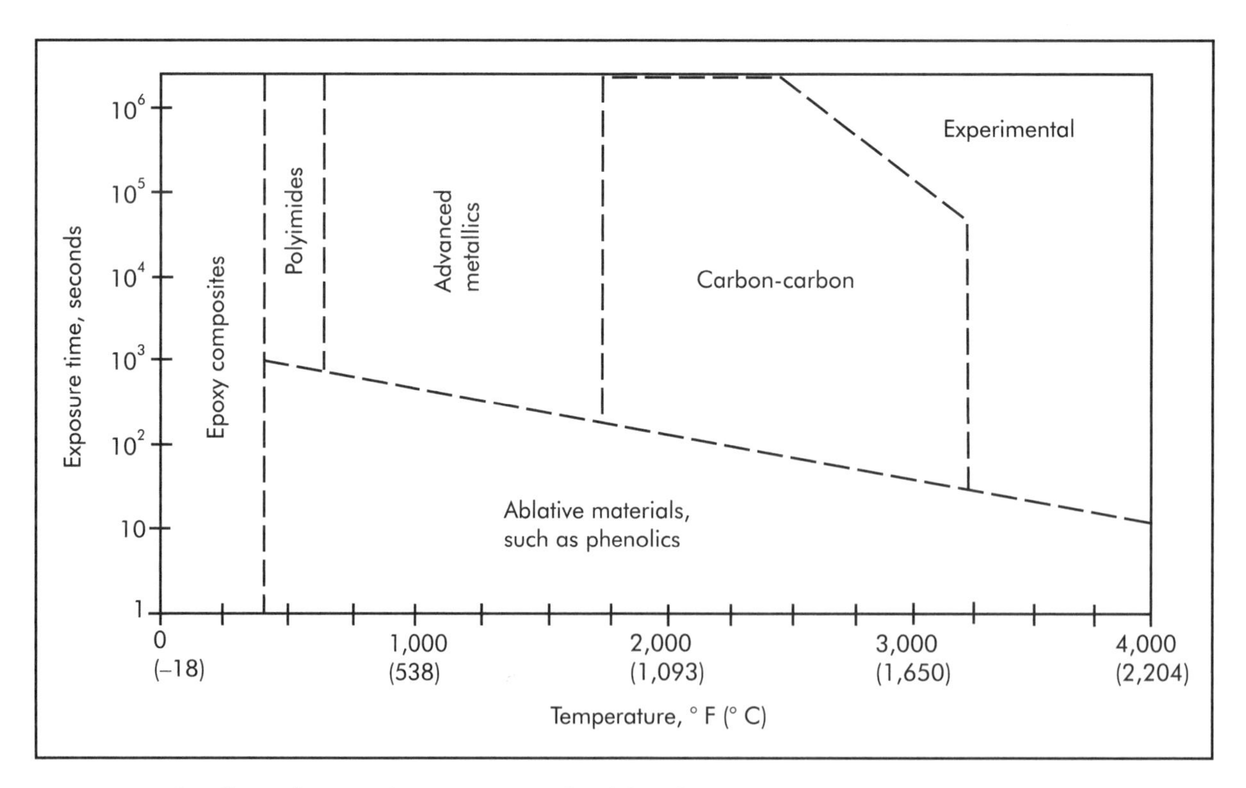

Figure 5-10. The effects of time and temperature on the ability of various composites to retain properties.

Table 5-2. Thermal stability tests.

Test	Comment
Thermogravimetric analysis (TGA)	• Monitors weight loss during a programmed temperature rise • Typical reported value is when 10% weight loss is experienced • Depends on the atmosphere used with inert values higher than air (oxygen) • Often is the first test used because it requires only a small sample and a short time
Isothermal weight loss	• Measures properties for long durations at a set temperature (usually the temperature of expected use) • Typically mechanical properties are also monitored • Can be a costly test because of the times involved and the large number of samples needed
Thermal mechanical analysis (TMA)	• Measures changes in mechanical properties (usually flexural or penetration) with a programmed temperature rise • Can be done quickly with only a small sample
Differential scanning calorimeter (DSC)	• Detects thermal transitions during a programmed temperature rise • Determines T_g, T_m, and T_d • Can also measure induction time to onset under isothermal conditions
Heat distortion temperature (HDT)	• Measures flexural rigidity during a programmed temperature rise • Useful in detecting use temperature
Gas chromatography (GC)	• Detects off-gases with time and temperature
Mass spectroscopy	• Analysis of off-gases

with the other specialty resins used in high thermal applications.

Applications

Carbon matrix composites have applications similar to many of the phenolic applications but with much higher standards of performance. The most important use of the carbon matrix composites is in high-temperature applications. Rocket nose cones and exit throats are two common applications. Whereas phenolic matrix composites are ablative, that is, the material is eroded away by the hot gases passing around the composite, carbon matrix composites have little ablation but, rather, simply withstand the temperatures of that environment. When critical dimensions need to be maintained, such as in the throat leading to the exit nozzle, or in various rocket valve bodies for vector control, or when the nozzle itself must be lighter weight and cannot carry the extra material required of an ablative, carbon matrix materials are used.

Most of the applications of carbon matrix composites involve high thermal stability and dimensional stability. Therefore, both the matrix and reinforcement should be capable of withstanding high temperatures. As a result, the most commonly used reinforcement for carbon matrix composites is carbon fibers, which are more thermally stable than any other common reinforcement. When carbon matrix materials are used with carbon reinforcements, the resultant composites are often called **carbon-carbon composites** (CCC).

Another application for the high-temperature stability of carbon matrix composites is in gas turbine engines. Some specific parts using carbon matrix composites include flame barriers, ducting, exhaust systems, combustors, turbines, and self-lubricated bearings. Carbon matrix composites are also used in traditional internal combustion engines for pistons and other high-temperature parts where the lighter weight and lower frictional losses of the composite are premium over the traditional metal material.

In some applications, direct exposure to flames will cause the carbon matrix composite to oxidize, and possibly burn, when oxygen is present. That possibility can be largely prevented by coating the carbon composite with a ceramic coating, often silicon carbide. Oxidation protection also has been improved by adding oxidation inhibitors to the matrix during the manufacturing process for parts.

The use of carbon matrix composites in brake linings for aircraft, trucks, racecars, and some luxury passenger cars has become a major market. The brake lining application takes advantage of the high thermal stability of the carbon matrix and the high abrasion resistance of the composite. Possibly even more important is the high thermal energy absorption of carbon matrix composites, which have the highest specific heat capacity of any known material. Both of the opposing brake linings are usually made of carbon matrix and carbon reinforced composites. These materials replace brake linings reinforced with asbestos and are, therefore, much safer for the environment.

An interesting application for carbon-carbon composites is as a tooling material. These tools will not be as durable as metal tooling. However, the low coefficient of thermal expansion (CTE) of carbon-carbon composites, which is close to the CTE for carbon fibers themselves, makes this tooling ideal for the manufacture of other composite materials, especially when high numbers of items are not expected. The thermal stability of the carbon-carbon composites means that resins that cure or process at high temperatures can be molded in the carbon-carbon tooling with little worry about changes in dimensions from expansion and contraction.

The major drawback to the use of carbon matrix composites is cost. This cost, which is a function of the difficult and time-consuming manufacturing process, dictates that carbon matrix composites be used only in high-value applications.

Manufacturing Process

A process that illustrates the major concepts of manufacturing of carbon matrix composites is diagramed in Figure 5-11. The creation of the carbon matrix begins with the combination of reinforcement and resin into an uncured preform. As previously indicated, the most common raw materials used to make this preform are phenolic resin and carbon fibers. This preform, when in its uncured state, is sometimes termed a **green preform**. Because of the frequent use of carbon fibers, the designation of carbon matrix composites as carbon-carbon composites is widespread. However, strict definition of the terminology would allow this name only when the reinforcement is carbon fibers.

The preform material is placed into a mold, shaped to fit the mold contours, and then cured. For most composites, this curing step and then a final finishing step

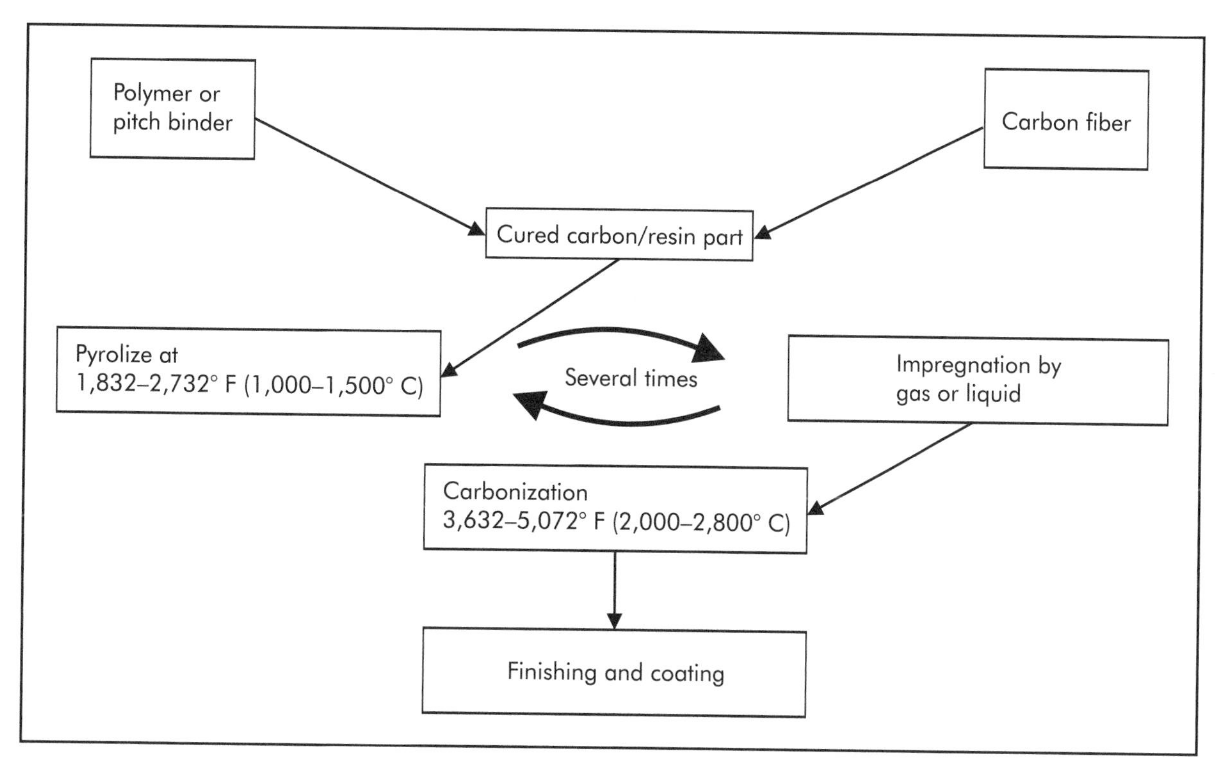

Figure 5-11. Production flow chart for carbon-carbon composites.

would complete the manufacturing process. However, for carbon matrix composites, the molded material must be converted first, and the steps required are unique to this material. As will be seen, the steps are time consuming and expensive, thus making carbon-carbon composites among the most expensive of all composite materials.

The next step in the process is a **pyrolysis** step. Pyrolysis involves heating the part in the absence of air (oxygen) so that the resin chars but does not oxidize (burn). The temperature of this process is usually about 1,832° F (1,000° C). During pyrolysis, the resin becomes nearly all pure carbon through the evolution of the oxygen, hydrogen, and other atoms contained therein. This evolution of non-carbon atoms leaves behind a porous structure that is held together by the strength of the residual matrix and the reinforcing fibers. If no further step were done, the thermal properties of the part would be good, but the mechanical properties would be inferior. Each of the pores would be a point for crack initiation, thus reducing greatly the tensile, flexural, and compression properties.

To improve the mechanical properties of the composite, the pores are filled by adding additional resin to the part. This step, called **impregnation**, can be done by infusing the porous carbon matrix material with additional resin, which can be either a gas or a liquid. When the liquid is in the gaseous state, the impregnation process is called **chemical vapor deposition (CVD)**. In this process, the impregnating resin or some other carbon-rich material, such as acetylene, is vaporized and then, with pressure or vacuum, caused to infiltrate the porous carbon matrix composite. When the impregnation is done in the liquid state, the porous carbon matrix composite is covered with the liquid resin, which is allowed to seep into the pores of the composite. This process can be speeded up with pressure. However, even with pressure or vacuum, both of processes take considerable time, often on the order of days or weeks.

When impregnation has proceeded so that the pores of the carbon matrix composite are filled with new resin, the composite is again pyrolyzed. This second pyrolysis carbonizes the new resin by driving off the oxygen, hydrogen, and other atoms and will, therefore, create a few more pores in the newly impregnated areas. It also drives off additional non-carbon atoms in the remainder of the carbon matrix material, thus creating some porosity in the bulk material as well.

The composite material is then subjected to another impregnation step, in which either gas or liquid resin is infused into the pores of the bulk material. Hence, the process of impregnation and pyrolysis is repeated several times.

When impregnation is completed, the material is subjected to a higher temperature. This is called **carbonization** and occurs at about 4,532° F (2,500° C). By this time, the number of non-carbon atoms is few and so the porosity is minimal and can be ignored.

The carbon matrix composite may then be coated. This step is not done for all parts. The applied coating is non-flammable so that the carbon matrix will not burn when heated in the presence of oxygen. The coating is almost always a ceramic, such as silicon carbide.

Finally, the part is finished. This step would include sanding, trimming, drilling attachment holes, or other similar processes that allow the carbon matrix composite to fit into an assembly, if required.

The entire process to make carbon/carbon composites takes many weeks and may even take as long as six months. Obviously the costs of such a process are high.

Properties

Because carbon matrix composites are generally intended for high-temperature applications, the performance of the composites at various temperatures is of great importance. Table 5-3 lists the performance of a carbon-carbon composite at room temperature (70° F [21° C]) and then at three higher temperatures: 1,500° F (816° C), 3,000° F (1,649° C), and 3,500° F (1,927° C). The excellent retention, and even increase, in some properties at high temperatures is due to the stability of both the resin and the reinforcement in carbon-carbon composites. Modulus, which reflects the softening of the material, seems to be the mechanical property most negatively affected by the rise in temperature.

The presence of the carbon fibers gives the carbon-carbon composite good crack-stopping capability. That is, when a crack begins within the carbon matrix, the crack grows until it hits a carbon fiber and then it terminates. This greatly improves the toughness of the carbon-carbon composites relative to other highly rigid materials.

POLYIMIDES AND RELATED POLYMERS

The extremely long processing times and high cost of carbon-carbon composites have led to efforts to develop other polymers that have good thermal stabilities at less cost. Such materials may also possess some additional benefits. A set of performance goals, partly idealistic and partly realistic, have been identified for polymer candidates that could offer high thermal stability and favorable economics. These polymers, which might be called *thermally stable structural composite polymers*, would have several important performance properties including:

- excellent thermal stability, usually defined as being substantially better than the best epoxies. The minimum performance would be greater than 90%

Table 5-3. Mechanical properties of a typical unidirectional carbon-carbon composite at various temperatures.

Property	Temperature, ° F (° C)			
	70 (21)	1,500 (816)	3,000 (1,649)	3,500 (1,927)
Tensile strength—warp/ longitudinal direction, psi (MPa)	48.0×10^3 (331)	54.0×10^3 (372)	60.0×10^3 (414)	62.0×10^3 (427)
Tensile strength—fill/ transverse direction, psi (MPa)	40.0×10^3 (276)	45.0×10^3 (310)	50.0×10^3 (345)	—
Modulus—warp/ longitudinal direction, psi (GPa)	16.5×10^6 (114)	16.5×10^6 (114)	15.0×10^6 (103)	12.5×10^6 (86)
Modulus—fill/transverse direction, psi (GPa)	15.8×10^6 (109)	15.8×10^6 (109)	14.4×10^6 (99)	12.0×10^6 (83)
Compressive strength—warp/longitudinal direction, psi (MPa)	26.0×10^3 (179)	33.0×10^3 (228)	40.0×10^3 (276)	—
Compressive strength—fill/transverse direction, psi (MPa)	26.0×10^3 (179)	30.0×10^3 (207)	35.0×10^3 (241)	—

retention of properties with long-term exposure to 600° F (316° C), usually implying that the T_g is greater than 600° F (316° C).

- good oxidative stability, defined as being better than carbon-carbon and at least as good as a high-performance epoxy.
- excellent toughness that is equivalent or superior to high-performance epoxies and superior to phenolics.
- good mechanical properties, equal to high-performance epoxies.
- acceptable processing ability that is equivalent to or better than phenolics.

Based upon previous experience with chemical structures, highly aromatic compounds, especially those with high stiffness from internal bonding as from crosslinking or high crystallinity, are ideal candidates for polymers having high thermal stability and other desirable properties.

Highly aromatic polymers with high internal stiffness, polyimides have been used for many years in applications requiring thermal stability. These polymers have proven to be valuable as composite matrix resins for such applications.

Polyimides can be either thermoplastics or thermosets. The thermoplastics will be discussed in the chapter on thermoplastic composites. Two general types of thermoset polyimides have had considerable commercial success: *end-capped polyimides* and *bismaleimide (BMI)*, discussed separately in this section.

Structure and Crosslinking Reactions

The polyimides are characterized by the presence of the imide group in the polymer backbone. This group is illustrated in Figure 5-12a. The imide group is a ring structure containing nitrogen and two carbonyl groups (carbons that are double-bonded to oxygen molecules). Rings having more than one type of atom, such as the imide ring, which contains both carbon and nitrogen atoms, are called **heterocyclic**, a general name that can be applied to imide rings, among many other chemical ring groups. The heterocyclic groups or molecules are often highly stable and can be aromatic themselves or coordinate their electrons with an aromatic molecule to achieve even more stability. When linked with an aromatic molecule, as illustrated in Figure 5-12b, the molecule can be called an **aromatic heterocycle**.

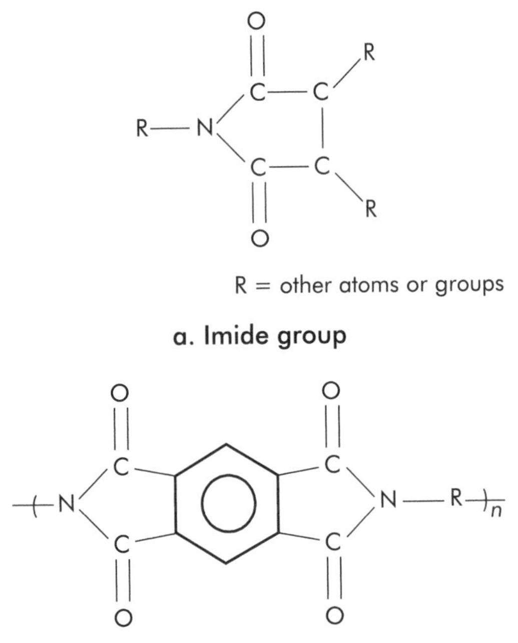

R = other atoms or groups

a. Imide group

b. Imide group in polymer

Figure 5-12. The imide group and its inclusion in a polymer.

The imide group can bond to other atomic groups and, therefore, can be included as part of a polymer backbone, as in Figure 5-12. The aromatic-heterocyclic polymer chain is *highly aromatic*. It conjugates the electrons from the aromatic ring with the electrons in the oxygen-carbon bonds of the imide group. *Very stiff* and generally planar with few backbone bonds around which rotation can occur, it is *composed of only strong bonds*. Therefore, it can be understood why this polymer has high thermal stability.

End-capped Polyimides

The end-capped polyimides are synthesized by combining *three monomers*—two that react to form the polyimide chains and another that attaches to the ends of the chains to form end caps. The most common of these polyimide types is called PMR-15. PMR stands for polymerization of monomer's reaction, and the 15 is an abbreviation for 1,500 formulated molecular weight, which will be explained later in this section. Several different monomers could be used in making polyimides. Those chosen for PMR-15, the most widely used end-capped thermoset polyimide, are readily available, relatively low cost, and have acceptable processing and properties.

The PMR-15 synthesis is complicated. It can be summarized as a reaction to form the imide ring (called imidization) and then a chain-lengthening step followed by an end-capping step. The length of the polymer chain is controlled by the end-capping step. In actual practice, the number of polymer units formed is typically between two and four. This results in a final molecular (or formulated) weight of about 1,500. Therefore, the molecular weight is controlled by the relative concentrations of the three monomers.

The cure time for PMR-15 and other polyimides is long. Another drawback is that a solvent is often used to impart some flexibility to the polymer during cure. When cured, the solvent must be removed or bubbles and surface blemishes might remain on the part. The removal process requires that the polyimides be cured under vacuum, usually in an autoclave. However, some molders have developed an alternate molding method that eliminates the problems of solvent removal during molding. These molders lay up the prepreg so that a preform is made. That preform can be placed in an oven at 165–400° F (74–204° C) for 1–2 hours to evaporate the solvent. Then, when desired, the preform can be placed in a mold (usually a compression mold) and heated to 600° F (316° C) for a relatively short period of time (a few minutes for small parts to a few hours for thick parts). Although pressure is still required to get the end groups to crosslink properly, the use of an autoclave is not required.

Although the PMR-15 polymer has properties that have been good enough to make it a commercial success, many believe that improvements can still be made. The high aliphatic content of the most common end-capper and other groups within the polymer can be points of vulnerability in highly thermo-oxidative environments. Further, the high crosslink density at a 1,500 molecular weight leads to rigidity and brittleness. Therefore, much research continues on polyimides made from different monomers in attempts to eliminate the problems with PMR-15 and yet not diminish its good properties.

Properties and Applications

The selection of 1,500 formulated molecular weight for the most common PMR resin is a compromise. Higher molecular weights would give improved strength and stiffness but would result in a glass transition temperature that would be above the temperature at which the crosslinking occurs. Therefore, they would have much slower crosslinking reactions since the crosslinking would occur, at least in part, in the solid state. The typical crosslinking window is 480–520° F (249–271° C). Lower molecular weights would give improved processing, including improved wet-out and lower voids, but would also result in higher brittleness because of the higher crosslink density that would occur with shorter chains. Experience has shown that 1,500 formulated molecular weight works for most applications, but when needed, the molecular weight can be easily adjusted by varying the relative concentrations of the monomers.

Thermal stability is clearly the most important reason to use polyimides. They can typically be used at 550–600° F (288–316° C) for continuous service. For short durations, temperatures even as high as 900° F (482° C) can be tolerated, although some surface oxidation occurs. Thermal data indicates that weight loss is relatively fast at 800° F (427° C), moderate at 700° F (371° C), and relatively slow at 600° F (316° C). Some studies have reported a significant loss in polyimide properties under hot/wet conditions. While this has not been fully characterized, good design practice would suggest that some consideration be given to this possibility.

Some of the major uses of polyimides are as bearings and seals in high thermal environments. Polyimides also have been used for ducts on engines and airplane engine blades. Recently built fighter aircraft employ polyimides for firewalls. Some high-speed developmental aircrafts require the thermal stability of polyimides for their skins. This is also true of some missiles, especially fins and radomes.

Bismaleimide (BMI)

BMI resins have gained popularity as a material that has better thermo-oxidative and hot/wet properties than epoxies and easier processing than other polyimides. These resins can be used in many automated and semi-automated composite manufacturing methods such as resin transfer molding

(RTM), filament winding, and pultrusion, as well as in hand lay-up and vacuum bagging using prepregs.

Structure

Figure 5-13 illustrates the structure of a typical BMI monomer. The name comes from the combination of two (bis) maleic groups (four carbon units containing a carbon-carbon double bond) with imide rings. Crosslinking occurs at the double bond in the terminal imide rings. The crosslinking reaction can be carried out simply by heating the monomer. Because of the low molecular weight of the monomer, the resulting thermoset has high crosslink density and is brittle.

Properties and Applications

Much effort has been given to toughening BMI and several routes are available. One is to extend the length of the monomer (called extending the chain) by the reaction of the monomer with a diamine, usually methylene diamine. Another toughening technique is to co-react the BMI monomer or extended chain monomer with additives based on inherently tough materials such as diallyl bisphenol A. All of these methods result in thermoplastic BMI materials that can be crosslinked simply by the addition of heat, usually at temperatures of about 250° F (121° C). These conditions are similar to those used for high-performance epoxies. Higher cure temperatures will, however, result in higher crosslinking and improved thermal properties. To obtain the best properties, curing is often done at about 482° F (250° C). Several cure accelerators are available for some of the BMI formulations that will allow the crosslinking to proceed at lower temperatures.

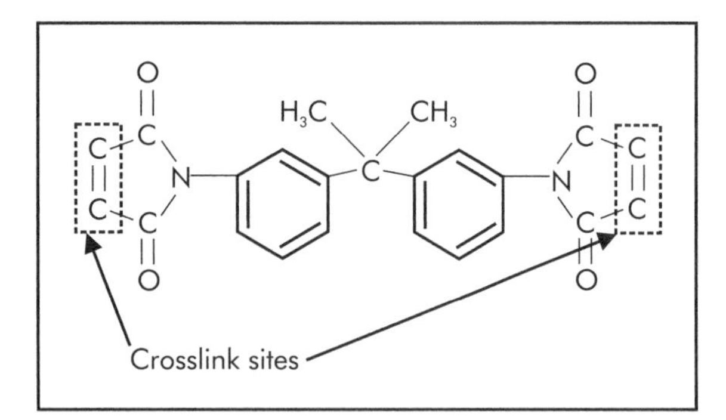

Figure 5-13. Bismaleimide (BMI) structure.

Crosslinked BMI can achieve a T_g as high as 572° F (300° C) with an oven treatment of 6 hours. This property is substantially better than high-performance epoxies and has become especially important in several high-performance aircraft because of their high speed and need for high heat distortion temperatures. For example, BMI is the resin of choice for the majority of the composite panels used in the F-22 Raptor fighter jet. The wings, empennage, fuselage, and understructure are all made with BMI composites. Other applications include the aft flap hinge fairing of the C-17, the inner core cowl for the Pratt and Whitney 4168 engine, the tail boom of the Bell Model 412 helicopter, and engine and exhaust parts for some Formula 1 racecars.

CYANATE ESTERS

Cyanate esters are another group of resins that have thermal and strength properties superior to epoxies. While not as good as BMI resins in thermal and strength properties, cyanate esters have found special applications because of their superior dielectric loss properties and low moisture absorption. The major uses for the resin are in radomes, skins covering phase-array antennae, advanced stealth composites, and space structures.

Many electrical applications require low dielectric loss properties so that the signals are not absorbed by the resin, which causes weaker transmission, return signals, and overheating of the components within the electronic enclosure. The cyanate esters have the lowest dielectric loss properties among the common and high-performance thermoset resins. Only two thermoplastics, polytetrafluoroethylene (PTFE) and polyethylene (PE), have more favorable dielectric properties among the common plastics.

Structure and Crosslinking Reactions

The structure of cyanate esters is depicted in Figure 5-14. The unique carbon-nitrogen ring is highly stable and stiff, much like aromatic rings.

The low dielectric properties of cyanate esters are attributable to the symmetrical arrangement of the electronegative oxygen and nitrogen atoms in its structure, resulting in weak dipoles. When the dipoles are weak, there is little tendency for the electrons to be transmitted through the structure and, therefore, dielectric losses are low. When reinforced with ultra-high-strength PE fibers or PTFE material, the electrical properties

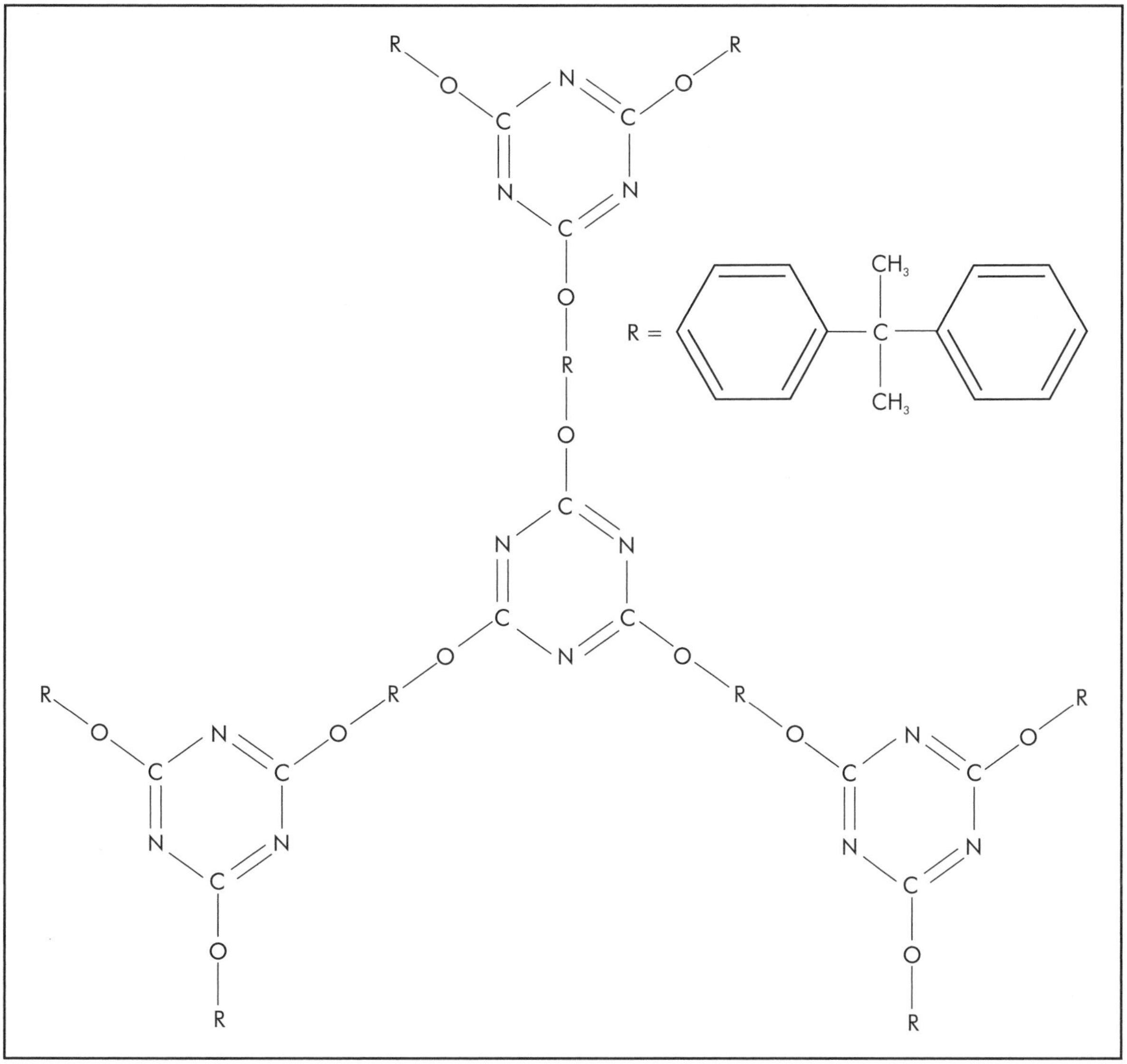

Figure 5-14. Cyanate ester structure.

of the composite can be even lower than for the **neat resin** (non-reinforced).

Specific properties can be improved by slightly varying the type of cyanate monomer. For instance, adding methyl groups to the rings will improve water absorption and dielectric loss but at the expense of T_g. Replacing the methyl hydrogen with fluorine gives a further reduction in the dielectric loss but also decreases T_g. Many such variations have been tried and some are commercially available.

Polymers with high elongation (usually thermoplastics) can be added to the cyanate ester monomers and prepolymers to give some improvement in toughness. In most of the cases where a thermoplastic is added to a cyanate ester, the T_g of the cyanate ester is reduced.

Cyanate esters are made by addition and rearrangement reactions, as opposed to condensation reactions, and without highly volatile solvents. Therefore, little tendency is seen for cyanate esters to out-gas in high vacuum. This property is especially favorable in space applications. Moreover, the density of cyanate esters is slightly lower than similar epoxy molecules, thus giving a weight savings—another valuable property for space applications.

POLYURETHANES

Several traditional thermoset resins not normally associated with composites are being examined as possible new resins for composite applications. Polyurethanes are one of these important groups. Several characteristics of polyurethanes make their use in composites quite logical. For instance, polyurethanes are known for their toughness. Therefore, they have great potential for applications in the automotive industry, such as automobile body panels, where toughness is a prime advantage. Furthermore, the rapid manufacturing possibilities of polyurethanes make it an attractive choice. The wide choice of technologies and formulations suggests that the properties can be suited to a variety of applications. As a result, some joint ventures have been announced to bring together polyurethane resin technology and fiberglass reinforcement technology.

Structure and Crosslinking Reactions

The basic chemistry for formation of a polyurethane is similar to a condensation reaction in that two monomers, each having at least two reactive groups, unite to form the polymer. However, in contrast to condensation reactions, no condensate is formed when the urethane bond is created. The urethane bond is formed from the reaction of a polyol with an isocyanate, as illustrated in Figure 5-15.

A **polyol**, which means multiple alcohol (OH) molecules, can have from two to many OH groups. Just as with a condensation polymerization reaction, the polyol monomer must have at least two reactive groups, in this case OHs, to polymerize. If three or more reactive groups are present, crosslinks can form. The polyol monomer depicted in Figure 5-15 is shown with two OH groups and some generic linkages between these groups. The linkages can be aliphatic or aromatic or mixed.

The other monomer in the reaction to form a urethane is an **isocyanate**, which is the double-bonded NCO combination of atoms as shown in Figure 5-15. This monomer must also have at least two reactive groups to form a polymer. The types of chemical compounds that have two isocyanate groups are called *diisocyanates*.

When a polyol reacts with an isocyanate, a molecular rearrangement occurs that creates a more stable molecular structure. The hydrogen on the polyol forms a bond with the nitrogen in the isocyanate. Further, the oxygen in the polyol forms a bond with the carbon in the isocyanate. Some previous bonds in the polyol and in the isocyanate break to allow these new bonds to form.

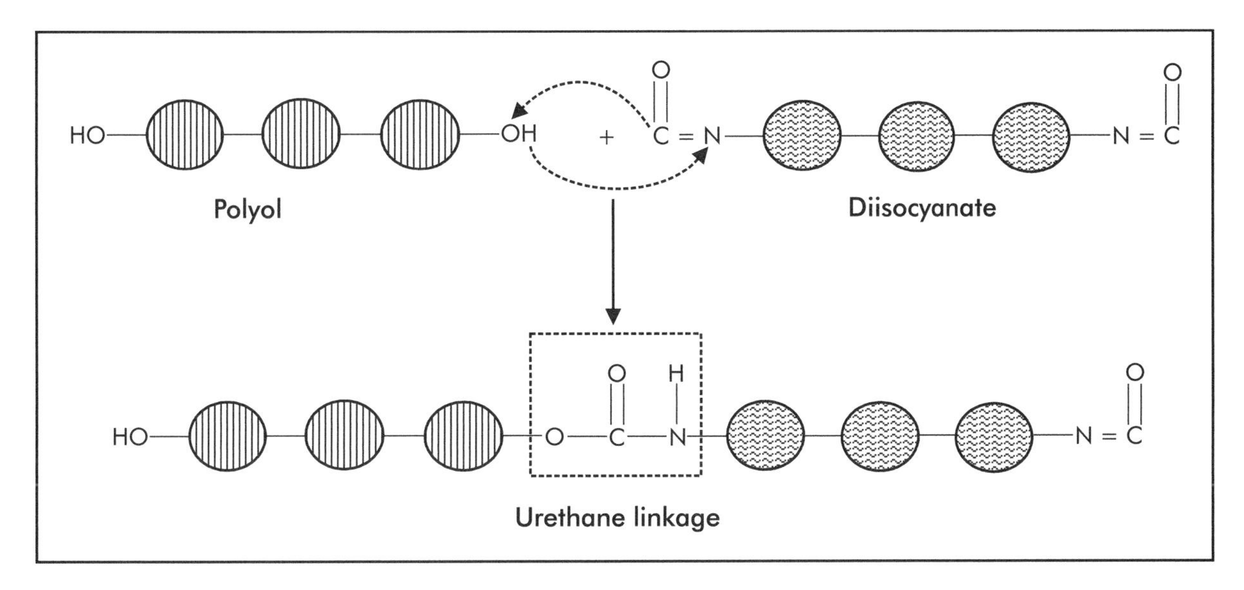

Figure 5-15. Urethane linkage formation.

Reactions by molecular arrangement have been previously encountered with epoxies and can give some useful comparisons to polyurethanes. The epoxy ring was opened by the reaction of a hardener and the bonding between the epoxy and the hardener involved a molecular rearrangement. The role of the polyol in polyurethane chemistry is like the role of the epoxy molecule in epoxy chemistry. The isocyanate role in polyurethanes is like the hardener in epoxy chemistry. Epoxy molecules generally have epoxy groups on the ends of the various branches of the molecule, much as polyols would have OH groups on the ends of the branches. Just as the epoxy molecules could have many different chemical groups or arrangements of atoms between the epoxy rings, the polyols also can have many different groups between the OH groups. It is this variation in the types and arrangements of atoms between the reactive groups that gives such variety in the types of epoxies and polyurethanes that can be made.

Although many types of linkage groups can be used in polyols, two general arrangements are common in commercial polyurethanes. Both are relatively short-chain polymers, one based on ether linkages and the other on ester linkages between the polymer units. The ether-based polyurethanes are generally more flexible than the esters. The principal use of ether-based urethanes is in foams. Ester-based polyurethanes have higher mechanical properties than the ethers. They are more often used in molded parts and coatings. The type of polyurethane (PUR) most likely to dominate composites has not yet been fully demonstrated.

Usually the linkage groups in the isocyanate monomer, the groups of atoms between the isocyanates, are large aromatic groups. Typical examples of commercial diisocyanates are toluene diisocyanate (TDI) and methylenediphenyl isocyanate (MDI). MDI is darker in color and has lower oxidative resistance and ultraviolet (UV) light stability than TDI. Even greater UV stability and increased toughness can be obtained by using an aliphatic diisocyanate. Excellent weather resistance and improved UV resistance can be obtained by using a triisocyanate.

Caution must be exercised in the handling of isocyanates. These chemicals cause respiratory distress and can be toxic. They will react rapidly with water, often with much heat being evolved. Therefore, extreme caution should be used in the storage and use of isocyanate materials.

Molding

Polyol and the isocyanate monomers are generally liquids when they are combined to form the polyurethane, which simplifies the mixing and metering of these materials. In most cases, the reaction between the polyol and the isocyanate is almost instantaneous, even without external heating. Manufacturing methods especially well suited to this mixing of reactive liquids and combining with a reinforcement are resin transfer molding (RTM), a closely allied process called reinforced reaction injection molding (RRIM), or structural reaction injection molding (SRIM). (These processes will be discussed in detail in the chapter on closed mold injection processes later in this textbook.)

In the closed mold injection processes, the two liquid components are pumped through a mixing chamber or tube and then into a closed mold into which a reinforcement preform has already been placed (assuming there is a reinforcement). The polyurethane polymer is formed inside the mold, usually with only moderate pressures required. Hence, the molds can be made of inexpensive materials and can be large assuming that some care is exercised to ensure that the entire mold is filled. When the polyurethane is formed (cured), the molds are simply opened and the part extracted. The closed mold injection process can be automated easily. The totally enclosed nature of the process has further increased its desirability.

Polyurethanes also can be molded by traditional thermoset processes such as compression molding, transfer molding, and open mold processes such as spray-up and lay-up, all of which will be discussed later in this textbook.

Properties and Uses

The major property of PUR that drives its use in applications is its toughness. This toughness comes from the high flexibility and elongation of the structure. Property comparisons between resins are hard to make because of the many grades available. However, a comparison of elongation of general casting polyurethane with 55% fiberglass reinforcement shows elongations from 5–55% whereas polyester SMC would typically have a maximum elongation of 5%.

PURs can be flexible or stiff. Some specialty grades of reinforced PUR have elongations as high as 600% and are, therefore, considered elastomers. Polyurethanes can have tensile strengths as high as 2,000–7,000 psi (14–48 MPa) and Shore hardness as high as 90D. These properties lead to an ability of some PUR sheets to deflect 6–8 in. (15.2–20.3 cm) at –20° F (–29° C) repeatedly. Polyurethanes are, therefore, especially useful for cryogenic applications.

Stiffness, generally measured by flexural modulus, is chiefly a function of the amount of glass reinforcement in polyurethanes just as it is in polyesters, although the matrix stiffness does have some effect. Polyester resins are generally stiffer than polyurethanes so the composites of polyesters are, on average, stiffer than polyurethanes. However, polyurethanes have a wider range of stiffnesses, including some that are quite flexible.

When it comes to thermal capability, the sag values of a PUR reinforced with 20–30% fiberglass are 0.05 in. (1.3 mm) at normal automotive painting temperatures of 325–360° F (163–182° C). The sag can be reduced significantly by additional crosslinking, but this results in more brittleness at room temperature. Therefore, reinforced PUR materials are not currently used for horizontal automotive exterior panels because of excessive sag. SMC would be the plastic of choice for these horizontal applications.

Polyurethanes are generally better adhesives than polyesters. Therefore, adhesion to the fiber reinforcements is better across a wide range of reinforcements. This adhesion is also an advantage when the reinforced part is to be bonded into an assembly or covered with another coating.

Many of the mechanical properties of PURs are competitive with polyesters and some much better. Further, the surface smoothness and paintability of PURs, which are so critical for exterior automotive parts, are not a problem. PURs can be easily molded to a class-A surface. Paintability, however, has been a little more difficult to achieve. After 1,000 hours in a QUV (a type of weatherometer), gloss is significantly lost, but can be regained to near original values with washing and waxing. Paint adhesion is somewhat less than SMC after 1,000 hours in the weatherometer.

Another property that is leading to sales of PUR is abrasion resistance. Reinforced PUR materials have Taber wearability that is far better than most other plastic materials. This advantage is one of the principle reasons PURs are used so widely in athletic shoes.

The most serious limitation to the use of polyurethanes is cost. Polyurethane grades suitable for use in composites cost about 10–20% more than polyesters.

Some authorities have suggested that polyurethanes have some problems with flammability, which are not present in polyesters. Two flammability issues have been raised. One is the possibility of formation of hydrogen cyanide (HCN), a highly toxic gas. Although this gas may be present in smoke from PURs, the levels detectable are small and generally lost amidst all of the other off-gases and combustion products. Another issue raised against PUR flammability characteristics is the low amount of internal oxygen (that is, oxygen bonded in the polymer molecule) compared to many other plastics resins, such as polyesters. The critics suggest that this lower oxygen content will lead to the formation of more carbon monoxide, which is poisonous, rather than harmless carbon dioxide. However, little evidence to support these charges has been widely distributed and accepted. Therefore, although some concerns have been raised with PURs' flammability characteristics, they have not been carefully demonstrated. PUR materials, however, will burn, as do many plastics. Some progress has been made in reducing the flammability of PURs with other polymer systems, but as of yet, no common reinforced PUR material can meet stringent flammability requirements.

In general, polyurethanes are not going to replace polyesters except where the unique properties or processing advantages are significant. The most important properties are flexibility, toughness, abrasion resistance, and good adhesion. The advantages in processing are shorter cure cycles, some unique processing methods, and the lack of volatile solvents or co-reactants. With these advantages, applications of reinforced polyurethanes will continue to develop. Some will be as replacements for polyesters, but the majority of applications are more likely to be as replacements for other materials or as totally new applications for composites.

Recently, some new applications for polyurethanes in composites have been of high interest. These applications use polyurethanes as a bagging material. It is sprayed onto the composite part to achieve an airtight bag, which is required to apply a vacuum during cure to remove trapped air and off-gases. These polyurethane or closely related polymer bags are much easier to apply than the traditional nylon bags. They are also much more durable (less ripping) and will therefore greatly increase the productivity of a molding operation that requires bagging. The use of these bagging materials also raises the possibility of entire bagging systems that are sprayed on rather than adhering layers of material. The importance of this innovation will be seen when bagging is discussed in the chapter on open molding of advanced composites.

SILICONES

Silicone resins are widely known but not necessarily as a resin matrix for composites. There are, of course, other uses of silicones in conjunction with composites, such as mold release and adhesives, but others may be a bit more hidden. Even those hidden uses are surprisingly important and can significantly affect operations. Even though this chapter is about resins as matrix materials for composites, it is useful to explore all the uses of silicones here so that its many uses can be appreciated and the best silicone for each use selected.

Structures and Crosslinking Reactions

A clear understanding of the basics of silicones requires that three terms, which are often confused, be clearly differentiated. They are:

- *Silicon* is the element (atom) characterized in the chemistry periodic table and listed just under carbon. It is often found in nature combined with oxygen and is the principal component of sand (silicon dioxide). Purified silicon is used in computer chips and transistors, but is not used directly in any plastics applications. The symbol for elemental silicon is given in Figure 5-16a.
- *Silane* is composed of small molecules of silicon combined with chlorine, hydrogen, small organic groups, and other atoms for special purposes. These small molecules are sometimes used as coatings, especially as coupling agents. However, their principal use is as monomers in making silicone polymers. A silane molecule is illustrated in Figure 5-16b.
- *Silicones* are polymers (made from silanes) characterized by a series of alternating silicon-oxygen bonds along the polymer backbone and various small organic groups bonded to the silicon atom. See Figure 5-16c. Depending upon how the polymerization is carried out, the silicones can take several forms: oils, rubbers, resins, and siliconates. Oils, rubbers, and resins have widespread use in the composites industry.

Properties and Uses

Silicone oils are polymers made by limiting the length of the polymer chain to less than 1,000 repeat units. These short-chain

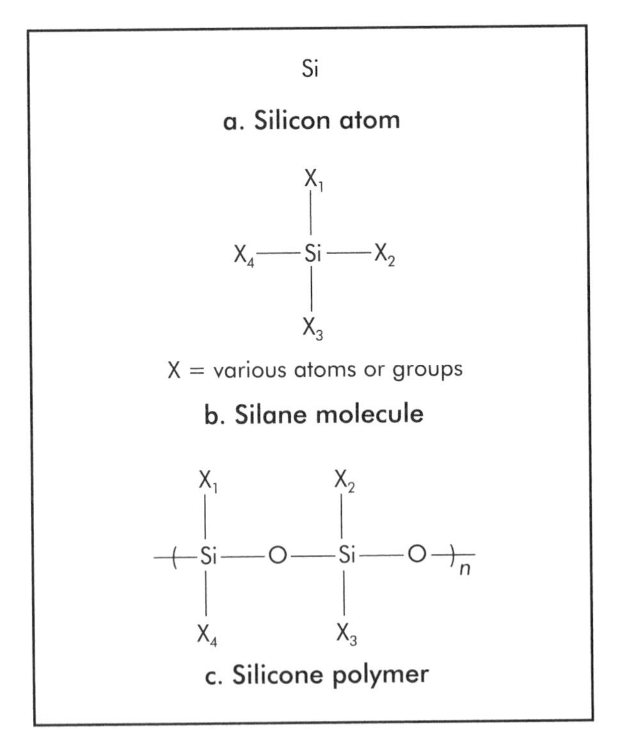

Figure 5-16. Various forms of silicon-containing materials.

polymers are usually liquids (oils) at room temperature. However, when the chain approaches 1,000 repeat units, the polymers become thicker, like greases or pastes. Some silicone greases or pastes can be made by blending short-chain-length silicone oils into fatty-like materials called thickeners. Silicon oils are also blended with low-viscosity fluids and used as sprays. They also can be dispersed into water to form emulsions, which are used as coatings, such as leather waterproofing agents.

Silicone oils are generally characterized by being water repellent, highly compressible, chemically and physically inert, and thermally stable. Silicone viscosity changes little with temperature and shear rate, and it has the ability to form a film that prevents things from sticking together.

Silicone elastomers probably are the most important segment of the silicone market. These materials are polymer chains of over 1,000 repeating units and can be as long as 100,000. The chains are linear, thus imparting high elongation and the elastic nature of rubbers. As with natural rubber, the silicone elastomers are crosslinked (also called curing or vulcanizing) to improve heat stability and mechanical properties. Like other rubbers, silicone elastomers can be compounded with fillers, pigments, and other additives to achieve a wide variety of viscosities, colors, pre-cure and post-cure mechanical properties, etc. The organic groups attached to the silicon atom in the polymer, labeled with Xs in Figure 5-16c, also can be selected to further change polymer properties.

The most common silicone polymer is dimethylsiloxane. This is the polymer formed when all of the organic groups bonded to the silicon atom in the siloxane polymer (the Xs in Figure 5-16c) are methyl groups. It has excellent flexibility, good water repellency, good low-temperature flexibility, and high chemical resistance, to name just a few prominent properties. If one or both of the methyl groups is replaced by a phenyl (aromatic) group, which is the most common substituent other than methyl, the properties are modified. The phenyl-substituted polymer has higher heat stability, better oxidation resistance, and improved toughness, but is stiffer.

The incorporation of fluorine into the organic groups attached to the silicon atoms in the siloxane polymer chain can improve the solvent resistance of the silicone material. This solvent improvement comes, however, with less heat stability and at a higher price. Other substitutions can be made to provide a wide range of properties for special applications.

Silicone rubbers can be dissolved in a variety of organic solvents and dispersed in water to provide a wide range of coatings. These coatings are generally applied to a substrate and then heated or simply allowed to stand to allow the liquid component to evaporate. The dry materials are then further heated to cure the silicone rubber.

Curing of silicone elastomers is usually done with peroxide initiators or curing agents (catalysts) but also can be done with sulfur (as is done with natural rubber) or other special crosslinking techniques somewhat unique to silicones. The groups of silicone elastomers that require high temperature to cure are generally called high-temperature-vulcanization (HTV) silicones. Those that can be cured at room temperature are called room-temperature-vulcanization (RTV) silicones.

Silicone resins can be liquids at room temperature or solids in solvent solutions such as lacquers. When applied as coatings, the solvents evaporate and the silicones crosslink to form a hard surface. The crosslinking usually is done at elevated temperatures with a peroxide initiator added to the silicone solution.

Silicone resins can be used as molding compounds, usually with fillers. The molded articles have excellent thermal stability and low flammability. However, the cure times are long (typically 30 minutes to 2 hours) and cure temperatures are high (typically 300° F [149° C]). The advantage of these coatings is high-temperature performance and high surface hardness. The coatings are sold as unmodified resins and with various fillers, such as aluminum, and pigments.

Siliconates are silane molecules where one of the substituents (Xs in Figure 5-16) is a metal oxide. Usually one of the other substituents is a short organic group, such as methyl, and the other two substituents are OH groups. These are not polymers but, rather, modified silanes. They are usually sold in solutions and used principally as waterproofing agents for masonry and wood.

Composite Applications

Matrixes for composites are the principal subject of this chapter. Silicone resins have long been used as matrix materials in composite parts. For example, a short-glass-fiber-reinforced silicone resin can be used as a molding compound to provide structural and protective characteristics in electrical applications. Silicone/glass laminates are commonly used in circuit boards where their good electrical properties are important. Those properties include fire and arc resistance, temperature stability, solvent resistance, and low electrical conductivity. The disadvantages over conventional organic resins include longer cure times, higher cost, and sensitivity to some solvents such as acetone, toluene, and carbon tetrachloride.

Mold releases are vital to most composites manufacturing operations to protect the tool and part from damage by unwanted adhesion. Releases are generally applied directly to the tool (mold) and then the composite part is built on top of the layer of release. Several materials besides silicones are frequently used as mold release agents including wax-based products, polyvinyl alcohol, fluorocarbons, and several proprietary mixtures.

Silicone mold releases are often applied in an aerosol spray form. While this is convenient, care must be taken to restrict the spray to the mold only, and not to allow the silicone to get on or in the components or layers of the composite part. If such contamination occurs, delamination of the composite or reduced properties can result. It is also important to be sure that the spray is isolated from other areas of production so there is no possible contamination of the composite material. There have been some reports that silicones were sucked into the ventilation system of a factory, which then subsequently contaminated parts.

Silicones as oils (liquids instead of sprays) are well adapted as mold releases. They are easily spread as a thin film over molds because of their low surface tension. They are also resistant to oxidation and thermal degradation, which gives them more life than many other mold releases in high-heat-cure cycles.

A technology related to mold release agents is the coating of paper and other substrates with silicone polymers to make release paper. This type of paper is used extensively in making composite prepregs. It also can be used as a release material for tool making and other applications where release from a flat surface is desired without use of mold release. A commercial application of this technology is in the baking industry where cookies and other products are placed on silicone paper to achieve release from the baking sheets.

Coupling agents are an important part of ensuring a good bond between the matrix and the reinforcement. Silane, the monomer form of silicone, is frequently used as a coupling agent in bonding organic polymers to fiberglass. Silane coupling agents used for this purpose have two different types of reactive end groups, which provide the ability to chemically bond to both the organic polymer and the inorganic reinforcements. The improvement in bonding from the coupling agents is especially effective when the composite is exposed to water or solvents. When a coupling agent is present, photomicrographs show improved wetting of the fibers by the resin. This wetting dramatically reduces the tendency of the fibers to pull out of the resin, thus transferring more load onto the fibers and making the composite stronger. In polyester/fiberglass systems, the dry flexural strength has been improved from 55,000–78,000 psi (379–538 MPa) when coupling agents are used. When wet, those same systems are improved from 20,000–80,000 psi (138–552 MPa).

Silane coupling agents are typically applied to the fiberglass by the fiberglass manufacturer along with the sizing (a protective coating). The type of coupling agent can be specified to ensure optimal compatibility with most of the common thermoset resins used in composite fabrication and with many of the thermoplastic resins to which fiberglass and mineral reinforcements (such as talc, clay, etc.) are added.

Coupling agents improve adhesion in sealants, coatings, and adhesives where the resin must bond to a metallic or other non-organic surface. In these cases, the coupling agent is added to the polymer mix.

Adhesives and sealants are other useful products in the composites industry. While some types of silicones are most noted for their excellent release properties, others are among the stickiest of materials, for example, being able to adhere to Teflon® under water. Properties that give silicone adhesives an advantage over other adhesives are temperature stability (over a range from –85 to 500° F [–65 to 260° C]); chemical stability (resistant to bases, acids, oils, many solvents, and most atmospheric pollutants); good weatherability (demonstrating excellent property retention over 20 years of life); and good electrical properties (low conductivity).

The same concepts that make silanes good coupling agents are used to make silicones good adhesives and sealants. Hence, bonding between composite materials and non-organic surfaces is excellent. In some especially difficult situations, a primer should be used to enhance adhesion. These primers are available from the manufacturers of the silicone adhesives. As with all adhesive materials, if using silicone adhesive and primer, apply them to carefully prepared, clean surfaces. Silicone adhesives tend to be somewhat more expensive than organic adhesives, such as epoxies. However, when difficult conditions are present, few materials can perform like silicones.

Flexible tooling (molds) is a rapidly growing application for silicones in composites. When the need arises to duplicate a complicated part, especially with undercuts and/or high-detail replication, silicone tooling is often the easiest and least expensive material for creating the mold.

The number of parts that can be made from a flexible mold is generally much fewer than from a hard mold, especially if the mold is heated. However, the cost and effort to make a new mold from silicone is considerably less than would be required to make a hard mold. Flexible tooling is, of course, less dimensionally stable than hard tooling. Therefore, if the dimensions of the part must be held to extremely tight tolerances, flexible tooling may not be appropriate. Also, if pressure is required during the molding cycle, flexible tooling will deform and the part dimensions will change.

The simplicity of making a mold using silicone is one of its major advantages. Silicone rubbers are generally sold as two-part systems, a resin and a curing agent. The resin is chosen to fit the application. Many resin types are available. The choice depends upon the complexity of the design to be reproduced, the extent of undercuts, and the compatibility required with the resin that will be cast in the mold to make the final parts. The curing agent is chosen to be compatible with the resin in light of other considerations, such as clear versus colored agents, which help ensure thorough mixing of curing agent and resin, working time required, and compatibility with the material of the master to be cast against and the resin to be used to make the final parts.

The mold is then made by first preparing the pattern. It must be clean and dry. The pattern is then mounted securely to the bottom of a rigid, open box that has an area and depth sufficient to give at least .25 in. (0.6 cm) clearance around and above the pattern. The pattern and inside of the box are then coated with mold release. The resin and curing agent are mixed and then poured or pumped around and over the pattern. Best results are obtained if the mix is de-aired. This is done by placing a top on the box and subjecting it to a vacuum of about 29 in. (74 cm) for a few minutes, or by placing the resin mix into a vacuum system before pouring into the mold. Automatic dispensing and de-airing equipment is available for high-volume production processes.

After curing for 24 hours at room temperature or less time if heated, the mold can be removed (stripped) from the pattern. The silicone mold is then placed in a frame to give it some dimensional stability and it is coated with a barrier coat to increase mold life and prepare the surface for proper casting.

Reusable vacuum bags have become an important application because of the high cost of vacuum bagging using disposable materials. Many composite applications, especially for advanced composites, require low void contents, typically <1%. Achieving low void contents usually requires applying a vacuum to the part during initial cure and then applying pressure during the latter stages of cure. Traditionally, this has been accomplished by creating a vacuum bag assembly over the part to be cured. This assembly consists of a sheet of nylon film secured to the upper face of the mold using two-sided tape to create a vacuum bag. Several sheets of release material, which could be made of silicone film, and absorbent materials are placed between the composite lay-up and the vacuum bag to collect any resin that may flow out of the part during evacuation. The entire vacuum assembly is then placed into an autoclave or oven and cured. The nylon bag and sealant materials are not reused.

Recently, several composite manufacturers have shown that silicone bags can replace the nylon sheet and sealant materials. Silicone bags vastly reduce the time needed to create the vacuum assembly and are reusable (from 30 to 100 cycles), thereby reducing long-term costs. Because the bags are conformable to the shape of the part, pressure from the autoclave is uniformly applied to the part through the silicone bag. Moreover, bag failures (breaks in the bag causing loss of vacuum), which are common with nylon bags, are almost eliminated with the thicker and more rugged silicone bags. Should the silicone bags become damaged, they can be repaired with room-temperature-vulcanization (RTV) silicone sealants. The benefits of using reusable silicone vacuum bags become even more pronounced at elevated temperatures (up to 550° F [288° C]) where nylon films must be replaced with the stiffer, more costly polyimide films.

Recent developments in using the thermal expansion of silicone elastomers in a controlled manner offer an economical alternative to achieve *pressurization of composite parts* during curing. The result is part definition superior to non-pressurized parts and, in the case of advanced composites, there is a reduced need for autoclaves. The technique is called trapped rubber molding (TRM). The composite lay-up is placed in one half of the mold and then a silicone rubber tool is placed on top of the composite. The other part of the mold or simply a platen is then placed on top of the silicone rubber tool and clamped or held shut in a press. As the assembly is heated, the silicone expands and creates pressure on the composite part. Silicone is well suited for implementation as a trapped rubber molding tool because of the following properties: it can be easily cast, it has high tear strength, it is temperature stable, and it exhibits long mold life.

An alternate system to trapped rubber molding is called the THERM-X™ process, invented by United Technologies. In this process, a tool and composite part are both placed inside a pressure vessel and then completely surrounded by compressible silicon rubber with a barrier, such as aluminum foil, between the silicone and the part for protection. The pressure is provided through the expansion of the silicone media that is heated, usually with an independent control from the heat supplied to the mold, to achieve curing of the composite.

Several silicone materials are used as **additives** in plastics for a variety of purposes. One application that has excellent promise is the use of a high-molecular-weight silicone powder as a flame retardant. Concentrations of only 1 to 5% can modify a plastic's burn characteristics, reducing the rates of heat release, and smoke and carbon monoxide evolution in halogen-free, halogenated, and phosphorus flame-retardant systems. In highly filled resin systems, silicone powders can reduce the amount of flame retardant filler required. In thermoplastic systems, the addition of the silicone powder can also improve processing by reducing the amount of energy required to extrude the material.

Other silicone additives improve dispersion properties and enhance resin compatibility. Some have been developed as crosslinking agents for thermoplastic resins.

A relatively new application is the use of silicones as low-volatile-organic-compound (VOC) solvents. These materials have proven to be invaluable in paint removal applications and as solvents for high-performance coatings. Some of the coatings are also made from silicone. With the continuing pressure to lower VOC levels, the use of non-polluting solvents, such as silicones, will continue to grow.

DICYCLOPENTADIENE (DCPD)

Occasionally a new resin emerges in the composites field. This occurs, of course, because the new resin has some distinct, compelling advantages over the materials currently being used. Dicyclopentadiene (DCPD) seems to be on the brink of becoming the newest star of the composite resin marketplace. The most important advantages of DCPD are improved toughness and chemical resistance over the traditional unsaturated polyesters.

Two important applications for DCPD have recently been commercialized. The first was a cover for chlor-alkalai cells. These covers are made by a liquid injection system into a closed mold. The other is corrosion-resistant pipe. Here the advantages of toughness combined with corrosion resistance better than vinyl esters suggest strong market potential.

DCPD has been used in composites for many years as an additive to improve the performance of traditional unsaturated polyesters. In these modified polyester resins, the DCPD reacts with maleic acid and results in a higher T_g and more rapid development of acceptable hardness. DCPD also improves the ability to hide the fibers so the visual characteristics of the reinforced polyester are better. The DCPD has low viscosity, thus allowing a reduction in the total amount of styrene, which reduces the amount of harmful emissions and shrinkage. DCPD also improves the cure of the resin when in an oxygen atmosphere, thus assisting in making the surface tack-free even when exposed to air.

Recently, some resin manufacturers have announced the inclusion of DCPD as a component in epoxy resins. The DCPD is polymerized in the backbone and epoxy groups are still on the ends for crosslinking. The addition of DCPD significantly improves the T_g of the epoxy, raising it from 320° F (160° C) at 22% concentration to 356° F (180° C) at 28% concentration.

Another modification to the formation of pure DCPD monomer is the copolymer between DCPD and butadiene. This gives a significant increase in toughness over what is already a tough polymer. Similar copolymers between DCPD and ethylene also have been commercialized. These copolymers, sometimes called cyclic olefin copolymers (COCs), are thermoplastics and compete against traditional thermoplastics such as polystyrene, polycarbonate, and acrylic.

Structure and Crosslinking Reactions

The structure of the dicyclopentadiene monomer can be depicted in several, equivalent representations. Because these can be confusing, the most common methods of showing the structure are given in Figure 5-17. Figure 5-17a is probably the most representative of the actual geometry of the molecule. Figure 5-17b is simply the same as Figure 5-17a, but without the atoms shown. Note that this representation assumes a carbon atom is at each node. Figure 5-17c is a planar representation of the molecule. Both Figure 5-17a and b emphasize the three-dimensional nature of the molecule and its similarity to norbornene, an important organic molecule. Note that the DCPD monomer is also similar to the nadic end-capper used in some polyimide resins.

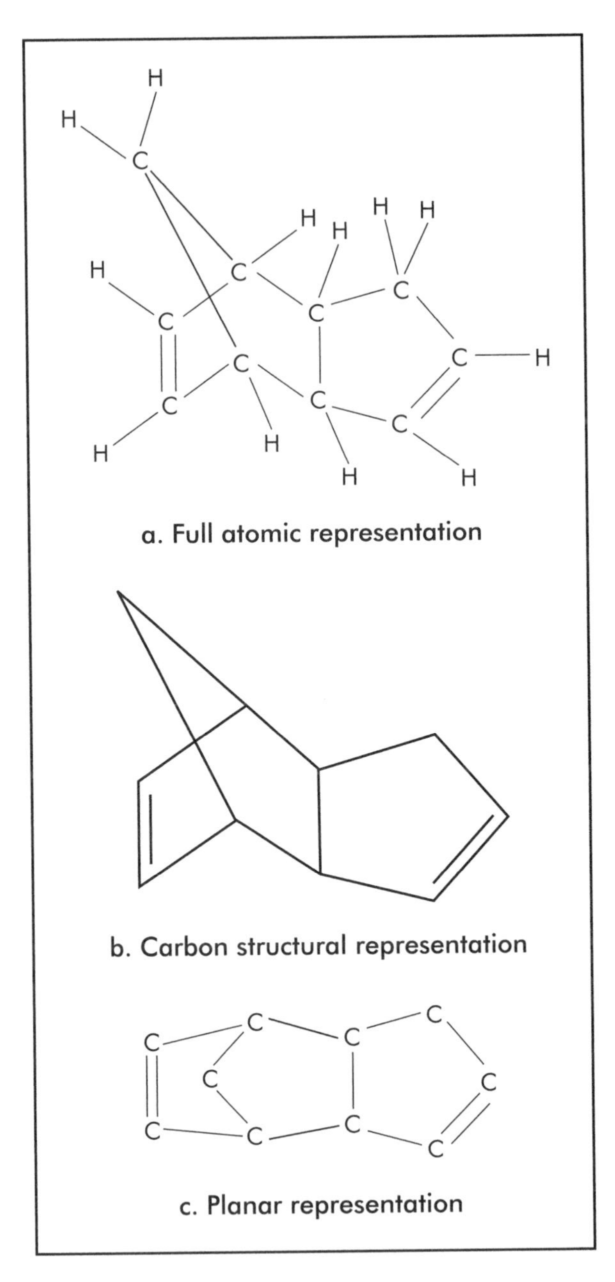

Figure 5-17. Alternate representations of dicyclopentadiene (DCPD).

The polymerization of DCPD begins with the opening of the six-member ring that contains the carbon-carbon double bond. This step results in a linear polymer in which the remaining bicyclic structure is pendant from the polymer chain. The chain backbone includes the carbon-carbon double bonds. This reaction mechanism has been called ring-opening-metathesis polymerization (ROMP). A complex organometalhalide catalyst is employed in polymerization. In the past, the sensitivity of this catalyst to the presence of oxygen and other contaminants was a major problem making the reaction difficult to carry out. In the last few years a new catalyst system based on ruthenium (Grubbs catalyst) has been developed that vastly reduces the complications associated with the polymerization. The reaction is highly exothermic so some control over the temperature is required. However, the use of the new catalyst along with a reaction modifier, triphenylphosphene (TPP), has made the polymerization quite routine. The reaction rate is usually fast.

Although it is theoretically possible to stop the reaction at the linear thermoplastic polymer stage, the practical matter is that the reactivity of the DCPD monomer results in crosslinking reactions proceeding at the same time the linear polymer is being formed. The high reactivity results from the many reactive sites on the DCPD monomer where crosslinking can occur. Crosslinking occurs at the double bond in the pendant bicyclic ring. Additional crosslinking can occur at the carbon-carbon double bond along the backbone. As a result, the polymeric structure is complex and has high crosslinking density. The interlocking structure combined with the aliphatic nature of the polymer results in extraordinary toughness.

Properties and Uses

The high reactivity of DCPD can be used in combination with other monomers and polymers to obtain some unique properties. As already discussed, DCPD has been used as an additive in relatively minor amounts with both unsaturated polyesters and epoxies. These products have been used to make automobile body filler, bath tubs/shower stalls, spas, sinks/vanities/counter tops, truck bodies/caps, boats, personal watercraft, decorative building columns, and in electrical applications. Many applications benefit from the superior surface finish that DCPD gives to traditional polyester parts as well as its superior physical properties. Of special note is the higher T_g that leads to improved thermal shock resistance in cultured marble applications.

The advantages of DCPD when compared to traditional unsaturated polyester and vinyl ester formulations are: improved toughness (sometimes initially as high as 100 times better but falling to about 10 times better over time), stiffness, corrosion resistance, reduction of styrene, and the ability to have a single-stage cure from a low-viscosity resin.

Processing

An important advantage of DCPD is the ability to use the resin in liquid injection molding processes, such as structural reaction injection molding (RIM) and resin transfer molding (RTM), which are discussed in detail in a later chapter. Some significant advantages are obtained in using DCPD versus other more traditional resins. Because of the reactivity of the DCPD polymerization, the need for expensive RIM metering equipment is largely eliminated. With DCPD, the only requirements are a simple mixing vessel, a feeding motor pump, and some temperature control over the mold. Impregnation of the fiber preform is rapid because of the low viscosity of the resin. Reaction time is fast.

Other typical composite manufacturing methods such as filament winding, centrifugal casting, pultrusion, and spray-up have been demonstrated to work at least as well with DCPD as with polyesters and vinyl esters.

DCPD also has some disadvantages versus traditional unsaturated polyesters and vinyl esters. For instance, its high toughness seems to decrease with time. This may be due to oxidative attack on the carbon-carbon double bonds over time. Hence, oxidation stability is also lower than in some high-performance vinyl esters and polyesters. At present, no appropriate coupling agents with fiberglass have been developed, so bonding to the reinforcement is less than optimal.

CASE STUDY 5-1

Rapid-curing Fiberglass Shin Guards

A student and an avid soccer player, Spencer Larsen has reported the following breakthrough. His report is used in part.

"In recent years a development in sports equipment has revolutionized the soccer world. A material that was initially used for making medical splints has now found its way onto the playing field. Now its job is to prevent some of the injuries that it so often aided in fixing. The product is a rapid-curing fiberglass shin guard. It is made from a proprietary, fiberglass reinforced polyurethane and seems to have just the right blend of manufacturability and properties.

"The material comes as a prepreg in a watertight bag. Several layers of fiberglass cloth are impregnated with a dry mixture of isocyanate and polyol that becomes reactive when brought into contact with water. In addition to solvating the materials so that they can react with each other, the water molecules react with isocyanate molecules, which are in excess concentration in the prepreg and form intermediate reactive groups. These groups further react with additional isocyanate to form a polyurethane.

"The instructions for the curing of the shin guards tell the user to wet the material by putting it under a running faucet or by dipping it into a container of water. The greater concentration of water molecules provides for a much quicker curing time, usually 10–20 minutes. Once the material has cured, further exposure to water does not affect the shin guards. They are even advertised as being machine-washable.

"Thanks to the combination of reactive resin and fiberglass, the guard can be custom fit to the user's shin. Before and immediately after being exposed to moisture, the material is flexible and can be shaped to the shin and held in place by an elastic wrap (like an ACE® bandage) or a tight sock. A few minutes after being wet down, the material is cured and the shin guard has the permanent shape of the user's shin.

"Several layers of fiberglass and plastic materials comprise the reinforcement and cushioning part of the shin guard. The inside layer is made of ethylene vinyl acetate that has a knitted covering and several holes punched for ventilation. The inner layers are several fiberglass layers of different weaves that provide thickness variation and strength as needed for optimal performance. The outer layer of the shin guard is a fabric capable of stretching to account for the shaping of the shin guard to the shin.

"The composite shin guards compete against several other designs and materials. The typical design of the fiberglass/PUR shin guard is better at dispersing impact forces than almost any other shin guard while being lighter than most of the others and about average in thickness. Overall, it is clearly the best choice."

SUMMARY

The specialty and high-performance thermosets are used when environmental conditions are severe. One of the most adverse environments for a composite material is when it is used as a pipe or tank for holding aggressive chemicals. For these applications, vinyl esters have proven to be both cost effective and durable. In general, vinyl esters are the resin of choice for corrosive situations or when most common solvents need to be transported or stored. Of course, vinyl esters also have their limitations. Some extremely aggressive liquids will attack vinyl esters. Therefore, care should be taken to test the materials and/or consult a reliable chemical compatibility table before using vinyl esters or any polymer matrix in liquids that might be detrimental to their performance.

When the environment may include the danger of a fire, the usual choice of matrix is a phenolic. These highly aromatic materials resist burning and have relatively little smoke. Moreover, the smoke emitted is generally of a lower toxicity than most other resins. Hence, critical transportation environments, such as on airplanes, ships, trains, and buses, as well as high-density public areas, require that phenolics be used as the resin matrix of interior components. Phenolics are also used in applications where a char needs to be formed so that the heat will not be transported throughout the material. Such an application would be the exhaust nozzle of a rocket. In this case, the phenolic is reinforced by carbon fibers. As the char is formed, the carbon fibers and char are slowly eroded away (called ablation) and the heat is retained in the char zone.

For situations where ablation must be much slower than can be obtained with phenolics, such as for aircraft brakes, and for high-temperature environments, the material of choice is carbon-carbon composites. The matrix is actually a char, usually of phenolic, which is then filled with additional resin and charred again. This process is repeated until the matrix is nearly without voids. The resulting material is largely pure carbon. The reinforcement is also carbon. Hence, carbon-carbon is the name of the material. The major drawback to carbon-carbon composites is high cost.

When a material must be found that has higher thermal stability than epoxies, easier processing than phenolics, and lower cost than carbon-carbon, the most logical choice is polyimides. These resins are highly aromatic and stiff—both characteristics typical of high strength and thermal stability. However, polyimides are quite difficult to process and are higher priced, at least in comparison to epoxies. Therefore, their use is usually restricted to only high-temperature environments.

Cyanate esters have especially good electrical properties and are therefore often used for housings over antennae and for radomes.

The polyurethanes are just now achieving high volumes as matrix materials for composites. They are tougher than most other matrix materials and have good processing capabilities. Therefore, it is expected that more polyurethane matrix composites will be used in applications where moderate damage might be a problem.

Silicones have been used in the composite industry for many years. Mold releases, vacuum bags, flexible molds, adhesives and, of course, matrices have all been made from silicones for many years. The matrix materials perform especially well in environments of both low and high temperature, although not as high as with polyimides.

DCPD is a relatively new matrix material that has some nice properties. It is tough and corrosion resistant. DCPD is a low-viscosity resin that cures quickly to give a nice surface and excellent properties.

LABORATORY EXPERIMENT 5-1

Solvent Resistance of Composite Resins

Objective: Determine the solvent resistance of specialty resins versus polyester thermosets.

Procedure:

1. Obtain samples of composites made from traditional polyesters and from several specialty resins. Some good specialty resins would include vinyl ester, phenolic, and cyanate ester, all with fiberglass.
2. Cut the samples into tensile bars. The shape can be found by looking at the tensile test in the ASTM procedures.
3. Divide the polyester samples and the specialty resins into at least two groups of at least two samples each. One of the sample groups of each type of resin is for the control and the other groups

are for exposure at different times, if possible.

4. Place the control samples in a place where they will not be subjected to chemicals or light.
5. Using goggles and protective clothing, carefully place the other samples in a pan containing sufficient amount of a chemical to cover the samples. You might choose one of the following as a possible chemical: nitric acid, hot water, acetone, bleach, etc.
6. Leave the samples in the material for a few days or, preferably, a few weeks. Make sure the test is clearly labeled with a warning indicating the nature of the chemical. Also make sure the test is in some location where the pan will not get bumped and spilled. If noxious odors are emitted from the chemical, the test should be run inside a chemical hood.
7. Using goggles and protective clothing, carefully remove some or all of the samples. If you have enough samples, you might take out some of them at two or more different times. Use gloves and tongs to remove them. Be careful not to spill or drip on your clothes.
8. Carefully wash the samples. Use safety equipment as previously outlined. Dry the samples with a cloth.
9. Obtain the control samples and then test both the control samples and the exposed samples using a tensile test machine. Note the tensile strength, modulus and, especially, the elongation.
10. Compare the elongation as a function of the exposure time for the polyester and specialty resins.

QUESTIONS

1. Indicate three advantages of vinyl esters over traditional unsaturated polyesters in composites.
2. What are three advantages and disadvantages of vinyl esters versus epoxies in composites?
3. Discuss an important application for phenolic composites. What are the molecular features that lead to its excellent performance in this application?
4. Why do phenolics make a good starting material for carbon-carbon composites?
5. What is meant by B-staging? What are the A and C stages? Explain how resins other than phenolics might be B-staged.
6. Why do carbon-carbon composites take so long to make?
7. What is the difference between pyrolysis and oxidation? Why is pyrolysis used in making carbon-carbon composites?
8. What are two key criteria for a thermally stable composite?
9. What molecular features suggest that polyimides will have good thermal stability?
10. What are the unique properties of PMR-15? How are they related to its molecular structure?
11. Give two key properties of polyurethane composites. In which applications are these properties important?
12. List four different uses for silicones in the composites industry.
13. What causes the high reactivity of DCPD?
14. Explain how a self-healing composite works.

BIBLIOGRAPHY

Brown, E.N., Sottos, N.R., and White, S.R. 2001. "Fracture Testing of a Self-healing Polymer Composite." Submitted for publication in *Experimental Mechanics: An International Journal*.

Cull, Ray A. 1980. "Characteristics of High-temperature Polymers." Personal communication, quoted from Cassidy, P. E., *Thermally Stable Polymers*. New York: Marcel Dekker, Inc.

Goodman, Sidney H. (ed.). 1998. *Handbook of Thermoset Plastics*, 2nd Edition. Westwood, NJ: Noyes Publications.

Larsen, Spencer. 1989. "Composites in Medical Applications." Report for MFG555.

McDowell, Enoch. No year. "A Closer Look at OSi® Fiberglass Technology." Report for plastics materials and manufacturing class. Quoted (in part) by permission.

Peters, S. T. 1998. *Handbook of Composites*, 2nd Edition. London: Chapman and Hall.

Reardon, Joe, Manager, Applications Engineering for HyComp, Inc. 2002. Private communication in January.

Society of Manufacturing Engineers (SME). 2005. "Composite Materials" DVD from the Composites Manufacturing video series. Dearborn, MI: Society of Manufacturing Engineers, www.sme.org/cmvs.

Strong, A. Brent. 1995. *Fundamentals of Polymer Resins for Composite Manufacturing*. Arlington, VA: Composites Fabricators Association.

Strong, A. Brent. 1996. *Composites Fabrication*, October, pp. 41–47.

Strong, A. Brent. 2006. *Plastics: Materials and Processing*, 3rd Ed. Upper Saddle River, NJ: Prentice-Hall, Inc.

White, S.R., Sottos, N.R., Geubelle, P.H., Moore, J.S., Kessler, M.R., Sriram, S.R., Brown, E.N., and Viswanathan, S. 2001. "Autonomic Healing of Polymer Composites." *Letters to Nature* 409, 794–797.

Woodson, Charles S. 1999, 2000, 2001. Cymetech, LLC, personal communications.

6

Thermoplastic Composites

CHAPTER OVERVIEW

This chapter examines the following topics:

- Review of thermoset and thermoplastic composites
- Engineering thermoplastic composites
- High-performance thermoplastic composites

REVIEW OF THERMOSET AND THERMOPLASTIC COMPOSITES

As discussed briefly in previous chapters, all resins are either thermosets or thermoplastics. The pre-cured thermosets are relatively low-molecular-weight resins in which bonds are caused to form between the molecular chains during molding—a process called **crosslinking** or **curing**. Crosslinking forms a polymer network of extremely high molecular weight, which results in significant changes in many properties. The thermoplastics, which are not crosslinked, have molecular weights higher than the uncured thermoset resins but not as high as the crosslinked (cured) thermosets.

Thermoset polymers are formed, therefore, when crosslinks are formed. The crosslinking bonds are created through chemical reactions between active sites on the resin molecules, sometimes employing the use of non-polymeric molecules, such as styrene and various initiators, to assist in the process. The crosslinking process is somewhat different for each of the polymer types, as described in previous chapters. The net result of the crosslink formation in all thermosets is a dramatic solidification of the resin, an increase in strength and stiffness, an increase in the glass transition temperature, and an increase in the melting point. The melting point is raised high enough that it is usually above the decomposition temperature of the polymer and the crosslinked polymer can no longer be melted.

Thermoset resins have distinct advantages in composites. The uncured thermoset resin is mixed with the fibers before the crosslinking reaction while it is still low in molecular weight and, therefore, low in viscosity. These low-viscosity resins (often having viscosities similar to thin motor oil) will easily wet the fibers. As will be seen, the higher molecular weight of the thermoplastics makes fiber wet-out much more difficult. Another advantage is that the crosslinking reaction can be interrupted and suspended at the point where the fibers are wetted but full cure has not yet occurred. This is called **B-staging**, and the resulting material (called a **prepreg** or **molding compound**) can be stored until the actual molding operation is carried out at a later time. The suspension of the curing process is overcome by heating the resin (and occasionally by adding an activating chemical), which is done during the molding operation. Usually special (cold) storage conditions are required for prepregs and molding compounds as the thermoset resin will slowly continue the crosslinking reaction even though it has been temporarily slowed by cooling. With all of these advantages, thermosets are clearly the preferred resin for making composites.

Thermosets have some disadvantages too, which have led to the use of thermoplastic resins in composites as well. Some of the disadvantages of thermosets include the following:

- required mixing of chemicals to effect the cure,
- potentially harmful vapors and chemical handling,
- limited shelf life,
- multi-step processing,
- longer times for molding the part (cure time),
- inability to reprocess (reshape) the material after it has been cured, and
- poor toughness.

The last of these—poor toughness—needs some explanation. Because the crosslinks tie the polymer chains together, the overall rigidity of the polymer system is usually increased as curing proceeds. This rigidity reduces elongation and generally reduces impact toughness in thermosets.

Thermoplastics do not have crosslinks. The molding process is simply one of melting the thermoplastic so it will flow into and fill the mold. Then the material is allowed to cool and solidify in the molded shape. No change in molecular weight occurs during the processing of thermoplastic resins. Therefore, whatever molecular weight is needed in the final product must exist in the thermoplastic resin before it is molded. All composite materials are solids at their use temperatures and generally at room temperature, regardless of the use temperature. So thermoplastics are solids at room temperature, and after being melted during molding, they return again to their same solid condition after cooling.

The absence of change in molecular weight of a thermoplastic during processing gives rise to a major compromise of properties, which is inherent in the formulation of all thermoplastic resins. In general, *most mechanical and physical properties of polymers are improved* as their molecular weight increases. However, increasing molecular weight makes *processing of the thermoplastic composite much more difficult*. This is because resin melting points are increased, wet-out of the fibers is more difficult with the higher viscosity resins, and the polymers are more difficult to dissolve should solvent processing be used to wet the fibers.

Much research and effort has gone into the task of improving the physical properties of thermoplastic resins and yet not making the resulting polymer extraordinarily difficult to process by melting or, less often, by solvation. Even though compromises on molecular weight and processing are inevitable, some important properties of thermoplastics are superior to thermosets. This has led to increased use of thermoplastics in composite applications.

To fully appreciate the impetus to use thermoplastics for composite applications, it is appropriate at this point to step back and look at the *entire plastics and composites marketplace*. This includes all the polymer composite products employing thermoset and thermoplastic resins, as well as the many *non-reinforced* plastics used. This marketplace is, of course, immense and getting bigger every day. In applications too numerous to even conceive, plastics are replacing wood, metal, ceramics, paper, leather, and other traditional materials. Moreover, they are being used in a multitude of unique applications for which no other material has ever been used. When viewing the entire market for plastic and composite materials, that is, all products employing polymeric resins, thermoplastics represent about 80% of the total. Thermosets represent the remaining

20%. Why are thermoplastics so dominant in the total marketplace? Generally, thermoplastics have the advantage over thermosets because of much faster molding times (simple cooling of thermoplastics to solidify versus the chemical reactions required for thermosets) and the overall adequacy of the properties that can be obtained.

Still putting the big picture into perspective, it should be understood that the vast majority of plastics materials do not use reinforcements and are not, therefore, composite materials. Hence, the inherent problems of fiber wet-out that come with thermoplastic resins do not apply. The market for just reinforced materials, that is, the composite materials, is about 20% of the entire plastics and composites marketplace. Within this more narrow composites market, thermosets represent about 80% of the total material used, just the reverse from the entire marketplace.

In summary, thermoplastics are the dominant plastic materials overall and especially in non-reinforced applications. Thermosets are used in non-reinforced applications for specific purposes where they have an advantage because of some unique property. Further, within the reinforced or composites marketplace, which is much smaller than the total, thermosets dominate and thermoplastics are used only in applications where their unique advantages are important. This chapter discusses the characteristics of thermoplastic composites that present those specific advantages.

Differences Between Engineering and High-performance Composites

The world of composites can be separated into two groups by resin type, thermosets and thermoplastics, and then into two additional groups, the engineering or industrial composite materials, and advanced or high-performance composite materials. Table 6-1 summarizes the differences between these materials.

A brief summary of the characteristics of the engineering and advanced thermoset and thermoplastic composites is given in Table 6-2. Engineering thermoset composites (FRP) might be used in automobiles and boats. Engineering thermoplastic composites could be used to strengthen traditional plastic parts such as valves and football helmets. The advanced thermoset composites might be used in helicopters and airplanes, whereas advanced thermoplastic composites could be used in submarines and tanks. These uses are, of course, merely representative of common applications that utilize these various materials.

For thermosets, the division into engineering and advanced composites is based on resin type, reinforcement type, precision placement of the fibers, and control over fiber-resin content during the molding process. Generally, the resins used for engineering composites are polyesters and vinyl esters. The reinforcement used with engineering composites is almost always fiberglass. An alternate name for the *engineering thermoset composites* is **fiberglass reinforced plastics (FRP)**. For this group of composites, the placement of the fibers during molding is often quite random as occurs in such processes as spray-up and wet fiber lay-up. Even when more precise methods such as filament winding, pultrusion, and resin transfer molding (RTM) are used, control over the resin-fiber concentrations is generally less precise than would be the case when using these same processes with advanced composites. The emphasis in engineering composite processing is often on throughput. Engineering composites are characterized by their relatively low price rather than for any particular outstanding performance property (except for corrosion resistance for vinyl esters).

As indicated in Table 6-1, the *advanced thermoset composites* are distinguished from the engineering thermoset composites by the

Table 6-1. Characteristics of engineering and advanced, high-performance thermoset and thermoplastic composites.

	Engineering Composites		Advanced Composites	
	Thermosets	Thermoplastics	Thermosets	Thermoplastics
Resin types	Low cost (for example, unsaturated polyester and vinyl ester)	Traditional commodity and low cost (for example, polypropylene [PP], nylon, polycarbonate [PC], polyethylene terephthalate [PET], and acrylonitrile-butadiene-styrene [ABS])	Epoxies and specialty resins (for example, phenolics, polyimides, polyurethane [PUR], silicones, and cyanate esters)	High mechanical property resins, usually highly aromatic (for example, polyetheretherketone [PEEK], polyphenylene sulfide [PPS], polyarylsulfone [PAS], liquid crystal polymers, and thermoplastic [TP] polyimides)
Reinforcement types	Fiberglass (chopped or woven)	Fiberglass (short)	Carbon fiber, aramids, other high-performance fibers, and specialty fiberglass (long fibers)	Carbon fiber, aramids, other high-performance fibers, and specialty fiberglass (long fibers)
Concentration of reinforcement	Up to 50%	Up to 40%	Above 50% and up to 65%	Above 50% and up to 65%
Processing methods	Traditional thermoset curing (for example, spray-up, lay-up, resin transfer molding [RTM], filament winding, pultrusion, and compression molding)	Traditional thermoplastic processing (for example, injection molding, extrusion, thermoforming, and blow molding)	Precision placement methods (for example, prepreg lay-up, filament winding, pultrusion, and RTM)	Adapted precision methods for thermosets and specific methods for thermoplastics only

Table 6-2. Comparison of advanced and engineering thermoplastics and thermosets.

Advanced Thermosets	Advanced Thermoplastics	Engineering Thermosets	Engineering Thermoplastics
High cost	High cost	Low cost	Low cost
High-temperature capabilities	Solvent resistance	Excellent fiber wet-out	Standard thermoplastic manufacturing
High strength	High toughness	Moderate strength	Short fibers
High modulus	Poor wet-out	Brittle	Moderate strength
Good fiber wet-out	High strength	Brittle	Good toughness

types of resins, types of reinforcements, and precision of the processes used to control fiber placement. The types of resins used in these materials have been discussed in previous chapters and the types of reinforcements and molding methods will be examined in detail in subsequent chapters. So, in brief, the high-performance thermoset resins include epoxies, phenolics, polyimides, etc., and the high-performance fibers include carbon/graphite, aramids, etc.

A similar separation into engineering and high-performance composites can be made for thermoplastic composites, also shown in Table 6-1. *Engineering thermoplastic composites* are made from relatively low-cost thermoplastic resins. They are almost always reinforced with short fiberglass strands. These composites are usually molded using traditional processes such as injection molding, extrusion, and thermoforming, which were designed for *non-reinforced materials*. The resins are introduced into the processing machines as pellets—small pieces of solid thermoplastic resin shaped like short pieces of spaghetti. These pellets are heated in the process until molten, pressed or injected into the mold, and then allowed to cool and solidify in their molded shape. The normal use of these traditional thermoplastic molding processes would start with non-reinforced thermoplastic resins, which are called **neat resins**.

Molders who use injection molding and other traditional thermoplastic molding machines would rarely have the capability to mix reinforcements into the plastic as this is usually done in an extruder. The addition of fibers is often done by the resin manufacturer or an intermediate resin **compounder** who might also add colors, UV stabilizers, and other minor components during the same extrusion operation. The actual addition is usually done by adding the short fibers through an inlet port in the barrel of the extruder. The reinforcements are mixed into the molten resin by the action of the extruder screw. This action will often break the fibers, especially if they are long. Therefore, the fibers used in engineering thermoplastic composites are short—small enough to survive the extrusion operation and pass through the many small clearances typically present in molding machines. The fiber lengths in these pellets are usually only a few millimeters and, by analogy, are often called **whiskers**. Difficulties in processing of the resin and fibers in a molding machine usually limit the maximum concentration of fibers to about 40%.

Even though the processes used to mold engineering thermoplastic composites are relatively common, some minimal machine modifications must be considered when molding with reinforced resins. The most obvious change is that all small restrictions in the processing stream need to be opened up as much as possible to allow for passage of the fibers. Those restricted openings might be in the molding machine or in the mold, such as the gates. The molder should also realize that long-term running of fiber reinforced resins will increase the abrasive wear along the flow path of the resin. This is due, of course, to the higher abrasive character of the reinforcement versus a neat resin. Therefore, wear should be expected and machines and tools should be examined often as wear could change part dimensions. The processes for molding thermoplastics will be discussed in greater detail in a later chapter in this book.

The *advanced (high-performance) thermoplastic materials* are characterized chiefly by the type of resins used, the types of reinforcements, and the types of processes, as shown in Table 6-1. The resins used in advanced thermoplastic composites will be discussed later in this chapter. The advanced thermoplastics typically include high-performance reinforcements, such as carbon fibers, aramid, etc., which are used at the longest possible length and highest practical concentration. As is the case with high-performance thermosets, the fiber lengths in high-performance thermoplastics are often the full size of the part. The fiber concentrations may be as high as 65% of the total in both thermoset and thermoplastic advanced composites.

Many of the precision fiber lay-down methods used for high-performance thermosets can be used for thermoplastics, provided that appropriate changes are made. These modified processes, as well as some processes used exclusively with thermoplastic composites, will be discussed later in the chapter on thermoplastic processes. In general, however, the high melting point of high-performance thermoplastics must be allowed for in processing these materials.

ENGINEERING THERMOPLASTIC COMPOSITES

Dozens of engineering thermoplastic resins are available and, as previously discussed, they dominate the world of non-reinforced plastics. Most of these resins also can be obtained with short fiber reinforcements. The reinforcements are generally added to obtain improvement in some specific mechanical property. A few of the most common thermoplastic resins, neat and reinforced, along with their properties, are listed in Table 6-3. Note that a wide range of properties can be obtained in just the neat resins (0% fiber). This variety has been developed over many years as polymer chemists have created new resins to fill in the gaps and expand the available mechanical, physical, and chemical properties of plastic materials.

The choice of resin often depends on many factors, which might include mechanical properties, but can also include chemical resistance, thermal resistance, resistance to ultraviolet light, resistance to environmental stress cracking, electrical properties, or others. Resin choice might be dictated by one of these non-mechanical properties. However, at the same time, some improvement in mechanical properties might be required over what can be obtained from just the neat resin. For example, a plastic valve may require the inherent abrasion resistance and machinability of nylon but could also require the tensile strength to be greater than the 5.5–12 ksi (ksi is the abbreviation for 1,000 pounds per square inch) (38–83 MPa) obtainable with neat nylon resin. A logical method of getting increased tensile strength in nylon would be to add fiberglass reinforcement. As seen in Table 6-3, tensile strengths of 20–27 ksi (138–186 MPa) are possible in nylon with 30% fiber content.

Table 6-3. Effects of fiberglass content on mechanical properties of various engineering thermoplastic resins.

Resin	Tensile Strength*, ksi (MPa)		Flexural Modulus*, ksi (MPa)		Notched Izod Impact*, ft-lb/in. (J/mm)	
	0% Fiber	30% Fiber	0% Fiber	30% Fiber	0% Fiber	30% Fiber
Acrylonitrile-butadiene-styrene (ABS)	4.8–7.5 (33–52)	13–16 (90–110)	290–400 (1,999–2,758)	900–1,100 (6,205–7,584)	2–6 (0.11–0.33)	1–1.5 (0.05–0.08)
Acetal	6.5–10 (45–69)	14–22 (97–152)	400–500 (2,758–3,447)	1,000–1,400 (6,895–9,652)	1.4–1.5 (0.06–0.08)	.7–1.8 (0.04–0.10)
Acrylic (polymethyl methacrylate [PMMA])	5.1–10.5 (35–72)	17.5 (121)	189–270 (1,303–1,862)	1,800–2,000 (12,410–13,789)	.4–1.8 (0.02–0.10)	1.0 (0.05)
Nylon (polyamide)	5.5–12 (38–83)	20–27 (138–186)	40–575 (276–3,964)	1,000–2,000 (6,895–13,789)	.6–1.7 (0.03–0.09)	1.5–3.8 (0.08–0.20)
Polycarbonate (PC)	9–10 (62–69)	14–19.5 (97–134)	320–335 (2,206–2,310)	680–1,250 (4,688–8,618)	12–15 (0.65–0.80)	1.8–2.0 (0.10–0.11)
Polyethylene	2.3–4.3 (16–30)	3.4–6.1 (23–42)	50–310 (345–214)	438–585 (3,020–4,033)	.4–4.0 (0.02–0.20)	1.2–1.3 (0.06–0.07)
Polyethylene terephthalate (PET)	7–15.2 (48–105)	16–23 (110–159)	350–450 (2,413–3,103)	1,050–1,730 (7,239–11,928)	5.2 (0.38)	1.2–2.2 (0.06–0.12)
Polypropylene	4.2–5.4 (29–37)	8.1–8.5 (56–59)	250 (1,724)	750–899 (5,171–6,198)	.3–.6 (0.01–0.03)	1.1–2.0 (0.06–0.11)
Polystyrene	6.5–7.2 (45–50)	7–11 (48–76)	435–450 (2,999–3,103)	920–1,300 (6,343–8,963)	.4 (0.02)	1.0 (0.05)
Polyvinyl chloride (PVC)	5.9–7.2 (41–50)	8.5 (59)	300–378 (2,068–2,606)	500 (3,447)	1.5 (0.08)	16 (0.87)

* Values shown are typical for several grades of each polymer. (Source: IDES database.)

In every thermoplastic resin, the addition of fiberglass increases the tensile strength and the flexural modulus. Typical increases with 30% fiber addition are 120–200% in tensile strength and 200–400% in flexural modulus. This results because the fibers carry much of the load when tensile and flexural stresses are applied. Note that the increases in tensile strength and flexural modulus are not uniform; that is, some resins are improved more than others. This variation is due to several reasons, among which is the ability of the resin to bond to the fibers and the ways in which the fibers interact with the internal crystalline structure of the resin.

Examination of the notched Izod impact data shows a much different result when reinforcements are added. Impact measures the *toughness* of the material, and toughness is a combination of both strength and elongation. Hence, if the elongation and strength are both changing, toughness can be difficult to predict. As already discussed, the strength of an engineering thermoplastic resin increases greatly with the addition of reinforcement. The elongation, however, will generally decrease because the fibers hold the composite structure firmly in place. Because the elongation is such a strong determining factor of toughness, the usual (but not exclusive) pattern for thermoplastics would be for the Izod toughness to decrease with fiber content. This trend is seen in the notched Izod impact data where the toughness of the reinforced material is generally lower than the toughness of the same resin without reinforcement.

A graphical presentation of the mechanical properties of a particular nylon resin with several loadings of fiberglass gives some additional insights into the effects of fiber content on various mechanical properties. As can be seen quite readily in Figure 6-1, tensile strength increases linearly with fiber content as does the flexural modulus. However, the elongation drops precipitously with even a small amount of fiberglass reinforcement. Hence, the Izod impact toughness, which is a combination of strength and elongation, has only a slight overall increase because the effects of strength and elongation more or less offset each other. Because the molecules are higher apart at high temperature (this is, after all, the reflection of an expansion), properties that depend on interactions between the molecules, like tensile strength, are decreased at higher temperatures. Hence, the decreases in properties with temperature are greater for non-reinforced than for reinforced plastics because the reinforcement holds the molecules in place better.

An interesting application of reinforced engineering thermoplastics is for wear resistance. When the resin is reinforced with aramid fibers, such as DuPont's Kevlar®, wear capability increases tremendously over both non-reinforced and even fiberglass reinforced materials. For example, the wear rate of non-reinforced nylon is 22.9×10^{-5} in./hr (58.2×10^{-5} cm/hr). When the nylon is reinforced with 20% aramid, the wear rate decreases to 6×10^{-5} in./hr (15×10^{-5} cm/hr). Similar improvements in durability are seen in other resins.

As discussed, the lengths of the reinforcement fibers used in the materials shown in Table 6-3 are almost like whiskers, typically .005–.010 in. (0.13–0.26 mm). If the fiber lengths are increased, the mechanical properties increase significantly. For instance, a process called *long fiber reinforcement* increases the fiber length by about 15 times and increases the tensile strength by 10%, flexural modulus by 40%, and notched Izod impact by over 100%.

In the process used to make the long-fiber-reinforced thermoplastic resins, reinforcement strands are impregnated with the resin. After impregnation, these spaghetti-like strands are chopped to normal pellet length—about .10 in. (2.5 mm). Note that, in contrast to the whisker-length reinforce-

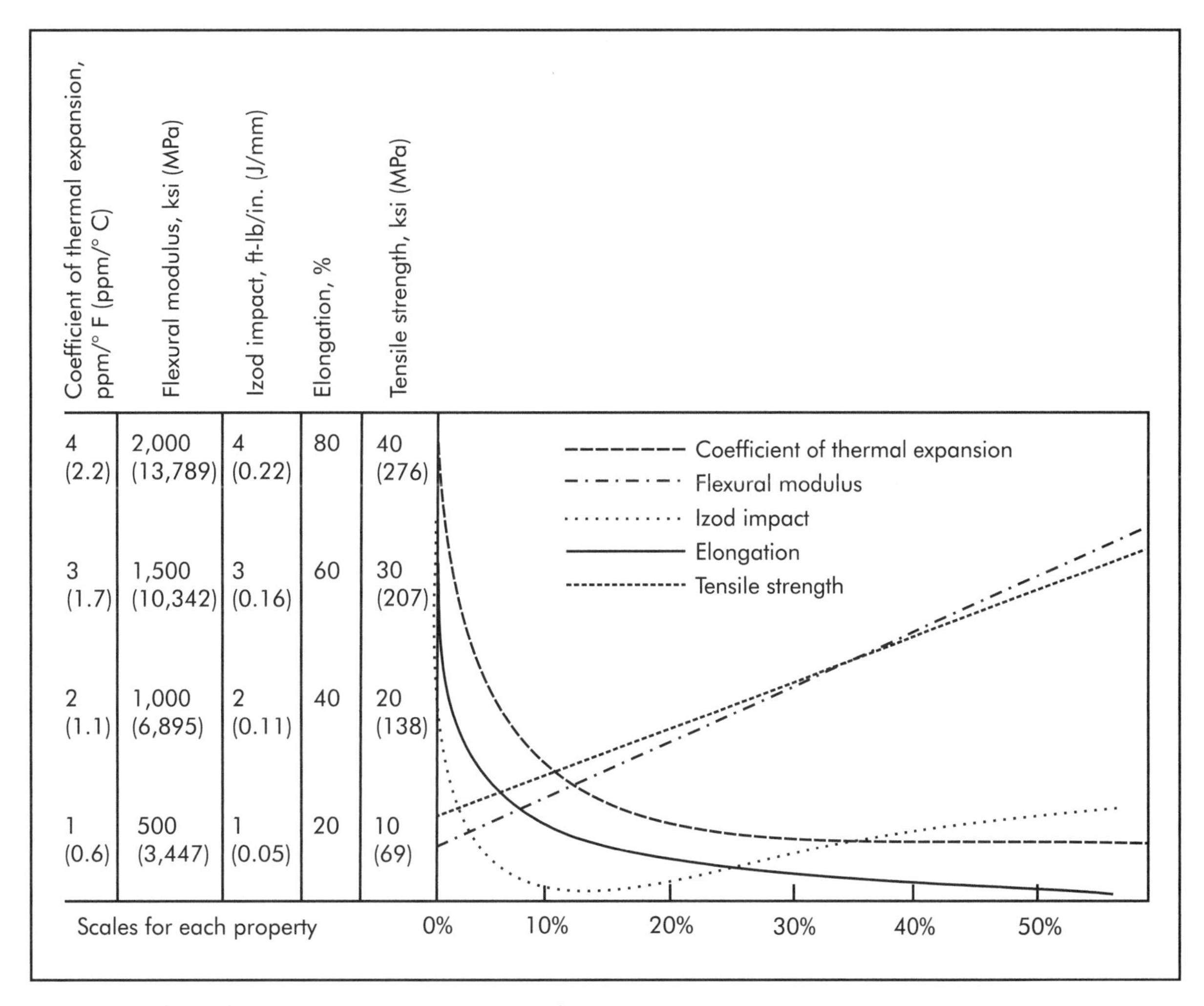

Figure 6-1. Effects of fiber content on the properties of nylon.

ments of common thermoplastic engineering resins, the long-fiber reinforcements will be the entire length of the pellet. The lengths of the long fibers are about the maximum that can be used in typical thermoplastic processes. Even with great care, many of the long fibers are broken during processing, but the net result is still longer fibers than the whiskers and, therefore, improved mechanical properties.

Even greater fiber lengths will generally result in even greater improvements in properties, although the rate of increase in properties with length diminishes. For most resins, the optimal length is about 3 in. (7.6 cm). Of course, when fiber lengths are longer than those used in pellets, different manufacturing processes must be used. The methods used to process thermoplastic resins where the fibers are long, often the entire length of the finished part, will be discussed in the chapter in this book on thermoplastic processing.

One resin has shown to have a strong entry into the marketplace—continuous reinforced polypropylene. Sheets of polypropylene and fiberglass are sold as rolls. In general, the reinforcement fibers have been fully wetted with the polypropylene, although one manufacturer sells fiberglass mat and fabrics in which the fiberglass strands are intermixed with polypropylene strands. Consolidation and wetting of the fibers takes place when the product is molded. The key properties for these polypropylene-fiberglass composites are high impact strength, low cost, and ease of molding. The markets for the polypropylene-fiberglass composites are as metal replacements (substituting for aluminum in ladders) and as replacements for bulk molding compound (BMC) or sheet molding compound (SMC) in automotive panels (such as non-dent door panels).

Although not shown in Table 6-3, other mechanical and physical properties are often affected by the addition of fiber reinforcements to engineering thermoplastics. In general, *wear resistance* is dramatically improved by the addition of fiber reinforcements. This occurs because the fibers are stiff and hard—highly desirable properties, which improve upon the wear resistance of resins that are often inherently good in wear resistance anyway. *Shear strength*, a property important when mechanical fasteners, such as rivets, are used to attach composites, increases with fiber content (see Table 6-4) suggesting a stiffening and internal strengthening of the material because of the fibers.

In spite of the significant improvements in the properties of engineering thermoplastic resins when fiber reinforcements are added, some serious limitations in properties still exist. The most important limitations are related to thermal stability, solvent sensitivity, and toughness. These limitations are directly addressed in the next section.

HIGH-PERFORMANCE THERMOPLASTIC COMPOSITES

Whereas the engineering thermoplastic composites could easily be treated as a general group because of the overall familiarity of most readers with the neat resins and the effects of the addition of reinforcements, the high-performance thermoplastic composites will be discussed by major family groups. These materials are less familiar because of their limited and specialized use.

Table 6-4. Coefficient of thermal expansion (CTE), ppm/° F (ppm/° C) for various thermoplastic composites at different fiber content levels.

Resin	Fiberglass, % by Weight		
	0	20	35
Acetal	4.5 (2.5)	3.5 (1.9)	1.9 (1.0)
Nylon	5.5 (3.1)	2.1 (1.2)	1.1 (0.6)
Polycarbonate	3.9 (0.50)	1.8 (1.0)	1.2 (0.7)
Impact polystyrene	2.2 (1.2)	2.2 (1.2)	1.7 (0.9)
Acrylonitrile-butadiene-styrene (ABS)	4.1 (2.3)	2.0 (1.2)	1.6 (0.9)
Polyphenyleneoxide	3.0 (1.6)	2.0 (1.2)	1.0 (0.6)
Polyethylene terephthalate (PET)	5.3 (2.9)	3.3 (1.8)	2.4 (1.3)

High-performance thermoplastics are occasionally used in non-reinforced applications where their high strength, high stiffness, solvent resistance, low flammability, and thermal stability are key properties. The remainder of this chapter will look at the major families of high-performance thermoplastic resins—comparing their advantages and limitations. In most cases, the structural formula will be shown as a key to understanding the basis of the overall properties of the resin.

A general characteristic common to all high-performance thermoplastics is a high degree of **aromaticity**, which adds stiffness, strength, flame-retardant properties, and high thermal stability to the backbone of the polymer. Therefore, these same characteristics are imparted to the resin in composite applications. In many cases, the stiffness of the backbone is modified (either by making it stiffer or more flexible) by the addition of some other molecular group or groups that link the aromatic sections together along the backbone. These other groups will impart unique characteristics to the molecules. Therefore, they are the focus of the comparisons between the various types of high-performance thermoplastic resins.

Thermoplastic Polyimides and Related Polymers

Polyimides were introduced in a previous chapter of this book on specialty thermosets. Thermoset polyimides are seen to have the highest thermal resistance of all the thermoset resins in common use. This excellent thermal stability is also characteristic of thermoplastic (that is, non-crosslinked) polyimides. The polyimides, both thermoset and thermoplastic, also have good mechanical properties (especially high stiffness), excellent electrical properties (high resistivity), excellent solvent resistance (no known reaction to organic solvents for some), and low flammability. A typical thermoplastic polyimide is pictured in Figure 6-2.

The high degree of aromaticity of the polyimide molecule and the accompanying stiffness of the planar imide rings leads to high stiffness in the backbone of the molecule and, therefore, the high mechanical stiffness, thermal stability, and other desirable properties mentioned. However, these same characteristics lead to high melting points that are near the decomposition temperatures, thus making thermal processing difficult. The inherent resistance of these materials

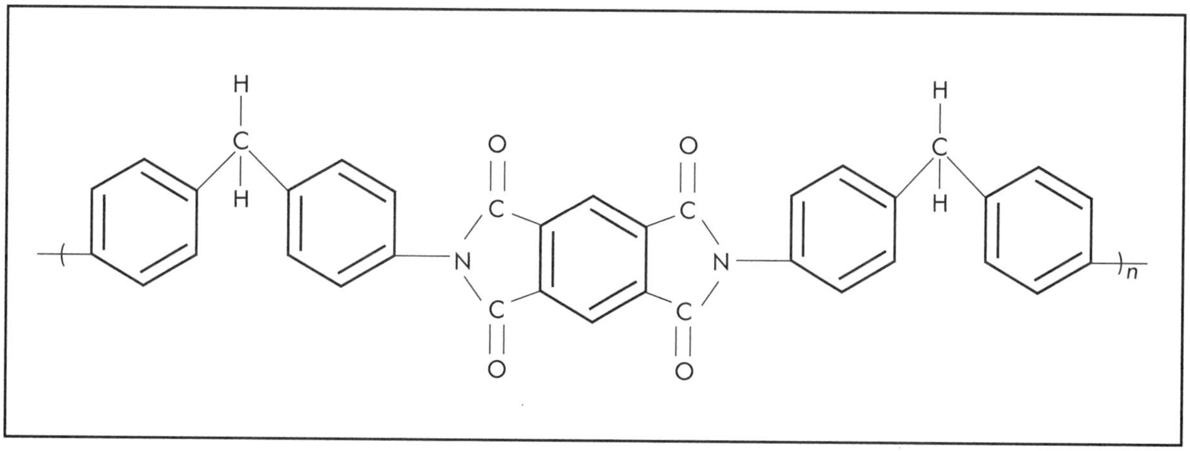

Figure 6-2. Thermoplastic polyimide.

to solvation by organic solvents also makes solvent processing difficult. Therefore, thermoplastic polyimides are difficult to process by either of the common methods that bring a thermoplastic material into the liquid form—melting or solvation.

A closely related thermoplastic polyimide has similar processing problems as discussed above but, because of its excellent electrical resistance, has found applications as an insulating material and as a substrate for circuit boards. The excellent ultraviolet resistance and mechanical strength of this material have led to its use in space for solar array panels. The material, in its film form is Kapton®, a DuPont trade name, and in its formable plastic/composite form is Vespel®, also a trade name of DuPont. The plastic/composite resin is used extensively for small washers, inserts, rods, and in other applications where the combination of thermal stability and solvent resistance is especially important. The plastics/composites are formed by **sintering**, a process wherein a powdered material (resin, metal, or ceramic) is heated to a temperature just *below its melting point* and held for some period. This causes the powdered particles to join together and form a solid mass. The sintered thermoplastic materials, often reinforced with short fibers or fillers, are formed into standard shapes and then machined to specific sizes and shapes. For high-volume applications, they also can be sintered directly into the shape desired.

The difficulty in processing thermoplastic polyimides has led to several attempts to modify them. NASA Langley Research Center has developed a material called LARC-TPI that has greater flexibility in the backbone, thus lowering the melting point sufficiently to allow melt processing of the material. LARC-TPI has been successfully used as a matrix for high-performance composites but making thick parts has proven to be difficult. However, LARC-TPI is used extensively as a high-performance structural adhesive and as a film.

Another important modification of the basic thermoplastic polyimide is the combination of the imide with an amide group to form **polyamide-imide (PAI)**. As with other modified polyimides, the PAI structure has an imide group bonded to an aromatic group, which results in a stiff and structurally rigid backbone. To give some flexibility and toughness, this aromatic group is bonded to an amide, which acts as a bridge to another aromatic group. The amide gives greater flexibility than the basic polyimide structure where the aromatic groups are joined directly to imide rings without any intermediate atoms.

One brand name of PAI is Torlon®, a registered trade name of Solvay (formerly owned by BP Amoco). This resin is the highest performing, **melt-processible resin**. It is able to maintain most structural properties up to 500° F (260° C) for extended periods (T_g = 537° F [281° C]). Parts and shapes can be made by extrusion, injection molding, and compression molding. These can be reinforced or non-reinforced. As with other polyimide-related materials, standard shapes are usually made and then machined to particular part specifications.

Polyetherimide (PEI) is another polymer in which an imide/aromatic group is separated from another aromatic group by a linking group. In this case, the linking group is an ether. A common brand name for PEI is Ultem® (GE Corporation). As with other modified polyimides, PEI has excellent thermal stability, resistance to most solvents, inherent low flammability and smoke emission, and low electrical conductivity. Several medical applications use PEI because parts can be made by injection molding, but will still have good thermal stability under high-temperature sterilization. PEI is often less expensive than the other high-performance polyimide materials.

PEI is available in numerous grades, both non-reinforced and reinforced. Some of the reinforced grades have as high as 40% fiberglass, aramid, or carbon fiber content. In some applications where wear resistance is needed, up to 10% polytetrafluoroethylene (PTFE) (Teflon®, a DuPont trade name), silicone, or molybdenum disulfide ("moly") can be added, thus imparting slipperiness to the resin's natural abrasion resistance. Typical uses for the wear-resistant grades include: gears, bearings, slides, cams, and pump components.

It is possible to achieve the backbone stiffness and other properties characteristic of the polyimide family of materials without actually employing the imide ring. The most common example of such a molecule is **polybenzimidazole (PBI)**. The PBI molecule is structurally similar to a polyimide except that the five-member ring is altered. However, the PBI ring structure is still flat and stiff, thus giving it many of the same properties as polyimides.

The PBI resin is nonflammable in air and emits little or no smoke up to a temperature of 1,000° F (538° C). PBI, in common with most polyimides, also offers excellent resistance to acids, bases, organic solvents, and fuels. It can be made into *fibers* with excellent fire-retardant properties. PBI fibers are also used in high-temperature protective gloves, firefighter clothing, and other protective gear; blended with other heat- and fire-resistant materials; and even coated with aluminum for enhanced thermal resistance from its reflectivity.

PBI resin can be molded, but with some difficulty, as is the case with thermoplastic polyimides. Typically, a proprietary molding process is used by a contract molder who makes standard shapes and then machines the parts to the particular shapes required. Typical applications are seals, o-rings, and valves. The outstanding properties of the molded parts include the highest compressive strength of any unfilled resin, excellent tensile and flexural properties, good fatigue resistance, excellent hardness, low friction, a relatively low coefficient of thermal expansion (CTE), high electrical resistivity, and good thermal stability. PBI also has the advantage that most other polymers will not stick to it, even when the other polymer is melted. Hence, PBI is often used as a bearing or seal in machines used to melt other thermoplastics. The T_g for PBI is over 800° F (427° C). Thus temperatures up to about 800° F (427° C) can be withstood for long periods of time and, for short bursts, the material can withstand up to 1,400° F (760° C).

In contrast to most thermoplastic polyimides, PBI is available in a solution. It can be used for film casting, encapsulation, or coating. It is hydrophilic, which gives it some improved comfort in fabrics but also increases its sensitivity to some polar solvents in comparison to thermoplastic polyimides. PBI can be doped with acid, thus increasing its hydrophilic character and allowing it to be used for membranes for chemical separations in aqueous solutions.

Polyetheretherketone (PEEK) and Related Polymers

Perhaps the most widely used of the high-performance thermoplastic resins is polyetheretherketone (PEEK), whose structure is shown in Figure 6-3. In this polymer several aromatic groups are separated by various linkages. The name of a polymer is derived from its linkages and, in the case of PEEK, the linkages are ether, ether, and ketone. Related polymers are PEK, with one ether linkage and one ketone linkage, and PEKK, which has one ether and two ketone linkages. The properties of all these materials are similar, so all will be discussed together, generally with specific reference to PEEK, the most common of the group. PEEK and its related polymers are sometimes called polyaryletherketones, where "aryl" indicates

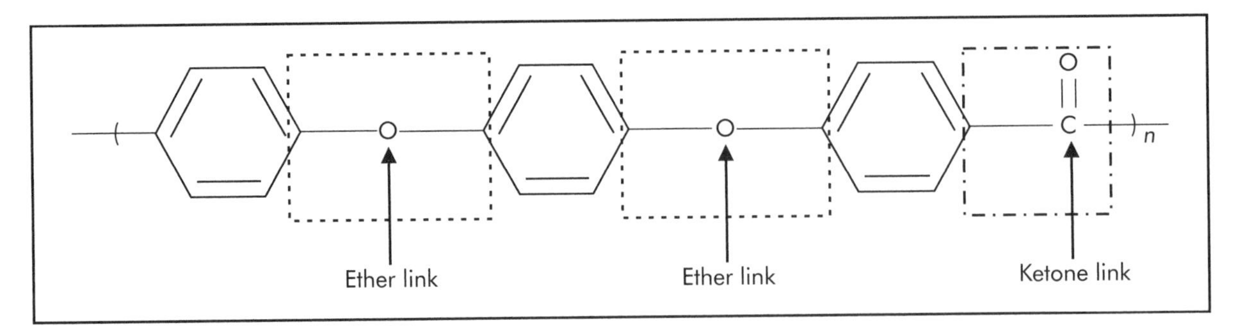

Figure 6-3. Polyetheretherketone (PEEK).

a high amount of aromatic content. Since PEEK and other related polymers are highly aromatic, they have increased T_g, which gives good thermal stability. As with the polyimide-related thermoplastics, PEEK has low flammability, low electrical conductivity, high solvent resistance, high strength, and excellent toughness.

Another polymer that is of the same general family is polyphenylene ether (PPE), which is also called polyaryl ether. The most common brand name for PPE is Noryl®, a trade name of GE, who offers the material in a 50:50 blend with polystyrene.

The regularity of groups along the polymer backbone leads to a high degree of crystallinity. Thus PEEK is often referred to as a **semi-crystalline** material. Structural rigidity, high strength, and high modulus come from the formation of these crystalline regions. The crystal structures also improve the solvent resistance of PEEK since the solvent molecules must penetrate the tightly bonded crystalline structure to have any effect on the polymer.

PEEK was one of the first reinforced high-performance thermoplastics to be qualified for use in aerospace applications. As a result, much more service data is available for reinforced PEEK than for most of the other high-performance thermoplastics. Also, the familiarity of aerospace engineers with the performance of PEEK has led to its use in several large aerospace parts, thus increasing the total volume of PEEK resin being used over most of the other high-performance thermoplastics.

PEEK is often used in composite applications because of its excellent blend of thermal stability and ease of processing characteristics. The thermal stability of PEEK is only slightly less then the polyimide and polyarylimides, but it is far easier to process. Furthermore, PEEK has excellent resistance to solvents, ultraviolet light, stress cracking, and low flammability.

Polysulfones and Related Polymers

Polysulfones and related polymers are highly aromatic molecules that use sulphur or sulphur-containing groups as links between the aromatic groups. The structure of the most common of this group, polysulfone (PSU), is given in Figure 6-4a. Another commonly used polymer in this group is polyphenylene sulfide (PPS), which is shown in Figure 6-4b. Possibly the most important feature of the sulphur-containing groups is their relative ease in melt processing compared to the other high-performance thermoplastics.

The overall properties of the other members of this sulphur-containing group of thermoplastic polymers are similar to PSU. Some of the more common brand names are: polysulfone—Udel® (Amoco);

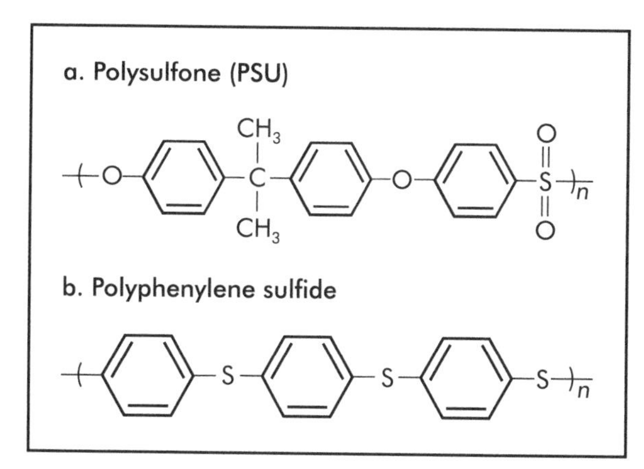

Figure 6-4. Sulphur-containing advanced thermoplastics.

polyarylsulfone—Astrel® (3M); polyethersulfone—Radel® (Solvay); and polyphenylene sulfide—Ryton® (Chevron Phillips).

Liquid Crystal Polymers

Liquid crystal polymers generally have long repeating units, which lead to a high degree of orientation or crystallinity even in the liquid state. When molded and solidified, these materials tend to be highly directional. A representation showing the three-dimensional nature of the crystalline and amorphous areas is given in Figure 6-5.

One of the unique properties of liquid crystal polymers (LCPs) is they rapidly change their viscosities with temperature. At certain temperatures, the liquid state changes from having areas of structure to being totally random with a corresponding drop in viscosity. This property allows LCPs to be used as lubricants of high or low viscosity, depending on the temperature.

In the solid state, LCP materials have some self-reinforcing characteristics. The high orientation of the linear units is much like the reinforcement obtained from fibers. As a result, the strength of LCP polymers, even without external fibers added, is high. One polymer of this type is Kevlar® by DuPont. It is used extensively as a fiber in composites and other applications where strength and toughness are important. This material will be discussed further in the chapter on fibers.

Several LCP matrix materials have been commercialized. The most common of these are: Vectra® (by Hoechst) and Xydar® (by Solvay). They are melt-processible with ordered sections in the melt state and high orientation in the solid state. The highly aromatic nature of these materials leads to excellent strength and stiffness, and good thermal stability. Because of their orientation, the materials have low thermal expansion and low shrinkage. Generally, the LCPs have good stress-crack resistance, solvent resistance, electrical resistance, and flame-retardant properties.

Fluoropolymers

The polymers that have a backbone of carbon to which fluorine atoms are pendant, sometimes called fluorocarbons or fluoropolymers, are unique among all the polymer materials. These unique properties

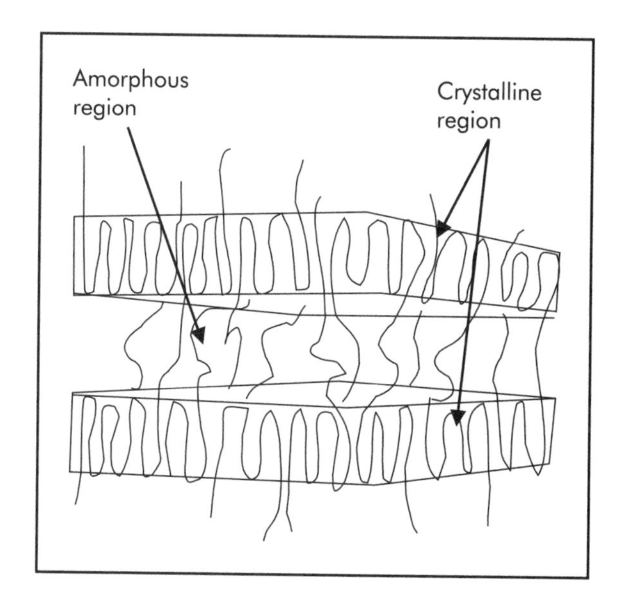

Figure 6-5. Three-dimensional representation of a liquid crystalline polymer.

result from the strong attraction of the fluorine to the electrons in the fluorine-carbon bond. This causes the bond to be short and strong, thus giving high density to the fluorocarbons. Furthermore, the electrons are held so tightly by the fluorine that electron activity, such as attraction or bonding with any other atom, is low. This results in a high resistance to chemical attack and generally poor bonding of fluorocarbons to any other material. The poor bonding of these materials is, of course, detrimental to their use as a matrix in composites. However, the addition of some fluorocarbon material to other matrices has given some advantages of solvent resistance, electrical resistance, flame-retardant properties, and thermal stability without major sacrifices in matrix-fiber bonding.

The first of the fluoropolymers made was polytetrafluoroethylene (PTFE), which is most commonly known by the trade name Teflon® (DuPont Company). Later, because PTFE was nearly impossible to mold using conventional melt-processing methods, a derivative molecule, called polyhexafluoropropylene (PHFP), was produced. In PHFP, one of the fluorine molecules is substituted with a carbon and three fluorine molecules, thus giving slightly higher carbon branching and flexibility to the polymer. A mixture of PTFE and PHFP, called fluorinated ethylene propylene (FEP), has a good blend of properties with ease of processing.

Later, fluorinated polymers containing less fluorine and retaining some of the hydrogen molecules on the carbons were made. The most common of these materials has two fluorine molecules and two hydrogen molecules on the carbons in the basic unit. This material is called polyvinylidene fluoride (PVDF or PVF_2). The most common brand name is Kynar® (Atofina). It is used as an additive to improve chemical resistance in other matrix materials. Like the other fluoropolymers, PVDF has excellent flame-retardant properties, electrical resistance, and low friction properties.

CASE STUDY 6-1

Conversion of the 737 Cargo Hold Smoke Detector to Thermoplastic Composites

After the crash of Value Jet 737 into a Florida swamp in 1996, the Federal Aviation Administration (FAA) and the Boeing Company agreed that all 737 passenger aircraft should have a cargo hold smoke detection system.

Originally, designers housed the smoke detectors in a four-ply, glass-reinforced, thermoset polyester, pan-shaped shell. With airplane production rates reaching 28 per month and with 10 parts per airplane, Boeing's Spokane plant personnel soon found themselves overwhelmed with orders. In addition, spares orders skyrocketed since the current fleet had to be retrofitted.

The original process made the pan using hand lay-up on an aluminum tool that was vacuum bagged and oven cured for 4 hours. The part was then trimmed on a three-axis router and assembled by hand to finish the part. Eventually, eleven lay-up molds had to be made and each tool had to be turned twice a day to keep up with production needs.

In 1999, a 50-ton thermoplastic press was installed and production started using Ten Cate Advance Composites' Cetex® material. This material is a woven, glass-reinforced Ultem® (polyetherimide) resin product. The new material had to comply with all the FAA's requirements for smoke, toxicity, and vertical burn. Also, since the part was to be installed in the ceiling of the cargo hold, the material had to pass the Class E cargo hold burn test, which requires an 1,800° F (982° C) flame to be applied 4 in. (10 cm) from the part for 5 minutes without having a burn through. The part passed all these tests and, moreover, its weight decreased by 5% while its strength and durability increased.

Because of the ductility of the material, the part could be trimmed using a stamping press, which provided additional cost savings. In the end, production costs were reduced by 80%, saving Boeing over $500,000.

SUMMARY

For years, injection molders have been using short fibers (whisker length) as reinforcements in thermoplastic resins. The properties of these materials are excellent for some applications with significant increases possible in tensile strength and flexural stiffness, sometimes achieving a 300% increase over the non-reinforced thermoplastics. Typical fiber contents range up to 30%, although 40% and even 50% fiber contents are available for some resins.

In every case where the reinforcements are added to the neat fibers, the tensile strength and flexural modulus increase with fiber content. This is expected because the fibers, which are stiff and strong, carry much of the load. The Izod impact results are not as clearly understood as the strength and modulus properties. Impact measures the toughness of the material, and toughness is a combination of both strength and elongation. Hence, if the elongation and strength are both changing and often changing in different directions, the toughness can be difficult to predict. However, the usual pattern for thermoplastics would be for the Izod toughness to decrease with fiber content since elongation will probably decrease faster than the strength will increase.

Some questions arise immediately when considering the relative values of reinforced thermoplastics and thermosets. First, if thermoplastics can have such good mechanical properties, why do thermosets dominate the reinforced plastics market? Another question then arises. Where do thermoplastics fit and where are thermosets more appropriate? Also, is there likely to be strong competition between the two materials?

One answer is simple—the higher cost of the thermoplastic materials effectively limits the scope of their use. Most of the thermoplastics that would be stiff and strong enough to compete with thermosets cost considerably more than basic unsaturated polyesters. Not only are the thermoplastic resins themselves more expensive, but the cost of including the fiberglass adds to the price of the reinforced thermoplastic, whereas with thermosets, the molder is able to add the fiberglass as part of the molding process.

If part performance requirements exceed those easily obtained using epoxies, the resin costs may actually favor thermoplastics even though they are considered high-performance or advanced materials. For example, many of the aromatic thermoplastic resins are inherently flame retardant, have low smoke emissions, and low heat release. They also have service temperatures higher than most epoxies. A good example of the value of these thermoplastic properties is in cargo container parts for aircraft. The excellent flammability, smoke, and heat properties combined with the natural toughness of high-performance thermoplastics makes them ideal for this application.

Another major advantage of high-performance thermoplastics is their general resistance to most solvents. For instance, some of these materials are resistant to airline hydraulic fluids (Skydrol) and have found application where contamination from that fluid is likely.

Some aerospace applications have been converted from aluminum to thermoset parts because of the weight savings that can be gained from using thermal welding instead of rivets. There are also lower labor costs, which also result from use of the welding technology.

Thermoplastics seem to have special advantages for the small part that can be injection molded from short, fiber-reinforced engineering thermoplastics or for the high-performance part that needs special properties such as flammability, solvent resistance, or extra weight savings. Considering the advantages in toughness, cycle time, and finishing (drilling, cutting, etc.), the future for thermoplastics appears to be bright.

Finally, and potentially the most important implication for long-term views, thermoplastic materials can be recycled into applications and retain the same value as the original resin. Thermosets cannot be remelted and so they can only be recycled into filler materials. Several countries have mandated recycling of all materials, which strongly favors the use of thermoplastic composites. In countries where such laws have been passed, the use of thermoplastic composites is rising at twice the rate of thermosets.

LABORATORY EXPERIMENT 6-1

Toughness of Thermoplastics versus Thermosets

Objective: Compare the toughness of thermoplastic composites with thermosets and examine the effects of fiber length.

Procedure:

1. Obtain samples of flat panels (at least 8 × 8 in. [20.3 × 20.3 cm]) of the following materials: thermoset reinforced plastics (polyester, vinyl ester, or epoxy with glass fibers); thermoplastic resin with long, continuous, fiberglass reinforcement; and thermoplastic resin with short-fiber reinforcement. All the samples must have the same fiber content (%).
2. Cut samples for impact tests. These tests can be Izod, Charpy, or falling dart impact, depending on the type of testing apparatus available. The sample sizes and other testing criteria should be obtained from the American Society for Testing and Materials (ASTM). At least three samples of each resin should be tested.
3. After testing, compare the toughness results of thermoplastics versus thermosets and the effects of fiber length. Also compare the results with toughness data for the neat resins (obtained from the resin manufacturer) to assess the effect of reinforcement content.

QUESTIONS

1. What is the key difference between the structure of thermoplastics versus thermosets and what property differences will be caused by this structural difference?
2. Identify and explain three differences between engineering thermoplastics and advanced thermoplastics.
3. Give three advantages of an advanced thermoplastic composite over an advanced thermoset composite. Give three advantages of the advanced thermoset over the advanced thermoplastic.
4. Why do engineering thermoplastics usually contain short fibers?
5. What are four key properties of composites that depend strongly on the nature of the resin? Indicate whether thermosets or thermoplastics would generally have the advantage in each of the four properties.
6. Describe the differences in the nature of recycling thermosets versus thermoplastics.

BIBLIOGRAPHY

Society of Manufacturing Engineers (SME). 2005. "Composite Materials" DVD from the Composites Manufacturing video series. Dearborn, MI: Society of Manufacturing Engineers, www.sme.org/cmvs.

Strong, A. Brent. 1993. *High Performance and Engineering Thermoplastic Composites*. Lancaster, PA: Technomic Publishing Co.

Strong, A. Brent. 1995. *Fundamentals of Polymer Resins for Composite Manufacturing*. Arlington, VA: Composites Fabricators Association.

Strong, A. Brent. 2006. *Plastics: Materials and Processing*, 3rd Ed. Upper Saddle River, NJ: Prentice-Hall, Inc.

7

Ceramic and Metal Matrix Composites

CHAPTER OVERVIEW

This chapter examines the following:

- Non-polymeric matrix materials
- Properties and uses of ceramic matrix composites
- Manufacturing of ceramic matrix composites
- Properties and uses of metal matrix composites
- Manufacturing of metal matrix composites

NON-POLYMERIC MATRIX COMPOSITES

The past several chapters have discussed the most prominent polymeric matrix materials for composites. The differences between these polymeric materials have been examined and their various advantages and disadvantages pointed out. Standing back now and taking a look at all the polymeric matrix materials as a group, some common characteristics can be seen when they are compared to non-polymeric matrix materials (ceramics and metals), which will be discussed in this chapter.

In comparison to ceramics and metals, the polymer matrices have *advantages* by being easier to mold, less expensive, lighter weight, and can wet out the reinforcements more easily. The polymer matrices' *disadvantages* are lower thermal stability, lower stiffness, and lower hardness, which affect the amount of frictional wear. These overall differences, and others that will be discussed later, can be understood, at least preliminarily, by examining the differences in the types of bonds that exist in ceramics, metals, and polymers.

There are different ways in which electrons behave in the bonds formed between the elements (atoms) of the periodic table. The elements can be divided into two major groups—metals on the left and non-metals on the right. Some elements along the intersection of the periodic table can behave as both metals and non-metals, depending on the circumstances, but that technicality will be ignored for this discussion.

The separation of the elements into metals and non-metals is based on the tendency of the element to give or take one or more electrons when forming bonds with other elements. The metals give electrons and the non-metals take electrons. All the elements are electrically neutral before bonding with other elements; that is, the number of protons in the nucleus equals the number of electrons surrounding the nucleus. So, when the metal atoms give up electrons they become positively charged. The non-metals become negatively charged. Any charged element, whether positive or negative, is called an **ion**.

There are three possible ways that the elements, metals and non-metals, can combine. First, a metal can combine with a non-metal. Second, a metal can combine with another metal. Third, a non-metal can combine with a non-metal. These combinations result in three different bonds.

Ionic Bond—Metal Atom with Non-metal Atom

When a metal atom comes near a non-metal atom, one or more electrons will move from the metal to the non-metal because a more stable state is formed in both atoms when the electrons transfer. This causes the metal to become positively charged and the non-metal to become negatively charged. These positive and negative ions will then attract each other strongly, much like north and south poles of magnets attract. The ions can be said to bond together. This type of bond is called an **ionic bond**. Most of the time many positive and negative ions join together in a three-dimensional lattice of alternating ions, where each of the positive ions is surrounded and attracted by many negative ions and vice versa. The lattice structure is usually called a *crystal*.

If the entire structure is a continuous lattice structure without boundaries, it is called a *single crystal*. Most crystal structures are not that uniform. There are crystals that are finite in size and that meet other crystals, which usually have different orientations. These materials are called *polycrystalline*. It is also possible to form ionic-bonded materials from atoms that do not form crystals. These latter structures form a rigid, disordered (amorphous) network that is termed a *glass*. Regular window glass is such a material.

Ionic bonds are the most important bonds in *ceramic materials*. These bonds are strong and therefore require much energy to break. That high bond energy is responsible for the high melting point of ceramics. Further, if the ceramic is placed under force, the layers in the crystal will resist movement relative to each other because any movement would result in the positive ions coming near other positive ions—a highly unfavorable situation since like charges repel. Hence, the crystal structure resists movement. This is demonstrated by the brittleness of ceramics. The resistance to atomic movement and the strong bonds between atoms also explains the hardness of ceramics and their wear resistance. Hence, the major properties of ceramic materials can be ascertained from knowledge of the characteristics of ionic bonds.

Metallic Bond—Metal Atom with Another Metal Atom

When several metal atoms come into close proximity, they all want to give up one or more electrons to form a more stable state. This is possible because the donated electrons can move freely in the spaces between the positively charged metal ions. The freely moving electrons cancel repulsive charges between the atoms and, therefore, bind the ions together. The highly mobile group of electrons is called the **sea of electrons**. When positive ions are held together by a sea of electrons, the bonding is called a **metallic bond**. Metals and most metal alloys are characterized by this type of bonding.

The metal ions arrange themselves into a three-dimensional lattice structure that has some similarity to the lattice structure of ceramics; both are called crystals. However, the metal structure is not as brittle as the ceramic because the metal layers can move more easily, facilitated somewhat by the sea of electrons. When the layers of the metal crystal move relative to each other, the positive-positive repulsions of the metal ions are cancelled to some degree by the sea of electrons that moves between the ions. (The sea of electrons is like a lubricant that forms a slick coating between moving parts.) Therefore, metals are more ductile than ceramics. This ductility gives metal atoms greater ability to move when impacted, thereby absorbing impact energy. Hence, metals are less easily broken on impact and tougher than ceramics. The greater mobility of metal ions, which is facilitated by the sea of electrons, means that less energy is required to separate the individual ions in the lattice. Therefore, the melting points of metals are generally lower than the melting points of ceramics.

The mobility of the electrons can be actually seen in metals. They reflect light and give metals their shiny appearance. The electrons' mobility also can be noted by the high electrical and high thermal conductivity of metals. The highly mobile electrons are carriers of the electrical charge and heat energy through the metal. In ceramics, the electrons are held strongly by the non-metal ions. So, they do not conduct electricity or thermal energy well (except when in solutions where the ions have been broken out of the crystal lattice and the ions themselves carry the electrical charge).

Covalent Bond—Non-metal Atom with Another Non-metal Atom

The third possibility for bonding is when a non-metal atom comes into close proximity with another non-metal atom. Both of the non-metals want to take electrons. This is done by sharing some of their outer-most electrons to achieve stable states for both atoms. The type of bond in which atoms are shared is called a **covalent bond**.

Many common small molecules are held together by covalent bonds. Some examples are water, carbon dioxide, methane (natural gas), and ammonia. These molecules are mostly gases and liquids and, therefore, not important structural materials. Some non-metal atoms, especially carbon, have the ability to bond with several other atoms to form long chains. These chains are called **polymers**. When a polymer gets to be about 20 units long, it is a solid and can begin to have some structural properties. As has been discussed in previous chapters, the polymeric matrix materials used in composites are long polymers, sometimes made of thousands of units.

Because polymer units consist of molecules that are long chains, the atoms are not able to pack as individual atoms or ions as they do in ceramics and metals. However, under certain conditions, the polymer molecules can align together and achieve some increased stability by packing together in a molecular structure similar in some ways to the crystal structures of the ceramics and the metals. When this occurs, it can be said that the polymer is *crystalline* or, more accurately, *semi-crystalline*, since in polymers, this packing never involves all of the molecules so there are always some areas that are non-crystalline. These areas are called **amorphous regions**. Some polymers form almost no crystalline regions and these are commonly called *amorphous polymers*. (The concepts of crystalline and amorphous were previously discussed in Chapter Two.)

The overall linear nature of the *uncross-linked* polymer molecules gives polymers some distinct differences from ceramics and metals. In amorphous regions, there are no primary forces between the atoms in adjacent polymers. Even in the crystalline regions, the forces attracting the molecules together are small when compared with the crystalline forces of ceramics and metals. Polymers will entangle, a characteristic of long chains, and the energy for untangling the chains is much less than the energy of breaking the ionic and metallic bonds. Therefore, the energy required to separate polymeric molecules from each other is relatively small compared to the energies required to break ceramic or metal crystals. As a result, the melting points of linear (thermoplastic) polymers are much lower than the melting points of ceramics or metals. Remember that melting a polymer does not mean that the atoms are separated out of the polymer chain; it merely means that the chains can move independent of other chains. The separation of the atoms from each other out of a chain is the process of decomposition, and that requires much higher temperatures than polymeric melting (although still not as high as many ceramic and metal melting points).

The relative openness of the polymer structure also means that polymer atoms can move quite easily and, therefore, absorb impact energy. Hence, polymers are tougher than either metals or ceramics. This movement, however, means that polymers are not as strong or as stiff as ceramics and metals. As will be seen in much more detail later in this text, the inherently lower strength and stiffness of polymers is one of the main reasons that they are reinforced with strong, stiff fibers to make composite materials. It is, therefore, obvious that polymers are made into composite materials to achieve better strength and stiffness. But if ceramics and metals are inherently strong and stiff, why are they reinforced to make composite structures? The answer to that question and the ways that these materials are reinforced will constitute the remainder of this chapter.

PROPERTIES AND USES OF CERAMIC MATRIX COMPOSITES

The most inviting applications for ceramic matrix composites (CMCs) are where polymer matrix composites are least likely to be used—in high-temperature environments and where high stiffness is required. The high-temperature capabilities of ceramic matrix composites are summarized in Table 7-1. With the exception of carbon-matrix composites, which must be protected from oxidation by ceramic coatings, ceramic matrix composites are clearly the best choice for high-temperature applications.

The problems of high temperatures can be summarized in the following generalized laws.

Table 7-1. Temperature limits for composites with various matrices.

Composite Type	Approximate Maximum Use Temperature, ° F (° C)
Organic matrix	800 (427)
Metal matrix	1,800 (982)
Ceramic matrix	3,000 (1,649)
Carbon matrix (oxidation protected)	4,000 (2,204)

- Everything reacts with everything at high temperatures.
- Everything reacts faster at high temperatures.
- The reaction products may be anything.

While these rules are obviously an oversimplification and meant to give some humor, their intent is clear: withstanding high temperatures is extremely difficult for materials. Much effort has been invested to develop products that can withstand the temperatures demanded by our high-tech world.

The aerospace industry gives some examples of the increasing need for high-temperature applications. Over the last several decades, the temperature requirements of materials have gone from a little over normal temperature on a hot summer day to approaching 3,000° F (1,649° C) as airplanes have increased in sophistication. When time at temperature is considered, the requirements for performance at high temperatures are even more dramatic. In some applications, such as the space shuttle, the technology has not fully met the requirement and, as is sadly evident, some mishaps have occurred. When a material's technology has not reached the required levels of performance for temperature and time, time has to be carefully controlled so that the material's ability to perform is not severely compromised.

As mentioned, non-reinforced ceramics inherently possess the high bond energies that allow them to be used in high-temperature/high-stiffness applications. However, inevitably, the brittleness and lack of impact toughness of these non-reinforced ceramic materials seem to limit the applications in which they can be used. Simply put, the structural applications for non-reinforced ceramic materials are severely limited.

Fortunately, it has been found that by adding a reinforcing material to a ceramic, the toughness of the material can be substantially improved. Moreover, the material's temperature capability is not adversely affected and, in many cases, is actually improved. These reinforced ceramics are, of course, ceramic matrix composites.

Toughening Mechanisms

To understand the mechanisms by which the reinforcements toughen the matrix, this section will examine the nature of toughness in detail. In previous chapters it has been pointed out that toughness is related to the ability of a material to absorb the energy created when the material is impacted. By absorbing impact energy, the material resists local concentrations of energy that are sufficient to initiate and propagate cracks. Therefore, **toughness** also can be defined as the ability of a material to resist the initiation and propagation of cracks.

The following toughness mechanisms in ceramic matrix materials have been proposed, either acting alone or in concert.

- Usually, if cracks form from local stress concentrations that exceed local strengths, they naturally propagate since the energy that encourages this behavior is often higher than the energy absorbed in that propagation. However, if the growing crack encounters a boundary of a strong and rigid material, such as a reinforcement, it will be unable to simply move through the reinforcement material. Instead, the crack is deflected and runs along the surface of the reinforcement where it must break interface bonds between the reinforcement and the matrix and, possibly, induce pullout of the reinforcement from the matrix. This deflection, breaking of the bonds, and pullout require energy (absorb energy). That absorbed energy is often sufficient to stop the growth of the crack. In other words, the reinforcement has arrested the growth of the crack. Some call this phenomenon **crack pinning** or **crack deflection**.
- Another method used in composites to stop crack growth is by **load transfer** to the reinforcement. This occurs when the reinforcement has a higher strain to failure than the matrix. In this case, when the crack encounters the fiber, the load is transferred to the fiber from the matrix. As the fiber elongates, it absorbs the energy of the crack along its length, and the energy then dissipates through internal atomic motions.
- Yet another method used to arrest the growth of cracks in ceramic matrix composites is **transformation toughening** in which the movement of the crack induces a phase or crystalline transition in the reinforcement (toughening) material. When such a transformation increases the volume of the phase or crystal, the crack is squeezed by the enlarged matrix and the energy required for further growth is increased. Thus, the crack stops growing. This method is especially important in zirconia-toughened alumina.

- If the growth of the crack creates **microcracks** that radiate from the leading point of the growing crack, the microcracks can absorb sufficient energy to stop the growth of the main crack.

These toughening mechanisms are probably not completely independent and might occur simultaneously and synergistically, although one mechanism may dominate in a particular system.

Reinforcement Shapes

Several types of reinforcement shapes have proven to be successful in improving the toughness of ceramic matrix composites. Reinforcement shapes for polymer matrix composites will be discussed in a later chapter. However, the shape effects of ceramics reinforcements are unique enough that some consideration should be given to them here. The shape effects are unique because the purposes of reinforcements in ceramics differ from polymer composites. In ceramics, the reinforcements have a primary purpose of giving toughness whereas in polymer composites the reinforcements carry the load to improve mechanical properties like strength and stiffness. The most common shapes to reinforce ceramics are fibers and whiskers. Flakes and particles are also used but have less significance in changing properties and will not be discussed here.

Since continuous or even long-segmented reinforcements force deflection of a crack to travel along the fiber for a longer distance, these long fibers are more likely to absorb sufficient energy to arrest crack propagation than are particles or whiskers. But, these longer reinforcements may result in greater difficulty in processing. However, some other advantages of longer fibers also have been recognized and exploited in various applications. Continuous fibers, often but not exclusively carbon, can be tailored to concentrate their strength and stiffness into particular orientations. The reinforcements can be unidirectional fibers, fabrics, or preforms. Sometimes further lateral or transverse toughness can be achieved by including some whiskers in the matrix material. The use of long-fiber composites has proven to be highly beneficial for applications such as turbine blades.

Whisker reinforcements do not impart the same toughness as continuous fibers, but whiskers are much easier to process. Moreover, whiskers impart sufficient toughness for many applications. They are, therefore, the dominant reinforcement shape for CMCs. A comparison of fibers and whiskers in a ceramic matrix is given in Figure 7-1.

Matrix Materials

Many materials can serve as ceramic matrices. They are chosen, in part, on their ability to withstand the environment and meet the specific requirements of the application. Most ceramic matrices are comprised of: alumina (Al_2O_3), borosilicate (B_2O-SiO_2), mullite (mixed Al_2O_3 and SiO_2), magnesium oxide (MgO), cordierite (mixed MgO, Al_2O_3, and SiO_2), zirconia (ZrO_2), lithium aluminosilicate glass (LiO_2-Al_2O_3-SiO_2),

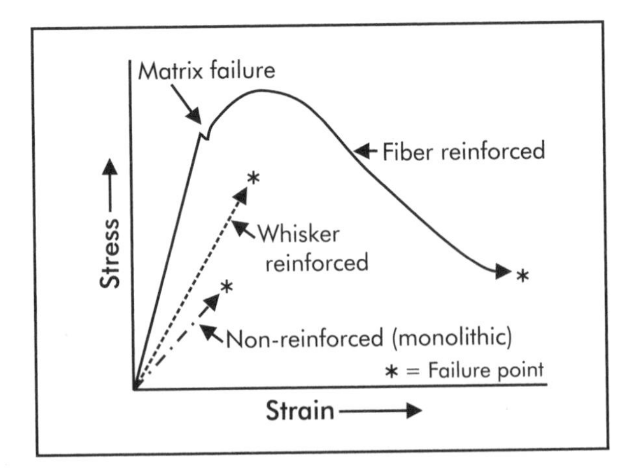

Figure 7-1. Stress/strain characteristics of various reinforcements for ceramic matrix composites.

silicon carbide (SiC), boron carbide (B_4C), silicon nitride (Si_3N_4), aluminum nitride (AlN), boron nitride (BN), zirconium boride (ZrB_2), titanium boride (TiB_2), molybdenum disilicide ($MoSi_2$), and sialon (a complex compound of Si, Al, O, and N). Some of these materials are well known ceramics while others are chosen for specific applications. $MoSi_2$ is an example, where its high melting point (3,660° F [2,016° C]) allows applications that are rare even for CMCs.

Applications

In many applications, especially those requiring good thermal and wear properties, CMCs compete with carbon-carbon composites. However, some experts classify carbon-carbon composites as CMCs. This is reasonable because the carbon matrix has properties similar to those of the CMCs that have been discussed. However, because the basis of the carbon-carbon matrix is a polymer, this text has treated this type of composite with the polymeric group. Clearly, however, the classification could be either way.

As noted previously, the ability of CMCs to withstand high-temperature environments has led to their use in many aerospace applications. Some of the most common uses include engine parts, especially turbine blades and in areas that confine the exhaust gases. Skin areas in the path of exhaust gases are also typically made of CMCs. As the speeds of aircraft increase, especially for planes and hypersonic vehicles, frictional heating will require higher-temperature materials on surfaces such as wing leading edges and nose tips.

Space applications include heat shields for space re-entry vehicles, nose cones, engine parts, rocket nozzles, and general space structures where thermal exposure is a critical issue. Strategic and tactical missiles increasingly share this need to withstand high temperatures. Sensors are being placed on these structures in key locations to monitor the heating and, increasingly, the stresses that arise during flight. Those sensors need to be especially protected from the heat and impact damage. Therefore, "sensor windows," covers that protect the sensors, are increasingly made from CMCs. Similarly, most of these structures have small parts that serve as fasteners (such as bolts and nuts) and fittings that have both high thermal and structural (often isotropic) requirements. These, too, are candidates to be made of CMCs.

CMCs are generally more transparent to radar, microwaves, and other telecommunication frequencies than other composites. Thus some applications of CMCs include radomes, space structural elements on optical devices, and stealth components.

Just as advanced aerospace applications are increasing toward higher temperature requirements, so, too, are automotive applications. Automotive engines operate more efficiently at higher temperatures, so there is a strong incentive to increase operating temperatures, which are progressively out of the range of metals. Initially a few parts were made from CMCs, such as piston heads and exhaust valves. As the trend to higher-temperature engines continues, it is likely that more components will be converted to CMCs, possibly even entire automotive engines.

Non-automotive power generators also have been shown to operate more efficiently at higher temperatures. Therefore, CMCs are being used as components for gas-turbine engines, oil- and coal-fired power plants, and nuclear power generation systems.

Another non-automotive application for CMCs is in heat exchangers. Heat exchangers are increasingly important as engines and other devices operate at ever-higher temperatures. As with most of these other devices, the efficiency of a heat exchanger increases with operating temperature. In this application, thermal conductivity and resistance to heat shock are important. While ceramics do not generally have good thermal conductivity, the presence of reinforcements with high thermal conductivity has made CMCs highly effective as heat exchangers. This is especially true when the reinforcement is a fiber that is continuous in the CMC and oriented so that the direction of the fibers is the direction that the heat is to move. Carbon fibers have proven to be especially good as reinforcements in this application.

The abrasion resistance of CMCs has led to their extensive use as cutting tools for the metal-cutting industry. These tools have demonstrated far longer wear lives than traditional tool-steel cutters. In addition to their abrasion resistance and general hardness, CMCs are dimensionally stable, thus improving on the accuracy of cutting over traditional steel cutting tools.

A closely related application for CMCs is tooling for extrusion dies. These tools are used to shape materials as they are pushed through a small opening (die) that gives shape to the final part. Metal extrusions include such common items as aluminum window frames, automotive trim, and wire, where the diameter of the wire is reduced as it is drawn through the die. In plastic part manufacturing, heated plastic melt is pushed through the die and then cooled into the shape desired. Some familiar plastic extruded parts include PVC pipe, beverage drinking straws, and synthetic fibers.

Some tooling, such as the cavities in an injection mold for plastics, is made by electrical discharge machining (EDM). This process uses a spark to form the shape required in the tooling by eroding away the unwanted material. The EDM process requires that the material being eroded away (the workpiece) be conductive. Hence, CMCs have not been able to be formed by this method. Recently, however, conductive CMCs have been made by adding conductive filler to the matrix material. This capability has opened the pathway for many more tools to be made of CMCs.

One application for CMCs that does not depend on thermal or wear capability is armor. Some armor consists of layers of CMC material between metal plates. CMC armor has been shown to offer unique protection against high-explosive, anti-tank (HEAT) rounds. These rounds had been a serious problem since the waning days of WWII when they were first used. The principle on which the metal-CMC-metal sandwich works uses the expansion in volume that occurs when the CMC is penetrated by the HEAT round. As revealed in high-speed photography, as the CMC is penetrated, it shatters and thus greatly expands volume. This expansion is directed outward through the penetration hole, which pushes the metal jet of the HEAT round with it. Thus, a counterforce is applied that helps to protect the second layer of metal and, therefore, the integrity of the tank. This technology has proven to be so effective that advanced tanks in the United States, Great Britain, and other nations use the technology. A similar technology is used for inserts placed into the pockets of bullet-proof vests for individual protection.

MANUFACTURING OF CERAMIC MATRIX COMPOSITES

Many ceramic matrix materials are solid powders before they are formed into composite materials. For whisker and particle reinforced composites, the whiskers or particles are usually blended with the powder and then pressed to form a preform. The preform is then heated, often under pressure, to a temperature slightly lower than the melting point of the ceramic. (This assumes the reinforcement has a higher melting point than the ceramic matrix.) The ceramic particles consolidate with each other and with the reinforcements in a process called **sintering**.

The whisker reinforced composites have higher strength and higher toughness compared to the non-reinforced ceramics. Some of the whiskers proven to be especially beneficial are SiC (silicon carbide) in a matrix of Si_3N_4 (silicon nitride) and SiC in a matrix of Al_2O_3 (aluminum oxide). This blending/sintering process can be done with short fibers rather than whiskers or particles. However, the short fiber reinforced composites generally have lower strength and higher toughness than the non-reinforced matrix itself.

While deceptively simple in concept, the process of making CMCs from powders using whiskers as the reinforcements is actually quite complicated as it involves many different variables. Therefore, the properties of the whisker reinforced CMC are dependent on these variables: whisker production method, whisker size, whisker loading, matrix particle size and shape, matrix powder/whisker blending, whisker/matrix compatibility, forming conditions, use of a sintering aid, and sintering temperature, pressure, and time.

The whisker, particle, or continuous reinforcements can be mixed with a slurry of the ceramic powders in either a water or organic (usually alcohol) solvent carrier. Wetting agents might be added to assist in wetting the fibers. Organic binders also might be added to help hold the materials together in a preform after the carrier has been evaporated. The slurry is filtered and then heated under pressure to eliminate the liquid and shape the ceramics and reinforcements into a convenient form for subsequent operations. This preform is then heated to eliminate any binder, and then further heated at higher temperature and pressed to sinter the matrix and entrap the reinforcement.

Another manufacturing method is to create a preform of continuous fibers using weaving, braiding, or knitting techniques. These fiber preforms are then infiltrated with the powder material, often by the slurry technique. The impregnated fibers are then heated in stages to remove the liquid and sinter the matrix. During the heating, pressure is applied using a mold that has the shape of the preform and, of course, the final part. Alternately, the fiber preform can be infused with melted ceramic. The high viscosity of the melt limits this method somewhat, but it can work if the fibers are sufficiently wetted and the infusion is done slowly. The ceramic matrix also can be infused into the fiber preform using chemical vapor deposition (CVD), also called chemical vapor infusion (CVI). CVD/CVI is a process in which the ceramic matrix is vaporized and then enters the preform and coats the fibers. Further gaseous ceramic material is infused and this material reacts with the ceramic coating already on the fibers and eventually fills the pores in the matrix. This is a slow process, but the results can be good because bonding of the matrix to the fibers is usually excellent.

A newly emerging technique for manufacturing CMCs is to create the reinforcement phase by direction reaction, which occurs during the consolidation of the matrix. This process, called **in-situ reaction sintering**, employs two or more constituents in a compact preform, which react during sintering to form new phases. The method has the potential to improve the uniformity of the distribution of the reinforcing materials since they are formed throughout the matrix material. In addition, some new composite structures may be possible with this method. The CMCs made with this technique include aluminum nitride with titanium boride, titanium carbide with alumina, titanium nitride with alumina, titanium boride with alumina, titanium nitride with silicon nitride, boron nitride with alumina, and titanium nitride with sialon. At this time, this manufacturing method has been demonstrated in the laboratory but has not been widely used on commercial products.

PROPERTIES AND USES OF METAL MATRIX COMPOSITES

Metal matrix composites (MMCs) consist of high-performance reinforcements in a metallic matrix. The first production use of MMCs was as components of the space shuttle (boron reinforcement in aluminum matrix to make tubes, 1974). Today, the most common matrix materials are: aluminum, titanium, magnesium, copper, nickel, and various alloys of these metals. The reinforcements can be in the form of particles, whiskers, or fibers. The most common reinforcements are boron fibers, graphite/carbon fibers, and silicon carbide fibers, whiskers, and particles.

Just as fiberglass reinforcements are routinely added to engineering thermoplastics to improve their strength or stiffness, so too are reinforcements routinely added to metals to improve strength and stiffness. Therefore, MMCs are common where part specifications dictate high performance.

Since metal matrices are usually inherently ductile and tough, the problem in ceramic matrix composites with brittle fracture is of little importance with MMCs. Rather, the MMCs are analogous to polymer matrix composites in that the reinforcement is stronger and stiffer than the matrix. Therefore, the composite is stronger and stiffer than the non-reinforced matrix.

Property Comparisons

The MMC properties of most importance are: strength and stiffness improvement over non-reinforced metals, higher temperature stability than polymer matrix composites, wear resistance, damping control of vibrations, electrical conductivity, thermal conductivity, and control of the coefficient of thermal expansion.

Examples of the changes in properties of MMCs with various reinforcements can be seen in Table 7-2. From the table it is obvious that particulate reinforcements will raise the modulus and strength significantly over the non-reinforced material. Note also that the 7090 alloy has a slightly higher strength and modulus than the 6061 alloy, but both are roughly the same when compared to having no reinforcement. Because whiskers are more difficult to incorporate into the matrix than particles, the whiskers are shown with only 20% loading. Even so, the modulus and strength data are nearly the same as the 30% particulate data. This suggests that whiskers are more efficient than particles at increasing strength and modulus. This is, of course, as would be expected because the whiskers have a more distinct length/diameter ratio, which allows better transfer of forces to the whiskers than can be done with the particles. The much higher strength and modulus values for the fibers in aluminum indicate that fibers are better than whiskers as reinforcements.

An advantage of whiskers over fibers is their ability to easily reinforce in all directions. Typically, the transverse and longitudinal strengths of SiC whiskers are within 5% of each other. Hence, whiskers are more effective than fibers when the application requires multidirectional reinforcement.

The remaining data in Table 7-2 allow comparisons between different metal matrices. When the strength or modulus of the matrix increases, the strength or modulus of the composite also increases. This demonstrates that the matrix and reinforcement in MMCs both contribute to the strength and modulus. This is somewhat different from polymer matrix composites where the fibers clearly dominate the strength and modulus values. The reason for this difference is that, in metals, the strength and modulus of the matrix are much closer to the values of the reinforcement, so the

Table 7-2. Properties of selected metal matrix composites.

Matrix	Reinforcement	Reinforcement Type	Density, lb/in.^{3} (g/cm^3)	Modulus, $\times 10^6$ psi, (GPa)	Strength, $\times 10^3$ psi (MPa)
Al 6061	None	—	0.097 (2.7)	10.0 (69.0)	35 (241)
Al 6061	30% SiC	Particulate	0.105 (2.9)	17.55 (120.7)	80 (552)
Al 7090	30% SiC	Particulate	0.108 (3.0)	18.5 (127.5)	112 (772)
Al 2124	20% SiC	Whisker	0.104 (2.9)	18.0 (124.1)	108 (745)
Al 6061	Graphite	Unidirectional fiber	0.089 (2.5)	48.6 (335.1)	120 (827)
Mg AZ91C	Graphite	Unidirectional fiber	0.072 (2.0)	52.2 (359.9)	117 (807)
Cu	SiC	Unidirectional fiber	0.230 (6.4)	30.0 (206.8)	150 (1,034)
Al	SiC	Unidirectional fiber	0.103 (2.8)	30.0 (206.8)	212 (1,462)
Ti	SiC	Unidirectional fiber	0.140 (3.9)	31.0 (213.7)	240 (1,655)

combined values in the composite are more of an average.

Plots of specific strength versus specific stiffness of MMCs and various resin-reinforced materials are shown in Figure 7-2. The reinforced metals are, in general, both stiffer and stronger than non-reinforced analogues. The exceptions would be when the reinforcement does not increase the specific strength or stiffness because it is heavier or in some other way adds weight disproportionately affecting the strength or stiffness. (Note that specific strength and specific stiffness are the strength and stiffness divided by the density. Therefore, heavy components tend to decrease these values.)

Metal matrix composites do not have the best combination of specific strength and specific stiffness when compared to carbon fibers in resin, as shown in Figure 7-2. However, they do have good specific strength and stiffness compared to most other materials, along with other properties that are of particular advantage for some applications. One obvious advantage over both polymeric and ceramic matrix materials is conductivity. Even with highly conducting fillers, polymeric and ceramic materials do not reach the conductive levels of metals. Another advantage to MMCs is their lack of out-gassing in space. Polymeric materials have a tendency to release trapped gases when in a low-pressure environment like space. Metal matrix composites have little chance of trapping gases. Further, even if gases are present, the low diffusivity of metals (especially compared to polymers) makes MMCs highly valued in space applications.

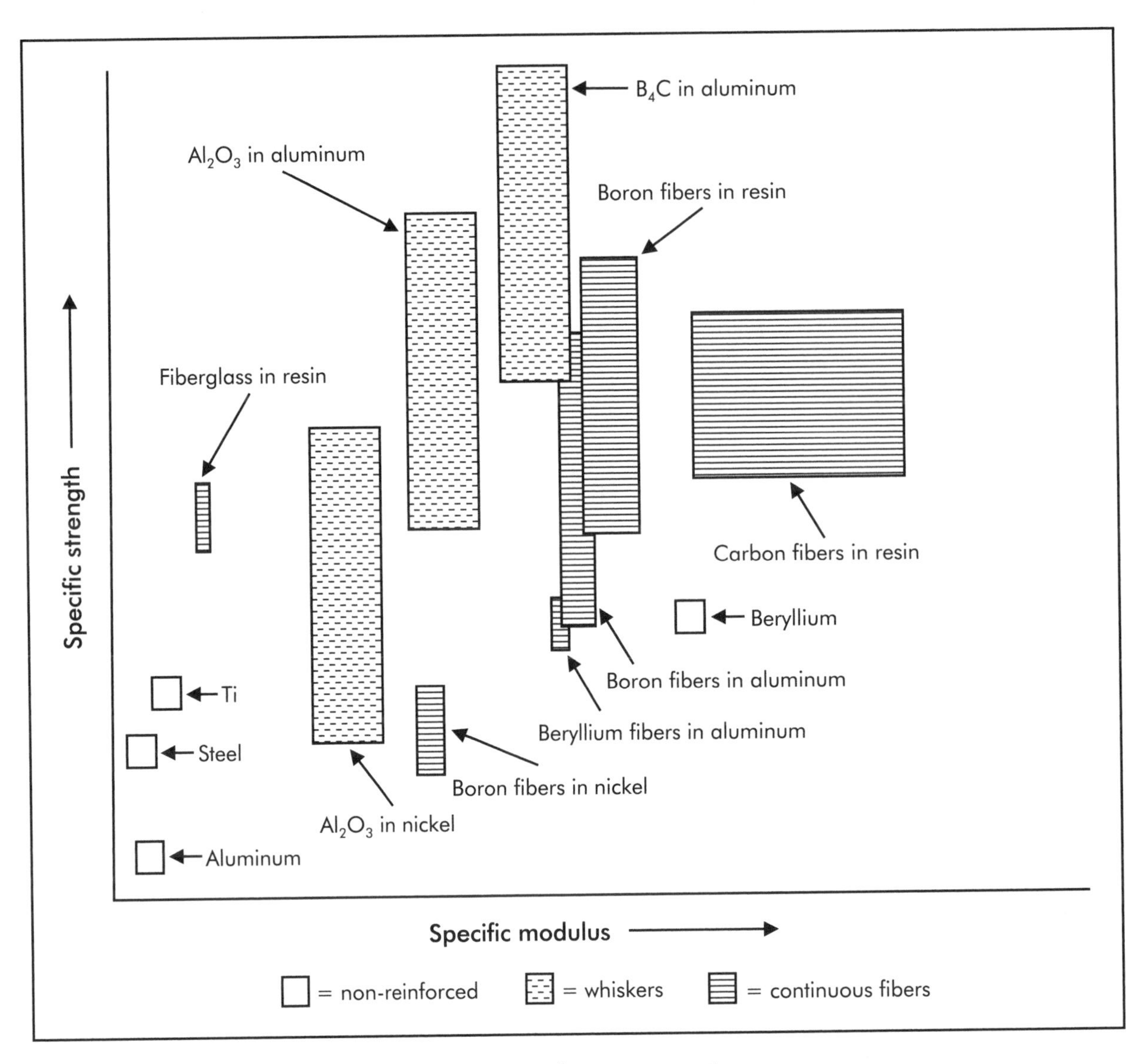

Figure 7-2. Specific strengths versus specific stiffnesses of various composites.

A particular advantage of MMCs is post-fabrication formability. Polymer matrix composites are usually thermosets and, therefore, cannot be reshaped after curing. Ceramic matrix composites are high melting and therefore cannot be shaped with ease. Metal matrix composites have a lower melting point than ceramics and can usually be reshaped using conventional metal shaping processes such as casting, forging, cutting, milling, EDM, and cold forming.

The incorporation of a second phase (the reinforcement) into a metal significantly affects the propagation of pressure waves through the material by acting as sites for scattering and attenuation. Measurements on boron fiber reinforced aluminum have shown a five-fold increase in dynamic damping capacity. In another example, the high thermal conductivity of a metal matrix, combined with the low thermal expansion of the composite, allows the design of high-performance heat sinks for electronics applications. Carbon reinforced copper is of particular interest for this application.

A problem that must be acknowledged in MMCs is the corrosion that arises when aluminum is reinforced with carbon/graphite fibers. In an environment in which the composite might be wetted by water or, even worse, seawater, the corrosion between carbon fibers and aluminum is rapid. Water-based corrosion of aluminum reinforced with SiC is not as severe as with carbon/graphite, but it still must be dealt with in the design of parts. The corrosion with SiC is usually restricted to pitting rather than the catastrophic failure that can occur with carbon/graphite. Boron reinforced aluminum also corrodes and corrosion can be severe if the parts are not carefully processed.

Applications

MMCs have been used in automotive, aerospace, sports, and numerous other applications. Some automotive applications include: piston ring inserts, pistons, connecting rods, impellers, brake calipers, sway bars, and critical suspension components. One automobile company is using MMCs as the crown of the pistons. It reports a seven-fold increase in piston life for its diesel engines. Another auto manufacturer is using an aluminum MMC block in its engines without the normal steel liners.

In aerospace, MMCs have been used for applications such as structural rods for space platforms, antenna masts for satellites, aircraft stiffeners, actuator rods, missile body casings, and various joints and fittings. Sports applications include tennis racquets, golf club heads, and golf club shafts. General manufacturing applications include cutting tools and bearings.

A pump company reported a 200% increase in pump life when it changed from a steel housing to one made of MMC. The application was pumping sandy and dirty water.

In general, anything that can be cast in aluminum can be cast in an aluminum matrix MMC. The cost increase is 30–40%. Typical industries using cast aluminum MMCs include defense, automotive, and sporting goods. The range of tolerances, surface finishes, production rates, and casting sizes are about the same as with non-reinforced aluminum. However, part complexity is reduced somewhat because of the higher viscosity of the reinforced melts. Shrinkage is about 1.5% in MMCs. Machining of MMCs causes more tool wear than with non-reinforced metals. Polycrystalline diamond tools are preferred for cutting MMCs. Typically, finishing costs for MMCs are about twice as much as for non-reinforced analogues.

MANUFACTURING OF METAL MATRIX COMPOSITES

As with other composite materials, the manufacturing methods used to make MMCs depend on the natures of the matrix and reinforcements. When reinforcement orientation is not important, as with particles and whiskers, the matrix and reinforcements simply can be blended and then molded. Conceptually the reinforcements could be blended into a molten metal. However, in practice, the mixing of the fibers and the melt is done in a machine like an extruder. The extruder melts the metal in a heated barrel and a screw applies pressure and forward movement to the molten metal and the reinforcements, which are added in the barrel. The reinforcements must be small enough to pass through as the screw turns and to clear other small tolerance zones within the extruder. The final parts may or may not be pressed while still hot. If pressed, the mechanical properties are better. Melting is usually done under an inert gas atmosphere to prevent formation of unwanted metal oxides and metal carbides.

A more common method of incorporating particles or whiskers into a metal is to blend the reinforcements with a metal powder. In this technology, called *powder metallurgy*, the metal powder is thoroughly blended with the reinforcements. The mixture is then degassed to remove any volatile components and trapped air. The powder-reinforcement mixture is then pressed into a preform that is shaped appropriately for later work. An organic binder might be added to help retain the shape of the preform. The preform is then heated to remove the binder and further heated at a higher temperature while being pressed to consolidate the metal powder and bond with the reinforcement. This is a sintering process and is similar to the sintering process discussed for ceramic matrix composites. In the case of metals, however, some experts have suggested that the processing temperature must be above the solidus temperature of the metal (that is, the temperature at which a liquid phase begins to form) to achieve proper wetting of the fibers. Sometimes a subsequent shaping process, such as pultrusion, can be done to reform the powder metallurgy billet into the final form.

Mixing long fibers with either a powder or a melt would be difficult because the orientation of the fibers would not be preserved and the fibers might be broken in the mixing process. However, it is possible to coat the fibers with the metal powder and then carefully place them in the orientations desired. The fibers can then be heated and pressed to sinter the metal powders. The problems with this method are the difficulty in retaining the metal powder on the fibers as they are oriented and the problem of disturbing the orientation while the materials are compressed.

A pressure infusion method has proven to be more convenient than the powder method and still preserves the orientations of the fibers. In this method, thin sheets of the metal matrix material (called foils) are placed under and on top of a layer of reinforcement, thus making a sandwich structure. Multiple layers of reinforcement and foils can be done. The fiber layer is laid down carefully in the proper orientation on top of a metal foil and then covered with another metal foil. The metal-fiber sandwich is then compressed (the foils help retain the fibers in place) and heated to above the melting point of the metal. This heating with compression causes the metal foils to melt and infuse with the fibers, thus wetting the fibers and distributing the metal throughout the structure.

Rather than infuse the metal using foils, it is also possible to melt the metal and cast it onto the fibers that have been arranged in the casting mold. This is especially appropriate when the shape of the part is complex. It is also a good way to make MMCs that have thick sections. A vacuum or pressure assist is often used to help with full wetting of the fibers. Some reinforcements, especially carbon fibers, are not wetted well by molten metals, so special fiber coatings are used to improve the bonding between the fibers and the metal matrix.

Just as with ceramics, vacuum diffusion can be used to infiltrate the fibers with the metal matrix. In fact, this process is simpler with metals than with ceramics. Fiber volumes of 50–60% have been achieved using diffusion techniques. (Fiber volumes above 50% are characteristic of advanced polymeric composites.)

An unusual method has been employed to make boron carbide reinforced metal matrix composites. In this method, the boron carbide and the metals that form the alloy matrix are combined. But, instead of combining them as molten liquids, they are combined as microscopic particles, which are then blended in an advanced jet mill and compressed at high pressures to form a billet. The billet is then shaped to form parts, such as bicycle frames, where the composite competes well against traditional aluminum alloys. The modulus of the new MMC is more than three times greater than the best modulus of aluminum. The new MMC is also about 38% stronger.

CASE STUDY 7-1

Diamond Reinforced Abrasive Pads

Diamonds are not usual reinforcement materials for MMCs, but when the composite is used as an abrasive pad, diamonds have a special value because of their hardness and wear resistance. One particular application that uses diamond-reinforced pads is the smoothing of silicon disks (wafers), which will be used to make integrated circuits. For this application, the metal matrix needs to be able to bind the diamonds securely. Hence, an alloy of several metals with nickel as the principal component is used because the nickel will sinter well and retain the diamonds.

The MMC composite is made by mixing the metal powder with an organic binder and then shaping the resulting mixture into a flat sheet about .06 in. (1.5 mm) thick. The sheet is positioned on top of a metal base that provides a solid backing and is an attachment to the grinding mechanism. After the sheet is properly placed, it is sprayed with an organic adhesive and the diamonds are set in place. The diamonds are usually in an array that not only facilitates their use in abrasion but also makes quality control easier as each diamond must be securely attached when the product is completed.

After the diamonds have been placed on the surface of the matrix sheet, they are pressed into the sheet slightly to insure good contact between the diamond and the sheet. The metal base, matrix, and diamond unit is heated to a moderate temperature to remove the organic binder in the preform material. Then, the unit is transferred to a heating press where the entire assembly is sintered. The diamonds are bonded into the top surface of the metal matrix when the process is completed.

Similar grinding wheels are made by mixing the diamonds (or diamond dust) throughout the matrix. When these materials are fully sintered, they provide excellent grinding action, presenting diamonds to the grinding surface regardless of the amount of wheel wear. Of course, other materials besides diamonds (such as SiO_2 and Al_2O_3) also can be used. These have lower hardness than diamonds but are, of course, less expensive.

SUMMARY

Whereas ceramic matrix composites are usually associated with specific high-temperature or high-stiffness applications, metal matrix composites might be chosen to simply upgrade a part that is otherwise made of the metal matrix material without the reinforcement. Therefore, the choice to upgrade to MMCs is viewed as routine and simple, much like the choice of fiberglass reinforcements is an upgrade to traditional engineering thermoplastics.

Ceramic matrix composites, on the other hand, can fulfill applications where traditional ceramic materials would not work because of some inherent property problem, usually brittleness. Therefore, CMCs have a unique role that is, in some applications, compelling. MMCs are less compelling but, interestingly, more common.

Both CMCs and MMCs are increasingly used in commercial applications. CMC research and development activity is especially strong in the United States. However, Europe, Japan, and China are rapidly achieving parity. Still, some technological gaps continue to limit even wider adoption of CMCs. These gaps include the following: reliable and low-cost forming technology, non-destructive testing capabilities, improvement of matrix/reinforcement bonding, joining technology, thermal fatigue, complex structure forming, quality assurance criteria, machining, and an extensive design database.

In the case of MMCs, the development effort in the United States seems focused on high-performance applications while in Europe and Asia it seems concentrated on commercial uses, especially automotive. As with CMCs, some technology gaps are limiting even more rapid acceptance of MMCs in commercial use. These gaps include the following: casting technology, non-destructive testing techniques, knowledge of the effects of defects, fracture properties, low-cost fabrication techniques, joining technology, complex structure forming, and quality assurance criteria.

Undoubtedly the gaps in both CMCs and MMCs will be overcome. Much work is underway and the promises of the superior performance of these materials are sufficiently rewarding that the development efforts should continue with vigor.

LABORATORY EXPERIMENT 7-1

Wear Properties

Objective: Discover the wear capability of metal or ceramic matrix composites.

Procedure:

1. Purchase a cast iron log. (This is a solid piece of cast iron shaped in a long cylinder.)
2. True the log by mounting it on a lathe and cutting off the outer layers until the cuts are uniform around the surface.
3. Use a newly purchased tool steel (M42 or M2) cutting insert and cut a 5-in. (12.7-cm) linear pass on the log. Set the turning speed, feed speed, and feed depth according to the suggestions of the cutting tool manufacturer. After cutting, remove the cutting tool and measure the wear on the tool flank (or relief angle). (See Figure 7-3 for an explanation of the wear measurement.)
4. Repeat cutting of 5-in. (12.7-cm) linear passes on the log, along with inspection of each pass until the wear reaches .015 in. (0.38 mm).
5. Using a newly purchased MMC or CMC cutting insert, duplicate the cutting of the log as was done with the tool steel. After cutting for the same number of cutting passes as the tool steel, compare the wear on the MMC or CMC.
6. Report the difference in wear as both mils of wear and as a percentage of wear where the tool steel is the standard.

QUESTIONS

1. Give a major advantage of a ceramic matrix composite over a non-reinforced ceramic. How does this advantage contribute to the use of CMCs for structural applications?
2. What are three advantages of using metal matrix composites versus metals alone?
3. Why do MMCs reinforced with whiskers have higher strength and stiffness when compared to the same matrix reinforced (to the same percentage content) with particles?

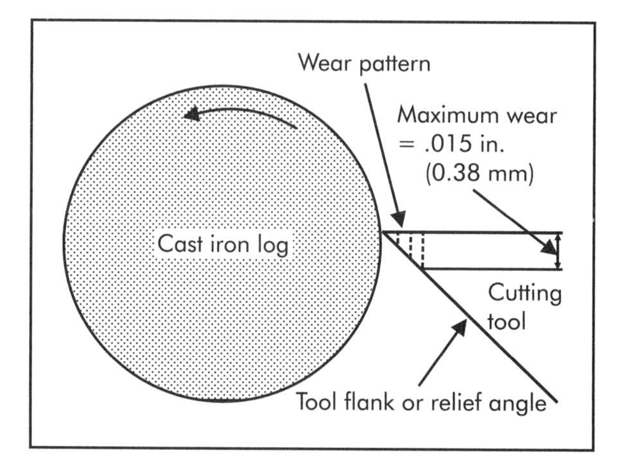

Figure 7-3. Wear pattern of a cutting tool.

4. What is the problem associated with using carbon reinforcements in an aluminum matrix?
5. Why don't ceramics conduct electricity as well as metals? What is the reason that metals conduct electricity so well?
6. Which material—ceramics or metals—would you expect to be a better thermal conductor? Why?
7. If the melting points of ceramics are generally higher than the melting points of metals, what can you conclude about the relative strengths of ionic versus metal bonds, in general?
8. If the melting points of ceramics are generally higher than the melting points of polymers, what can you conclude about the relative strengths of ionic versus polymeric bonds, in general?
9. How does the decomposition temperature of a polymer relate to the strength of the covalent bond? How does that relate to the melting point?
10. What physical phenomena are associated with the decomposition temperature and the melting points of polymers?

BIBLIOGRAPHY

ASM International. 1987. *Engineered Materials Handbook, Vol. 1: Composites*. Metals Park, OH: ASM International.

Bowles, R. R., Mancini, D. L., and Toaz, M. W. 1987. "Metal Matrix Composites Aid Piston Manufacture." *Manufacturing Engineering*, May, pp. 61–62.

Chawla, K. K. 1993. *Ceramic Matrix Composites*. London: Chapman and Hall.

Ginburg, David. 1991. "Scaling Up Metal-Matrix Composite Fabrication." *Composites in Manufacturing*, Second Quarter, Vol. 7, No. 2.

Lee, Stuart (ed.). 1990. *International Encyclopedia of Composites*. New York: VCH Publishers.

Morgan, Peter. 2005. *Carbon Fibers and the Composites*. Boca Raton, FL: Taylor & Francis.

Schuldies, John J. 1992. "Ceramic Composites—Emerging Manufacturing Processes and Applications." *Composites in Manufacturing*, Fourth Quarter, Vol. 8, No. 4.

Schwartz, Mel M. 1997. *Composites Materials*. Upper Saddle River, NJ: Prentice-Hall, Inc.

Society of Manufacturing Engineers (SME). 2005. "Composite Materials" DVD from the Composites Manufacturing video series. Dearborn, MI: Society of Manufacturing Engineers, www.sme.org/cmvs.

Strong, A. Brent. 2006. *Plastics: Materials and Processing*, 3rd Ed. Upper Saddle River, NJ: Prentice-Hall, Inc.

Warren, R. 1992. *Ceramic-matrix Composites*. New York: Chapman and Hall.

8 Reinforcements

CHAPTER OVERVIEW

This chapter examines the following concepts:

- General fiber characteristics
- Glass fibers
- Carbon/graphite fibers
- Aramid and other synthetic organic fibers
- Boron, silicon carbide, and other specialty fibers
- Natural fibers
- Fiber-matrix interactions

GENERAL FIBER CHARACTERISTICS

Reinforcements for composites can be fibers, whiskers, and/or particles. Each has its own unique applications, although fibers are the most common and have the greatest influence on the mechanical properties of the finished part. In fact, the influence of fibers, especially long fibers, is so dramatic that composite materials containing long fibers might be considered a new class of solid materials beyond the traditional materials—ceramics, metals, and polymers. Long fiber composites have contributed to so many major innovations in our modern world. Composites with whiskers or particles are more likely to be thought of as just improvements over the non-reinforced matrix material. Therefore, the emphasis of this chapter will be on the long fibers with only a minor treatment of whiskers and particles.

Fibers are materials that have one long axis (length) compared to the other dimensions (width and height). The other dimensions are often circular or near circular. This relationship between length and the other dimensions is often expressed mathematically using a term called the **aspect ratio,** which is simply the ratio of the length to the diameter of a fiber. Fibers are materials that have high aspect ratios. This means that loads can be transferred onto the fibers along the long axis. Therefore, the fibers can be efficient in giving reinforcement to the composite structure.

In addition to the natural length that characterizes fibers, their internal molecular structure is often oriented in the long direction as well. This gives the fibers exceptional strength and stiffness in the long direction moreso than the other directions. This orientation of internal molecules is usually produced by *drawing* the fiber. This is accomplished by physically pulling the solid but pliable (plastic) fiber in the long direction. This process orients the molecules along the length of the fiber. A tension load on the fiber pulls against the molecular chains themselves, stretching their polymeric backbones, rather than against a mere entanglement of the chains. In some cases, merely forming the fiber by forcing it through a die also will have the effect of aligning the internal molecules in the long direction. However, drawing is usually more effective than mere flow alignment.

Fibers are available in many diameters as seen in Figure 8-1. The diameters shown are those typical of each fiber type. Some are smooth around the perimeter since they are made from melts. Others have rough perimeters since the fibers are heat treated after being formed and the treatment causes some degradation of the surface.

Fibers of non-circular cross-section, such as a tri-lobe design that looks somewhat like a cloverleaf, have been investigated for reinforcement of composites. Such fibers are well known and frequently used for non-composite applications such as sleeping bag filler, thermal clothing, and even carpets where the advantages include improved thermal retention and soil hiding. However, these properties have little value in composites. What was hoped is that the non-circular cross-sections, which have higher perimeters than circular ones, might have improved adhesion to the matrix materials and, therefore, improved properties in the composite. Careful choice of the shape might also result in a stiffer fiber, possibly through a structural effect like that seen in I-beams. Some improvements were initially reported, but these non-circular cross-sectional fibers

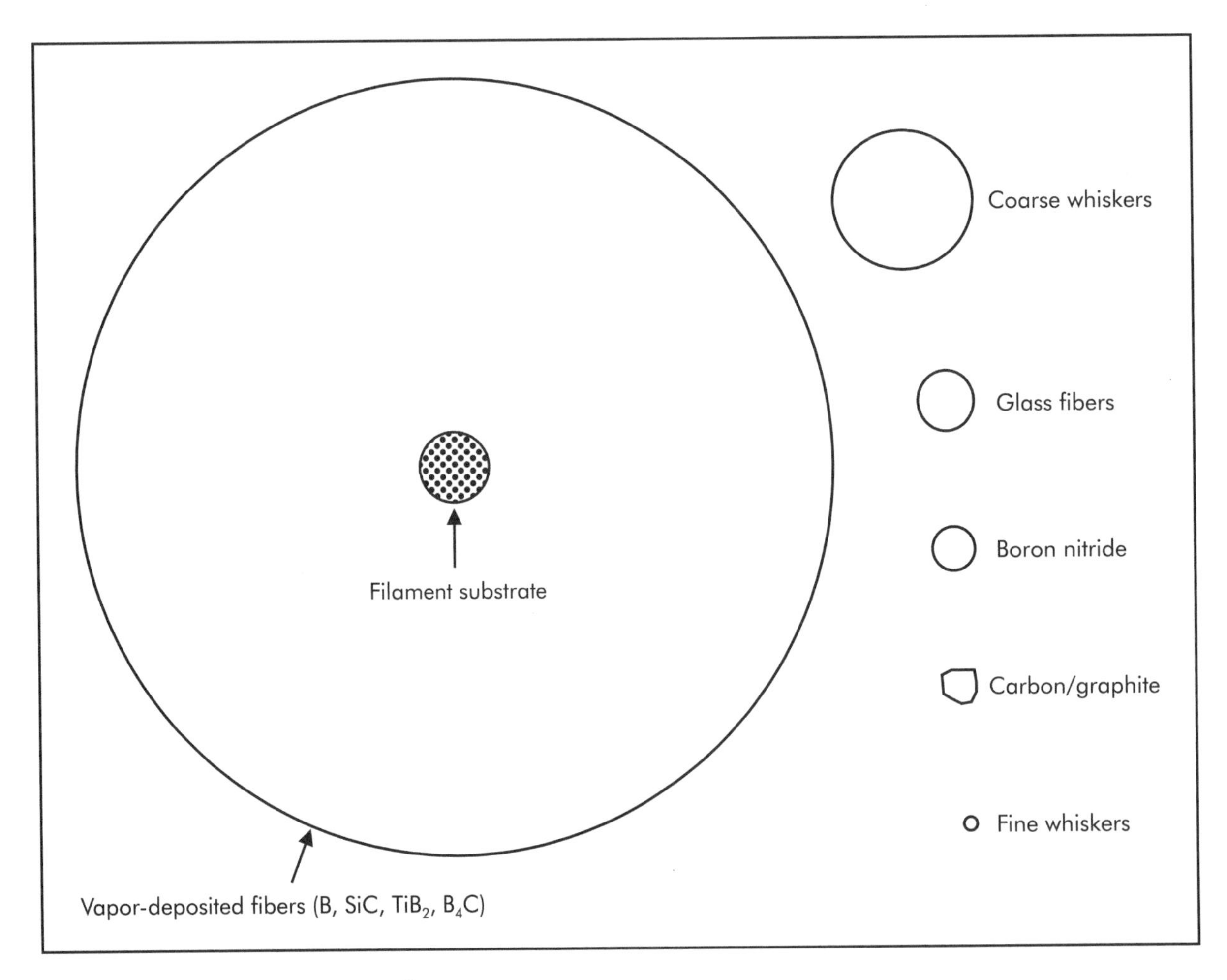

Figure 8-1. Relative size of common fibers.

have not been widely used and are not now commercially available.

Fibers are generally made in continuous lengths that can be used as is or chopped to the length desired. They also are available in many oriented forms such as woven cloth, mat, knits, and braid (all discussed in a later chapter).

Studies have been conducted to determine the relative importance of various fiber parameters on the properties of the composite into which they have been incorporated. The fiber length and concentration have the strongest effects on composite properties. As expected, the following composite properties increased with longer fiber length: tensile strength, modulus of elasticity, flexural strength, and elongation. The rate of increase rises sharply until the fibers are about 3 in. (7.6 cm) and then the rate of increase in properties rises less sharply. However, in some applications, the relatively minor increases in properties obtained with long fibers are needed for part performance. Also, some manufacturing methods are more favorable with long fibers. Of course, the upper limit of fiber length is the length of the part. In these situations, the fibers are referred to as **continuous**.

Increased fiber loading has the following effects on composite properties: increased tensile strength, increased tensile modulus, increased flexural strength, and decreased elongation. A limit of fiber loading has been found to be about 60–70%. Above that level there seems to be insufficient resin in most composites to fully wet the fibers.

Before discussing each of the fiber types individually, a comparison of the properties of several fibers commonly used in composites is useful. That comparison is given in Table 8-1 along with data for some common metals that might compete against composites in some applications. Note the wide variation in all the properties listed. As might be expected, the tensile strengths of the fibers most commonly used in composites (glass, carbon, aramid) are quite high, but not so high that other materials (boron, SiC, and even steel) could not compete. The same can be said of modulus—the traditional composite fibers are generally higher in value, but other materials are within the same general range. With elongation, the metals are clearly higher than the traditional composite fibers. So, why do glass, carbon, and aramid dominate the composites market? The answer to this question will be a focus of this chapter.

Another consideration has proven to be of overwhelming significance in choosing the materials that are to be reinforcements for composites. That is the material's mechanical properties and density, which is a combination of the advantages of strength, stiffness, and light weight. In aerospace applications where the cost of launching a satellite can be as high as \$10,000/lb (\$22,046/kg), low weight while still maintaining mechanical strength and stiffness of a material cannot be overemphasized. While aircraft costs per lb/kg are less than space structures, the premium for weight savings is still compelling. Even in sporting goods, the weight and mechanical properties combination in a material is important. For example, there is increased hitting speed that results when a tennis racquet or golf shaft is lightweight, stiff, and strong. Even for applications like wind turbine blades, lightweight composites dramatically improve the efficiency of operation.

The mechanical strength, stiffness, and density properties combination is called the **specific strength** and **specific stiffness** of a material. Normally, these are calculated by dividing the tensile strength and tensile modulus of a material by its density. This has been done for the materials in Table 8-1 and the resulting data are given in Figures 8-2 and 8-3. In these charts, the reason for using traditional composite fibers is clearly

Table 8-1. Comparison of properties for several key fiber types (representative values for each fiber type, bare fibers).

Fiber Type	Density, g/cc	Tensile strength, ksi* (MPa)	Tensile Modulus, Msi** (GPa)	Elongation to Break, %
Glass (E-glass)	2.5	500 (3,447)	10 (69)	4.9
Glass (S-glass)	2.5	665 (4,585)	12 (83)	5.7
Carbon/graphite (standard modulus)	1.8	600 (4,137)	33 (228)	1.6
Carbon/graphite (intermediate modulus)	1.8	780 (5,378)	40 (276)	1.8
Carbon/graphite (ultra-high modulus)	1.9	500 (3,447)	64 (441)	0.5
Aramid (high toughness)	1.4	523 (3,606)	12 (83)	4.0
Aramid (high modulus)	1.4	580 (3,999)	19 (131)	2.8
Aramid (ultra-high modulus)	1.5	494 (3406)	27 (186)	2.0
Ultra-high-molecular-weight polyethylene (UHMWPE) (standard modulus)	0.97	375 (2,585)	17 (117)	3.5
UHMWPE (high modulus)	0.97	435 (2,999)	25 (172)	2.7
Boron (on tungsten)	2.5	500 (3,447)	56 (386)	0.9
SiC	3.0	500 (3,447)	60 (414)	1.8
Flax	1.5	116 (800)	9 (62)	2.5
Spider silk	1.3	145 (1,000)	14 (97)	35
Steel	7.8	145 (1,000)	29 (200)	30
Aluminum	2.8	70 (483)	10 (69)	20
Titanium	4.5	166 (1,145)	25 (172)	30
Magnesium	1.8	40 (276)	7 (48)	15

*ksi = psi × 10^3, **Msi = psi × 10^6

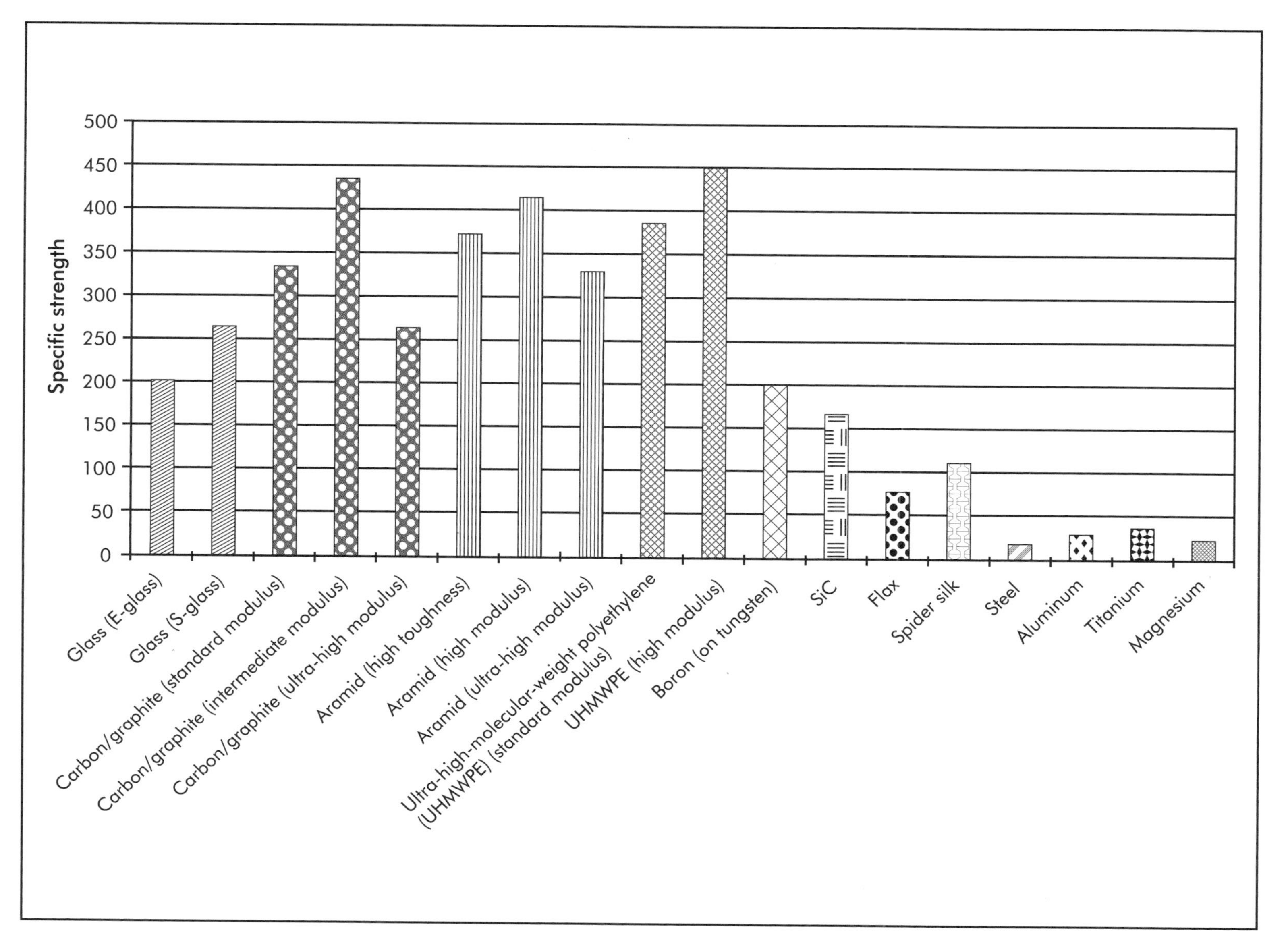

Figure 8-2. Specific strengths of various materials.

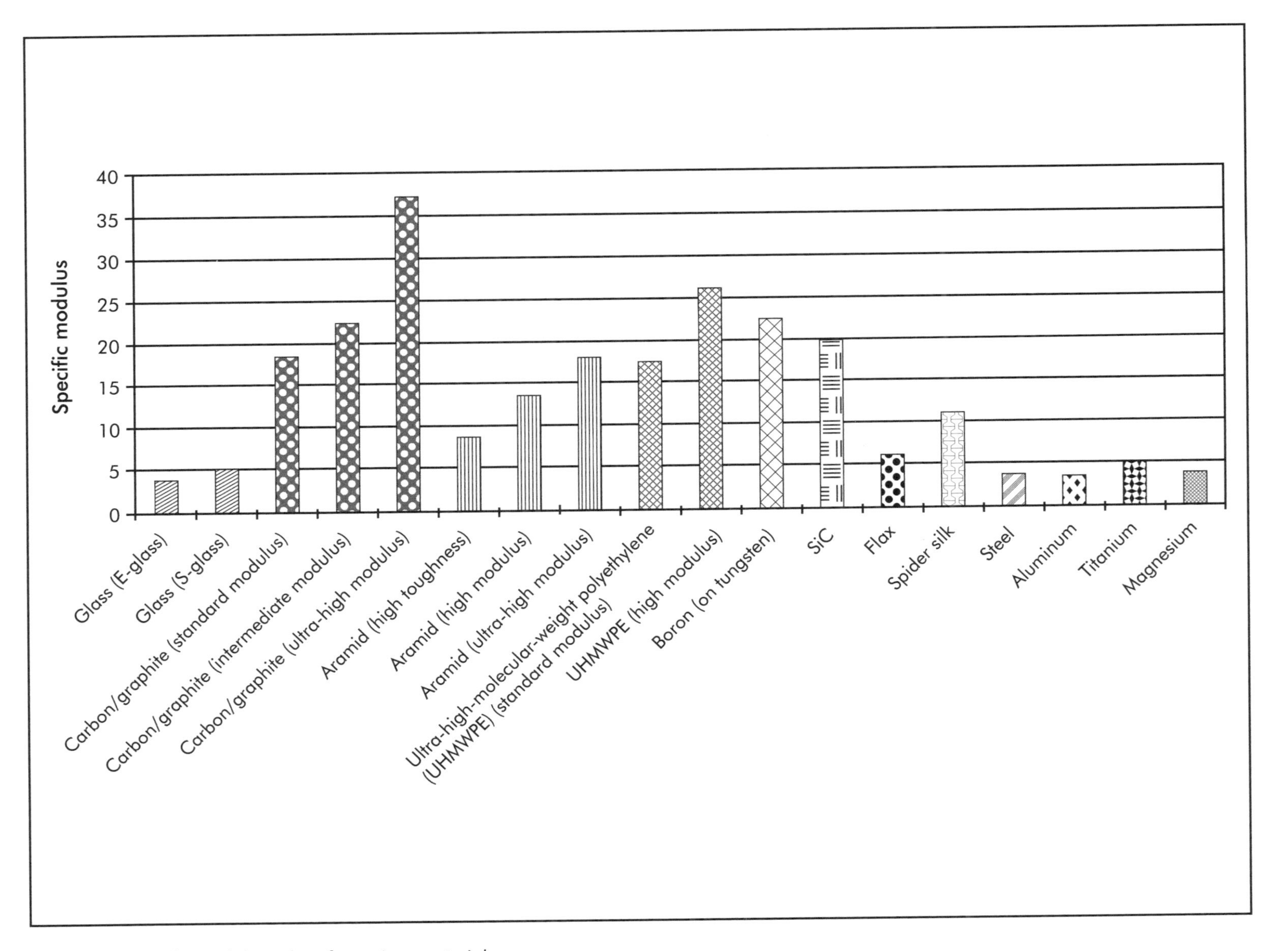

Figure 8-3. Specific modulus values for various materials.

evident. The specific strengths of the traditional composite fibers are about an order of magnitude greater than the metals. Similarly, the specific stiffness values for the advanced fibers (carbon/graphite, aramid, and ultra-high-molecular-weight polyethylene [UHMWPE]) are also about an order of magnitude greater than the metals. The specific stiffness of fiberglass is not as high.

An interesting property comparison related to specific strength can be seen by considering a length of suspended cable that can support its own weight. An example would be a tether of material used for a great height. In this scenario, carbon/graphite can support a length of about 140 miles, aramid about 150 miles, and steel only about 20 miles. The value of the specific strength for the advanced materials is obvious from these data.

The advantages and disadvantages of each fiber reinforcement type will become evident as each is discussed individually in this chapter. (The Composite Materials video [2005] from the Society of Manufacturing Engineers, which complements this section, offers excellent overviews of the reinforcements used in composites materials.)

GLASS FIBERS

Glass fibers as reinforcements have been known and used for centuries. Renaissance Venetian artisans incorporated glass strands in web-like configurations to strengthen their fine, thin-walled glass objects such as vases, pitchers, and decorative pieces. Commercially important continuous-filament glass fibers resulted from the joint Owens-Illinois and Corning Glass research, which culminated in the formation of a manufacturing facility in 1937.

Fiberglass was originally intended as an insulation material and that application continues to be important. During WWII when shortages of strategic materials forced manufacturers to seek alternate materials, fiberglass was combined with resin to create the first modern composites. These materials had a few wartime applications, but really increased in volume with the tremendous growth in the economy that occurred after WWII. Fiberglass was used to reinforce airplane parts, automobile bodies, recreational boats, and numerous other consumer products. Almost always made with polymeric resins as the matrix, the entire class of materials became known as **fiberglass reinforced plastics (FRP)**. That name persists today as a way to describe parts made with fiberglass and, usually, polyester thermoset resin.

Fiberglass is, by far, the dominant reinforcement material in terms of usage. This dominance is due to the combination of its low price (roughly an order of magnitude less than carbon fibers) and excellent properties. As can be seen from Table 8-1 and Figure 8-3, the strength and specific strength of fiberglass are quite good. Therefore, the use of this material in many consumer products is widespread, especially when seen in the context that for several years following WWII fiberglass was the only commercial reinforcement for composites. Early designs of many products were made around fiberglass and those products have served well. The high volume of use has contributed to the continuing low price of fiberglass. Therefore, for most consumer products, there has been little incentive to change to another reinforcement. This situation does not apply to aerospace where performance is highly valued and there are strong incentives to change to the more recently developed and higher-performing materials like carbon and aramid fibers.

Manufacturing Processes

The raw materials of glass fibers are silica sand and various subcomponents, such as limestone and boric acid, and some minor ingredients, such as clay, coal, and fluorspar. These are dry mixed and melted in a high-temperature refractory furnace. The temperature of this melt varies for each glass composition, but is generally about 2,300° F (1,260° C). Two similar processes are used for the manufacture of glass fibers from the molten mixture of sand and the subcomponents. These processes are called the marble process and the direct-melt process. The marble process is shown in Figure 8-4. As can be seen, after the glass is melted, marbles are formed, which allow for ease of transport of the glass and subsequent reheating as shown in the latter part of the figure. The direct-melt process simply removes the marble formation and couples the melt chamber directly to the filament formation stage.

The first process developed was the marble process. In this process, the raw materials are added to a blender where they are thoroughly mixed. The blender then feeds the materials to a furnace where they are melted. The molten glass exits the furnace into a machine that cools the glass and forms it into small glass marbles. These marbles

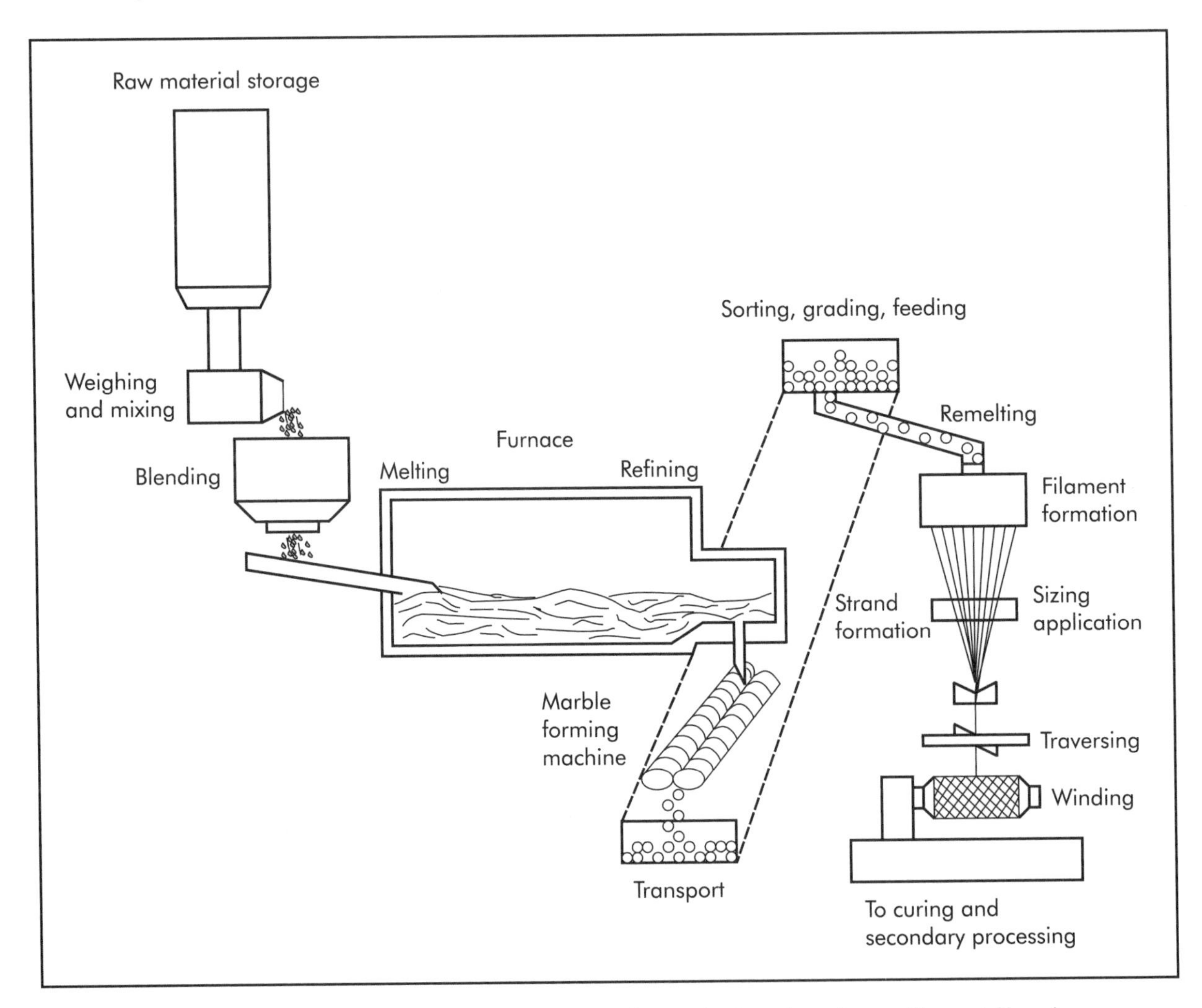

Figure 8-4. Schematic representation of marble melt process for production of continuous filament fiberglass.

are graded and inspected and then stored until needed for conversion into fibers. That conversion begins by feeding the marbles into a remelting furnace where a glass melt is again created. The fibers are formed by directing the melt into the formation bushings, which are low-corrosion metal (such as platinum) plates that have a multiplicity of tiny holes (typically 200 to 1,200). The molten glass flows through the bushing holes and forms continuous strands called **filaments**. Then, the filaments are cooled (quenched) by spraying with water or simply with air. Thereafter they are gathered together. The diameter of the fibers is controlled by the hole size in the bushing, temperature, viscosity of the melt, and cooling rate and method. A protective coating or sizing is applied so that the fibers can pass easily through mechanical equipment without breaking or abrading. A sizing may be applied before the fibers are cooled. The sizing may contain a coupling agent and other chemicals to enhance the bonding of the fibers to the matrix when they are later mixed together to make composites. These additives will be discussed in more detail later in this chapter. The fibers are drawn (mechanically elongated) and then wound up on rolls. They can then be transported to a curing step and other secondary processing before they are sent to the customer.

The direct-melt process is similar to the marble process in raw material mixing and melting. However, after the melting furnace, the molten glass flows directly into the fiber-forming bushing and plate without being made into the marbles. The downstream steps after the fibers have been formed are the same as with the marble process.

Originally, the use of the marbles resulted in improved control over the quality of the fibers. However, the direct-melt process has been sufficiently improved that it is now the major process used for making glass fibers.

Types of Glass Fibers

Glass is an amorphous material that consists of a silica (SiO_2) backbone with various oxide components to give specific compositions and properties. Several types of glass fibers are manufactured but only four are used in composites: **E-glass**, **S-glass** (and its variation S2), **C-glass**, and **quartz**.

E-glass fibers have a calcium aluminoborosilicate composition with maximum alkali content of 2%, similar to Pyrex®. They were originally used when strength and high electrical resistivity were required. E-glass fibers are still used in various electrical devices. Because of their good strength properties and low cost, E-glass fibers are the most common type of fiberglass used in composites. (The electrical properties have little importance in most composite applications.)

S-glass is approximately 35% higher in strength than E-glass and has better retention of mechanical properties at elevated temperatures as shown in Table 8-2. S-glass is preferred in advanced composite applications where fiberglass is used rather than one of the other high-performance reinforcements like carbon or aramid.

Other properties for E-glass, S-glass, and C-glass are given in Table 8-3. These data point out the advantages of E-glass in electrical applications and of C-glass in acid solutions.

Quartz fibers are made from mineral quartz, which is mined chiefly in Brazil. It is then made into rods that are drawn into fibers. The strength of quartz is approximately 25% lower and the density is only slightly less than fiberglass. Quartz is used only in applications where its superior softening point temperature and/or superior electrical signal transparency are necessary. Applications for quartz include its use in ablatives, thermal barriers, antenna windows, and radomes.

Table 8-2. Effects of temperature on the strength of glass fibers.

Glass Type	Temperature, ° F (° C)			
	–310 (–190)	72 (22)	700 (371)	1,000 (538)
	Tensile Strength, ksi (MPa)			
E-glass	770 (5,309)	500 (3,447)	380 (2,620)	250 (1,724)
S-glass	1,200 (8,274)	665 (4,585)	645 (4,447)	350 (2,413)

Applications

Fiberglass reinforced plastics (FRP) have become so widespread in use that it is hard to imagine what society would be like without them. Rather than attempt to list the applications or even the markets for FRP, a few applications will be discussed just to show the breadth of fiberglass composites use. More detailed discussions of applications are in the last chapter of this book.

When most people think of fiberglass, they envision a fiberglass composite boat. Certainly this is a major application for fiberglass and the reasons are obvious. Compared to other materials that have been used for boats, like wood and metal, fiberglass is easier to shape (mold), lighter, not susceptible to water logging or corrosion, more easily colored and, often, does not require painting. Furthermore, the overall cost of manufacture of FRP is often less than the other materials.

If boat hulls are thought of as panels, other major panel-like applications for FRP can be envisioned such as light rail cars, building roof structures, housings and cabinets for appliances, tub and shower units, wall and bathroom panels, and even automobile bodies.

Automobiles are a major market for FRP and not just for body panels. Many parts in automobiles are made of fiberglass composites. Some examples are the ducts for the air conditioning and heating, housings of many types throughout the car, and various small parts that are injection molded using fiberglass reinforced thermoplastics. In a similar product line, FRP has now become the standard material for entire truck cab bodies. In addition to its light weight and low corrosion properties, FRP allows a significant reduction in the number of parts. This saves on original assembly costs and improves durability because there are simply fewer parts and therefore fewer places for different sections to rub against each other and wear.

Thinking more about the injection molding of fiberglass composites, tremendous variety exists of thermoplastic parts reinforced with fiberglass. To strengthen parts, one of the first options considered is to make them out of the same resin but add fiberglass reinforcements. This decision usually allows the same mold to be used and, probably, the same resin manufacturer. Hence, the use of fiberglass is an easy way to improve part strength or stiffness. Some consideration should be given to toughness and elongation, but for many applications the reductions in these parameters is not critical.

Some sporting goods use FRP. Examples would be vaulting poles, bows and arrows, surfboards, snowboards, and skateboards. Snowmobile housings are usually FRP. Even the walls of swimming pools are usually strengthened with fiberglass.

Many fiberglass products are not readily observable and yet are critically important to everyday life. Some examples include the

Table 8-3. Properties of glass fibers.

Property	Type of Glass		
	E-glass	S-glass	C-glass
Coefficient of thermal expansion, 10^{-6} ° F (10^{-6} ° C)	2.8 (5.2)	3.1 (5.7)	4.0 (7.3)
Specific heat at room temperature, BTU/lb•° F (kJ/kg•° C)	0.193 (0.810)	0.176 (0.737)	0.188 (0.787)
Softening point, ° F (° C)	1,555 (846)	1,778 (970)	1,382 (750)
Dielectric strength, V/mil (kV/cm)	262 (103)	330 (130)	—
Index of refraction	1.562	1.525	1.532
Weight gain after 24 hours in H_2O, %	0.7	0.5	1.1
Weight gain after 24 hours in 10% HCl, %	42	3.8	4.1
Weight gain after 24 hours in 10% H_2SO_4, %	39	4.1	2.2
Weight gain after 24 hours in 1% Na_2CO_3, %	2.1	2.0	24

pipes that bring out water, especially the mains that run under the street, and pipes used in plants where many chemicals are processed. Storage tanks, especially for hazardous materials, are frequently made of FRP. Even the circuit boards in computers and other electronic devices are likely to be reinforced with fiberglass.

CARBON/GRAPHITE FIBERS

The demand for reinforcement fibers with strength and stiffness higher than those of glass fibers has led to the development of **carbon** or **graphite fibers**. Although carbon fibers were used by Thomas Edison for his first successful electric light bulb, the development of high-strength/high-modulus carbon fibers did not occur until the 1950s. By the 1960s, fibers based on rayon and polyacrylonitrile (PAN) had been patented and offered for commercial sale. In the late 1960s, pitch-based carbon fibers were developed and commercialized. Since then, the mechanical properties of carbon fibers have steadily increased through improvements in starting materials and manufacturing methods. Today, carbon/graphite fiber has among the highest specific strength and highest specific modulus of any material.

When originally developed, there was a difference between carbon and graphite fibers. The graphite fibers were made by processing carbon fibers through an additional high-temperature step. (The process for making carbon/graphite fibers is discussed later.) Today, the demands on performance have dictated that all carbon fibers meet the higher performance capabilities of graphite fibers, so there is no difference between the two types. Therefore, this text will generally use the term "carbon fiber" to indicate both fiber types unless the graphite fiber designation is required, such as when marketing and some processing issues are discussed. Today the designations of carbon fibers focus on the modulus and strength specifications of each variety.

Manufacturing Processes

The current preferred technologies for making carbon fibers use **polyacrylonitrile (PAN)**, **pitch**, and, to a lesser extent, **rayon**. All of these materials have high carbon concentrations relative to other atoms and all can be converted into a graphite material. Each process offers distinct advantages and disadvantages in terms of both fiber cost and properties. PAN-based fibers have good properties with relatively low costs for standard modulus products. Pitch-based fibers have high modulus and good thermal conductivity. Rayon-based fibers were the first carbon fibers made, but are now made only for longstanding customers for which they were specified many years ago.

The PAN-based and pitch-based fiber manufacturing processes are summarized in Figure 8-5. PAN is a commercially available textile fiber. It is therefore shown as a ready-made fiber for the starting material of the PAN-based carbon fiber process. PAN fibers are usually produced at a different facility than the carbon fibers and often by a different manufacturer. However, PAN fibers used to make carbon fibers are a different grade from those used in traditional textiles. The textile fibers are often referred to as acrylic fibers, but those intended for carbon fibers are simply called PAN. The PAN fibers are formed as strands and then gathered into a group called a **tow**, and wound onto a spool.

Pitch is the material that remains at the bottom of the distillation tower used to separate the components of coal tar or petroleum. It has high free-carbon content, thus making it a suitable starting material for making carbon fibers. In addition to this, pitch is used as a binding material for asphalt road surfaces and fuel briquettes. Pitch is an amorphous mass that must be formed into fibers. In the conversion process, pitch goes through a meso-phase in which the polymer chains are somewhat oriented even though the material is liquid—a liquid-crystal phase. This orientation is responsible for the ease of consolidation of the pitch-based product into a carbon/graphite form.

Both PAN- and pitch-based fibers must be stabilized by thermosetting (crosslinking and rearranging) so that the polymers do not melt in subsequent processing steps. This **stabilization** step, sometimes called the oxidation step, also must be accompanied by stretching of the fibers, which involves holding the fibers at a constant length against their inherent shrinkage as they become stabilized. Thermosetting is done with moderate heat and in an air atmosphere. The molecular structure of a PAN-based fiber during the stabilization step is shown in Figure 8-6. Two equivalent representations of the consolidation process are shown. Notice that rings are formed during the stabilization step. These rings are critical to the later formation of graphite. In the pitch-based process, a similar ring-based structure is formed. Tows of fibers from many spools are directed into the stabilization oven through a series of combs that keep the fiber tows separate but adjacent to one another as they go through the process. These tows are spread (flattened) by passing them over a bowed roller. Spreading allows equal access to the heat for all of the fiber strands.

The next step in the process is **carbonization**. (This step is sometimes also called pyrolization.) During this step the fibers are heated to higher temperatures in a nitrogen atmosphere and the tension on the fibers is maintained. Hydrogen molecules attached to the interconnected rings are removed by the heat and the chains of rings merge together to form plates. This process is depicted on the molecular scale in Figure 8-7. The material on the left is a product of the stabilization process. The molecules within a carbon strand are stretched and aligned. Two of these molecules are shown in the first part of Figure 8-6. One is reversed relative to the other, but they are otherwise equivalent. The heat and stretching pressure causes the molecules to join together by the elimination of hydrogen (dehydrogenation). The new structure emerges and is seen to have additional six-member rings at the junctions of the two previously separated molecules. This results in the formation of a plate-like structure. The carbon content after carbonization is 80–95%. Because of material loss and shrinkage during carbonization, the fiber diameter is reduced to about one-half the original diameter.

To achieve the high strength and stiffness properties demanded in today's marketplace, the fibers are further heated, with continued tension, under a nitrogen atmosphere in a step called **graphitization**. In this step, the nitrogen atoms present in the rings are eliminated and the ring structure is further consolidated. The graphitization structure is shown in Figure 8-8. Again, the product of the previous step is shown on the left side of the figure along with another, equivalent plate. The heating and tension during graphitization eliminates the nitrogen and new rings are formed that join

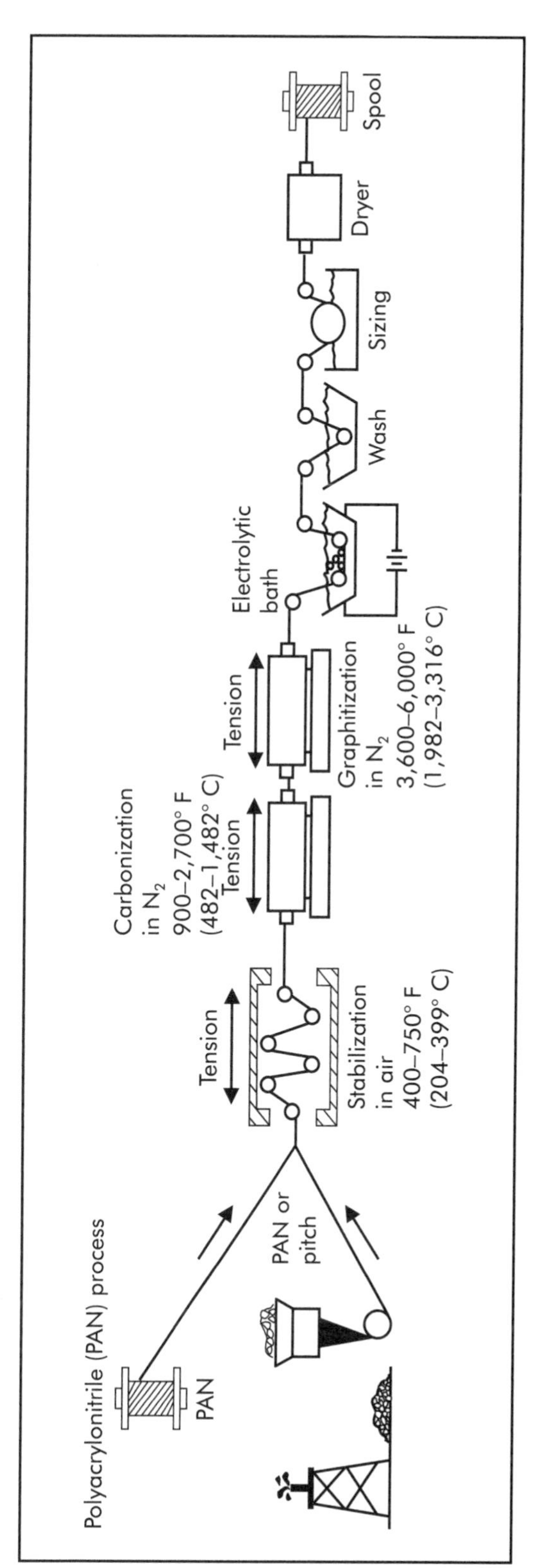

Figure 8-5. Manufacturing process for making carbon/graphite fibers.

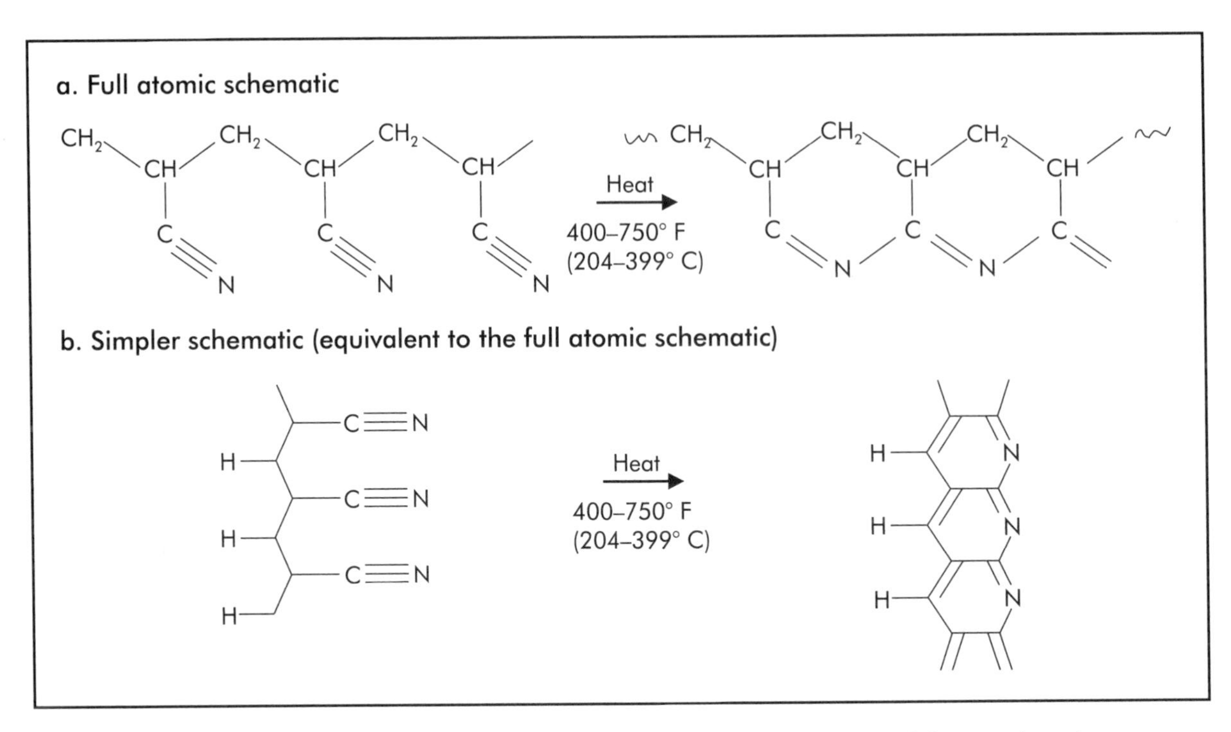

Figure 8-6. Structural changes of PAN-based carbon fiber during the stabilization step of the manufacturing process.

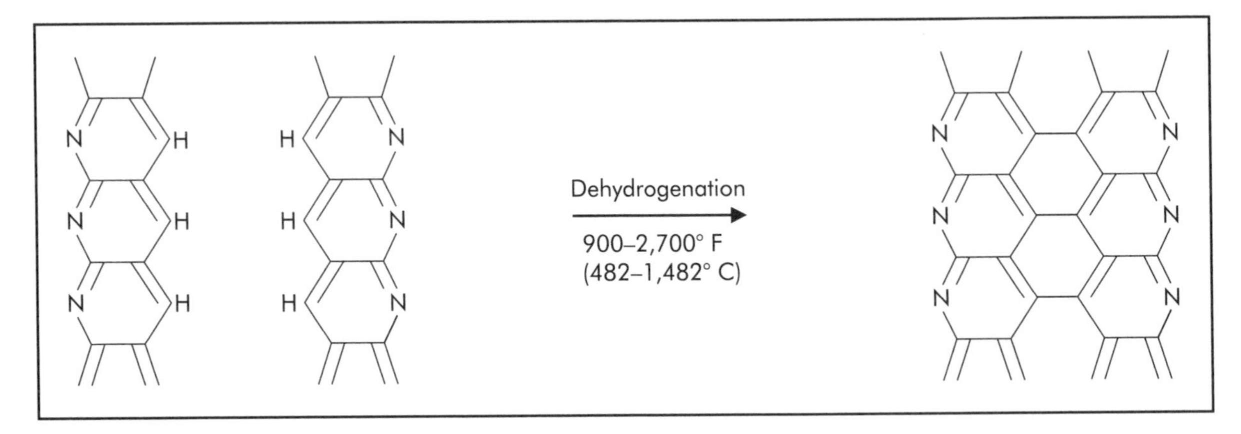

Figure 8-7. Structural changes of PAN-based carbon fiber during the carbonization step of the manufacturing process.

the previously separate plates. The result is a larger plate with only a few nitrogen atoms on the edges. The carbon content of the fibers after graphitization is generally above 99%. The differences in modulus and strength of the fibers are largely dictated by the conditions present during graphitization. Higher-modulus fibers are processed at higher temperatures and for longer times.

After graphitization, the fibers are cleaned in an electrolytic bath. This bath removes debris that might have been formed during the heating processes and oxidizes the surface of the fibers. The oxidation of

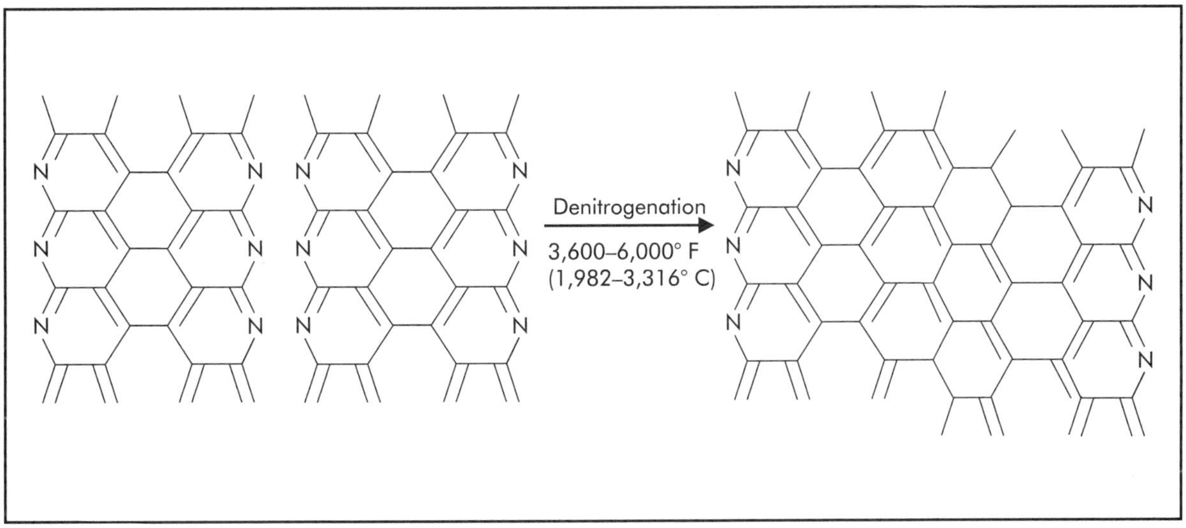

Figure 8-8. Structural changes of PAN-based carbon fiber during the graphitization step of the manufacturing process.

the surface adds chemical groups, like hydroxyls, to the carbon surface and improves bonding, which is needed for subsequent surface finishes.

Some manufacturers and, especially laboratory researchers, have investigated non-electrolytic surface treatments of carbon fibers. Gas-phase oxidations and radio frequency (RF) plasma treatments have both been demonstrated but fiber properties are lower than with the electrolytic process. Liquid oxidization can be done by simply pulling the fibers through a bath of some convenient oxidizing agent such as nitric acid, potassium permanganate, or sodium hypochlorite. However, the electrolytic process is still preferred because it can be run continuously whereas with the solution processes the oxidizing agent must be added intermittently.

Some non-oxidizing treatments also have been demonstrated in the laboratory. One method, called whiskerization, grows ceramic whiskers (such as SiC, TiO_2, or Si_3N_4) on the surface of the carbon fiber. The whiskers are oriented perpendicular to the surface of the fibers. While these whiskers greatly increase the ability of the carbon fibers to bond to the matrix, the whiskerization process is expensive and thus is not used commercially. Another non-oxidative method, called pyrolization, consists of vapor-phase deposition of pyrolitic carbon on the fiber surface. The method shows good increases in bonding to the matrix, but is also expensive.

The effectiveness of the surface treatment of the carbon fibers is often measured by examining the wettability of the fiber. The extent of the wetting of the fiber by water is directly proportional to the number of chemical groups attached and, therefore, the effectiveness of the surface treatment. Wettability is most often measured by looking at the contact angle of the fiber/water interface.

After washing the electrolyte solution off, the fibers are introduced into a bath where an appropriate polymeric finish is applied. Usually, this finish is an epoxy so that good compatibility with the eventual matrix resin can be achieved. After drying, the fibers are then wound back onto spools. It should be noted that each tow of fibers that started through the process is retrieved at the end as a separate tow. Therefore, each spool of original fiber is again spooled at the end of the process.

Throughout the process several changes have occurred in the fibers. Some typical values for these changes in a PAN-based fiber are:

- The density increased from 6.94 to 1.03 oz/in.^{3} (1.20 to 1.78 g/cc) with changes occurring more or less uniformly throughout the process.
- The diameter decreased from 550 to 291.4 μin. (13.97 to 7.40 μm) with changes occurring more or less uniformly throughout the process.
- The resistivity dropped from 454 to .0008 ohm-in. (1,154 to 0.002 ohm-cm) with most of the change occurring during graphitization.
- The tensile strength increased from 36 to 600 ksi (248 to 4,137 MPa) with most of the change occurring during graphitization.
- The modulus increased from 1.5 to 33 Msi (10 to 228 GPa) with most of the change occurring during graphitization.
- The elongation to failure dropped from 4.8 to 1.6% with most of the change occurring during oxidation.
- The cost of the fiber has increased significantly from $3.88 to $8.30 per pound ($8.55 to $18.29 per kilogram) with costs spread more or less evenly over the entire process.

The manufacturers of the fiber add new plant capacity as the volume increases and, little by little, the price drops. A major innovation to reduce the price has been the use of larger tows, that is, tows with more strands. The standard for many years was 4,000–12,000 strands per tow. Some manufacturers have begun making 50,000 strands per tow in an effort to more efficiently use the production equipment. For some applications, these larger tows seem to work well, although process problems associated with handling the larger number of fibers remain a problem for the molder. Some applications, such as aerospace, do not use the larger tows. This is because of the greater possibility that some of the strands, especially those in the center of the tow, are not fully exposed to the thermal treatments of the manufacturing process and therefore have lower physical properties. The conversions of pitch-based and rayon-based fibers are conceptually similar.

Structure

As shown in Figure 8-9, the plates stack upon each other to form a three-dimensional structure. They are oriented within the fiber so that the long axis of the fiber is the same direction as the long axis of the plates. Hence, when the fibers are pulled in tension, the force is pulling against the ring structures of the plates. These rings are formed by strong covalent bonds so the resistance against the tensile force is strong. The ring structure is also dimensionally stable, thus adding even more strength in the fiber axis direction and significantly improving the stiffness (modulus) of the fibers.

In the cross-fiber direction there are bonds. These bonds, however, are relatively weak Van der Waals bonds that form to hold the stacked rings in place. The weakness of these bonds in traditional bulk graphite (not fibers) is attested to by how easy it is to write using a pencil. Remembering that the bulk graphite of a pencil is made so that there is little alignment of the graphite plates, it can be extrapolated that the pencil mark is formed when one layer of graphite is separated from the stack and deposited on the paper. However, the direction of the plates within a fiber dictates that the sloughing off of the layers, were that to occur, would not be at the end of the fiber but, rather, off the sides.

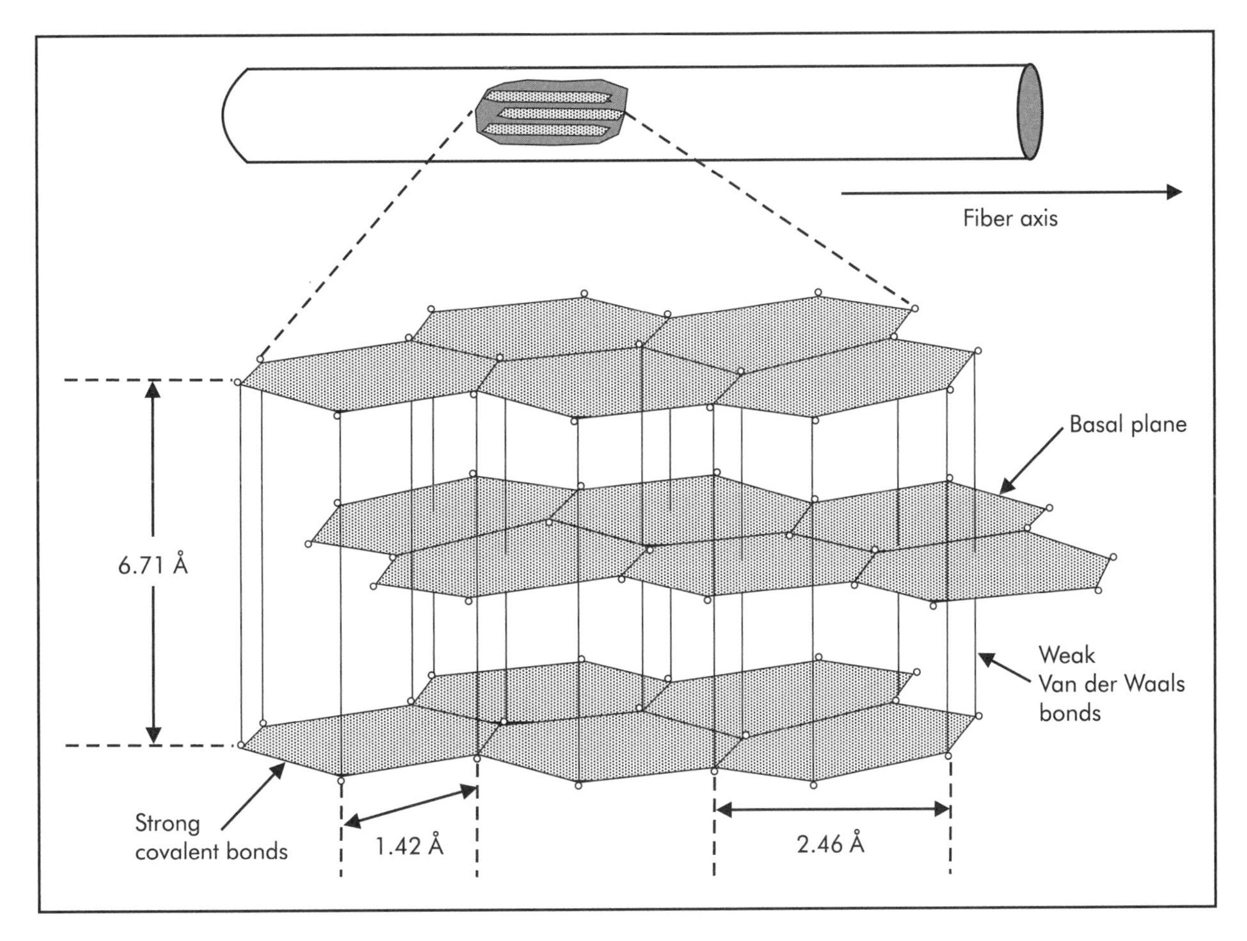

Figure 8-9. Three-dimensional representation of the carbon fiber structure.

Because carbon fibers are made under tension, the rearrangements of the atoms and molecules tend to eliminate the inevitable imperfections that occur along a plate and the non-planarities that can occur between plates. The rearrangements improve the natural stacking of the plates and result in a significant increase in the number of Van der Waals bonds. This stabilizes the cross-fiber direction and increases the size of the plates so that the surface of the carbon fibers has little sloughing off of the layers of graphite. Still, carbon fibers have a tendency to abrade along their surface, so surface cleaning via electrolytic treatment and addition of polymeric sizing are important to reduce the tendency for surface abrasion.

Properties

Carbon fibers are available in a variety of grades from many different manufacturers. These grades are generally grouped according to the modulus of the fibers. However, the presentation of property data can be misleading because of the large number of grades and different points of optimization for the various fibers. The existence of three major types of carbon fibers (PAN-based, pitch-based, and rayon-based) further complicates the picture. Nevertheless, property data for a variety of carbon fibers are presented in Table 8-4.

Analyses of the data from Table 8-4 tests reveal that, for PAN-based fibers, both strength and stiffness increase going from

Table 8-4. Properties of various carbon fibers available from manufacturers.

Fiber Type	Tensile Strength, ksi (MPa)	Tensile Modulus, Msi (GPa)	Elongation to Break, %
PAN-based Fibers			
Standard modulus	500 (3,447)	33 (228)	1.6
	653 (4,502)	34 (234)	1.8
	512 (3,530)	33 (228)	1.5
Intermediate modulus	780 (5,378)	40 (276)	2.1
	880 (6,067)	42 (290)	2.1
	924 (6,371)	43 (297)	2.2
Ultra-high modulus	500 (3,447)	64 (441)	0.5
	640 (4,413)	64 (441)	1.0
	554 (3,820)	85 (586)	0.7
Pitch-based Fibers			
Standard modulus	276 (1,903)	55 (379)	0.5
Intermediate modulus	305 (2,103)	75 (517)	0.4
Ultra-high modulus	525 (3,620)	105 (724)	0.5
	527 (3,633)	128 (883)	0.4
Rayon-based Fibers			
	110 (758)	6 (42)	—
	119 (821)	5 (35)	—

standard modulus to intermediate modulus materials. However, when going from intermediate modulus to ultra-high modulus, the strength goes down and the modulus goes up. These trends confirm what has been seen in tensile measurements taken on fibers at various stages of the manufacturing process under a variety of conditions. These data suggest that both strength and modulus initially increase with treatment temperature but that after a certain point, a maximum in strength is reached. Beyond that point a trade-off occurs—optimizing for a higher modulus will decrease the tensile strength.

The optimization of modulus is generally a function of the temperature of the graphitization stage in the process, although time and other factors also can have effects. The increase in temperature and time during graphitization increases the carbon content of the fiber. This suggests that as the non-carbon atoms are driven off, the fiber continues to consolidate and form a more orderly structure—both are factors that tend to increase modulus. However, the decrease in strength suggests that defects are being created in the fibers and these defects are points where failure can begin. The trade-off between strength and modulus is illustrated in Figure 8-10. For pitch-based fibers, the trade-off is not as dramatic, at least within the range of products currently in production. However, the elongation for the pitch-based fibers is generally lower than for PAN-based fibers. Therefore, pitch-based fibers are being used in applications where both strength and modulus are critical, and the lower elongation is not a problem. In general, however, PAN-based fibers are more commonly sold in the marketplace.

The decrease in strength that occurs when defects are present also can be seen in the effect of fiber length on tensile strength. This trend reflects the essentially brittle nature of the carbon fiber. In such a material, and similarly to ceramics, whenever a defect is present, the defect serves as a failure initiation point. As the length of the test fibers increases, there is a greater chance of having a defect within the test sample and, therefore, the tensile strength decreases. The tensile strength also decreases with fiber diameter for a similar reason—the more material present in the test fiber, the more likely the chance of having a defect in the sample.

Although mechanical properties like tensile strength and modulus are the most important for most applications of carbon fibers, thermal properties are becoming increasingly important. One of those thermal properties is the coefficient of thermal expansion (CTE). For almost all materials, the CTE is positive because the increased motion of the molecules at higher temperatures

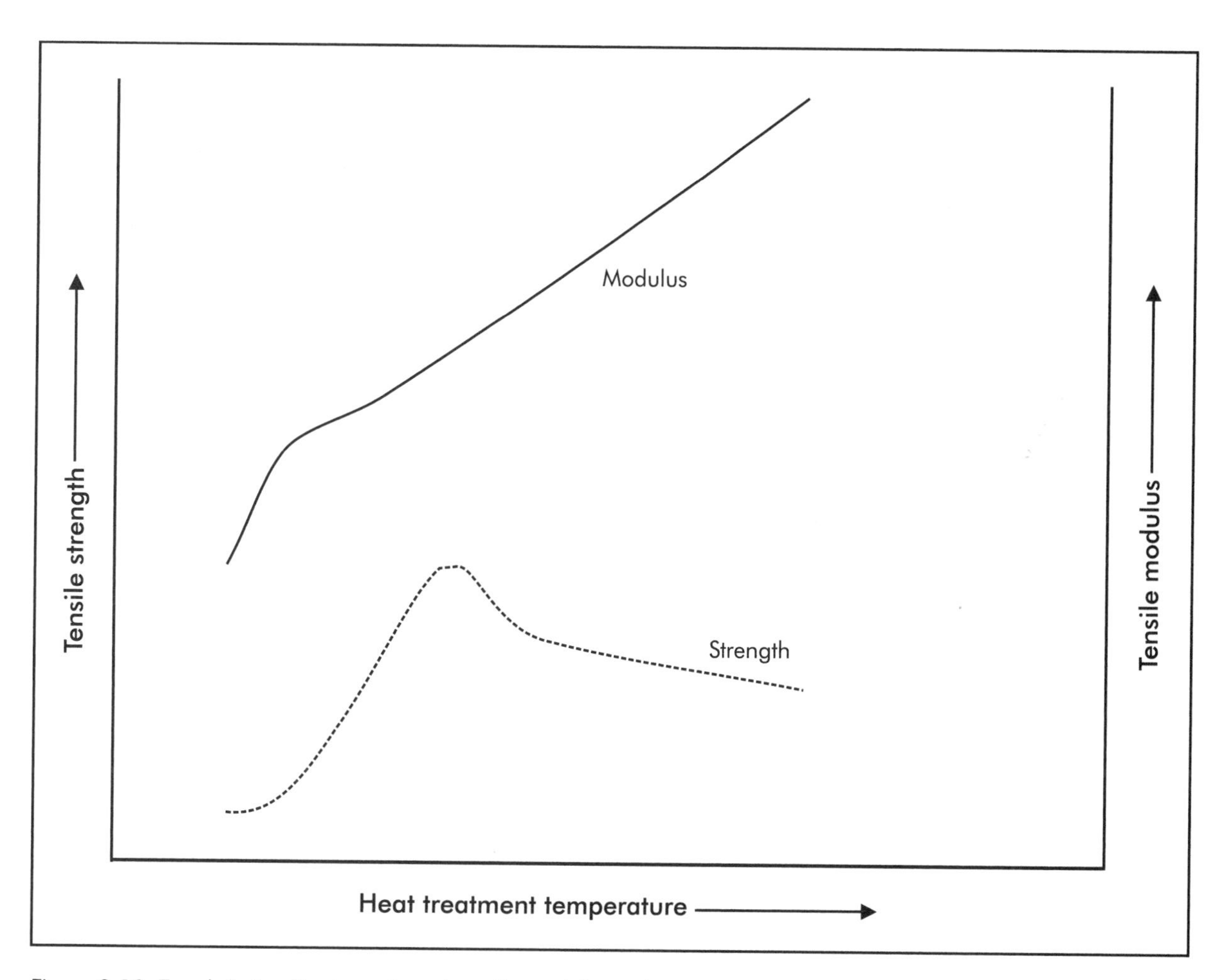

Figure 8-10. Trends in tensile strength and tensile modulus with temperature for PAN-based carbon fibers.

will increase the size of the sample. Carbon fibers are the rare exception to this rule. They have a negative CTE indicating that as the temperature goes up, the sample gets smaller. (For the carbon fibers, the CTE is measured in the longitudinal direction, that is, along the major axis of the fiber.) The contraction is stronger in ultra-high modulus than in standard-modulus fibers. The explanation given is that the carbon fibers are contracting to achieve better symmetry and uniformity in the molecular structure (toward the idealized perfect graphite crystal).

Another thermal property of great interest and increasing importance is thermal conductivity. Pitch-based carbon fibers have high thermal conductivity, surpassing that of most metals. These fibers can be used as radiators (heat pipes) to transmit heat away from the heat source. The thermal properties of various materials are shown in Figure 8-11.

Carbon fibers are moderately good conductors of electricity. This is assumed to arise from the natural conductivity of graphite and the orientation of the rings that lie parallel to the axis of the fiber.

Carbon fibers are elastic to failure at normal temperatures (meaning that they do not deform before failure), which renders them creep resistant and not susceptible to fatigue. They are chemically inert except in strongly oxidizing environments or when in contact with certain molten metals. Carbons have excellent damping characteristics as well.

However, carbon fibers have some shortcomings. They are brittle, have low impact resistance and low break extension (elongation), relatively low compressive strength, and a small coefficient of linear expansion, which is a benefit except that it might complicate processing. They are also expensive, although the price has continued to decrease with increased production.

Applications

The ideal engineering material would have high strength, high stiffness, high toughness, and low weight. Carbon fibers, combined with high-performance matrices, meet these criteria more closely than any other material. The uses of carbon fiber composites obviously depend upon the unique properties available with these materials. Although there are some obvious overlaps between categories, it is convenient to examine applications according to the primary property utilized.

Representative applications based upon *strength*, *stiffness*, and *low weight* include the following:

- aircraft control surfaces and, increasingly, full fuselages;
- helicopter rotor blades and wind turbine blades;
- aircraft structural parts such as doors and landing gear assemblies;
- automotive drive shafts and leaf springs;
- racing car bodies and frames;
- spacecraft, rockets, and missiles; and
- high-precision tooling.

These applications are based on *thermal properties*:

- heat shields for missiles and rockets;
- brakes;
- aerospace antennas (because of the low coefficient of thermal expansion);
- space structures such as telescope mounts; and
- housings for computers, small motors, and electrical control panels.

The following applications are based on *chemical inertness*:

- storage tanks (especially when weight is a consideration such as the waste tanks on airplanes);

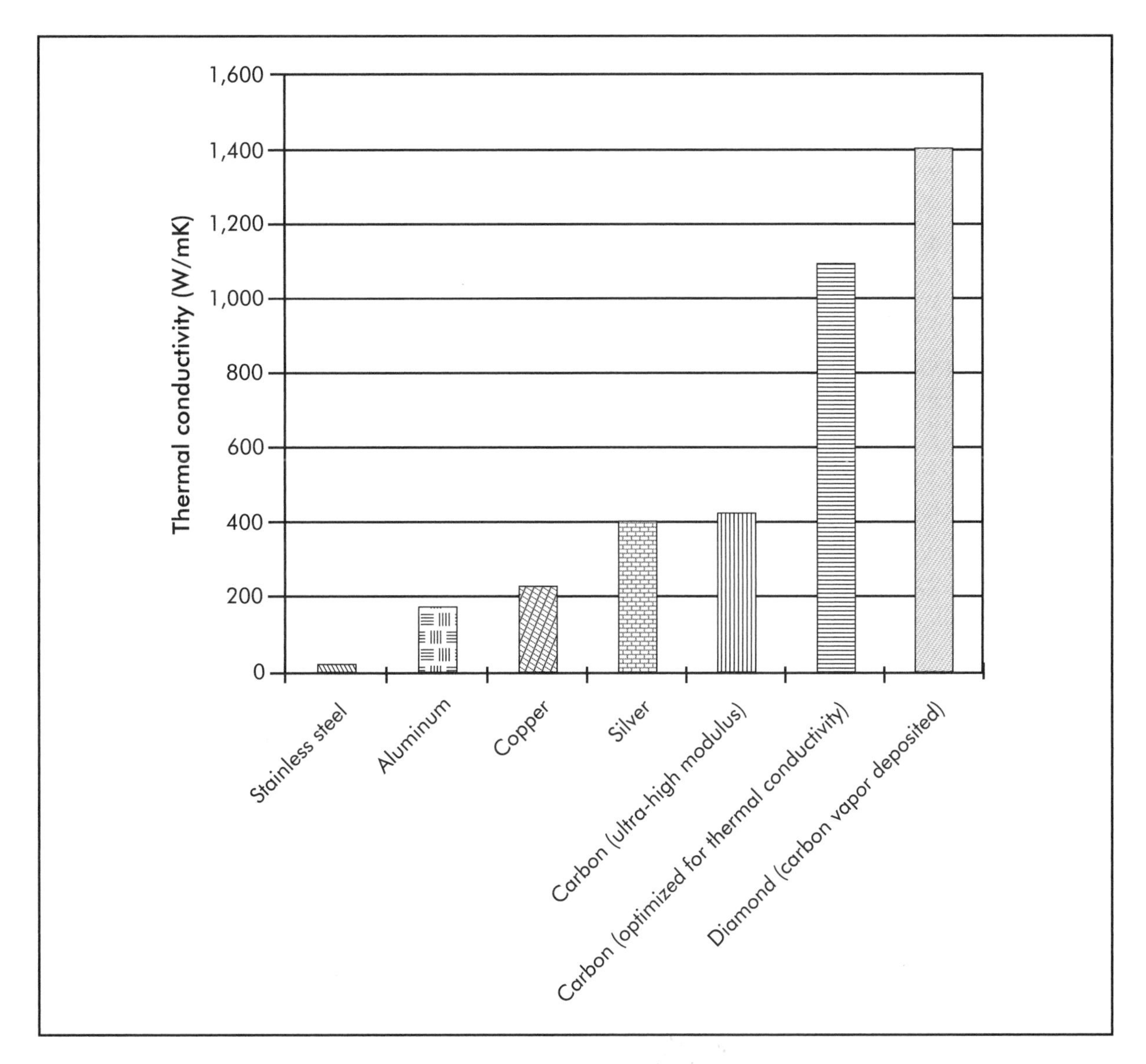

Figure 8-11. Thermal conductivities of various materials.

- bridge structures (which will not corrode and have good seismic resistance); and
- uranium enrichment centrifuge in the nuclear industry.

Applications based upon *rigidity* and *good damping*:

- musical instruments;
- audio speakers;
- rollers for industrial processing such as in the paper industry; and
- arms for mounting the heads to read computer storage devices.

These applications are based on *electrical properties*:

- shields against radio frequency interference;
- circuit boards; and

- touch switches.

Applications based upon *biological inertness* and *x-ray permeability*:

- artificial joints;
- heart-valve components; and
- x-ray tables and mounting arms.

Applications based upon *fatigue resistance* and *self-lubrication*:

- textile machine components;
- air-slide valves;
- compressor blades; and
- artificial limbs.

Most of these applications for carbon fibers are growing rapidly and new applications are constantly being created. Therefore, the volume of carbon fiber usage keeps increasing and the price keeps dropping. There is probably a minimum price based on the cost of the raw material, but even that price is subject to the normal price-volume relationships. So, the price of carbon fiber may continue to drop. This should further accelerate the number of uses as it becomes ever more competitive.

ARAMID AND OTHER ORGANIC FIBERS

The most common organic fibers used for reinforcements are the aramids, with Kevlar®, a DuPont fiber, currently the major brand. Other aramids are available but, as yet, have not penetrated the market as fully. Another high-performance organic fiber is made of UHMWPE. This fiber is sold in the United States as Spectra® and in Europe as "Dynema." Other organic fibers might be used as reinforcements, but generally not in high-performance applications. For example, nylon fibers might reinforce tires or conveyor belts. These materials are certainly composites, but they are not generally considered with the aramid and UHMWPE fibers. This is because their properties are much lower and therefore the improvement in properties that they impart is less significant. These other organic fibers are, as a group, usually referred to as **textile fibers**. These are not discussed in this text except in passing.

Aramid

Manufacturing Process

In 1971, **aramid fibers** were introduced in commercial products. They were originally used as reinforcements for tires, belts, and other rubber-related goods. After a few years, they became widely used for ballistic protection and then as a reinforcement for high-performance composites. The fibers have also found applications in high-strength cloth, such as that used for sails on racing boats and protective clothing and gloves.

The synthesis and structure of the aramid polymer are shown in Figure 8-12. The polymer is a polyamide having the same chemical linkages (amides) along the backbone as nylon. However, in the case of nylon, these linkages join segments that are aliphatic, whereas with aramid, the segments are wholly aromatic (that is, they contain the benzene ring). Therefore, as with other highly aromatic molecules, especially when the aromatic rings are in the backbone, the strength and stiffness of aramids are high. Note also that the polymer is highly polar (as are all polyamides) and so there is significant intermolecular bonding as illustrated in Figure 8-12. These intermolecular bonds further increase the strength and stiffness of the polymer.

After the polymer is formed it must be spun into fibers. This is done by filtering, washing, drying the polymer, and then dissolving it in an appropriate solvent as illustrated in Figure 8-13. Because of high regularity, high aromaticity, and intermolecular bonding, the polymer is difficult to dissolve. That difficulty can be appreciated by noting that the solvent used is hot, concentrated sulfuric acid. Even in solution the polymer

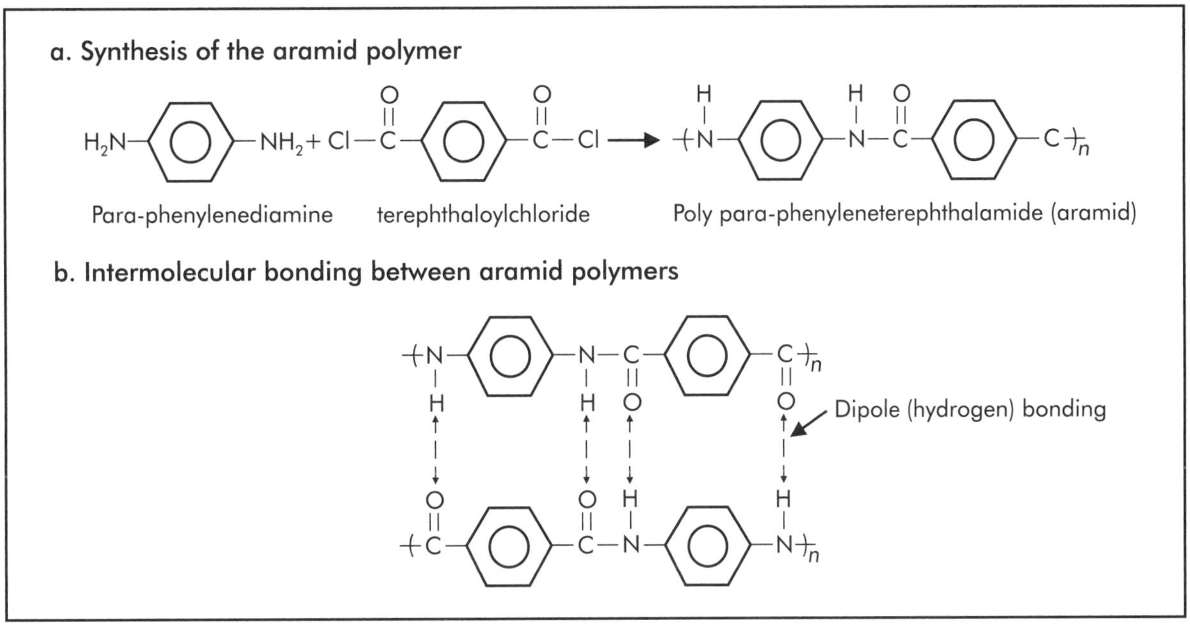

a. Synthesis of the aramid polymer

Para-phenylenediamine terephthaloylchloride Poly para-phenyleneterephthalamide (aramid)

b. Intermolecular bonding between aramid polymers

Dipole (hydrogen) bonding

Figure 8-12. Synthesis of aramid and its intermolecular bonding.

maintains some intermolecular bonding and forms rod-like crystals, similar to the liquid-crystal thermoplastics discussed previously. The solution is then forced though a plate (**spinneret**) into which many small holes have been drilled. The material emerges as continuous fibers. The fibers are then washed, collected, dried, and wound onto spools. As indicated by the inherent rod-like nature and high crystallinity of the fibers, the molecules are already aligned in the long direction as they are formed. Therefore, aramid fibers are not drawn.

In contrast to glass fibers and carbon fibers that have sizing coated onto their surfaces, aramid fibers often do not require sizing. This is possible because aramid fibers are tough and not, therefore, subject to the same incidental processing damage that occurs with glass and carbon fibers. However, if the fibers are to be subjected to extensive mechanical operations, such as during braiding or some weaving, a sizing can be applied. As will be discussed later in this chapter, a coupling agent may be applied to improve the bonding between the fibers and matrix. However, in the case of aramids, there are some applications, such as for ballistic protection, where the bond between matrix and fiber is intentionally weak so it can be broken during impact and, thus, absorb more energy.

Properties

Three grades of aramid fibers are available from DuPont and similar aramids are available from other companies. These grades are designated as Kevlar 29 (high toughness), 49 (high modulus), and 149 (ultra-high modulus). The differences between the grades are determined by the processing conditions, which generally give more crystallinity to the higher-modulus fibers. (The mechanical properties for these fibers are given in Table 8-1 and the specific strengths and specific modulus are given in Figures 8-2 and 8-3.) Aramid fibers are less dense than fiberglass and carbon fibers, and intermediate in strength

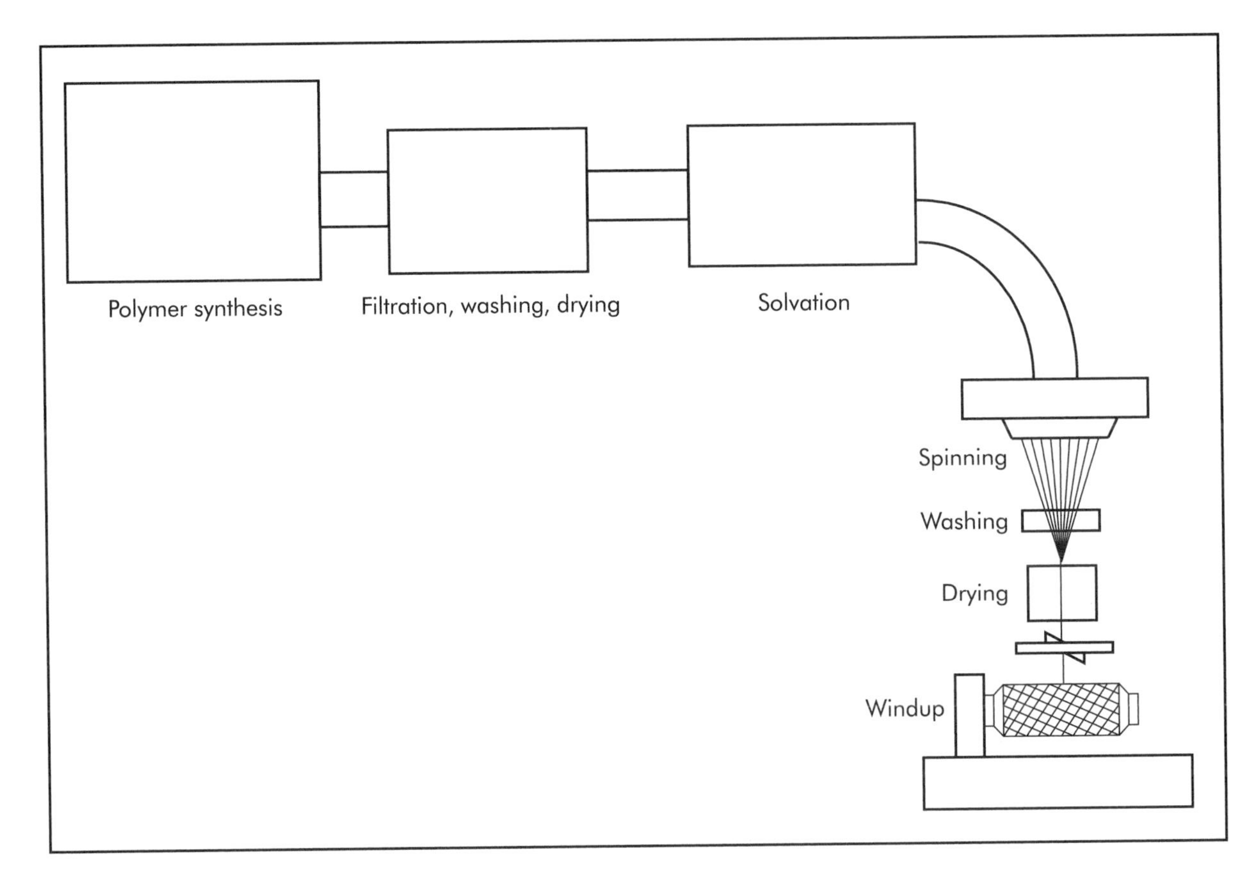

Figure 8-13. Manufacturing process for aramid fibers.

and stiffness between glass and carbon fibers. Therefore, the specific strength of the aramids is quite high, roughly comparable to the specific strength of carbon fibers. The specific modulus of aramids is higher than glass but not as high as the outstanding specific modulus of carbon.

The elongations of aramid fibers are much higher than either fiberglass or carbon fibers. That elongation, plus the excellent tensile strength, results in high toughness for aramid fibers that the other fibers do not offer. This toughness is one of the most important properties and leads to the majority of applications in which aramid reinforced composites are used.

The failure of aramid fibers is unique in that when they fail, the fibers break into small fibrils, which are like fibers within the fiber. These fibrils arise from the rod-like structure of the liquid crystals as they are spun into fibers. Tensile strength failure is believed to initiate at the ends of the fibrils and is propagated through the fiber by shear. Therefore, aramid fibers fail by a series of fibril failures rather than a brittle failure like carbon and glass. Analogous to this would be the difference between the failure of a cable consisting of many fibers versus failure of a solid rod. Even though the size of the cable and rod might be the same, the cable is tougher because it has many failure modes and is far less susceptible to formation of a single defect leading to brittle failure. The unique failure mechanism of aramids is responsible for their high strength and high toughness. However, the fibrils do not have the same resistance to compression forces as do the monolithic glass and carbon fibers. Therefore, in general, aramid fibers are rarely used when compressive force resistance is important.

The effect of aramid toughness can be seen in Figure 8-14. Pressure bottles made of aramid fiber composite and carbon fiber composites (the same matrix resin) were subjected to impacts and then tested for retention of original pressure-holding capability. Clearly the retention of pressure strength was much higher for the aramid-reinforced bottles.

Aramid fibers are made by a traditional polymeric synthesis, which is modified to allow for its unique characteristics. Since aramids are thermoplastics, they possess many of the same fundamental characteristics. (Although carbon fiber also starts as a thermoplastic polymer, PAN, the heat treatments of carbonization and graphitization change it into a ceramic-like material.) The thermoplastic nature of aramids means that, among other things, the fibers are sensitive to heat. That sensitivity is seen in Figure 8-15. The decrease in tensile strength seen when the aramid fibers are exposed to a high temperature (482° F [250° C]) for extended periods is not unexpected in light of the thermoplastic nature of the fibers. However, the negligible decrease in tensile strength at lower temperatures (up to 302° F [150° C]) is encouraging and indicates that aramid fibers have good resistance to heat when compared with other thermoplastic materials. This characteristic is consistent with the high melting point/decomposition temperature of aramid fibers (932° F [500° C]).

Another characteristic of thermoplastic resins that should be considered with aramids is sensitivity to solvents. As already indicated, aramid is resistant to most solvents. This is confirmed in data presented in Table 8-5. Retention of tensile strength is shown for aramid fibers exposed to several aggressive liquids. Only in some of the strongest acids,

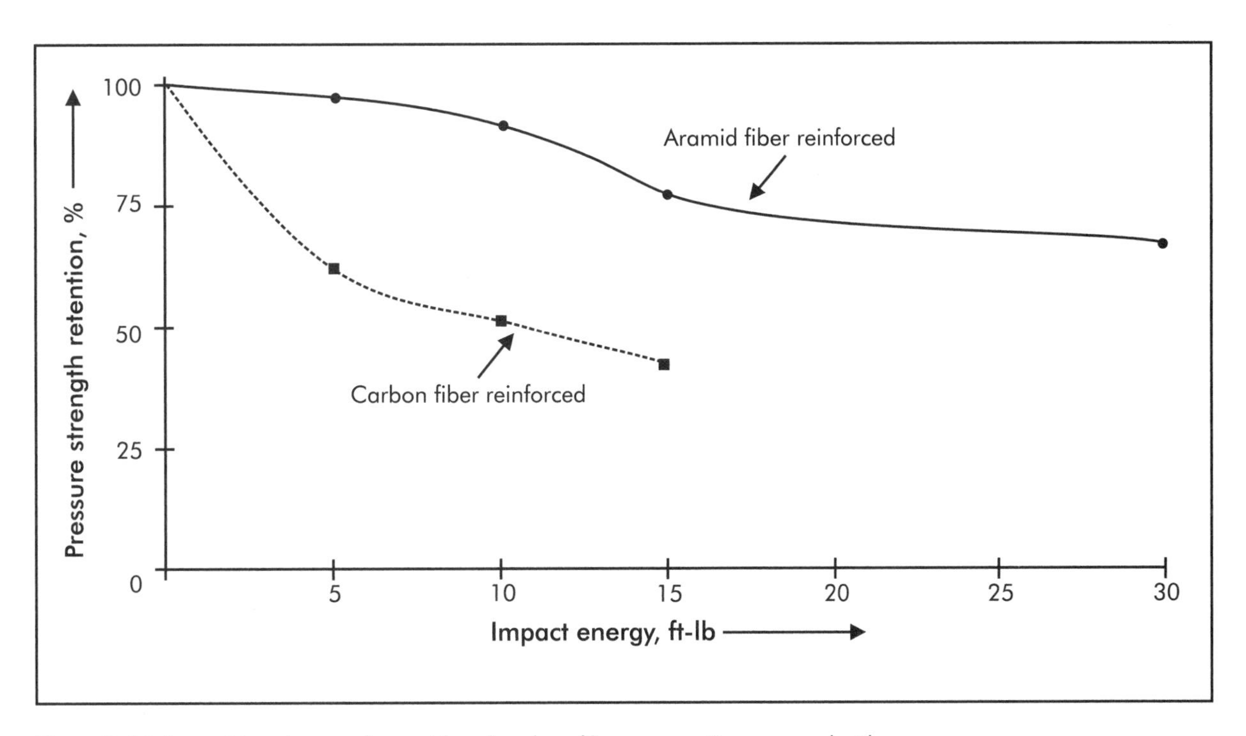

Figure 8-14. Impact toughness of aramid and carbon fiber composite pressure bottles.

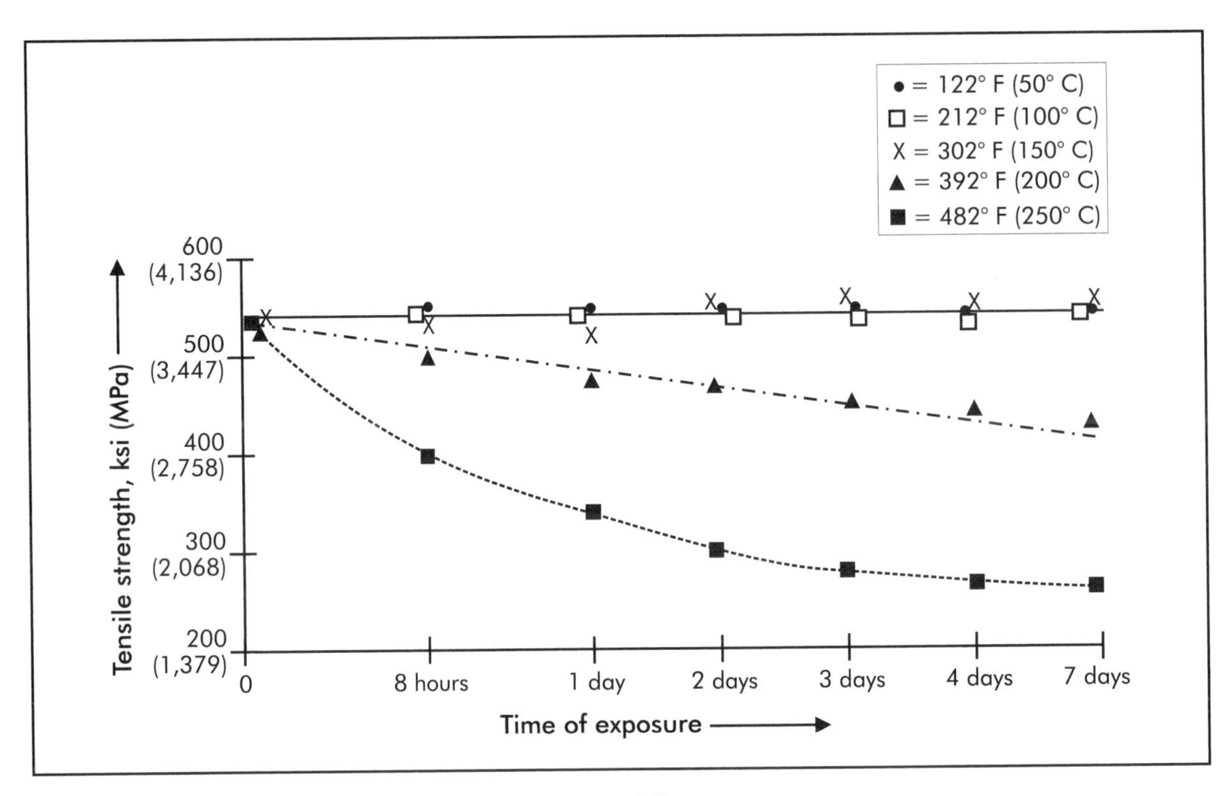

Figure 8-15. Effect of temperature on a high-modulus aramid fiber.

such as hydrochloric, nitric, and sulfuric, are aramid fibers strongly affected.

Polymeric materials are affected by sunlight to a greater extent than are ceramics and metals. Therefore, it is prudent to investigate the effect of outdoor exposure on aramid fibers. For example, two aramid products—a rope of .5 in. (1.3 cm) and a plain-weave fabric were exposed to sunlight for various exposure periods. As would be expected, the rope gives better protection against the sunlight because it is thicker and therefore protects the inner fibers better (it is actually the ultraviolet [UV] light of the sun doing the damage). The fabric, with its inner fibers only within the woven small fiber yarns, shows rapid deterioration. Because tests of this type often take 50% loss in a property as the end-point of the test, the fabric test was terminated after only 5 weeks of exposure. By contrast, the rope held up well for over 24 months losing only 69% over that period.

In light of the above somewhat negative results, the positive aspects of aramid fibers should be re-examined. The most important positive property is impact damage prevention and this can be seen in the data presented in Figure 8-16. For this test, prototype surfboards were constructed. Each of them was a 1-in. (2.5-cm) thick structure with a layer of composite material on the bottom, a layer of polyurethane foam on top of that, and a test layer on top of the foam. The test layer was made of 6.0 oz/yd^2 (203 g/m^2) fiberglass cloth and polyester resin in the first prototype, of 5.6 oz/yd^2 (170 g/m^2) aramid cloth in the second prototype, and of 3.5 oz/yd^2 (119 g/m^2) in the third prototype. Weights were dropped onto the sample from different heights to show the effects of increasing impact. The data show clearly that

Table 8-5. Chemical resistance of aramid fibers.

Liquid in which Fibers are Soaked at 70° F (21° C)	Tensile Strength Retention, %
Acids	
Acetic (99.7%)	100*
Hydrochloric (37%)	36*
Hydrofluoric (48%)	94*
Nitric (70%)	13*
Sulfuric (96%)	0*
Bases	
Ammonium hydroxide (28%)	92**
Sodium hydroxide (40%)	97*
Solvents	
Acetone	100**
Carbon tetrachloride	100**
Benzene	100**
Other Chemicals	
Gasoline	100**
Salt water (5% solution)	99.5*
Water (boiling)	98*

* = 100 hours exposure
** = 1,000 hours exposure

the fiberglass prototype sustains much more damage than either aramid prototype. In fact, even the second aramid sample (about half as heavy as the fiberglass), is superior in impact damage protection.

Applications

Several uses for aramids take advantage of its desirable combination of light weight, strength, modulus, and especially toughness. One use is in ballistic protection. Not only is aramid fiber used for bullet-proof vests, but it is also used as armor for ships and motorized combat vehicles such as tanks and personnel carriers. Battlefield shelters made of aramid fibers are lightweight while retaining strength and antiballistic properties.

Other uses include leading edges of aircraft wings and other structures where impact damage might be expected. For example, the leading edges of the Voyager aircraft that flew nonstop around the world were made of aramid. Damage control applications have also included protecting parts while still in the fabrication stage. This is accomplished by making an aramid-reinforced shield for the part that can be removed when it is ready for use.

Aramids have found a special niche in high-performance pressure vessels. In these applications, not only is the hoop stress high, but the ability to withstand impact damage is important because they may be subjected to incidental damage and must not leak.

Ultra-high-molecular-weight Polyethylene Fiber Synthesis and Manufacture

People familiar with plastics are often amazed that a high-performance fiber could be made from polyethylene. Traditional polyethylene is the highest volume plastic and is used for a wide variety of applications ranging from trash bags to milk jugs to toys for children. Traditional polyethylene is not, however, used as a textile fiber because of durability (cracking) problems. Therefore, the use of a polyethylene polymer for composites suggests that something special has been done to the polyethylene.

One obvious change in UHMWPE fibers from traditional polyethylene is the molecular weight. As the name suggests, UHMWPE fibers have molecular weights that are orders of magnitude higher than traditional polyethylenes. This higher molecular weight improves almost all physical properties. The techniques for polymerizing

Impact data for various test layer configurations

Drop Force, in.-lb (N-M)	Fiberglass, 6.0 oz/yd² (203 g/m²)	Aramid, 5.0 oz/yd² (170 g/m²)	Aramid, 3.5 oz/yd² (119 g/m²)
20 (2.3)	Resin cracks	Slight indent	Slight indent
30 (3.4)	Resin cracks, fabric ruptured	Slight indent	Slight resin cracking
40 (4.5)	Resin cracks, fabric ruptured	Slight resin cracks	Slight resin cracking
50 (5.6)	Severe resin and fabric damage, core exposed	Slight resin cracks, no fabric damage	Large indent, resin cracks, no fabric damage
60 (6.8)	No test	Large indent, resin cracks, no fabric damage	Large indent, resin cracks, no fabric damage
70 (7.9)	No test	Large indent, resin cracks, no fabric damage	Severe resin and fabric damage, core exposed

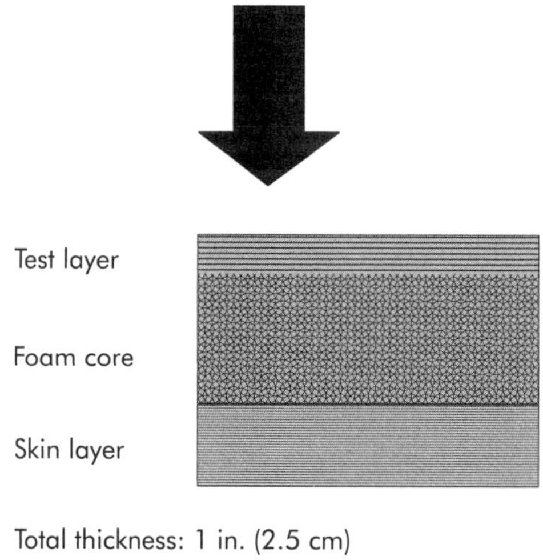

Figure 8-16. Impact studies on prototype surfboard.

UHMWPE have been known for many years. Powders of UHMWPE have been molded into articles such as cutting boards, snowboard bottoms, and other products where high strength and scuff resistance are important.

But just increasing the molecular weight does not explain all of the performance characteristics of UHMWPE fibers. The other change is that the manufacturing process of UHMWPE fibers stretches out the molecules in the fiber direction so that the normal folding of the polyethylene molecules does not occur. For this reason, UHMWPE fibers are also called **extended chain polyethylene (ECPE) fibers**. Extending the molecules creates rod-like structures. Somewhat similar in shape to the aramid and liquid-crystal polymers, these rods have a high degree of molecular orientation (95–99%) and high crystallinity (60–85%).

The starting material for the fibers is UHMWPE powder with a molecular weight of 1–5 million. The powder is dissolved in a solvent so that the normally intertwined molecules are free to move completely independent of other molecular chains. The solution is then forced through a plate with small holes in it (spinneret) similar to what is done when making aramid fibers. The molecules are aligned as they pass through the spinneret holes. The solvent is removed and then the fibers are dried. Later, the fibers are drawn.

Properties

The mechanical properties of two of the most common UHMWPE fibers in the United States are presented in Table 8-1 and Figures 8-2 and 8-3. From Table 8-1 it can be seen that the tensile strength and modulus are good but not significantly better than fiberglass or aramid fibers. However, the specific strength is as high as any other material and the specific modulus is higher than anything except carbon fibers.

In addition to the high strength and stiffness, UHMWPE fibers are tough. This can be anticipated from the elongation data in Table 8-1. That toughness has led to these fibers being used extensively for impact and ballistic protection. In those markets, UHMWPE competes against aramids. But UHMWPE has one major advantage—it weighs less for the same level of protection.

The basic polyethylene nature of UHMWPE fibers brings up the question of the effectiveness of the bond between the fibers and the matrix. Bonding to standard polyethylene is extremely difficult. But poor bonding may actually improve impact resistance. This is because slippage along the matrix-fiber bonds uses energy and therefore improves impact strength. When used in applications where the fiber-matrix bond must be good, UHMWPE fibers can be treated to oxidize their surface, usually with a corona or plasma treatment. In these cases, the bonding to the matrix is as good as the aramid-matrix bond.

UHMWPE fibers are thermoplastics and therefore have some of the same advantages and disadvantages that were noted with aramid fibers. The resistance of the fibers to solvents and most acids is better than with aramid. This resistance reflects the inherent chemical inertness of polyethylene.

In light of all its excellent properties, why aren't UHMWPE fibers used more extensively in composite applications like those described for fiberglass or carbon fibers? The answer is simply that the UHMWPE fibers are not able to withstand the temperatures of curing used in many composite applications. The melting points of the fibers differ slightly between grades but are approximately 290° F (143° C). Also, UHMWPE fibers have much higher creep, that is, movement under a load, which is less than the load required to break the fibers. Hence, structural applications are usually not an option for these fibers.

Applications

The major uses for UHMWPE fibers are similar to those for aramid with the exception of the structural aramid applications. Hence, UHMWPE fibers are used for sailcloth and other fabrics and applications, like ropes, where toughness and possibly light weight are important. A related application is protective clothing and, especially, protective gloves. The high toughness of these fibers in gloves protects hands from cuts and abrasions.

By far, however, the most important application of UHMWPE is for ballistic protection. Light weight and excellent bullet-stopping capability give these fibers a strong position in the ballistic protection market. Some manufacturers of bullet-proof vests have noted that when UHMWPE fibers are combined with aramid fibers in adjacent layers, the performance of the entire part is improved. It is speculated that additional energy is used when dissimilar materials rub against each other and, therefore, ballistic protection is improved.

Other Organic (Synthetic) High-performance Fibers

A few other thermoplastic fibers have properties high enough that they have been used as reinforcements in structural composites. One approach to making these fibers is to follow the lead of aramids and UHMWPE and form rod-like polymers that exist as crystals in solution and then spin fibers from the solution. One example of such a fiber is Vectra, a polyetherimide (PEI). Just as a PEI can be an excellent matrix (as was discussed in the chapter on special matrices), it also can be a good reinforcement. To overcome the high cost of this material, researchers have developed a method to reinforce polyester by properly coupling it with PEI and epoxy. All of the integration of the reinforcement is done at the time the polyester is crosslinked in a sort of self-reinforcing step. It was found that the strength of polyester doubled with only 0.7% PEI content. A matrix material and textile fiber with excellent flame retardant properties, poly-p-phenylenebenzobisoxazole (PBO), also has been suggested for use as a reinforcing material.

Another rod-like thermoplastic is based on chemistry similar to polyimides. This material, termed M5, is targeted at aramids and has properties that, at least in the laboratory, are superior.

Just as UHMWPE can be made into high-performance fibers by special processing, it has been found that another commodity thermoplastic, polypropylene, can be made into ballistic protection materials. This is done by forming the polymer into tapes of unidirectional fibers and then compacting by a patented process. The result is a flat sheet material that can be thermoformed to make products of relatively simple shape but with high ballistic protection capabilities.

BORON, SILICON CARBIDE, AND OTHER SPECIALTY FIBERS

Boron and silicon carbide (SiC) are high-modulus reinforcements. These materials were originally developed in the 1960s at Texaco and United Technology under the auspices of the Air Force Materials Laboratory. While the total production volume of these materials is small compared to the other fibers discussed, boron and SiC have excellent properties suited to some specialized and important applications.

Both boron and SiC fibers are made by the chemical vapor deposition (CVD) process. In this process, illustrated in Figure 8-17, a substrate filament passes through a cleaning oven in which the fiber is heated to remove any lubricants or other surface contaminants. The filament is typically tungsten in the case of boron fibers and carbon in the case of SiC, although both fibers have been used with both deposition materials. The substrate filament then passes into a chamber into which a deposition material has been added as a gas. The typical deposition gas for boron fibers is BCl_3 and H_2, and for SiC it is a mixture of silanes with H_2. The deposition gas decomposes when it touches the substrate filament, which is heated by resistance from a variable direct current (DC) power supply. The coated fibers are then wound onto a reel as they exit the chamber.

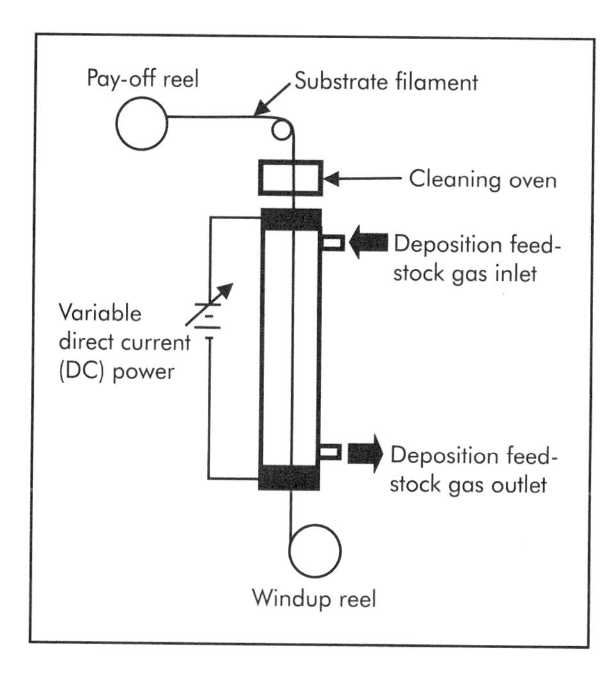

Figure 8-17. Chemical vapor deposition (CVD) process for making boron and SiC fibers.

As can be seen in Figure 8-1, the diameter of the fibers made by the CVD process is many times larger than the other commonly used fibers. The time required to make the vapor-deposited fibers is usually long because there needs to be time for the deposition buildup to occur. Typical reaction time inside the deposition chamber is generally from 30 seconds to a few minutes. Therefore, to achieve the high output required for efficiencies and cost, a company manufacturing vapor-deposited fibers will usually employ many CVD chambers.

CVD process conditions for making boron fibers can be modified to improve temperature resistance, tensile strengths (by eliminating defects), and wettability by a metal matrix.

A minor change in the gas flows inside the chamber when making SiC fibers creates a different SiC crystalline structure on the outside of the fiber. This change improves the adhesion of the SiC fibers to metal matrices. Such changes have allowed several grades of SiC fibers to be made and sold commercially.

Applications

Boron fibers are unique among the high-performance fibers because they combine the usual good tensile properties with good compressive properties. The typical compressive strength for boron fibers is 1,000 ksi (6,895 MPa).

Most boron fibers are sold as prepreg tapes with epoxy as the matrix. These prepregs have been used to make the following aerospace products: F-14 horizontal stabilizer; F-15 horizontal stabilizer, vertical stabilizer, and rudder; B-1 dorsal longeron; F-111 wind doubler; Mirage rudder; and the Hawk and Sea Stallion helicopters' horizontal stabilizers.

Some recent applications have used a combination of boron fibers with conventional carbon fiber reinforced epoxy prepreg. These products have the advantages of both carbon fiber and boron fiber reinforcement. The tensile modulus and compressive strength are both high for this hybrid reinforcement product.

Boron fibers also have been used in sports and recreation products such as rackets for tennis, racquetball, squash, and badminton; fishing rods; skis; and golf club shafts.

Boron fibers have been combined with metals to create metal matrix composites. One principal application for boron fibers in a aluminum matrix is as tubular struts in the frame and rib truss members and frame stabilizing members in the space shuttle.

Silicon carbide fibers are often sold as reinforcements for metals and ceramics. The primary property required for this application is the ability of the SiC fibers to retain their bonding strength to the matrix even during high-temperature processing. Some applications for SiC fibers in an aluminum matrix include aircraft wing structural elements, elements for movable military bridges, and missile body casings. SiC fibers have been placed in a titanium matrix for drive shafts to give increased stiffness over other candidate materials. Fan blades also have been made of titanium with SiC reinforcement. Ceramic armor reinforced with SiC fibers has become a major product and has a bright future.

Specialty Fibers

Several fibers have a unique property or properties that direct their use toward specific applications. For example, carbon fibers have been exposed to fluorine to produce carbon/fluoride fibers. Electrical resistivity for these fibers ranges from 10^2 to 10^1 ohm-cm, making them either insulators or semiconductors. In addition, these fibers conduct heat. The combination of insulating against electricity and conducting heat is especially useful in applications such as heat sinks for electrical devices.

Alumina fiber, with a high modulus, 40–55 Msi (276–379 GPa) in diameters of 3–20 microns, has been explored as a reinforcement for aluminum. This fiber remains stable in molten metals and will not corrode. It is transparent to radar.

Nitrides are made from organosilicon polymers. They have been shown to resist thermal shock better than SiC. The coefficient of thermal expansion of the nitrides is lower than alumina. Nitrides retain their stiffness and strength at temperatures over 2,200° F (1,204° C).

Fibers made from basalt (an inorganic mineral) have high modulus and reasonably good strength. Basalt fibers generally compete against carbon fibers. These fibers have been commercialized using a process developed in Russia and later licensed to firms in other countries. While still small in volume, the opportunities for these fibers seem to be growing, especially when carbon fibers are in short supply.

One of the problems with aircraft reinforced with carbon fiber composites is their inability to conduct electricity away from critical electronic components when the aircraft is struck by lightning. This problem has been remedied by installing a metal mesh in the outer layers of the aircraft skin. However, this mesh has several problems including the added weight, complexities in manufacturing, difficulty to repair, and potential for reducing the stealth of the aircraft. An alternate method of giving aircraft lightning strike protection is to use nickel-coated carbon fibers. These have been shown to protect the aircraft and reinforce as normal carbon fibers, thus simplifying the manufacture of the parts and their repair. The carbon fibers are coated with nickel in a process that is roughly analogous to that used to make boron fibers.

NATURAL FIBERS

The properties of some natural fibers are given in Table 8-1. As can be observed, the tensile strength of flax and spider silk are less than the strengths of synthetic fibers such as fiberglass, carbon fibers, and aramids. Likewise, the modulus of these fibers is somewhat lower than the more traditional reinforcement fibers. However, costs of these and other natural fibers can be much lower than synthetic fibers. Further, there is a certain appeal from an environmental viewpoint for using natural fibers. Therefore, natural fibers have been used for some composite parts, especially in countries where the more traditional fibers are largely imported, and in applications where the strength of the natural fibers is sufficient.

Some of the properties of natural fibers cited include low cost, low density, biodegradability, sustainability, health benefits due to reduced dermal and respiratory irritation, and reduced machine wear during manufacturing operations. While these properties are laudable, some negative factors associated with natural fibers include variations of properties because of climate, harvesting techniques, plant variety, maturity, extraction technology, and general fragility. In spite of these problems and because of the advantages, some countries have pursued a vigorous program of development of natural fibers for composites. One example is New Zealand and their work with locally grown hemp. The fibers proved to be adequate with unsaturated polyester resins for some applications in which strength and stiffness were not critical such as for tub and shower units. The fibers were also used with polypropylene and exhibited acceptable properties for non-critical applications.

Efforts in India and Latin America have been similarly successful with hemp and a variety of other natural fibers such as flax and cotton. Comparisons of these materials with traditional high-performance reinforcements show that the synthetic materials have higher properties but, as with the New Zealand case, some applications do not need high properties and so natural fibers are acceptable.

Spider silk is an interesting natural fiber that has the potential of acquiring high-performance properties. The difference with spider silk is that it has a unique molecular shape, which may be key to making an imitation synthetic fiber with properties that may be at least as good as some of the high-performance fibers. Even though the resultant silk material is synthetic, environmentalists would be happy with the projected manufacturing method since it is done at room temperatures and in aqueous solutions. This is in contrast to the high temperatures and hazardous solvents used in fiberglass, carbon fiber, and aramid manufacturing. Other material advantages of spider silk are its ability to absorb dye, luster, extensive softness, and environmental stability.

The advantage of spider silk over silkworm silk is that a spider has the ability to synthesize a variety of silks for different applications, each from one of its six different silk glands. The uniqueness of the structure of spider silk results in extraordinarily high elongation (35% in draglines and up to 200% in other silks) and therefore high impact toughness. It is helpful to think of the forces involved on the thin fiber web when struck by a fast-flying insect. This invites the use of spider silk as an armor material and as a rope or net. Medical devices such as wound closure systems, vascular wound repair devices, hemostatic dressings, patches, glues, sutures, ligament prosthetic devices, and other bio-related products are especially appealing as applications for silk.

While the appeal of natural fibers is certainly high, reality requires that some high-volume application in a major market needs to adopt natural fibers before they can be taken seriously as a part of the composites market. That situation may actually be close to happening. The major application is automobiles and the marketplaces are Europe and the United States. The movement into using natural fibers as a reinforcement for sheet molding compound (SMC) in automobiles is based on their proven and long-term use in composites with thermoplastics. They are used in applications such as synthetic lumber, decking, furniture, and a few automotive components. Most of the natural fibers are wood products and so they are mostly seen as fillers rather than strength reinforcements. However, about 4% of the 400-million-pound (180-million-kilogram) market comprises natural reinforcement fibers such as flax, kenaf, hemp, and jute. These fibers are often a replacement for fiberglass. The major impetus for the use of these fibers is that they are recyclable. This is especially important in Europe but is also increasingly a motivation in the United States.

Mercedes-Benz introduced jute-based door panels in its E-Class vehicles in 1994.

German automakers continue to use natural fibers, especially those imported from developing countries as a way of assisting their economies. Some developmental products have been created in the United States for interior door panels and similar non-critical automotive applications.

The ultimate environmental combination may now be possible for composites. John Deere has made parts for its combines from natural fibers and a soy-based resin. Obviously both are natural and recyclable. While the environmental reasons for their adoption will continue to be strong, these all-natural composites will improve in properties as experimenters study how to make them better, potentially competing in all ways with synthetic resins and fibers.

FIBER-MATRIX INTERACTIONS

The bonding between the reinforcement and matrix is critical to the performance of the composite material. Usually a strong bond between reinforcement and matrix is desired so that loads can be transferred efficiently from the matrix to the reinforcement. This text concentrates on the more common methods of strengthening the reinforcement-matrix bond and some of the important considerations involved in the modification and optimization of that bond.

Although not considered in detail here, the toughness of a material, which is the ability of the material to absorb energy, is sometimes increased by a weak bond between the matrix and fibers. This allows slipping between matrix and fibers, which absorbs energy.

Sizings and Finishes

Sizings and finishes were discussed briefly when the major fiber types were introduced. A narrow distinction was made between sizings and finishes. **Sizings** are used chiefly for their ability to protect the fibers from mechanical damage. **Finishes** enhance the bonding of the fibers to the matrices and impart desirable fabric qualities. However, in practice, one material often serves both purposes and so the distinction between sizings and finishes has largely disappeared. Therefore, this larger view will be taken here. Materials coated onto the fiber surface will be referred to as both sizings and finishes with no attempt to distinguish them from one another precisely except where obvious or necessary. Another term that is sometimes confusing is **fiber surface treatment**. This is a step in the manufacture of the fiber in which the fiber surface itself is modified as opposed to a coating being applied as with a sizing or finish. The electrolytic treatment of carbon fibers is an example of a fiber surface treatment. However, it is possible that during the fiber surface treatment some coating could be added, thus complicating the terminology.

Sizings have been used for many years in the textile industry. When textiles were first used in automated equipment during the Industrial Revolution of the 18th and 19th centuries, manufacturers found that the fibers were being damaged and their strength drastically reduced. They discovered that starch could be applied from an aqueous solution to protect the fibers. Then, when the fibers had been converted into cloth and the processing was complete except for sewing, the starch could be removed by washing the fabric in hot water. Over the years, other materials, most notably polyvinyl alcohol, have been found to impart a protective coating to fibers, which is easily removed when manufacture of the cloth is complete.

In high-performance composites where the fibers are so critical to the performance of the part, sizings continue to be important. This protection is especially important to carbon fibers, fiberglass, and other ceramic-like fibers that are brittle and highly sensitive to surface defects. Today, sizings are typically polymeric materials that remain on the fibers. They deliver dual-purpose sizing and finish.

The finish chosen should be compatible with both the reinforcement and matrix. To accomplish this, finishes are often comprised of molecules that have one type of functional group on one end and a different type on the other. These types of finishes are called **coupling agents**. For example, a finish for fiberglass and polyester might have one end made of silicon-containing groups, which are compatible with the glass, and the other end composed of an organo group, which is compatible with the polyester. Figure 8-18 illustrates such a coupling agent and the polar and non-polar interactions it would have with fiberglass and a resin matrix. This is analogous to soap molecules where the non-polar end attaches to the grease and the polar end attaches to the water, thus forming a coupling of the water to the grease and allowing the grease to be washed away.

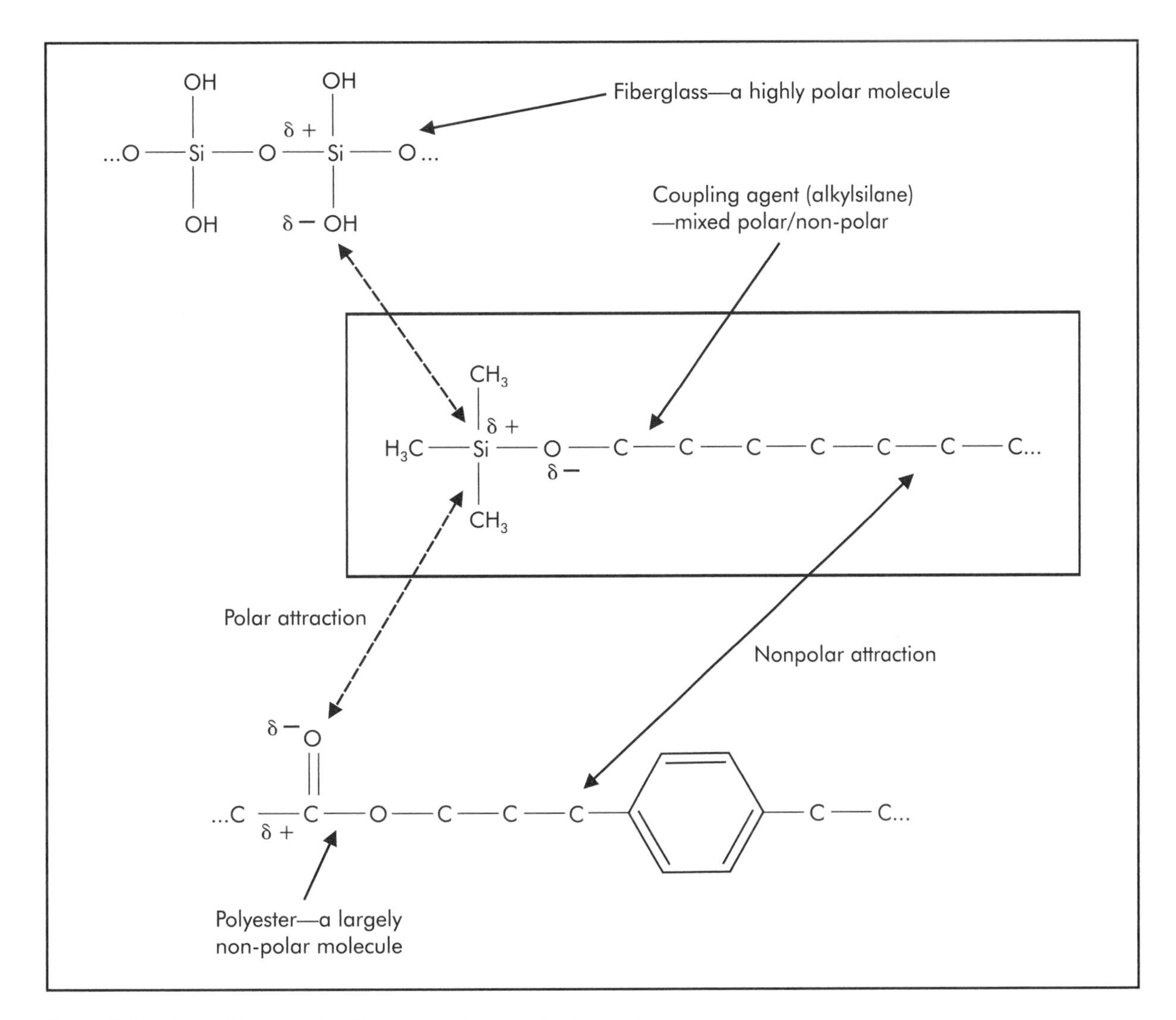

Figure 8-18. A coupling agent with polar and non-polar interactions.

Some recent research suggests that actual molecule-to-molecule coupling does not exist, but rather, the coupling takes places as groups of molecules associate with others through a network of forces. Regardless of which model is chosen, the concept of the molecule-to-molecule coupling provides a simpler picture, which has led to choices of coupling agents that have given the desired improvements in mechanical properties of the composites.

In some cases, as with carbon fibers, the chosen finish is often the same type of polymer as will be eventually used as the matrix. Although this would seem to solve the compatibility problem, it often results in only fair reinforcement/finish compatibility since the finishes of the matrix resins often have only a marginal interaction with the fiber. For this reason, fibers are often treated with electrolysis, plasma, or corona. This activates the surface of the fiber and makes it more compatible with the finish resin.

The moisture sensitivity of the composite is often compromised by the presence of the finish. Finishes are often polar to obtain good compatibility with the fiber or interaction with the matrix; but, as polar molecules, they are subject to water absorption. Therefore, the use of a finish is sometimes a necessary evil—needed for strength and modulus, but detrimental to other properties.

For metal matrix composites, the use of coupling agents is not as common as with polymeric matrix composites. However, surface treatment of the fibers is often done to improve adhesion. For example, to improve the bonding of silicon carbide fibers with aluminum, the fibers can be treated in a silicon-rich environment. This increases the amount of silicon on the fiber surface and enhances the bond between the fibers and the aluminum matrix. Nickel coatings are also known to aid the wetting of ceramic reinforcements by liquid aluminum alloys. Another approach to improving bonds in metal and ceramic matrix composites is to introduce metal ions by infiltration, which alloy with the matrix and bond to the silicon carbide. Examples of such metals are magnesium and lithium.

The development and optimization of a surface treatment for fibers can be arduous and time consuming. Therefore, fiber companies often guard the exact nature of their coupling agents and finishes with great secrecy. Further, the difficulty and development cost of good coupling agents have proven to be barriers to the introduction of several promising matrix materials. Nevertheless, fiber manufacturing companies continue to optimize current surface treatments and develop finishes that are compatible with new matrix resins.

Interphase Theory

The theory of molecules touching other molecules to form bonds was introduced as the **interface theory** in a previous chapter. This is the basis of coupling agent concepts. An alternate view of how molecules interact is called the **interphase theory**. In this theory, the nature of reinforcement/matrix interactions involves the recognition of a flexible, three-dimensional interphase. This is thought to exist between the matrix and the coupling agent and, possibly, even between the coupling agent and the fiber. The interphase is a polymer network formed by the coupling compound or sizing and into which the matrix molecules can penetrate. An analogy of the interphase would be a briar patch formed by the coupling agent. The network may have occasional chemical attachments (bonds) to the fiber surface. However, its chief purpose is to provide a lattice that the matrix molecules can penetrate and then be held in close proximity to the fibers. Therefore, the interphase is a region where sizing (finish) and matrix have diffused into each others' domain. There also may be selective absorption of the resin onto the fiber. The interphase would be responsible for transferring the load from the matrix to the fibers.

Although the interphase probably has a lower modulus and lower strength than the fiber or the matrix, true cohesive failure in tension (that is, failure between similar parts of the interphase) is probably rare within the main part of the interphase zone. The rationale behind this statement is simply that tension failure usually occurs because of a critical flaw. The probability of encountering a critical flaw is much greater in the bulk phases of the matrix and fiber than in the interphase zone because the bulk phases are much larger.

Even if a crack is directed specifically at the interphase region, as with the peel test (discussed later in this chapter), the failure will not necessarily be at the interface. The possible failure mechanisms are complex but have been studied in adhesion bonding technology. Determining whether fiber/matrix interphase failure has occurred can be difficult.

Scanning electron microscope (SEM) studies of the fracture surfaces of composites often reveal fibers that appear to be clean of adhering matrix. These observations have been interpreted as interfacial (adhesive) failure when, in fact, the failure has actually been close to, but away from, the boundary and slightly inside the interphase region (cohesive failure). However, some confirming evidence is often needed to affirm these theories.

Adhesion Theory

Predictions of the bond strength between the matrix and reinforcement or between the interphase and reinforcement can be obtained by applying the general principles of surface chemistry. The attractive forces on a molecule in the bulk are essentially equivalent in all directions because it is completely surrounded by other molecules. However, the forces on the surface molecule are unbalanced because it does not have molecules of the same material on one side. This unbalance gives rise to a **surface free energy**, γ, which is defined as the energy necessary to form a new surface of a unit area. It is the same energy as would be required to move a bulk molecule to the surface. If the surfaces are a liquid and a vapor, γ_{LV}, the energy is called the **surface tension**.

The energy associated with bonding at the surface is defined as the **work of adhesion**, which is the energy required to separate two particles. In other words, it is the energy of the bond between the particles. It is possible to define the work of adhesion in terms of the surface free energy between two surfaces, A and B, as:

$$W_{Adhesion} = \gamma_A + \gamma_B - \gamma_{AB} \quad (8\text{-}1)$$

where:

$W_{Adhesion}$ = work of adhesion for the two particles
γ_A = surface free energy of particle A
γ_B = surface free energy of particle B
γ_{AB} = surface free energy of the interaction between particles A and B

The work of adhesion can be approximated using the peel test. In this test, one material is pulled from the other. However, the work of adhesion assumes equilibrium conditions, which in this case means that the pulling rate is infinitely fast and there is no effect of temperature. In reality, of course, the pulling rate is some finite number and the effect of temperature is present. However, within the limits of a fast pulling rate and low temperatures, the conditions that would determine the work of adhesion can be approximated. But the work of adhesion should not be equated to the adhesive bond strength since the bond strength (as observed in any real peel strength test) also contains the energy of dissipative processes such as viscoelastic deformation, plastic deformation, local micro-cracking, and so on. Hence, when these energies are included, the actual curve of the peel strength is much higher than just the work of adhesion.

As noted previously, if a surface has high energy, it is far more likely to bond than if it has low energy. Using the concepts of surface tension, the surface energy for a given surface can be determined by measuring the angle between a liquid and the surface. The angle to be measured, θ, is shown in Figure 8-19. The work of adhesion in terms of liquids, solids, and vapors is expressed as:

$$W_{Adhesion} = \gamma_{LV} + \gamma_{SV} - \gamma_{SL} \quad (8\text{-}2)$$

In terms of the angle, θ, the work of adhesion can be derived, resulting in the following equation:

$$W_{Adhesion} = \gamma_{LV}(1 + \cos\theta) \quad (8\text{-}3)$$

Since the surface tension between the liquid and the vapor and the contact angle can be determined experimentally, the work of adhesion can be found. Then, if the surface energy, γ, is known for the liquid (usually water) and for the gas (usually air), the surface energy of the solid relative to the liquid can be calculated. This quantity is a relative measure of the energy of the surface. Therefore, by observing whether the angle of the liquid on the solid is high or low, the surface energy of the solid can be evaluated.

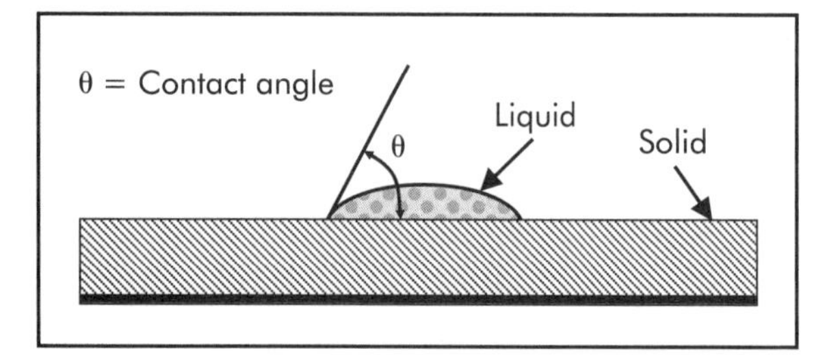

Figure 8-19. Surface energies determined by contact angle.

The above phenomenon is often referred to when discussing the energy of the surface. The surface of the solid is said to be "wetted" by the water when the angle is low and the water readily spreads across the surface. This happens when the surface energy of the solid is high. In contrast, when the water does not spread on the surface, it tends to bead up. This is indicative of low surface energy. An analogy is given to car waxes. When wax is present, the water beads because the wax has low surface energy.

Measurement of Fiber/Matrix Bond Strength

Several methods have been used to determine the fiber/matrix bond strength but none of the methods gives unambiguous results.

Single Filament Pullout

The single filament pullout test shown in Figure 8-20 uses a modified tensile test machine to measure the force required to dislodge a single drop of resin from a single filament. The test uses a filament to which a single drop of matrix has been applied. After the resin has cured, the filament is threaded through a hole in a plate and then attached to a tensile machine. The hole must be smaller than the resin drop. After attaching the lower end of the filament to the other tensile jaws, the fiber is pulled by the tensile test machine and the force to cause the fiber to separate from the resin drop is measured. This test will not work unless the fiber is stronger than the interface. This condition is generally met, but for any particular fiber the condition may not be met because of fiber flaws. Some of the flaws could come from handling during the test. Therefore, a large number of samples must be run to obtain reproducible results. This test is sensitive to fiber sample length since the probability of encountering a flaw increases as the length of the fiber

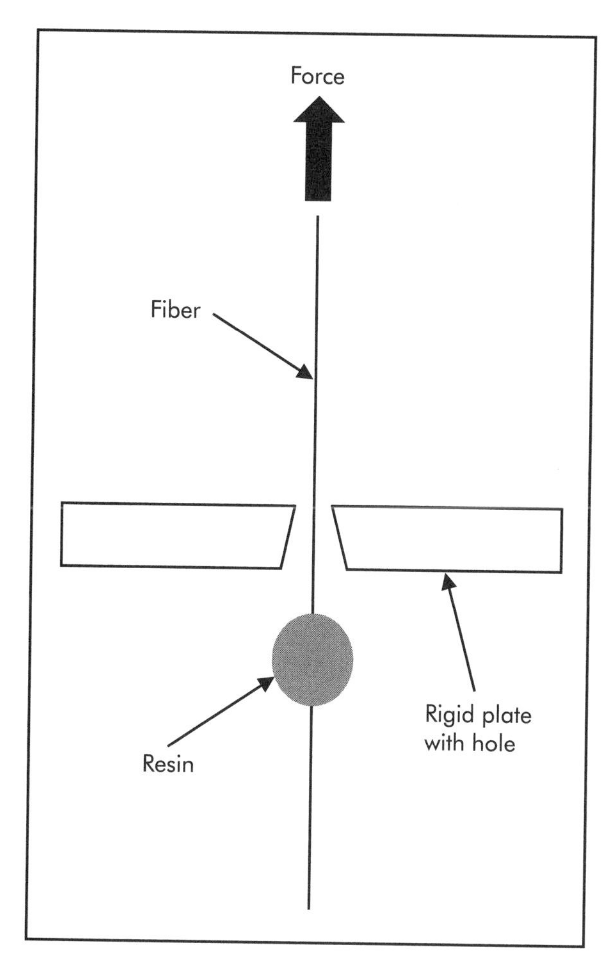

Figure 8-20. Fiber pullout test.

increases. It is also tedious because of the fragility of the fibers and the difficulty in getting a good resin droplet.

Transverse Tensile Test

Figure 8-21 shows the transverse tensile test in which a sample containing unidirectional fibers is oriented so that the pull is perpendicular to the direction of the fibers. The major difficulty with this test arises because of the large differences in the moduli of the matrix and the fibers, which may focus failure into the interfacial region but not exactly at the interface of the fiber and the matrix. In this test, aligning the direction of the fibers in the sample to be perpendicular to the pull direction is critical to prevent shear forces.

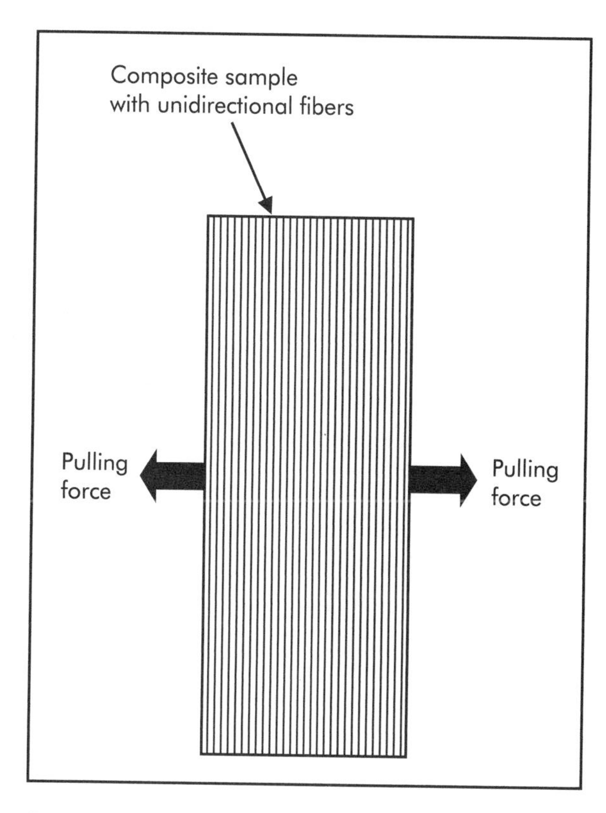

Figure 8-21. Transverse tensile test.

Short Beam Shear Test

The short beam shear test is the classical test used most frequently to determine fiber/matrix bond strengths. The test apparatus is illustrated in Figure 8-22. The sample is reasonable in size. Typical dimensions for the test are: specimen length = 1.5 in. (3.8 cm); specimen width = .25 in. (0.6 cm); specimen thickness = .25 in. (0.6 cm); and the support span = 1.0 in. (2.5 cm). Specimens are, therefore, easy to prepare. Also, equipment to run the test is generally available. The test is much like a flexural strength test except that the sample is much shorter and thicker than the flexural sample. Therefore, whereas the flexural sample bends when the

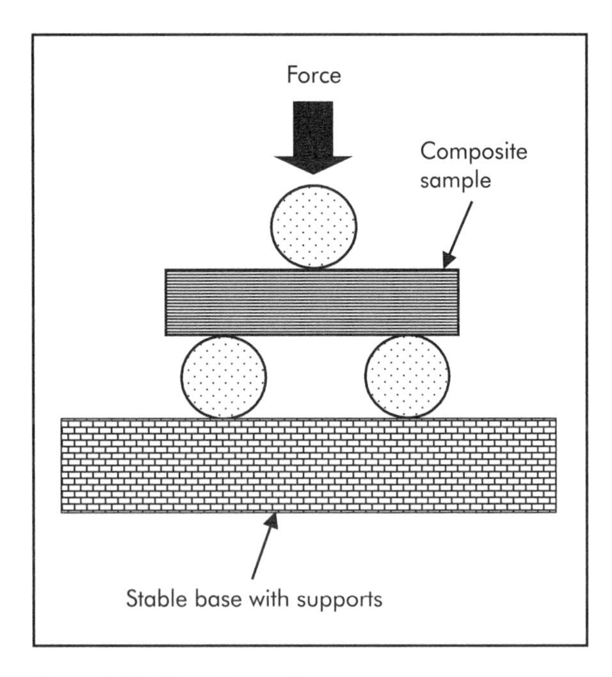

Figure 8-22. Short beam shear test.

force is applied, the short beam shear sample resists bending and, instead, shears. This shearing action is understood by noting that in all flexural samples, including the short beam shear, the top of the sample is in compression and the bottom is in tension. This situation requires, therefore, that a shear plane be located within the sample at the place where compression changes to tension. By being short and thick, the short beam shear sample forces that shearing action to be the major force within the sample.

There are many failure modes associated with the short beam shear test and, therefore, it is difficult to predict theoretically. However, when comparing materials that are substantially the same, this test is excellent to determine the relative strengths of the fiber/matrix bonds.

Embedded Single Filament Test

The embedded single filament test depends on the determination of the critical fiber length. The critical fiber length is the minimum length at which the matrix cannot transmit enough tensile stress to break the fiber. The equation relating the critical length and the strength between the matrix and the fiber is:

$$\tau_c = \frac{\sigma_c d}{2l_c} \qquad (8\text{-}4)$$

where:

τ_c = strength between the matrix and the fiber (shear strength)
σ_c = strength of the fiber (ultimate tensile strength)
d = diameter of the fiber
l_c = critical length

The shear strength is inversely proportional to the critical length so small critical lengths indicate good adhesion.

The critical length is determined by encapsulating a single fiber in a matrix sample, which is prepared as a micro-tensile test specimen. That is, the sample, usually about 3 in. (7.6 cm) long and .5 in. (1.3 cm) wide, is much smaller than the standard tensile test specimen. The embedded fiber test specimen is illustrated in Figure 8-23. After the matrix has fully cured, the specimen is slowly pulled in small increments and inspected after each pull. The fiber will begin to break into small segments as load is transferred to the fiber by the matrix. Eventually, the fiber segments become so small that the matrix cannot transfer sufficient force onto them to cause them to fracture. When all the segments have reached this length, the test is concluded. The length of the fibers at this point is the critical length. The typical critical length for carbon-epoxy fibers is .0016 in. (0.041 mm).

Because of the nature of fiber strengths, a statistical version of equation 8-4 is required. In Equation 8-5, the critical strength, σ_c, is measured, summed, and averaged:

$$\tau_c = \frac{d}{2l_c} \sum_i \sigma_c^i \qquad (8\text{-}5)$$

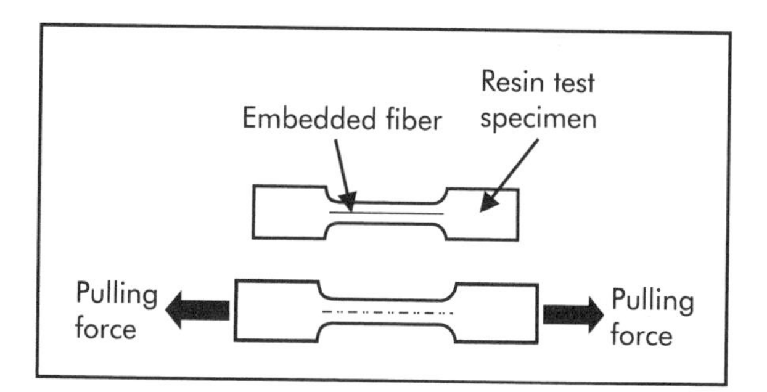

Figure 8-23. Embedded single filament test.

Observation of the lengths of the fiber segments and the point at which they are no longer breaking can be facilitated by using optical birefringence, which arises because of the stresses induced in the sample by the pulling. The birefringence is observed by viewing the sample between crossed-polarizing filters and using monochromatic light. Once the fiber begins to fracture, a characteristic birefringence pattern develops at the broken fiber ends. It looks like the fiber segment is sheathed in light. When the fiber fracture is complete, the birefringence pattern develops into a non-nodal pattern; that is, the sheath appears to surround all the fiber segments in a single, continuous light pattern. The intensity of the fiber pattern (sheath) seems to be proportional to the bond strength; so, when the bond strength is low, the development of the sheath is rapid suggesting separation of the bond between the matrix from the fiber.

Micro-indentation Test

Another method recently developed to measure the fiber/matrix bond strength is the **micro indentation test**. This method allows calculation of the interfacial shear stress and takes into account factors such as: local geometric distribution of the fibers, shape and size of the indenter, friction between the indenter and the fiber, and residual stresses due to thermal loading. As a result, the test method allows greater freedom in sample selection and reduces standard deviation in the testing.

Failure Modes of the Fiber/Matrix Bond

The nature of the bond between the fiber and the matrix results in several different failure modes, which depend on the effectiveness of the bond and the relative strength and stiffness of the matrix and the reinforcement. The most common types of failures for polymer matrix composites with high-modulus fibers (such as fiberglass, carbon, and aramid) are shown in Figure 8-24. Three different failure modes are depicted.

In the first, Figure 8-24a, the breakage points of the composite are at the fibers themselves. This is the preferred failure mode for maximum strength and indicates that the bond between the fibers and the matrix is strong. The matrix has transferred the load onto the fibers and has stayed in contact with the fibers to the stress level at which the fibers break.

The second failure mode is referred to as matrix-fiber separation (Figure 8-24b). In this case the matrix separates from the fiber. That is, the bond between the matrix and the fiber breaks. This is an undesirable circumstance and reflects a weak fiber-matrix bond. The matrix recedes from the fiber because it has greater elongation than the fiber. During tensile pulling it stretches and when the bond to the fiber fails, the matrix pulls back elastically.

The third failure mode is fiber pullout (Figure 8-24c). In this case, the bond between the matrix and the fiber is poor. Little or no energy is transferred to the fibers and, therefore, the composite fails in its basic purpose.

The nature of failures for metal and ceramic matrix composites is somewhat different from the polymeric matrices. In metal and ceramic matrix composites the moduli of the matrix and the reinforcements are often nearly the same. The failure

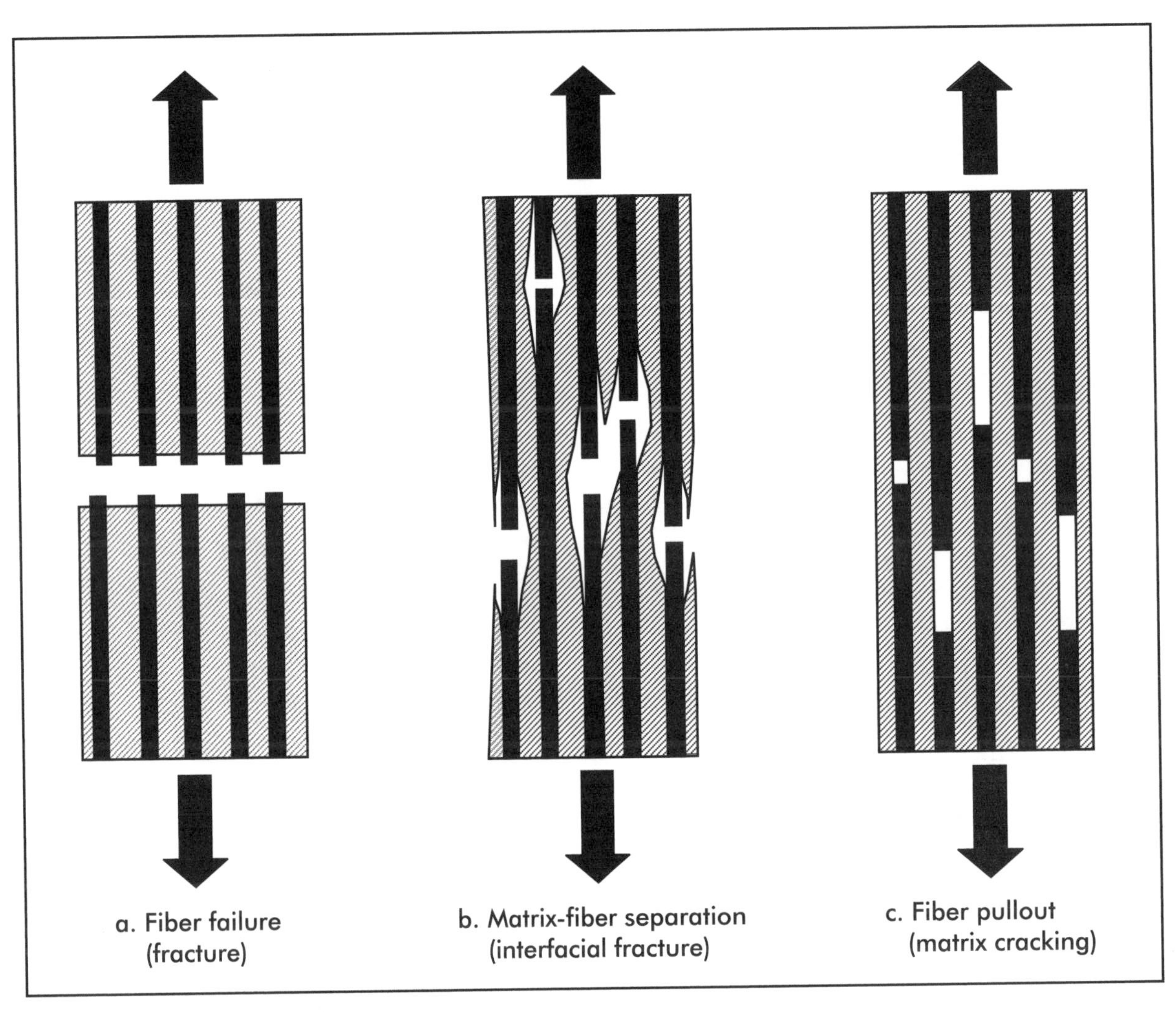

Figure 8-24. Common failure modes for polymeric matrix composites.

modes of these materials are illustrated in Figure 8-25.

Figure 8-25a shows the fiber failure of a metal matrix composite in which the bonds between the matrix and the reinforcement are good. Note the random nature of the fiber failure and the lack of failure of the matrix. This failure point is the situation just prior to the failure of the matrix. Figures 8-25b and 8-25c illustrate ceramic matrix composite failures. In Figure 8-25b the matrix and fiber fail simultaneously. This is the case most desired for structural applications. In Figure 8-25c the matrix fails before the fibers. While this composite can absorb high amounts of energy and can be of benefit for ballistic protection, it would not be chosen for structural applications.

Influence of Coupling Agent

The most obvious and desired improvement of properties from the inclusion of an effective fiber surface treatment is an increase in mechanical properties. In particu-

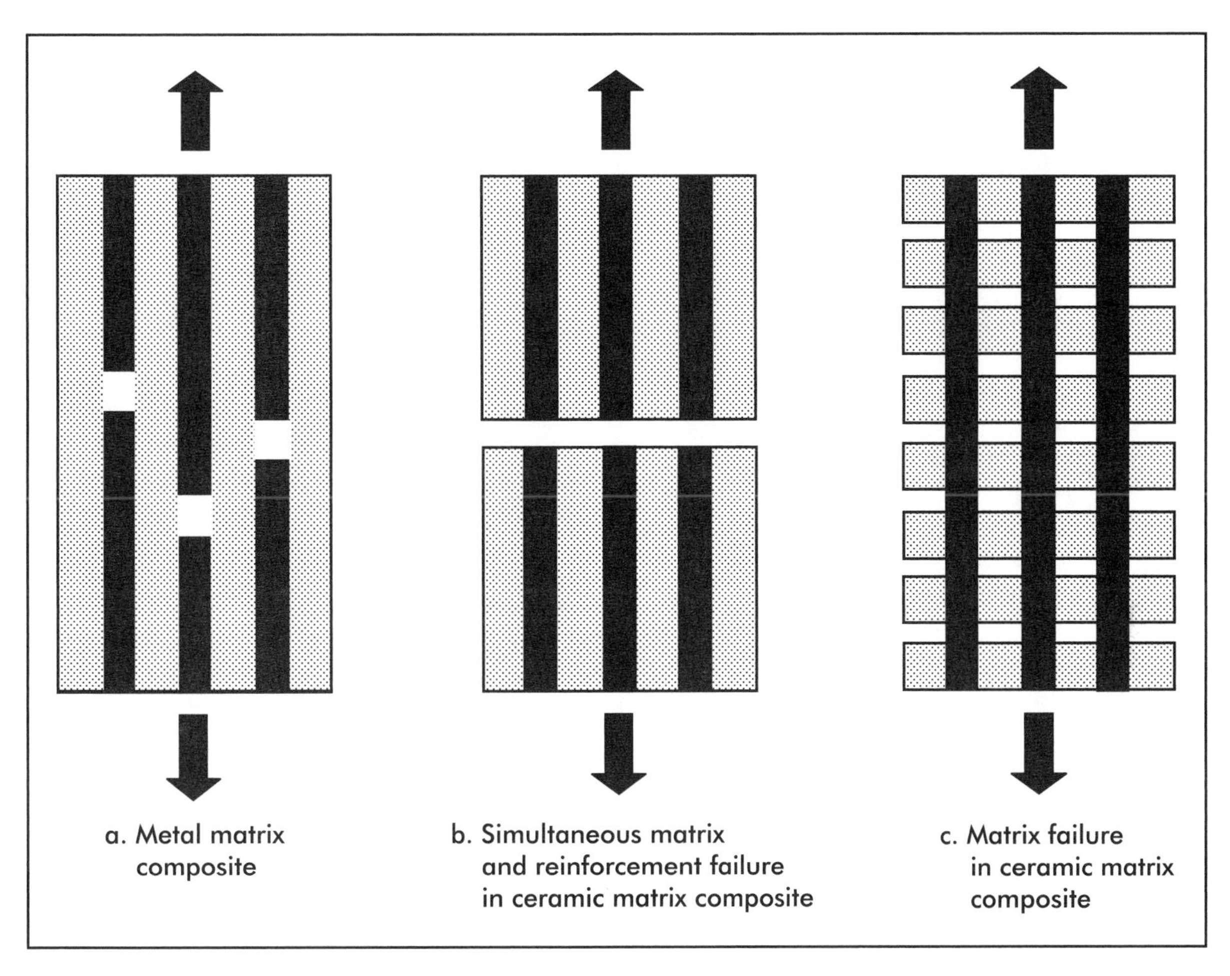

Figure 8-25. Failure modes for metal and ceramic matrix composites.

lar, it is hoped that both tensile strength and modulus will be improved by the addition of a coupling agent that enhances the bond between the matrix and the fiber surface. As expected, the addition of the sizing greatly improves the strength and the modulus of the composite. The effect of sizing is roughly to increase the strength and the modulus of the composite by about 50%.

Another interesting trend is the effect of **annealing**. Annealing is exposing the composite to a moderately hot temperature above room temperature but not close to the melting point or decomposition temperature. When a composite is annealed (or post-cured), the strength and modulus are improved. However, the effect of annealing is not as great as the improvement achieved with sizing.

Another property that can be improved with sizing is the resistance to cyclic fatigue. A composite without sizing can be compared to a composite with sizing. The sized composite is substantially superior to the composite without sizing in its retention of properties (applied stress) and that improvement increases with the number of cycles.

CASE STUDY 8-1

Bullet-proof Vests

The ability of a material to serve as anti-ballistic personal armor (bullet-proof vests) is dependent on several factors. The most obvious factor is that the anti-ballistic material must withstand the penetration of the bullet (or other ballistic device, such as shrapnel). This capability is related directly to the strength of the material. Many materials, from steel to aramid cloth, will satisfy this requirement. In fact, almost any material, if layered thick enough, would suffice. However, realistically the personal armor must be reasonably thin and pliable so that the individual wearing the vest does not feel like a medieval knight encased in hard steel armor. This restriction leads to other requirements of the anti-ballistic material. One requirement is that the material must be strong enough that a reasonable thickness will stop the bullets. In this regard, the thickness required is directly proportional to the flexural strength of the material. Therefore, materials like aramid (Kevlar®) or similar strong fibers will work. But so will steel plates. Hence, the second restriction must be noted—the material must be flexible. Thirdly, the material must be light weight.

It is important to note that strength, flexibility, and light weight are not the only requirements. These will prevent the bullet from penetrating, but another requirement is that the impact of the bullet must not do major bodily harm. That is, the force of the bullet must not be so intense that it breaks bones, injures internal organs, or causes other bodily trauma. This requirement is usually stated in terms of energy dissipation. The material must dissipate the energy sufficiently and in a short enough time that the body will not be harmed. Aramid and UHMWPE fibers have been shown to possess this property. They are naturally tough materials and, therefore, absorb energy. Moreover, they have been shown to spread the impact energy laterally with great efficiency. Other fibers, such as carbon or fiberglass, have strength sufficient to stop the projectiles, but they are not tough and do not spread the energy. Carbon and glass fibers tend to break when they are hit. This results in the transfer of much of the impact energy of the projectile and, therefore, the potential of bodily harm.

Most commonly, the fibers used in bullet-proof vests are not encased in a matrix. They are sheets of fibers (usually woven into cloth) stacked to the thickness desired and then sewn into cloth pockets. This gives the flexibility desired and the thickness required to stop the bullets. The sliding of the layers over each other also helps to spread the impact energy. The number of layers is chosen to specifically counteract a particular threat. For instance, the threat may be a .22 rifle bullet or a .30-06 armor-piercing round. Both threats are measured by the weight of the projectile and its velocity of travel. Then, through experimental testing, the number of layers to stop each particular threat is determined. When a bullet-proof vest is purchased, the threat level is identified so that the user will be aware of the product's limitations. Sample threat levels are illustrated in Table 8-6.

Table 8-6. Ballistic protection.

Threat Level	Number of Layers	Ammunition Stopped
2A	22	9 mm
2	32	44 magnum, 357 magnum
3A	40	Wad cutter, 240 grain bullet

Other armor types may, and often do, contain a matrix. For example, helmets are shaped and rigid. They need the matrix to impart these properties. Armor for vehicles is also shaped and rigid. Since the threat for a vehicle is often more severe, the types of armor are often very elaborate, often consisting of several layers, each of which has a different capability. Some of the layers would be non-matrix material, some might be polymeric matrix materials (especially those that are shaped), and others might be ceramic matrix materials (to stop high-impact projectiles). While the technology to protect vehicles has already reached a high level, the needs for armor and the increasing sophistication of weapons suggest that further developments will continue.

SUMMARY

When fiber reinforcements are put into composites, the overall properties are decreased because the fiber does not occupy all the volume. Thus the composite properties are a combination of the fiber and the matrix. Therefore, the overall strength, modulus, and so on, are modified by the effect of the matrix. Still, because of their importance in determining the properties of the composite, the physical properties of the fibers are closely examined during the design process. Putting it simply, the better the physical properties of the fibers, the more efficient the composite structure.

Some of the key learning points in the study of fibers are as follows:

- Reinforcements can be fibers, whiskers, or particles. Fibers are the most important and have the largest effect on composite properties.
- Glass fibers are the most widely used of all reinforcements. The major types of glass fibers for composites are E-glass, S-glass (including the improved S2-glass), and C-glass, which are all specifically formulated for their electrical, strength, and corrosion-resistant properties. However, because of cost, E-glass dominates in composite applications except where there is a high priority on performance. Glass fibers are mainly used for engineering applications (as opposed to high-performance applications).
- Carbon fibers are the most prevalent fibers for critical or high-performance applications. They are made by pyrolizing while the fibers are held under tension or by spinning directly from a pitch (carbonaceous mass). Today, carbon and graphite fibers are essentially the same. Carbon fibers have high specific strength and specific stiffness.
- The most common organic fibers are aramid and UHMWPE. These fibers are tougher than those made of glass or carbon. Because of their high toughness and ability to spread energy laterally, these fibers are used for ballistic protection.
- Boron and silicon carbide continuous fibers are made by chemical vapor deposition using other high-modulus fibers, such as tungsten and carbon, as the substrates. The high cost of the boron and SiC fibers limits their use even though they have high moduli and strengths. They are used for metal and ceramic reinforcements and occasionally with epoxy.
- Fiber strength decreases greatly in the presence of surface defects and moisture absorption. The surface of fibers can be treated with a sizing, finish, or coupling agent to improve mechanical properties.
- The bond between the matrix and reinforcement is critical to the performance of the composite. An interphase region formed by the sizing material, between

the fiber and the matrix, has been postulated and described.

- Coupling agents, sizings, finishes, and surface treatments improve shear strength and most other mechanical properties of composites.
- Tests to determine the strength of the bond between the matrix and fiber include fiber pullout, transverse tensile, short beam shear, and embedded single filament.

Costs of fibers are an issue in most composite applications. Because the total costs of fibers are determined by the market and the costs of production, they are constantly changing. However, some generalizations are probably helpful in understanding the general relationships of the prices of the various fibers.

The least expensive fibers are glass. They usually have a selling price in the vicinity of $1.00/lb ($2.20/kg). Carbon fibers sell over a wide range of prices depending, in part, on the amount of certification and testing that must be done. Fibers for aerospace and military applications require far more certification and testing than those sold for use in sports, medicine, and industrial applications. Military/aerospace fibers are about $15.00/lb ($33.07/kg) and general fibers are about $8.00/lb ($17.64/kg). Aramid fibers cost about the same as the high-end carbon fibers, that is, around $15.00/lb ($33.07/kg). Spectra fibers are about the same price as aramids.

LABORATORY EXPERIMENT 8-1

Fiber Length Effects

Objective: Discover the differences in the tensile strength of fibers as a function of fiber length.

Procedure:

1. Select 15 single filaments from a tow of fiberglass or carbon fibers. Be careful to handle them carefully.
2. Cut 10 of the filaments to 10-in. (25 cm) lengths.
3. Cut 10 of the filaments to 6-in. (15 cm) lengths.
4. Cut 10 of the filaments to 4-in. (10 cm) lengths.
5. Prepare a fixture for the jaws of the tensile test machine that will allow gripping of each filament individually so that there is no slippage.
6. Carefully clamp each filament into the jaws and run a tensile test.
7. Note the ultimate tensile strength of each filament.
8. Note the range of results for each length.
9. Average the tensile strengths for each length.
10. Plot the average tensile strength versus length for the fibers.
11. If available, repeat the test with aramid fibers and compare the results.

QUESTIONS

1. Distinguish between a finely extruded rod and a fiber in regard to properties and the additional processing done to a fiber.
2. Describe the differences expected in the tensile strength of glass fibers that are short (less than 1 in. [2.5 cm]) versus those that are long (over 10 in. [25 cm]). Compare the range of the results for each fiber length. Why would you expect these differences?
3. Compare the differences that would be expected in the tensile strength of glass fibers of different lengths with aramid fibers of the same length. Compare the range of values with the range for fiberglass filaments. Why would you expect these differences?

4. State the purposes for sizings, finishes, and coupling agents.
5. Describe the differences between pitch-based and PAN-based carbon fibers.
6. When would aramid fibers be used in place of carbon fibers?
7. Describe the method of manufacture for boron fibers.
8. Describe a possible application and a manufacturing method using magnetic whiskers as reinforcements.
9. Describe the concept of an interphase between the fiber coupling agent and the matrix as compared to an interface.
10. Describe the advantages of three types of tests for fiber/matrix bond strengths.
11. List three advantages and three disadvantages of natural fibers.

BIBLIOGRAPHY

Al-Ostaz, A., Drzal, L.T., and Schalek, R.L. 2004. "Fiber-Matrix Adhesion Measured by the Micro-Indentation Test: Data Reduction Based on Real Time Simulation." *Journal of ASTM International*, Vol. 1, Issue 3.

ASM International. 1987. *Engineered Materials Handbook*, Vol. 1, "Composites." Materials Park, OH: ASM International.

Bascomb, W.D. and Drazl, L.T. 1987. "The Surface Properties of Carbon Fibers and Their Adhesion to Organic Polymers." NASA Contractor Report 4084, July.

Bascomb, W.D. and Jensen, R.M. 1986. "Stress Transfer in Single Fiber/Resin Tensile Tests." *Journal of Adhesion*, 19, pp. 219–239.

Callister, William D., Jr. 2000. *Materials Science and Engineering*, 5th Ed. New York: John Wiley and Sons.

Carley, James F. (ed.). 1993. *Whittington's Dictionary of Plastics*, 3rd Ed. Lancaster, PA: Technomic Publishing Co., Inc.

Delmonte, John. 1981. *Technology of Carbon and Graphite Fiber Composites*. New York: Van Nostrand Reinhold.

Donnet, Jean-Baptiste and Bansal, Roop Chand. 1984. *Carbon Fiber*. New York: Marcel Dekker, Inc.

George, Andrew. 2002. "Spider Silk, Exploring Its Mechanical Properties and Reviewing Progress Towards Its Marketplace Presence." Report for MFG 355, August 12.

Lee, Stuart M. (ed.). 1991. *International Encyclopedia of Composites*. New York: VCH Publishers.

Lubin, George, (ed.). 1982. *Handbook of Composites*. New York: Van Nostrand Reinhold.

Miall, Mackenzie. 1961. *A New Dictionary of Chemistry*, 3rd Ed. New York: Interscience Publishers, Inc..

Mohr, J. Gilbert and Rowe, William P. 1978. *Fiber Glass*. New York: Van Nostrand Reinhold.

Morgan, Peter, 2005. *Carbon Fibers and their Composites*. Boca Raton, FL: Taylor & Francis Group (CRC title).

Society of Manufacturing Engineers (SME). 2005. "Composite Materials" DVD from the Composites Manufacturing video series. Dearborn, MI: Society of Manufacturing Engineers, www.sme.org/cmvs.

Strong, A. Brent. 2006. *Plastics: Materials and Processing*, 3rd Ed. Upper Saddle River, NJ: Prentice-Hall, Inc.

Wheeton, John W., Peters, Dean M., and Thomas, Karyn L. 1987. *Engineers' Guide to Composite Materials*. Materials Park, OH: American Society for Metals.

9

Reinforcement Forms

CHAPTER OVERVIEW

This chapter examines the following concepts:

- Filaments, strands, tows, rovings, and yarns
- Woven and knitted fabrics (cloth)
- Non-woven fabrics (mat)
- Prepregs
- Braided, stitched, and three-dimensional laminates
- Preforms
- Hybrids
- Whiskers

INTRODUCTION

Reinforcements are sold and used in composites in many different forms. Some of these forms are essentially available as produced by the fiber manufacturer. Others are converted into cloth, mat, or a variety of other forms. These technologies have been developed over the centuries by the textile industry and the composites industry has shared them. Still other forms have been developed specifically for use in composites.

Because of the importance of the reinforcement in the final properties of the composite, the form in which that reinforcement is used makes a major difference in the characteristics of the composite. Even more important, however, is the influence of the form of the reinforcement on the process by which the composite is made. Therefore, as the various types of reinforcement forms are explored in this chapter, keep in mind the properties of the composite that might be affected by the form and the type of process that might be used to make a composite structure using that particular type of reinforcement form. (The Society of Manufacturing Engineers [SME] produces videos that complement this text. They offer an excellent review of the forms of reinforcements used in composites [SME 2005].)

FILAMENTS, STRANDS, TOWS, ROVINGS, AND YARNS

As discussed in Chapter Eight, reinforcements are made as single **filaments** in a process called **spinning**. For efficiency, the filaments are usually made by passing the liquid raw material through a metal die with many holes in it where each hole forms a continuous filament. In the case of carbon fiber, precursor fibers are formed in this way and then converted. Despite the fact that most fiber reinforcements are formed as single filaments, they are almost never used as single filaments in either composite applications or textile applications. Filaments are simply too fragile and small to be used individually. To emphasize the solitary nature of filaments, these single fibers are sometimes called **monofilaments**.

A few comments are appropriate here about the word "**fiber**." This term is used in both precise and general terminology. The precise definition refers to a single filament in which the length is substantially greater than the diameter. Hence, a fiber has a high **aspect ratio**. However, the term "fibers" is used generically for any single or group of filaments that have a high aspect ratio. Practically, fibers are usually considered to have a length of at least .20 in. (0.5 cm). Below that length, the material is called **whiskers**.

As indicated previously, when synthetic fibers are made, they are usually continuous. This continuous nature facilitates subsequent processing of the fibers in processes such as weaving and knitting. However, occasionally it is desired to have fibers that are not continuous but longer than whiskers. So, the fibers are chopped into shorter lengths, usually in the range of .50–4.00 in. (1.3–10.2 cm). These chopped fibers are called **staple fibers**. In the textile industry, these fibers are often blended with cotton or wool fibers, which are naturally in the length of staple fibers. Having all the materials in the blend the same length simplifies blending and co-processing. In the composites industry, staple fibers are used in mat (to be discussed later in this chapter) and in molding compounds (discussed briefly in the polyester chapter and will be discussed further in the one on compression molding).

When formed, hundreds and sometimes thousands of filaments emerge from the many holes in a single spinneret. They are gathered together into a group and then wound onto a spool or processed further (possibly chopped into staple fibers). The group into which the fibers have been gathered has a specific name. In the case of fiberglass, the group of filaments is called a **roving**. For advanced fibers (carbon, aramid, ultra-high-molecular-weight polyethylene [UHMWPE], etc.), the group of filaments is called a **tow** (pronounced "toh"). The filament group also can be called a **strand**, which applies to both advanced and glass fibers. If the strand of fibers is twisted, which helps keep the fibers in the bundle, the group is called a **yarn**. Figure 9-1 depicts the various forms of continuous fibers.

The diameter of the filaments and the number of filaments in a strand varies greatly depending on the specifications of the purchaser, the manufacturing concept, and the capabilities of the fiber manufacturer. Fiberglass manufacturers have developed a system to identify the critical parameters for their products (BGF Industries, Inc. 1997). The identification system in the United States would identify a yarn as, for example, ECG 150 3/2 3.0S, where this code represents the following:

- E = the type of glass (the example represents E-glass but others could use a C [C-glass] or S [S-glass] identifier).
- C = the type of filament (the example is a continuous filament but others could be staple fibers and textured fibers).
- G = the filament diameter according to the standard designation given in Table 9-1. Fibers of diameters larger than K are rarely used in composites because of their tendency to break.
- 150 = the filament count (example is 15,000 yards/lb). Rather than a specific count of the filaments, the fiberglass

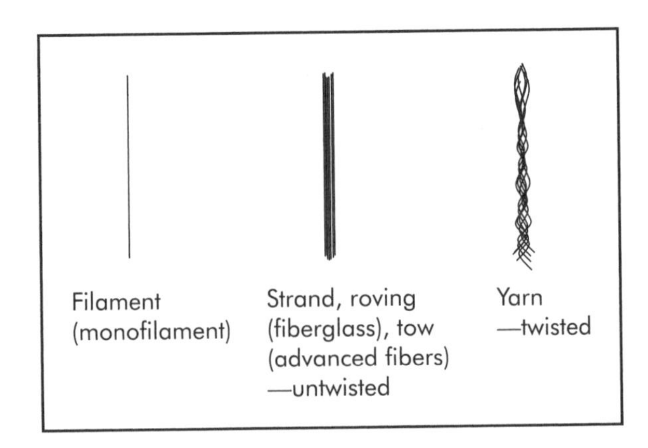

Figure 9-1. Various continuous fiber forms.

Table 9-1. Selected diameter designations for fiberglass products (U.S. system)*

Designation Code	Nominal Filament Diameter, in. × 10^{-5} (micron)
B	14 (3.5)
C	18 (4.5)
D	23 (5)
E	28 (7)
G	35 (9)
H	39 (10)
K	51 (13)
U	98 (25)

* This table and the example in the text were derived from BGF Industries, Inc. brochure (12/97)

system uses a number easily determined by weight.

- 3/2 = the number of strands twisted together/the number of plies in the total bundle (the example is two plies of a 3-strand construction (total of 6 basic strands).
- 3.0S = the number of turns per inch in the twist of the final yarn and the direction of the twist (example is 3 turns per inch in the S direction).

Advanced fiber manufacturers typically identify the following parameters for their products:

- Fiber types, unique to each manufacturer, carry designations such as AS4, IM7, P-55S, K-1100, T-650/35, 34-700, and HR40.
- To indicate the umber of filaments, the actual count is given (typical values are 3,000, 6,000, 12,000, and 50,000).
- Filament diameters are typically in the 20 to 30 × 10^{-5} in. (5–7 micron) range.

Fiberglass and advanced fiber manufacturers may also designate the dimensions of the spool or creel onto which the fibers have been wound. The spool's outside diameter, weight, core inside diameter, core length, fiber length on the spool, and number of spools per box are some of the values that would typically be used. These values help the purchaser match the material with the actual manufacturing process.

Occasionally, especially with standard textile fibers such as nylon and rayon, another term, "denier," is used to identify the fibers. A **denier** is the weight, in grams, of 9,000 meters of fiber. A 7-denier fiber is thinner than a 12-denier fiber. This measure is similar to the filament count measurement used to designate fiberglass strands.

Strands are used in many composite part manufacturing processes. The strands may be fed into systems where they are chopped and coated with resin (spray-up) or coated in a bath of resin and then laid down as continuous strands onto a mandrel (filament winding). These and many other processes will be discussed in later chapters in this book. The most important use of strands is in weaving and knitting fabrics and in continuous fiber processes such as filament winding and pultrusion.

WOVEN AND KNITTED FABRICS (CLOTH)

Some directional conventions have been developed to simplify the discussion of fabrics. Dating from the early days of weaving, these conventions are based on the way fabrics are woven on a loom. In the weaving process, fibers are arranged in a parallel array on a loom between two supports. The fibers are aligned in the **warp**, **longitudinal**, or **machine direction**. The machine direction is usually 0°.

A means is provided for other fibers to be interwoven above and below the warp fibers. In primitive weaving, the means is simply manually interlacing the crossing fibers. In more complex looms, the interweaving process is highly automated. These interwoven, crossing fibers are called the **weft**, **woof**, **fill**, or **transverse direction** fibers. These fibers are arranged at 90° to the machine direction.

Another direction is one that cuts across the fibers at a 45° angle. This is called the **bias**. Generally the bias direction toward the right of the machine direction is called positive bias and to the left is negative bias.

The directional conventions for woven fabric also pertain to rolls of fabric. These are given in Figure 9-2 where a roll of material is illustrated. Because fabrics are often sold on rolls and are planar, they are sometimes called **broad goods**, a term from the textile industry. Note that the warp fibers are in the long direction of the roll and the fill or weft fibers are in the cross or short direction of the roll. The edges of the roll are often treated to prevent unraveling. These edges are called the **selvage**. In general, when fabrics are made for composites, no selvage edge is created. The warp count and fill count are simply the number of strands

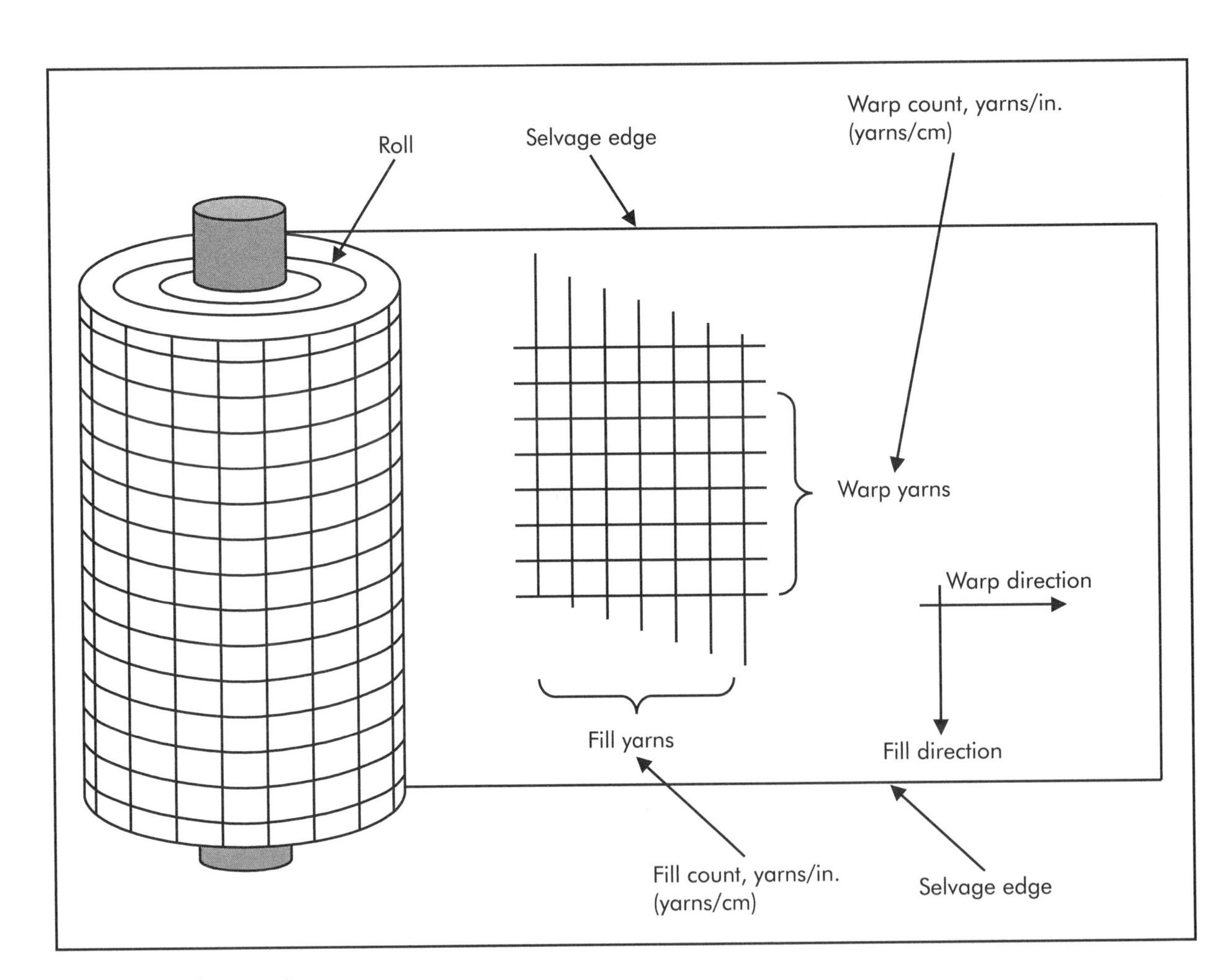

Figure 9-2. Roll terminology.

in the warp and fill directions respectively. These are usually expressed as yarns or ends per inch (or centimeter).

Types of Weaves

Textile manufacturers have developed fabrics with various handling and mechanical characteristics such as openness, drape, and strength. In fabrics, **openness** is a relative description of the space between the parallel fibers. In other words, it refers to how tightly packed the fibers are in the array. Openness is inversely related to the warp and fill counts. **Drape** is the ability of a fabric to hang without creases and to form graceful folds when used as draperies, shower curtains, and the like. In composites, drape is the ability of the fabric to conform to the shape of the mold. These characteristics are not just dependent on the fiber's properties but on the way in which the fibers are formed into the cloth. Figure 9-3 illustrates some of the most common weaving patterns.

- **Plain weave** (Figure 9-3a), the simplest of all the weaving patterns, is made by interlacing strands in an alternating over-and-under pattern. In other words, there is one warp fiber for

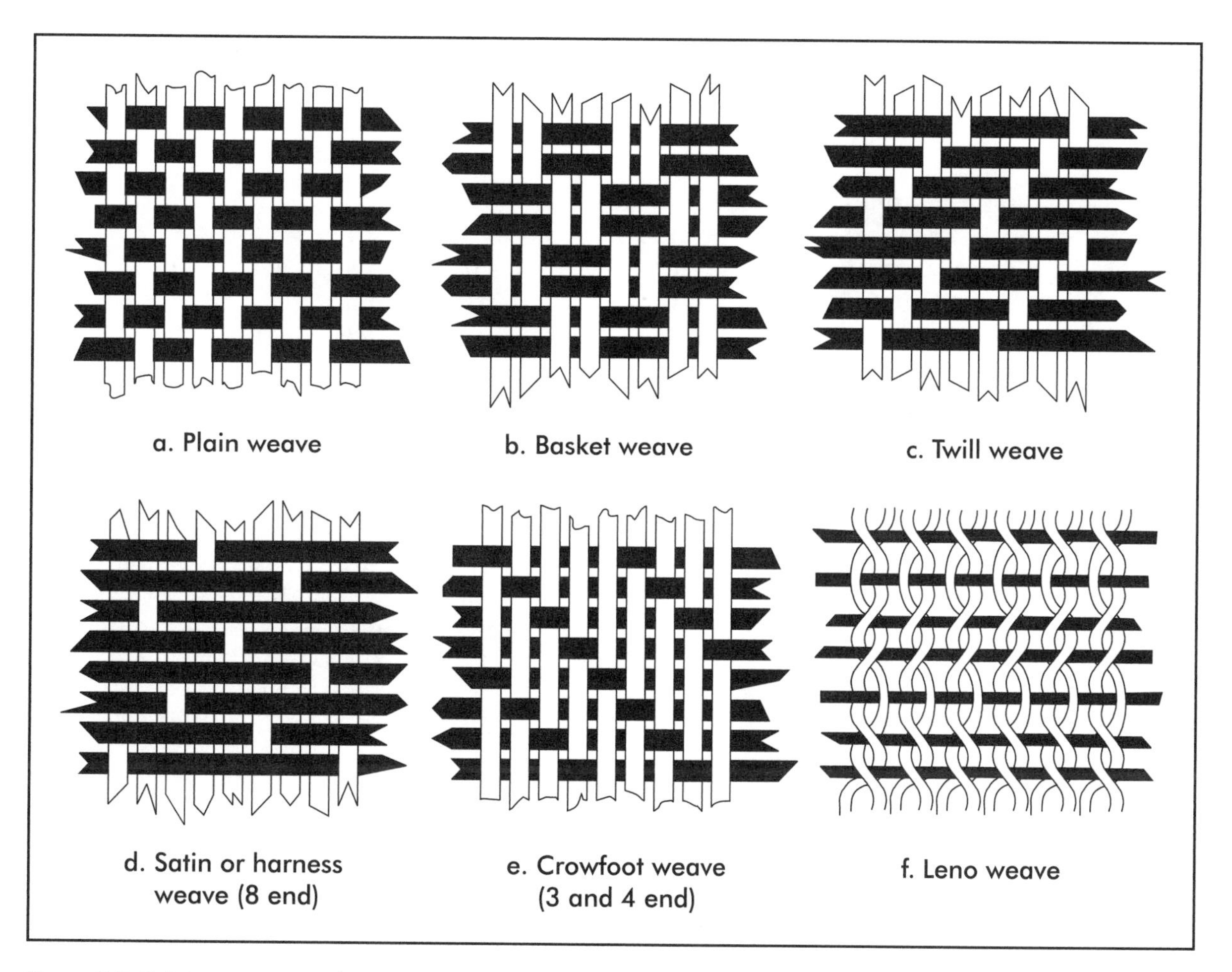

Figure 9-3. Fabric construction forms.

one fill fiber without skipping. Maximum fabric stability and firmness with minimum strand slippage results from this weave. The pattern gives uniform strength in both warp and fill directions when strand size and count for warp and fill are the same. This weave type is the most resistant to in-plane shear and is, therefore, considered to be a rather stiff weave. Because the weave is stable, it is usually left moderately open so resin penetration and air removal are fair to good. Plain weave fabrics are used for flat laminates, printed circuit boards, narrow fabrics, tooling, and covering wood structures such as boats.

- **Basket weave** (Figure 9-3b) has a pattern similar to plain weave except that two warp strands are woven as one over and under two fill strands. This weave is less stable than plain weave so it is more pliable. The fabric is flatter and stronger than an equivalent weight and count of plain weave. The additional strength comes because there are fewer crossover points between the warp and weft fibers. Crossover (crimp) points can pinch the fibers and cause defects that diminish the strength of the fabric. Basket weave uses are similar to those for plain weave but with better drape on mild contours.

- **Twill weave** (Figure 9-3c) has the fill fibers pass over one and under two or more warp fibers in such a way that the fabric has the appearance of diagonal (bias) lines. This weave has superior wet-out and drape compared to plain weaves with only a small reduction in stability. There is reduced crimp and so the strength of twill fabric is slightly higher than plain weaves. The surface of twill-based composites is smoother than plain weave composites.

- **Satin** or **harness weave** (Figure 9-3d) is similar to the twill weave but has one warp strand weaving over four or more strands (4 harness, 5 harness, 8-harness, etc.) then under one fill strand. This weave has a high degree of drape and stretch in all directions. A high strand density is possible. Satin weaves are flat. Less stable than plain weave, satin weave is also less open than most other weaves so wetting and air removal can be a problem unless a vacuum is used. Because the fabrics are non-symmetrical, care must be taken in the design of the part to ensure that asymmetric lay-ups do not cause warping of the part. (There is more on this subject in the chapter on design.) Satin weave is used extensively in the aircraft industry where complex shapes are common. It is also used for housings, radomes, ducts, and other contoured surfaces.

- **Crowfoot weave** (Figure 9-3e) is a satin weave in which the stagger pattern is not strictly repeated. This produces a weave with improved unidirectional stability and more strength in the warp directions than a plain weave. The crowfoot and satin weaves can incorporate a higher strand count than plain or basket weaves. The fabric is like other satin weaves, highly pliable, and can conform to complex contours and spherical shapes. The crowfoot weave is used for fishing rods, diving boards, skis, and aircraft ducts.

- **Leno weave** (Figure 9-3f) is produced by having two parallel warp strands twisted around each fill strand, providing a locked effect. This weave reduces distortion of low-count, open-weave fabric. It also provides heavy fabrics for rapid buildup of plies. The leno weave is used as an inner core for support of thin coatings and for tooling and repairs, most often in conjunction with other fabrics.

In addition to the conventional woven products, other fabrics have special applications in composites. These fabrics are illustrated in Figure 9-4.

- The looping pattern of **weft knitting** (Figure 9-4a) gives fabrics much greater drape and stretch than can be obtained in woven fabrics. Hence, the ability to contour these fabrics is exceptional. The contouring capability of knits has been useful in forming parts with complex geometries such as conduits for aircraft electrical wiring and air conditioning. These conduits are rarely straight for more than a foot or so and therefore require a highly complex mandrel for filament winding or hand lay-up. The knitted fabric can be applied over the mandrel as a sock, thus giving a well-fitted reinforcement with little time expenditure. Moreover, the transfer of energy within the fabric is excellent and so knitted fabrics are often used in ballistic devices, especially when the device is highly shaped (such as a helmet). This desirable property is enhanced by the superior crack-stopping capability of knitted fabrics.
- The method for **triaxial weaves** (Figure 9-4b) moves the warp fibers into the bias directions while leaving the fill fibers in the 90° directions. These fabrics obviously have high strength in the bias directions. In applications such as drive shafts, where the loading is chiefly along the 45° axis, triaxial weaves are highly efficient reinforcements. They are also highly effective when mixed with traditional woven fabrics to obtain a composite with high strength in all planar directions.

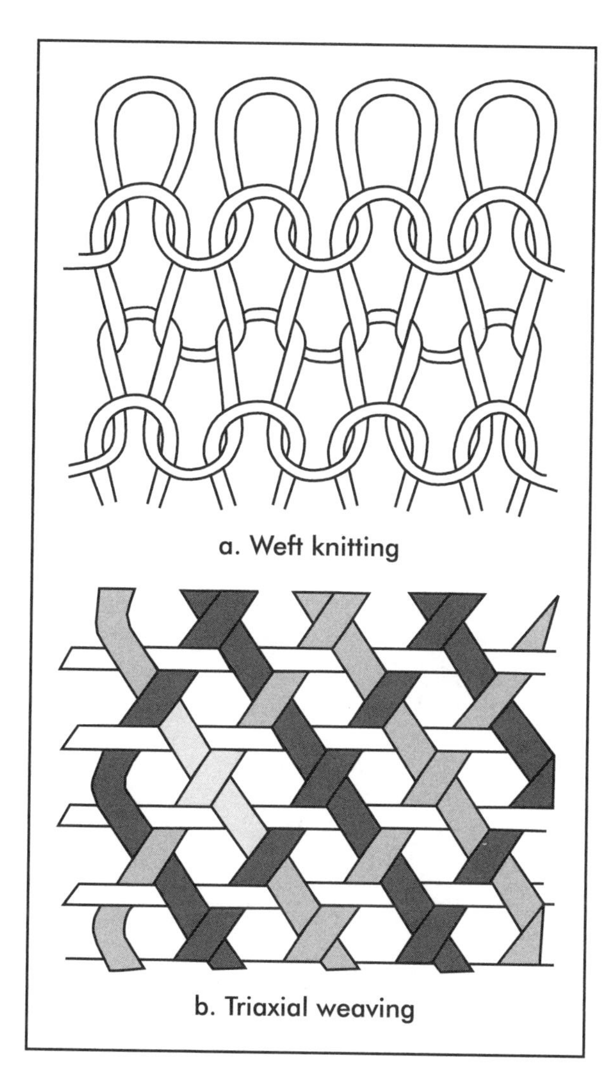

Figure 9-4. Knitted fabrics.

Properties

Fabric properties can depend upon a host of parameters including the following: type of weaves, areal weights, fabric thicknesses, thread counts, variety of fiber diameters, other fiber properties for both warp and fill, and directional breaking strengths. This represents an incredible variety, especially because each manufacturer has a separate group of fabric combinations. Therefore, extensive lists of fabrics by various categories and style numbers are published to assist users in obtaining the type of fabric desired. Additionally, most weavers will make special orders if the volume is sufficient to justify the setup time.

The wide variety of fabrics can be illustrated by mentioning two fiberglass fabrics used widely in composites—one thin and the other thick. The thick material is called **woven roving**. It is simply a large-count roving material loosely woven into one of the patterns discussed previously. The looseness is needed to ensure proper wet-out of the fibers. This material is used when a rapid buildup of reinforcement weight is desired. This can occur when an area of the composite part needs special reinforcement because it is subject to high stresses. The thin fabric is **veil**. It is comprised of a simple, tight weave made from fine strands so that it appears to be a fine, uniform fabric. Veil is often applied behind a gel coat to mask the underlying composite material and give a smooth surface. It also serves as a physical barrier to fiberglass strands that may become prominent because of the natural movements of the composite. (Note that some veils are made by non-woven methods as explained in the next section.)

NON-WOVEN FABRICS (MAT)

Some composite applications do not require the sophistication of a woven reinforcement. For example, a manufacturer of boats may be currently using chopped fiber spray-up (discussed in a later chapter) but would like the convenience of a sheet reinforcement. While changing to a woven fabric would likely result in higher physical and mechanical properties, it also would be more costly than the chopped spray-up method. An economical compromise for the molder would be the use of **chopped strand mat**. This material, shown schematically in Figure 9-5a, is a sheet product made by chopping fiberglass into similar lengths as would be typical for a spray-up operation (1–3 in. [2.5–7.6 cm]) and directing these chopped fibers onto a belt. A light coating of binder is applied and dried to create a material that can be handled as a sheet. The chopped strand mat gives the same quality of reinforcement that would be obtained in spray-up but with the convenience of a lay-up fabric and only a small increase in cost.

Chopped strand mat is not as rugged as a woven fabric. The fibers are bound

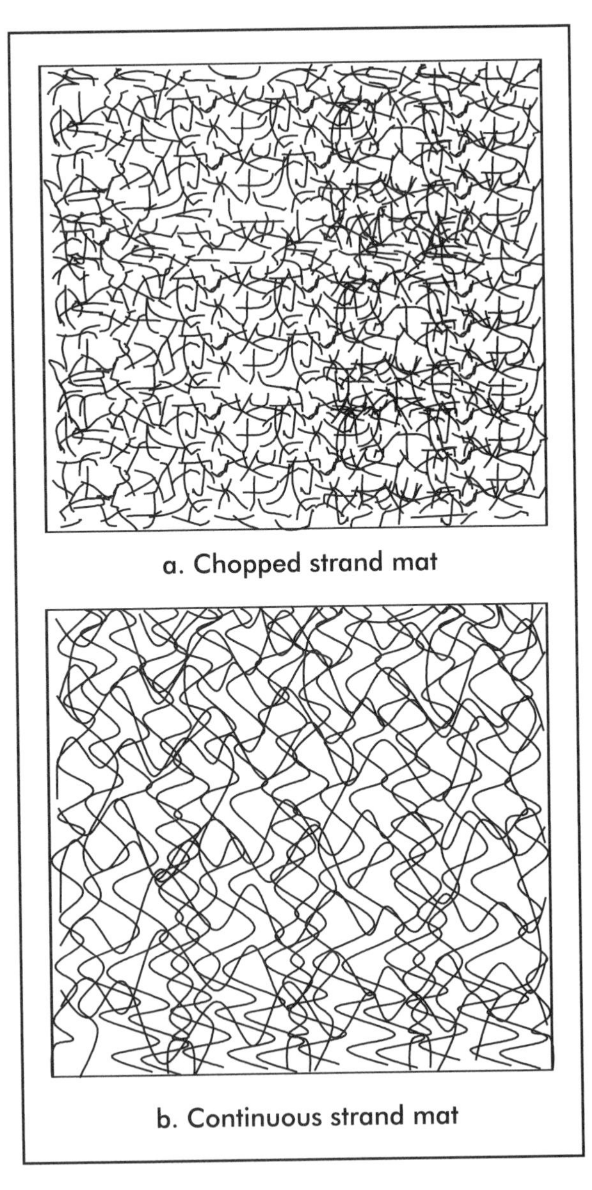

Figure 9-5. Types of mat.

loosely and so when handled, some of the fibers may fall from the mat. Therefore, some care should be taken in handling the product so that the integrity of the mat can be retained.

Some applications require slightly better properties than can be obtained with chopped strand mat but may not need the properties or cannot bear the costs of a woven fabric. A product that fits this niche is **continuous strand mat,** which is illustrated in Figure 9-5b. This material has longer strands than the chopped mat and, therefore, improved mechanical properties. The mat is also more rugged and can be handled without the loss of fibers, which is common with chopped mat. However, the continuous strand mat is not as durable as woven fabric. The product is made by swirling continuous strands of fiber onto a moving belt and then, just as in the case of chopped strand mat, applying a light coat of binder, which is then dried.

The composite designer should recognize that laminates made with mat are only 33–50% as strong as fabric laminates of comparable thickness. This drop in properties is because of lower fiber content in the mat, shorter fibers (in chopped mat), and more randomness in fiber direction, which results in some fibers being oriented in directions that are not required in the loading pattern of the part. However, the low cost and generally acceptable properties for many parts (especially those in which loads are moderate and/or multidirectional) has led to widespread use of chopped strand mat in the fiberglass reinforced plastics (FRP) marketplace.

The low cost of mats has led to many veil materials being made of mat rather than woven fabrics. This is logical because veils have no structural properties and so, in most cases, a mat veil would be just as acceptable as a woven veil.

It is possible, of course, to make mats out of materials other than fiberglass. One material that has found widespread use in non-composite markets is non-woven polyethylene. In this product (Tyvek®, by DuPont), polyethylene fibers of two different molecular weights are deposited onto a belt in much the same way that continuous strand fiberglass mat is made. However, with non-woven polyethylene, no binder is applied. Rather, the entire sheet is pressed between heated rolls. The temperature of the rolls is sufficient to melt the lower-molecular-weight polyethylene fibers, which then melt and bond the higher-molecular-weight materials together. This material is used as a wrap for buildings and in non-tear envelopes, among many other uses. While it is not used extensively in composites, the technology is inviting for other non-woven products.

PREPREGS

Advanced reinforcements like carbon and aramid fibers are often made into woven fabrics but are rarely made into either chopped or continuous mats. These advanced fibers usually require more precise control of the direction of the fibers than can be obtained with mats. In many cases, even the relatively fixed geometry of fabrics results in some sub-optimal arrangement of the fibers. For example, there may be an application wherein 80% of the load is in one direction. Therefore, the designer wants to orient 80% of the fibers in that direction so the design can be as efficient as possible. Fabrics with 80% of their fibers in one direction would rarely be made and would likely not be stable enough for practical applications.

Another method has been developed that allows designers to optimize the fiber orientation and yet still have the convenience of a sheet material for ease of manufacture. The method is to supply a sheet of fibers that are all oriented in one direction. Since the bare fibers are not woven, they are loose and, therefore, difficult to handle. The fibers could be sprayed with a light binder coating, as is done with fiberglass mat products. However, another solution to holding the fibers in place is now preferred. This method is to coat the unidirectional fibers with the matrix resin and use the same amount desired in the completed composite. So, no additional resin needs to be added to these materials. The resin holds the fibers together in a sheet form. These materials are called **prepregs** because they are fibers *pre-impregnated* with the matrix resin. Prepregs in which the fibers are all in the same direction are called **unidirectional tapes**.

The resin coating the fibers in prepreg sheets is usually epoxy. It is partially cured (B-staged) so it is easy to handle does not run off the fibers.

Because the resin is already in place on the fibers, no additional materials can be added to the prepregs. Therefore, the resin coating must include all of the hardening agent and other constituents required for full curing. Since the resin coating contains everything needed for curing the resin, the prepregs must be kept refrigerated to prevent them from curing prematurely. Typically, prepregs are kept at 0° F (–18° C) for up to 6 months. Beyond that time, most prepregs must be recertified as acceptable or simply discarded. Because the amount of time that a prepreg is exposed to room temperature is critical to its shelf life, when a roll of prepreg is removed from the freezer, the time it is out of the freezer is carefully noted. This is called the roll's **out-time**. If the entire roll is not consumed during one out-time, the remainder is placed back in the freezer so that the shelf life can be preserved. Some new developments in resin technology have resulted in epoxy resins that have much longer shelf lives. Some of these claim that even after a year without refrigeration no significant curing has resulted.

Prepregs are made by carefully collimating fibers in a single direction; that is, the fibers are passed through a device that aligns them carefully into columns and prevents the fibers from crossing over one another. Typically, a comb-like device is used to collimate the fibers. Then the columns are carefully drawn together into a sheet so that there are no gaps between the fibers. Next, the sheet of fibers is directed onto the surface of a paper sheet to which a thin layer of resin has been applied. This resin is the matrix. The thickness of the resin is carefully controlled so that the amount is equivalent to the final resin content desired in the prepreg. If a viscous resin is to be used, it can be applied as a solution. However, it is important to evaporate the solvent so that there are no (few) residual volatiles that might complicate the final cure of the resin or compromise the properties of the final part.

After the fibers are properly placed on the resin-coated paper, another sheet of resin-coated paper is applied on top. The sandwich of paper, resin, and fibers then passes through rolls where the resin is pressed into the fibers to achieve wet-out. The rolls may be heated to facilitate wet-out but the heating must be carefully controlled to prevent excessive curing of the resin. The fully wetted prepreg is then wound up and shipped to a cold storage facility. The paper used in this process is coated with silicone so that when the prepreg is formed, the resin separates from the paper and adheres to the fibers. Usually the top sheet of paper is removed when the prepreg is rolled up at the end of the prepreg line. The other piece of paper on which the prepreg sits is wound with the prepreg to prevent the rolled up layers from touching each other and sticking together. This final paper layer is removed when the prepreg is used.

Prepregs are supplied in rolls of convenient widths, typically 12–24 in. (30–61 cm) wide. However, they are available in widths as narrow as 3 in. (7.6 cm) and as wide as 72 in. (183 cm) or the maximum width of the machine on which the prepreg is made. The lengths are usually specified so that the weight of the roll is not excessive for easy handling.

Properties

Prepregs have a degree of flexibility. This permits shaping a prepreg to fit a mold of complex shape. The property of flexibility is called *drape* (using the same term as is used to describe the flexibility and conformability of textiles). The prepregs also have a sticky surface because the resin is not fully cured. This stickiness is called **tack**. The methods of lay-up and curing of these laminar prepreg products will be discussed in a later chapter on vacuum bag lay-up methods.

The specifications for prepreg materials are stringent. Some of the requirements include the following: fiber weight per unit of width, volatile content (if the resin was applied from a solution), resin content, tack, tack retention, flow of the resin (that is, viscosity curve with temperature), workmanship (drape), alignment (no fiber crossovers), gaps/spacing, width, length, splices, uniformity, and storage and shelf life.

Traditionally the resin content of the prepreg is given as a weight percentage, whereas the resin content of the finished part is given as a volume percentage. The explanation for this is simply that when the prepreg is made, the resin-weight percentage is easy to measure and control. However, the resin- (or fiber-) volume percentage is preferred for finished laminates because it is related directly to the mechanical properties through laminate theory (discussed in Chapter 11). Historically, the trend has been toward lower resin content so that the specific strength of the composite is increased. Because consistent removal of large excesses of resin has become a costly problem, prepregs are now made with near-net resin contents, allowing only a small excess of resin to accommodate the resin bleed process that occurs in bagging (discussed in Chapter 14).

If a more dimensionally stable sheet is desired, prepregs can be made with woven fabrics. These have some obvious handling advantages and, depending on the type of weave, can have improved drape over the unidirectional material. However, the crimping (crossover) of fibers in woven goods diminishes the strength of the prepregs compared to unidirectional tapes. Therefore, when the optimum performance in a laminar product is desired, unidirectional prepreg tapes are used.

Prepregs also can be made with tow. These pre-coated materials are used in automated operations where the application of a wet resin would be impractical. As with other prepreg material, prepreg tows have the advantage of carefully controlled resin-fiber content and, therefore, higher precision in this property than most wet impregnation processes.

BRAIDED, STITCHED, AND THREE-DIMENSIONAL LAMINATES

An irony (and difficulty) of making composite products is that they are usually three-dimensional but the raw materials (reinforcements) used to make them are generally one- and two-dimensional linear or sheet materials. Therefore, in making composite parts, a designer must be able to envision three-dimensional outputs from two-dimensional inputs. Thus reinforcements are layered to get three-dimensional products, which are often built-up laminates.

The efficiency and uniformity of the physical conversion from raw materials to the final product are important to the success of subsequent manufacturing processes. Composite manufacturers struggle with uniformity in thickness and other issues related to this dimensional conversion. Thankfully, some of the raw material suppliers have been inventive in the forms they supply and have already made some of the dimensional conversion. This pre-conversion to higher dimensions simplifies the manufacturing process for some composite parts and has been shown to result in improved properties.

Braided Reinforcements

Braids are strictly another form of fabric although they are usually perceived as three-dimensional materials. In braiding, several strands are alternated over and under each other to create either a flat or tubular fabric (like wrapping a maypole or, of course, braiding hair). The direction of crossover between the strands, which is affected by the number of strands being woven together, gives a characteristic angle for the braid. The direction of maximum reinforcement strength is in the direction of the fibers, which are at the characteristic angle of the braid. If desired, some strands in the braid can be oriented in the 0° direction to give longitudinal strength as well.

Using methodology developed over the centuries, braids are now made on highly automated equipment. The inclusion of computer controls for this equipment has led to some amazing options, which are being explored by composite manufacturers. Some braids are made on a core or mandrel, thus giving a shape to the material not generally possible with woven or knitted fabrics. Braiding need not be linear. Therefore, curved mandrels can be covered with braided material, thus allowing easy buildup of fibers for oddly shaped products.

Braids can be built up to create a pseudo-three-dimensional structure. This is called "pseudo" because it is still largely laminar in that no fibers are actually oriented in the direction perpendicular to the plane of the braid. However, the thickness yields a strong product that has many characteristics of a truly three-dimensional product.

When a braided structure is put in tension, the shorter fibers experience the load first and then straighten, thus putting load on the longer fibers. This occurs even with the relatively stiff matrix limiting the dimensional change of the composite structure. Therefore, most of the fibers participate in sharing the load. This phenomenon leads to good mechanical properties and highly energy-absorbent structures. The energy-absorbing nature of the composite is further enhanced because when fibers do start to fail, the lost load is again transferred to other fibers. Therefore, the failure mechanism of a braided composite is progressive breakage, which absorbs ever-greater energy and can often give indications of imminent failure. Because of the many crossover points, braids also have excellent crack-stopping capability. Braided structures have the advantage of bias fiber directions, resulting in improved performance when they are subjected to shear forces. A potential problem, often alleviated by good braid design, is the loss in properties because of crimping.

Some of the composite products that use braided reinforcements include hockey sticks (see the case study at the end of this chapter), prosthetics, canoe oars, aircraft horizontal stabilizers, automotive parts, furniture, motorcycle seats, turbofan guide vanes, water skis, and bicycle frames.

Creating fiber forms that already have some three-dimensionality is often highly desirable. These higher-dimensional materials can usually be handled more efficiently than lower-dimensional fibers or sheets and can, therefore, reduce manufacturing time. The higher-dimensional materials may also ultimately improve the mechanical properties of parts.

Stitched Laminates

Composite manufacturers discovered that if they stitched through a lay-up of reinforcement fabrics, some additional beneficial properties could be obtained. These beneficial properties arise because of the orientation of the stitches perpendicular to the plane of the composite layers. When at least 5% of the total fiber volume is oriented perpendicular to the plane of the plies, the laminate is called a **trans-laminar reinforced composite**.

The benefits in properties realized from stitching include improved compression after impact, increased fracture toughness, and severely restricted size and growth of impact damage and edge delamination. Clearly, there is an improvement in delamination since the layers are sewn together. Also, the stitching of the laminate improves the ease of handling the composite, thus reducing the time for manufacture.

Instead of stitching with conventional fibers, an alternate approach is to use discontinuous pins or rods to hold the layers together. This is called **Z-fiber construction**.

Stitching and Z-fiber construction work because the normal laminate depends on the relatively weak matrix to hold the laminates together in the perpendicular direction. Because the laminate is brittle (because it is crosslinked), the construction is especially susceptible to impacts and shear. Hence, reinforcements in the perpendicular direction have a major effect on these two properties in particular.

Two approaches are common with stitched fabrics. The first is called **comprehensive stitching**. It involves stitching across the entire laminate. This is generally done to improve the general properties of the laminate as already mentioned. Clearly, increasing the amount of stitching will increase the properties. Larger stitching threads and higher stitching densities (stitches per unit area) are both effective in improving the interlaminar shear and toughness properties. Unfortunately, both larger stitching threads and higher stitching densities lower the in-plane tensile strength, that is, the tensile strength in the plane of the laminate. This is not surprising because the stitching creates discontinuities in each layer of the laminate.

The second approach to stitching is called **selective stitching**. This involves stitching only in selected areas of the laminate. The most common area where such stitching might occur is at a joint. In this case, the advantages are that the area where shear forces are most likely, the joint, is strengthened, while other areas are not compromised in their other properties.

The improvement in laminate handling with stitching is intuitively obvious. The layers do not slide around and the laminate stays neatly layered. Less obvious is the improvement in other processes when the laminates are stitched. One example is pultrusion. This process, which will be discussed later in this text, involves drawing fibers or fabrics through a resin bath and then into a mold where the laminate is cured. If the fabrics are stitched, more complex arrangements of the plies can give improved properties in the part. Also, stitching has been shown to reduce the amount of stray fiber accumulation, which is always a problem at the entrance of the curing die.

Three-dimensional Laminates

With the advent of computer-controlled weaving machines, some three-dimensional woven fabrics have become practical as reinforcements for composite parts. These products are like traditional woven fabrics except that a weave is also introduced in the vertical direction. In making 3-D woven fabrics, multiple layers of fabric, which are woven at the same time in parallel sheets, are simultaneously interlaced with strands of fibers in the perpendicular direction, that is, in the thickness or z direction. This is like having weft fibers across the warp and perpendicular to the warp. The result is an integrated stack of fabrics bound together in a three-dimensional array.

The three-dimensional woven product has another advantage that can be critical for some parts. That advantage is a dramatic increase in formability (the ability to be draped) over equivalent 2-D stacks of material. Using this technology, a helmet was formed from a 3-D woven fabric without pleating. The total manufacturing time for the helmet was reduced from 45 minutes using 2-D technology to only 3 minutes with the 3-D weave and the vacuum-assisted resin transfer molding (VARTM) process. This rapid time was assisted by the much higher resin infusion rate of 3-D fabrics when compared to equal thicknesses of traditional 2-D lay-ups.

Knitting also has been adapted to making preforms for composite structures. However, because the stiff fibers used in composites, especially carbon fibers, do not adapt well to knitting, a different approach has often been taken. In this new approach, the precursor fibers, such as polyacrylonitrile (PAN), are knitted into the three-dimensional shape and then the entire structure is carbonized to convert the PAN fibers to carbon fibers.

Just as non-woven products, such as mats, have some unique advantages over woven, braided, and knitted fabrics, non-woven three-dimensional lay-ups have some advantages over interlaced three-dimensional products. The major advantage of the non-woven products is the lack of crimping. This advantage is even more important in three-dimensional products. Comparisons of the mechanical properties of composite parts have shown that 3-D products can have a 30% increase in strength and modulus over otherwise equivalent crimped constructions.

A major project at Boeing (and previously at McDonnell Douglas, Long Beach) used 3-D materials for fabrication of a prototype aircraft wing. The entire wing was made from a stitched preform, which was shaped with appropriate stiffeners (such as T-sections). The near-net-shape preform was made using an automated stitching and lay-down machine as a single unit. Many different fiber orientations were layered and then stitched together. The layers included mat, unidirectional sheets, woven fabrics, and knitted fabrics. Thicknesses varied from over 1.5 in. (3.8 cm) at the wing root to less than .75 in. (1.9 cm) at the tip. Resin was infused into the preform using a VARTM technique. The completed wing was shown to have superior mechanical properties and, of course, greatly reduced manufacturing costs when compared with traditional 2-D manufacturing methods.

All forms of 3-D materials allow easy inclusion of various types of reinforcements, such as glass, carbon, and aramid fibers together in one integrated preform. This permits optimization of the structure.

PREFORMS

The application of advanced textile technologies like three-dimensional weaving and stitching has clearly improved the manufacturing ease of making composites by creating broad goods that are pre-layered and held in place. There is one more textile-manufacturing concept related to the form of the fibers, which has further improved composite manufacturing. That additional concept is the **preform**.

The concept of a preform is to make a reinforcement package in the shape of the final product (or nearly so). The preform lacks only the resin and curing to be fully completed. Said another way, the reinforcements are arranged into a form that is the same as the finished product will be when the resin is added and cured. Because they are made to be in their final shape, such preforms are called **net shape**. The advantages are, of course, the elimination of shaping and finishing after molding and generally improved uniformity of fiber distribution. Also, in some closed-mold operations like resin transfer molding (RTM) and compression molding, preforms are needed to ensure that the layers do not move around in the mold during resin injection, mold closing, or curing, especially where deep draws and complex shapes are required.

The use of preforms depends on issues such as quality, repeatability, manufacturing efficiency (cycle time), and economics. Sometimes preforms are dictated because, as cited previously, movement of layers during the molding cycle would prevent good parts from being made. In other cases, the reason for preforms is simply cycle time. In some situations, shorter cycle time will easily justify the increased costs of making a preform because of the high value of the product, the high value of the molds (and the need to have more parts made on each mold), or because the production demands are high and parts are needed in large volume.

Some molders have long practiced a method of improving efficiency by precutting and assembling stacks of material so that at the molding stage the operator does not have to worry about getting the materials in the proper lay-up. This process is called **kitting**. Some molders request that material suppliers do the kitting for them. This is often more efficient because the material suppliers have sophisticated cutters and other material-handling equipment.

A preform can be thought of as two- or three-dimensional, based on the nature of the materials from which it is made. In both cases, of course, the final preform is three-dimensional because all real, finished parts are three-dimensional.

Two-dimensional preforms are made by laying up traditional two-dimensional materials like fabrics, knits, mats, and prepreg sheets, and then cutting to the final shape of the part. The layers are then secured in some fashion to hold them in place and make the preform. Alternately, the layers of material can be cut and then laid up in the final shape and secured, although this is less common. Some common methods of securing the layered materials would be with a lightly applied adhesive, with light stitching, and with pinning.

Another way of making a two-dimensional preform is by laying up materials like mat that already have a light binder coating and then thermoforming to create the shape. The light binder in the mat must be compatible with the thermoforming process. After thermoforming, the preform is trimmed to net shape. It is then ready to be used in a process that will add the resin and cure the final part. The disadvantages of the thermoforming method are the cost of the extra step and the usually higher cost of the thermoforming-compatible binder.

Another method of forming a preform using two-dimensional materials is by *directed fiber performing*. In this method, a skilled operator uses a chopper gun to direct chopped fibers and resin into a perforated mold shaped like the part. A strong suction behind the mold assists in getting uniform coverage, compacting the fibers, and holding them in place until the binder cures.

In a more automated method, fiberglass roving is directed into a chopper located at the top of a chamber. The chopped fibers are blown into the chamber and directed onto a perforated screen shaped like the desired preform. The fibers are deposited on the screen in a way that gives uniform coverage. As the fibers are directed onto the screen, they are sprayed with a binder so that, after drying or curing, the fibers are held in the preform shape. Some models of this equipment have several preform molds that are rotated under the stream of fibers and binder. The multiple forms speed production. The drawbacks of this system are the high cost of the equipment and the poor compaction of the preform, which can be a problem. Another is that the preform is made with chopped fibers and that could be a problem where strengths need to be optimized.

Three-dimensional preforms are not made from layered two-dimensional materials but, in contrast, are made directly using one of the 3-D processes discussed previously. For preforms the processes would not just make 3-D weaves, they would make 3-D weaves to a particular net shape—the preform. This requires sophistication not present in the broad goods products. For example, when making a three-dimensional preform by 3-D knitting, the machines must stop knitting on some needles and continue knitting on others, all the while maintaining an even tension on all yarns. When the 3-D weaving or knitting is completed, the preform is ready for resin addition and molding.

Three-dimensional I-beams, T-stiffened panels, and similar shapes can be created automatically using 3D techniques. These shaped preforms are useful in processes such as pultrusion and RTM where the final part can be made by simply infusing resin into the preforms and then curing. As the thickness of the part increases, the value of the preform also increases because labor can be dramatically reduced. Preforms are required for the RTM process and because of the manufacturing efficiency they afford, it is a process that is growing rapidly (RTM is discussed in detail in a later chapter).

HYBRIDS

Materials that combine two or more types of reinforcements are called **hybrids**. Typical examples would be carbon with aramid, glass with carbon, and glass with aramid. The advantage of hybrids is that the superior properties of the various reinforcement types can be utilized to optimize the part.

A highly practical application is in an I-beam that will likely be subjected to impact. The hybrid lay-up is illustrated in Figure 9-6. Aramid fibers, usually as a prepreg sheet, are placed on the top and bottom of the I-beam in the areas where impact is expected. The rest of the I-beam is made of carbon fibers. If the resins are the same, full integration of the aramid section with the carbon materials should be possible.

Other hybrids are made by laying a sheet or fabric of one material on top of others in a laminate. This has improved energy-absorbing properties because energy dissipation is increased across layers of dissimilar materials. There is, of course, some worry of delamination, but that can be minimized if the resins are the same in layers of different reinforcements. An application of this method of using hybrids is in armor. Layers of aramid and UHMWPE are alternated to give the highest performance in bullet-proof vests and rigid armor applications.

Still other hybrids are made by weaving or knitting different fibers into the same fabric. This has the advantage of improving the average properties of the fabric according to the average of the two materials. Another advantage of this technique can be achieved by weaving together reinforcement fibers

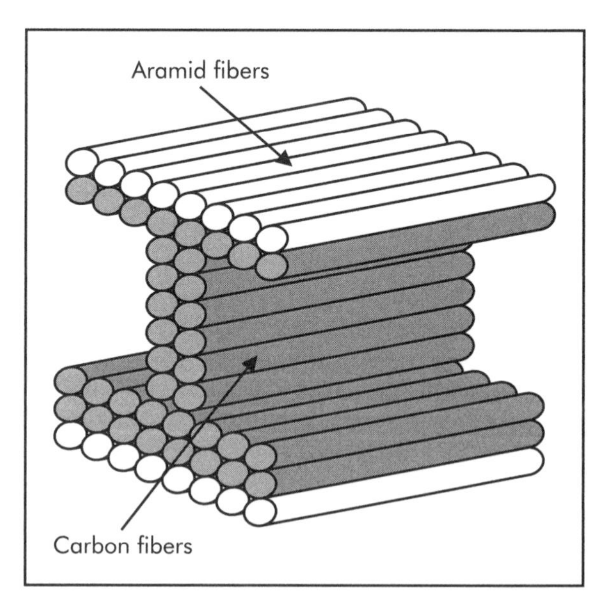

Figure 9-6. Selective placement of fibers within hybrid composites.

with thermoplastic fibers that can be later melted to form the matrix. This eliminates the need for a separate matrix introduction and, of course, the weaving pattern guarantees that the matrix is well distributed throughout the reinforcements (for example, making the reinforcement the warp and the matrix fibers the weft).

WHISKERS

Whiskers differ from fibers in that whiskers are shorter. Practically, fibers are usually considered to have a length of at least .2 in. (5 mm). Below that length the material is considered to be a whisker. (If the material does not have a high aspect ratio, that is, high length to radius ratio, it is called a *particle*. Particles are not usually considered reinforcements in the same sense as fibers and whiskers.)

Because whiskers are short, the probability of encountering a defect is much lower than with conventional fibers. Hence, on average, individual whiskers have higher tensile properties than individual fibers. However, in a composite, the higher properties are often not fully realized because of end effects (weakening of the composite because of problems at the ends of the fibers). In addition, the whiskers are so much closer to the critical length that full transfer of force from the matrix to the reinforcement is not sustained for as long a time as with conventional fiber lengths.

Most of the time whiskers are sold in one of two different forms. The first is **milled fibers** sold as loose whiskers. These are usually made by chopping and then milling regular fibers. Milling is an operation where the fibers are placed between rollers and squeezed as they are rolled. This reduces the length of the fibers. The milled fibers are often used to make repairs to composites and fill in areas in the pre-cured part that are too small or narrow for fibers to be effectively placed. Many fiber manufacturers make whiskers from spooling remnants. These are the fibers left on the ends of spools, which are slightly longer than the average and, therefore, not processible with regular length spools. The other common method of selling whiskers is as a paste in matrix resin. This has the advantage of ease of application.

CASE STUDY 9-1

Braided Preform Use in Hockey Sticks

Although composites have proven to be excellent materials for both the shaft and the blade of hockey sticks, the blade has been a problem because its curved shape has been difficult to match with conventional, flat prepreg sheets. Hockey stick manufacturers reported some fiber buckling in the curve and, therefore, loss of strength and reduced "feel" for the hockey player. However, those problems have been solved through the use of a braided preform.

The preform for the blade is made from fiberglass and is braided over a mandrel. The fibers used are softer than conventional roving fiberglass so they can be handled more easily in the braiding process. They also give improved performance in the part. The braider ships the preforms to the molder.

The hockey stick is made of two parts—the shaft (handle) and the blade. The handles are usually made of wood, metal, or composite. After the handle is made, the blade core (made of white ash) is attached to the handle. Then, after final sanding of the blade, the molder slips the preform over the wooden blade core much like a sock is slipped over a foot. This process is much faster and more reliable than the old hand lay-up process. In fact, the preform process has been semi-automated so that the time to install the sock is less than 15 seconds.

The next step is to apply the resin. This is done by hand and the resin is rolled onto the fibers to achieve good wet-out. The resin is usually a toughened polymer like polyurethane. Curing is usually done at room temperature in about an hour, although a post-cure of about 24 hours is common. Then the blade is trimmed and finished.

SUMMARY

The form in which the reinforcements are used makes a difference in the properties of the composite. The form also makes a difference in the methods of manufacturing. Therefore, examining the nature of the reinforcement is an important aspect of understanding composites.

Many composites are made from the simplest of reinforcement forms—fibers in bundles. These bundles, called strands, tows, rovings, or yarns (depending on the type of fiber and whether it is twisted), are used directly in many processes such as filament winding and pultrusion. The fiber bundles also can be chopped into shorter lengths so that they can be blended with natural fibers or sprayed into a mold along with the resin matrix.

Fibers can be woven into cloth. But, the cloth is not all the same. The properties are manipulated by varying the over-and-under pattern of the fibers as they are crossed to make the cloth. Some of the patterns are chosen for their desirable properties but with others the major focus is on the shape of the part and the ease of conforming the reinforcement to the part's shape.

Sometimes a broad goods shape is desired, as in a woven fabric, but other characteristics are lacking. Hence, knits and braids can be made. These are generally more complex than just woven broad goods. The fibers also can be made into a broad goods form without weaving, knitting, or braiding. These nonwoven materials are called mats or sheets. The mats are generally randomly oriented and the fibers are held in place with a light application of a binder. The matrix is added in a later step. Prepregs are sheets comprised of unidirectional fibers held together by the matrix itself.

Real composite parts are three-dimensional. Actual parts can be made by assembling layers of essentially two-dimensional material. However, this results in little fiber being oriented in the direction perpendicular to the plane of the broad goods. Therefore, these built-up materials have little strength in the perpendicular direction. But some textile techniques like perpendicular stitching and 3-D weaving allow reinforcement assemblies to have a significant number of fibers oriented in the perpendicular direction. These materials, called three-dimensional reinforcements, have improved perpendicular properties, including toughness.

Both two-dimensional and three-dimensional reinforcements can be cut and shaped so they are in the final shape of the molded part. These cut and shaped assemblies are called preforms and generally can be made into composite parts merely by adding the resin. Preforms are becoming increasingly important because of the rise in the use of RTM as a manufacturing process, although other processes also use preforms advantageously.

Whiskers, while not really a different form of reinforcement, are mentioned in this chapter because they can be delivered as a paste and used as a form of reinforcement.

LABORATORY EXPERIMENT 9-1

Fabric Openness

Objective: Determine the relative wet-out characteristics of open and tight fabrics.

Procedure:

1. Make two lay-ups using fabrics that are vastly different in openness. The lay-ups should be at least four layers thick.
2. Place the lay-ups over a container and then clamp the edges of the reinforcement assemblies so they do not move.

3. Pour a set amount of resin, such as 10 oz. (300 ml), into the middle of the reinforcement assembly so the resin flows through it and then into a container. (If preferred, corn syrup or motor oil can be used in place of the resin.)
4. Time the flow of the resin through the reinforcement assembly. (If preferred, the container into which the liquid flows can be calibrated and the timing measured as the time to reach a certain volume in the container.)
5. Discuss the flow rate versus the openness of the fabrics.

QUESTIONS

1. Discuss the differences in weaving pattern and physical properties for a plain weave and a satin weave.
2. Distinguish between a cloth and a mat.
3. What are the differences between a random chopped mat and a continuous strand mat in terms of construction and performance?
4. Which type of satin weave would be expected to have more drape—8-harness or 5-harness? Why?
5. Give three advantages of a stitched fabric.
6. What are two advantages of a 3-D weave over a stitched fabric?
7. Explain why preforms are gaining in popularity. What is a disadvantage of preforms?
8. Why do whiskers have more strength than fibers?
9. List three products for which a knitted fabric would be better than a woven.
10. Explain why crimping diminishes the properties of a fabric.
11. Distinguish between a prepreg and a mat.

BIBLIOGRAPHY

A&P Technology. 1998. "Why Braided Reinforcements should be Part of Your Vocabulary and Part of Your Composite!"

BGF Industries, Inc. 1997. Brochure, December.

Carley, James F. (ed.). 1993. *Whittington's Dictionary of Plastics*, 3rd Ed. Lancaster, PA: Technomic Publishing Co., Inc.

Mohr, J. Gilbert and Rowe, William P. 1978. *Fiber Glass*. New York: Van Nostrand Reinhold Company.

Salamone, Joseph C. (ed.). 1999. *Concise Polymeric Materials Encyclopedia*. Boca Raton, FL: CRC Press.

Society of Manufacturing Engineers (SME). 2005. "Composite Materials" DVD from the Composites Manufacturing video series. Dearborn, MI: Society of Manufacturing Engineers, www.sme.org/cmvs.

Strong, A. Brent. 2006. *Plastics: Materials and Processing*, 3rd Ed. Upper Saddle River, NJ: Prentice-Hall, Inc.

10

Quality and Testing

CHAPTER OVERVIEW

This chapter examines the following concepts:

- History of composite materials testing
- Quality control principles
- Component materials testing
- Mechanical testing
- Thermal and environmental testing
- Flammability testing
- Non-destructive testing and inspection

HISTORY OF COMPOSITE MATERIALS TESTING

In general, the properties of metals, ceramics, and plastics are known because of the large databases for these materials compiled over the years. The same is not true for composites. The lack of an accepted composites database is due to many factors, including the following:

- the complexity of composites,
- the relative newness of composites as materials,
- the continuous development of new composites, and
- the scarcity of standardized tests for composites.

The complexity of composites arises because the reinforcement and the matrix are such different materials. This is further complicated by the fact that the reinforcement can be oriented in specific directions. As will be discussed in more detail in the chapter on design, these characteristics make composites very different from metals, ceramics, and plastics.

The emergence of composite materials has been relatively recent, generally dating from the 1950s. The initial developments were largely made by aerospace companies who developed tests specific to their own applications, thus using their own criteria for acceptance. Although these tests were not usually proprietary, the companies rarely worked together to develop standardized tests. Hence, the database of composites data was fragmented and inconsistent from company to company. Further, new materials were continually being developed—true carbon fibers in the mid-1960s, aramid fibers in the 1970s, and many advanced resins in subsequent years. Even the development and approval of new designs made the accumulation of standard databases difficult because each design might be made with new and occasionally surprising results. In essence, the innovations in composites design and manufacturing were occurring so rapidly that uniform testing was not able to keep up.

Some attempts at standardization have been made by the major testing and professional societies of the world. In the United States these groups include: American Society for Testing and Materials (ASTM),

Military Handbook Committee (MIL-HDBK-17), Suppliers of Advanced Composite Materials Association (SACMA), National Aeronautics and Space Administration (NASA), American National Standards Institute (ANSI), National Institute of Standards and Technology (NIST), Underwriters Laboratory (UL), ASM International, Society of the Plastics Industry (SPI), Society for the Advancement of Materials and Process Engineering (SAMPE), and the American Society of Mechanical Engineers (ASME). Non-U.S. groups have also worked to standardize tests. These include: Deutsches Institut fur Normung (DIN), Association Francaise de Normalization (AFNOR), British Standards Institute (BSI), Japanese Industrial Standards (JIS), and the International Organization for Standardization (ISO).

Because of the lack of an accepted database for composites and the uniqueness possible with each design, testing of each composite structure or design is far more important than is typical for metals, plastics, and composites. It is therefore important to understand the overall nature of composites testing and quality control. There are specific tests that have become common within the composites industry. Thus this chapter begins with a discussion of the fundamentals that underlie quality control and all composites testing. It then proceeds to discuss specific tests for material components—resin and reinforcements—and then tests for the composites themselves. The composites tests will be arranged according to the type of property being examined: mechanical, thermal and environmental, and flammability. The chapter concludes with a look at non-destructive testing and inspection.

QUALITY CONTROL PRINCIPLES

The fundamentals of quality are straightforward and amazingly simple. Whether the process to improve is a unique lay-up operation of large parts or a compression molding process for thousands of pieces a day, the basic approach is similar. In the end, the goal is to describe a principled approach to quality using the tools, methods, and techniques that make sense and to illustrate the development of a logical plan for improvement.

Quality is a process not a destination. This statement is a bit trite from overuse, but it does describe the true nature of quality improvement. However, in the process of improvement, it has been found that most people, especially company executives, give themselves more credit for implementing good quality practices than they may actually deserve. A study was done to evaluate the familiarity and use of quality methods and techniques in small manufacturing operations. It was found that almost without exception, company managers reported that they were more familiar with and more involved in using quality programs than was actually observed during inspections of their facilities. Clearly, guilt or self-deception should not be allowed to cloud the evaluation of a company's position in terms of quality, which is so critical to long-term success.

As part of a self-evaluation on quality, it should be understood that use of a quality tool or system does not guarantee its correct use. For example, the fact that a company uses control charts in the tracking of its manufacturing process does not mean the control charts are being used appropriately or that the use is sufficient. Therefore, reporting or believing that control charts are implemented correctly within a company may be an overly optimistic judgment by managers, yet understandable. However, if a company gets caught in this over-estimation of quality performance, the danger can be that it will fail to invest the time and effort needed to really use the quality tools that will help make a difference. A company truly focused on continuous improvement will not get caught up in measuring *how many* or *which* tools it is using or not using. The focus will be on *getting better*, realizing that the tools are simply the means to continue on the course of improvement. Thus assessments of the effectiveness of the quality program will take precedence over merely verifying that some quality tool is being used. If a company measures its quality program by how many or what kind of tools are being used, it is likely focused on the wrong thing.

To begin to make some sense of the variety of methods for quality control, there are five primary quality functions that quality tools are to address. These are:

1. Documentation and certification programs
2. Analysis and statistical tools
3. Improvement methods
4. System integration strategies
5. Management philosophies

All are important to quality and the improvement of quality. The degree of importance of any one depends on the company's particular stage of progress.

Documentation and Certification Programs

Documentation and certification of quality (like ISO 9000) is being forced upon many companies by competitive pressures. Although companies may not like it, they will benefit from those pressures in the long run. In the short run, however, there is confusion and anxiety.

Certification programs, such as **ISO 9000**, are popular and the intent of these programs is good. There are some interesting issues with these efforts, however, which are not necessarily compatible with improvement-focused quality plans and systems. First, ISO 9000 and the QS 9000 are, for this discussion, essentially the same; that is, they provide a way to document, evaluate, and certify a company's quality system. ISO 14000 is similar in intent but focused on environmental issues.

ISO 9000

Originally, ISO 9000 was to replace scores of national and international quality standards with one single family of standards that would be recognized and accepted worldwide. This, for the most part, has occurred. But at the same time there has developed a dangerous idea that if a company has achieved certification then it is a company of high quality. *Certification does not guarantee excellent quality.* In short, this is because it does not require much more effort to document a poor process than it does a good one. ISO 9000 has made an admirable attempt to include continuous improvement in the certification process; it is just not as easy as it sounds.

What is ISO 9000? It is a set of international standards that outlines stringent requirements for documenting the components of a company's design and production system. Certification is granted based on a company's ability to adhere to the requirements set forth in the standards. ISO 9000 is actually composed of four certification sets: 9001 being the most comprehensive and 9002, 9003, and 9004 addressing lesser sets of requirements of the organizational system.

ISO 9001 is a quality assurance model made up of 20 sets of quality system requirements. This model applies to organizations that design, develop, produce, install, and service products. ISO 9001 expects organizations to apply this model and meet the requirements by developing a comprehensive quality system. The primary components of this system include all management responsibilities, quality system requirements, product design requirements, process control requirements, and training and service requirements.

ISO 9002 is made up of 19 sets of quality system requirements. It is essentially the same as 9001 except that product design is not required. Thus it focuses on management, quality, process, and training and service requirements.

ISO 9003 is made up of 16 sets of quality system requirements and is essentially the same as 9001 except it leaves out product design, purchasing, process control, and servicing requirements.

ISO 9004 primarily focuses on the critical elements of a strong quality system—all aspects of management responsibilities, quality planning, improvement, and customer interactions and service.

As already indicated, the ISO certification process presupposes but does not really require good quality. That task is left largely to the company. However, the task of obtaining good quality is ultimately much more important to the well-being of a company than compiling documentation for the various standards to which it might be certified.

Analysis and Statistical Tools

The "basic tools" are so basic that you might not even consider them to be statistical in nature or related to quality. This list of tools includes check sheets, flow charts, correlation plots, and plots of occurrence frequencies called *histograms*. Most of these are intuitive. For example, check sheets are used to ensure that all of the required steps are being done in a process. To fully understand a process, it is smart to begin with a flow chart showing each of the steps in the order they occur.

Correlation plots have to do with running the process. To find out if changing one variable in a process, such as cure temperature, would cause a change in another variable, such as hardness, the cure would be run at various temperatures and the hardnesses measured. Then, the hardnesses would be plotted versus the temperatures. Then a correlation analysis would be performed to determine how closely the two variables were related. Correlation plots can be done using most spreadsheet computer programs (such as Excel®) with little special training needed. Such a plot is shown in Figure 10-1. The correlation coefficient, R^2, is shown along with the equation of the line.

A histogram is shown in Figure 10-2. In this example, hardnesses were measured for 30 days and then plotted according to their general values. The data show that the plotted hardnesses represent roughly a bell curve (normal distribution). When the data are normally distributed, several statistical methods can be applied such as statistical process control (SPC), capability analysis, and design of experiments (DOE). A Pareto chart is a special type of histogram (or bar chart) used for quality purposes. This chart lists the causes and numbers of product rejects, and orders them from highest to lowest. Figure 10-3 shows an example of a Pareto chart.

To effectively improve a process, it must first be determined how well the process or system is functioning. The most important of the tools for monitoring a process is *statistical process control* (SPC). It uses average values of important process variables, such as temperature or hardness, and plots them over time. Upper and lower control limits are established to help monitor when a variable is not within the desired range for smooth and reliable production. SPC helps to identify when a process is drifting out of control.

Capability analysis is a method of statistically analyzing a process to determine how good it is relative to the process specifications. In other words, is the process, as it is running, capable of staying in statistical control within the specification limits set for it?

Design of experiments (DOE) then comes into play to help reduce the common cause variation that occurs in statistically stable processes. DOE, with its strong relationship

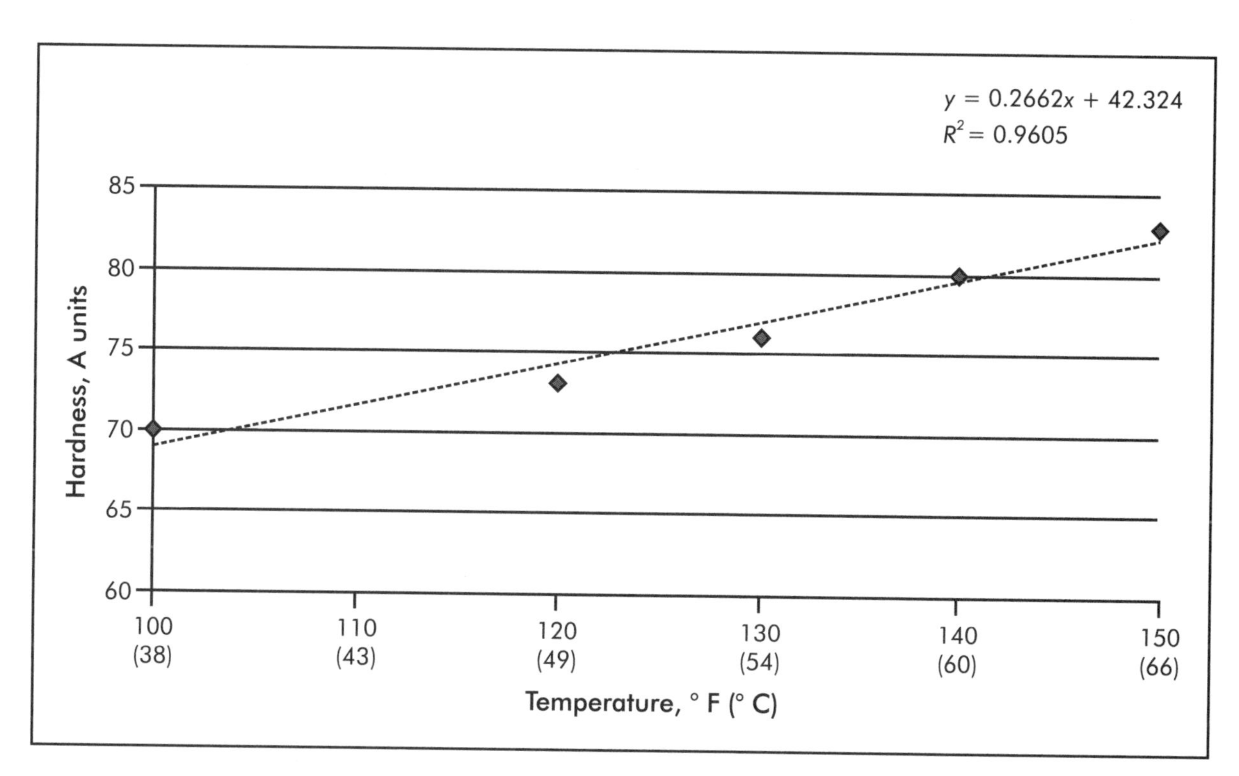

Figure 10-1. Correlation plot of hardness versus temperature.

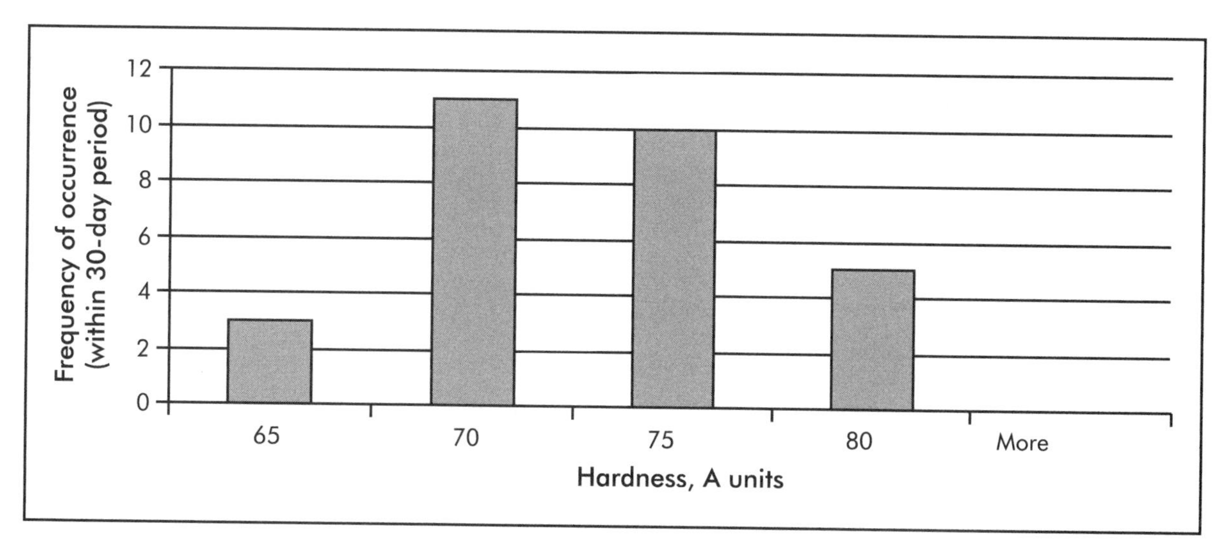

Figure 10-2. Histogram.

to improvement methods, supports progression to higher stages of quality. In essence, DOE is a method of performing experiments and analyzing the process when more than one variable might be changing, as is the typical case in actual operations.

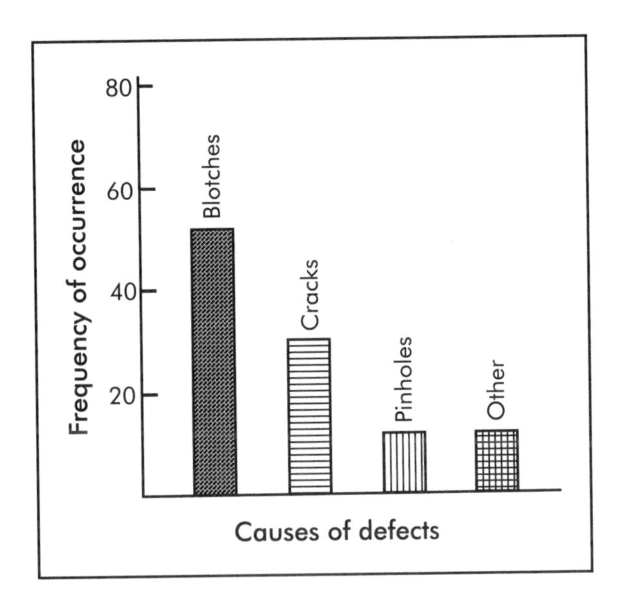

Figure 10-3. Pareto chart.

Improvement Methods

The entire idea behind the most powerful and long lasting of the quality strategies is the idea of continuous improvement. This is what Ed Deming preached and, in fact, what his 14 points are based on. In his view, the purpose of the analysis tools is to determine how and where the process should be improved for most immediate effect. Then quality efforts should continue the analysis so that new areas for improvement are identified and worked on for improvement.

Failure mode and effects analysis (FMEA) is a technique that identifies the foreseeable failure modes of a product, service, or process and plans for the elimination of those failure modes. The hoped for result is, of course, the elimination of failures. The FMEA technique involves three steps:

1. Recognize and evaluate the potential failure of a product, service, or process and the effects of that failure.
2. Identify actions that could eliminate or reduce the chance of the potential failure occurring.
3. Document the process.

Another tool that is helpful in identifying the causes of a problem or defect is the *cause-and-effect diagram* (sometimes called the Ishikawa or fishbone diagram).

Quality function deployment (QFD) is a system that focuses first on the desires of the customer and then sets the product or process parameters in accordance with those expectations. When problems arise, changes to the process are examined in light of potential customer satisfaction.

Most companies find that improvement efforts need to be quantified or at least compared to some standard. The most common of those standards is the competition or some industry standard. This comparison process is called *benchmarking*. However, even though benchmarking is laudable, it is difficult for a company that is just starting to apply quality control methods. Used properly, benchmarking requires an excellent understanding of the company's processes, its weaknesses, and how to best apply the ideas and methods obtained from other companies. If this prerequisite understanding does not exist, a company will find itself trying to imitate the operation it is benchmarking rather than intelligently implementing learned methods.

Another method for establishing a quantified standard is *six sigma*. In this improvement scheme, a standard of process deviation is set such that the standard deviation, σ, is reduced to the point that the specifications are at $\pm 6\sigma$. In other words, 99.9999998% of the product or service will be between specifications, and the nonconformance rate will be 0.002 parts per million. This standard is difficult to achieve and should not be attempted by companies that have not already worked for a considerable time at improvement.

A company might seek an alternative standard for improvement that is appropriate for their status but does not depend upon either industry standards (benchmarking) or an arbitrary value (six sigma). Possibly the best alternative is to employ the *Taguchi loss function*. In this method, the process is run under SPC but, rather than accept the process whenever the critical values are within the acceptance limits, the process is continually centered toward the mid-point of the specifications. In other words, the process is *continually refocused*. This method has proven to be highly effective and applicable to companies regardless of their state of quality implementation.

Total quality management (TQM), an excellent concept that lost luster falling victim to the same end as other management fads, was founded on the idea of continuous improvement. Unfortunately, its falling was more the fault of improper implementation by impatient and naïve managers than it was a fault of the philosophy. Improvement programs are, as they should be, overflowing with milestones and goals, but they do not have an ending point. In attempting to use TQM, many managers focused on endings and final results and thereby fell into one of two fallacies. First, having accomplished a set goal to improve quality and having once achieved it, they stopped their efforts, usually far short of what the company needed to claim success. The other fallacy involved companies who did not accomplish the goals they had set, and having failed to do so, blamed TQM, though the fault lay within their methods, implementation procedures, or style.

Improvement strategies must be directed by transformational leaders and accompanied by improvement in the leader, the company, and the process. By virtue of its title and purpose, done any other way, it is not an improvement strategy.

System Integration Strategies

At its root, quality is truly a system concept and quality improvement is a system problem. It is affected by, and affects all aspects of the design and production system. For example, at the root of quality is the reduction of variation. In fact, Deming said, "If I had to reduce my message to a few words, I would say that it all had to do with reducing variation." Regardless of how good a quality strategy is or how broad and precise the view is of the system, if processes and parts are not stable and ever striving to achieve low variability, quality will never be good. In the book *Running Today's Factory*, Standard and Davis claim that "the core strategy of lean manufacturing is to achieve the shortest possible cycle time by streamlining production flow" and "that factory performance is largely determined by variability." Thus, quality improvement and lean manufacturing are brought together in such a way that there cannot be impact on one without impacting the other.

The system's overall performance depends on the interdependence in its entire structure, not a compilation of the sum of its parts. The structure and organization of the system determine the behavior, which in turn dictates various events. Usually, only the events are seen and therefore acted on. That is why, especially early on in improvement efforts, actions are shortsighted and reactionary whereas later they become more objective and wise. Early on the broader scheme of things is not considered, but eventually experience forces this to be the first consideration. Too often it is thought that by breaking up the system, all that will have to be dealt with are its parts. Systems theory is a reminder that if you segment an elephant you may get more "manageable" pieces, but you will not gain an understanding of how the elephant as a whole works.

Today's world requires astute management of the supply chain. In the past, the supply chain for any given company usually involved only suppliers that were nearby or within the same country. Now the supply chain is likely to be international. With this broader focus, quality issues become even more important and, often, costly.

Therefore, a good management system for the supply chain is imperative in today's global economy.

Management Philosophies

The employees' quality of life has a lot to do with the quality of a company's product. How management sees employees and the systems and priorities of the company are critical factors in the quality improvement process. Nearly every quality guru has a set of principles (for example, Deming's 14 points) that are counted as fundamental to the particular program or method. Total quality management, considering the term "management" as part of its name, depends on sound management for success as has already been discussed. Some of the other management philosophies seek to instill a special kind of quality awareness, sometimes without the direct reference to quality embodied in Deming's 14 points or TQM.

Lean manufacturing has been, and continues to be, an extremely important concept in reducing waste throughout the manufacturing enterprise. One of the important aspects of lean manufacturing is the 5S (sort, straighten, shine, standardize, sustain) program, which helps to put everything in the plant in good order. During lean implementation, the company will do a value stream map of its process(es), in which the various capacities and production systems are described in terms of downstream demand. The company then attempts to improve the current process, possibly using a technique called *kaizen*. A "kaizen" is an event whose outcome is a future state map, which embodies the change system and path for continuous improvement (Mika 2006). All together, the lean manufacturing method yields great results and improved profits because the entire system is considered together in an organized and disciplined way. This overall view will, of course, favorably impact the quality of the products or process. Other systems of good management will similarly give positive impact on quality.

COMPONENT MATERIAL TESTING

Strategy

Testing can be viewed from two different perspectives. One view is a determination of the *fundamental* properties of the composite material. The other is to investigate the *in-use* properties of the composite by monitoring changes in the composite's properties over time or under various environmental conditions. This information is used to verify that a composite part has the properties expected over its useful life.

The conditions under which testing occurs can often affect the test results, sometimes dramatically. Therefore, if a standard test is to be performed, the specifications of sample conditioning, the type of equipment, and the operation of the equipment (such as the speed, temperature, etc.) should all follow that standard precisely.

If non-standard tests are to be performed, the conditions of the test should be carefully noted and justified. Usually, a standard with known properties or a comparison material might be tested at the same time, thus ensuring that the conditions are the same for both samples.

Raw Material Testing

The manufacturers of reinforcements and resins usually conduct extensive tests on their products and, upon request, furnish the results of these tests to their customers. It is generally good practice for molders to obtain these test results so that they are aware of the specifications of the reinforcements and resins. This is useful information to the design process and flags possible conditions that may need monitoring.

Because reinforcements do not change appreciably over time and because the tests to verify their properties are tedious and require specialized equipment, few molders independently verify the properties of the reinforcements. Some of these tests are discussed in Chapter Eight. On the other hand, resin is constantly changing because of its reactive nature. Therefore, it is recommended that molders conduct some verification of resin properties to ensure that the batch is appropriate for use.

Testing for viscosity is simple and can be done using relatively inexpensive equipment (see Figure 10-4). The measured viscosity can be compared with the viscosity of the resin as indicated by the manufacturer. If the viscosity has dramatically increased, it is an indication that the resin has begun to cure and may not be appropriate for use.

More extensive tests (requiring more testing sophistication) include Fourier transform infrared (FTIR) and various chromatography methods. FTIR examines the frequencies of the molecular motions in the polymer and thereby creates a spectrum, which is unique for each polymer. Chromatography separates various components of the resin. In one chromatography method, a small amount of solvated or liquid resin mixture is injected into a small-diameter tube in which a gelatin material (in the most common method) has been coated on the inside walls. A flow of gas is injected and the tube is often heated. Under these conditions, the components of the resin separate. Some of the components are absorbed more strongly by the gelatin and therefore they take longer to traverse the length of the column. The retention is a function chiefly of size since the smaller materials penetrate the gel and are absorbed more strongly. Another factor affecting the permeation is chemical affinity for the gel. Chromatography can be combined with mass spectrometry to identify the nature of the components that make up the resin mixture.

For prepreg samples, the viscosity of the resin cannot be determined easily. Therefore, tests that measure the flexibility and the stickiness of the prepreg are used as

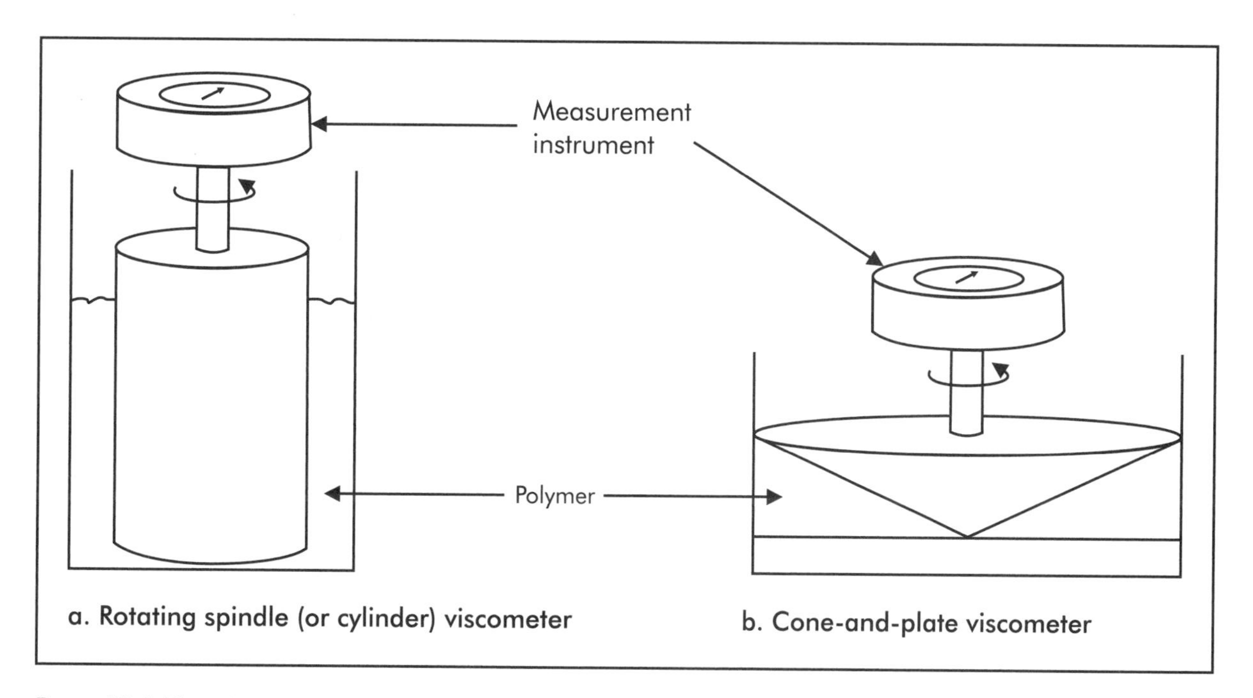

Figure 10-4. Viscosity measuring equipment.

reasonable approximations of the resin's cure state. The test for flexibility is called the **drape test**. In this test, a sample of the prepreg is placed over a solid mandrel of a set diameter and the amount that the sample drapes over the mandrel under its own weight is measured. If the cure of the resin has advanced excessively, the prepreg will be stiff and the drape will be less. The stickiness is determined using the **tack test**. In this test, the prepreg is pressed against a standard surface and the amount of force needed to lift the prepreg is measured, usually by hand or by a force scale. The tack will be less over time as the cure of the prepreg advances (usually as it continues in storage).

The acceptability of the initiator (catalyst) is best measured by mixing it with resin in the correct concentration and then running a test for gel time and peak exotherm as discussed in Chapter Three. If the times and temperatures are similar to those specified by the manufacturer and data collected by the molder, the initiator can be assumed to be acceptable.

Most of the other components in a composite can be accepted from reliable manufacturers as shipped unless some specific problem arises.

Curing Tests

It is important to ensure that the cure has occurred properly in a composite. The gel and exotherm tests are directly related to the cure and can be performed after assurance that the resin and the initiator are themselves acceptable (as discussed previously). If problems with the cured composite are evident or suspected, other tests are available to determine whether the part cured properly.

A hardness test is possibly the easiest way to investigate the extent of cure on a cured sample. This test uses a hand-held device, which has an indenter that is pressed into the part. The force to indent is the hardness, which is measured on a scale determined by the type of hardness tester and the nature of the indenter. One common device is the Barcol hardness tester. In general, the higher the hardness value, the more complete or advanced the cure. Some caution must be exercised in using hardness tests. The variation in these tests can be high. Also, the tests are usually not appropriate for gel coats because the sample thickness for good hardness measurements is thicker than that recommended for gel coats. (See ASTM D2583.)

Another test to evaluate the extent of cure uses **differential scanning calorimetry (DSC)**. This test requires a large test machine that is quite costly and, therefore, is not generally owned by molders. However, standard testing laboratories or resin suppliers often have such machines. The test measures thermal transitions in the resin. Of particular interest is the transition called the glass transition temperature, T_g. As the amount of crosslinking increases, the T_g also increases until a maximum is reached when the amount of crosslinking is essentially complete. If data is available from previous samples known to be properly cured, that data can be compared with the T_g of the sample in question, and a determination of the extent of cure can be made.

In the DSC test, a small sample of the composite is placed in a sealed packet and then inserted into the heating chamber of the differential scanning calorimeter where a known standard material is housed. The temperature of the chamber is increased and the amount of heat taken up or given off by the test specimen is compared with the standard material. The information is displayed in a plot of heat absorbed (or given off) versus temperature. The increases or decreases in the plot are endotherms (heat taken up) or exotherms (heat given off) and correspond to phase changes and other molecular rearrangements in the sample. One of the transitions measured is the T_g.

Other methods used to determine T_g are dynamic mechanical analysis (DMA) and thermo-mechanical analysis (TMA). In the DMA method, the sample is loaded with a stress and the displacement is observed. Alternately, the sample can be displaced and the stresses observed. The glass transition temperature is taken as the point where key properties associated with the stiffness of the composite change. In the TMA method, changes in the length or thickness of a sample as a function of temperature are monitored.

After curing, it is sometimes desirable to determine the concentration of fiber and resin in the composite. If no fillers are present, the percentage of reinforcement can be determined by placing a weighed sample in a tared crucible and heating to burn off the resin. The sample is then reweighed and the percentage of reinforcement and resin determined. This method works because the reinforcements are not affected by the temperatures at which the resin is burned off. However, the method cannot be used for reinforcements such as aramids and UHMWPE.

If a furnace is not available for burning off the resin or if the reinforcement will not withstand the temperature required to burn off the matrix, a solvent extraction method can be used. In this method, the weighed composite sample is placed into a solvent bath, which is usually heated and equipped with a reflux condenser to return the hot solvent back into the heating vessel. After digestion of the resin, the reinforcements are dried and weighed to determine the percentage of composition. A test method also exists for determining the fiber content of a composite through electrical measurements. This method only applies to unidirectional conducting fibers in a non-conductive matrix.

If a filler is present, it will interfere with the calculation of the resin-reinforcement percentage. Fillers has to be separated from the reinforcement and from the resin. This is often a complicated process that depends on the type of filler and other components in the composite.

One test to help sort out the presence of filler from the resin is the thermo-gravimetric analysis (TGA) test. This test involves the weight lost, expressed as a percentage, as a function of increasing temperature. The temperature increase is controlled at a fixed rate and a purge gas sweeps any evolved material out of the system. TGA can be used for organic and inorganic systems or for a mixture of both. Moisture is generally lost at about 212° F (100° C) or higher depending on whether moisture is bonded in the system. Organic systems are lost below 932° F (500° C) depending on the nature of the material. Inorganic systems may have transformations that can be used for identification. TGA also can be used for stability studies to determine the rate of decomposition at a fixed temperature.

The *void content* of the composite can be determined from a combination of the percentages of component contents and the densities of the composite, fiber, resin, and any other components that might be present. The fiber and resin densities are usually obtained from the manufacturers. The composite density is obtained by the buoyancy test. The procedure is to calculate the theoretical density of the composite based on the densities and percentages of the components and then compare the theoretical density with the actual density determined in the laboratory. The difference is assumed to be the void content.

MECHANICAL TESTING

Mechanical properties are determined by testing the composite material when subjected to various forces. As depicted in Figure 10-5, five forces are important in composite structures: tensile, compressive, flexural, torsional, and shear.

All of the force types are important to determine the behavior of composites in different circumstances. Some of the important properties are determined under standard testing conditions (room temperature, slowly applied forces, etc.). Determination of other properties involve these forces but with various changes to testing conditions. For example, impact tests involve forces that are added quickly. The creep tests are like tensile tests in which the forces are static but applied for a long time. The fatigue tests involve flexural loads that are applied repeatedly over extended periods. The vibration and damping tests measure the response of the composite to loads that are intermittent.

Some mechanical properties for a typical aerospace composite, carbon-epoxy, are given in Table 10-1.

Table 10-1. Mechanical properties of a typical carbon-epoxy composite.

Properties for 62% Fiber Volume Composite	Test Value
Tensile strength	315 ksi (2,172 MPa)
Tensile modulus	21 Msi (145 GPa)
Flexural strength	260 ksi (1,793 MPa)
Flexural modulus	19 Msi (131 GPa)
Shear strength	18 ksi (124 MPa)

Tensile Properties

Tensile properties are determined using procedures such as those given in ASTM D638 and D3039. The former test is for all materials in general whereas the latter is specifically for resin-matrix composites reinforced by oriented continuous or discontinuous high-modulus fibers.

The standard sample for test method D638 is shaped so it will break between the gripping regions of the tensile test machine. The

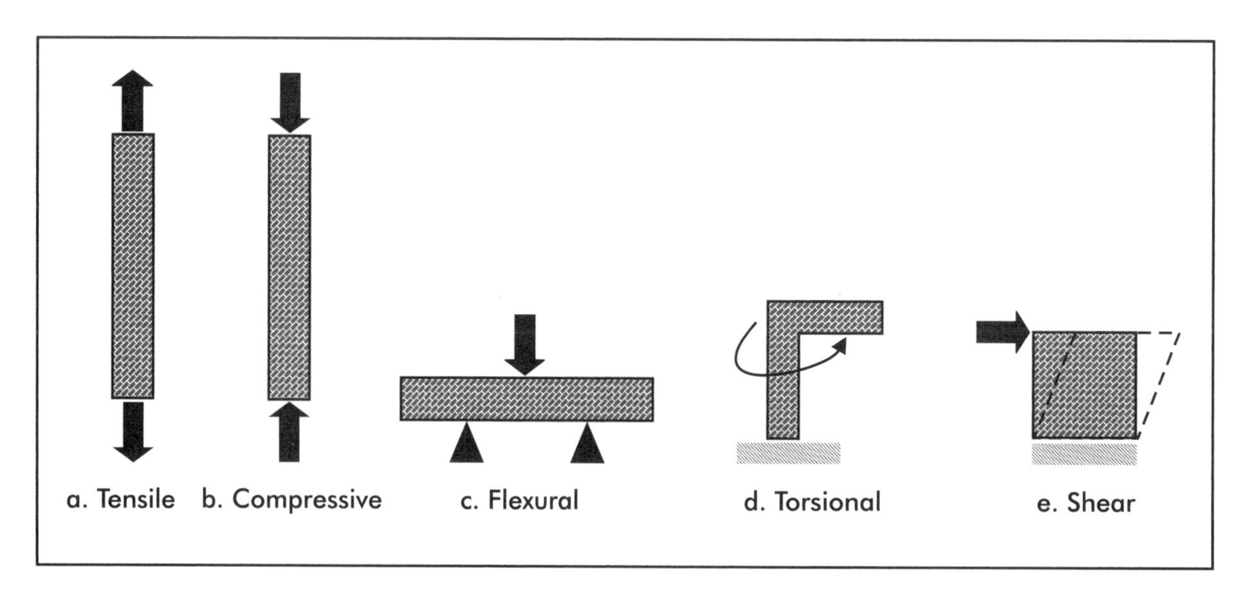

Figure 10-5. Types of forces.

most common shape for non-composite plastic test specimens or composites reinforced with short fibers is a "dog-bone" shape, illustrated in Figure 10-6a. The increased width of the specimen in the gripping zones ensures that the specimen will break within the failure zone between the grips.

The standard specimen for composites (test method D3039) is illustrated in Figure 10-6b. Note that this specimen achieves breakage in the desired zone by reinforcing the ends with tabs bonded to a flat, rectangular sample. This latter specimen avoids the problem of cutting the specimen into a dog-bone shape, which requires some of the fibers in the specimen to be cut short. Since this affects the properties of the composite, the tabbed specimen is superior for continuous reinforcements. Note that the tabs in specimen 10-6b are tapered at

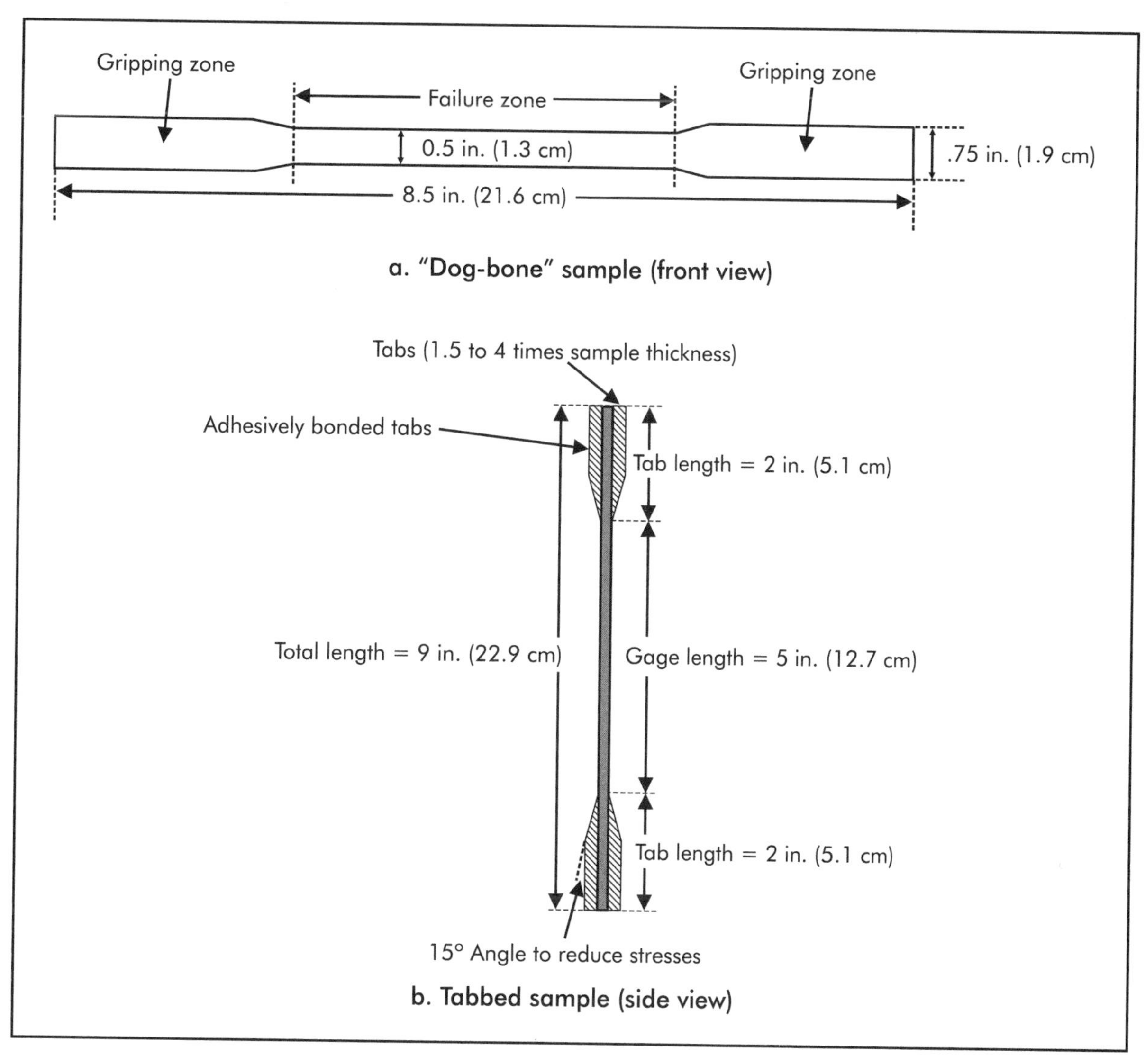

Figure 10-6. Tensile test specimens.

the point where the tabs meet the body of the specimen. This taper avoids the stress riser (causing early failures) that can occur when a sharp edge is present. Care should be taken in sample preparation so the tabs are properly aligned and bonded at a temperature that will not distort or otherwise affect the sample. Generally, the samples are cut from a laminate sheet after the tabs have been bonded onto the laminate. The tabs are more likely to be parallel and, therefore, the adhesive is less likely to run over the edges of the samples. To prevent fiber fraying along the edges and to give the most precise width control, diamond cutting blades with waterjet cutters have proven to be the best way to cut the samples.

Some data indicate that the thickness of the sample is important in giving less spread in the data. It is suggested that the samples be in the thickness range of .2–.4 in. (5–10 mm). Other data indicate that the surface finish (either from lamination or from machining the edges) can cause differences in properties. Surface ply orientation makes a significant difference in properties too. For example, in two samples that have the same number of plies in the same direction, the order of the plies makes a difference in the tensile test. An orientation in which the surface fibers are parallel to the pull direction yields different results from a sample having all the same plies in which the surface is oriented in the perpendicular direction.

The crosshead speed is generally run at 1–2.5 in./minute (25–64 mm/minute). Because of the brittle nature of the samples, especially if they are carbon fiber reinforced, care should be taken to protect the eyes and skin against flying fragments. The break is often sudden with a large amount of shattering of the sample.

When a specimen is pulled in tension, there is a resistance to movement. If only small deformations are considered and the solid material returns to its original shape when the force is relieved, then the deformation is called **elastic**. In these deformations, all of the mechanical energy put into the material by the applied force to cause the deformation is held within and then used to cause the material to return to its original shape and position when the stress is relieved. A common example would be a spring that is deformed slightly, thus imparting potential energy, which is then available within the spring to cause it to return to its original shape. Another way of saying this is that energy is returned or recovered. Energy is always recovered in elastic deformations.

The force divided by the area is called the **stress** (σ). So that materials of different sizes can be directly compared, the force is usually divided by the area of the sample, yielding units of pounds per square inch or pascals (newtons per square meter). (The area is the width times the thickness. The length dimension is not important in the calculation because it is assumed that the applied force is evenly distributed over the entire length of the sample between the pulling forces.)

The displacement (movement) of the material is called the **strain** (ε). Normally the strain is given as the change in length (Δl) divided by the original length (l_o); the units are dimensionless.

The **elongation** is a measure of the strain when the force is tension. Elongation is usually expressed as a percentage increase in length compared to the original length of the test specimen.

Plots of the stress versus strain such as the type shown in Figure 10-7 are called stress-strain curves or stress-strain diagrams. Elongation is difficult to measure because of the stiffness of the samples. Accurate determination of elongation requires strain gages, extensometers, or optical interferometers. These are attached to the sample during setup of the test.

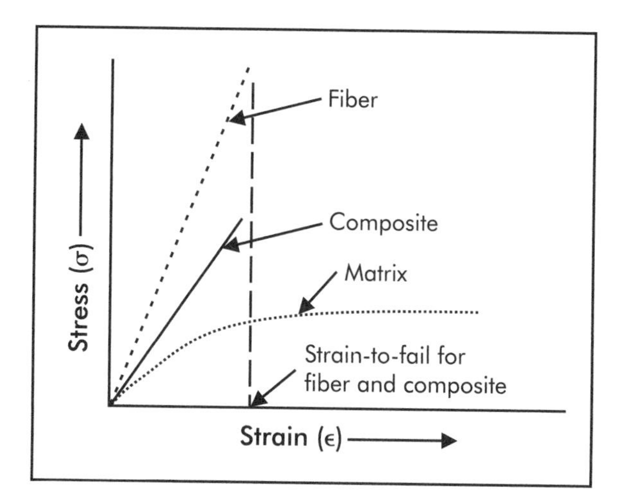

Figure 10-7. Relationships of the tensile strength of fiber, matrix, and composite.

Note in Figure 10-7 that the stress-strain curve for the fiber is linear. This indicates that the stress-strain behavior is elastic. The curve for the matrix begins linearly but then starts to curve. Hence, the stress-strain behavior of the matrix is elastic and then non-elastic. The composite curve is elastic because it is dominated by the characteristics of the fibers.

The relationship between stress and strain in the elastic region can be shown by,

$$\frac{F}{A} = \sigma = E\varepsilon \qquad (10\text{-}1)$$

where:

F = force
A = cross-sectional area
σ = stress
ε = strain
E = proportionality factor, called the **modulus**, and sometimes referred to as **Young's modulus** for the tensile stress case. The modulus is the slope of the stress-strain curve. If the modulus is large, corresponding to a steep angle of the curve, the material resists deformation strongly. Such materials are said to have **stiffness**.

It is important to remember that the properties of the individual components of a composite are not the same as the properties of the composite itself. For example, the strength of the composite is not the same as the strength of the fibers even though they carry most of the load. Among the reasons for this difference is some of the load, even though small, is carried by the matrix; not all fibers are exactly arranged as predicted; and most importantly, the tensile test procedures require that the load applied to the part being tested be divided by the part thickness. This normalizing of the thickness allows for comparisons of data for many parts of many thicknesses. However, in real composite parts, the thickness is not normalized and, therefore, the actual load-carrying capability will vary from the reported tensile strength. For composite materials, this factor will generally result in lower values for the tensile strength of composites versus the tensile strength of the fibers. This is because, in the composites, the thickness includes the matrix, which carries little of the load.

In theory, the tensile properties of a pressure vessel can be determined by cutting a ring from the vessel and then pulling on the ring with pins attached to the tensile machine. However, in practice, a more common method of determining the tensile properties of a pressure vessel is to pressurize the vessel (almost always with water) and simply measure the pressure at which the vessel bursts (loses pressure). This is called the **burst test**.

A modification to the tensile test, called the **open-hole tensile test**, evaluates the effect that shortening the length of the fibers has on part performance. In this test, a .25-in. (0.6-cm) diameter hole is drilled in the center of the tensile specimen. Then, a tensile test is run to determine the effects of the hole.

Compression and Compression-After-Impact Properties

Compressive force yields information about the strength and stiffness of a columnar sample supported on its sides to prevent buckling when it is pressed on its ends. There are several methods for supporting compression test specimens and these have given rise to various test methods. Perhaps the most prevalent of the methods is the Boeing modified test method, which is similar to the procedure and fixture given in ASTM D695. It consists of metal plates between which the tabbed sample is placed. In this case, the tabbing extends along both sides of the sample except for a small region (.50 in. [1.3 cm]) in the middle. The support fixture prevents the sample from buckling but does not prevent the non-tabbed region from failing. Pressure is applied only on the end of the sample. Other compression test systems include those developed by Celanese, Illinois Institute of Technology Research Institute (IITRI), and the National Aeronautics and Space Administration (NASA). These methods differ principally in the way the sample is supported by the fixture.

The compressive properties can be quite different from the tensile properties because of the difference in the ability of the composite to support a columnar load versus a pulling load. The fibers, in particular, have a tendency to buckle within the composite, especially when voids are present, and this greatly diminishes the compressive properties.

An important test to assess the loss of properties after impact is called the compression-after-impact test. In this test, several flat-panel specimens are made and compared in compressive strength. The special fixture supports the specimens on their sides and bottom. A cap is placed on the top of the sample where the compression machine presses. A control specimen is tested for flat-panel compressive strength. Then, other specimens are impacted with a known energy and then tested for flat-panel compressive strength. The compressive strengths are compared to evaluate the effect of damage to the composite sample.

Flexural and Shear Properties

The flexural or bending forces are determined by placing a rectangular sample across two supports and then pressing on the top of the sample. Most commonly the pressing is done with a single point of contact as shown in Figure 10-8. This is called the standard flexural test or the three-point bend test. Occasionally, although rarely in composites, two points of pressure are used. When this is done, it is called the four-point bend test.

During the test, the top of the sample under the loading force is in compression and the bottom opposite the loading force is in tension. Clearly a transition in type of force must occur in the middle of the part. At this transition plane pure shear occurs. This situation was discussed previously when short-beam shear testing was discussed as a method of determining the fiber-matrix bond strength.

While flexural testing is simple and easy to perform, the mixture between tensile and compressive forces means the flexural test

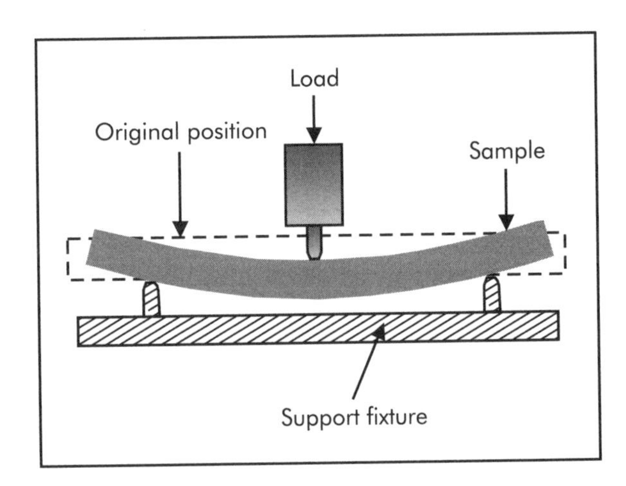

Figure 10-8. Flexural test.

cannot determine basic material properties that are useful for theoretical calculations in design. Nevertheless, it is a commonly used test and is especially useful when flexural forces are those that are likely to be encountered in the use of the composite part. An example when such forces would be important is in a tub and shower unit where the principal forces are the flexural forces from a person standing on the bottom of the tub.

The short-beam shear method is probably the simplest shear test to run, but it is not the only test procedure used for composites. Other shear tests include: Iosipescu (v-notch), two rail, and three rail. Test fixtures for these and other test methods can be obtained commercially.

Shear testing is important for samples bonded by adhesives. Examples would be sandwich panels and certain joints. The most important of the tests for these materials view shear as a peel phenomenon (shearing along an adhesive plane). A simple test can be done by simply turning the ends of a sample or molding a sample with turned ends so it can be pulled in a tensile machine. An example of this simple peel test is illustrated in Figure 10-9. A peel test that gives more reliable data, the climbing drum peel test (ASTM D 1781), measures the force required to peel a layer from a sandwich structure. If a more severe peel test is desired, the rolling drum peel test (ASTM D3167), depicted in Figure 10-10, can be used.

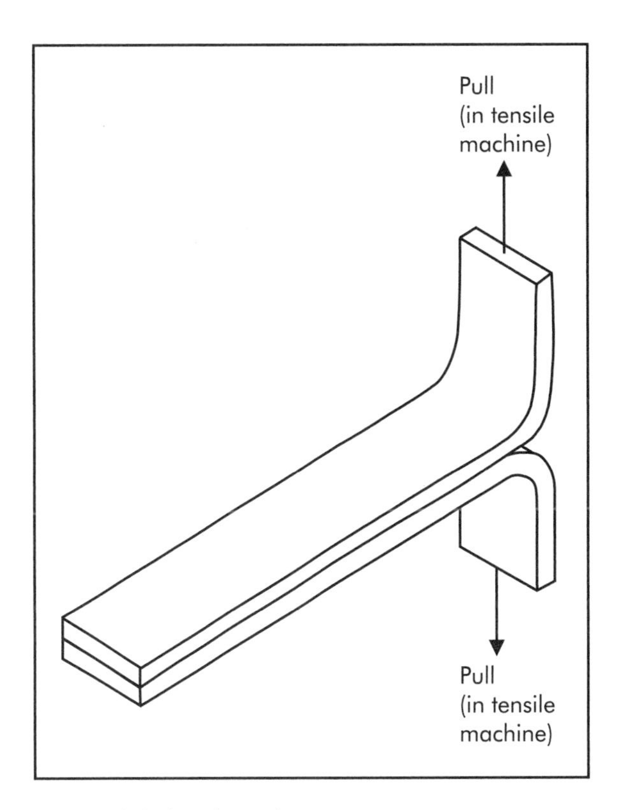

Figure 10-9. Simple peel test.

Impact/Toughness Properties

The ability of a material to absorb energy without breaking (rupture) is called **toughness**. There are two types of toughness—equilibrium toughness and impact toughness. **Equilibrium toughness** is so named because the speed of the tensile test is usually so slow that equilibrium conditions can be assumed. It is related to the area under the stress-strain curve such as that developed when a tensile test is performed because energy absorption is the summation of all the force resistance effects within the system. Using calculus, this summation is accomplished by integrating the equation describing the stress-strain curve. This type of toughness only approximates actual toughness performance, but it is useful in comparing similar materials.

Generally, the toughness of a material is much more important when the force is applied suddenly in an impact rather than over a relatively long period of time, as is done in the stress-strain experiment. Therefore, a property called **impact toughness** (sometimes called **impact strength**) is defined as the energy absorbed by a material upon sudden impact.

Impact toughness is strongly dependent upon the ability of the material to internally move or deform to accommodate the impact.

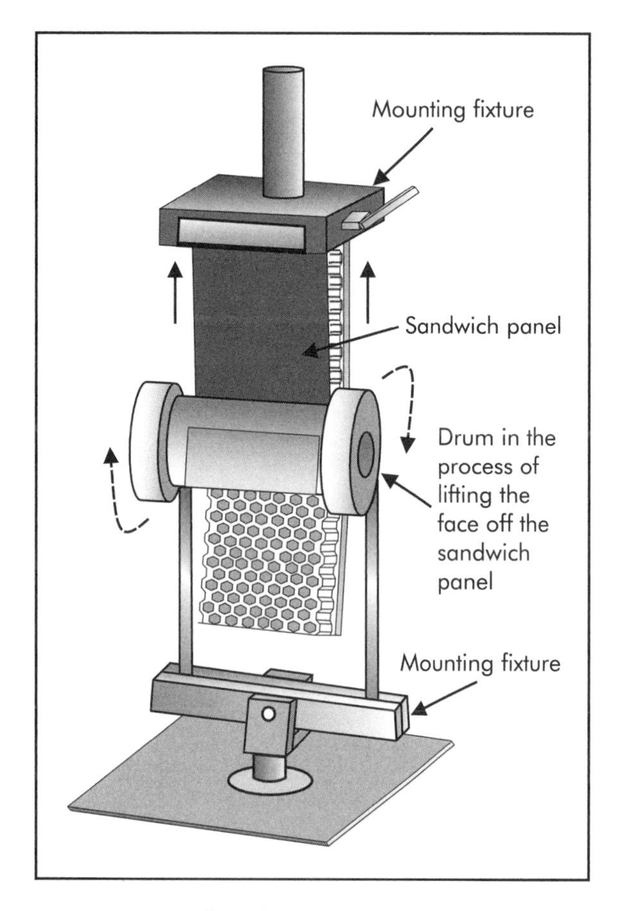

Figure 10-10. Rolling drum peel test.

This movement is related to elongation or strain. Therefore, materials that exhibit high elongation are often tough, especially if they also have good strength. However, the modulus is usually low in these materials (low stiffness). Hence, materials with high elongation and low modulus are called tough, and materials with low elongation and high modulus are called brittle (the opposite of toughness).

High molecular weight favors high toughness. This results from the combination of higher strength and better sharing of the impact force along the polymer chain, which causes more atoms to rotate, vibrate, and stretch to absorb the energy of the impact. Crystallinity gives higher strength but lower toughness unless the nature of the backbone changes. For example, even though nylon is crystalline, the ability of this material to absorb energy by passing the energy from one molecule to another through intermolecular forces makes it tough. Crosslinking of a brittle polymer will usually decrease toughness because of the increased limitation of motion within the polymer mass caused by the intermolecular bonds. Alternately, crosslinking a flexible polymer will increase impact toughness because of the greater ability of the polymer to transfer the energy.

The Izod and Charpy tests (ASTM D256) utilize a pendulum-impact testing device (illustrated in Figure 10-11). The sample is clamped in the holder vertically (for Izod) or laid horizontally (for Charpy) and may have a notch cut in it to initiate the rupture. In the Charpy and Izod tests, the pendulum will generally break the sample and continue its path beyond the impact point. Some of the energy in the pendulum will be absorbed in the breaking of the sample; therefore, it will not swing as high after the impact as before the impact. The reduction in height (potential energy) can be related directly to the energy absorbed by the sample. A pointer, which is pulled by the pendulum and then stops at the maximum height, records the maximum height of the swing. The energy absorbed from the impact is measured from the height of the pointer. The energy absorbed is a measure of the toughness of the sample. The results are usually expressed as impact toughness (Izod or Charpy) in units of ft-lb/in. (J/m).

Another test to determine impact toughness is the falling dart test (ASTM D5628). In this test, the sample is placed under a column that guides a metal dart, tup, or other impacting device onto the sample. The impactor has an accelerometer attached that measures the energy absorbed on impact and thus determines the toughness of the sample. The weight of the impactor, the drop height,

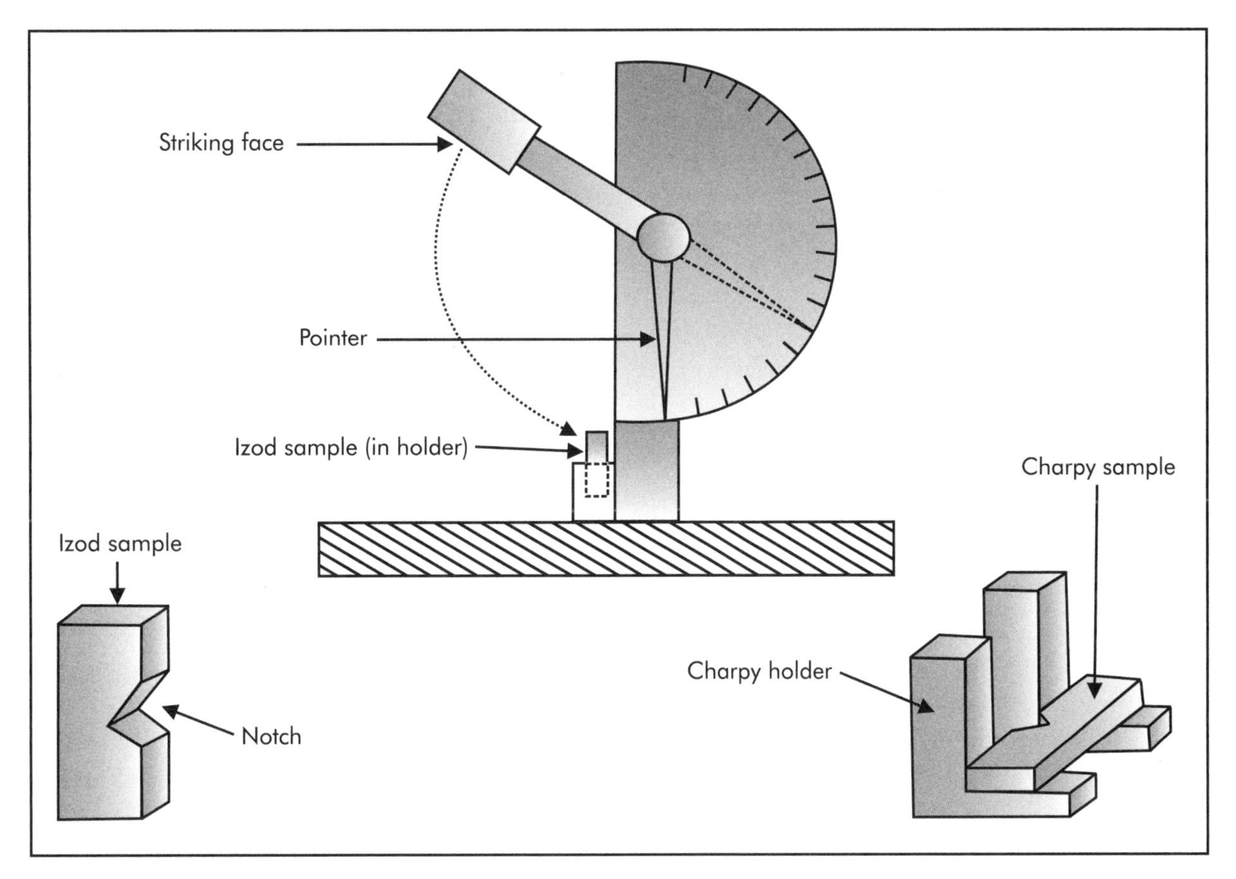

Figure 10-11. Izod and Charpy impact tester and specimens.

and the size of the impactor point can all be varied to accommodate different materials and different sample thicknesses.

Some tests lack the accelerometer and measure the toughness by noting the drop height and weight of the impactor that causes sample rupture to occur. An iterative process of raising or lowering the height to obtain a range of "just fail" and "just pass" heights determines the results of the test. In some cases, the results are given as the height at which 50% of the samples break. This method is called the stair-step procedure.

Creep

The tests discussed up to this point all take a relatively short period of time to measure results (even tensile tests are completed in a few minutes). However, polymeric matrix materials will often deform over time under a continuously applied load that is less than that required to give immediate or near immediate failure. This property is called **creep**. Creep is illustrated in Figure 10-12 for a simple beam. A load is applied over a long period of time causing the material to deform. When the load is removed, a small amount of the deformation will be immediately recovered (the elastic or short-range portion of the deformation). However, some of the deformation is not recovered. This deformation is called **permanent set**.

In addition to an obvious dependence on the nature of the composite, the amount of creep is strongly dependent upon the amount of the load, the length of time the load is

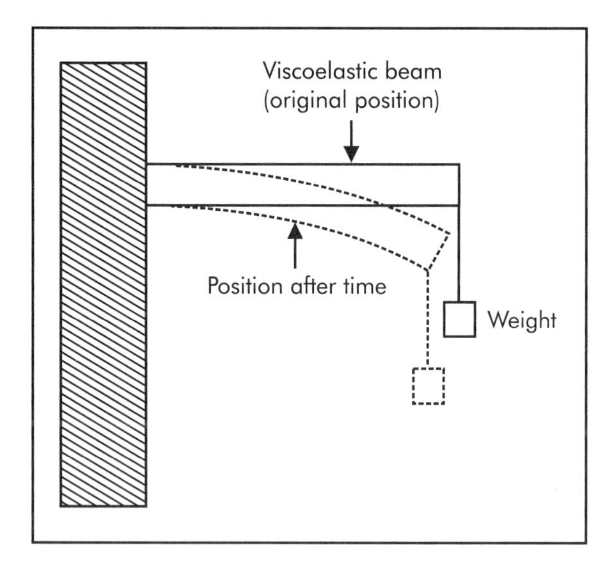

Figure 10-12. Creep test.

applied, and the temperature of the material. Increases in any of these parameters will cause creep deformation to be greater. These relationships can be expressed by the modified stress-strain equation,

$$\sigma_o = E(t,T)\varepsilon(t,T) \quad (10\text{-}2)$$

where:

σ_o = applied stress (usually a constant over the course of the experiment)
E = creep modulus
ε = amount of deformation (strain due to creep)

Both the modulus and the deformation are functions of time and temperature. Further, increases in the amount of stress (σ) from experiment to experiment will also increase the amount of deformation.

Significant creep is generally observed in ceramic matrix and metal matrix composites only at high temperatures and for long duration. However, non-reinforced polymers and polymer matrix composites are considerably different and can exhibit pronounced creep at temperatures of 100–300° F (38–149° C) where operational loads can cause long-term creep. Some of the methods that can be used to reduce the amount of creep include the following:

- Place the fiber orientations in directions of high loads to reduce the creep loading on the transverse or resin-dominated direction. In other words, let the fiber, not the resin, be the principal load-bearing member of the structure.
- Increase the fiber content and correspondingly reduce the content of the creep-prone resin matrix.
- Select stiffer fibers such as high-modulus carbon or use hybrid designs to achieve a higher stiffness.
- Reduce the stress level.
- Reduce the temperature.
- Utilize initial loading cycles, which sometimes relieve the residual stresses that contribute to the creep. It has been observed that initial load cycles relieve the stresses, causing initial micro-cracks and flaws to arrest at fiber-matrix interfaces and creating a more stable substructure within the material.

If a material is loaded for a long period of time, at a high temperature, with a substantial load, the sample will eventually break. This is called *creep rupture*, and it is useful in developing allowable stress limits in applications. If the stress-to-rupture test is performed at several loads, the times for rupture will be different assuming constant temperature. A plot of the stress level versus the time to failure will give the allowable stresses for design considerations with the additional allowance for safety factors.

Creep can be especially troubling for joints. Adhesives are generally not reinforced and therefore will creep just as other plastic and resinous materials do. The creep for these materials is higher than for fibers, thus the non-reinforced materials will creep more than the reinforced composites. To minimize the amount of creep in non-reinforced materials, the adhesive should be as highly crosslinked as possible without becoming excessively brittle. Creep is also a problem in mechanical joints. The composite material will often creep under the loading of a bolt or rivet. This can be minimized by using the highest possible fiber concentration in the composite, reducing the temperature or level of stress, or by spreading the stress over the largest practical area.

Some residual stresses in a material may cause creep rupture to occur more rapidly than in a non-stressed sample. In these cases, the residual stresses combine with the stresses induced by the creep (migrating through the polymer structure) and result in early rupture.

Fatigue

Composites and almost all other materials are negatively affected by intermittent loads applied over long periods. This phenomenon is called **fatigue**. The fatigue test is performed by measuring a critical property in a composite, such as tensile strength, and then subjecting the sample to intermittent loads over long periods of time. The critical property is re-measured after set numbers of cycles. In actual practice, several identical samples are prepared; all are exposed to the load with a few being withdrawn for testing at each interval. Figure 10-13 shows a chart used to monitor the fatigue of various materials. These types of charts are occasionally called S-N charts (indicating the stress and the number of failures).

One of the major advantages of carbon-epoxy composites in aircraft applications is their improved fatigue performance versus aluminum as illustrated in Figure 10-13. Many aircraft are designed with a maximum service life of around 10^6 to 10^7 cycles. Their fatigue endurance limit, the stress level below which no significant failures will occur, is low in this cycle range. In fact, the fatigue endurance limit is on the order of one-half the static ultimate strength for conventional aluminum alloys. This means that after about 10^6 cycles, the aluminum can only safely withstand 50% of the static load it could originally withstand. In most cases, the limit of performance is taken as the point when the retention falls below 50%.

Fiber reinforced composites, especially carbon fibers, appear to be better than their metal counterparts. In fiber-reinforced composites, the loss of a few failed fibers is not noticeable to the overall strength of the composite because the load is redistributed among the other fibers through load sharing.

Composites tend to stabilize early in fatigue loading through the following mechanisms, each of which absorbs or redirects the energy to other parts of the structure:

- matrix micro-cracking, which absorbs energy by breaking matrix bonds but does not lead necessarily to catastrophic breaks in the fibers;
- blunting of cracks at the fiber surface, which reduces further crack growth and transfers the load to the fiber;
- delaminations between layers, which may relieve internal cure stresses through bond breakage and additional flexibility;
- stress redistribution and load sharing that exists within the overall composite structure; and
- energy dissipation resulting from the viscoelastic effects of the matrix (that is, movements among the polymer chains). This is also called *internal damping*.

Thus, composites have several mechanisms through which fatigue can be reduced or at least significantly retarded, whereas metals do not.

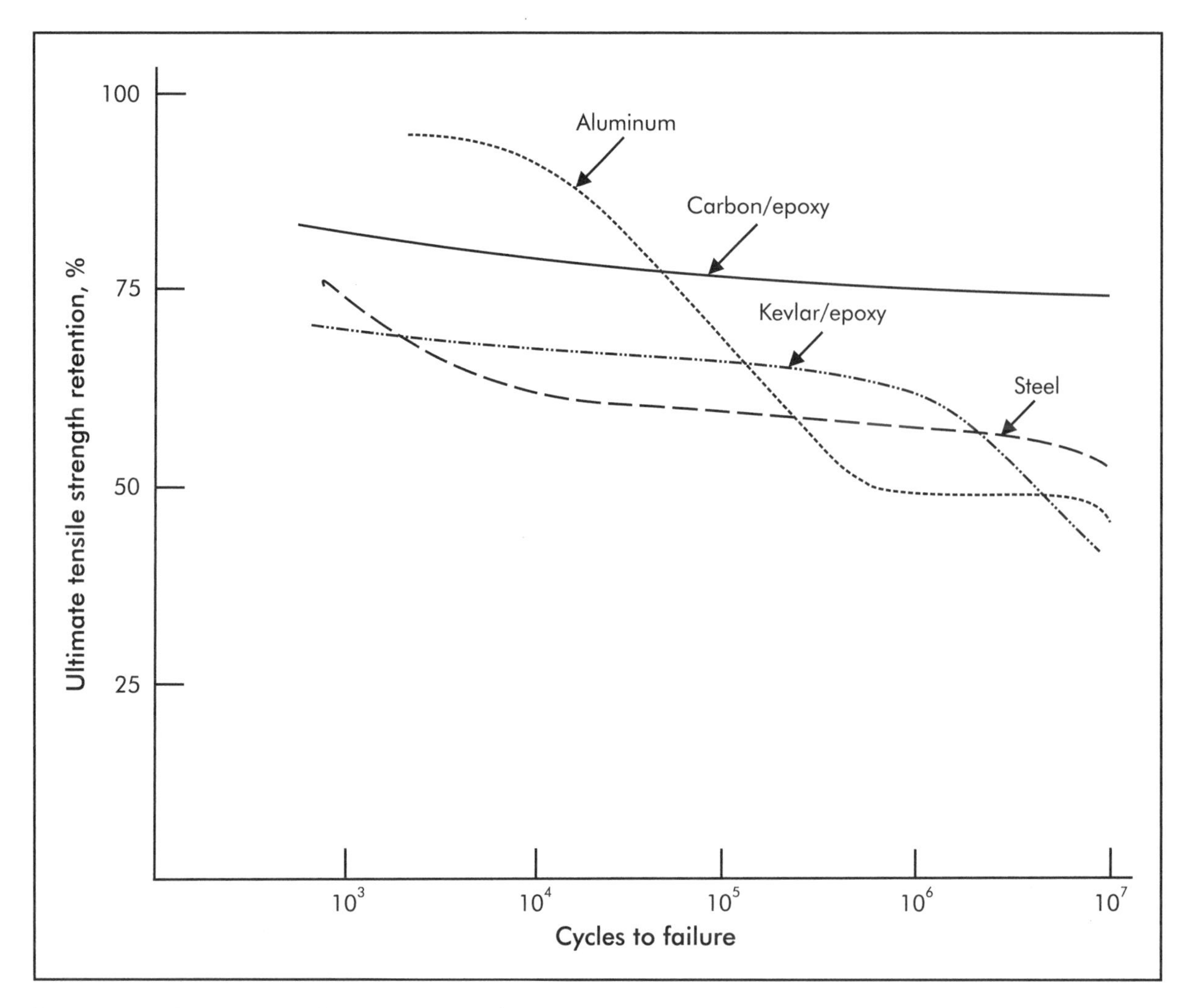

Figure 10-13. Fatigue testing for various materials.

Vibration and Damping

When a composite material receives a sudden impact, it vibrates. That is, it oscillates about the original position in cycles with ever decreasing amplitude. The speed with which the vibrations end is called the **damping** of the composite. Vibrations are related directly to the stiffness of the composite, so the stiffer the material, the longer the vibrations will persist.

In general, vibrations in a composite are to be avoided. Vibrations cause internal movements, which can result in fiber-matrix bond breakage and delaminations, especially over long periods. The vibrations might also initiate micro-cracks, which could later cause failures. They can also affect external attachments, bonds, and performance such as in an airplane wing. Vibrations can cause distortions of function such as misalignments in space telescopes. In many cases, they are also objectionable because of discomfort. (Think of the vibrations that occur in an airplane floor or in a golf club shaft.) Therefore, damping is highly desired.

Compared to most metals, composites have excellent damping characteristics. The many fiber-matrix bonds absorb energy (even without rupturing) and thus cause the vibrations to decrease. But some additional features can be added to composites to further improve damping. One such feature is called *constrained layer damping* and it is done by bonding a sheet of another material, often a metal, to the surface of the composite. This creates an interlayer that absorbs considerable energy and therefore dampens the structure. Another method is called *active damping*. In this method, a piezo-ceramic device is incorporated in the composite. This device senses the movement and sends an electrical current that drives a motor to counteract the vibrations. Still another method is to bond the composite together with an adhesive layer that has high elongation. This type of material absorbs much of the energy and therefore dampens the entire structure. Another method is to orient the fibers in a damping sheet, which is incorporated onto the composite. The fibers in the sheet are at an angle (usually 45°) with other fibers in the same sheet, which are oriented exactly opposite (–45°). The layers are bonded together with a high-elongation adhesive. When vibrations occur, these fibers attempt to align with the direction of vibration, usually trying to rotate to 0°. The two layers are, therefore, acting in opposite directions. This causes a shearing action in the adhesive and that absorbs energy and damps the vibration. This latter method has been called *stress-coupled activated damping* (SCAD) and has been shown to be several times more effective than the other common types of damping.

As a result of their improved damping characteristics, composites provide excellent properties for aircraft and missile control surfaces where fast, rigid response is needed. This characteristic is important in controlling *aero-elasticity*, the interaction between the structure's elasticity and its need to maintain a rigid aerodynamic surface. In other words, the composite provides stiffness and improved "feel" in the aircraft. Proper selection of resin and fiber combinations, along with fiber and layer orientations, can provide that control.

THERMAL AND ENVIRONMENTAL TESTING

Temperature Effects

Metals and ceramics have a response to temperature that is uniform in all directions. Therefore, the **coefficient of thermal expansion (CTE)**, often represented by α, is independent of location within the metal and ceramic. Further, the CTE for metals is usually constant over the entire range of temperatures for the solid metal. For plastics, on the other hand, the CTE is different below and above T_g. The rubbery state above T_g expands from two to four times more than the glassy state below T_g, and the CTE in both states is substantially higher than the CTE for a metal or a ceramic.

Composites, which are a mixture of a plastic matrix and a ceramic-like fiber, have a complicated dependence on temperature. As a result, they have different CTEs in the direction of the fibers and in the transverse direction. These differences are noted in Table 10-2 where the CTE in the fiber direction is seen to be at least an order of magnitude larger than the transverse CTE. Clearly, the fibers are restraining matrix expansion in the fiber direction, which is the direction in which the bonding of the matrix and the fibers would logically be most effective.

Table 10-2. Representative coefficients of thermal expansion for epoxy-matrix composites.

Reinforcement	Direction	Coefficient of Thermal Expansion, 10^{-6}, ° F (° C)
Glass fibers	Parallel to fibers	8 (15)
	Transverse to fibers	30 (55)
Carbon fibers	Parallel to fibers	–0.4 to 0.6 (–0.7 to 1.0)
	Transverse to fibers	4 to 6 (7 to 10)
Aramid fibers	Parallel to fibers	–1 (–2)
	Transverse to fibers	33 (60)

Though the values in Table 10-2 are merely representative, the range of values for carbon indicates real variation in the CTE depending on the nature of the carbon fiber itself. As the carbon fibers are converted from polyacrylonitrile (PAN) to graphite, the structure of the fibers goes through a dramatic change. This change in structure results in a significant change in CTE as the fibers move from a plastic-like structure to a ceramic-like structure. Therefore, it might be expected that the CTE changes in a similar way, which is exactly what occurs. It can be followed by noting the CTE as a function of fiber modulus. This is possible because the fiber modulus is a good indicator of the extent to which the fiber has been converted to graphite. Also in that figure are some CTE values for metals and ceramics, which indicate that for them too, as the modulus increases the CTE decreases. Hence, stiff materials tend to expand less with temperature.

The negative CTEs for carbon and aramid fibers are highly unusual and deserve some discussion. For most materials, the CTE reflects the greater activity of the atoms at higher temperature. This activity results in the atoms moving away from each other to allow for the greater movement (vibrating, twisting, rotating, and translating). Such activities are also occurring in carbon fibers and in aramid fibers. However, in both of these materials, the strong stretching (drawing) of the fibers during the fiber-making process built in some residual stresses that, upon heating, are relieved. This causes an initial shrinkage in the fibers with heating and, therefore, a negative CTE. The negative CTE has a benefit in allowing the creation of a composite material that is essentially a zero CTE material. This is done by carefully selecting the directions of the fibers used in the composite. Such a material has great benefits in applications such as space telescopes where the thermal changes experienced are great and yet the need for precise alignment of various optical components must not be compromised. These can both be achieved with a zero CTE material.

In addition to the expansion that most materials experience with increasing temperature, they also experience a decrease in important mechanical properties such as strength and stiffness. Figure 10-14 shows the effect of temperature on the specific strength of various metals and composite materials. While the polyimide-carbon fiber composite has high specific strength at room temperature, its high-temperature range is limited. On the other hand, iron has low specific strength but has good retention

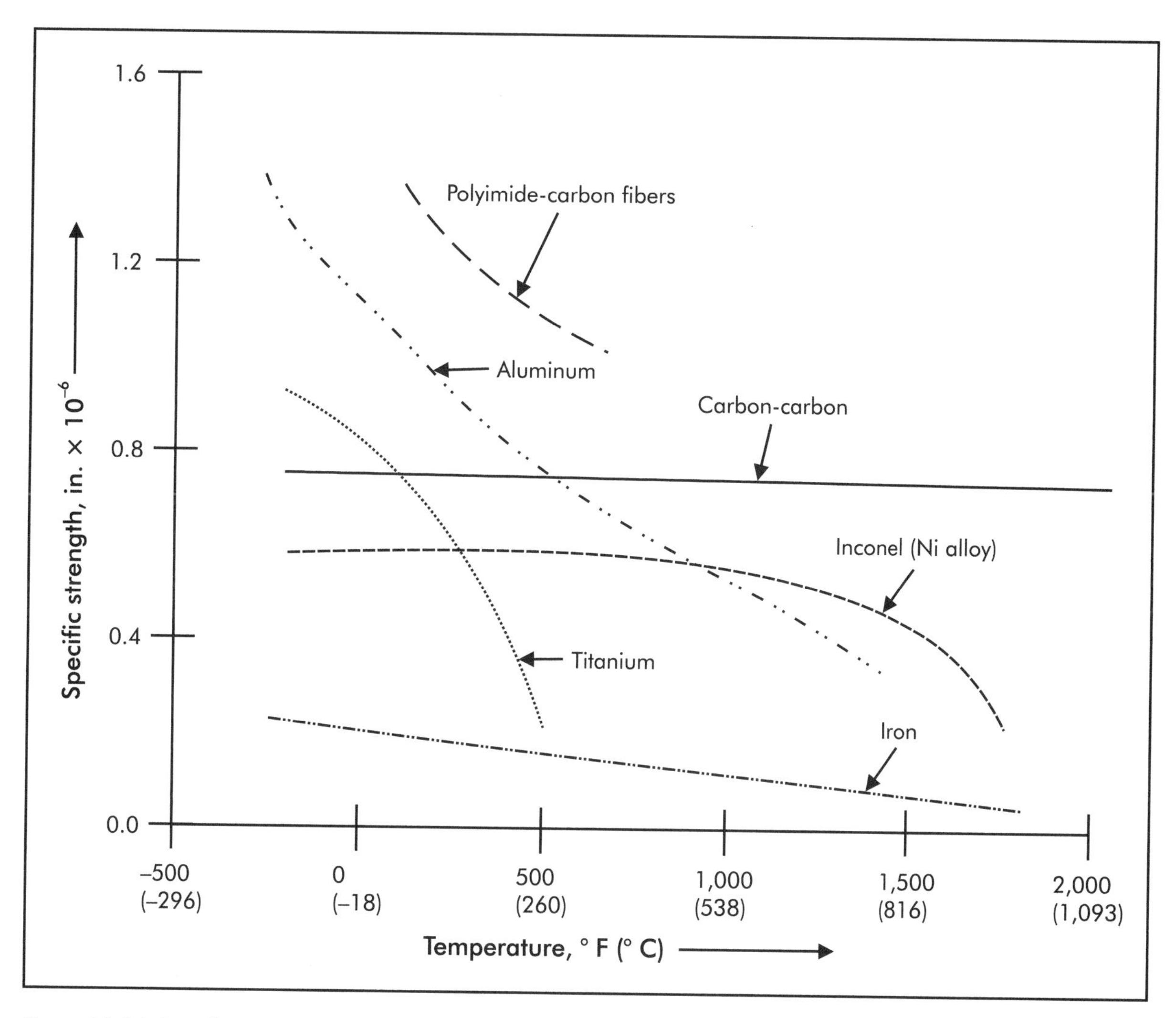

Figure 10-14. Specific strength as a function of temperature.

of properties through a wide temperature range. Other materials fall between these extremes. The best material for both high specific strength and retention of specific strength is carbon-carbon.

Moisture and Solvent Effects

Composites absorb moisture through the matrix, the fiber, the fiber-matrix interface, and porous regions or areas where micro-cracking or delaminations have occurred. Table 10-3 indicates the degree to which the polymer matrix can absorb water under equilibrium conditions. The goal within the industry has been to develop matrix materials that do not take up much moisture. A general rule-of-thumb has been to target 3% as an upper range. However, it is obvious that values for existing resin systems are often greater than this goal.

Fiberglass and aramid fibers can absorb considerable moisture, although it is typically less than most matrix materials. Carbon fibers and ultra-high-molecular-weight

Table 10-3. Moisture absorption for neat (non-reinforced) resins.

Polymer Type	Moisture Absorption at Saturation, %
Epoxy	5.0
Epoxy, aliphatic toughened	4.0
Epoxy, rubber toughened	7.0
Epoxy, rubber toughened at 250° F (121° C)	9.4
Polysulphone	1.0
Polyetheretherketone (PEEK)	0.5
Polyamide-imide	0.3

polyethylene (UHMWPE) fibers absorb little moisture. Consequently, a good approximation of the amount of moisture absorbed by a composite is the amount that would be absorbed by the resin matrix. However, when using moisture absorption data for neat resins (without fibers), it should be remembered that the fiber is a major, non-moisture absorbing part of the structure. Therefore, at a typical fiber content of 60%, the resin moisture pickup will be about 40% of the values shown in the moisture absorption data for the resin. For example, consulting Table 10-3, the amount of moisture that could be absorbed in a graphite composite made with epoxy, with the fiber volume at 60%, would be 2% (40% of the maximum value of 5%).

As shown in Table 10-3, there are several factors that influence water absorption. Toughening epoxy by adding aliphatic material to the chain decreases water absorption slightly. However, adding rubber increases water absorption. The increase in water from the rubber toughening is due to the presence of a second phase wherein absorption occurs on the phase boundaries. Note the low water absorption rates of the thermoplastic materials (polysulfone, polyetheretherketone [PEEK], and polyamide-imide). The high-performance thermoplastics are well known for their low moisture absorption and resistance to other solvents.

Solvent sensitivity can be a major problem with some composite materials. The polyesters are sensitive to many common solvents such as acetone and methyl ethyl ketone (MEK). Epoxies also have sensitivity to many common solvents including the hydraulic fluid used in many aircraft. This sensitivity is a problem because of the occasional spills that inevitably occur during aircraft maintenance. Therefore, the excellent solvent resistance of the high-performance thermoplastics, such as polysulfone (PSU), PEEK, and polyamide-imide (PAI) allows polymer composites to be used in applications where these and other solvents are present.

Polymer materials are sensitive to ultraviolet (UV) light. Therefore, in outdoor applications, the decrease in polymer capability needs to be determined. Some environments of high sunlight are obviously more severe in this regard than others.

A similar problem exists in near space with atomic oxygen. Just as UV light degrades the matrix materials, so does atomic oxygen. Studies of spacecraft placed in orbit for several years and then retrieved indicate that most polymers have a significant loss of properties. Hence, in both outdoor and space applications, a significant safety factor might be required or, alternately, a surface coating employed on polymeric matrix composites. Even in earthly conditions, the long-term effects of simple environments can result in a significant loss of properties over long periods. Figure 10-15 shows data from a study in which a fiberglass-epoxy composite was exposed to several high-moisture environments in an effort to gage the effect of moisture and attempt to correlate its performance in real time with accelerated tests. Of course, the value of the accelerated tests is that the performance of the composite can be anticipated in far less time than would be required with real-time monitoring.

Figure 10-15 represents data for pressure tanks that were burst (taken to failure) after exposure to three conditions: boiling water, high humidity and elevated temperature (95% relative humidity and 174° F [79° C]), and actual service in the humid and hot conditions of Panama. Comparison of the data indicates that the boiling water is more severe than the other conditions. Therefore, even though it was a short test, it may not correctly reflect the condition of the part after actual service. The temperature-humidity conditioning, on the other hand, seems to agree well with the conditions of actual service in Panama. The time frame of the temperature-humidity testing is in the order of weeks (rather than days for boiling water) but the weeks are better than experiencing the length of years of actual service in Panama. For example, the accelerated test indicates that 40 days in the temperature-humidity environment is equivalent to nearly 4 years in Panama.

Aging and Service Life Considerations

Few composite products have anticipated short-term service lives or planned obsolescence built into their designs. Most structures are designed for a minimum of 3–5 years service and a maximum of approximately 10–25 years. However, as has been seen in this chapter, composite properties degrade over time. Therefore, some rational basis for predicting the service life

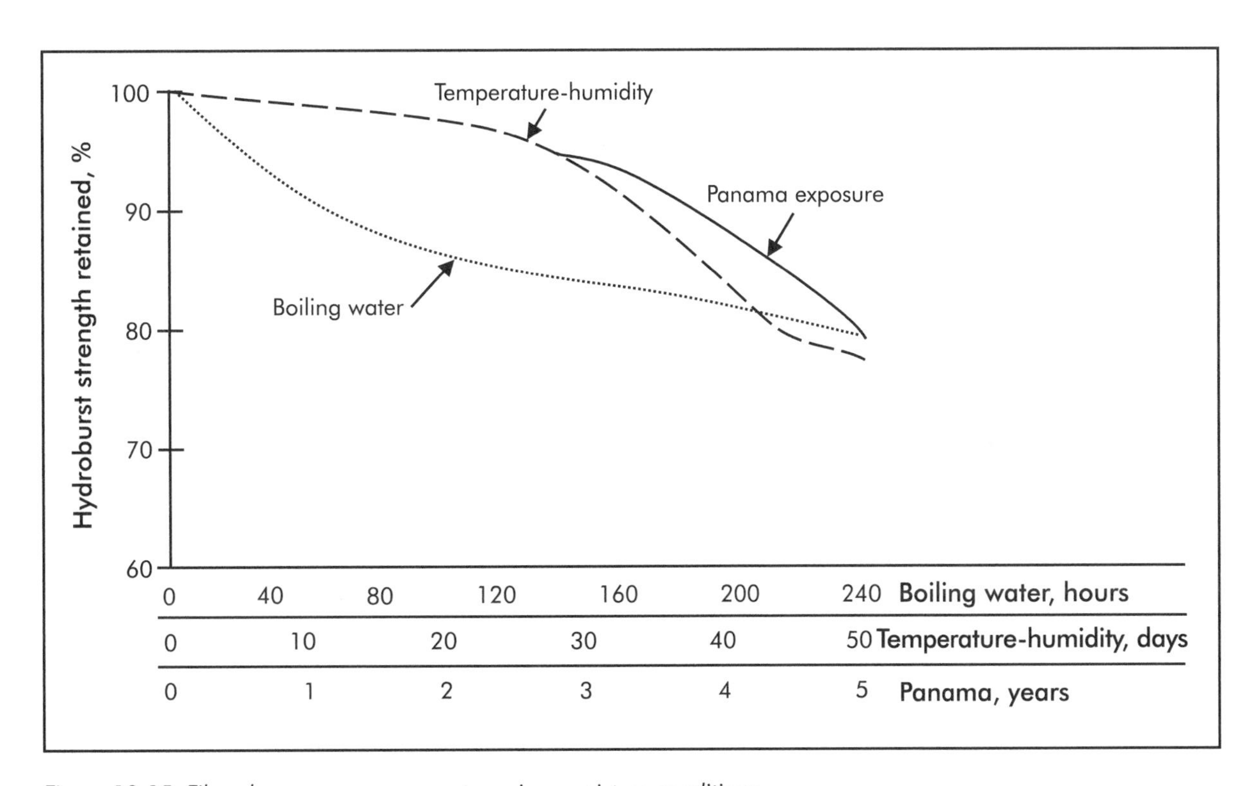

Figure 10-15. Fiberglass-epoxy exposure to various moisture conditions.

of parts needs to be devised. Following are some suggestions for establishing a basis for predicting service life.

- Define the service environment in terms of exposure time, temperature, humidity, and loading conditions. Concentrate on identifying the worst combination of conditions.
- After selecting the materials and constituents, review the database for those materials to see what is already known about them.
- Conduct accelerated aging tests on the materials and components to develop trends in performance where the existing data is inadequate or where performance records are unknown. These accelerated tests usually consist of putting samples into elevated temperature and severe humidity conditions. If the anticipated environment is repeated loading, a cyclical loading test would be appropriate.
- Accelerated aging analysis usually requires modeling the data taken from several environmental conditions and including some real exposure data so that an extrapolation to real operational environments can be made.
- Verify the data by subjecting the part to real service conditions. Keep good records over long exposure times so the real performance can then be placed in a database to give credence to future models and extrapolations.

Thermal aging is an important way to accelerate composite behavior. Tests for thermal aging depend upon the phenomenon called the **time-temperature transposition**. This principle states that increasing the temperature of a sample is the same as extending the time for the sample. That is, to see how a sample behaves over time, the temperature of the sample can be raised. Aircraft companies and others who expect their products to be used for many years often conduct these tests. These companies may, for example, compare the baseline (unheated) sample with materials that have been exposed to a set time, such as 5,000 hours, at two temperatures to study the loss of a particular property. Properties often examined include bearing strength, compression after impact, flexural strength, open-hole tension, and open-hole compression.

FLAMMABILITY TESTING

Understanding How Composites Burn

The study of how composites and plastics burn has been, and continues to be, a major area of research at universities, in industry, and at various national, military, and private laboratories around the country. No one has yet been able to sort out all of the factors that occur during the combustion of composites and plastics. Even less, no comprehensive model of combustion exists for these materials. Some aspects are clear, however, taken by themselves. However, when several effects combine (as occurs in real fires), only gross approximations are possible. Even so, there is a surprising amount of information that can be learned just by making some reasonable observations.

What is most clear is that under the conditions that normally exist for fires, fiberglass does not burn and does not, therefore, contribute to the combustion of the material. The other common fiber reinforcements—carbon fibers, aramid fibers, and UHMWPE fibers—will burn under normal fire conditions. However, their contributions to the fire are generally much smaller than the resin matrix and will be ignored for the purposes of this discussion. Some minor contributions may arise from the orientation effects of the fibers, such as when the outer layers of fiberglass insulate the inside from the flame, but these effects are small and will be ignored.

Most inorganic fillers do not burn, but they can decompose in a fire and strongly affect the combustion process. One obvious example is aluminum tri-hydrate (ATH), which releases water molecules when heated. These water molecules tend to reduce the flammability of the composite, affecting both flame and smoke. Other fillers that can strongly affect combustion include those containing halogen atoms such as fluorine, chlorine, bromine, and iodine. These fillers reduce flammability, but they may increase smoke density. Many minor additives, such as antimony oxide, can affect the behavior of the combustion process. Sometimes even color pigments can have an effect. To understand the effects of fillers, the best practice is to run some flammability tests.

The resin matrix is, by far, the most important aspect of the composite in determining the combustion characteristics. The reason is simple—most organic resins will burn. To reduce that burning, some modifications will usually have to be made to the resin or one of the few inherently flame-resistant resins will have to be used. Other chapters have addressed the changes most often made to make the resins less combustible so they will not be treated here. In brief summary, the most common methods of modifying the resin are to add halogens or increase the amount of aromatic (benzene) groups within it.

Combustion Characteristics of Organic Resins and Plastics

The following properties are usually those considered when examining the burning of composites and plastics.

- Ease of ignition/flame resistance—how readily does the material ignite? What kind of ignition source is required?
- Flame spread/fire-retardant properties/surface flammability—how rapidly does the fire spread? Is it different in different directions?
- Heat release rate—how much heat escapes from the material during combustion and at what rate?
- Fire endurance—how rapidly will fire penetrate a barrier of the specific material? What are the orientation effects on fire penetration?
- Ease of extermination—how easily will the fire go out?
- Smoke emission/toxicity—how much smoke is released? What is the rate of smoke release? Is the smoke toxic or corrosive?
- Physical characteristics—does the material drip, droop, or sag as a result of burning?

Types of Flammability Testing

The most obvious and prevalent reason for testing is to satisfy the requirements of some agency that controls the specifications for use of the product. Other reasons might include improvement of the composite materials, study of the nature of the combustion process, establishment of a specification, and assessment of overall design for fire safety.

Flammability tests can be divided into three groups, generally reflecting the purpose for performing the test. These categories are:

- official tests or those performed to meet some official requirement,
- laboratory tests or those usually done within a manufacturing facility for informal studies, and
- full-scale tests used to simulate actual use conditions.

Official Tests

For numerous reasons, various market sectors have developed their own flammability requirements and tests. The test procedures have been carefully specified so that all parties performing the tests do them in precisely the same way. The agencies that control the specifications usually change the procedures only after considerable study and experimentation, thus guaranteeing continuity over time in interpreting the test results. The procedures are also studied to ensure that variations from lab to lab are kept statistically small. In many cases, equipment used in the tests has also become standardized, often requiring purchase from only approved sources. In light of these requirements, places where official tests can be run are usually limited to two groups—commercial testing laboratories and major research facilities such as those owned by resin manufacturers.

The major market or use sectors for composite products and their associated fire requirements are summarized in Table 10-4. Note that each test procedure is designated by a alphanumeric code in which the agency controlling the test procedure is designated, such as the American Society for Testing and Materials (ASTM), federal motor vehicle safety standard (FMVSS), Federal Railroad Administration (FAR), International Organization for Standardization (ISO), National Bureau of Standards (NBSIR, now the National Institute of Standards and Technology), and the military standard (MIL-STD). Details about the tests can be obtained from these agencies.

A brief summary of the most important of the official tests follows.

- The cone calorimeter (ISO 5660/ASTM E-1354) test, which is only about 10 years old, appears to be the single, most comprehensive test that might satisfy most of the marketing/use sectors. Several advantages of this test include: the sample size is modest (4 × 4 × .25 in. [10.2 × 10.2 × 0.6 cm]) so that preparation costs are low; costs are usually modest (in the $200–$500 range at commercial labs); and most of the fundamental combustion characteristics (ease of ignition, rate of heat release, weight of sample as it burns, temperature of sample as it burns, rate of weight loss, rate of smoke release, and yield of smoke) can be determined under a wide range of heater and ignition conditions. As a result of the vast amount of data available from this test, a model of the combustion of a material might be developed, thus enabling an estimation of the potential effects of a fire on surrounding areas and occupants.
- The principal result of the radiant panel test (ASTM E-162) is a flame spread index. This index is the product of a flame spread factor and a heat evolution factor, which are determined as the material is subjected to a radiant heat source. The sample size is not large (6 × 18 in. [15.2 × 45.7 cm]) and the costs of the test are modest ($200–$300). Further, this test has been used extensively for many years and so a wealth of comparative data is available. However, results are often confusing, especially when different materials are compared, thus limiting the broad applicability of this test.
- Combustion research has shown that many fire fatalities occur in rooms where the flames never reach. Hence, the nature of the smoke generated in a fire is important. The smoke chamber test (ASTM E-662) is designed to measure the density of the smoke generated as a function of time, under both flaming and smoldering conditions. Because this test was originally developed by the National Bureau of Standards (NBS), it is sometimes referred to as the NBS smoke density test. The test apparatus is a closed box (smoke chamber) approximately 3 × 3 × 2 ft

Table 10-4. Summary of fire requirements for composite materials in the United States.

Sector	Component	Property	Test Procedure	Criteria
Surface transportation (cars, trucks, and buses)	Panels	Flame spread	FMVSS 302 (federal motor vehicle safety standard)	Rate of flame spread, 4 in./minute (10 cm/minute)
Surface transportation—mass-transit vehicles (buses, light rails, and passenger trains)	Seat materials	Flame resistance	FAR 25.853 (Federal Railroad Administration)	Flame time ≤ 10 seconds, burn length ≤ 6 in. (15 cm)
	Panel, partition, wall, ceiling	Flame spread	ASTM E162 (American Society for Testing and Materials)	Flame spread index ≤ 35
	Floor structure	Fire endurance	ASTM E119	Nominal evacuation time ≥ 15 minutes
	Seat, panels, walls, partitions, ceiling, floor	Smoke emission	ASTM E662	Smoke density, D_s (1.5) ≤ 100 D_s (4.0) ≤ 200
	All materials	Toxicity	NBSIR-82-2532 (National Bureau of Standards)	As appropriate
Air transportation (commercial aviation aircraft)	Cabin and cargo compartment materials: seat, panel, liner, ducting	Flame resistance: vertical, horizontal, 45°	FAR 25.853 (a–b), FAR 25.853 (b2–b3), FAR 25.855	Flame time: 15 seconds Flame time of drippings: 3 seconds Burn length: 6 and 8 in. (15 and 20 cm) Rate of flame spread: class b2 is 2.50 in./minute (6.4 cm/minute); class b3 is 4 in./minute (10 cm/minute)
	Cargo compartment liners, seats	Fire endurance	FAR 25.855, FAR 25.853	No flame penetration of liner, peak temperature at 4 in. (102 mm) above specimen, ≤ 400° F (204° C); mass loss and flame spread criteria for seats
	All large area cabin interior materials	Smoke emission	FAR 25.853 (a-l)	D_s (4.0) ≤ 200
		Heat release rate (HRR)	FAR 25.853 (a-l), ASTM E906	Peak HRR in 5 minutes: 65 kW-minute/m^2 Total HRR in 2 minutes: 65 kW-minute/m^2

Table 10-4. *Continued.*

Sector	Component	Property	Test Procedure	Criteria
Marine transportation (life boats, rescue boats, and small passenger vessels)	Main structure, hull	Resin/laminate flammability (fire-retardant properties)	MIL-R-21607 (military standard) or ASTM E-84	Resin qualified under MIL-R-21607 or laminate tested to ASTM E84, flame spread index ≤ 100
Marine (high-speed craft, fire-restricting materials)	Bulkheads, wall and ceiling linings (surface materials)	Heat release rate, surface flammability, smoke production	ISO 9750 (International Organization for Standardization) room/corner test	Average HRR ≤ 100 kW-minute/m^2, maximum HRR ≤ 500 kW-minute/m^2, average smoke production ≤ 1.4 m^2/second, maximum smoke production ≤ 8.3 m^2/second, flame spread ≤ 20 in. (0.5 m) from floor, no flaming drops or debris
	Furniture frames, case furniture, other components	Ignitability, heat release, smoke production	ISO 5660 cone calorimeter	Criteria currently under development
Military (U.S. Navy submarines)	Structural composites inside pressure hull	Peak heat release (PHR), smoke, etc.	MIL-STD-2031 (SH)	Flame spread index ≤ 20, maximum smoke Dm ≤ 200, 25 kW/m^2: PHR ≤ 50 kW/m^2 Tign ≥ 300s • 50 kW/m^2: PHR ≤ 65 kW/m^2 Tign ≥ 150s • 75 kW/m^2: PHR ≤ 100 kW/m^2 Tign ≥ 90s • 100 kW/m^2: PHR ≤ 150 kW/m^2 Tign ≥ 60s

(0.9 × 0.9 × 0.6 m). The smoke chamber contains an optical light source and detector to measure the reduction of optical density from the smoke generated by the burning of a sample placed inside. The sample is small, usually 3 × 3 × .25 in. [7.6 × 7.6 × 0.6 cm], and the cost is moderate ($300–$500). Results are usually expressed as smoke density, D_s, after 1.5 or 3.0 minutes.

- The Steiner tunnel test (ASTM E-84) measures the flame spread and smoke generated by the burning of a sample within a large (25 × 2 ft [7.6 × 0.6 m]) chamber. The principal use of the test is for building materials. The sample is attached to the top of the chamber and a fire is started at one end. The rate and extent of burning are reported. The cost of performing this test (about $500 per run) is relatively high, as is the cost of the materials because of the large sample size.

Laboratory Tests

Sometimes official tests are either too costly or too time consuming to run. Such a case might occur when a company is trying to adjust the formula or design of a product. The company might go to an outside testing agency (commercial laboratory or resin supplier) and have an official test run on a candidate formula or design and then use inside laboratory tests to monitor the effectiveness of changes. In this way, nearly immediate results can be assessed. Also, because the nature of the materials are usually similar, the laboratory tests will general give the same ranking of materials as would be obtained in official tests. Laboratory tests are not, however, as useful as official tests in ranking the combustion characteristics of widely different materials.

Some of the most common laboratory tests are listed next.

- The limiting oxygen index (LOI) test (ASTM D-2863) is probably the most accurate of the laboratory tests. In the LOI test a sample is suspended vertically inside a closed chamber (usually a glass or clear plastic enclosure) as shown in Figure 10-16. The chamber is equipped with oxygen and nitrogen gas inlets so the atmosphere within it can be controlled. The sample is ignited from the bottom and the atmosphere is adjusted to determine the minimum amount of oxygen required to sustain burning. This minimum oxygen content, expressed as a percentage of the oxygen/nitrogen atmosphere, is called the oxygen index. Higher numbers are associated with decreased flammability.
- In the vertical burn test (ASTM D-568 and D-3801), the sample is suspended vertically so that it can be ignited at the bottom. After ignition, the ignition source (usually a Bunsen burner) is

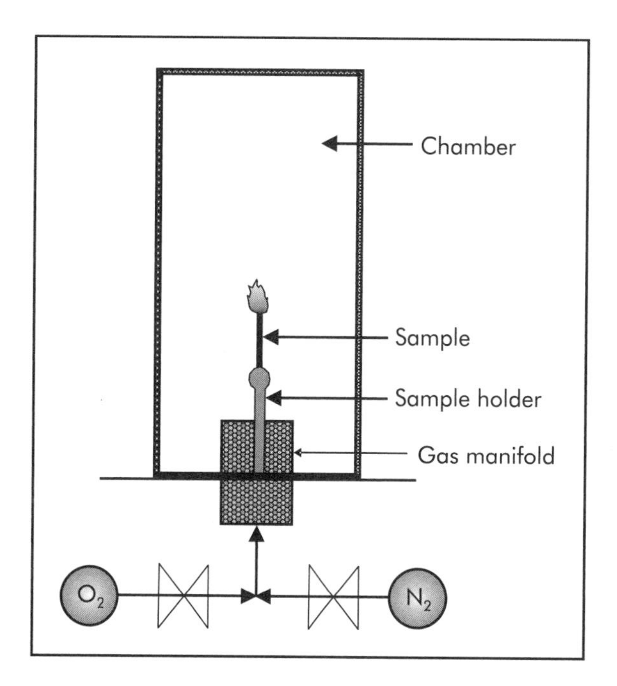

Figure 10-16. Limiting oxygen index test.

withdrawn and the length of burn of the sample in a set period of time (10 seconds) is measured. If the sample does not burn for the entire time, the time to extinguish is noted. Materials that burn while the ignition source is in contact but go out quickly when it is removed are termed self-extinguishing. This test is most useful in measuring the burning characteristics of similar materials. Dripping of the resin should be noted.

- The horizontal burn test (ASTM D-635) is similar to the vertical burn test except that the sample is supported vertically. This test is less stringent and used when the vertical test cannot distinguish between materials. Only self-supporting materials should be tested in the horizontal test.

Full-scale Tests

The major problem with the results from official tests and laboratory tests is that performance in an actual combustion situation can only be speculated. In fact, the standard procedures often caution against using the results of the test for any determination of performance in actual burning conditions. The best that can be hoped for in the official and laboratory tests is a correct ranking of materials as to their performance in real burning conditions. In most of the tests, failure is considered "bad," but passing cannot necessarily be extrapolated to being "good."

To determine actual performance, full-scale tests should be run. Full-scale tests are only rarely controlled by standard test procedures. Therefore, the company or agency requiring the results is usually free to configure the test. Often, because of the complexity of combustion, companies will request assistance from private testing organizations (often the same as those who run the official tests). Some of these testing organizations have even established full-scale test facilities (rooms, vehicles, etc.) that can be used as a framework on which the test materials can be attached.

Full-scale tests often examine temperatures in and around the test enclosure, smoke generation in and leaking out of the enclosure, and heat generation in and around the chamber. Ignition often simulates a likely fire hazard, such as a burning waste basket, or a controlled source such as a gas burner or an assembly of wooden slats.

Flammability testing will continue to be important in assessing composite behavior. The use of this type of testing in specifications will likely continue to rise as more agencies see the value in trying to protect the public and limit liability through testing. Therefore, it is prudent for manufacturers of composite parts to assess the combustion characteristics of their products so that, should reformulations be necessary, they can be anticipated and worked on before a crisis situation exists.

NON-DESTRUCTIVE TESTING AND INSPECTION

Most of the tests that have been described in this chapter are destructive, that is, the sample is no longer fit for use after the test has been conducted. As a result, real parts are rarely tested using destructive tests. In most instances, proof parts are used to confirm performance in simulated actual use. An example is aircraft. Before an airplane is certified for flight, an entire airplane must be subjected to a series of proof tests that simulate and exceed actual service conditions. However, once the certification is successfully completed, real production parts are no longer tested in destructive tests. Instead, coupons and samples from trim areas, which are assumed to correctly reflect the conditions of the real parts, are tested in the destructive tests.

To gain further confidence in the integrity of the parts, several **non-destructive tests (NDT)** have been developed to investigate the nature of the product, including the effectiveness and reliability of the manufacturing process. Some have divided NDTs into two groups—non-destructive evaluation (NDE) in which the part is changed, and non-destructive inspection (NDI) in which no change occurs in the part. NDT is applied far beyond the field of just composites. Some of the most important applications include: weld inspection, concrete curing, electrical conductivity, printed-circuit-board inspection, wooden beam inspection, medical investigations (like x-rays and magnetic resonance imaging [MRI]), and many others too numerous to list.

In composites, non-destructive tests assume that the correct design for the part has been made. That is, the correct fibers and matrix have been chosen and the arrangement (direction) of the fibers in each of the laminates and the number of laminates required has been worked out. However, problems during manufacturing can always occur and these need to be monitored. As discussed previously in the section on quality, the most important monitoring of the manufacturing process is done by controlling the machines. This is called statistical process control (SPC). Even with superb SPC procedures, occasional verification of the actual parts is useful. The most important method of doing this verification is with ultrasonic scanning.

General Principles

Most NDT techniques involve sending an energy signal (wave) through or onto the part to be inspected and then detecting changes in that signal when compared to some standard. In general, the changes are associated with defects in the part. By increasing the frequency of the wave, the energy of the wave is increased, and the higher the energy, the greater the penetration capability of the wave. Wave penetration is also affected by the density of the material and its uniformity (as well as some other factors that are beyond this discussion). Sometimes penetration is beneficial, as with dense materials (like metals) or thick materials. However, on other occasions penetration can be viewed as a drawback because the waves may be so energetic that they pass through everything and cannot distinguish the flaws from the good material. Another problem with high penetration waves is that, in general, the higher the energy of the wave, the more difficult it is to control and interpret. Furthermore, high-energy waves can be physically harmful.

The secret of good NDT is to find the frequency and wave type that will reveal the most about the material and the potential defects that are likely to be found. Each frequency range has its own equipment, methods, and characteristics. This section examines some of the most common. Some of the methods use electromagnetic signals. Others, like ultrasonic and acoustic emissions, use non-electromagnetic signals. Still others, like eddy current, use imposed electrical current.

Ultrasonic Testing

Ultrasonic signals are not really electromagnetic. They are, instead, acoustic and, even though they have frequencies and wave properties, they are fundamentally different from electromagnetic waves. Ultrasonic frequencies (20–100 MHz) are just right for examining carbon-based, resinous materials because waves of these frequencies are highly affected by changes in the density of the material being examined. This is because the wavelengths are about the same size as common defects. When the ultrasonic waves encounter changes in density, such as a void space within the composite layers, the signal changes are dramatic, thus facilitating easy detection. Since the rate of signal travel is dependent on the stiffness of the material, ultrasonic measurements can be used to determine the modulus of a material, assuming that the density is known. Ultrasonic signals are the most commonly used of the NDT techniques for composites.

Ultrasonic through-transmission testing is probably the simplest of the ultrasonic methods. In this type of testing, the part is placed between an ultrasonic transmitter and a receiver. The pulsed signals are monitored as the transmitter and receiver devices move (as a coupled pair) across the area of the part to be investigated. Defects are detected as losses in the transmitted signal. This method is relatively easy to automate and simple to interpret. To minimize losses in signal strength due to air gaps between the transmitter and receiver, a couplant material that has high ultrasonic signal transmission strength is placed between the transmitter and the part and the receiver and the part. Water and acoustic jelly are common couplants. When the area to be examined is large, the entire part, along with the transmitter and the receiver can be immersed in the couplant. The couplant in this case is usually water.

An obvious problem with the through-transmission method is that both sides of the part must be accessible. This is often the case with parts before they are assembled, but is not the case when field inspections are required.

Ultrasonic pulse-echo testing requires access to only one side of the part. The ultrasonic signal is transmitted into the part and then the signal reflected off the backside of the part is detected on the same side as the transmission occurred. Defects are detected as signals that return faster than the signals coming from the opposite side of the part. This method is commonly used for field inspections and, because of other advantages, is gaining popularity in all stages of composites manufacturing.

One advantage of the pulse-echo method is that defects at different depths in the part can be distinguished from one another because of the differences in return times from these various depths. The output of commercial ultrasonic scanners gives a picture of the internal structure of the composite and shows where defects such as delaminations, contaminants, and similar defects are located. Another advantage of the pulse-echo method is the increased sensitivity it gives to detecting many types of defects, such as foreign material inclusions, which might be missed by the through-transmission method. Pulse-echo devices also allow for precise measurement of thickness and, therefore, can detect changes in thickness from processes such as corrosion, wear, or manufacturing fluctuations.

Recent improvements in ultrasonic testing have taken great strides in reducing, or even eliminating, the need for a couplant. One improvement uses sprayed water, thus allowing much greater portability. In addition, machines are now available that work well without any couplant material beyond just air.

Acoustic Emission

In the **acoustic emission** method, the sounds emitted by a sample placed under stress are detected and compared to other samples of similar parts. A difference in the sound pattern indicates that a sample is structurally different and, presumably, it has a defect. An example of a similar phenomenon is the audible sounds made by wood just before it breaks.

A problem with acoustic emission lies in the requirement for a standard emission pattern from a known good part. Quantitative results are also difficult with this method. Still, the method is gaining popularity because of its simplicity and ease of performance. The stress does not need to be mechanical. In actual practice, thermal stresses are the most common. Therefore, the test can be done by simply placing the acoustic detection equipment on the sample and then shining a light or other heat source on the sample.

Acoustic emissions are sensitive to many different phenomena in composite manufacturing. For instance, the curing speed, consistency of fiber distribution, impact damage, finish or gel coat uniformity, fatigue failures, and stiffness are just some of the most common properties that can be detected.

One non-composite application that is interesting is the use of acoustic emissions as a way to monitor vibrations on computer hard disks as they come under environmental jarring. Evidently, acoustic emission testing has a response rate and sensitivity that are especially good for this application. Another interesting application that has relevance for composites and other products is the ability of acoustic emissions to detect minute fluid flows, such as leaks. Valve leaks are routinely monitored using portable acoustic emission devices. In composites, when a honeycomb sandwich material has been exposed to water, any absorbed water can be detected as the sandwich is heated. If water has seeped into the honeycomb cells, the water will escape back through the entry crack when the water is vaporized, and that leaking sound can be detected.

Radiography

Radiography is a general term for the inspection of a material by subjecting it to penetrating irradiation such as x-rays. These high-energy waves are the most familiar type of radiation used in medical practice and in inspecting metals. Radiography has an important role in the inspection of metal-matrix composites. However, most polymer composites are nearly transparent to x-rays and, therefore, defects cannot be easily detected in these materials. In some instances, an opaque penetrant can be used to detect defects. The high energy of x-rays is a further deterrent to their use as these rays require special precautions to avoid excessive exposure.

Thermography

Thermography (sometimes called infrared [IR] thermography) detects differences in the relative temperatures of the surface of a sample and, because these temperature differences are affected by internal flaws, this can indicate the location of those flaws. If the internal flaws are small or far removed from the surface, however, they may not be detected. Two modes of operation are possible—active and passive. In the active mode, the sample is subjected to a stress (usually mechanical and often vibrational) and then the emitted heat is detected. In the passive mode, the sample is externally heated and the thermal gradients are detected.

Optical Holography

The use of laser photography to give three-dimensional pictures is called *holography*. This method can detect flaws in composite samples by employing a double-image method where two pictures are taken with an induced stress in the sample between the times of the pictures. In the past, this method has had limited acceptance because of the need to isolate the camera and sample from vibrations. However, a new phase-locking technique seems to eliminate this problem. The stresses imposed on the sample are usually thermal. If a microwave source of stress is used, moisture content of the sample can be detected. This method is especially useful for detecting bond defects in thick honeycomb and foam sandwich constructions.

A related method is called **shearography**. In this method, a laser is used with the same double-image technique as in holography with a stress applied between exposures. However, in shearography, an image-shearing camera is used. The camera takes the signals from the two images and superimposes them to reveal interference, thereby indicating the strains in the samples. Because strains are detected, the size of the pattern can give an indication of the stresses concentrated in the area and, therefore, a quantitative appraisal of the severity of the defect is possible. Shearography has gained positive acceptance in a relatively short period of time because of this attribute, its great mobility over holography, and the ability to apply stress with mechanical, thermal, and other methods.

Microwave Frequencies

As in microwave cooking, microwave energy is especially useful in exciting water molecules, which vibrate strongly during microwave irradiation to heat the food. Therefore, moisture content is readily detected using microwaves.

Microwaves can also detect changes in the dielectric properties of a composite. Areas in which the resin is especially rich or poor can be seen since the dielectric properties would be different in those areas, especially if an air gap were present. One other obvious application for microwave technology is, therefore, the measurement of the dielectric nature of radomes and other devices that need to transmit electrical signals.

When a resin is liquid, it can carry charge much more readily than when it is solid. Therefore, the extent of cure can be measured by following the changes in the dielectric properties of the resin. This also may be useful in following the degree of crosslinking.

The microwave spectrum also can be used to detect defects, especially those that are small (about .04–.39 in. [1–10 mm]). These are shown as decreases in transmitted signal.

Eddy Current

An **eddy current** is an induced electrical current within a conductive material. It is not part of the electromagnetic spectrum. This technique is, therefore, limited to composites that are conductive, including those with carbon fibers. A coil induces an alternating current in the material. These currents are affected by variations in conductivity, permeability, mass, and homogeneity. The changes in these currents are detected by a magnetic detector and compared with spectra from other samples. Surface defects are the most commonly detected because the penetration depth of this technique is limited.

Visual Inspection

Nearly everyone uses visual inspection as a method for detecting some defects in composites, especially those on the surface or within the gel coat. A wide range of visually detected defects have been classified according to their depth and nature and are a measure of the severity of the damage sustained. Some of the types of defects classified include: scratches, abrasions, dents, and cuts. Other defects that are often visually detected are blisters, wrinkles, porosity, orange peel, pitting, and discoloration.

If the composite is translucent, defects within the structure and reinforcement uniformity can be detected visually. In general, visual frequencies are not able to penetrate highly filled composites because of light interference with the opaque fillers.

Visual inspection is especially helpful in color determinations and can be enhanced by instrumentation that detects color matching, fading, loss of gloss, and other spectral characteristics. Sometimes viewing under oblique angled light is useful in detecting surface irregularities, especially subtle problems such as color irregularities and surface waviness. Microscopes also can be helpful in enhancing visual detection of flaws.

Penetrating dyes can be used as a visual technique to detect surface cracks and other imperfections that result in small, otherwise undetectable surface defects. The visibility of these dyes can sometimes be enhanced by the use of special lights.

CASE STUDY 10-1

Leaf Springs

The principal properties for most composite parts are related to tensile properties. However, an application based on the flexural properties of composites may prove to be one of the largest applications of composites. This application is leaf springs—those springs that are made from stacks of relatively flat strips of a resilient material.

The principal applications of leaf springs are in automobiles and trucks. These springs have the advantage of reduced vertical space required when compared with the more familiar coiled springs. Therefore, leaf springs are especially useful in cushioning the rear wheels of the car. The springs are placed on the underside of the car in the very restricted space between the rear axle and frame.

Composite leaf springs have been used in several vehicles including the Corvette® and Ford Aerostar®, as well as several trucks and trailers. The main materials used in leaf springs are E-glass and epoxy resin. The needed load-bearing capability is obtained by stacking the springs.

The composite leaf spring originated in the racing circuit out of a need for a more durable product than the metal leaf springs then in use. The punishment encountered during racing caused premature sagging (permanent set) or fracture of the metal leafs. The failures forced replacement of the springs nearly every race, thus adding to the costs of the cars and requiring additional time of the already pressed mechanic. Composite springs proved to have about 20% longer fatigue life, which eliminated the need for such frequent replacement. The composite springs also gave a significant savings in weight over the metal springs—a significant competitive advantage in racing. Many racing teams also found that the use of composite springs eliminated the need for sway bars, thus further reducing maintenance, cost, and weight. The weight savings in some cases was 3% of the total weight of the car. (The weight savings in just the springs was about 20%.)

The ability to vary many parameters in the composite springs allowed designers to alter spring rates, flex locations, specific areas of stiffness, and even make progressive resilience within one spring.

Fabrication is done by cutting prepreg tape into the desired shape and then stacking the layers within a matched metal mold. Pressure and heat are applied to consolidate and cure the layers in the spring. The cured part is then trimmed using a diamond cutting wheel.

SUMMARY

Testing and quality assurance are not only good business practices; they are mandated by many government and commercial customers. The success of any effort will depend primarily on three factors.

First, management's understanding, support, and involvement are integral to every successful quality improvement effort. Success is directly related to the level or attitude of quality within the organization, but primarily within management. Quality is not just a task within the production system, consigned to one area that is primarily involved in sampling and inspection. If this is the primary view of quality by management and the employees, then any efforts to improve it will cost more than they are worth, and the real quality of the process will not improve. Management must have an attitude that quality begins with them, be willing to invest in the education of employees to further this idea, and be involved themselves.

Second, success is directly related to the time and resources committed to quality. Interestingly, the commitment of management time and other personnel is more important than money.

Third, success follows the actual content and fundamental validity of the strategy, program, or method being applied. Fads are not new to the business environment and quality has not been immune to its share of faddish sales talk. For any operation or organization the correct strategy is critical.

The careful manufacturer will understand the standard test methods and the specific tests that might be applied to the actual use conditions of parts. Testing is the basis for improvement of a part's design and manufacturing process and should, therefore, be practiced routinely and as an integral part of every product-improvement project. Key learning points related to testing include the following.

- Test results for composites have a much wider distribution of data than do results from single component materials like simple metals, plastics, and ceramics.
- Many standard test procedures are carefully prescribed. The American Society for Testing and Materials (ASTM) is the dominant body in the United States for defining tests.
- Incoming fibers are rarely tested by the molder, but the release data of the fibers should be monitored.
- Incoming resins are often tested by the molder to ensure quality and shelf life. These tests generally monitor viscosity and cure conditions.
- Fiber volume and void content should be monitored occasionally to ensure that the manufacturing method is working properly.
- Tensile tests provide key design and manufacturing information about composite laminates. Tensile strength, tensile modulus, and elongation are all important properties.
- Compression testing is important in design and in determining the presence of delaminations and the extent of damage from impacts.
- Flexural tests are not as valuable as the tensile and compression tests are for design, but they may be more important in actual use testing.
- Toughness is usually determined using impact tests (Izod, Charpy, dart drop). These tests have high variability and so many samples need to be run.
- Moisture and water effects are important environmental considerations for composites. Special applications also may consider the effects of other environmental factors such as solvents, thermal cycling, and creep.
- Important non-destructive tests for composites include: ultrasonic, radiography, tomography, and acoustic emission.
- In-use testing in both small and full-scale sizes is important to determine durability and fatigue of composite structures. The failure mechanisms are investigated to verify that design considerations were proper and that manufacturing and quality assurance were properly done.

LABORATORY EXPERIMENT 10-1

Alignment of Fibers in Tensile Tests

Objective: Determine the differences in tensile properties when fibers are not aligned.

Procedure:

1. Using unidirectional prepreg material, lay up a sheet that is roughly 12 × 12

in. (30 × 30 cm). The sheet should be six layers thick with all of the fibers carefully oriented in the 0° direction.

2. Using the same prepreg material, lay up another sheet in which the middle two layers are oriented at + and – 90°.
3. Using the same prepreg material, lay up another sheet in which the middle two layers are oriented at + and – 60°.
4. Cure all of the sheets.
5. Apply tabbing material to each of the sheets. This tabbing material can be additional composite material or a metal, but should be of the proper length and properly tapered.
6. Cut at least five tensile specimens from each sheet.
7. Test the specimens in tension and record the data. Average the data for each type of specimen.
8. Compare the data and plot the results indicating changes in ultimate tensile strength as a function of lay-up angle.

QUESTIONS

1. Why are the mechanical properties of composite materials often lower than the mechanical properties of the bare fibers?
2. Why is viscosity a good measure of uncured resin quality?
3. How can the fiber volume fraction be determined for a composite part made from hand lay-up without destroying the part?
4. Compare the use of five different NDT methods for determining whether a void is present in a thick, 1 in. (2.5 cm) part.
5. What differences are to be expected in the following tensile tests and what information can be obtained from each one?
   a. run in the fiber direction
   b. run in the direction perpendicular to the fiber direction
   c. run in the 45° direction
6. Why is the compression test more meaningful in determining voids than the tensile test?
7. Describe creep for both thermosets and thermoplastics. In testing for creep, what are the effects of temperature, fiber content, and direction?
8. What type of test is normally performed to measure fatigue life?
9. What is the difference between tensile strength and flexural strength in both testing and actual use? Discuss the differences in tensile and flexural modulus.
10. What mechanical properties will change as a polymer is post-cured?
11. Describe the elongation of a material.
12. What happens to the tensile strength of a material if the resin does not bond well to the fibers? What happens to the toughness?
13. Why are test procedures specified by ASTM or some other standards organization used when performing some tests? When would they not be used?
14. Draw a stress-strain curve for both a stiff material and a material with lower stiffness. Explain the differences in the curves. What would be the expected differences in ultimate tensile strength in comparing the two materials?

BIBLIOGRAPHY

American Society for Testing and Materials (ASTM). 1987. *Standards and Literature References for Composite Materials*. Philadelphia, PA: ASTM.

Composites Applications Guide, 10th Edition. 2005. Kansas City, MO: Cook Composites & Polymers.

Mika, Geoffrey. 2006. *Kaizen Event Implementation Manual*, 5th Edition. Dearborn, MI: Society of Manufacturing Engineers.

Society of Manufacturing Engineers (SME). 2005. "Composite Materials" DVD from the Composites Manufacturing video series. Dearborn, MI: Society of Manufacturing Engineers, www.sme.org/cmvs.

Sorathia, Usman, Lyon, Richard, Ohlemiller, Thomas, and Grenier, Andrew. 1997. "A Review of Fire Test Methods and Criteria for Composites." *SAMPE Journal*, July/August, p. 28.

Standard, Charles and Davis, Dale. 1999. *Running Today's Factory: A Proven Strategy for Lean Manufacturing*. Dearborn, MI: Society of Manufacturing Engineers.

Strong, A. Brent. 2006. *Plastics: Materials and Processing*, 3rd Ed. Upper Saddle River, NJ: Prentice-Hall, Inc.

11

Composite Design*

CHAPTER OVERVIEW

This chapter will present the following:

- Introduction
- Methodology of composite structure design
- Basic stress types
- Laminate theory
- Rule of mixtures
- Modeling and finite element analysis
- Lay-up notation
- Symmetric and balanced equations
- Cracking in composites
- Vibration and damping
- Smart structures
- Fatigue
- Composites versus metals
- Residual stresses

INTRODUCTION

Composite materials are inherently more complicated in their structures than single-phase, one-component materials such as most metals, plastics, and ceramics. Even when these one-component materials are filled with particles, their structures are still far less complex than most composites. The principal reason for the relative simplicity of metals, plastics, and ceramics is that the structures of these traditional materials are the same in all directions whereas in composites there is a specific directionality. When the structure of a material is the same in all directions, it is called **isotropic**. Because the fibers in most composites are not equally oriented in all directions, composites are, in general, *non-isotropic* or, synonymously, **anisotropic**.

If fibers are oriented in the vertical and planar directions, as in a 3D weave, the fibers are not really the same in all directions. This is because some of the directions, especially those off-axis, do not have fibers. However, the structure is a reasonable approximation of one that is isotropic. Therefore, materials reinforced along the principal x, y, and z axes are called **pseudo-isotropic**.

Because composites are often laminar structures, the fibers are often located in planes. When the fibers are directed in all directions (random planar orientations), the material is called **orthotropic**. If the fibers are oriented in specific directions in a plane, for example, at 0°, 90°, and ±45°, the material approximates orthotropic and is called pseudo-isotropic. (The example given is not the only way to get a pseudo-isotropic material. For example, instead of ±45° fibers, ±30° and ±60° fibers could be used.)

* The principal author of this chapter is Rikard Benton Heslehurst, Ph.D., Australian Defence Force Academy, School of Aerospace, Civil and Mechanical Engineering, University College, The University of New South Wales.

The design of composite structures is complex. As such, this chapter will provide:

- an understanding of the principle characteristics and issues so that good structures and components will result;
- an understanding of the influences and issues of the individual constituent materials (fiber and resin) so good designs will be produced;
- emphasis on the attention to detail required in design and manufacture to achieve the desired outcomes in behavior and performance;
- an understanding of how the manufacturing (fabrication) processes are closely tied to good composite structure design; and
- an awareness that there are many variables in the design of composite structures that need to be understood and addressed.

METHODOLOGY OF COMPOSITE STRUCTURE DESIGN

The basis of the composite structure design methodology can be undertaken by a series of development procedures. These are outlined in the flow chart shown in Figure 11-1. A brief description of each step follows.

1. Design specification is the most important of the steps in design. The most appropriate type of design is a functional specification. It is based on what the part must do in actual use (function). The development of a thorough and detailed functional specification will always save time and money. The end result is a product that works the first time and meets customer requirements. Design specification is particularly crucial for composite structures as there are many more ways of doing the design in the wrong way than, for example, with traditional materials, such as metals. A good approach to developing a design's functional specification is the quality function deployment (QFD) method. This method is discussed elsewhere in numerous design methodology books.
2. Choosing the material and manufacturing is like opening Pandora's box. There are many fiber and resin materials to select from, as well as fiber/resin system combinations. Most material vendors have a preferred list of the fiber/resin system combinations from which to choose, but the first requirement is choice of fiber type, fiber form, and resin for the structure. Typically, the fiber choice is either a glass or carbon type depending on the application and, of course, cost. Other fibers should be investigated where necessary. The selection of unidirectional (tape, rovings, etc.), woven cloth, mat, or some other form is next. The fiber form selection is generally based on either fabrication convenience or shape complexity. Note that there is a major impact on the strength-to-weight ratio and stiffness-to-weight ratio when selecting a cloth over a tape form. Resin selection is typically an operational environment issue. The operational environment, including vapor/fluid absorption characteristics, temperature range (hot and cold), and the corrosive condition dominate resin selection. Also with respect to the environment, attention should be given to fiber/resin interfacial issues. The other part of the step is choice of manufacturing method. This usually comes down to what a company already has available and the time/cost of manufacturing the part. Also impacting the manufacturing process selection is quality control of the product's performance require-

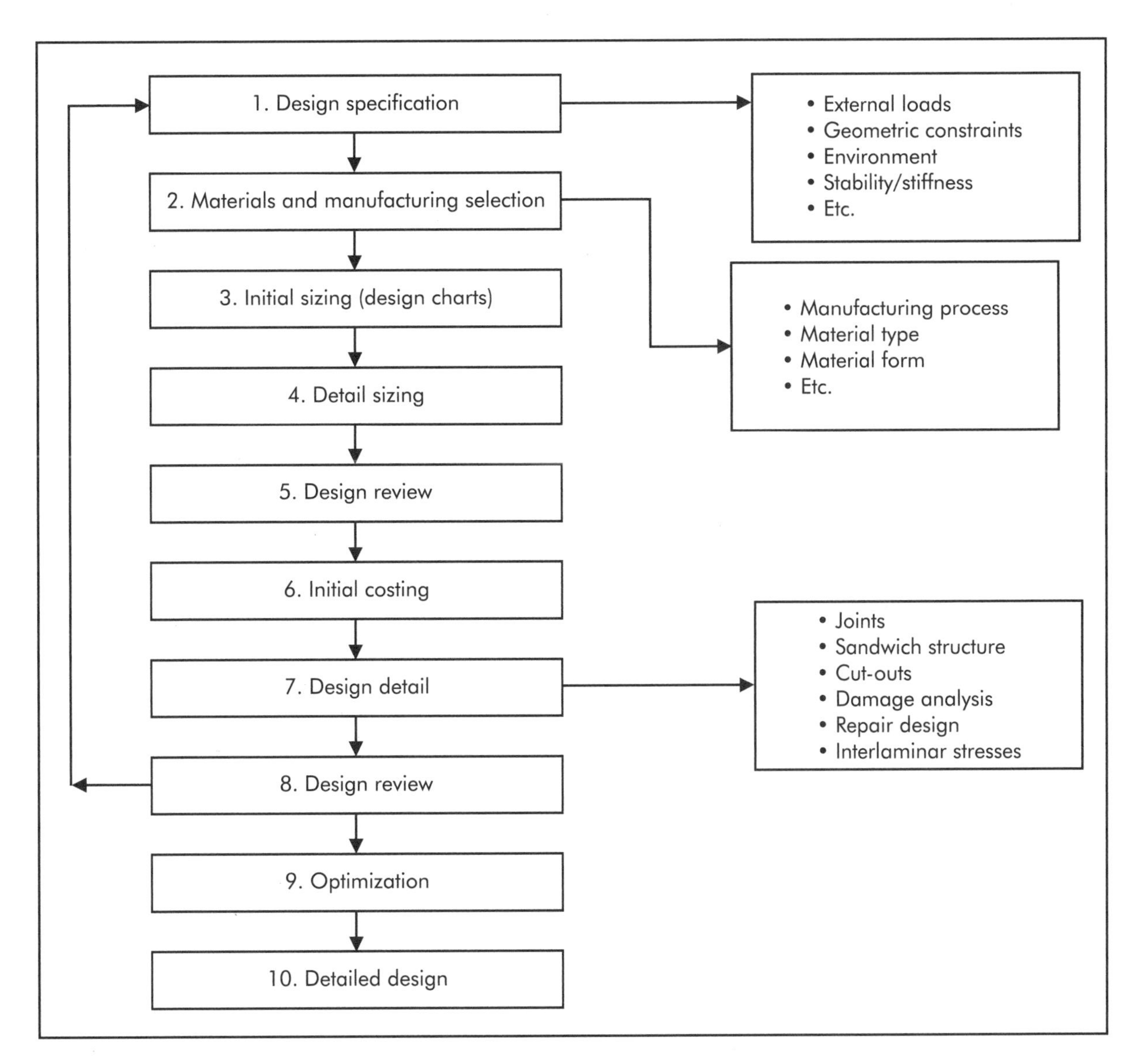

Figure 11-1. Composite design methodology flow chart.

ments and the number of parts to be produced.

3. Initial sizing of the composite structure is used to determine the number of plies required in the various orientations. This step can be done using simple sizing charts such as the one shown in Figure 11-2. For example, when the laminate's axial stiffness or strength requirement is known as a ratio of the single-ply equivalent property, then the percentage of plies in the 0°, ±45°, and 90° directions can be determined. However, do not rely solely on these results to finalize a design as other factors need to be considered when the final configuration is set. Figure 11-2 is just one of several types of design

charts that can be used to initially size the composite structure.

4. Detailed sizing of the composite structure requires a computer program to evaluate the engineering properties of the composite laminate. To determine the in-plane properties, a relatively simple program can be used; some will even run as a spreadsheet. But this step must be done to evaluate the complete in-plane stiffness properties and estimate the strength of the laminate in the principal directions. These simple programs also will estimate many of the other physi-

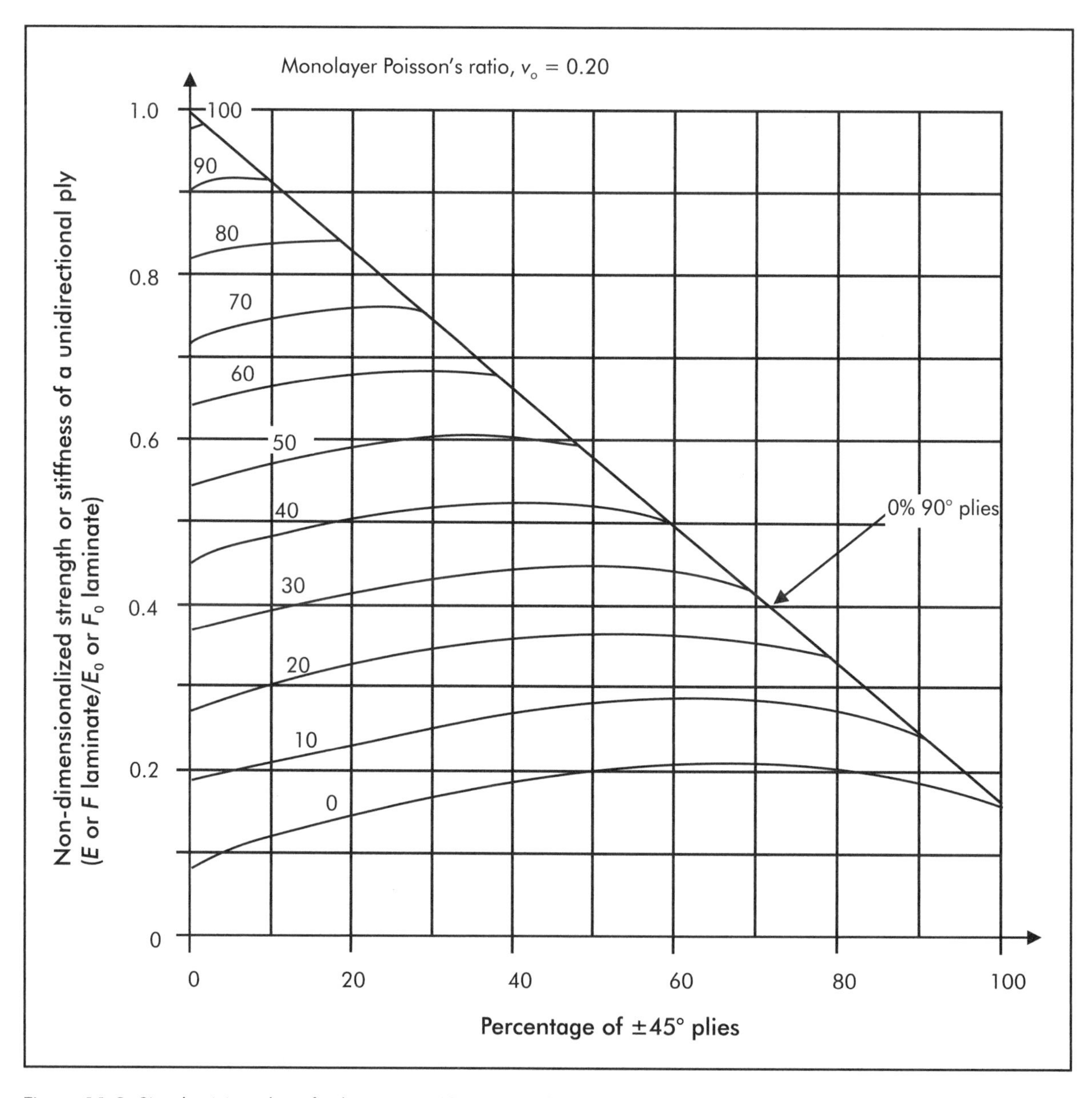

Figure 11-2. Simple sizing chart for laminates. (Courtesy John Hart-Smith, Boeing, 1989)

cal and mechanical properties, such as Poisson's ratio and the coefficient of thermal expansion. The analysis also should give an estimation of the initial failure condition, referred to as first-ply failure, and the ultimate strength of the structure. In detailed sizing, the stacking sequence of the plies becomes important, particularly in estimating the flexural rigidity (bending) resistance and strength characteristics of the laminate.

5. The first design review is used to check on the development of the laminate's design. The outcomes of the design, the mechanical and physical properties, are compared against the design's functional specification. Here the rationale of meeting, exceeding, or falling short of the requirements guide the designer in changing either one or all of the materials used, the manufacturing process selected, or the actual ply orientation and stacking sequence chosen.
6. Initial costing of the product can now give a good idea if the project is financially worthwhile. The previous steps defined the volume of the structure, the materials to be used, and the manufacturing method employed. With this information, knowledge of the consumables to be used, and the production quantity, the cost per part can be estimated. This is compared with other competing products and the functional specification to determine project continuation.
7. The design detail step covers a multitude of activities such as joint design, inclusion of holes and cut-outs, flexural stiffening with cores, investigation of potential interlaminar stress areas, and the effect of damage and reparability of the design. Such activities will lead to local ply buildup or thinning where required. Good stress analysis techniques, applicable to composite materials, must be used to complete this phase of the project design.
8. The second design review considers the detailed design work just completed. This review considers improvements to the design of the composite structure to help it meet the design goals with cost reduction. Typical cost reduction is achieved through reduced material usage, leaner manufacturing steps, and faster part fabrication.
9. Optimization of the structure involves making small, incremental changes to the individual ply materials, material form, ply orientation, and stacking sequences, etc. It is strongly recommended that only one aspect be changed at a time and the analysis re-run. Making too many changes at once will not indicate which parameter change was contributing to the overall improvement or degradation of structural performance.
10. The detailed design step typically involves performing some form of complex stress analysis such as finite element analysis. Detailed design analysis will look for further structural improvements without cost gains. Note that detailed stress analysis of composite structures requires significant processing time. This is especially so if the analysis is to be done thoroughly.

BASIC STRESS TYPES

The basic stress types to be considered in composite structure design are shown in Figure 11-3. They are usually designated as follows:

1. Axial and transverse in-plane tensile stresses (1–2 plane);
2. Axial and transverse in-plane compressive stresses (1–2 plane);

3. In-plane shear stresses (1–2 plane); and
4. Interlaminar or through-the-thickness stresses (1–3 and 2–3 planes).

Axial and Transverse In-plane Tensile Stresses

The in-plane tensile stresses in a composite structure act in the principal directions of the laminate. These stresses are predominately reacted by the fibers in the 0° and 90° laminate directions. The axial stresses are primarily supported by the 0° plies or the warp fibers in a woven cloth ply. The 90° plies or the fill direction for woven cloth support the transverse stresses. These stresses are designated σ_1 and σ_2.

Axial and Transverse In-plane Compressive Stresses

The in-plane compressive stresses in a composite are essentially the same as the in-plane tensile stresses. However, while the fibers take the compressive stresses, the resin plays an important role in compression. The fibers have a tendency to micro-buckle and it is the role of the resin (matrix) to support the fibers and resist this failure mode. The compressive strength properties are therefore controlled by the resin. When the resin is operated at a high temperature or has a degree of absorbed moisture, it becomes more rubbery and the maximum compressive strength reduces. The compressive stresses are designated $-\sigma_1$ and $-\sigma_2$.

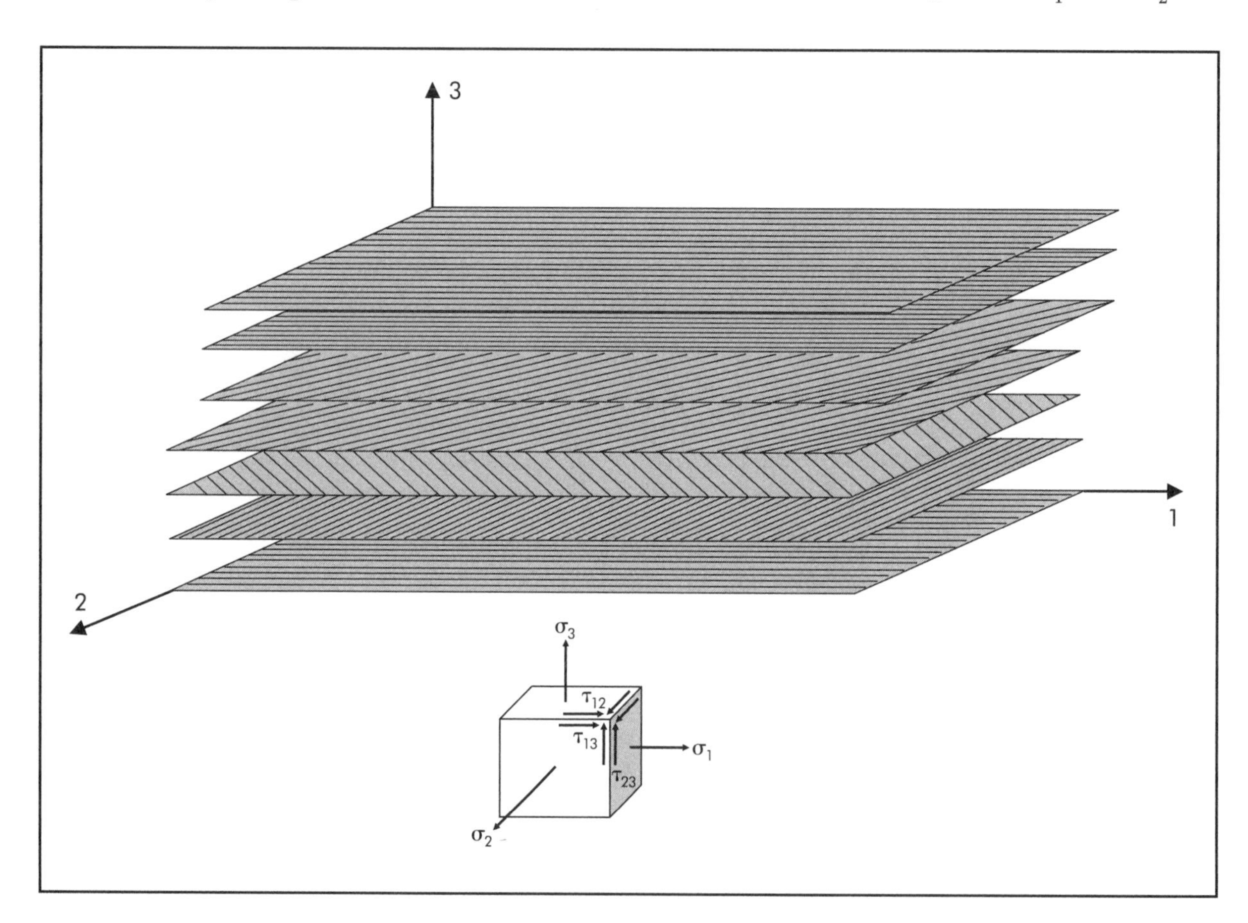

Figure 11-3. Stress designation.

In-plane Shear Stresses

The in-plane shear stresses are designated τ_{12}. Based on the well-known Mohr's circle of stress transformation, a pure shear stress field can be transformed into the biaxial principal tensile and compressive stresses along the ±45 degree lines in the 1–2 plane. Hence, it can be seen that the ±45 plies play a crucial role in supporting the shear stresses in a composite structure.

Interlaminar Stresses

The interlaminar stresses are also known as the out-of-plane or through-the-thickness stresses. There are three interlaminar stress components: the interlaminar normal stress σ_3, and the two interlaminar shear stresses, τ_{13} and τ_{23}. For conventional two-dimensional composite structures with the fibers in the 1–2 plane, the interlaminar stresses are typically taken by the resin. Since the resin is a relatively weak material in the composite structure, there is much effort expended to design the laminate so the interlaminar stresses are reduced as much as possible. Interlaminar stresses are typically concentrated at free edges and where there are ply variations as shown in Figure 11-4.

LAMINATE THEORY

The basic theory of laminate analysis is referred to as *Classical Laminate Plate Theory*.

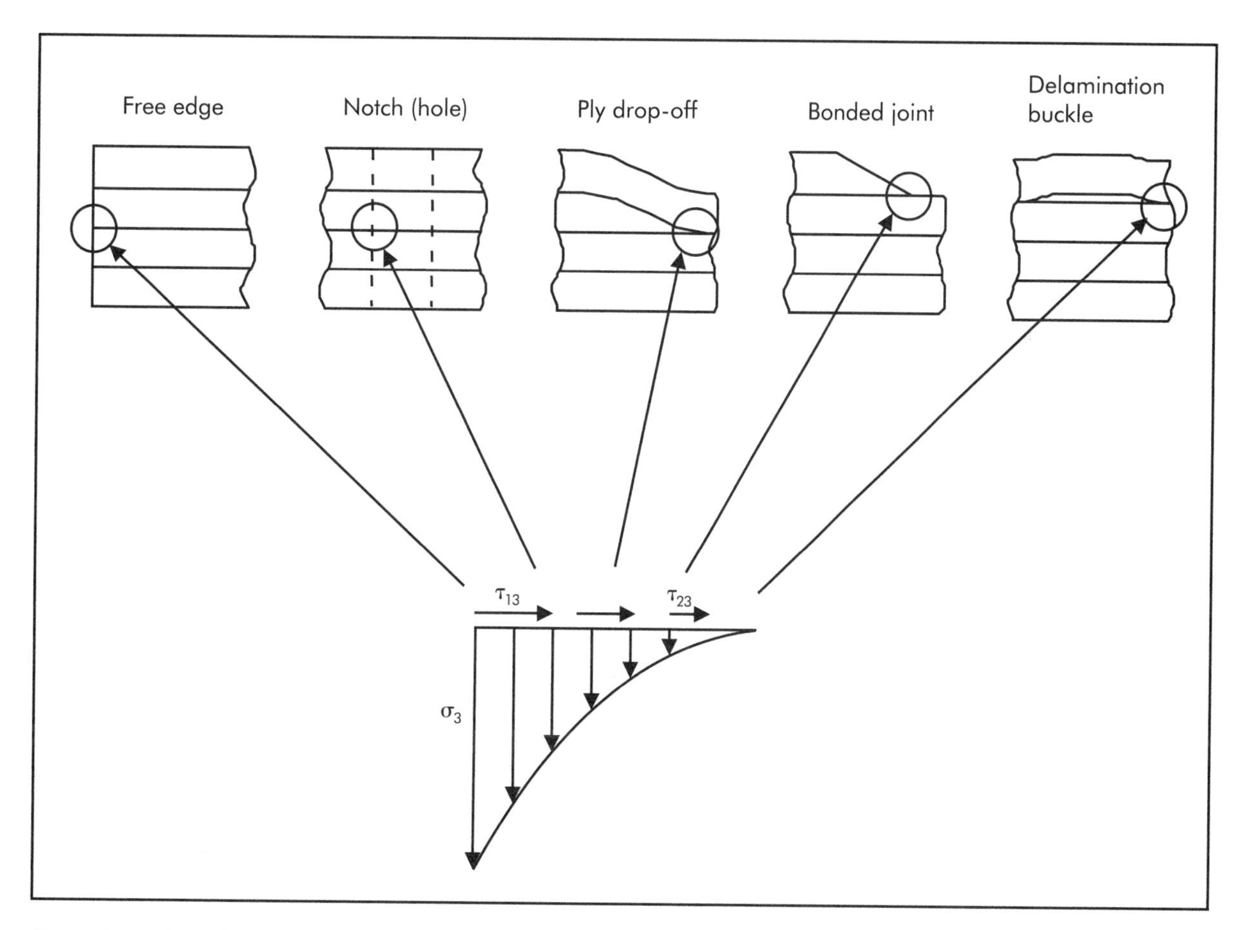

Figure 11-4. Out-of-plane stresses induced in composite structures.

The theory is essentially based on the development of the engineering and physical properties of a ply based on the fiber angle to the principal loading direction. Then, the total resistive force is obtained by simply summing the properties in a particular direction with respect to the ply thickness and dividing by the total thickness. If this assumption is made for a high-performance, carbon-fiber unidirectional composite, then the ply transverse and shear properties are approximately 10% of the axial ply properties as shown in Figure 11-5. Hence, the more 0° plies, the greater is the contribution to strength and stiffness in the axial direction.

Generally speaking, a unidirectional ply or woven cloth with the principal loading direction aligned to the ply fiber direction behaves in an orthogonal manner. That is, the material behavior has mutually perpendicular degrees of symmetry. When the ply axis of symmetry is not aligned to the loading axis, the ply behaves anisotropically. This means that the deformation behavior of the ply is influenced by both in-plane extensorial and flexural forces. When anisotropically behaving plies are subject to in-plane loads, the composite will twist and bend, as well as have in-plane deformations.

By laminating with specific ply orientations, the in-plane properties are uniform in all directions. Uniform properties in a material mean it has isotropic behavior. When the in-plane properties are uniform, the laminate is referred to as a *quasi-isotropic laminate*. Special quasi-isotropic laminates have equal numbers of plies in specific directions, for example, $\pi/3$ laminate [0°/+60°/–60°], $\pi/4$ laminate [0°/+45°/90°/–45°], $\pi/6$ laminate [0°/+30°/–30°/+60°/–60°/90°], etc. The order

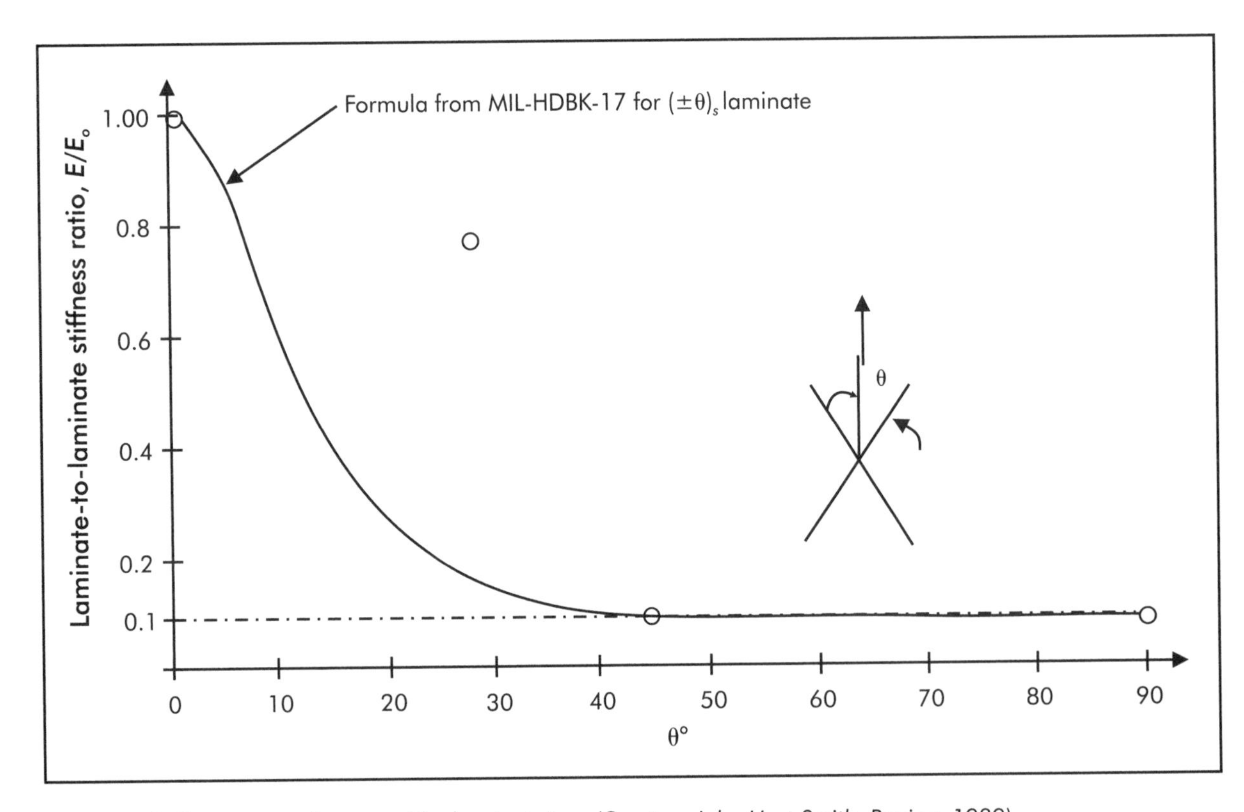

Figure 11-5. Ply property changes with ply orientation. (Courtesy John Hart-Smith, Boeing, 1989)

is not important for the in-plane properties; as long as there are an equal number of plies in each direction, the laminate's behavior will be quasi-isotropic.

The overall in-plane behavior of a structural laminate is a function of the in-plane stiffness and the flexural stiffness properties. In some laminate configurations there is coupling between in-plane and flexural behavior and this results in non-zero terms in the B matrix, B.

The in-plane or extensorial stiffness matrix relates the in-plane strains (axial, transverse, and shear) of the laminate to the load per unit width:

$$\begin{Bmatrix} N_1 \\ N_2 \\ N_6 \end{Bmatrix} = \begin{bmatrix} A_{11} & A_{12} & A_{16} \\ A_{21} & A_{22} & A_{26} \\ A_{61} & A_{62} & A_{66} \end{bmatrix} \begin{Bmatrix} \varepsilon_1 \\ \varepsilon_2 \\ \varepsilon_6 \end{Bmatrix} \qquad (11\text{-}1)$$

where:

N = load per unit width
A = in-plane stiffness
ε = strain

The flexural behavior relates the radii of curvatures to the D matrix to obtain the unit moments as:

$$\begin{Bmatrix} M_1 \\ M_2 \\ M_6 \end{Bmatrix} = \begin{bmatrix} D_{11} & D_{12} & D_{16} \\ D_{21} & D_{22} & D_{26} \\ D_{61} & D_{62} & D_{66} \end{bmatrix} \begin{Bmatrix} k_1 \\ k_2 \\ k_6 \end{Bmatrix} \qquad (11\text{-}2)$$

where:

M = edge moment per unit width
D = flexural stiffness
k = radii of curvature

The extensorial/flexure couple stiffness relates the radii of curvatures with unit loads or the in-plane strains with unit moments by:

$$\begin{Bmatrix} N_1 \\ N_2 \\ N_6 \end{Bmatrix} = \begin{bmatrix} B_{11} & B_{12} & B_{16} \\ B_{21} & B_{22} & B_{26} \\ B_{61} & B_{62} & B_{66} \end{bmatrix} \begin{Bmatrix} k_1 \\ k_2 \\ k_6 \end{Bmatrix}$$

and

$$\begin{Bmatrix} M_1 \\ M_2 \\ M_6 \end{Bmatrix} = \begin{bmatrix} B_{11} & B_{12} & B_{16} \\ B_{21} & B_{22} & B_{26} \\ B_{61} & B_{62} & B_{66} \end{bmatrix} \begin{Bmatrix} \varepsilon_1 \\ \varepsilon_2 \\ \varepsilon_6 \end{Bmatrix} \qquad (11\text{-}3)$$

where:

B = extensorial/flexural coupling stiffness

A total representation of the in-plane/flexural behavior for given deformations follows (note that diagonal symmetry is assumed):

$$\begin{Bmatrix} N_1 \\ N_2 \\ N_6 \\ M_1 \\ M_2 \\ M_6 \end{Bmatrix} = \begin{bmatrix} A_{11} & A_{12} & A_{16} & B_{11} & B_{12} & B_{16} \\ \ldots & A_{22} & A_{26} & B_{21} & B_{22} & B_{26} \\ \ldots & \ldots & A_{66} & B_{61} & B_{62} & B_{66} \\ \ldots & \ldots & \ldots & D_{11} & D_{12} & D_{16} \\ \ldots & \ldots & \ldots & \ldots & D_{22} & D_{26} \\ \ldots & \ldots & \ldots & \ldots & \ldots & D_{66} \end{bmatrix} \begin{Bmatrix} \varepsilon_1 \\ \varepsilon_2 \\ \varepsilon_6 \\ k_1 \\ k_2 \\ k_6 \end{Bmatrix} \qquad (11\text{-}4)$$

(For greater detail in the application of the classical laminate plate theory refer to Tsai and Hahn 1980, Tsai 1988, and Jones 1975.)

RULE OF MIXTURES

The rule of mixtures approach is a useful method in determining the engineering and physical properties of plies in a laminate. It is based on knowledge of the properties and volume of fibers and resin that make up a ply. The principal term used in the rule of mixtures approach is the **fiber-volume fraction** or fiber-volume ratio (V_f). The total expression is:

$$V_f + V_m + V_v = 1 \qquad (11\text{-}5)$$

where:

V_f = fiber-volume ratio, which is the physical volume of fibers to the total ply volume
V_m = matrix or resin-volume ratio

V_v = void volume ratio, which is the volume of voids and porosity as a result of the process used

The following equations result in the engineering and physical properties that can be determined using the rule of mixtures approach. (A more extensive listing can be found in Donaldson 2001). In using these equations, please note the following:

1. The void content is assumed to be zero.
2. Voids reduce the fiber-volume ratio.
3. Voids are a result of the manufacturing method and the cure process used.

In some cases the fiber or resin property cannot be easily determined. However, by measuring the ply property and determining the fiber- and resin-volume ratios, the individual fiber or resin properties can be estimated. A complete listing of the various testing methods for composites materials is available (American Society for Testing Materials 1990).

Density:

$$\rho = V_f \rho_f + V_m \rho_m \tag{11-6}$$

where:

ρ = total density
ρ_f = density of fibers
V_m = matrix volume ratio
ρ_m = density of matrix

Longitudinal stiffness (modulus):

$$E = V_f E_f + V_m E_m \tag{11-7}$$

where:

E = Young's modulus of a ply or single layer (laminate)
E_f = Young's modulus for fibers
E_m = Young's modulus for matrix

Poisson's ratio:

$$\upsilon = V_f \upsilon_f + V_m \upsilon_m \tag{11-8}$$

where:

υ = major Poisson's ratio of a ply or single layer (laminate)
υ_f = Poisson's ratio for fibers, which is difficult to measure
υ_m = Poisson's ratio for matrix

Shear modulus:

$$G = \frac{1}{\left[\frac{V_f}{G_f} + \frac{V_m}{G_m}\right]} \tag{11-9}$$

where:

G = shear modulus of a ply or single layer (laminate)
G_f = shear modulus of fibers
G_m = shear modulus of matrix

Tensile strength (longitudinal):

$$\sigma = V_f \sigma_f + V_m \sigma_m \tag{11-10}$$

where:

σ = axial stress strength of a ply or single layer (laminate)
σ_f = axial strength of fibers
σ_m = axial strength of matrix

Transverse modulus:

$$E_y = E_z = \frac{E_m}{\left[1 - \sqrt{V_f}\left(1 - \frac{E_m}{E_{f_y}}\right)\right]} \tag{11-11}$$

where:

E_y = Young's modulus in the y orthogonal direction of a ply or single layer (laminate)
E_z = Young's modulus in the z orthogonal direction of a ply or single layer (laminate)
E_{fy} = Young's modulus of the fibers in the y orthogonal direction

Interlaminar shear:

$$G_{yz} = \frac{G_m}{\left[1 - V_f\left(1 - \frac{G_m}{G_{f_{yz}}}\right)\right]} \quad (11\text{-}12)$$

where:

G_{yz} = shear modulus in the *yz* orthogonal plane of a ply or single layer (laminate)

$G_{f_{yz}}$ = shear modulus of the fibers in the *yz* orthogonal plane

Tensile strength (transverse):

$$\sigma_{yT} = \left[1 - \left(\sqrt{V_f} - V_f\right)\left(1 - \frac{E_m}{E_{f_y}}\right)\right]\sigma_{mT} \quad (11\text{-}13)$$

where:

σ_{yT} = axial tensile strength in the *y* orthogonal plane

σ_{mT} = axial tensile strength of the matrix

Generally, there is a difficulty in determining the measured value of the fiber-volume ratio in a fabrication process. Occasionally the weight ratios of the fiber and resin are used, but weight ratios cannot be used directly to determine the engineering properties. To determine the fiber-volume ratio from weight ratios, the fiber, resin, and ply densities must be known. The densities can be found in the manufacturing safety data sheet (MSDS) or materials data sheet (MDS) documentation. If the individual fiber and resin densities are not known, but the ply density and fiber-volume ratio are provided, then use Equation 11-6. The expression that determines the fiber-volume ratio from the fiber and resin weight ratios is (Heslehurst 2006):

$$V_f = \left[1 + \frac{1}{\left(\frac{W_f}{W_r}\right)\left(\frac{\rho_r}{\rho_f}\right)}\right]^{-1} \quad (11\text{-}14)$$

where:

W_f = weight of the fiber

W_r = weight of the resin system

ρ_r = density of the resin or matrix system

The properties of some common composite systems are shown in Table 11-1.

MODELING AND FINITE ELEMENT ANALYSIS

The modeling of composite laminated structures varies from the relatively simple to complex. All modeling of composite structures requires computer support due to the detail in the lamination of the individual plies.

Each ply has its own individual mechanical and physical properties in three dimensions. However, some simple assumptions allow for less complex analysis and initial sizing of composite structures to be performed. The assumptions made for ease of analysis are that the structure is orthotropic and homogenous initially and through-the-thickness effects are ignored. A slightly more detailed approach is to analyze the individual in-plane state of the laminate and use these results to interpret the total structure's performance. An expansion of this approach is to assume that the through-the-thickness performance is purely based on the resin properties. But for a complete analysis of the structure, the three-dimensional stress analysis of the plies must consider the anisotropic behavior of the laminate.

In-plane Orthotropic, Homogenous Laminate Analysis

The in-plane orthotropic, homogenous laminate approach is the same analysis method used with traditional isotropic materials. The basic assumption is that the entire laminate is homogenous. Thus

Table 11-1. Common composite system fiber/resin properties.

Property	Composite System					
	Carbon Fiber Reinforced Plastic (CFRP)	Boron Fiber Reinforced Plastic (BFRP)	Carbon Fiber Reinforced Plastic (CFRP)	Glass Fiber Reinforced Plastic (GFRP)	Kevlar® Fiber Reinforced Plastic (KFRP)	Boron Fiber Reinforced Aluminum (BFRA)
Fiber	T300	Boron (4)	AS	E-Glass	Kevlar 49	Boron
Matrix	N5208	5505	3501	Epoxy	Epoxy	Al
Fibers' relative density (density of the fiber system to the density of water)	1.750	2.600	1.750	2.600	1.440	2.600
Matrix relative density (density of the matrix to the density of water)	1.200	1.200	1.200	1.200	1.200	3.500
Density of fibers/ density of matrix, ρ_f/ρ_m	1.458	2.167	1.458	2.167	1.200	0.743
Void ratio, V_V	0.005	0.005	0.005	0.005	0.005	0.000
Fiber-volume ratio, V_f	0.700	0.500	0.666	0.450	0.700	0.450
Matrix volume ratio, V_m	0.295	0.495	0.329	0.545	0.295	0.550
Composite's relative density (density of the composite to the density of water)	1.579	1.894	1.560	1.824	1.362	3.095

the overall properties of the structure are smeared throughout the laminate. The individual ply stress state is ignored. However, the laminate does possess orthotropic properties in that its properties vary along the principal axes. Note that the material's in-plane axes of symmetry are aligned with the principal loading axes. Beware that the off-axis properties are not modeled well. This is because they are typically assumed to vary linearly between the axes of symmetry, and *not* like that shown in Figure 11-5.

In-plane Orthotropic Behavior of Individual Plies Analysis

The next level of complexity in modeling composite structures is to look at the in-plane orthotropic behavior on a ply-by-ply basis. One important aspect of this analysis is that the laminate as a whole is also orthotropic such that all angled plies come as a matched pair of positive and negative plies. This results in a balanced laminate that eliminates the in-plane anisotropic effects. Note that the out-of-plane deformations due to bending are not truly balanced (this is explained later in this chapter). The analysis interrogates the stress state of the individual plies and then combines the data from each ply to arrive at the total laminate performance. This approach requires detailed knowledge of the ply stacking sequence. The method also allows for the use of different fiber/resin layers or hybrid composite structures. One word of caution is that the laminate's properties are typically a weighted average over the thickness and this can lead to erroneous results when using the same method on, for example, sandwich structures.

Three-dimensional, Anisotropic Ply-by-ply Analysis

The most complex approach is to use three-dimensional, anisotropic ply-by-ply analysis. The difficulty in using this approach is more related to what material properties are known. For example, if the behavior of a laminate in three-dimensions was analyzed for a given applied strain, the following expression would be used. Evaluation of the stiffness matrix, C, is necessary. Assuming diagonal symmetry, that is, $C_{12} = C_{21}$, etc., the analysis required 21 tests to be performed to provide these constant of proportionality.

$$\begin{Bmatrix} \sigma_1 \\ \sigma_2 \\ \sigma_3 \\ \sigma_4 \\ \sigma_5 \\ \sigma_6 \end{Bmatrix} = \begin{bmatrix} C_{11} & C_{12} & C_{13} & C_{14} & C_{15} & C_{16} \\ C_{21} & C_{22} & C_{23} & C_{24} & C_{25} & C_{26} \\ C_{31} & C_{32} & C_{33} & C_{34} & C_{35} & C_{36} \\ C_{41} & C_{42} & C_{43} & C_{44} & C_{45} & C_{46} \\ C_{51} & C_{52} & C_{53} & C_{54} & C_{55} & C_{56} \\ C_{61} & C_{62} & C_{63} & C_{64} & C_{65} & C_{66} \end{bmatrix} \begin{Bmatrix} \varepsilon_1 \\ \varepsilon_2 \\ \varepsilon_3 \\ \varepsilon_4 \\ \varepsilon_5 \\ \varepsilon_6 \end{Bmatrix} \quad (11\text{-}15)$$

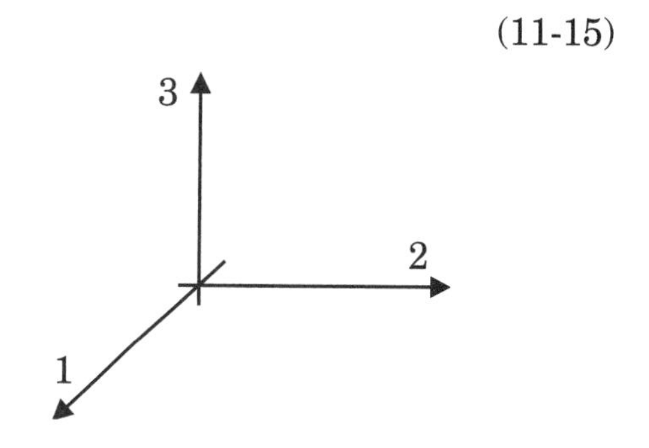

where:

C_{ij} = generalized stiffness
i = numerical values 1…6
j = numerical values 1…6

However, when the plane of symmetry is assumed through the thickness (mid-plane symmetry or the ply is symmetrical through the thickness), for example, parallel to the 1–2 plane, then the out-of-plane shear behavior is decoupled from the in-plane behavior and only 14 constants of proportionality need to be evaluated.

$$\begin{Bmatrix}\sigma_1\\\sigma_2\\\sigma_3\\\sigma_4\\\sigma_5\\\sigma_6\end{Bmatrix}=\begin{bmatrix}C_{11}&C_{12}&C_{13}&0&0&C_{16}\\C_{21}&C_{22}&C_{23}&0&0&C_{26}\\C_{31}&C_{32}&C_{33}&0&0&C_{36}\\0&0&0&C_{44}&C_{45}&0\\0&0&0&C_{54}&C_{55}&0\\C_{61}&C_{62}&C_{63}&0&0&C_{66}\end{bmatrix}\begin{Bmatrix}\varepsilon_1\\\varepsilon_2\\\varepsilon_3\\\varepsilon_4\\\varepsilon_5\\\varepsilon_6\end{Bmatrix} \tag{11-16}$$

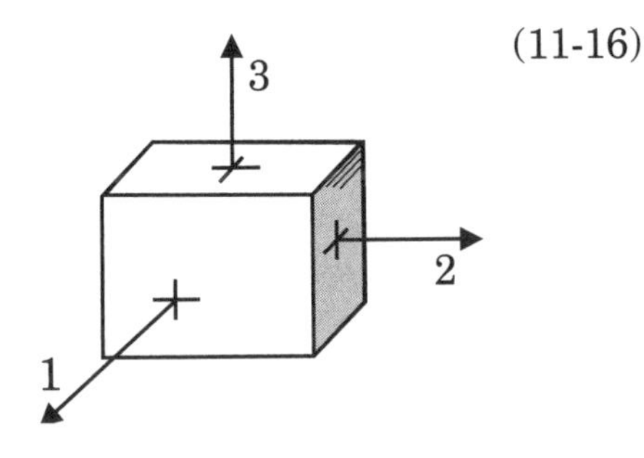

When there are three mutually perpendicular planes of symmetry, the shearing behavior is decoupled from the normal stresses. If the shear stresses in the three planes are independent, then only nine constants of proportionality need to be evaluated.

$$\begin{Bmatrix}\sigma_1\\\sigma_2\\\sigma_3\\\sigma_4\\\sigma_5\\\sigma_6\end{Bmatrix}=\begin{bmatrix}C_{11}&C_{12}&C_{13}&0&0&0\\C_{21}&C_{22}&C_{23}&0&0&0\\C_{31}&C_{32}&C_{33}&0&0&0\\0&0&0&C_{44}&0&0\\0&0&0&0&C_{55}&0\\0&0&0&0&0&C_{66}\end{bmatrix}\begin{Bmatrix}\varepsilon_1\\\varepsilon_2\\\varepsilon_3\\\varepsilon_4\\\varepsilon_5\\\varepsilon_6\end{Bmatrix} \tag{11-17}$$

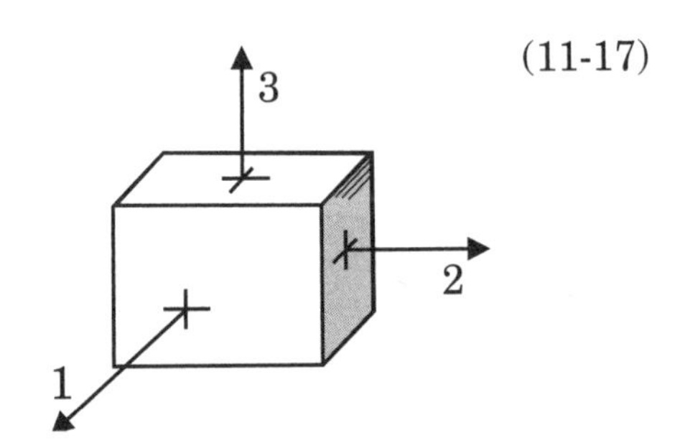

With in-plane (1–2 plane) behavior being the primary concern, the stress-strain behavior is further simplified to:

$$\begin{Bmatrix}\sigma_1\\\sigma_2\\\sigma_6\end{Bmatrix}=\begin{bmatrix}Q_{11}&Q_{12}&0\\Q_{21}&Q_{22}&0\\0&0&Q_{66}\end{bmatrix}\begin{Bmatrix}\varepsilon_1\\\varepsilon_2\\\varepsilon_6\end{Bmatrix} \tag{11-18}$$

where:

$\sigma_6 = \tau_{12}$, stress component 6 is equivalent to in-plane shear stress

$\varepsilon_6 = \gamma_{12}$, strain component 6 is equivalent to in-plane shear strain

Q_{11} = ply axial modulus, which is equivalent to the axial Young's modulus

Q_{22} = ply transverse modulus, which is equivalent to the transverse Young's modulus

Q_{66} = ply shear modulus

$Q_{12} = Q_{21}$, Poisson's ratio effect of transverse strain due to axial stress and vice versa

Thus, for the in-plane behavior, only four constants are required to be evaluated in tests: the axial and transverse Young's moduli, the shear modulus, and the Poisson's ratio.

Finite Element Analysis

Finite element analysis (FEA) allows detailed stress and deformation solutions to complex shapes. The analysis is based on dividing the structure into a large number of small elements and assigning each element the appropriate material properties. Then all of the equations are resolved simultaneously in the form of the load versus deflection relationship in terms of the known stiffness of the element. The element stiffness is based on the Young's modulus of the material and the element's cross-sectional area and length:

$$P = k\delta \tag{11-19}$$

where:

P = applied or internal load

δ = deformation

k = element stiffness = EA/L

and

E = Young's modulus

A = cross-sectional area

L = element length

The solution can be visualized as a plot of the stresses in a contour map of the structure showing deformation, as seen in Figure 11-6.

In composite materials, the main issue in the application of FEA is what material properties to use. The level of complexity is the ultimate choice. At the lowest level the structural element has global orthotropic properties. In other words, the structural components possess the laminate properties that have mutually perpendicular property variation. The through-the-thickness properties remain an issue, however. The next level uses the orthotropic properties of the individual plies and, again, the issue of through-the-thickness properties needs to be addressed. The subdivision of the plies requires a significant increase in the number of elements and processing time increases accordingly (more simultaneous equations to solve). The most complex approach requires more time for analysis and model building and includes the full 3D properties of the individual layers to allow for anisotropic behavior in the analysis. (For further reading, see Whitney 1987, Reddy and Miravete 1995, and Gibson 1994.)

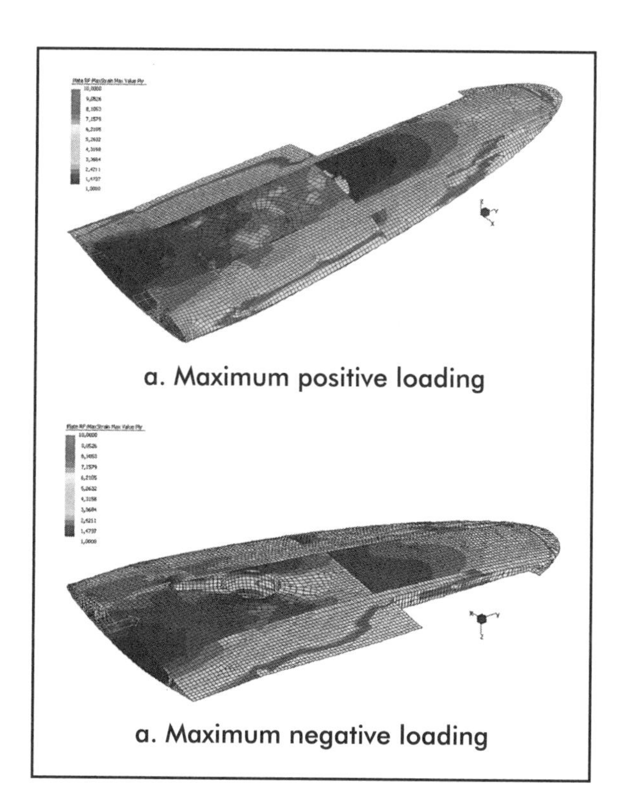

a. Maximum positive loading

a. Maximum negative loading

Figure 11-6. Finite element analysis of WWII fighter wing fabricated with composite materials.

Closed-form Solution Approach

The closed-form solution approach in the stress analysis of composite structures allows quicker results than with FEA. It is not as accurate as FEA, but in many cases is within 5% of the more accurate solution. The closed-form solution approach is excellent for undertaking the initial sizing stress analysis. It provides a good initial basis for further detailed stress analysis such as FEA. The only issue with the closed-form stress analysis approach is that it requires a good understanding of the behavior and anisotropic relationships of composite materials. The case study at the end of this chapter provides an example of a closed-form solution approach. (For further reading, see Whitney 1987, Reddy and Miravete 1995, and Gibson 1994.)

LAY-UP NOTATION

The basic principle behind lay-up notation is to ensure that the lay-up pattern achieved in fabrication of the composite laminate is the same as the engineering requirements. The engineering lay-up requirements are set from the engineering stress and stiffness calculations. The basic patterns of ply orientation can be illustrated with a simple diagram of the principal ply directions in a laminate. The ply lay-up pattern will aid in distinguishing between unidirectional plies and woven cloth. The basic reference for ply lay-up notation is taken from the Boeing Drafting Standard, BDS-1330. However, there are a number of ways to specify the ply lay-up patterns, which are also acceptable. In any case, a key to the notation should be provided and there should be logic and consistency in its use.

To assist in determining the difference between plies laid up at angles to the principal direction, the warp clock, shown in Figure 11-7, is used. The recognized industry standard for its use in notation states that the counter-clockwise (CCW) symbol be used on the main views in the manufacturing drawing to show the relative fiber or warp yarn direction, normally with the 0° axis parallel to the primary load direction of the part laminate or assembled structure.

Some basic rules apply for ply lay-up notation:

1. Laminate notation is written between square brackets. For example:

   [notation]

2. The notation is written starting with the first ply laid down, that is, at the tool surface.
3. Each ply is separated by a slash. For example:

   [ply1/ply2/ply3]

4. Plies of equal but opposite ply angle (plus/minus orientation) can be shown as:

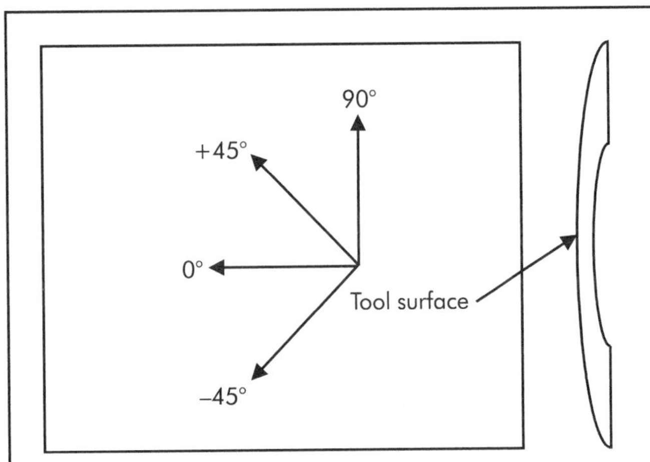

a. Clockwise (CW) warp clock

The clockwise warp clock is drawn from the *engineering design* standpoint, where the fiber angles are viewed from the outside of the structure, or from the tool surface, looking in. Notice that the 0° axis is parallel to the long axis of the rectangular shape and +45° is located by moving in a clockwise direction from 0° to 90°. This symbol is a mirror image of the CCW warp clock.

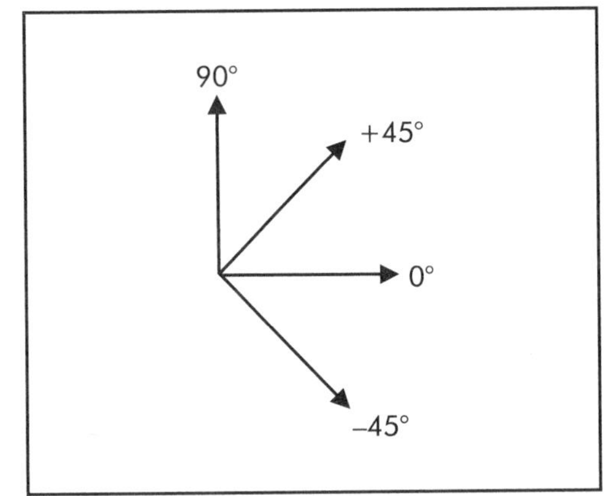

b. Counter-clockwise (CW) warp clock

The counter-clockwise warp clock is drawn from the *manufacturing* point of view, where the fiber angles are viewed from the inside of the panel looking toward the tool surface. This would be the case during lay-up. Notice that the 0° axis is parallel to the long axis of the rectangular shape and +45° is located by moving in a counterclockwise direction from 0° to 90°.

Figure 11-7. Directional conventions of the warp clock. (Courtesy Abaris Training, www.abaris.com)

[±45] → [+45/–45] and

[45] → [–45/+45]

5. Multiple (adjacent) plies can be shown with a subscript:

   $[0/0/0/90/90] \rightarrow [0_3/90_2]$

6. A lay-up with mid-plane symmetry uses an "s" subscripted outside the brackets:

   $[0/0/90/90/0/0] \rightarrow [0_2/90]_s$

7. Symmetry with an odd number of plies is shown as:

   $[0/0/90/0/0] \rightarrow [0_2/\overline{90}\ ]_s$

8. For clarification, a "T" subscripted outside the bracket designates that the entire (total) lay-up is shown:

   $[90/0/{-30}/{-45}]_T$

9. Combined tape and fabric is shown as the following, noting that the warp direction for the 45° fabric is to be orientated +45° to the laminate's 0° axis:

   [0/±45fabric/0] → [0/+45F/0]

10. When the orientations of the warp directions of fabrics to the laminate's 0° axis are optional:

    $[0/\pm 45F/90/(0 \text{ or } 90)F]_s$

11. When the warp direction of the angled fabric ply is essential:

    $[0/{+45F}/{-45F}/0]_s$ and

    with warp parallel to +45° → +45F and
    with warp parallel to −45° → –45F

12. For hybrid laminate lay-ups, the individual plies are coded with the material of that ply (FG = fiberglass and C = carbon):

    $[(0 \text{ or } 90)F_{FG}/(+45/0/{-45}/90)_C/(0 \text{ or } 90)F_{FG}]$

An example notation for a laminate lay-up sequence is shown in Table 11-2.

SYMMETRY AND BALANCED LAMINATES

The basic lay-up properties of a composite laminate are defined by the symmetry about the mid-plane and the balance of any angled plies in the laminate. The properties of the lay-up configuration have a major impact on the loading and deformation behavior of the laminate. First, here are the related definitions.

- A *symmetric laminate* has the ply orientations mirror imaged about its structural mid-plane as shown in Figure 11-8 with unidirectional tape.
- The *in-plane deformation balanced laminate* has a positively angled ply for every negatively angled ply. This definition does *not* include 0° and 90° orientations. By this definition, a 45° orientated plain weave cloth is effectively balanced, assuming the same number of filaments in the warp and fill directions. Note that other weave patterns can have slight variations in the warp and fill directions so they are not really balanced by themselves.
- To achieve a *flexural behavior balanced laminate*, the angled plies must be co-mingled to have equal distances from the mid-plane. This is because out-of-plane behavior is a function of ply position away from the mid-plane. Generally speaking, only the plain weave cloth's 45° orientation plies achieve this, but not in the purest sense as warp/fill direction ply counts can be different.

In terms of structural behavior, the following deformation phenomenon will occur.

- A *symmetric laminate* does not possess coupling between in-plane (extensorial) and flexural behavior. Thus symmetric laminates show the extensorial/flexural

Table 11-2. Example of lay-up notation system.

Meaning	Notation
45 Fabric (fiberglass) 0 or 90 (Carbon fabric) 0 Tape (carbon) 0 Tape (carbon) 90 Tape (carbon) +45 Tape (carbon) –45 Tape (carbon) +45 Fabric (carbon) +45 Fabric (carbon) –45 Tape (carbon) +45 Tape (carbon) 90 Tape (carbon) 0 Tape (carbon) 0 Tape (carbon) 0 or 90 (Carbon fabric) 45 Fabric (fiberglass)	$[45F_{FG}/((0 \text{ or } 90)F/0_2/90/\pm45/+45F)_C]_s$

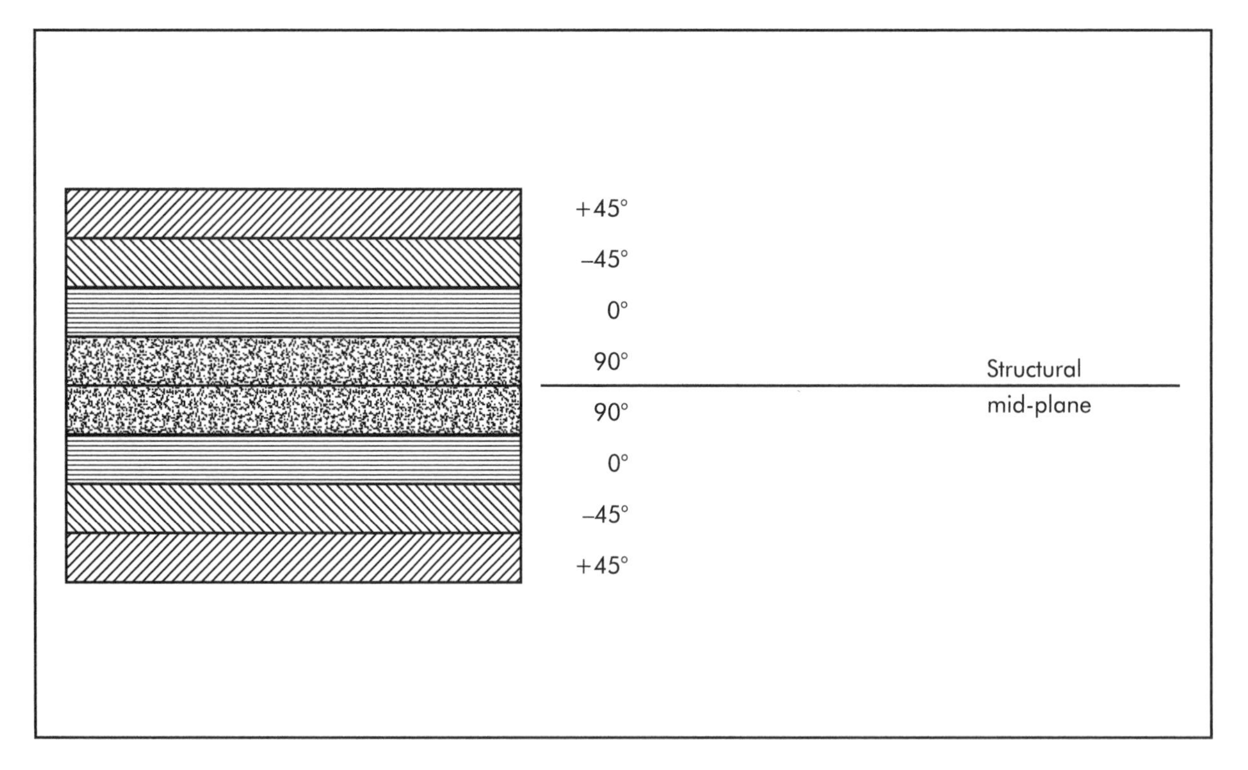

Figure 11-8. Symmetric laminate lay-up.

coupling matrix as $B = 0$. Asymmetric cross-plies (0° and 90°) result in $B_{11} \neq 0$, $B_{12} = B_{21} \neq 0$, and $B_{22} \neq 0$. With the angled plies in an asymmetric configuration, $B_{16} \neq 0$, $B_{26} \neq 0$, and $B_{66} \neq 0$.

- *In-plane balanced laminates* show no coupling between in-plane axial deformation and in-plane shear deformation. In other words, $A_{16} = A_{26} = 0$. However, this does not constitute flexural balance as $D_{16} = D_{26} \neq 0$.
- What is termed a *specially orthotropic laminate* has flexural behavior balance, so $D_{16} = D_{26} = 0$.

Figure 11-9 summarizes the behavior of the laminates based on mid-plane symmetry and angled ply balance. (Many of these issues are detailed in Tsai 1988.)

CRACKING IN COMPOSITES

There are three major categories of cracking in composite structures (as was illustrated in Figure 8-24):

1. Fiber fracture,
2. Fiber/resin interfacial fracture, and
3. Resin or matrix cracking.

Fiber fracture will most likely occur as final fracture in well-designed composite structures. If fiber fracture occurs early in the composite fracture process, then it is likely that the matrix is too compliant for the fiber system. For example, if carbon fibers were incorporated in a sealant material and loaded to failure then the fiber would more than likely fail first.

Interfacial fracture is usually due to poor material selection and fabrication process

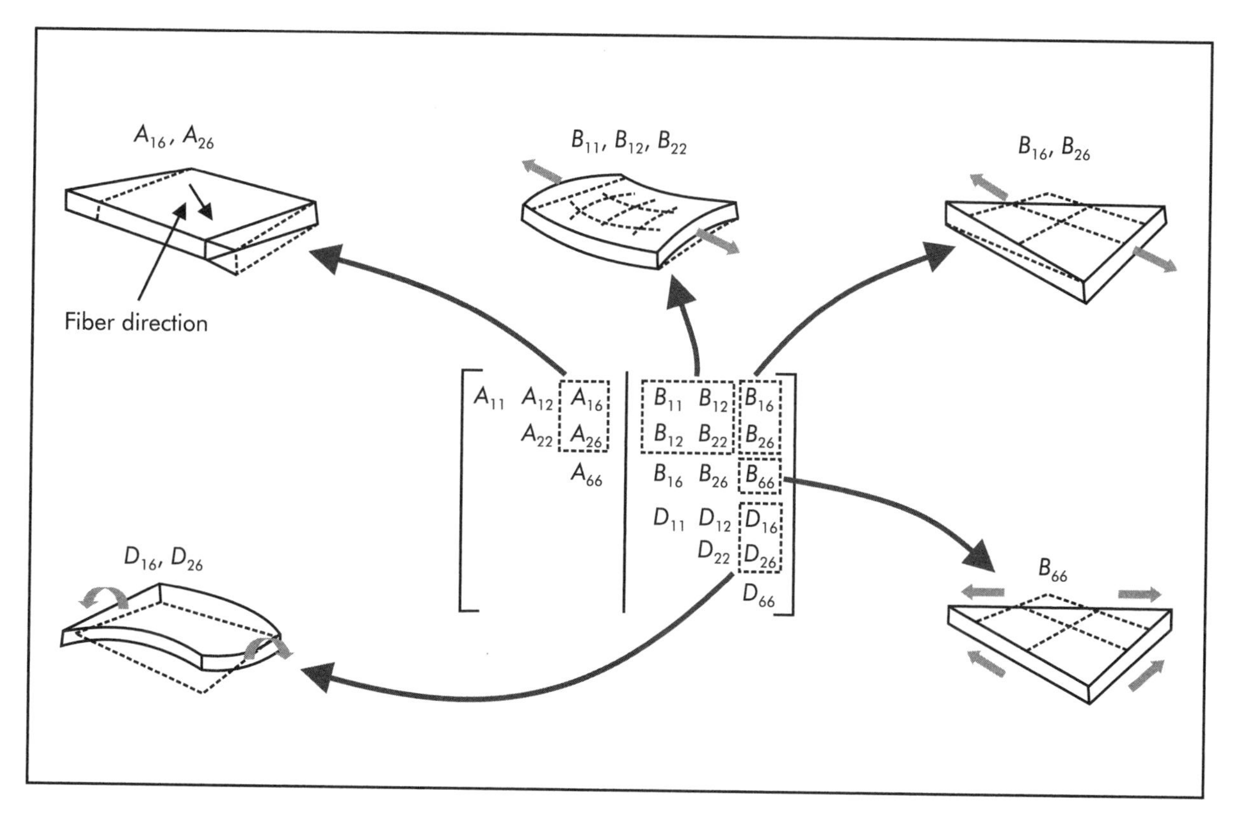

Figure 11-9. Stiffness coupling phenomenon (Davis 1987).

problems. Sizing the fiber for the resin system inhibits fracture between fiber and resin. Typically, interfacial failures are an indication of poor surface preparation of the fibers. Often the interfacial failure mode can be associated with fiber/resin interfacial degradation due to the absorption of some fluid, for example, moisture or hydraulic oils.

The most common low-load cracking mechanism in composite structures is *matrix cracking*. This is often what is termed **first-ply failure**, where the strength of the matrix is exceeded. Matrix cracking is not typically an ultimate failure mode but the early stages of a fracture process. There are basically two types of matrix cracking: intralaminar and interlaminar or delaminations. Figure 11-10 delineates the difference between **intralaminar cracking** and **interlaminar cracking** of the matrix. By definition, intralaminar cracks are within a ply in the matrix and interlaminar cracks are between adjacent plies.

Fracture Process

Well-designed composite structures will first experience fracture and cracking in the form of intralaminar matrix cracks. Intralaminar matrix cracks will typically appear in high concentration but are often localized as shown in Figure 11-11. These cracks appear parallel to the fibers in unidirectional composite plies. Intralaminar cracks generally initiate delaminations. The delaminations are the matrix cracks between the adjacent plies as shown in Figure 11-11.

The ultimate failure mode is fiber fracture as shown in Figure 11-11. Note that the various ply orientations define individual cracking mechanisms. The 90° plies will fail early by pure intralaminar cracks. The angled plies (in this case 45°) show all three failure modes of cracking in the sequences just described. The 0° plies will have intralaminar cracking and likely fiber fracture with little or no delamination occurring. The fabric composite structure will have limited widespread intralaminar cracking as the transverse fibers tend to minimize the matrix cracks. However, extensive delaminations and then final fiber fracture can occur. Figure 11-12 illustrates the fracture of a composite structure made with eight-harness satin glass cloth and epoxy resin. (Some discussion of the fracture process was presented in Chapter Eight where fiber/matrix interactions and the interphase were discussed.)

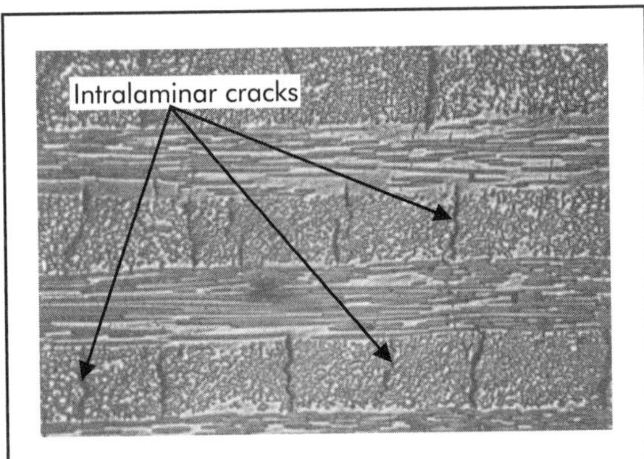

a. Intralaminar cracking

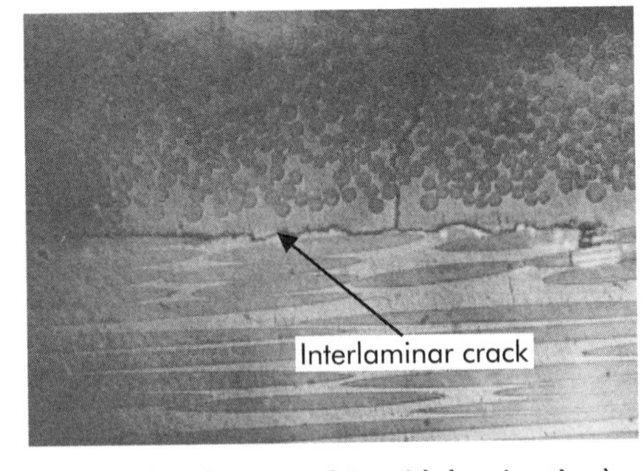

b. Interlaminar cracking (delamination)

Figure 11-10. Types of matrix cracks in composite structures (Agarwal and Broutman 1990).

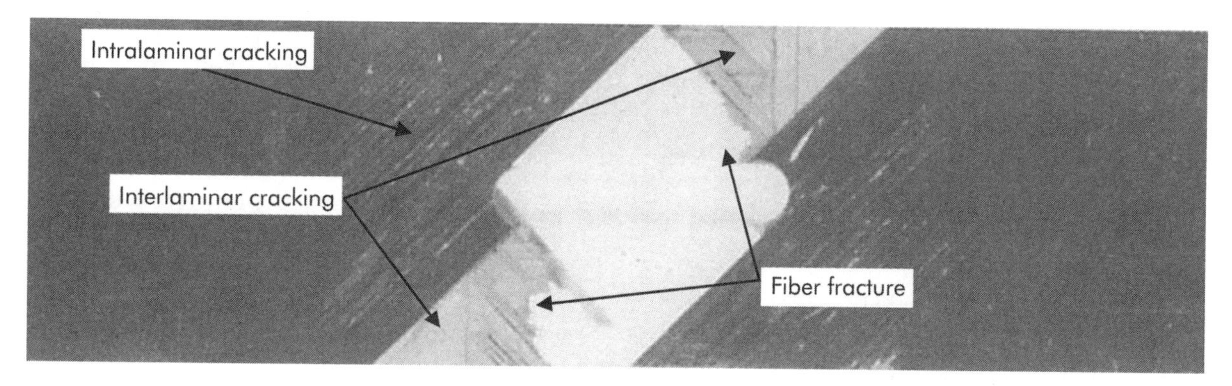

Figure 11-11. Composite structure fracture process.

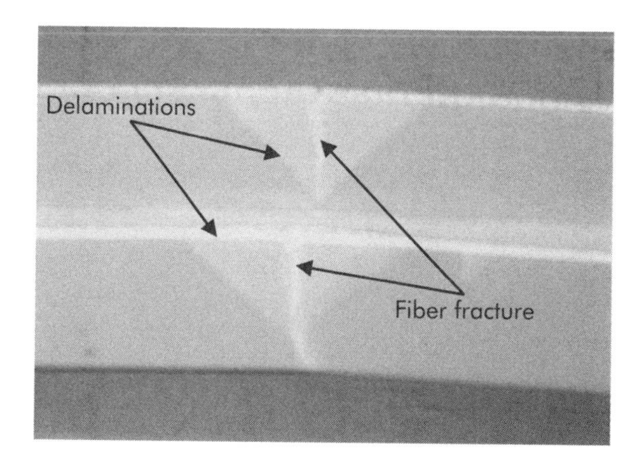

Figure 11-12. Fabric composite structure fracture.

VIBRATION AND DAMPING

The beneficial vibration and damping properties of composite structures give them yet another advantage over traditional engineering materials. The vibration performance is enhanced by the stiffness properties, which is a product of the fiber and ply orientation. The good damping characteristics are attributed to the polymeric matrix. This section addresses the vibration characteristics (natural frequencies and modes of vibration), and the natural damping behavior, particularly at high frequencies, of composite structures.

As a simple example, a flat rectangular plate is examined. The plate thickness, t, is substantially smaller than the plate's in-plane dimensions. The frequency response of the plate is an out-of-plane behavior and thus is influenced by three factors:

1. The plate aspect ratio, R,

$$R = a/b \tag{11-20}$$

where:

a = plate length

b = plate width

2. The plate density, which is influenced by the fiber and resin densities and the fiber-volume ratio, V_f.
3. The flexural stiffness of the plate, which depends on ply orientation and ply through-the-thickness position in the plate.

There are several modes of vibration in a structure. The natural frequency over a range of vibration modes for a specially orthotropic laminated plate (that is, the laminate is symmetric and specially balanced such that $A_{16} = A_{26} = D_{16} = D_{26} = 0$) is given by (Whitney 1987):

$$\omega_{mn} = \left(\frac{\pi}{Rb}\right)^2 \sqrt{\frac{1}{\rho}} \sqrt{\left(D_{11}m^4 + 2\left(D_{12} + 2D_{66}\right)m^2n^2R^2 + D_{22}n^4R^4\right)} \tag{11-21}$$

where:

ω = natural vibration frequency

ρ = plate density

D_{ij} = flexural stiffness of the D matrix of the plate; i and j are integers

m = longitudinal direction of the mode shape

n = transverse direction of the mode shape

Note for the first mode of vibration, $m = n = 1$. Hence, there are a lot of potential variables in the development of the plate's natural frequency. The panel aspect ratio plays a major role in determining the natural frequency of the structure, as does the panel density, which is controlled by the fiber/resin type and the fiber-volume ratio to some extent. However, the key factor is the flexural rigidity of the panel and thus the position of the contributing plies.

The damping of composite structures is more complex in analysis, but is fundamentally a function of the matrix (resin), fiber-volume ratio, and ply orientation. The natural damping characteristics of composite structures are of great benefit in design. Damping properties can be tailored, as are most composite properties, through variations in the fiber and resin selection, fiber-volume ratio, and ply orientation. The most significant feature of the composite structure that contributes to the damping properties is the viscoelastic behavior of the resin. Figure 11-13 illustrates the typical characteristic behavior of viscoelastic materials. The increasing load will store strain energy (area under the curve). As the load is removed, the deformation path is not similar to the load application path. Hence, the strain energy released is the area under the lower curve. The difference in energy stored and released is absorbed energy or energy loss. Understanding this concept allows some degree of analysis to be undertaken on the damping characteristics of composite materials.

The damping characteristic of materials is typically given as a loss factor, which is a function of the applied frequency. The loss factor is a ratio of the loss modulus to the storage modulus, both functions of frequency. The storage and loss moduli are derived from the relationship between the individual fiber and resin moduli, and the fiber-volume ratio. For example, the longitudinal loss factor of a unidirectional ply can be estimated by (Gibson 1994):

$$\eta_1(\omega) = \frac{E''_1(\omega)}{E'_1(\omega)} = \frac{E''_{f1}(\omega)V_f + E''_m(\omega)V_m}{E'_{f1}(\omega)V_f + E'_m(\omega)V_m} \tag{11-22}$$

where:

η = longitudinal damping loss factor

E'' = loss modulus

E' = storage modulus

E''_f = loss modulus of fiber

E'_f = storage modulus of fiber

E''_m = loss modulus of matrix

E'_m = storage modulus of matrix

V_m = matrix or resin-volume ratio

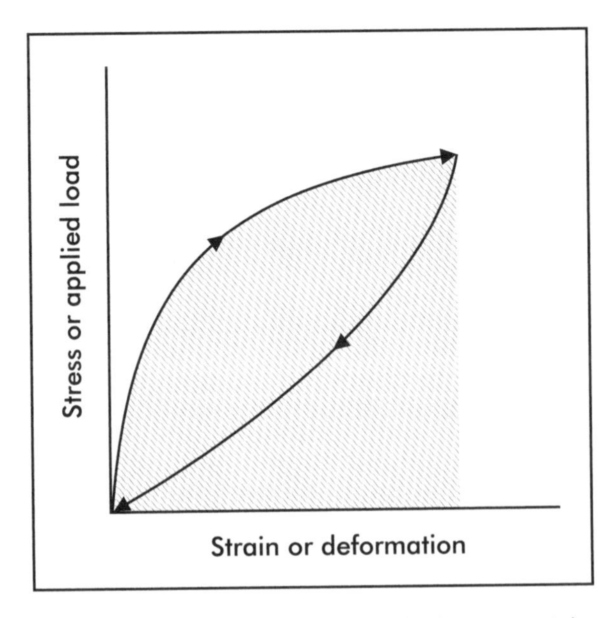

Figure 11-13. Viscoelastic behavior of polymer materials.

By using similar micromechanics expressions for the storage (stiffness) and loss moduli in the shear and transverse directions, the loss factor can be calculated for a variety of laminate configurations. Note, the loss factor of a unidirectional ply is typically dominated by the stiff fiber, but for matrix-dominated directions (45–90°) the loss factor increases significantly and is dominated by the viscoelastic matrix material. Therefore, when attempting to design a highly damped structure, a good proportion of ±45° plies is important.

SMART STRUCTURES

Due to the fabrication process, composite materials can easily allow the imbedding of other materials that make the structure inform or react in certain situations. With the imbedded devices the structure becomes "smart." Essentially, the smart structure can sense threats, heal itself, or adapt in form and/or function.

The application of smart structures falls into the following three categories:

1. Health monitoring of the composite structure.
   a. An imbedded sensor can identify local damage, whether it is caused by overload, extreme temperature, or removed material.
   b. Environmental degradation changes to mechanical properties can be identified.
   c. Changes to the physical properties through the absorption of fluids and gases can be identified by imbedded sensors.
2. Adaptive structures.
   a. With the use of imbedded actuators, the structural shape can be changed by external signals or reactive feedback.
   b. The various operation signatures (radar, infrared, visual, etc.) can be modified by the imbedded sensor to change or reduce the signature of the structure.
   c. The inclusion of electronics in the structural skin of a system will improve operational performance (such as conformal antennas).
3. Life cycle improvement.
   a. Active imbedded sensors can eliminate vibration through damping and thus extend the fatigue life of a structure.
   b. The smart repair patch will indicate if the patch adhesion has degraded thus allowing the repair to be changed.
   c. Matrix cracks self-heal—imbedded capsules break and fill the cracks in the resin material or re-bond the fiber and resin.

(A detailed discussion of smart structures can be found in Baker, Dutton, and Kelly 2004. Additional discussion of smart structures is presented in Chapter 21 where the damage prevention and detection aspects of smart structures are featured.)

FATIGUE

The fatigue damage associated with composite materials is either a cracking of the fibers, cracking at the fiber/resin interface (hereafter called interfacial cracking) seen as fiber pull-out, or cracking of the resin system (referred hereafter as matrix cracking). For any given material, these failure conditions are typically associated with the number of fatigue cycles. Low cycle fatigue (high stress levels) typically results in fiber fracture and interfacial cracking. High cycle fatigue (low stress levels) will more commonly result in matrix cracking. Figure 11-14 illustrates low and high cycle fatigue characteristics of composite materials.

Low cycle fatigue is dominated by the matrix stress levels in composite materials.

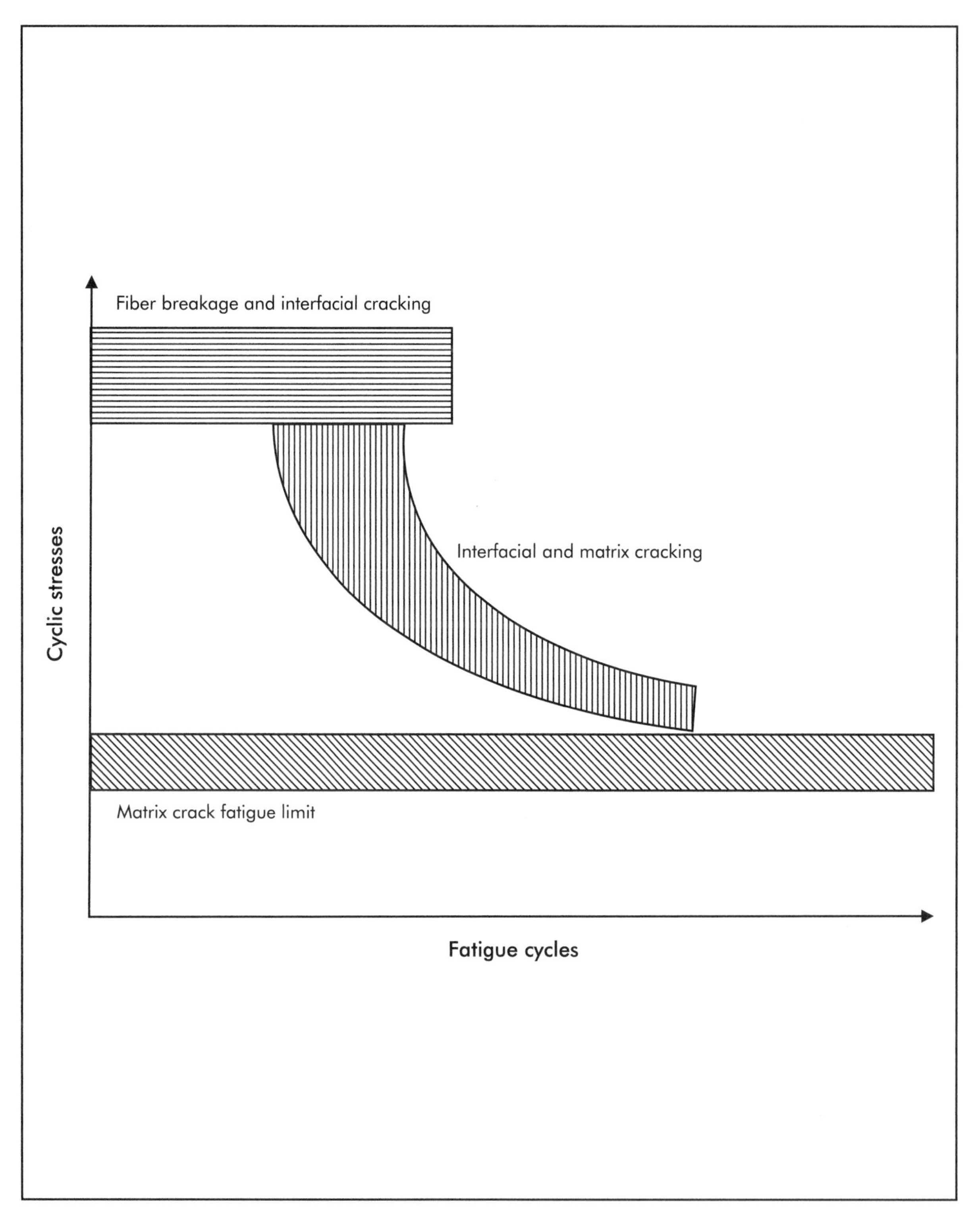

Figure 11-14. Low and high cycle fatigue characteristics of composite materials.

Thus fatigue crack inspection focuses on looking for small micro-cracking in the resin. The behavior of composite fatigue cracking is different from that of metals where small cracks can be critical in size. Well-designed composite laminate can tolerate large matrix cracks and thus provide a higher level of damage tolerance than metals. Figure 11-15 highlights the differences between damage size and inspection requirements for metals and composites. Note in particular the large damage size in a composite material before it becomes critical.

A further comparison of the fatigue behavior of composites versus metals is shown in Figure 11-16. The typical S-N curves are based on specific alternating stresses (alternating stress divided by material density). Note the typically lower strength loss of composites. Of particular interest is that boron fiber composites perform better than glass fiber composites. This is further explained in Figure 11-17 where the global stresses are proportioned to the fibers and the resin (matrix) based on the ratios of their individual stiffnesses.

The form of the fiber system also impacts the fatigue cracking of the matrix. Figure 11-18 illustrates this effect. Cracking in the matrix increases when the fibers are not aligned with the loading direction. Thus unidirectional fibers perform better that woven fabrics and random fiber forms tend to have more serious fatigue-cracking problems.

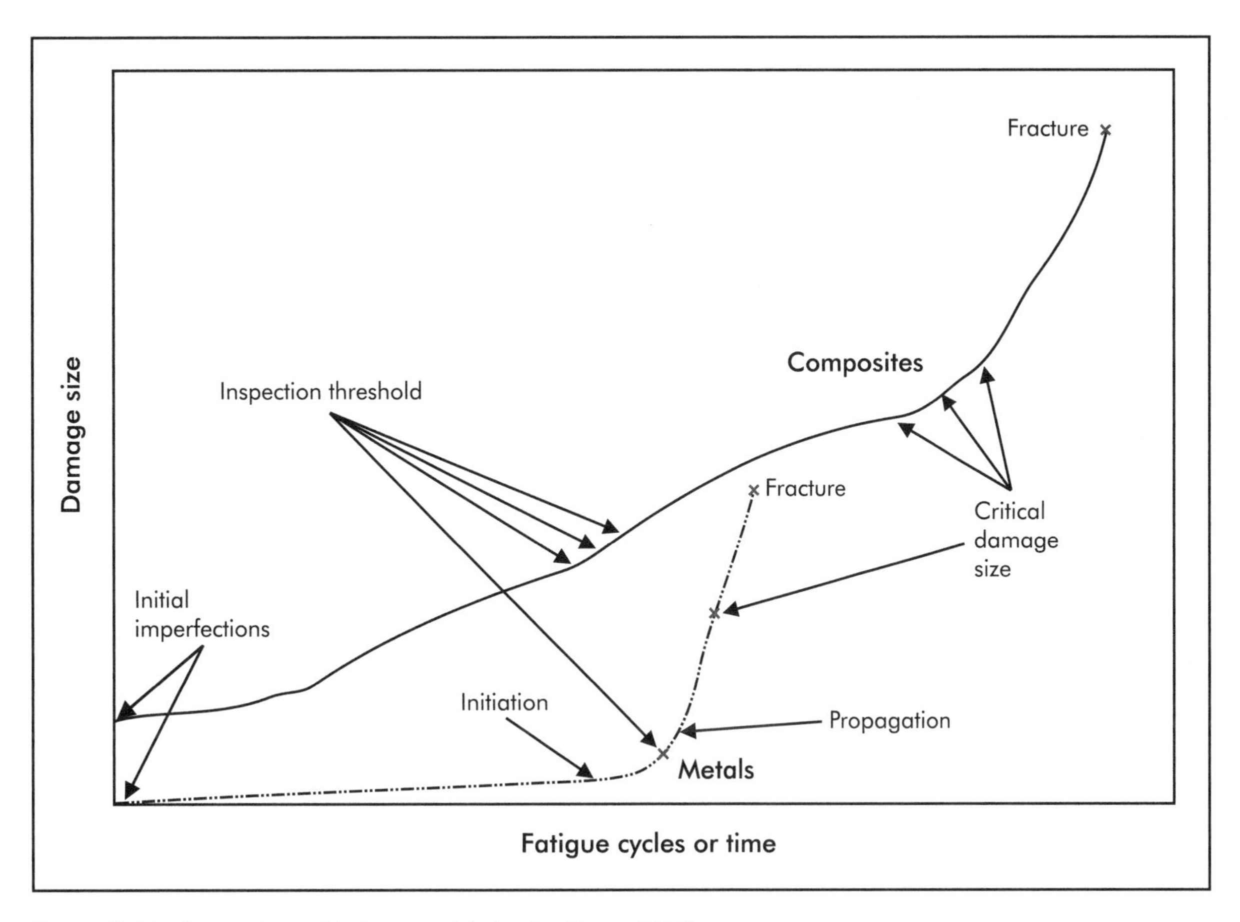

Figure 11-15. Comparison of fatigue crack behavior (Jones 1975).

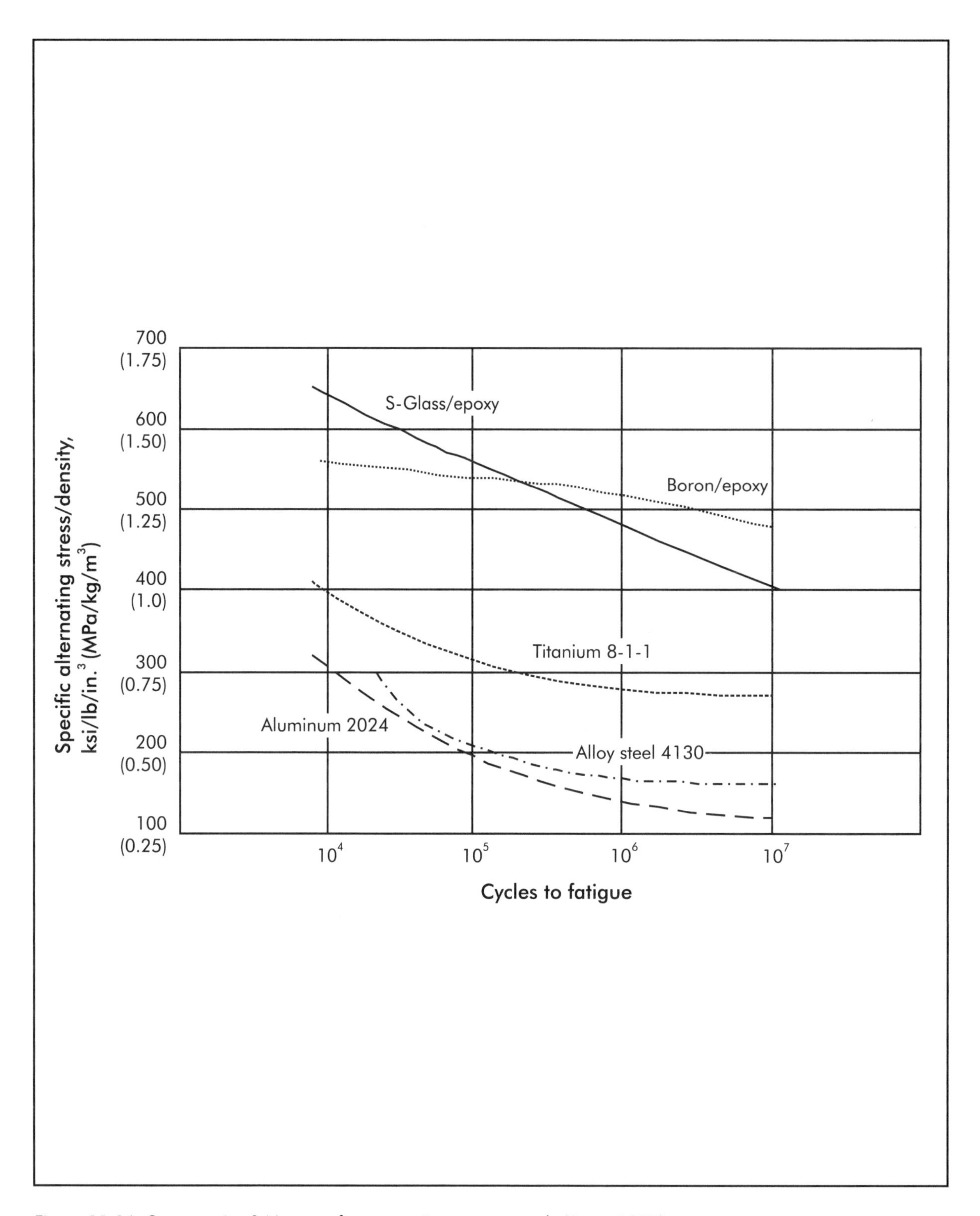

Figure 11-16. Comparative S-N curves for composites versus metals (Jones 1975).

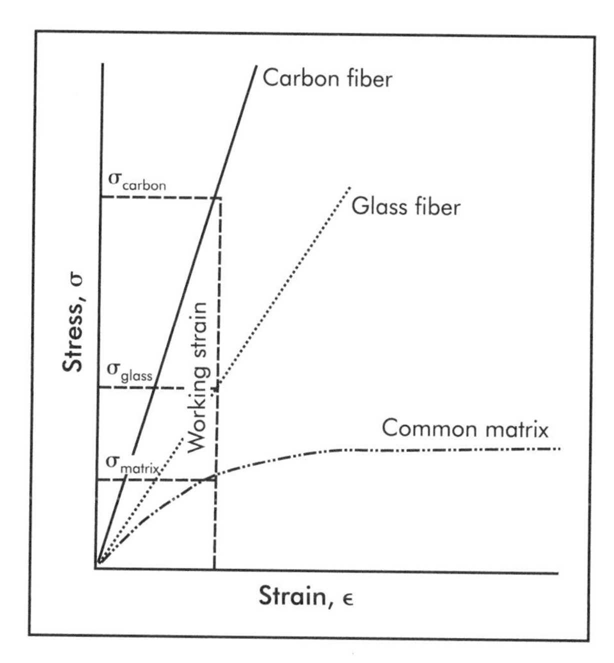

Figure 11-17. Stress distribution effects based on fiber stiffness versus matrix stiffness.

Included in this effect are the reduced fiber-volume ratios with fiber forms.

COMPOSITES VERSUS METALS

A direct comparison of composite properties to those of metals does not normally show the complete story. A better method of comparing the properties is to take advantage of the materials' densities. This shows the distinct advantage of composite materials over metals due to the lower densities of composite fiber/resin systems.

Another issue to be addressed in the comparison of composites to metals is that there is a wide range of composite material properties available with various fiber orientations. Comparative charts shown in Figures 11-19 through 11-24 plot the composite system with all plies oriented in the 0° direction and the quasi-isotropic laminate configuration. The unidirectional composite represents the upper limit of the composite properties and the quasi-isotropic laminate represents the other practical lower limit.

Figures 11-19 through 11-24 compare the properties of four typical composite systems with three common metals. The composite systems represented are a glass/epoxy, aramid (Kevlar®)/epoxy, graphite/epoxy, and boron/epoxy fiber/resin systems. Aluminum alloy 7075-T6, titanium alloy 6Al 4V, and stainless steel PH 17-4 are the common aerospace-quality metal alloys used in comparison.

Figure 11-19 indicates that there is a clear stiffness-to-weight advantage for the graphite and boron fiber/resin composite systems over metals for both the upper and lower limits of the material modulus property. Glass and Kevlar are on a nearly equal footing with metals with respect to specific tensile modulus.

In the specific shear modulus chart in Figure 11-20, composites do not do as well as the metals in comparison. The shear properties are at the lower limit with unidirectional fiber orientations and perform poorly against the metals. The quasi-isotropic laminates of graphite and boron are slightly better than metal in the shear modulus-to-weight ratio.

Figure 11-21 shows that the unidirectional specific tensile strength performance of the composite fiber/resins is clearly superior to the metals and, in particular, the aramid (Kevlar) and graphite fiber composites systems. In the quasi-isotropic laminate configuration, only glass, graphite, and boron preform better than the metals for the tensile strength-to-weight ratios, with graphite fiber systems being the best of the composites.

In Figure 11-22, there is reduced performance of glass and aramid (Kevlar) fiber systems against metals with regard to specific compression strength behavior. However, boron fiber composites have a clear advantage over the entire range of composites shown.

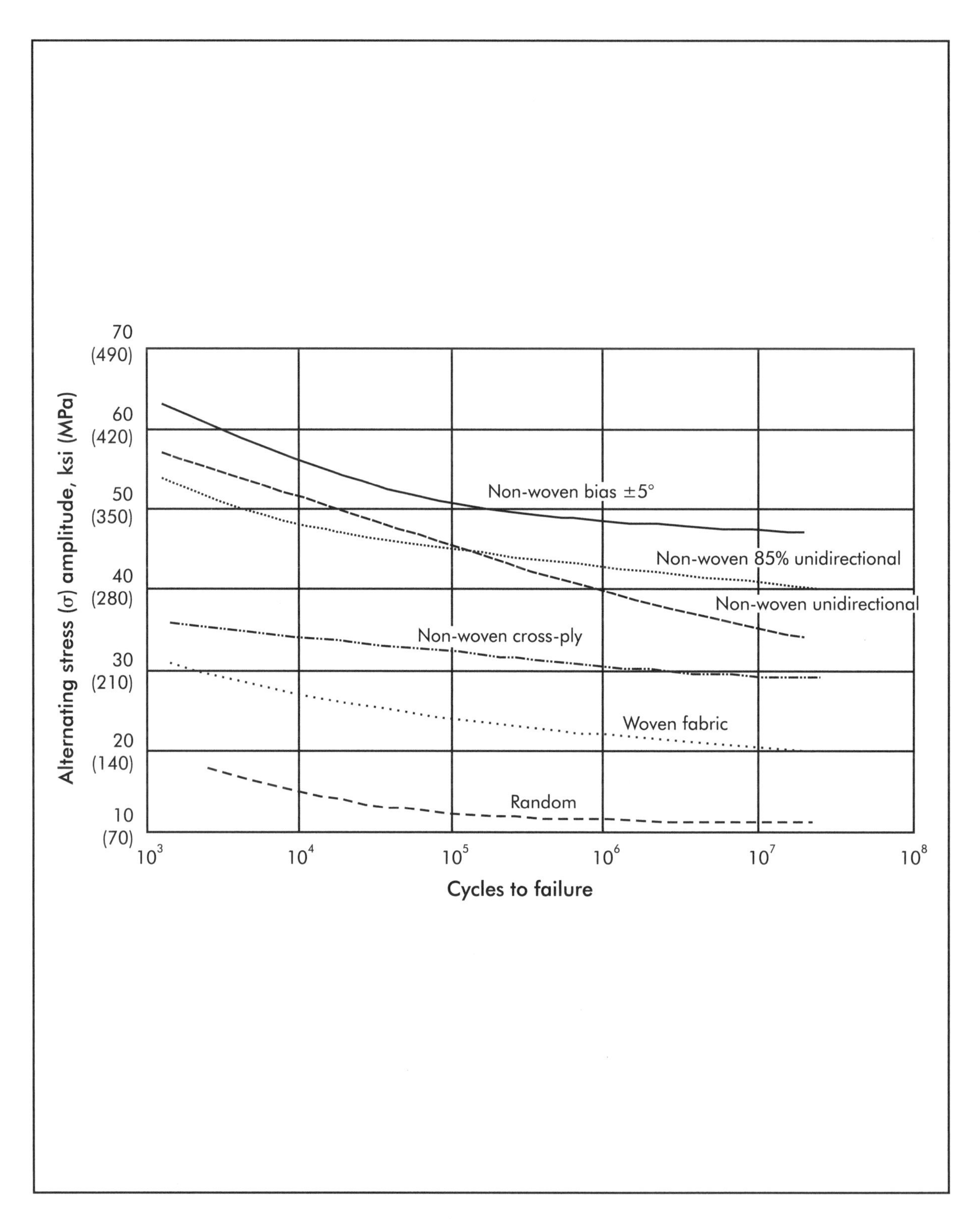

Figure 11-18. Fatigue stress behavior based on fiber form (Agarwal and Broutman 1990).

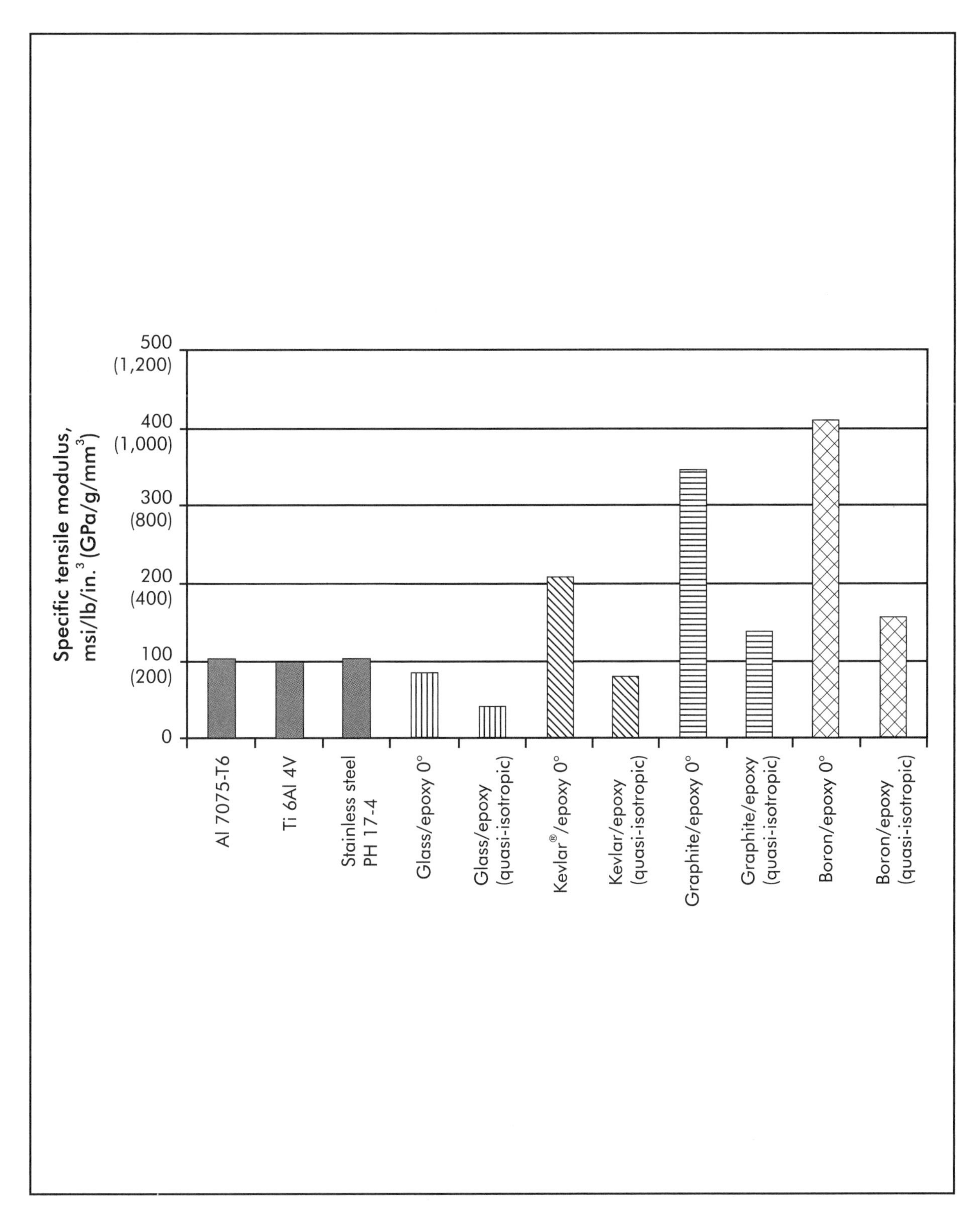

Figure 11-19. Comparison of the specific tensile modulus of four composites versus three metals.

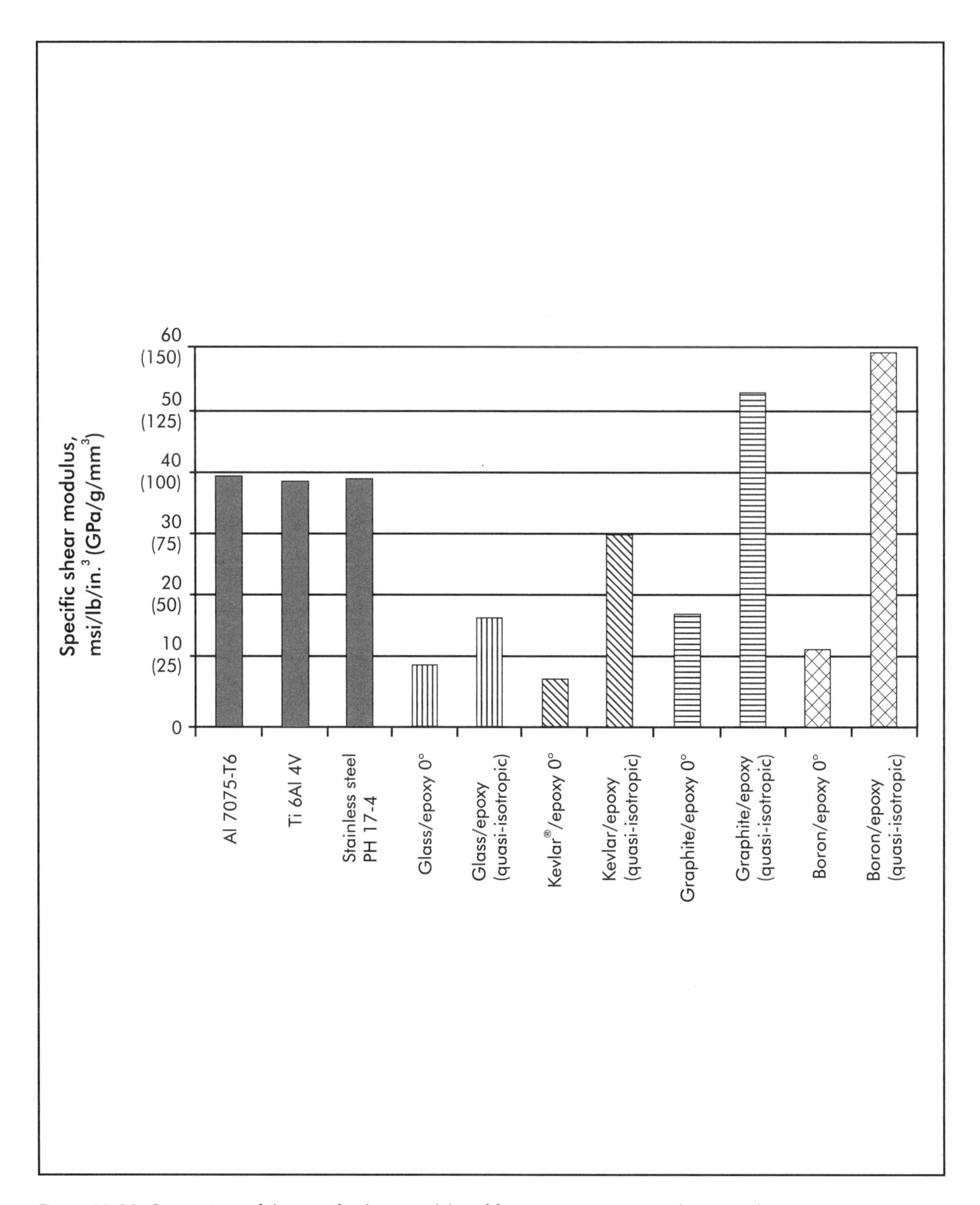

Figure 11-20. Comparison of the specific shear modulus of four composites versus three metals.

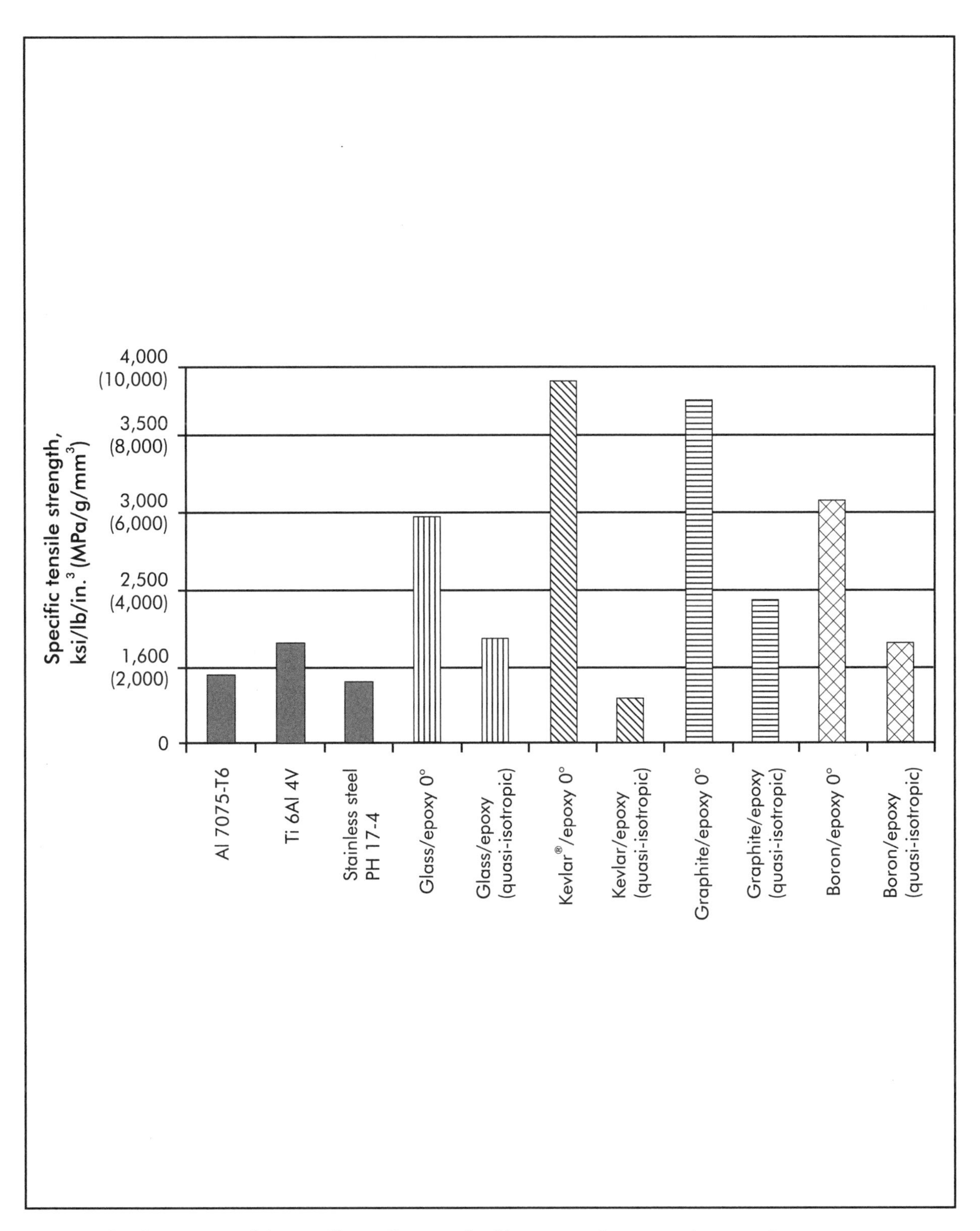

Figure 11-21. Comparison of the specific tensile strength of four composites versus three metals.

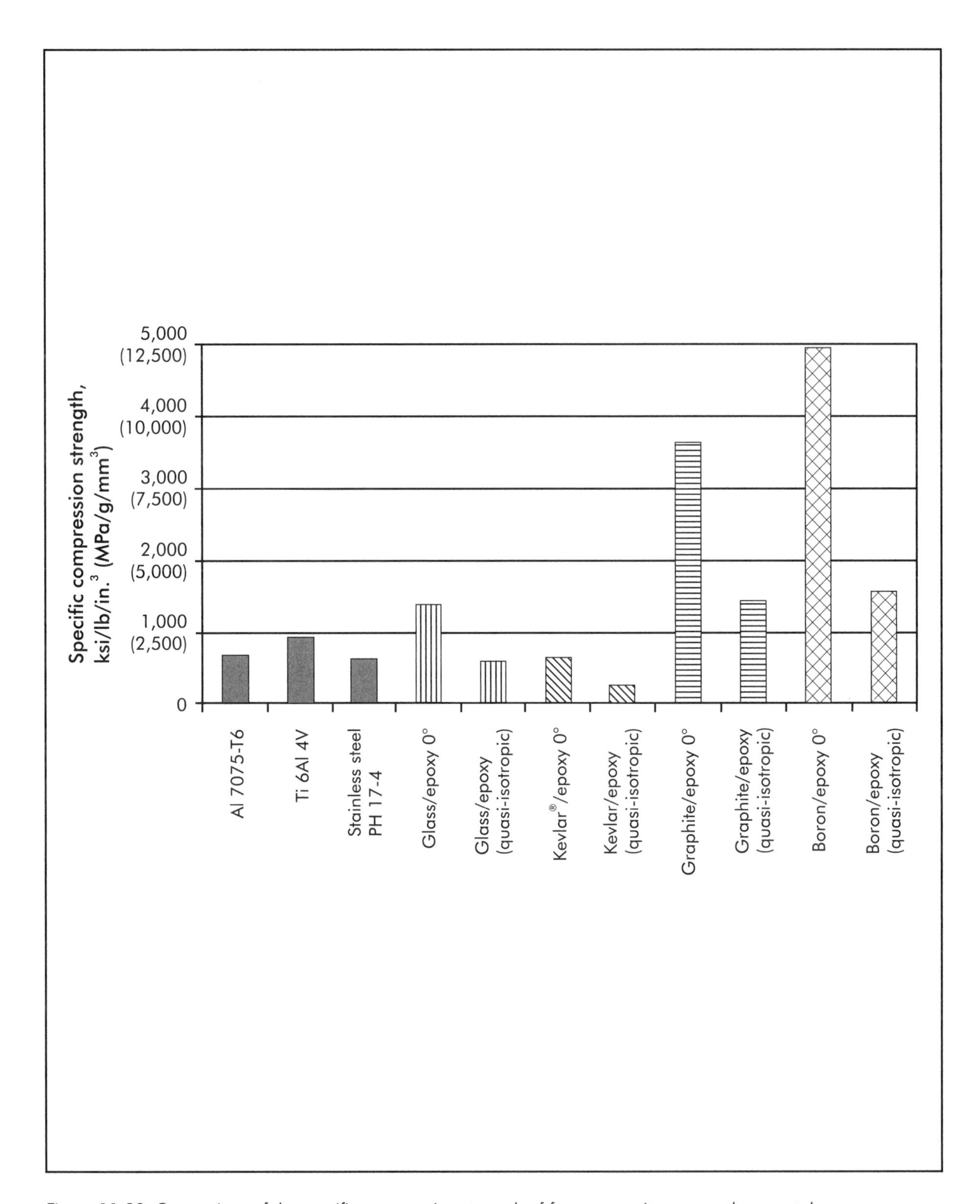

Figure 11-22. Comparison of the specific compression strength of four composites versus three metals.

The specific shear strength performance property of most composite laminates is poor between the unidirectional and quasi-isotropic configurations. Shear strength is only improved with more ±45° plies, with the pure ±45° angled ply configuration giving the best performance. Only graphite and boron quasi-isotropic configurations show an on-par performance with metals as shown in Figure 11-23.

While the metals typically have a higher coefficient of thermal expansion (CTE) value than most of the composite systems as shown in Figure 11-24, the extreme difference between the aluminum alloy and the composite systems should be noted. The major issue that arises with the differences between the CTEs of metals and composites is the residual stresses that arise in joint regions, particularly in heat-cured adhesive joints.

RESIDUAL STRESSES

Residual stresses in composite structures are observed at two levels. The individual ply can have residual stresses between the fibers and the resin (matrix), and there can be residual stresses between adjacent plies in the laminate. In both cases, the residual stresses typically result in some form of warping of the composite structure. Residual stresses originate from thermal variation between the cure and operating temperatures of the laminate, the differences between the coefficients of thermal expansion (CTE) of the fibers and resin, and each ply orientation. Moisture absorption also can lead to

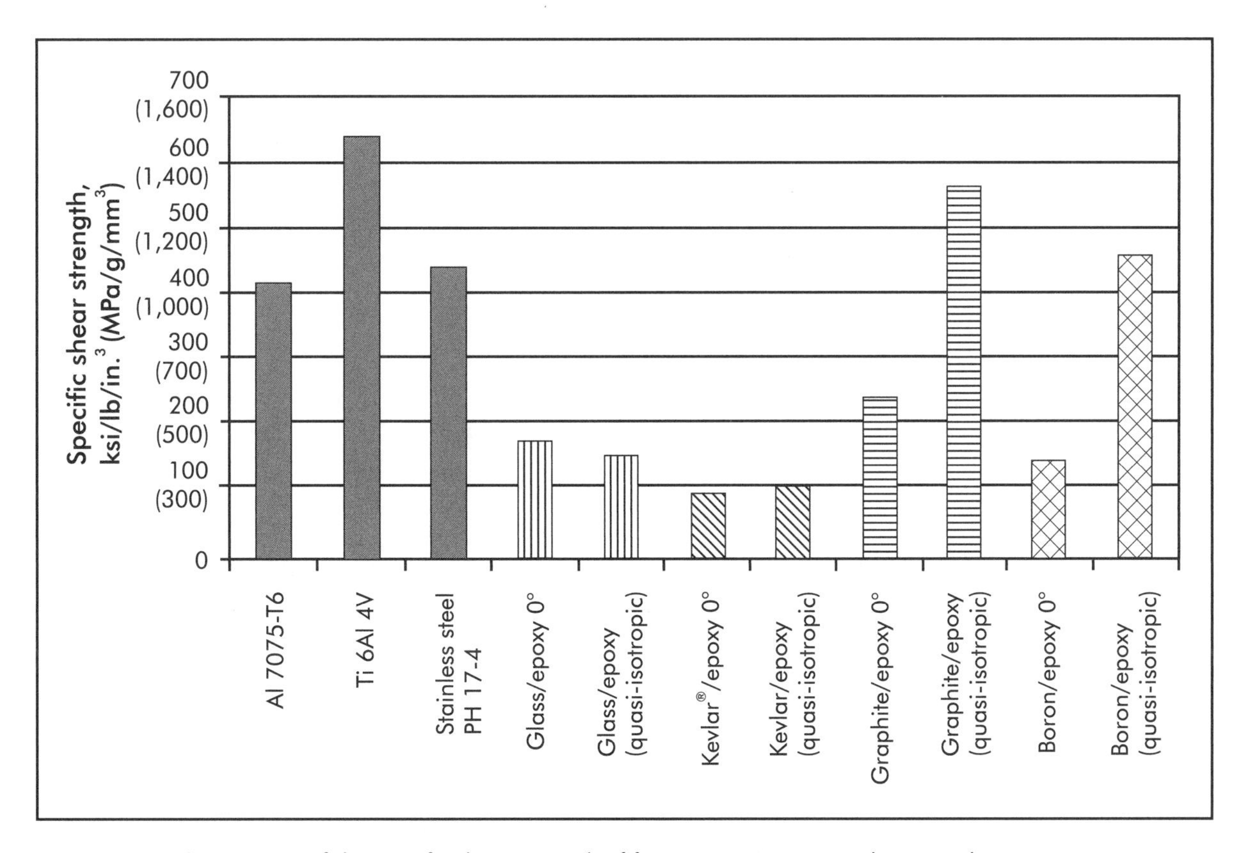

Figure 11-23. Comparison of the specific shear strength of four composites versus three metals.

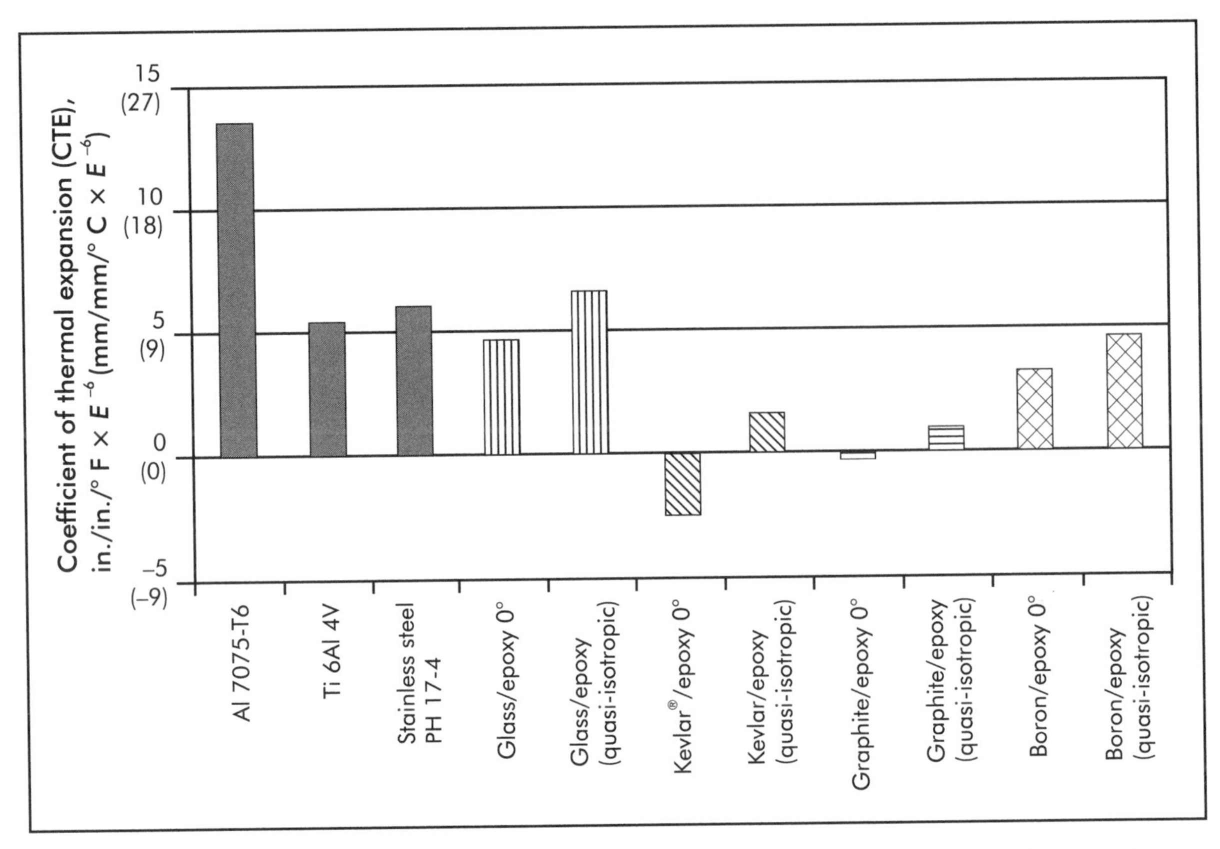

Figure 11-24. Comparison of the coefficients of thermal expansion (CTE) of four composites versus three metals.

expansion variations in the resin and thus residual stresses in the laminate. On some occasions the fibers are pre-stressed prior to infiltration and curing of the resin; this induces residual stresses between the resin and the fiber at the interface.

Residual Thermal Stresses

Residual thermal stresses arise because of the variation between the CTEs of the fibers and the resin, and the difference between the cure temperature and the operating temperature referred to as ΔT. Initially, the selection of the fiber/resin combination and the curing temperature needs to be considered so that thermal residual stresses do not cause fiber/resin interfacial shear failures. In a unidirectional composite ply, the fiber/resin CTE difference will produce in-plane direction variations of the ply's CTE. For example, looking at four typical unidirectional fiber/resin systems, the longitudinal (fiber direction) CTE is often much smaller than the transverse direction as shown in Table 11-3.

Large variation in the orthogonal plies' CTE values can have a significant effect on the laminate's post-cure residual deformation. When a laminate does not have mid-plane symmetry, room-temperature deformation of the laminate can be extreme, especially for thin laminates (fewer than 12 plies). As an example of this CTE variation, Table 11-4 gives the post-cure deformation results of a series of four-ply laminates made

Table 11-3. Common fiber/resin CTE properties (Tsai 1988).

Type	Boron Fiber Reinforced Plastic (BFRP)	Carbon Fiber Reinforced Plastic (CFRP)	Glass Fiber Reinforced Plastic (GFRP)	Kevlar® Fiber Reinforced Plastic (KFRP)
Fiber	Boron (4)	Graphite	E-Glass	Kevlar 49
Matrix	Epoxy	Epoxy	Epoxy	Epoxy
Longitudinal CTE per ° C × 10^{-6} (per ° F × 10^{-6})	6.08 3.38	–0.31 –0.17	8.60 4.78	–4.00 –2.22
Transverse CTE per ° C × 10^{-6} (per ° F × 10^{-6})	30.3 16.8	27.2 15.6	22.1 12.3	79.0 43.9
Resin/fiber CTE % difference	80%	101%	61%	105%

Table 11-4. Common fiber/resin system CTE properties.

Configuration	[0/0/0/0]	[0/90/90/0]	[0/90/0/90]	[0/0/90/90]	[0/0/90/0]	[0/0/0/90]
Symmetric	Yes	Yes	No	No	No	No
Post-cured shape	Flat	Flat	Slightly warped	Warped (oil canning)	Significantly warped	Grossly warped

from carbon/epoxy unidirectional prepreg tape cured at 350° F (177° C).

Though the symmetric panels are flat, they still have residual stresses within them. However, the residual stresses are balanced in the symmetric lay-up. If the laminate 0/90/90/0 is split along the mid-plane, the two panels are no longer symmetric and will be warped. Likewise, if the laminate 0/0/90/90 is split along the mid-plane, the two panels are individually symmetric and thus are flat. Figure 11-25 explains this behavior. Thermal stress behavior has a critical impact on repair using pre-cured or co-bonded repair patches.

CASE STUDY 11-1

Torque Tube

A torque tube is to be designed from graphite/epoxy composite materials at minimal weight. It is to be a maximum of 3 in. (76 mm) in diameter and 4 ft (1.2 m) in total length between the end constraints. The tube is centrally loaded with 1,350 in.-lb (152 Nm) ultimate design torque, as shown in Figure 11-26. The torsional deflection of the tube at design load is to be no greater than 1°.

The next task is to determine the following: tube wall thickness and ply configuration.

As this design project is a prototype, one of the simplest and cost-effective manufacturing processes available to manufacture it is by hand lay-up on a deflatable tubular mold.

As a general rule, the allowable shear strain for the design is restricted to 2,000 μstrain, which reduces the chance of first-ply

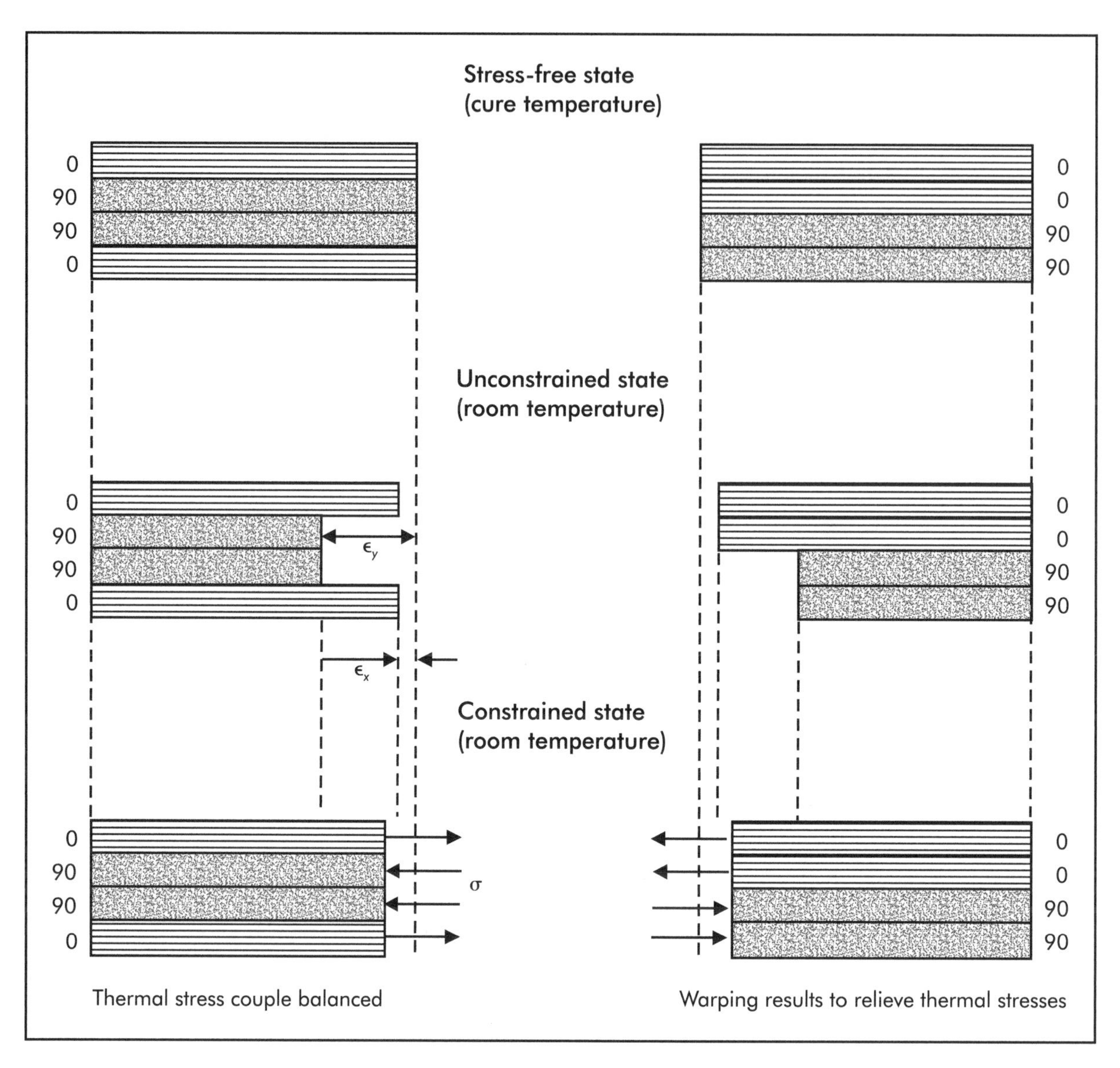

Figure 11-25. Effect of thermal stresses in symmetric and asymmetric laminates.

failure in the tube (intralaminar matrix cracking). The basic design equations are:

Maximum shear stress:

$$\tau_{allowable} = \frac{TR}{J} \leq G\varepsilon_{2,000\mu} \quad (11\text{-}23)$$

where:

$\tau_{allowable}$ = maximum allowable shear stress

T = ultimate design torque = 1,350 in.-lb (152 Nm)

R = tube radius = 1.5 in. (38 mm)

J = polar moment of area (see Eq. 11-24)

G = shear modulus

ε = strain = 2,000μ

Polar moment of area:

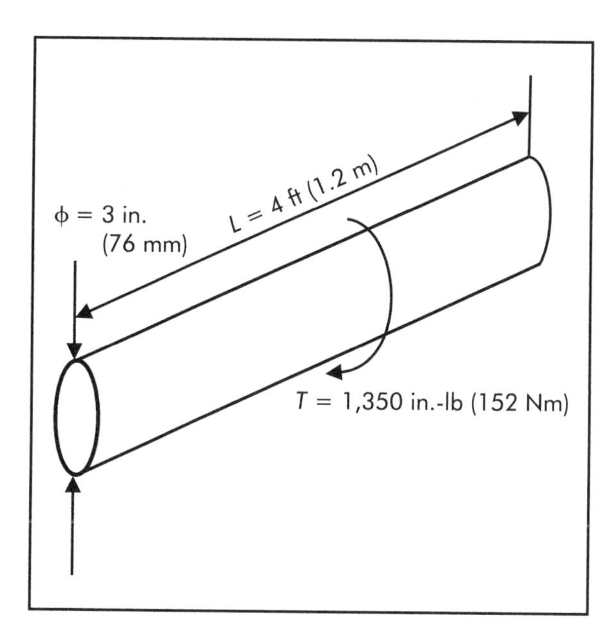

Figure 11-26. Torque tube geometric description.

$$J = \frac{\pi}{2}\left[R^4 - (R-t)^4\right] \quad (11\text{-}24)$$

where:

π = 3.14159, which is the ratio of a circle's circumference to its diameter

t = tube wall thickness

Angle of twist:

$$\phi = \frac{TL}{GJ} \le 1° \quad (11\text{-}25)$$

where:

ϕ = angle of twist

L = 4 ft = 48 in. (1.2 m)

So,

$$GJ = \frac{TL}{\phi} \le \frac{1{,}350x48}{1x(\pi/180)} = 3.713x10^6 \text{ lb.-in.}^2$$

(10.5 $KN\text{-}m^2$)

(11-26)

where:

x = orthogonal direction of plies

A major step in the design of the tube is to estimate the shear modulus, G. This can be done by first looking at the range of shear stiffness values for a given composite material. For a typical graphite/epoxy, unidirectional, prepreg composite material:

G_0 = 1.22 msi (8.91 GPa), the lowest value of shear stiffness for a 0° ply

$G_{\pm 45}$ = 7.55 msi (52.1 GPa), which is the highest value of shear stiffness for a ±45° ply

G_{QI} = 4.39 msi (33.3 GPa), which is the quasi-isotropic value of shear stiffness

A relatively large value for shear modulus is required for this design, but it is not good to have all 45° plies in the configuration as there are often axial loads that will need 0° plies for added strength. So the value of the shear modulus as a starting point is:

G = 6.0 msi (41.4 GPa)

The unidirectional axial Young's modulus is E_0 = 29.46 msi (203 GPa), which gives the ratio with the shear stiffness of the design:

$$\frac{G}{E_0} = \frac{6.0 \text{ msi}}{29.46 \text{ msi}} = 0.204 \quad (11\text{-}27)$$

Figure 11-27 reveals the percentage of ±45° plies is 76%.

From Eq. 11-24, the effective tube thickness is determined:

$$J = \frac{\pi}{2}\left[R^4 - (R-t)^4\right] = \frac{3.713x10^6}{G} = .619 \text{ in.}^4$$

($257{,}647 \text{ mm}^4$)

(11-28)

Therefore, with R = 1.5 in. (38 mm), then t = .03 in. (0.8 mm).

Assuming that a ply's thickness is .005 in. (0.13 mm), only six plies are required to achieve the design stiffness goal. Since 76% of the laminate is to contain ±45-degree

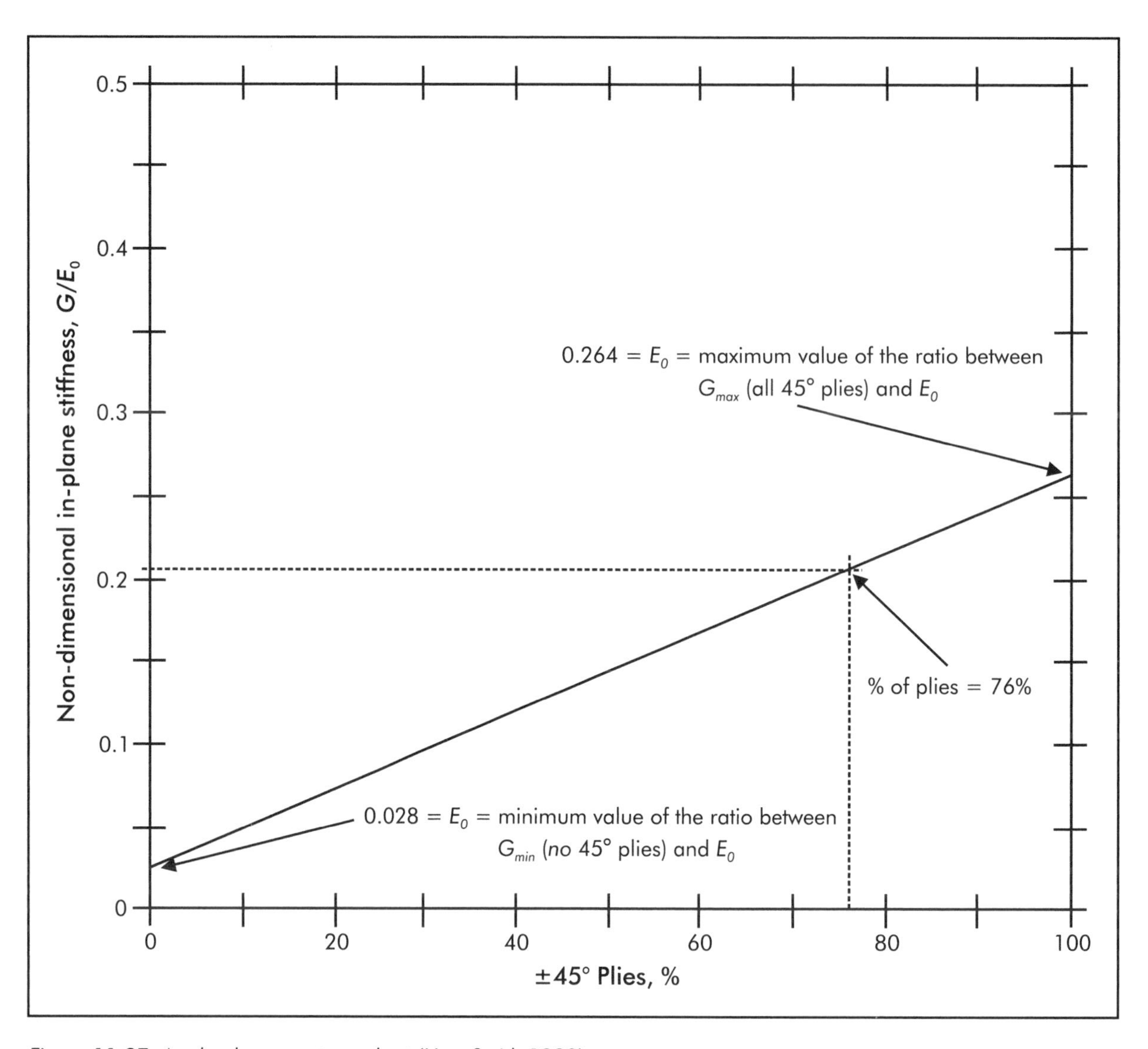

Figure 11-27. Angle ply percentage chart (Hart-Smith 1989).

plies, this means five of the six plies are required. However, as the angled plies need to be balanced, six plies will be used. As a general design rule, there also should be at least one ply in every direction of the [0/90/±45] combinations. The final laminate lay-up could be:

$$[\pm 45/0/\pm 45/90/\pm 45]_T$$

This ply configuration is based on the following reasoning.

- The ±45° plies on the outer surface will maximize the shear stiffness.
- The interspersion of the plies will minimize interlaminar stresses and reduce the risk of delaminations.

A detailed analysis of the laminate configuration produces the following results.

In-plane properties:

E_1 = 8.33 msi (57 GPa)
G_{12} = 5.97 msi (41 GPa)

$\nu_{21} = 0.4848$

$\alpha_1 = 0.8866 \times 10^{-6}/° F$ (0.4926/° C)

Flexural/torsion properties:

$E_1 = 9.13$ msi (63 GPa)

$G_{12} = 6.07$ msi (42 GPa)

Torsional strength at 2,000 µstrain:

$\tau_{12} = 11.94$ ksi (82 MPa)

$\tau_{ult} = 26.15$ ksi (180 MPa)

The torsional shear stress of 3.38 ksi (23 MPa) meets the design goals. Next, the results are checked against the design equations.

Geometry:

$R = 1.5$ in. (38 mm)

$t = 0.005 \times 8$ plies = 0.04 in. (1.02 mm)

$A = 0.372$ in.^{2} (240 mm^2)

$J = 0.8149$ in.^{4} (339×10^3 mm^4)

Shear stresses:

$$\tau_{applied} = \frac{TR}{J} = \frac{1{,}350x1.5}{0.8149} = 2.48 \text{ ksi (17 MPa)} \quad (11\text{-}29)$$

Margin of safety is:

$$\left(\frac{11.94}{2.48} - 1\right)\% = 380\% \quad (11\text{-}30)$$

Twist:

$$\phi = \frac{TL}{GJ} = \frac{1{,}350x48}{6.07x10^6 x0.8149} = 0.75° \leq 1° \quad (11\text{-}31)$$

Hence, for strength and performance the design is successful.

Other issues to consider further are:

- end fixtures and appropriate joint analysis,
- load application fitting and appropriate joint analysis,
- areas of local reinforcement,
- damage tolerance, and
- long-term manufacturing approach.

SUMMARY

The complex nature of composites makes designing composites parts complicated. Therefore, it is important to follow a general plan for design. The plan presented in this chapter has proven to be successful. It first requires, as do most other design systems, the establishment of the functional specifications of the part. Then, various steps in design and analysis are carried out.

Design analysis can be difficult, especially if there is little symmetry to the part. However, some assistance can be obtained by estimating the properties of the composite from the rule of mixtures. The assumptions of the rule of mixtures are reasonably applicable to composite structures and give good approximations of properties. The assumptions are as follows.

- The composite ply is macroscopically homogeneous and linearly elastic.
- The fibers are linearly elastic and homogeneous.
- The matrix is linearly elastic and homogeneous.
- Both the fiber and matrix are free of voids.
- The interface between the fibers and matrix is completely bonded, and there is no transitional region between the matrix and reinforcement that might add a complication of a third material.
- The mechanical properties of the individual constituents are the same whether they are made by themselves or made up within the composite.

With this in mind, the values of the stiffness, E_{11} (longitudinal modulus), υ_{12} (principal Poisson's ratio), and α_{12} (principal thermal expansion coefficient) can be expressed in terms of the matrix/fiber properties themselves and the volume fraction of the respective ingredients. These expressions are derived from the rule of mixtures as follows:

$$E_{11} = V_f E_f + V_m E_m \quad (11\text{-}32)$$

$$\upsilon_{21} = V_f \upsilon_f + V_m \upsilon_m \quad (11\text{-}33)$$

$$\alpha_{12} = V_f \alpha_f + V_m \alpha_m \quad (11\text{-}34)$$

where:

E_{11} = principal longitudinal modulus for composite ply
E_f = Young's modulus for the fibers
E_m = Young's modulus for the matrix
υ_{21} = major Poisson's ratio for composite ply
V_f = fiber-volume ratio
υ_f = Poisson's ratio for the fibers
V_m = matrix volume ratio
υ_m = Poisson's ratio for the matrix
α_{12} = principal coefficient of thermal expansion for composite ply
α_f = coefficient of thermal expansion for fibers
α_m = coefficient of thermal expansion for matrix

The ratio of the load carried by the fibers compared to the load carried by the entire composite structure can be found from the rule of mixtures. This load ratio is enlightening as it shows how the fibers dominate the load carrying within the composite structure. Figure 11-28 illustrates that when the modulus of the fibers, E_f, is much greater than the modulus of the matrix, E_m, so that the ratio of the moduli is a number over 10, and especially when the fiber volume fraction, V_f, is high, the fibers carry almost all of the load. For example, referring to Figure 11-28, when the ratio of the moduli is 10 and the fiber volume fraction is 0.6, the fraction of the load carried by the fibers is over 90%. This unequal distribution of load is a tremendous advantage in composites and is the reason that composites have such high specific strengths and specific stiffnesses compared to metals.

The unique nature of composites can be an advantage in damping and vibration applications. These advantages also relate to the long-term fatigue advantages of composites.

The manufacturing methods for composites allow the creation of smart structures that are able to monitor and even self-heal some cracks.

LABORATORY EXPERIMENT 11-1

Fiber-volume Fraction

Objective: Discover the effect of varying the amount of fiber in a composite.

Procedure:

1. Select a fiberglass roving for which the density of the material can be obtained from the manufacturer. (Alternately, the density can be determined experimentally.)
2. Select a resin for which the density is known.
3. Using the densities, calculate the weights of the fiberglass and the resin that will result in two different fiber-volume fractions—40% and 60%.
4. Weigh the fiber and resin components for making the composite panels (of 40% and 60% fiber-volume fraction).
5. Make the two panels. Typically, they can be made most easily using the hand lay-up technique. The fibers should be carefully laid up so they are all unidirectional. Be sure to wet the fibers thoroughly.

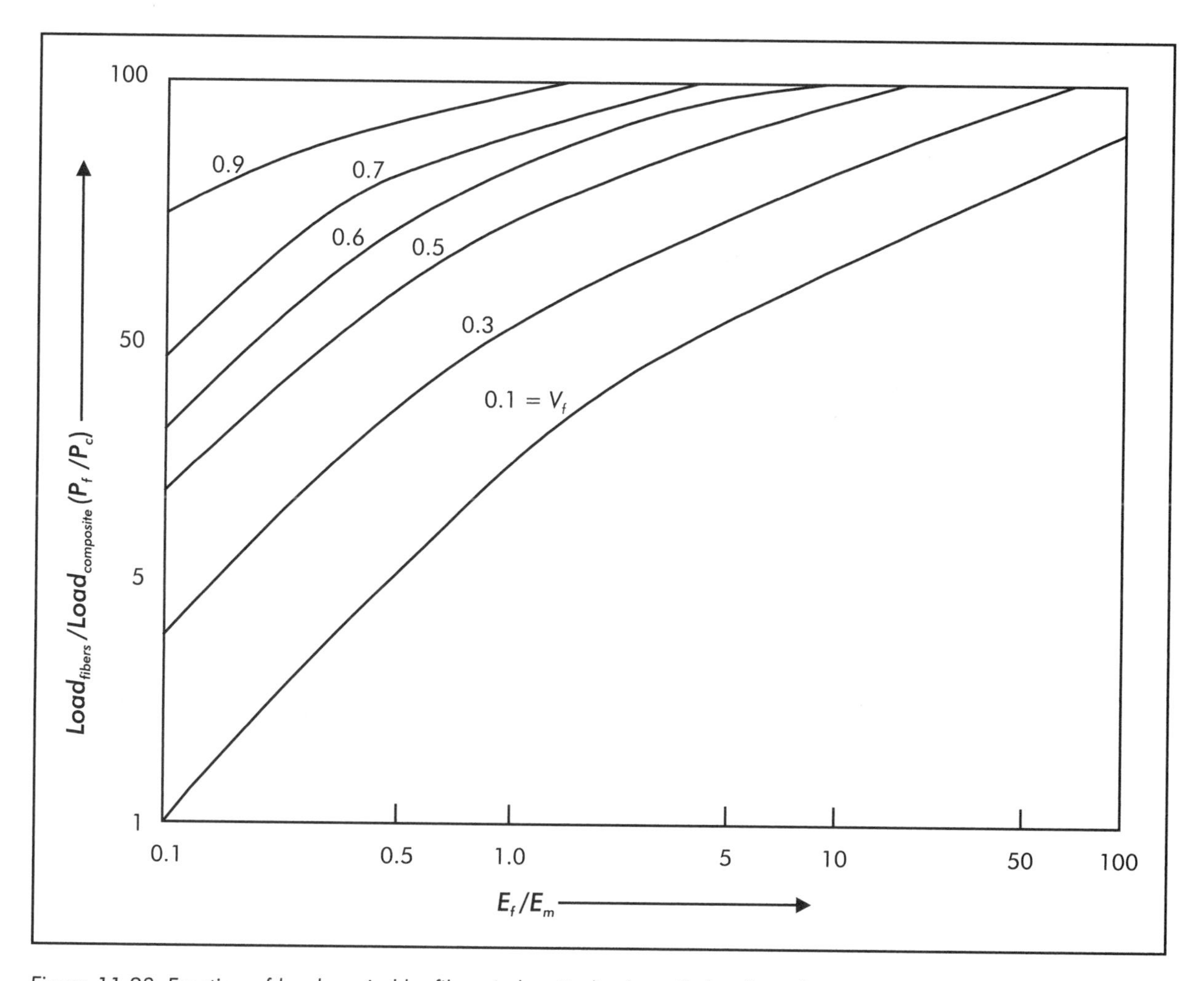

Figure 11-28. Fraction of load carried by fibers in longitudinal tensile loading of a continuous-parallel-fiber laminate.

6. After fully curing the panels, test them for tensile properties. Make a note as to which of the panels has the higher tensile strength and stiffness.

QUESTIONS

1. Describe the effects of the fiber volume in composite parts.
2. Describe the effects of fiber orientation in composite parts.
3. Discuss the differences between isotropic, anisotropic, and homogeneous materials.
4. Define the following terms:
   a. unidirectional
   b. specific tensile strength
   c. specific tensile stiffness
   d. Poisson's ratio
5. Why would you conduct accelerated aging tests on materials and components and not just wait to run real-time aging tests on the final product?
6. What techniques can be used to reduce the creep in a composite material with a polymeric matrix?

7. Calculate the values of E_{11}, υ_{12}, and α_{11} using the rule-of-mixtures relationships for a unidirectional, graphite/epoxy composite with the following constituent properties and 65% volume loading of fiber:

   $E_f = 43 \times 10^6$ psi (296 GPa)
   $E_m = 0.5 \times 10^6$ psi (3.4 GPa)
   $\upsilon_f = 0.2$
   $\upsilon_m = 0.4$
   $\alpha_f = 1.5 \times 10^{-6}/°$ F ($2.7 \times 10^{-6}/°$ C)
   $\alpha_m = 40 \times 10^{-6}/°$ F ($72 \times 10^{-6}/°$ C)

8. Draw a diagram showing the layers for the following composites and discuss whether they are balanced and symmetrical:

   a. $[45_2/{-45_2}/0]$
   b. $[0/45/90]_s$
   c. $[\pm 45, 90]_s$

BIBLIOGRAPHY

Agarwal, B. D. and Broutman, L. J. 1990. *Analysis and Performance of Fiber Composites*, 2nd edition. NY: John Wiley & Sons, Inc.

American Society for Testing and Materials. 1990. ASTM Standards and Literature References for Composite Materials. Philadelphia, PA: American Society for Testing and Materials (ASTM).

Baker, A., Dutton, S., and Kelly, D. 2004. *Composite Materials for Aircraft Structures*, 2nd Edition. Reston, VA: American Institute of Aeronautics & Astronautics (AIAA).

Davis, M. J. 1987. "Mechanics of Composite Materials." International Conference and Workshop on Composites in Manufacturing. Melbourne, Australia: April.

Donaldson, D., ed. 2001. *ASM Handbook, Composites,* Vol. 21. Materials Park, OH: ASM International.

Gibson, R. F. 1994. *Principles of Composite Material Mechanics*. NY: McGraw-Hill.

Hart-Smith, L. J. 1989. "A New Approach to Fibrous Composite Laminate Strength Prediction." Douglas Paper 8366. Palm Beach, FL: MIL_HDBK-17 Committee Meeting, Singer Island.

Heslehurst, R. B. July 2006. "Composite Structures Engineering Design vs. Fabrication Requirements." Invited paper. ACUN-5: International Composites Conference. Developments in Composites: Advanced, Infrastructural, Natural, and Nano-composites. Sydney, Australia: UNSW, July 11–14.

Jones, R. M. 1975. *Mechanics of Composite Materials*. Washington, DC: Scripta Book Company.

Reddy J. N. & Miravete, A. 1995. *Practical Analysis of Composite Laminates*. NY: CRC Press.

Society of Manufacturing Engineers (SME). 2005. "Composite Materials" DVD from the Composites Manufacturing video series. Dearborn, MI: Society of Manufacturing Engineers, www.sme.org/cmvs.

Strong, A. Brent. 2006. *Plastics: Materials and Processing*, 3rd Ed. Upper Saddle River, NJ: Prentice Hall, Inc.

Tsai, S. W. and Hahn, H. T. 1980. *Introduction to Composite Materials*. Lancaster, PA: Technomic Publishing Co.

Tsai, S. W. 1988. *Composite Design*, 4th Ed. Dayton, OH: Think Composites.

Whitney, J. M. 1987. *Structural Analysis of Laminated Anisotropic Plates*. Lancaster, PA: Technomic Publishing Co.

12

Sandwich Structures, Joints, and Post-processing Operations

CHAPTER OVERVIEW

This chapter examines the following concepts:

- Sandwich structures—concept and design
- Core material
- Other z-direction stiffeners
- Joints
- Post-processing operations

SANDWICH STRUCTURES—CONCEPT AND DESIGN

Up to this point, this book has focused on laminate structures (such as that shown in Figure 11-3). Composite structures are composed of a series of sheets or plies that have a particular fiber orientation. Some have unidirectional fibers and some, like cloth, can have two directions of fiber orientation. Still other materials, like mat, can have fibers oriented in many directions but all of them are in a plane because the mat is, itself, planar. All the fibers are held in their orientation by a resin matrix.

The use of flat sheets to make composites is logical because, largely, composites are manufactured by using various methods to arrange fiber reinforcements in flat sheets that are placed into molds. The composites industry calls the plane in which these fibers are laid the x-y plane (also referred to as the 1-2 plane in Figure 11-3 and all of Chapter 11). If the force that is to be resisted is to act along the x-y plane, the fibers can be oriented in the direction of the force to increase the strength of the composite. If one layer or ply of composite is not sufficient to give the strength or stiffness desired, more layers of composite can be bonded to those already present and the total strengths of the fibers act together in the x-y plane. So, by increasing the number of layers and by orienting the fibers in the directions of the forces, forces in the x-y plane can be resisted.

But what if the force applied to the composite is in the direction perpendicular to the x-y plane? (In Figure 11-3, this direction is designated as 3 but in other places it is called the z direction.) In laminate structures, if the force is perpendicular to the direction of the fibers, the fibers have little ability to resist it. In essence, the z-direction strength and stiffness of the laminar composite is determined by the strength and stiffness of the matrix. The advantages of a composite structure (that is, the synergy between fiber and matrix) largely disappear in the z-direction.

The most common method of addressing this z-direction problem is by making a **sandwich structure** such as that shown in Figure 12-1. It consists of two face sheets, a core, and adhesive materials. The face sheets (top and bottom of the structure) are traditional composite laminates. The nature of the core materials will be discussed later in this chapter. The adhesives (also discussed later in this chapter) are present to hold the structure together.

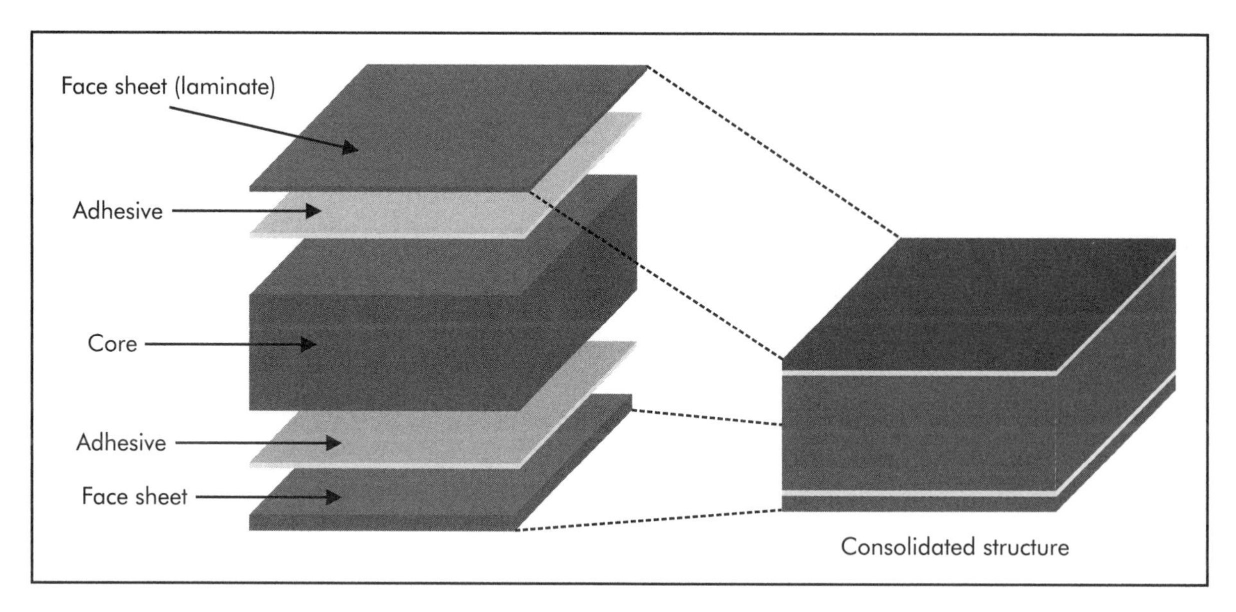

Figure 12-1. Components of a sandwich structure.

The term **double composite structure** describes the concept of a sandwich structure. The double composite's nature is simply that the face sheets are composites and then the entire sandwich structure is also a composite. Both consist of disparate materials joined together to create something that has better properties together than the components would have separately. In other words, it is a composite structure overall and contains composite components within it.

Thinking of these sandwich structures as double composite structures, it is clear that if the structure is to work properly, all the components must work together. On the other hand, the overall composite structure might have some of the same critical design and manufacturing issues as do the simpler composite structures. For instance, in designing and making fiber-resin composites and using them in sandwich structures, the following must be considered:

- trade-offs in the amount of component materials for performance and cost,
- careful selection of materials for optimum structural properties,
- environmental conditions during use,
- manufacturing of the sandwich structure, and
- various other design and use issues, especially as related to the specific application.

These considerations, so critical in the design and manufacture of all fiber-resin composites, are critical as well to the performance of sandwich composites. Each of these design/manufacturing considerations will be examined so that the material choices and compromises can be better appreciated and understood. Just as with resin-fiber composites, a particular component may be ideal in one area of consideration, such as properties, but not good in another, such as cost or manufacturing capability. Therefore, all the considerations must be weighed before the materials and manufacturing methods are chosen.

Trade-offs to Maximize Performance and Cost

The trade-offs between component materials in a sandwich composite go right to the basic issue of why a sandwich composite should be made at all. A structure quite different from a traditional laminate is the result when core materials are used to make a sandwich structure. The greater thickness of the core materials yields tremendous increases in bending/flexural strength and, especially, stiffness to an applied z-direction force. This effect is illustrated in Figure 12-2. Doubling the thickness at essentially the same weight as the resin-fiber composite and adding only the light core material gives a three-and-a-half-fold increase in strength and a seven-fold increase in stiffness. If the thickness is doubled again, still maintaining essentially the same weight, the increase in strength doubles to tenfold over the single skin and the stiffness increases to 37 times that of a non-sandwich composite.

Structural engineers have long known that increases in thickness will give large increases in strength and stiffness. This is the principle on which I-beams work. As illustrated in Figure 12-3, the face plates of the I-beam are the upper and lower surfaces and these correspond to the resin-fiber laminates. The web of the I-beam corresponds to the core of the sandwich composite. Note that the web of the I-beam is relatively thin and, therefore, lightweight. Likewise, the core materials are generally much lighter weight than the resin-fiber laminates. The increase in stiffness of the I-beam and of the sandwich composite are principally functions of the distance between the face plates and not of the nature of the web, provided certain overall parameters are

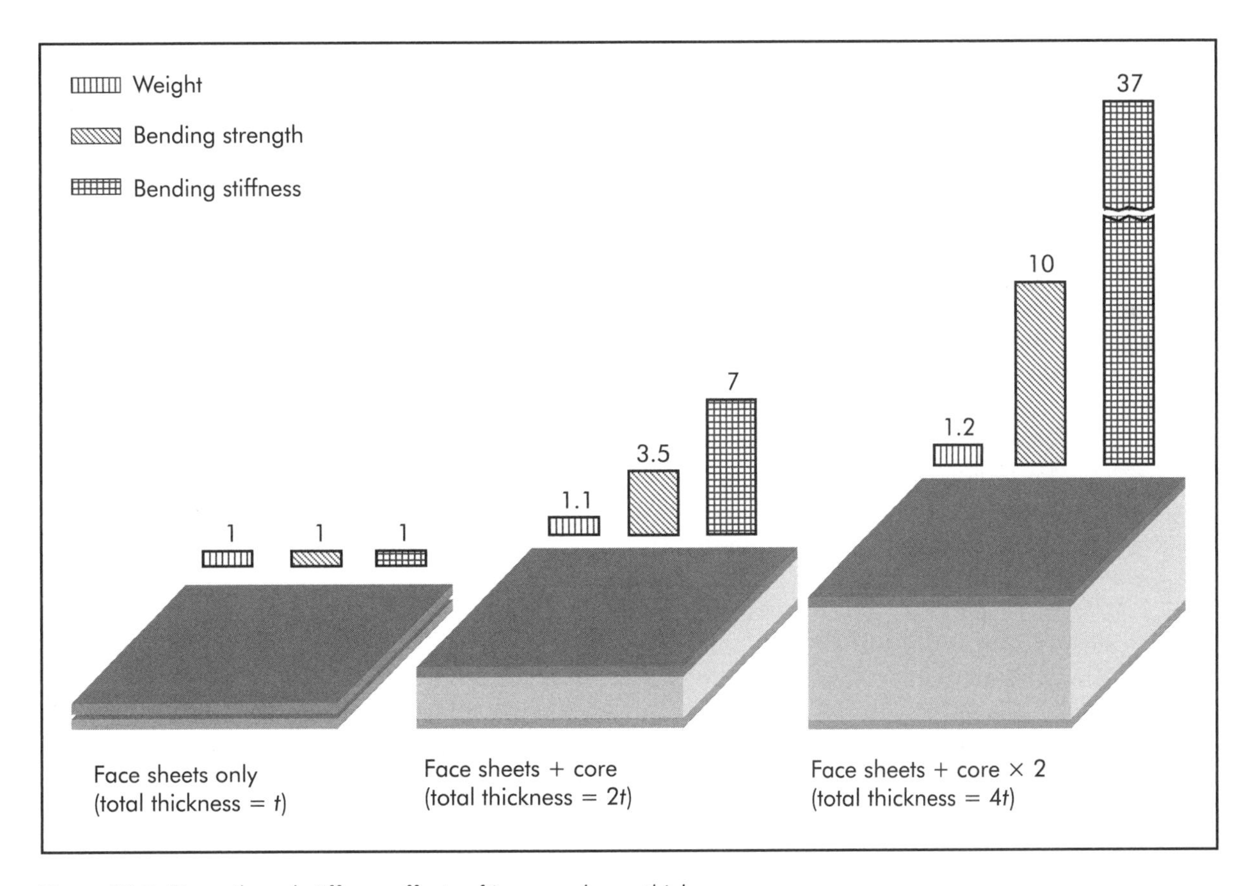

Figure 12-2. Strength and stiffness effects of increased core thickness.

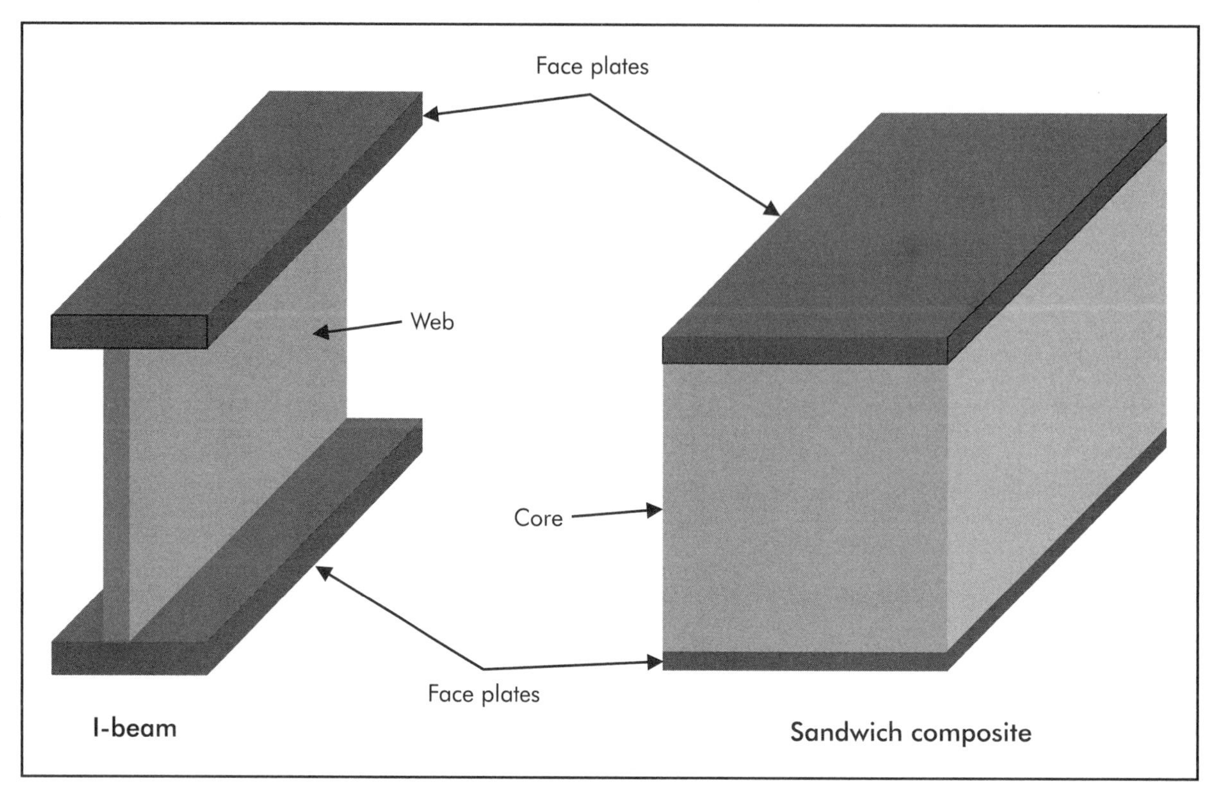

Figure 12-3. Comparison of an I-beam to a sandwich composite.

maintained. However, there is an important difference between I-beams and sandwich structures. When the sandwich is subjected to a z-direction force, the force tends to be spread all across the surface of the core, thus diminishing the force in any single location. The I-beam is far less capable of spreading the force. Moreover, because the force is spread in a composite, the core material need not be as rigid or strong as required in an I-beam. In other words, the combination of the face sheet and the core material, if bonded properly together, will have greater stiffness and strength than would either material by itself. Truly the sandwich is a synergistic composite structure.

The design question then becomes, how much resin-fiber composite should be used for the face sheets and how much for the core? If it is assumed that a constant overall thickness is required in the sandwich as might be used in just the laminate structure, then clearly using core material will reduce weight. In general, the use of more core material in place of face sheet will reduce costs since cores are usually less costly than face sheets. However, at constant overall thickness, increasing core thickness and reducing face sheet thickness will result in less strength and stiffness in the x and y directions. The optimization of component thicknesses is clearly a compromise that must be decided for each application. Typically, the laminate portions (face sheets) are designed for the proper x-y plane strengths and stiffnesses and then the amount of core is chosen for z direction stiffness and strength. This may result in some overbuilding of the composite (since less laminate might be acceptable), but this is reasonable for safety.

CORE MATERIAL

Properties

Up to now, this chapter has discussed sandwich structures in general, with little reference to specific core materials. Some core properties are critical to overall sandwich structure performance. Others are not as critical but might be a bonus in the choice of material for some specific applications. Table 12-1 lists the common core materials and some of their important characteristics.

The most common core materials can be divided into four general types: balsa wood, foams, honeycombs, and stitched/compressed materials. Within each category there are several specific types and all are distinct in properties. Each has inherent advantages and disadvantages that should be considered for every application.

The nature of the core is important in determining the crush strength of the sandwich structure. Clearly, the core material must be able to withstand the force applied in the z direction. (This phenomenon can be envisioned if you think of the deflections that would occur if the web or core was made of rubber.) Hence, it is important that the core materials be strong and stiff in the z direction. The common core materials are generally acceptable in stiffness and strength. However, a reasonable designer would make up sandwich composites for testing and subject them to the anticipated maximum z directional loading to be sure.

Another critical structural property is the ability of the core material to withstand x- and y-direction loads. Because the entire sandwich structure is bonded together, forces from the x and y directions will place the core into potential shear and buckling failure modes. These generally become worse as the core material is increased in thickness. Examples of some of the failure modes of core materials are shown in Figure 12-4.

In some materials the orientation of the core is important. For example, balsa wood is a common core material because of its light weight, but it is oriented so that the direction of the wood fibers is in the z direction. This is characteristic of end-cut balsa. The natural strength of the wood cells helps give support to the core. A similar situation exists in honeycomb materials. They are used with the cells oriented in the z direction because that is their direction of strength.

A property of immense importance in some applications is impact damage resistance or impact toughness. Here the high stiffness of a core material is often a disadvantage because high stiffness usually comes with high brittleness. As a result, when composites made with highly brittle cores are impacted, the core materials may shatter. The core materials best in resisting impact damage are those with slightly higher resilience. This property is generally associated with thermoplastic materials such as polypropylene, non-crosslinked polyvinyl chloride (PVC), or styrene-acrylonitrile copolymer (SAN), and with metallic honeycomb cores. If the composite core is damaged, the repair method usually involves insertion of a plug of core material and then re-establishment of the laminate layer. (These methods are discussed in Chapter 21.)

Another property of interest is the resistance of the core material to crack propagation. Again, high stiffness and the resulting brittleness tend to result in high crack propagation. Still another property that is related to stiffness/brittleness is fatigue resistance. This property is sometimes discussed in terms of withstanding high dynamic loads.

Testing

Some comments are appropriate regarding the nature of composites testing. The structural properties of composites and other materials are usually determined using standard tests (such as those outlined by the American Society for Testing and Materials [ASTM]), many of which were

Table 12-1. Common core materials and their properties.

Core Material	Characteristics and Benefits
Balsa wood (end grain)	Good shear strength, high fatigue endurance, low cost, easily bonded, easily finished, good temperature range
Polyvinyl chloride (PVC) foam (crosslinked)	High strength, high stiffness, low cost, easily bonded
PVC foam (linear)	Low cost, easily bonded, good impact resistance
Polymethacrylimide (PMI) foam	High dimensional stability under heat, excellent mechanical properties, solvent resistance, low thermal conductivity, high strength and stiffness
Polyetherimide (PEI) foam	Low water absorption, high thermal stability, high strength, fire resistant, good dielectric properties
Styrene-acrylonitrile copolymer (SAN) foam	No outgassing, high stiffness, high impact and fatigue strength, no environmental problems with resin or recycling
Ceramic foams	Unsurpassed thermal resistance, excellent thermal insulation, solvent resistance
Paper honeycomb	Low cost, easily bonded, strong for weight
Polyimide paper honeycomb	High strength to weight, corrosion resistant, good thermal insulation, fire resistant, easily shaped, excellent dielectric properties, easily bonded
Polyolefin honeycomb	Rigid and elastic, high toughness, sound and vibration dampening, explosion containment vessels, scrim cloth available, high strength to weight, corrosion resistant, fungi resistant, can be thermoformed, recyclable
Engineering plastic honeycomb	Tough, relatively high-temperature tolerant, excellent dielectric properties, good thermal insulator, fire resistant, fungi resistant, highly variable cell sizes and densities
High-performance honeycomb	Carbon fiber reinforced, carbon-carbon, aramid, quartz, superior strength, superior thermal resistance
Metal honeycomb	Aluminum, titanium, stainless, nickel available, no outgassing, high-temperature tolerant, fire resistant, fungi resistant, high thermal conductivity
Stitched/compressed	Excellent drape, needs to be fully wetted, high impact resistance, reduces cracks, absorbs resin for added strength

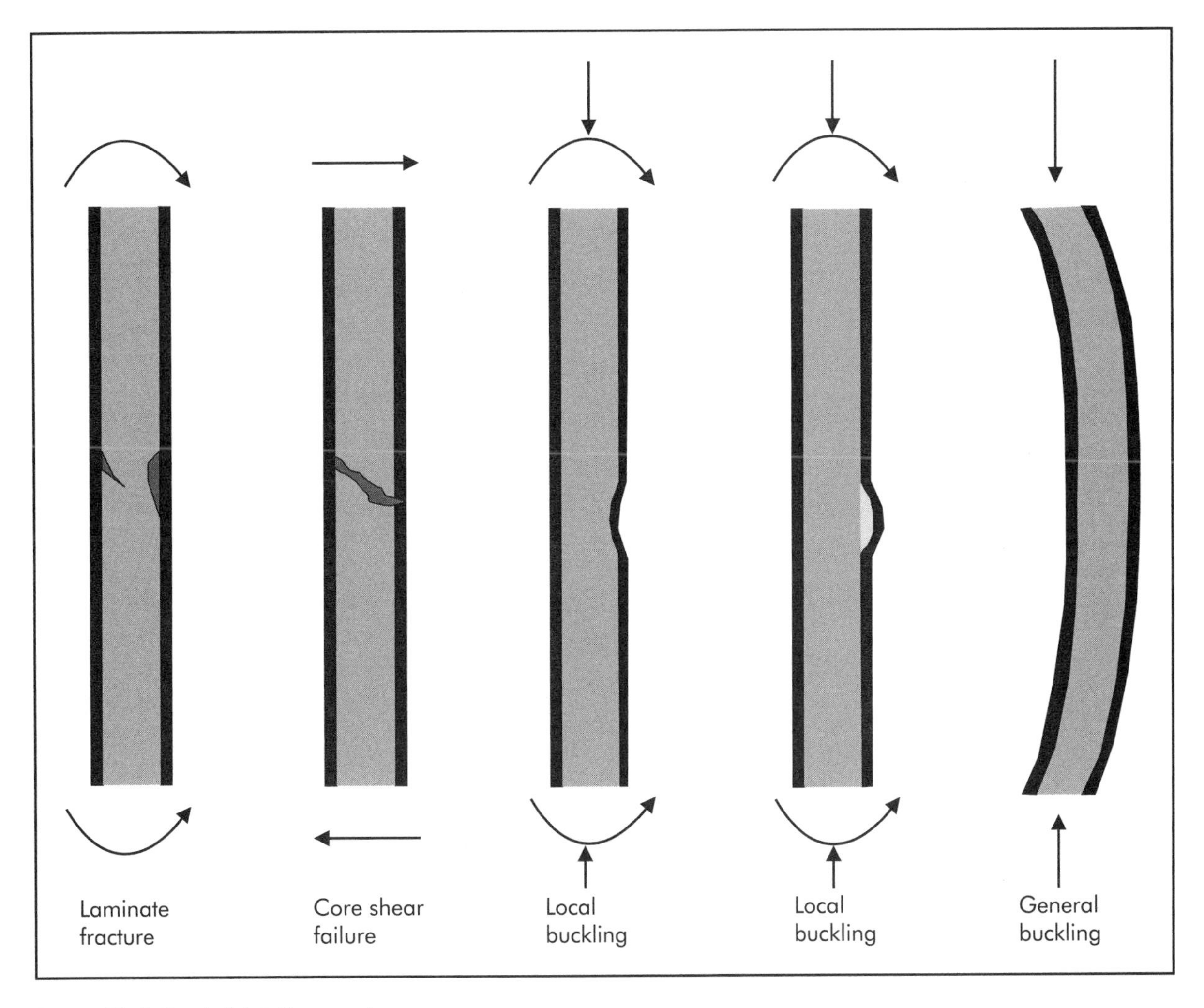

Figure 12-4. Sandwich failure modes.

developed for metals. Many of those tests have been modified to give comparable results for traditional resin-fiber composite laminates, but few have been modified for sandwich materials. Therefore, sandwich materials are sometimes at a disadvantage when compared to other materials if just the test data are compared. If the property of interest corresponds directly to the test and is important in the performance of the structure, this disadvantage is justified. But, if the specifications are written around material properties that may have little relationship to actual performance, then some alternate test might be devised that allows sandwich materials and metals or laminate composites to be compared fairly.

Some tests have been shown to be especially appropriate for sandwich materials and many of the most common applications. These include the long beam flex test, the roller cart test, the core shear test, the climbing drum peel test, falling dart impact test, flatwise tensile/compression test, vertical flame test, the National Institute of Standards and Technology (NIST, formerly the National Bureau of

Standards [NBS]) smoke chamber test, and the edgewise compression test.

Environmental Conditions During Use

Clearly the core material must perform many functions well. It has structural requirements and a myriad of other properties required by the conditions of each particular use. Some have described the core material in sports terms. It is like the athlete who competes in the decathlon—it is able to do many things well.

Some environmental conditions suggest special care in selecting the core material. These include the following:

- *Water absorption*: some composite materials have a tendency to absorb water and so might be avoided in marine environments. A related problem is entrapment of water, even when the core material itself does not absorb water. The water can adversely affect performance by adding weight and may diminish strength. The problem of water entrapment was especially important in aircraft for service on aircraft carriers. The U.S. Navy found that water was being trapped in the honeycomb material and was adding so much weight to the aircraft that take-offs were compromised. In addition, some composite materials could be subject to degradation in moist environments.
- *Temperature*: the face plates of a sandwich material are usually good thermal insulators so the core is often protected from direct contact with high heat. Nevertheless, thermal degradation of the core over time is a possibility that should be considered whenever high thermal conditions exist. Thermal stability and toughness are often difficult to achieve at the same time. An advantage of most core materials is their thermal insulative capabilities. Consequently, sandwich composites are widely used when thermal insulation is required. A consideration in these applications should be given to the differences in the coefficients of thermal expansion (CTE) between the composite and the face plates. If the CTE difference is too large, debonding can occur and most of the structural properties of the sandwich structure will be severely deteriorated.
- *Off-gassing*: some applications are particularly sensitive to off-gasses that may be emitted by the core materials. Of course, the resin in the face also might have a problem with off-gassing. Some core materials, such as metals and a few of the polymeric types, have low off-gassing and are therefore preferred when sensitivity to this problem is important, as might be the case in panels used in clean rooms.
- *Flammability*: clearly some applications, such as panels for trains, planes, and ships have strict flammability requirements. These often also include requirements on smoke generation and toxicity. Some of the core materials are especially good in resisting both burning and smoke generation.
- *Sound insulation*: most core materials are excellent sound absorbers. However, when sound is a key part of the application, the best policy is to test the entire sandwich structure for sound suppression.
- *Vibration damping*: for shipping and other transportation applications, the ability of the material to absorb vibrations can be critical. The natural frequency of the core is a key component in absorbing these vibrations. However, the entire structure has unique natural frequencies that must be determined

for the whole composite. Testing the entire structure is usually required.

Manufacturing of the Sandwich Structure

The most important part of manufacturing a sandwich structure is the bonding of the core material to the face plates. This is done with an adhesive film that is placed between the face plates and the core. That adhesive must be strong enough to transfer the load from applied forces onto the face sheets and to resist debonding from the surface of the core. Experience has shown that tough adhesives work best because they can elongate slightly to accept small displacements without cracking.

The adhesive must be thick enough to give a good bond over the entire surface but not so thick that it becomes a point of failure. The adhesive is, after all, a non-reinforced resin layer that has less strength than the resin-fiber laminate.

When the core is of an absorbent material like balsa wood, some foams, some honeycomb materials, and some stitched/compressed materials, care must be taken to prevent excessive resin infusion into the core layer. If the resin were absorbed or flowed into the core material, this could starve the interface and greatly weaken the bond. Therefore, if the core absorbs resin, the amount of adhesive must be increased accordingly or some measure must be taken to prevent absorption of the resin. With honeycomb, experience has shown that the correct amount of adhesive will wick slightly into the honeycomb cells to give a good bond to the walls but will not run excessively. To achieve this moderate wicking, the viscosity of the adhesive material must be carefully controlled at the temperature used for curing. Some core materials have a scrim sheet (that is, a thin mat material) bonded to the top of the core to prevent excessive resin infusion.

In one circumstance resin flow into the core is desired. That is when weight is not a problem and the core material might be subjected to high vertical (z direction) loads. In this case, sufficient adhesive is applied to allow the resin to flow into the core and bond at the interface. The cells of the core material are designed to have resin within them so that they will be stronger in crush strength. The resin inside the cells tends to keep the cells parallel, thus increasing the crush strength significantly.

In almost all sandwich structure applications, after the core and face plates are bonded to make the sandwich, the ends of the sandwich material usually need to be sealed. This sealing can be done by several different methods. One is to apply a thick resin as a sealant, sometimes reinforced with milled fibers along the edge. The **milled fibers** are short and give some reinforcement to the resin. Another method of sealing the sandwich is to plug the exposed end with some material. Some typical plug materials are balsa, injection molded plastics, and shaped metals or ceramics. These materials are usually bonded in place. Yet another method is to shape the core so it tapers to a smooth finish and then bond the faces together or wrap one of the face sheets over the end. Several methods of closing the ends of the sandwich are shown in Figure 12-5.

OTHER Z-DIRECTION STIFFENERS

Several methods other than sandwich structures have been developed to improve the z direction strength and stiffness of composite laminates. One strategy (discussed in Chapter Nine) uses fiber preforms in which some of the fibers are actually oriented in the z-direction component. These fiber materials are usually layers of cloth that have been stitched through the x-y plane laminate layers to place fibers oriented perpendicular to the plane of the majority of the other fibers in the structure. The structure with z-direction fibers also could be made by knitting or weaving. These fiber methods

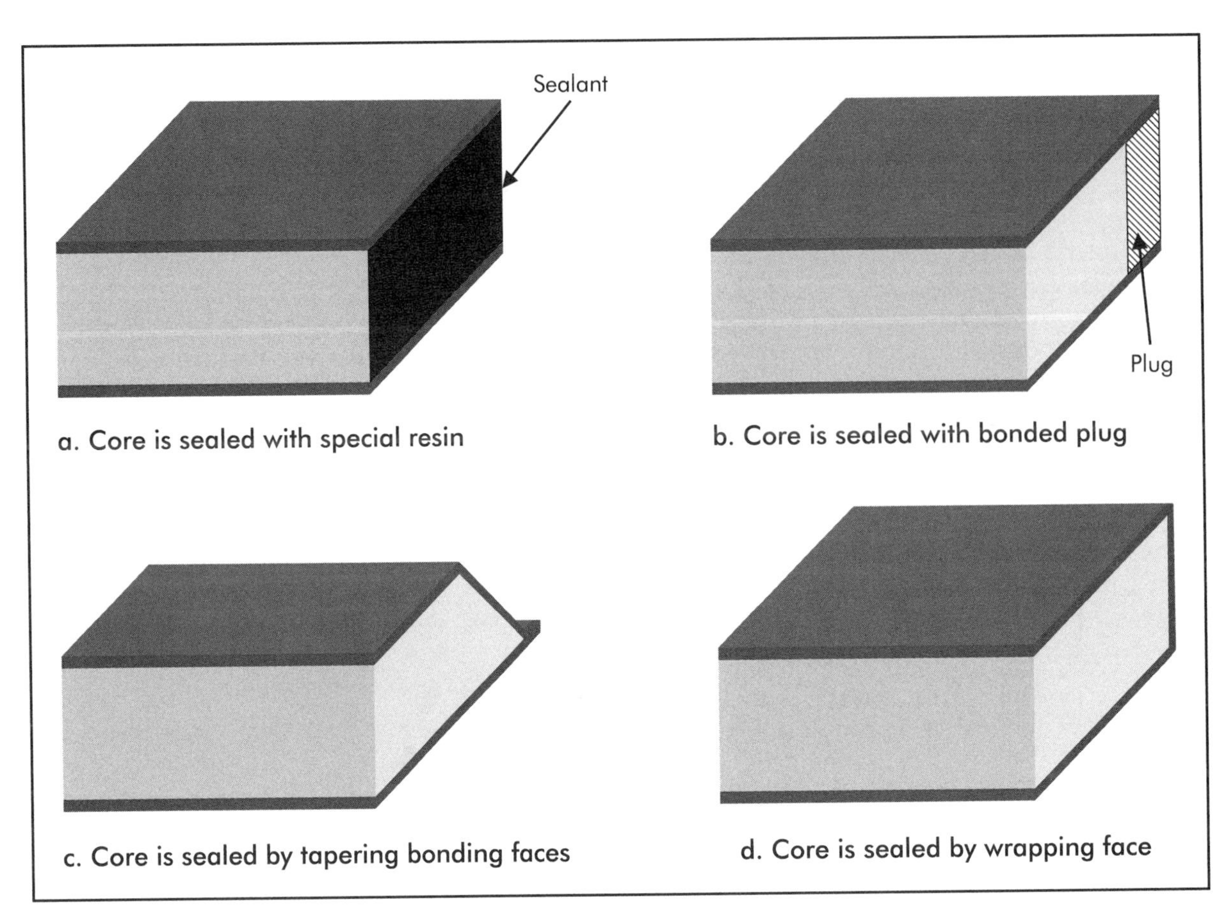

Figure 12-5. Methods for closing the edges of sandwich structures.

are valuable, especially for manufacturing methods that do not depend on laying up the fibers in sheets. But most manufacturing methods utilize laminar structures and therefore cannot use preforms that have z-directional fibers. (Some exceptions are the resin infusion processes discussed in Chapter 16 where fiber preforms are the norm.) A drawback is that these preforms are made on highly specialized textile fabrication equipment and, therefore, are more costly than traditional laminar structures. So, even though such z-directional preforms are available, they are not dominant in the composites market.

Other methods for increasing z direction strength and stiffness involve the addition of structural reinforcing materials or stiffeners. These stiffeners are often referred to by their cross-sectional shape. Typical examples are **hat stiffeners** and **J stiffeners**. The airplane industry uses these structures extensively. Many airplanes once used sandwich materials. But the problems of water entrapment, which are so common in sandwich structures on aircraft carriers, were also encountered in non-naval airplanes. Hence, sandwich construction is now rarely used in commercial airplanes. The alternate method is to use structural stiffeners in the hat and J shapes (and others that serve similar purposes).

When viewed from the top, that is, looking down the z axis, the pattern of intersection of the stiffeners can be of various types. The most common is **orthogonal**. In this pattern, the stiffeners intersect at right angles. Another pattern that has much interest, but not as much use, is called **isogrid**. This structure, composed of equilateral triangles, is actually stronger than the orthogonal, but is limited in actual application because the triangular structure is more difficult to manufacture. (A special structure based on the isogrid pattern is discussed further in Chapter 24.)

Some other types of structures that might give increased *z*-direction stiffness can be considered besides resin-fiber laminates and sandwich materials. For example, a structure can be built up using support struts between the laminate faces. This type of structure is usually shaped like an egg crate. Such structures can be stiff and strong, but are often difficult and expensive to build. Their most common use is for backing mold faces where access is relatively easy because only one composite face is used. However, egg-crate structures could be used for sandwich-like structures.

Another construction, especially common in the fiber reinforced plastics (FRP) marketplace, uses plywood or pressboard as the material between the composite laminate faces. This was an early construction and is still used in applications where weight is not a major consideration. The cost is also low for the wood products. Both plywood and pressboard are, of course, composite structures. Plywood is a laminate with its principle strength in the *x-y* plane. Therefore, it does not have optimized *z*-direction properties and is better thought of as a backing material than as a core material.

JOINTS

Just as the general design criteria and methods for composite laminates and sandwich materials are still developing, the design criteria and methods for joining composite materials to each other and to other materials are still developing. In fact, these joining methods are even more complex and, therefore, more tentative. The development of aluminum joining methods in the 1940s and 1950s could be a predictor of the future pathway of composites joining methods. The first methods for joining aluminum were riveting and bolting, then welding and diffusion bonding, followed by adhesive bonding, and most recently, stir welding. While some of these methods for joining aluminum may not be directly applicable to composite joining, certainly some of the steps in aluminum joining reflect the situation of composites. Thus there are two general types of joints in composites: mechanical, which is like riveting and bolting in aluminum, and chemical, such as welding, diffusion bonding, adhesive bonding, and stir welding.

Mechanical Joining

Several joint designs have been used to join composites to other composites and to non-composite materials (chiefly metals). The most common of these joint designs are shown in Figure 12-6. In each joint the composite materials are brought together and a hole is created through them. A mechanical joining device—rivet, bolt, screw, etc.—is placed through the hole and secured. While the straight lap is the simplest, its non-symmetry means that shear forces are almost always present in the joint. These forces can, of course, break the joining device and may even alter the nature of the way the laminate resists forces. Therefore, the offset lap has less inclination for shear and the double lap even less. The drawback for both of these joints is the additional material that must be used. The butt lap is a simple way of joining two laminates when the edges are to be aligned. This method is similar to a type of common repair (called doubler repair), which will be discussed in Chapter 20.

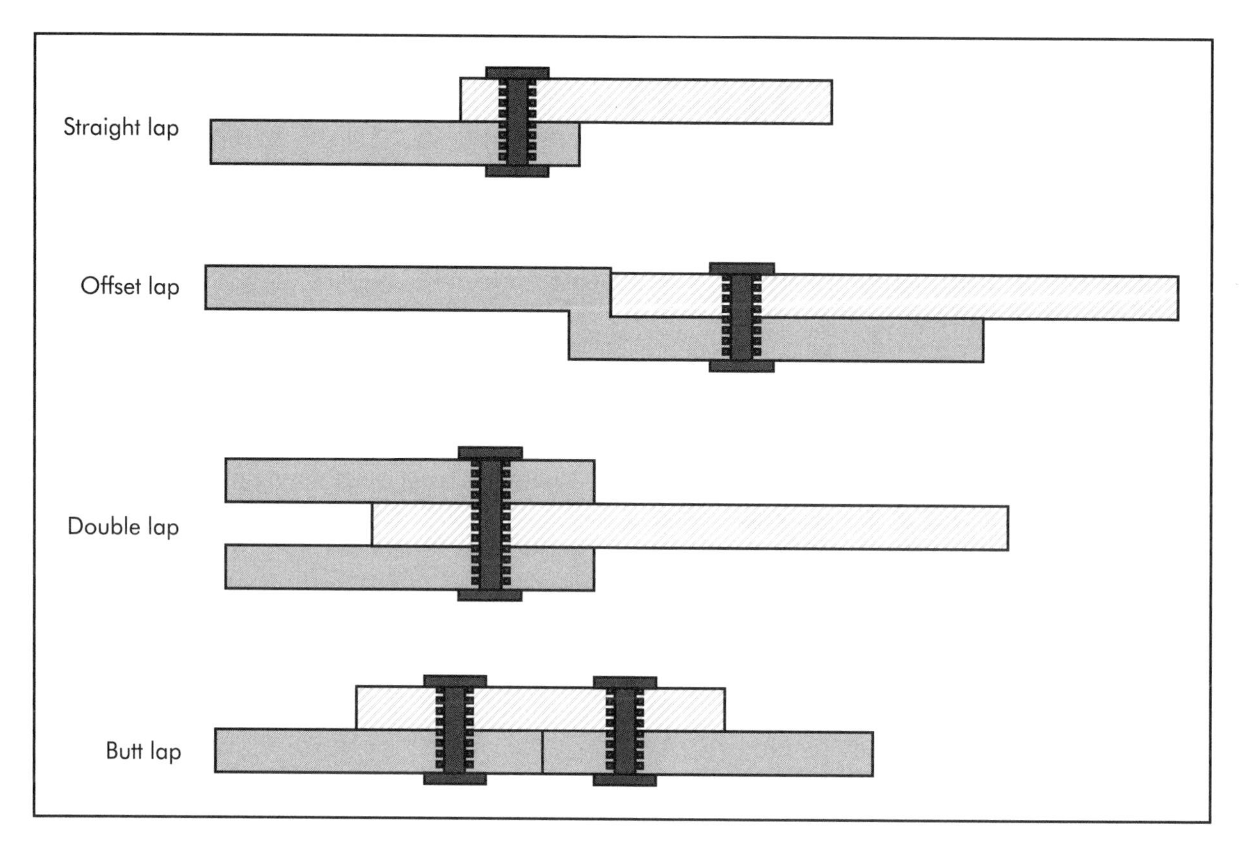

Figure 12-6. Mechanical joint designs.

Besides the design of the joint, other considerations in mechanical joining are:

- materials of the fastener,
- shape of the fastening device,
- strength of the fastener,
- hole size, and
- loss of strength because of the material removed in making the hole.

Materials of the Fastener

The choice of material for the fastener device is important, especially in joining carbon fiber reinforced composites. Because carbon is electrically conductive, galvanic corrosion occurs with many fasteners—aluminum and cadmium-plated being two of the worst. This problem is reduced by using nickel, titanium, or nonmetal (composite) fasteners, or by separating the fastener from the carbon with a glass-reinforced outer layer or sealant.

Shape of the Fastening Device

The physical shape of the fastening device is important, largely in ensuring that it and the process of joining do not damage the composite laminate. For example, the tightening process must not be so aggressive that layers of laminate are crushed. This type of failure, called a *bearing failure*, is illustrated in Figure 12-7a. To help prevent this from occurring, the head of the joining device and the end (nut, rivet end, etc.) must be large in diameter so the forces of closure are spread over a large area. It is also advisable to use a

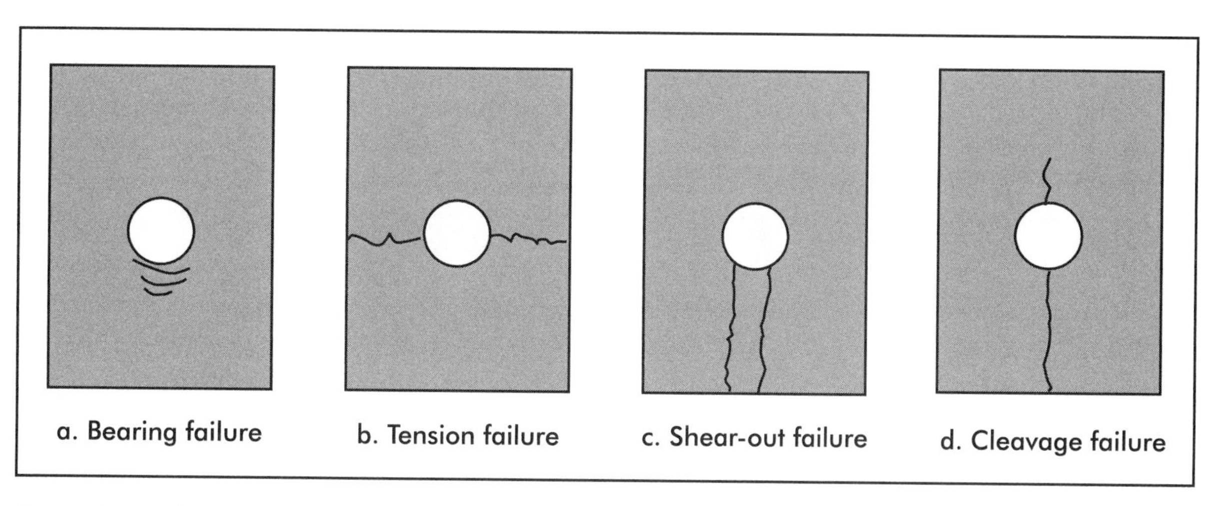

Figure 12-7. Failure types for mechanical joints in composites.

torque wrench or some other moderating device to ensure that the securing forces will not be excessive. Fasteners with small heads or ends also may have a tendency to pull out.

Hole Size and Loss of Strength

The fastener hole is often drilled to give an interference fit. If made larger than this, stress concentration points can develop at the ends of the hole when the joint is loaded in tension. This could result in failures of the type shown in Figure 12-7b. In addition, by making the hole large, more of the composite material is removed and, therefore, the composite is inherently weaker. The shear-out failure mode, shown in Figure 12-7c, occurs most readily when the fastener has a small diameter and the joint is subjected to shear forces. Another problem with a small-diameter fastener is, of course, failure of the fastener itself. The cleavage failure, Figure 12-7d, is from transverse shear forces.

Advantages

In spite of the problem with mechanical joining, it continues to be employed, largely because of one or more of the following *advantages*.

- Mechanical joints have little creep.
- Mechanical joints are stronger in peel loading than adhesive bonds.
- Mechanical joints are relatively easy to inspect.
- Mechanical joints can be designed for access, whereas adhesive bonds are permanently fixed.
- Little surface preparation is required for mechanical bonding.
- The mating surfaces of the materials to be joined is relatively unimportant.
- Mechanical joints are less sensitive to thermal, water, and other environmental degradation (except for the carbon-metal corrosion).
- Different materials, such as a composite and a metal, can be joined with little adjustment in the method.

Chemical Joining

Many composite joints are made by forming chemical bonds. This is in contrast to using an adhesive to bond the materials. For example, the bonds between layers of a composite laminate are, simply, bonds between the matrix of one layer with the matrix of another. Similarly, stiffeners, such as those discussed previously in this chapter, are often made from the same material as the laminate they are supporting and are cured at the same time as the laminate. This results in migration of the common matrix molecules across the boundary of the stiffener and the laminate so that the materials are integrally bonded. The same practice could be applied to two laminate sections that are joined together by curing them simultaneously. The process of curing laminate materials at the same time and thus causing them to bond together is called **co-curing**. Co-cured bonds are often the strongest chemical bonds that can be made between laminate structures. But, of course, co-curing requires that the composite materials be made of the same matrix resin.

Adhesive Joining

When different materials are to be bonded, the most common method is to apply an adhesive between the already cured materials and to cause the adhesive to harden in a separate step. The material to be joined or bonded to is called the **adherend**.

The effectiveness of adhesive bonding is dependent upon many variables, including:

- surface preparation materials and the method of surface preparation,
- shape of mating parts and joint design,
- adhesive composition,
- application procedure,
- tooling or holding jigs, and
- the curing process.

Surface Preparation

The preparation of the surfaces to be bonded is extremely important in adhesive bonding. It has a large effect on the permanence of the bond. The surface of the composite material must be free of grease and other contaminants. One contaminant that requires removal is mold release agent, which is often present on cured composites.

A typical surface cleaning process would include: wiping with a solvent or vapor degreasing, rinsing with water or an alkaline solution (if required to remove solvent), then abrading the surface by sanding or scrubbing with an abrasive pad, wiping again with solvent, rinsing with deionized water (if required to remove solvent), and drying in an oven (if water was required).

Sanding or scrubbing with an abrasive pad increases the surface energy and area (roughness), thus improving the bonding capability of the material. Placing a peel ply on the surface of the part during laminate curing can be done as an alternative to sanding since removal of the peel ply will result in a roughened surface.

When the matrix is polyester or vinyl ester, special attention must be given to determine whether the matrix cured completely (**cured hard**) to the outer surface. Oxygen inhibits polyester curing as does the evaporation of styrene at the surface. As a result, the outer layer of many polyester and vinyl ester materials is not totally cured even when the interior is properly cured and the normal cure cycle has been completed. This condition can be detected by a slight softening of the outer surface of the composite. If the condition exists and a bond is to be made, the incomplete cure may actually help to improve the bond between the laminate and the adhesive. This is because the uncured matrix molecules can migrate into the interphase of the adhesive molecules and vice versa, thus creating the same effect as when two laminates are co-cured.

Many polyester and vinyl ester formulations contain wax or some other internal agent that forms a coating on the top of the curing polymer and thus excludes the oxygen and retains the styrene. This results in the complete curing of the composite all the way to the outer surface. These materials are cured hard. But, the wax or other agent often prevents good bonding with an adhesive. Therefore, it is critical to remove these coatings as part of the surface preparation process. The most common method of removal of these coatings is by sanding.

Normally, adhesive bonding is not effective on a gel coat. The practice is to remove the gel coat completely by sanding before applying the adhesive.

If the composite is to be bonded to a metal, the surface of the metal must be carefully prepared. In the case of aluminum, the most commonly used cleaning method is phosphoric acid anodizing followed by a primer. Titanium, on the other hand, would be treated with a phosphate fluoride or by an anodizing method. Any metal that forms a surface oxide should be carefully treated to achieve a stable surface.

Shape of Mating Parts and Joint Design

Depending on the type of joint to be formed, some shaping of the mating surfaces may be required. The surfaces to be joined should be mated for shape and general smoothness (except for the slight roughness that occurs after sanding or abrading).

The most common chemically (adhesively) bonded joint designs are shown in Figure 12-8. Several of these are similar to the types of joints made with mechanical fasteners except that, of course, adhesive joints do not require holes or fasteners. The single-lap joint is simple but is asymmetric and, therefore, subject to shear. The double lap largely eliminates the asymmetry but requires more material. Similarly, the single- overlay butt joint is simple but asymmetric whereas the double-overlay butt joint is symmetric but requires additional material.

The scarf and stepped-lap joints are especially important in adhesive bonding. These designs allow forces to be transferred across the joint, thus minimizing their effect. (Not all the forces are transferred, but these joints are better than others in this respect.) Therefore, these two joint designs are used extensively in primary structures and for repairs. In the scarf and stepped-lap joints, the areas of the mating surfaces should be as large as practical and matched as closely as possible in shape. This not only helps to give a good bond, but it helps to transfer the forces.

The transfer of forces across a joint and the minimization of stress-riser points in the laminate itself can be accomplished by tapering the termination of the laminate rather than simply allowing all of the layers to stop at the same point. This tapering is illustrated in Figure 12-9. Note that layers are "dropped" in a staggered fashion, thus reducing the thickness gradually and smoothly.

Adhesive Composition

When applied, adhesives are usually liquids, pastes, or solid (uncured) films. In all cases, however, the adhesive goes through a liquid phase and wets the surface of the adherend or is somehow pressed against the surface of the adherend (or both). Most adhesives go through some chemical or physical change such that they become hard. (Exceptions to hardening adhesives would be those that are meant to stay flexible, such as rubber cement. In these cases, the only change, if any, is a thickening of the material, normally from solvent loss.) This hardening can result from loss of solvent, cooling of the adhesive from a melt to a solid, chemical reaction within the adhesive such as crosslinking or polymerization, or pressure against the adhesive that causes some interaction to occur with the adherend.

A wide range of adhesives are available to meet the many applications that require adhesive bonding. The adhesive is chosen for its ability to bond to the adherend and for its other properties, such as structural strength, cost, ease of application, and resistance to environmental effects. Structural adhesives

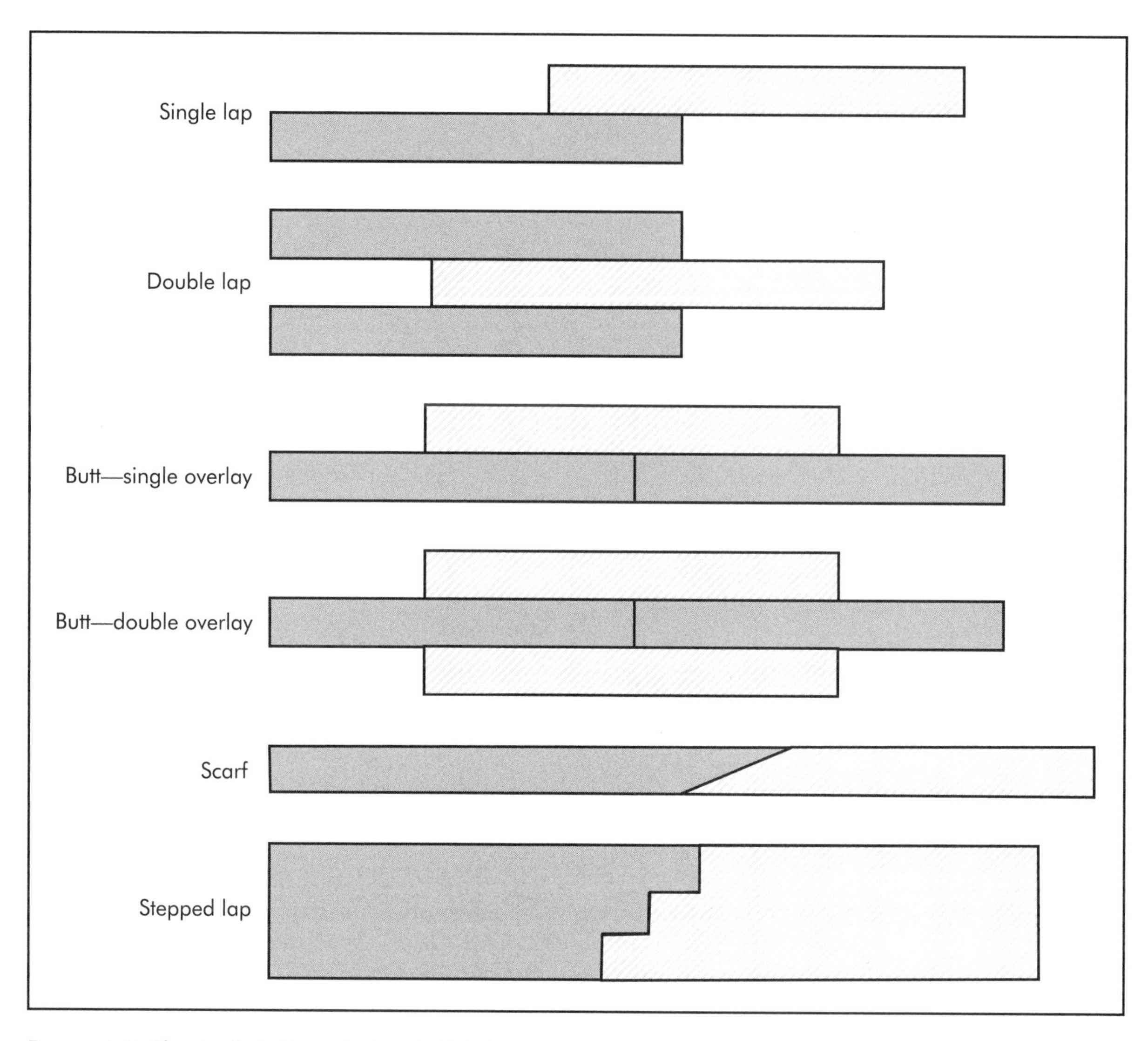

Figure 12-8. Chemically (adhesively) bonded joints.

are those used most often with composites. These can be stressed mechanically and can withstand a variety of difficult environmental conditions such as high temperature, adverse solvents, and creep. Requirements dictate that most structural adhesives be either one- or two-component thermoset systems. The most important polymeric groups within this class of adhesives are epoxies, polyurethanes, modified acrylics, cyanoacrylates, anaerobics, silicones, phenolics, and various high-temperature polymers. The following are brief descriptions of each type of structural adhesive.

- *Epoxies* are the most common of the structural adhesives. They are generally two-part adhesives, where one part is the epoxy and the other the hardener. (The chemistry of crosslinking of epoxies is discussed in Chapter Four.) Shelf

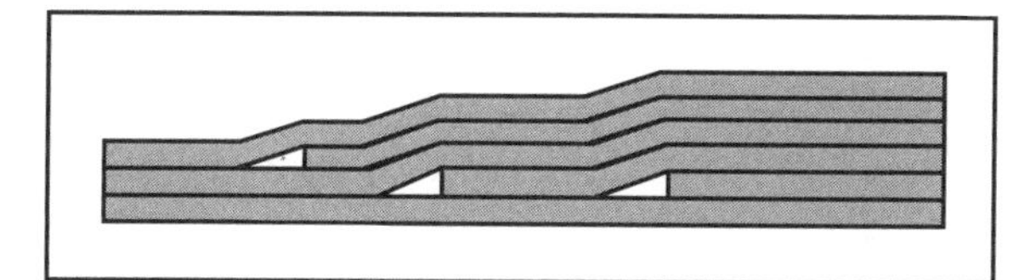

Figure 12-9. Tapering to end a section or make a joint.

life before mixing of the components is long (nearly infinite) but after mixing, the pot life is short.

- *Polyurethanes* are made by the reaction of a polyol and an isocyanate (as discussed in Chapter Five) and are, therefore, usually sold as two-component systems. These adhesives are often tougher than epoxies because of their higher elongation. However, the polyurethanes are not as strong as epoxies.
- *Modified acrylics* cure by free-radical polymerization reaction. They differ from standard acrylics in that rubber or other tougheners have been added. Typically, one component is applied to one adherend and the second component to the other adherend. The materials react when the adherends are pressed together. These adhesives can be used on most materials, even when the surface is slightly oily or improperly cleaned. Cure times are quite long. Impact resistance is improved over standard acrylics but is still only good (rather than excellent). Service temperatures are moderate.
- *Cyanoacrylates* are the materials used in superglues. They are single-component systems that cure when in the presence of even small amounts of moisture, as is present in most ambient atmospheres. These adhesives bond well to most materials, including skin. This can present a safety hazard but is also an advantage as they are used to suture skin together in surgical procedures. Cyanoacrylates are relatively expensive and strong but brittle.
- *Anaerobics* are single-component systems that cure by free-radical polymerization. The polymerization reaction is inhibited by oxygen, so no reaction will take place until oxygen is removed from the system. The air is excluded when the coated adherends are pressed together. Therefore, by mating the parts, the reaction is initiated and proceeds rapidly. Toughness and strength are moderate.
- *Silicones* are available as either one- or two-component systems. The one-component systems cure on contact with atmospheric moisture. Silicones can be formulated to cure at room temperature (room-temperature vulcanizing [RTV]) or high temperature (high-temperature vulcanizing [HTV]). They have good adhesion over a wide temperature range, –76 to 480° F (–60 to 249° C), and can go as high as 700° F (371° C) for some formulations. Silicones are tough and flexible with good solvent, moisture, and weathering resistance.
- *Phenolics* are low-cost thermosetting systems available with one or two components. They are generally strong but brittle. Phenolics are dark colored, which limits applications. These materials dominate the plywood and chipboard markets.

Application Procedure

After the surfaces are properly prepared, the adhesive is applied. The adhesive (if it is a one-component system) is generally kept at a low temperature before use so that its cure will not be advanced during storage. After warming, or in the case of two-component adhesives, after mixing, the adhesive is carefully applied to the treated surfaces.

The adhesive should be applied to the recommended thickness. It should cover the mating surfaces but not be excessively thick. The adhesive is a non-reinforced resin and is not, therefore, as strong as the laminates being joined. If the adhesive layer is too thick, the adhesive could become an important failure point for the entire structure. This is to be avoided unless done intentionally to create failure at a specific location or load rather than risk unpredictable catastrophic failure. But this latter situation is rare. Typical thicknesses of the adhesive layer are .004–.008 in. (0.1–0.2 mm).

Workholding and Curing

After applying the adhesive, the pieces to be bonded are placed in a holding tool, compacted to remove trapped air, and cured. The cure most often is done in an autoclave for advanced composites. However, ovens, presses, and even clamps for room-temperature cures are also used. If materials having different coefficients of thermal expansion are to be bonded, the cure temperature is often reduced until gel occurs and then the part is post-cured.

Advantages

Adhesive bonding has some drawbacks but also some advantages. The *advantages* of adhesive bonding compared to mechanical bonding include the following.

- Adhesive bonds have lower stress concentrations.
- Adhesive bonds eliminate potential delamination caused by the drilling required for mechanical joining.
- When the loading is first applied, adhesive bonds have less permanent set.
- Adhesively bonded joints weigh less than equivalent mechanical joints.
- Adhesive bonds permit smooth external surfaces at the joint.
- Adhesive bonds are usually less expensive.
- Adhesive bonds work well on even thin laminates.
- Adhesive bonding is less sensitive to cyclic loading.
- Bearing failures, pull-out, and other similar joint failures are not a problem.

Other Bonding Processes

With thermoplastic matrix composites, the materials can be fusion bonded (welded) as well as adhesively bonded. Welding is accomplished by cleaning the surfaces with solvent and/or water and detergent, heating the surfaces to the point of incipient melting, and then pressing the surfaces together. In addition to conventional heating, ultrasonic heating, friction heating, and induction heating (which requires the addition of some metal material) can be used. If the entire laminate gets hot enough for the resins of the two layers to intermingle, excess resin is usually not required. However, additional resin is often used in structural reinforced composites. A related process, hot-melt adhesion, uses a rod or sheet of thermoplastic resin that is heated and applied between the layers of laminate.

POST-PROCESSING OPERATIONS

Although composites are often molded to near-net shapes (that is, almost finished shapes), trimming, finishing, and assembly into larger structures may be required. In some of these processes, the potential for damaging or otherwise weakening the composite (delamination, fiber fraying, drill break-through, etc.) is great and care must be taken to maintain the composites' properties. (The post-processing operations commonly used with composites are outlined in the *Composites Manufacturing* videos that complement this text. The DVD entitled "Composites Post-fabrication & Joining" is especially valuable to enlarge on the information contained in this chapter [SME 2005].)

Conventional Metalworking Equipment

The equipment and procedures used in cutting, drilling, and machining of composites differ in some major considerations from similar processes for metals, wood, ceramics, or plastics. The abrasive nature and type of reinforcement material (aramid, fiberglass, carbon, etc.) make certain process modifications necessary.

Cutting, drilling, and other machining processes are largely dictated by the type of reinforcement with minor modifications to allow for the easily melted or degraded nature of the matrix. On the other hand, the processes of joining and painting are largely dependent on the matrix material since these are chiefly surface phenomena.

Composite materials can be trimmed and cut with conventional metalworking mills, lathes, circular saws, router cutters, and various abrasive tools. However, the machine settings (speeds, feeds, etc.) and cutting bits usually need to be modified. In most cases, the cutting speed (spindle speed) should be increased and the feed rate decreased relative to normal values used for metal machining.

Optimum feeds and speeds can vary significantly depending on the resin/fiber system and on the thickness of the materials. Typical cutting speeds for machining FRP materials are 600–1,000 ft/minute (180–300 mm/minute). Feed rates are .002–.005 in./revolution (0.05–0.13 mm/revolution). The following values have proven successful for epoxy/carbon composites approximately .25 in. (6 mm) thick in the routing process: cutter = .25 in. (6 mm) solid carbide (diamond-cut burr); cutter speed = 10,000–12,000 rpm; cutter lineal speed = 10–14 in./minute (250–350 mm/minute); and surface cutting speed = 600–800 ft/minute (180–250 m/minute).

Cooling or lubricating (preferably with fluorocarbon cutting fluid) may be necessary, especially for thermoplastic composites, to prevent resin build-up on the cutting tool and overheating of the part. If overheating of a thermoset composite part occurs, the damage can only be corrected by removing the damaged portion of the part.

The use of diamond- or carbide-tipped tools is recommended for composites. Care should be taken to ensure that the cutting tools are kept sharp to minimize delaminations. The brittle nature of most composites should be remembered in machining operations. For example, holding fixtures should provide for backing of the part to prevent delamination and backside breakout. In milling, both edges should be cut and then the center section can be cut, rather than starting at one edge and working toward the opposite edge. By following this sequence, the possibility of breaking off the last edge section is less likely.

Sanding should be done at 4,000–20,000 rpm or more using 80-grit aluminum oxide (dry) or 240 to 320-grit SiC (with water). Sawing should use a fine offset, high-strength blade. Good success has been achieved with a bandsaw blade with diamond grit on the cutting edge and a relief cutout (area without grit) at 2–4 in. (50–100 mm) intervals. In all cutting operations, an efficient vacuum system capable of removing cut material and dust will significantly increase (up to five times) cutter life.

Drilling of composites is important in almost every application, but especially in aircraft and electronic circuit board manufacture. Conventional metal-drilling equipment can be used, but the drills should be specifically tailored for composites. Drill bits should have a large positive rake angle without becoming so large that the cutting edge gets thin and dulls quickly. (The rake angle is the angle between the tool face and vertical surface of the part). A large positive rake angle, which means that the tool face is slicing into the part, will allow the drill to penetrate into the material with less force, avoiding the heat that arises from shear forces and friction. The positive angle also tends to push the fibers aside, whereas a cutter with a smaller or neutral angle would tend to pack the fibers in front of the tool, increasing the cutting forces (and heat) required to cut. The resulting chips are small and granular (sawdust-like) and should not fuse together, which would be an indication of too high a feed rate or pressure. Therefore, drills with high helix angles and wide, polished flutes are recommended to facilitate chip lifting and removal.

Carbide and, especially, polycrystalline diamond (PCD) inserts significantly increase drill cutter life compared to high-quality steel bits. PCD cutters often last more than twice as long as conventional cutters. When cutting composites (not combined with metals), a pointed drill-bit geometry can prevent delamination and fiber breakout because of the lower thrust forces on the exit of the hole. A drill proven successful for cutting carbon fiber and fiberglass has been used to make up to 40 quality holes per tool life cycle, with smoothly cut edges and without splintering or the use of backup material.

The problem in cutting aramid reinforced composites is quite different from cutting either carbon or glass fiber reinforced composites. With aramid, the fibers are tough and tend to shred rather than cut. Research has found, however, that if the fibers are preloaded (stretched) and then cut, cleanly cut fibers will result. For a rotating tool, such as a drill, this requirement dictates a C-shaped cutting edge that cuts from the outside toward the center. In this way, the fibers are engaged by the tool at the edge of the hole, stretched, and then cut by the following part of the cutting edge. The tools must be kept sharp for adequate performance with these materials.

With any of the reinforcement materials, peck drilling is recommended for deep holes and whenever dissimilar materials are used such as in sandwich structures. In this method, the drill is advanced into the hole a preset amount. It is then retracted to remove chips and dissipate heat. The insertion and retraction (like the pecking of a woodpecker) is continued until the hole is completed. Lubricants and coolants are not used in cutting deep holes since they tend to cause the chips to stick together and thus make removal more difficult.

Another problem occurs with the repetitive drilling of holes, which is often done in the manufacture of printed circuit boards. In repetitive drilling, the drill itself is gradually heated to a steady state with heat dissipating through the hole wall to the surrounding air and chips balancing the heat buildup in the drill. When the drill becomes dull, the frictional heat dominates the dissipation and the chips melt and stick to the drill causing hole enlargement, a phenomenon known as *drill smear*.

The requirements for drilling composites can be quite different from those for metals. In some cases, drilling of sandwich materials or parts with two different materials, such as a titanium bonded to a carbon reinforced composite, often requires major compromises in machining conditions.

Waterjet Cutting

Waterjet cutting is gaining strong support as a method to cut composite materials. This method utilizes a fine, high-pressure jet of water to impact and, therefore, cut or erode the material. In this process, intensifier pumps increase the pressure of filtered water to high levels (typically 60,000 psi [400 MPa]). The water is then pumped at low volumes (1–2 gal/minute [4–8 l/minute]) through an orifice having a diameter of about .010 in. (0.25 mm) and then onto the part to be cut. Abrasives can be added (usually entrained in the water flow after the orifice) to give additional cutting capability. These are especially useful for cutting sandwich materials.

Waterjet cutting can be initiated at any point on the part. It can cut complex patterns and can be numerically controlled for accurate and rapid cutting when the same shape is to be cut several times. However, because delamination can occur in the area of initial penetration, cutting is usually started at an edge or away from the finished area and then moved into the desired area. Waterjet cutting produces less cutting force than most of the mechanical methods and generally requires only clamping to support the part. It produces no noticeable heat-affected zone in the vicinity of the cut and can dwell in one spot for some time without widening the cut width.

The waterjet cutting process is quite noisy (requires ear protection), requires good filtration equipment for the water, and must be carefully monitored for jet wear. Water penetration between the fibers and laminations is a potential problem if the laminate has voids or if the water pressure builds up because it is not penetrating the part. This is especially problematic at excessive water pressures.

The cutting speed, which varies with the thickness of the workpiece and the type of material being cut, typically ranges from 60 ft/minute (18 m/minute) for .125-in. (3.18-mm) thin parts to 5 ft/minute (1.5 m/minute) for 1-in. (25-mm) thick parts. These cutting speeds are slightly higher than speeds for soft metals (aluminum, brass, etc.) and three to four times faster than for hard metals (steel, stainless, etc.).

The quality (roughness) of the cut depends on the hardness and strength of the fibers. For example, the cut is much rougher with boron fibers than with the comparatively softer aramid fibers.

Laser Cutting and Drilling

The use of lasers to cut composites is growing rapidly in those applications where the inherent problems associated with laser cutting are not serious. Lasers produce a coherent beam of light caused by the excitation and subsequent relaxation of electrons in specially chosen molecules, such as CO_2, which fill the excitation chamber of the laser. The light is focused onto the workpiece and then burns or melts the material away. Because no cutting force is exerted, fragile pieces often can be cut best using lasers. Lasers can be started at any point in the workpiece and are capable of cutting intricate patterns. They can be focused or collimated with apertures to drill holes that are too small to be effectively drilled mechanically. Therefore, they are used widely in the manufacture of printed circuit boards.

However, the thermal effects of laser cutting are sometimes serious detriments to its use. In some cases, the creation of a melted or charred edge or of a heat-affected zone surrounding the cut is a problem, which may be more severe because of the good heat conductivity of carbon fibers. The heating problem also may be worsened because of the need to heat glass fibers to a high temperature to melt them. Thick parts are usually problematic to cut with lasers because of the difficulty of removing the waste material, which may remain in the cut, dissipate the incoming light energy, and block further cutting. Therefore, laser cutting is generally limited to cutting those parts where the laser can break through before sufficient melted material accumulates to prevent further cutting, about .3 in. (8 mm). Cutting speeds depend on workpiece thickness, type of material, and the power of the laser. Typical values are 25–120 in./minute (635–3,048 mm/minute).

High-power lasers are often rather large machines that require careful alignment of various mirrors and focusing devices. Such machines are usually not portable. Therefore, the normal rule in using lasers is to move the workpiece under the laser rather than move the laser head.

Diamond Wire Cutting

The use of a wire saw with abrasives is old technology dating back to rock cutting in quarries in ancient Rome.

In the case of cutting composites, the wire is impregnated with diamonds. It is mounted in a cutting mechanism that moves the entire length of the wire, generally about 100 ft (30 m), first in one direction and then in the reverse (a sawing action) over the surface to be cut. The abrasive nature of the diamonds and their strength permit the cutting of essentially all material by this method. Though more time is required to make the cut using the diamond wire cutting method than with waterjet or laser cutting, the temperature rise at the cut is usually only a few degrees. The kerf (material lost in making the cut) is typically 10% greater than the diameter of the wire. The intricacy is reasonably good with a turning radius of .005 in. (0.13 mm) reported.

CASE STUDY 12-1

Comparison of Various Composite Constructions

Three different constructions are compared for making a canoe or kayak hull. The constructions are as follows:

1. All laminar construction of three plies of fiberglass cloth, Style 1800, 9.0 oz/yd^2 (300 g/m^2).
2. All laminar construction of three plies of aramid cloth, Style 500 cloth of Kevlar® 49, 5.0 oz/yd^2 (170 g/m^2).
3. Sandwich construction where the top and bottom faces are Style 500 aramid cloth and the core material is PVC foam.

The relative properties for the three composites are shown in Figure 12-10. The aramid laminate material is lighter weight than the others and has greater tensile strength and stiffness. This is characteristic of a laminate made from advanced materials (such as aramid or carbon fiber). However, the high flexural stiffness of the sandwich material illustrates the advantage of this construction with its favorable *z*-direction properties.

CASE STUDY 12-2

Designing a Platform Using Sandwich Construction

Assume that you want to make a maintenance platform for an industrial plant. The platform is to be 48 × 120 in. (122 × 305 cm). As long as large crowds of people or heavy machinery will not be moved over the platform, the standard pressure the design is required to withstand is 75 lb/ft^2 (.52 psi or 3.59 KPa). Experience indicates that a maximum acceptable deflection for a platform of this type would be equal to the span divided by 100. So, for a 48 in. (122 cm) span, the deflection would be .48 in. (12 mm). Assume a normal design safety factor of 2. Other required design criteria are:

- stiffness = 75,000 psi (517 MPa),
- flexural strength = 300 in-lb (34 J), and
- shear strength = 25 in.-lb (2.7 J).

It is decided that the platform is to be made of sandwich construction. Other design calculations considering toughness to resist dropped tools, abrasion resistance, and *x*-*y* plane forces have indicated that the exterior surfaces will be .10 in. (2.5 mm) thick. The laminates will be made from two plies of 1.5 oz/ft^2 (6.6 g/cm^2) chopped strand mat and 18 oz/yd^2 (618 g/m^2) knitted 0/90° fabric. This combination gives the thickness desired and has been shown to have the toughness and abrasion resistance required.

The core material needs to be chosen and the principal force to contend with is shear. The presence of a core material reduces the shear strength of the laminate by a factor of at least 10 and as high as 100

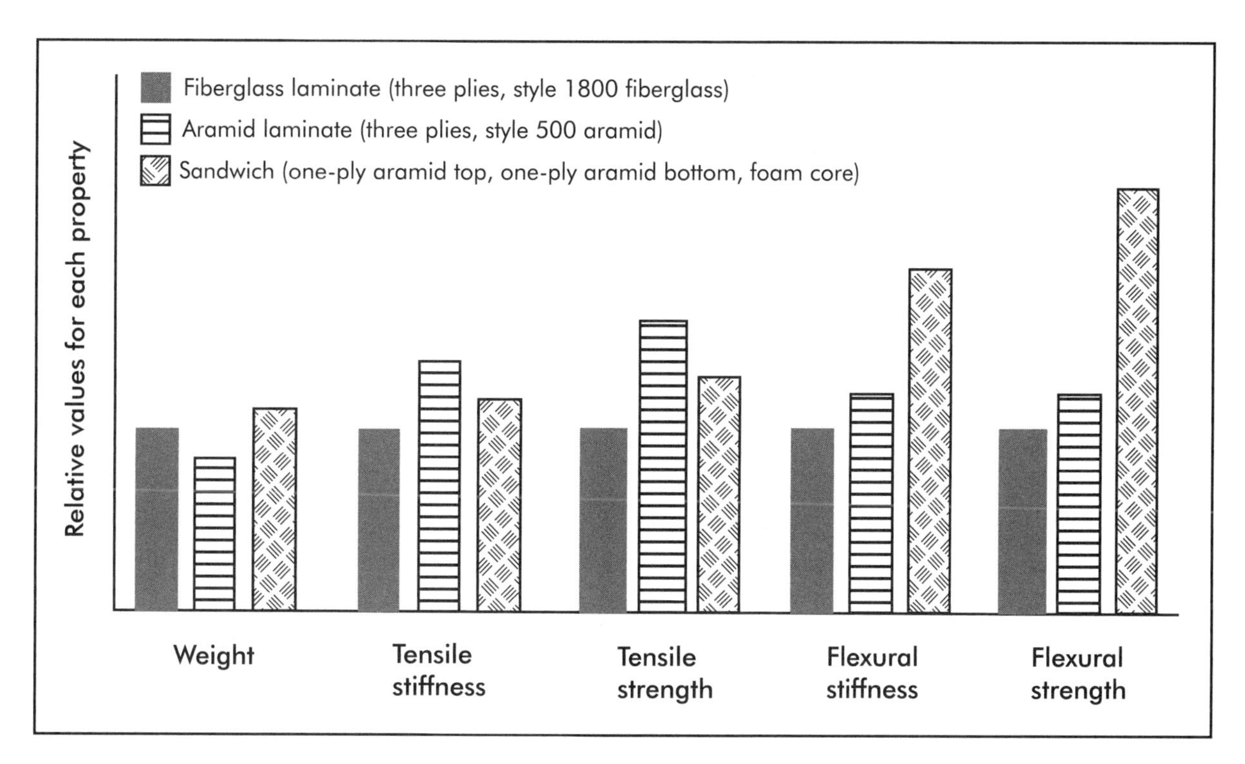

Figure 12-10. Comparison of laminate-only and sandwich constructions.

with some core materials. For example, solid fiberglass laminates have a shear strength ranging between 12,000–14,000 psi (83–97 MPa). Most common core materials have shear strengths between 70–360 psi (.48–2.5 MPa). Plywood and syntactic foams are the exception, having shear strengths of around 1,000 psi (6.9 MPa) (but at significantly higher weights than most other core materials).

To obtain the required 25 in.-lb (2.7 J) shear, the thickness of the core is calculated from reported (literature) values of the core material's shear strength. For example, PVC foam is reported by the manufacturer to have a shear strength of 87 psi (0.6 MPa). The overall shear strength of a cored panel is equal to the shear strength of the core times the thickness of the core plus one-half of the skin thickness. This means that the core plus one-half the skin thickness must be roughly a minimum of 25 divided by 87, or .29 in. (0.7 cm) thick. This is a rather thin core, but it meets the shear requirement. However, as will be seen later, a thicker core is needed to meet the strength requirement.

The flexural strength of the sandwich panel can be best assessed using a laminate stack computer program (as outlined in Chapter 11). In this case, the sandwich outlined easily meets the flexural strength requirement.

The stiffness requirement is established to prevent the feeling of walking on a surface that is too flexible. (If not stiff enough, it might feel like walking on a trampoline.) It is even possible that if the panel is not stiff enough, it might sag so much that it could slip out of its supports. The laminate stack program shows that the laminate proposed is only 8% of the required stiffness.

The stiffness deficiency can be corrected by increasing the thickness of the core. It is, after all, for stiffness that the sandwich construction was chosen. Through a series of trial and error calculations using the stack laminate program, the stiffness requirement can be met with a PVC core that is 1.125 in. (2.86 cm) thick. It also would be possible to increase the thickness (number of plies) in the face laminates, but this is generally a more expensive alternative to increasing the core thickness.

CASE STUDY 12-3

Ski Construction

The construction of a modern composite ski is shown in Figure 12-11. The incorporation of several different materials is of major interest. A sandwich of reinforced polymer and laminated wood core is at the center of the ski and provides the desired structural stiffness. To provide cushion, a layer of rubber is placed beneath the sandwich. Therefore, the stiffness felt by the skier is a combination of the sandwich and this rubber sheet.

The hard outer shell can be made of any durable material. Steel is common but so is composite. The bottom of the ski is a sintered running surface, usually ultra-high-molecular-weight polyethylene (UHMWPE). This material has a low coefficient of friction and thus is a good material to improve sliding. The phenolic spacer takes space, although there is some stiffness from it as well. The

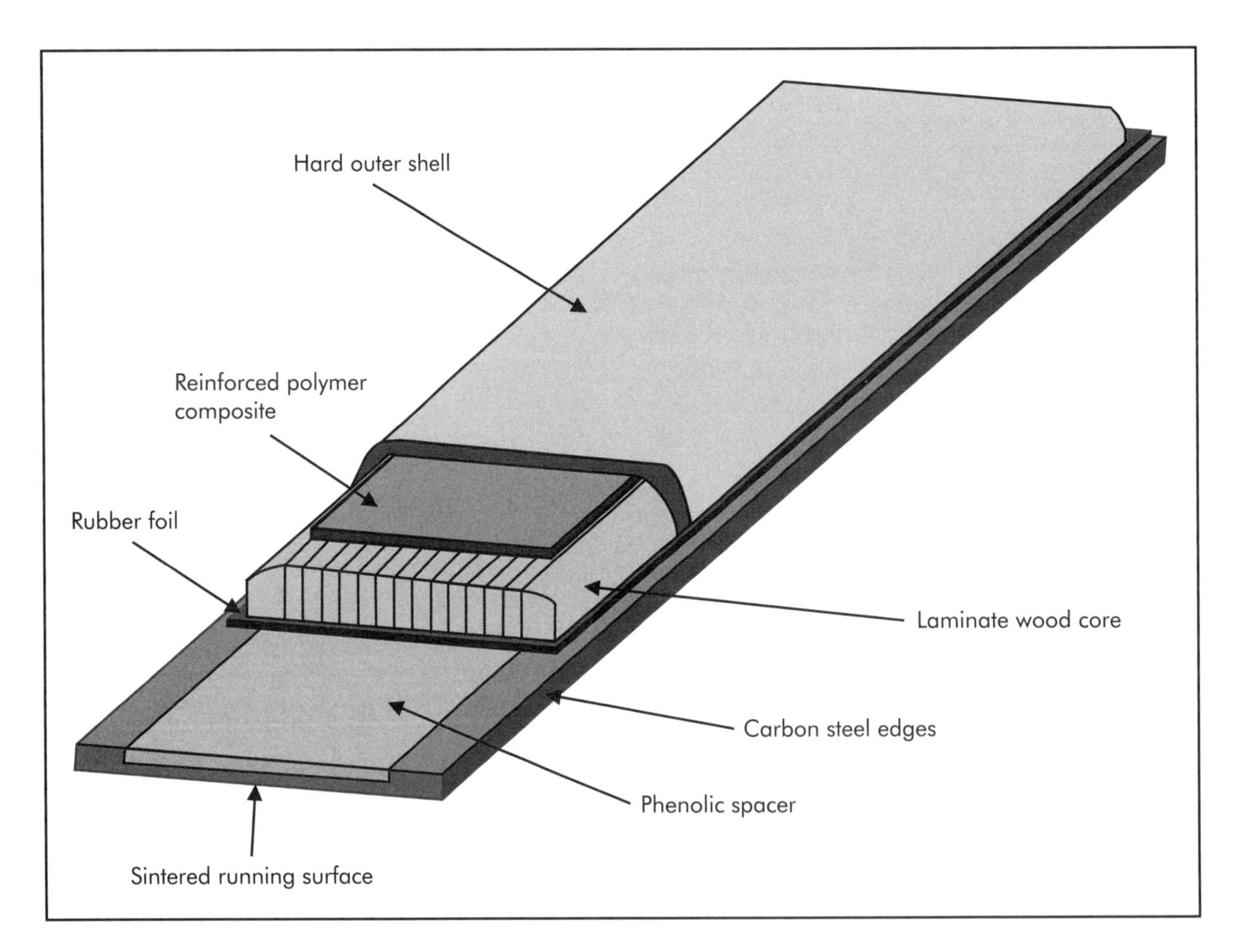

Figure 12-11. Ski construction.

top and the bottom are held together by steel edges. Clearly the construction is complicated and illustrates how various layers can be combined to achieve a unique combination of properties.

SUMMARY

The ability to give z-direction strength and stiffness to a composite structure greatly expands the capabilities of the material. Composite structures can be made lighter and at lower cost by forming a sandwich construction in which core material is bonded between composite laminate face plates. However, the complexity is also increased. There are more materials in a sandwich than in a traditional resin-fiber laminate and that means there are more things that can go wrong. The bonding must be excellent between the core and the face panels, for example. Also, the amount and type of adhesive used to form the bond must be carefully chosen. Further, the manufacturing process must be compatible with that choice. Finishing of the sandwich structure is more complicated than that for the average resin-fiber composite. However, in spite of these difficulties, the use of core materials is increasing greatly; cores are being used in an ever-wider assortment of applications.

Composites are joined by various methods, generally using standard joint design concepts. However, each joint is unique and often requires considerable design time. Manufacturing of the joints is also complicated and must be done with great care.

Composite joining methods have traditionally been dominated by mechanical fastening methods like rivets, bolts and screws. Increasingly, co-curing of parts and adhesive joining are becoming dominant. These methods, while still beset with difficulties, are proving to be more reliable, less costly, of lower weight, and longer lasting than mechanical joints.

Cutting, drilling, trimming, and other post-processing operations for composites are highly varied. Several methods are employed including modified metalworking methods and waterjet, laser, and diamond saw cutting. These methods all have some negatives, but with an understanding of the complex nature of composites and patience, post-processing of composites is becoming routine.

LABORATORY EXPERIMENT 12-1

Crush Strength of Core Materials

Objective: Discover the crush strength of various core materials.

Procedure:

1. Obtain a variety of core materials. Some suggestions are rigid foams, Nomex® honeycomb, aluminum honeycomb, balsa wood, and particle board.
2. Cut out three samples of each material in approximately 6 × 6 in. (15.2 × 15.2 cm) squares.
3. Weigh each sample and note the differences in weight.
4. Using a plate attachment on a tensile test machine, determine the compressive strength of each material.
5. Average the compressive strengths from the samples of each material and record.
6. Determine the ratio of the average compressive strength to the average weight for each material.

QUESTIONS

1. Give two advantages of using a sandwich construction rather than just a laminate.
2. Give three disadvantages of sandwich materials versus laminates alone.
3. What are the principal reasons why composites are more difficult to machine than metals?
4. Indicate three changes from traditional metal drilling that should be made when drilling composites.

5. Indicate the advantages and disadvantages of the following methods for cutting a thin composite part:
   a. waterjet
   b. laser
   c. diamond saw
6. What is co-curing and when can it be used most effectively for adding stiffeners to a composite part?
7. Indicate three advantages and three disadvantages of adhesive bonding versus mechanical bonding of composites.
8. Why was honeycomb material removed from naval airplanes?

BIBLIOGRAPHY

ATC Chemicals, Inc. www.atc-chem.com

Baltek Corporation. www.baltek.com

Benecor, Inc. www.benecorinc.com

Degussa Gmbh. Rohacell®. www.degussa.com

DIAB Core Materials. www.diabgroup.com

Hexcel Corporation. www.hexcel.com

Impact Composite Technology Ltd. www.impactcomposite.com

Lantor BV. www.lantorgroup.com

M. C. Gill Corporation. www.mcgillcorp.com

Nida Core. www.nida-core.com

Plascore, Inc. www.plascore.com

Schofield, R. A. 1997. "Rational Design of Core Materials." *Composties Fabrication*, September, 9–15.

Schofield, R. A. 1998. "Design of FRP Joints." *Composties Fabrication*, October, 12–19.

Society of Manufacturing Engineers (SME). 2005. "Composites Post-fabrication and Joining" DVD from the Composites Manufacturing video series. Dearborn, MI: Society of Manufacturing Engineers. www.sme.org/cmvs

Spheretex America, Inc. www.spheretex.com

Strong, A. Brent. 2006. *Plastics: Materials and Processing*, 3rd Ed. Upper Saddle River, NJ: Prentice-Hall, Inc.

Tricel Corporation. www.tricelcorp.com

Ultracore, Inc. www.ultracorinc.com

13

Open Molding of Engineering Composites

CHAPTER OVERVIEW

This chapter examines the following concepts:

- Process overview
- Gel coat considerations
- Lay-up molding
- Spray-up molding
- Tooling (molds)
- Quality control and safety

PROCESS OVERVIEW

The simplest technique to mold a composite part, probably the first used in modern times, puts fibers into an open mold and then adds resin to the fibers. After ensuring that the fibers are fully wetted by the resin, the resin is cured. Because the resin is added at about the same time the fibers are placed into the mold, these processes are called "wet."

The methods of placing the fibers into the mold can be divided into two general methods—lay-up and spray-up. (For a demonstration of the techniques and equipment used in these processes, see the *Composites Manufacturing* DVD series from the Society of Manufacturing Engineers 2005.) In lay-up, the fibers are in the form of broad goods, usually fabric or mat. In spray-up, the fibers are chopped and sprayed into the mold simultaneously with the resin. It is assumed that the fibers are placed or sprayed into the mold in essentially a manual process (as opposed to, for example, the automatic application of fibers to a mandrel, which is addressed in a later chapter).

The mold can be made of many materials but, to qualify for consideration in this chapter, the mold is single-sided (that is, open) and is not covered by a second rigid mold part during cure, although it might be covered by a flexible sheet.

In the lay-up and spray-up methods, a better outer surface of the part is created if a layer of special resin is applied to the inner surface of the mold before fibers are applied. (The inner surface of the mold becomes the outer surface of the part after the part is removed from the mold.) This first layer of resin is called the **gel coat** and has some special characteristics in manufacturing and operation that are quite different from the resin that forms the matrix for the composite layers. So, in this chapter, the gel coat is considered first and then the lay-up and spray-up processes to make the actual composite layers. These will then be followed by a more detailed discussion of the tooling (mold) in which the parts are made.

GEL COAT CONSIDERATIONS

The gel coat is intended to be largely a protective and decorative surface for the part. It contributes very little to the structural capability. Therefore, the formulation and application of the gel coat focus on the appearance of the part and its ability to continue to look good over time.

The gel coat is applied to the surface of the mold but, when the part is finally removed from the mold, the gel coat ends up as the outer surface of the part. Therefore, the gel coat must release from the mold. To allow this to happen, the mold is treated with a release material (discussed in detail later in this chapter). Another consideration with respect to the mold is the condition of the mold surface. The gel coat will replicate the surface precisely, so great care should be given to ensuring that the mold surface is clean and free of defects (like scratches, dents, or worn areas). After the mold is prepared, attention should be given to the gel coat materials, equipment, application, curing, and properties.

Materials

Almost all commercial gel coats are based on polyester or vinyl ester resins. Because gel coats must resist wear and weathering, the principal polyester resin in a gel coat is usually made with isophthalic acid as one of the components. (See the discussion in Chapter Three on various polyester resin types.) Styrene is the most common reactive diluent (sometimes called the **monomer** or **crosslinking agent**). The gel coat also might be formulated for specific properties such as color stability, processing ease, rapid cure, compatibility with other resins, ease of maintenance, and high elongation. To enable these variations, most gel coat formulations contain pigments, fillers, thixotropic agents to control viscosity, **accelerators** (also called **promoters**), **inhibitors** to extend shelf life, and specialty additives.

Gel coats are typically shipped in drums. They must be stored at moderate temperatures, ideally 72–78° F (22–26° C), to reduce the possibility of premature curing and low-temperature effects, which could be negative for resin formulation stability. The gel coat should be stirred or agitated gently so that solid materials stay in suspension. Normally, a promoter has been added to the gel coat by the resin manufacturer. Therefore, when the gel coat is to be used, the only ingredient that needs to be added by the molder is the **initiator** (sometimes called the **catalyst**). The initiator should be added just prior to using the gel coat.

Color is an important consideration in gel coat manufacture and use. Normally the color is created by a combination of pigments and dyes that are added to the gel coat by the resin manufacturer. The color formulation often can be matched to a specification supplied by the molder. The complexities of color perception and color matching, combined with the complexity of the gel coat formulation, suggest that color matching can best be done by the resin manufacturer.

Equipment

The most effective method of applying gel coats is to spray them into the mold. When properly done, this gives the most even coat with the best appearance. However, spraying properly requires good equipment that is adjusted and operated properly, good operating conditions, and a trained operator.

Several manufacturers make spraying equipment that is adequate for applying gel coat. In selecting the equipment to be used, some considerations are the method by which the resin is delivered to the spray gun, how the initiator is mixed with the resin, and the method of atomization of the resin. The choice of nozzle design and orifice size is also important. The molder should work closely with the manufacturer of the spray equipment to ensure that the equipment is working properly as a complete system, especially in light of the need to cut styrene emissions. The resins that are compliant with the maximum achievable control technology (MACT) generally have higher solids content and, therefore, are more difficult to spray properly.

The ambient environment for gel coat application should be a well-ventilated area with an air exhaust system and adequate methods to prevent overspray contamination. Temperatures should be within a few degrees of 77° F (25° C). After the gel coat has reached ambient temperature, the initiator concentration should be set (usually about 1.8%) and a small sample of initiated gel coat should be sent to the laboratory for gel time determination. If the gel time is not correct, the initiator concentration can be varied to achieve the time stipulated by the resin manufacturer. It is possible to adjust other parameters to achieve the correct gel time, but the initiator concentration (within the accepted range) is normally not only the easiest parameter to adjust but also the one most likely to give a good result.

Application

If the materials, equipment, and environment are correct, the operator is the determining factor for proper gel coat application. The operator has the responsibility of coating evenly without applying the gel coat too thickly. For most applications, the best gel coat thickness is 18 mils ±2 mils (0.46 mm ±0.05 mm). The thickness should be frequently checked. The device used to check the thickness is called a **mil gage**. It is simply a small piece of metal that has two posts that are pushed through the gel coat until the posts touch the mold. Between the posts are a series of fingers that extend toward the ends of the posts but are shorter than the posts by varying amounts. The thickness of the gel coat is measured by noting which of the extending fingers is not covered with gel coat. That finger is the first one above the thickness of the gel coat. A coating that is too thin may not cover or cure properly. A coating that is too thick could result in cracking.

Several manuals are available to assist the operator in developing a technique that will give uniform coating at the right thickness. In general, it is better to use a series of passes rather than a single concentrated pass. Difficult areas should be sprayed first, working outward to other areas. Most gel coats are formulated with thixotropic agents that will allow spraying on side (vertical) areas without resin sagging or running. Also, gel coats are formulated to flow together without pinholes or bubbles; however, this characteristic is affected by proper technique too.

Curing

Because gel coats are pre-formulated with all the required ingredients for curing except the initiator, curing begins immediately when the initiator is added just before spraying. However, the cure is not instantaneous; that is, the gel coat does not immediately become hard. Typical time for a gel coat to be ready for the next step in the molding process is about 45 minutes. This time should rarely exceed 8 hours since beyond that time it is so highly cured that a good bond between the gel coat and the laminate layer is difficult to achieve.

Properties

Many problems can occur with a gel coat. Some are seen immediately, usually because of poor application or problems with the conditions under which the application was done. Therefore, inspections should be routine and vigorous. Troubleshooting guides are available from resin manufacturers, equipment manufacturers, and professional or trade organizations that can assist in identifying the causes of defects.

Other problems are not detected until long after the gel coat has been applied. They may not, in fact, be discovered for many months or years. One problem that has been especially difficult for the composites industry is **cracking**. Cracking may occur simply because the gel coat is over-crosslinked. This could result from too much initiator being used or from excessively hot curing conditions. Therefore, the range of initiator volume and the conditions of cure should be carefully controlled.

As already discussed in previous chapters, cracking of a cured polymer occurs because energy within the material exceeds the strength of the material. That energy can be caused by the molding and demolding operation itself. Demolding is often difficult because the part sticks and, subsequently, stresses occur in trying to free the part from the mold. The energy to crack the gel coat also might come from impacts, although these are fairly easy to identify and obvious as to the cause.

More subtly, the energy could come from two or more causes that are, by themselves, insufficient to cause the cracking. For example, the part may experience stresses during curing because it has different laminate thicknesses or because of binding of the part within the mold. (Some areas may have thicker laminates and some may even have wood reinforcement added to the composite.) Stresses also might be generated by sharp edges at corners or around fittings or holes. Subsequently, when the product is placed in use, additional stresses may occur that, combined with the residual stresses in the part from molding, might result in sufficient energy to crack the gel coat. This is especially true if the gel coat is thicker than the manufacturer recommends. The additional stresses could come from general operation, such as the conditions a boat might experience in rough seas, or from extremes of temperature (the differences in the coefficients of thermal expansion [CTE] in various sections of the part).

To prevent cracks from occurring, the molder must test each design and each change in materials under a variety of normal and extreme operating conditions. These tests will provide the molder with the envelope of conditions within which the part can be operated. The molder must also monitor and control the manufacturing process so that:

- The gel coat thickness is within the correct range.
- Proper molding conditions exist (mixing of the gel coat in the drum, gel time, initiator level, shop temperature, spray-gun setup, spraying technique, time between application of gel coat and application of the laminate layers, etc.).
- Other potential stresses, such as demolding problems, are eliminated.

Sadly, all polymeric materials degrade over time. This will affect the appearance and the performance of the gel coat. Proper maintenance, such as applying protective coatings like waxes, and good cleaning can minimize the amount of degradation, but they cannot stop it. Therefore, some deterioration of properties is inevitable.

Because the gel coat is non-structural, it will be backed with layers of composite material (that is, resin reinforced with fibers). This backing material, called the structural layers or the laminate, might be seen through the gel coat, even when the gel coat is highly pigmented. Therefore, it is common practice to lay a thin fiberglass layer onto the gel coat before the composite material is applied. This thin fiberglass layer is called the **veil**. The veil not only visually hides the laminate layer, but also physically prevents fibers from pushing their way into or through the gel coat layer (a process called **blooming**). It also helps to smooth the final gel coat surface because it minimizes the effects of changes in laminate thickness.

LAY-UP MOLDING

Lay-up molding is one of two methods commonly used to apply the composite layers behind the gel coat. The other is spray-up, which is discussed later in this chapter.

Lay-up parts can be of almost any size and shape. In fact, the largely manual nature of lay-up molding makes it the method of choice for some parts that, usually because of high complexity of shape, can only be made by this method. The manual nature of the process also allows easy inclusion of inserts around which the composite material is placed. The wide variety of parts that can be made leads to a wide variety of curing conditions. Room-temperature and elevated-temperature cures are common. If only a few parts of any design are to be made, lay-up is usually the most effective method of manufacture.

Resin and Curing Agents

Due to the simplicity of the lay-up method and the generally manual nature of resin application, a wide variety of resins can be used. Although almost any polyester or vinyl ester resin is a candidate, the vast majority of applications use polyesters containing orthophthalic acid as one of the components. These are generally the least expensive polyesters and usually provide adequate properties for most applications. (These and other polyester resins are described in Chapter Three.) The lay-up resins usually have styrene as the reactive diluent. Fillers and other additives are most commonly added by the molder rather than by the resin manufacturer.

Lay-up resins are frequently shipped in drums; however, in large operations, the resins can be shipped in tank trucks or tank railcars. The resins should be stored within the temperature range specified by the resin manufacturer, which is usually moderately narrow—about 77° F (25° C). Because the resins are often shipped **neat** (containing no fillers or reinforcements), stirring the resin while in storage is less critical than with gel coats.

Just as there is a wide selection of lay-up resins, many initiators (catalysts) can be chosen for use with them. The initiator is usually chosen according to two criteria: the temperature at which cure will occur and the type of resin. Since most resins are ortho polyesters, which are compatible with most initiators, the cure temperature is the key parameter. But, if a resin other than an ortho type is used, the resin manufacturer should be contacted to check on the compatibility of the resin and initiator. To select the initiator based on cure temperature, the initiator manufacturer should be consulted. Usually written guides can be provided that list the available initiators and the effective temperature ranges. These guides also list acceptable accelerators and other additives for each initiator. Generally, the molder has the responsibility of adding these curing agents to the resin.

Reinforcements

Broad goods are the most common form of reinforcements for lay-up. These can be mats, woven cloths, knits, or any other of the common textile broad goods discussed in Chapter Nine. The choice of reinforcement, the orientation of fabrics (if any), and the number of layers are dictated by the design.

The reinforcements are generally simply laid into or onto the mold as indicated in Figure 13-1. They are placed on top of the gel coat, which may have a veil.

For ease of handling and speed of application when the same part design is to be made many times, broad goods are often precut before they are placed into the mold. Several cutting methods are available, ranging from simple hand-held devices to large automatic cutters that use a bed for supporting the broad goods and a gantry system to mount the cutting head. The actual cutting can be done with blades, dies, sonic-driven cutters, lasers, or waterjets, to name the most common. Sometimes a molder finds it easier and less expensive overall to use precut broad goods that have been bundled together. This

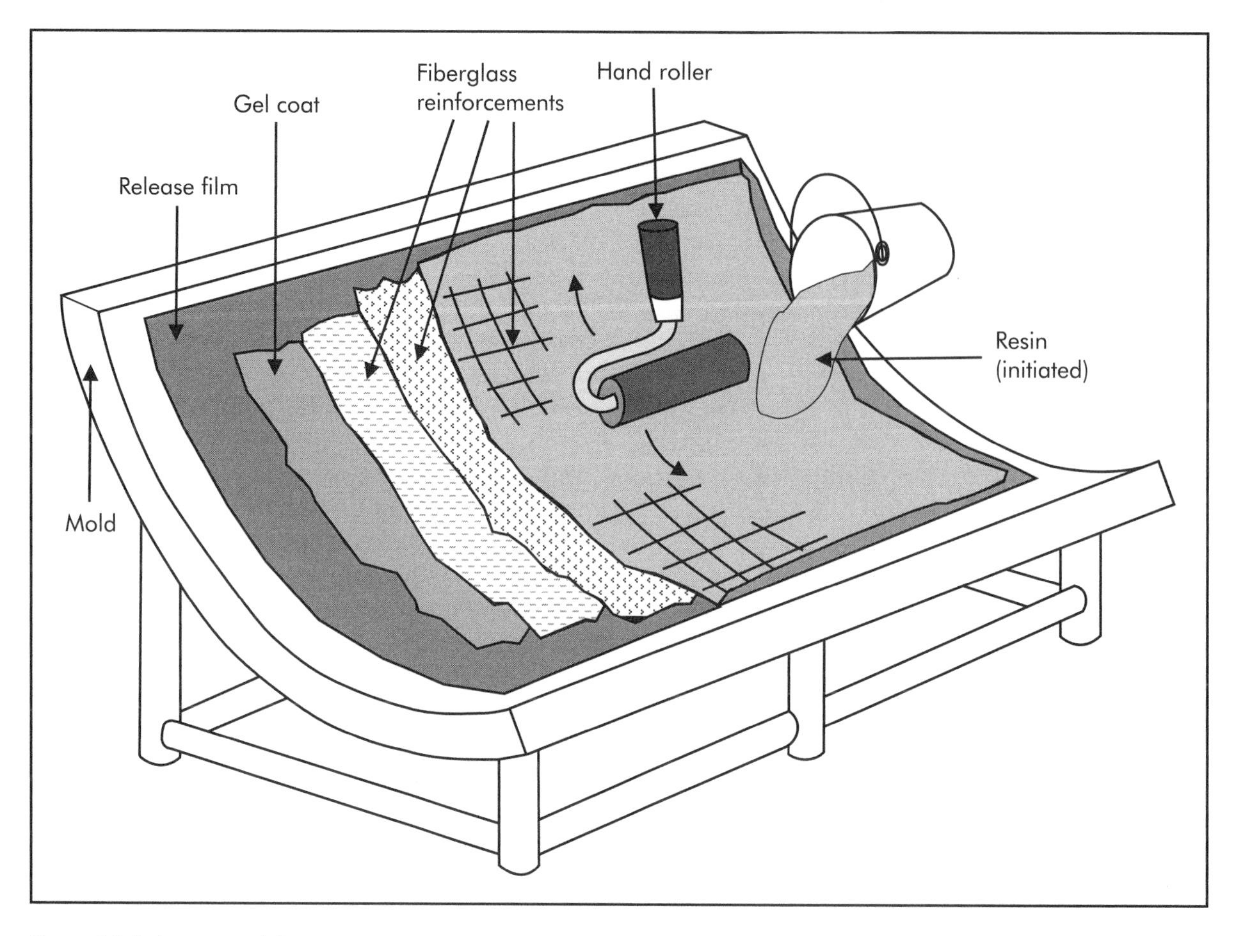

Figure 13-1. Lay-up molding.

is called **kitting**. This means the reinforcements have been cut and assembled so all that is needed is to simply place the reinforcement pack into the mold.

Resin Application, Wet-out, and Curing

In normal practice the resin is mixed with the initiator in a mixing vessel and stirred to ensure full dispersion. If fillers or other additives are used, they are also added in the same mixer. When fully mixed, the proper amount of resin to achieve the desired resin-fiber ratio is metered or weighed into a vessel (bucket) and transported to the molding area. To improve wetting, the upper surface of the gel coat can be pre-wetted with a thin layer of resin before the reinforcements are placed in the mold, but this step is optional. The reinforcements are placed in the mold and the remainder of the resin is poured to cover them as evenly as can be reasonably done. Alternately, the reinforcement can be wetted external to the mold and then transferred into the mold and smoothed.

Full wet-out and coverage of all the reinforcements is done by rolling the resin into the reinforcements. Small hand-held rollers with ridged surfaces have proven to be excellent tools for wetting out the surface. By rolling, the resin is distributed evenly over the fiberglass reinforcements. When the fiberglass is wetted, a change in color (translucency) is noted as the resin is absorbed, thus aiding in monitoring the wet-out process.

Curing can be done at room temperature or at elevated temperature (usually in an oven) as desired. The procedures and cautions discussed in Chapter Three should be observed in the curing operation. One important point that must be considered is the exotherm. Because exotherms can damage the part or cause safety issues, the composite layer cannot be too thick. Alternately, if the layer is too thin, proper curing may not occur. Therefore, experiments must be done to define the range of thicknesses for each layer so a proper cure is obtained. If one layer of reinforcement and resin is not sufficient for the structure of the final part, as is often the case, additional reinforcement layers and resin are added after the previous layer is cured. This is called **secondary bonding**. It is desired that chemical bonds form between the primary and secondary layers. This can best be accomplished when the resins in the two layers are the same. Also, adding the second layer to the structure before the first layer is nearly but not totally cured permits the softness of the primary layer to facilitate the formation of bonds between the layers. This is because the resin and styrene soften and soak into the primary layer. By use of the repetitive curing and application method, composites of any thickness can be made. However, the thickness of each layer should be controlled within the range defined for acceptable cure.

Care must be taken in the time between curing of the primary layer and application of the secondary layer so that the surface between them does not become contaminated. Some resins contain materials that will naturally contaminate the surface. An example is wax. This material is added to some resins to prevent air from reacting with the surface to prematurely quench the cure. The wax forms a layer that prevents air from reaching the surface. However, it also inhibits secondary bonding. Therefore, any resin having wax should be mechanically abraded (sanded) before applying the secondary layer. Other times that abrading should be done include when the primary layer has been cured at an elevated temperature, exposed to UV light, cured for more than 24 hours, or has a glossy resin-rich surface.

Additional structural materials, such as wood panels, might be added for increased stiffness. These are usually covered with a layer of uncured reinforcement and then cured to bond the wood into the laminate structure.

After the part has been completely cured, it is removed from the mold (as discussed later), trimmed or otherwise finished, and inspected for quality.

SPRAY-UP MOLDING

Spray-up molding is used when the part is of sufficient size and simplicity of design that spraying the resin and reinforcement can be accomplished easily and with good uniformity. Small and complex parts are usually not good choices for spray-up molding. However, for many composite parts, such as boats and tub/shower units, spray-up is the preferred method. The advantage of spray-up over lay-up for most parts is simply the speed with which the fiber and the resin can be applied to the mold.

In addition to the limitations already mentioned, the disadvantages of the spray-up method versus lay-up are the special spraying equipment required, the more limited choice of resins, the inability to control the direction of the fibers, high air pollution because of spraying the resin, and the higher skill level needed by the operator. The choices of resin are more limited because the resin's viscosity must be more carefully controlled to ensure that spraying can be done appropriately.

Equipment, Spray-up Technique, and Curing

The spraying equipment used for the composite material is similar to the type used for gel coats with some obvious differences such as the inclusion of a chopping mechanism on the spray gun. The fibers are brought into the chopper as roving and then chopped so that they fall into the stream of resin just after the nozzle. This means that the chopped fibers are entrained with the resin and together are sprayed into the mold. The spray-up method is illustrated in Figure 13-2.

The high degree of care needed in choosing the type of spray nozzle and the narrow operational limits required in gel coat spraying are not as important in spraying the composite materials. Nevertheless, some training for the spray-up operator is required, especially since the uniformity of fiber placement is critical to the performance of the part.

The operator sprays the resin and the entrained fibers into the mold. Coverage must be uniform and the thickness of the sprayed layer must be within the limits set for good curing.

A compromise is required in the length of chopped fiber. On one hand, the length of the chopped fiber should be short for ease of spraying and coverage into tight areas of

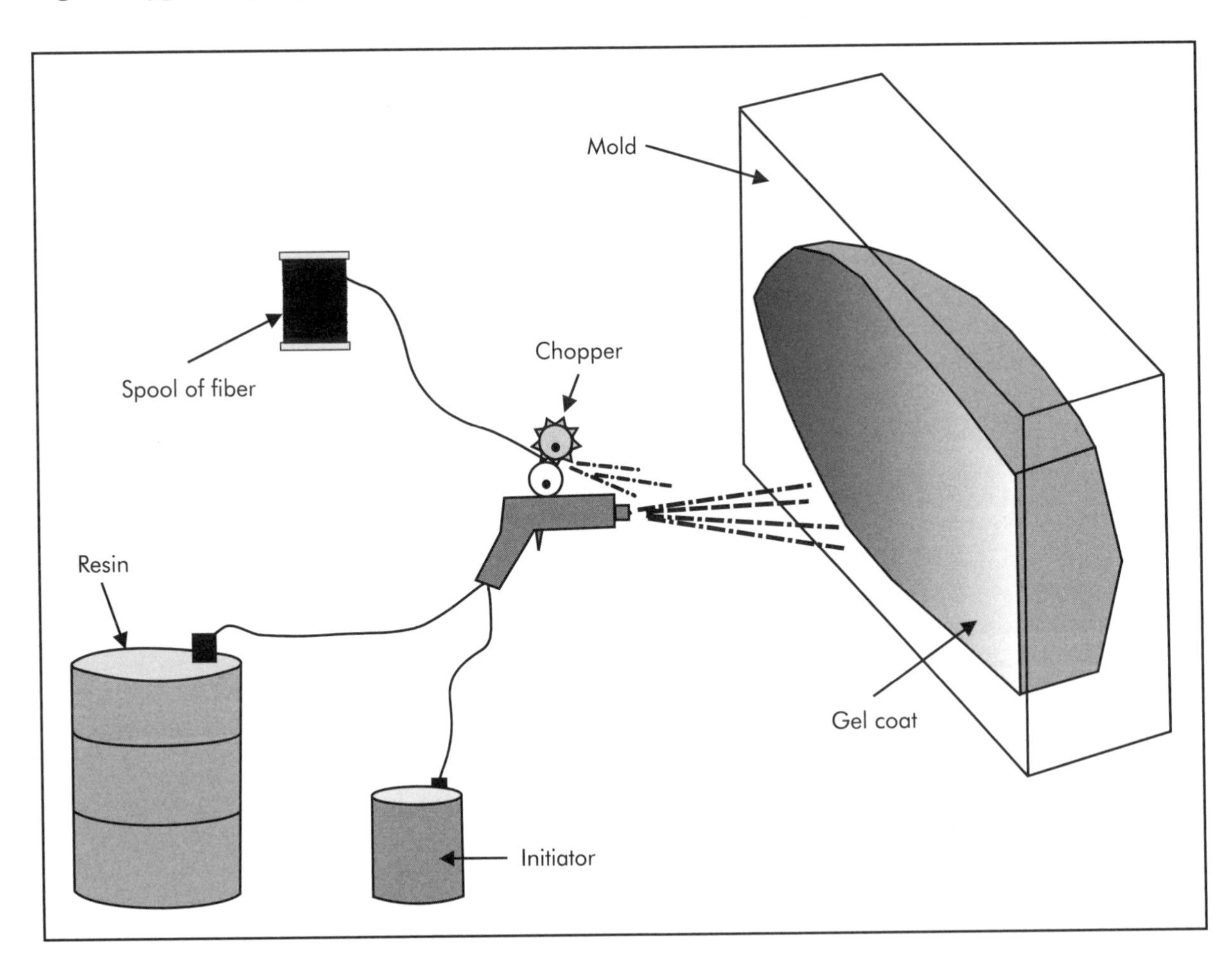

Figure 13-2. Spray-up molding.

the mold (including corners). But for best mechanical properties and ease of chopping, the length should be long. Most molders have found that fiber lengths in the 1–3 in. (2.5–7.6 cm) range are best.

In spray-up, wet-out of sprayed fibers is accomplished in the same way as with lay-up. After the fibers and resin are sprayed into the mold, a roller is used to make sure the resin fully wets the fiber bundles. Therefore, even spray-up uses manual labor as part of the process.

Curing in spray-up is identical to curing in lay-up. The addition of more layers for increased structural capability or wood or other panels for even more stiffness is also possible as with lay-up. However, these additional layers and the material placed around the panel are added by spraying rather than by hand. To make it easier to see where the second (or later) layers have been sprayed, many molders add a small amount of dye to the resin when the second layer is applied. This color difference allows the operator to see the areas that have been sprayed and to gage the thickness and coverage of the laminate.

TOOLING (MOLDS)

In both lay-up and spray-up, the mold is not only the definer of the shape of the part, but it provides the surface on which the materials are deployed prior to cure. Hence, the tooling in open-mold processes is important. Further, because cure times are often long and productivity requirements can be high, such as for tub and shower units, a company may be forced to have a large number of tools to meet production demands. Therefore, some attention should be given to making the mold and its use in the lay-up and spray-up processes.

Mold-making Method

Because the mold defines the shape of the final part, great care must be given to ensure that it has the exact dimensions and surface characteristics that will result in the type of part desired. Often, care also should be taken to ensure that the mold is durable so it will last for the production of several parts.

Many materials can be used for molds but the most common are metals and composites. The choice of material depends upon:

- expected wear (metal molds have lower wear),
- temperature of the cure (metal molds withstand higher temperatures),
- pressure exerted (metal molds withstand higher pressures),
- thermal expansion matching with the part (composite molds generally match better), and
- cost (composite molds are less expensive).

Since lay-up and spray processes usually require room-temperature cures at atmospheric pressures, the factors suggest that composite molds would be best and this is true in most cases.

It is conceivable to machine the surface of a mold directly and thus create the shape that will define the parts. However, several complications arise when this is attempted. First, the mold must be machined in the reverse image of the part. That is, because the part will be made against the mold, the part and mold must be reflections of each other. This is often difficult to do from a logistics standpoint as it requires thinking in the reverse and mistakes are common when this is attempted. Second, the cost of machining a mold is high and it usually requires the use of expensive machines. Third, the wear on the mold can be high. Therefore, it is better to find another way to make the mold. A common method for making composite molds is shown in Figure 13-3.

The requirements for mold performance are usually met by making the mold from

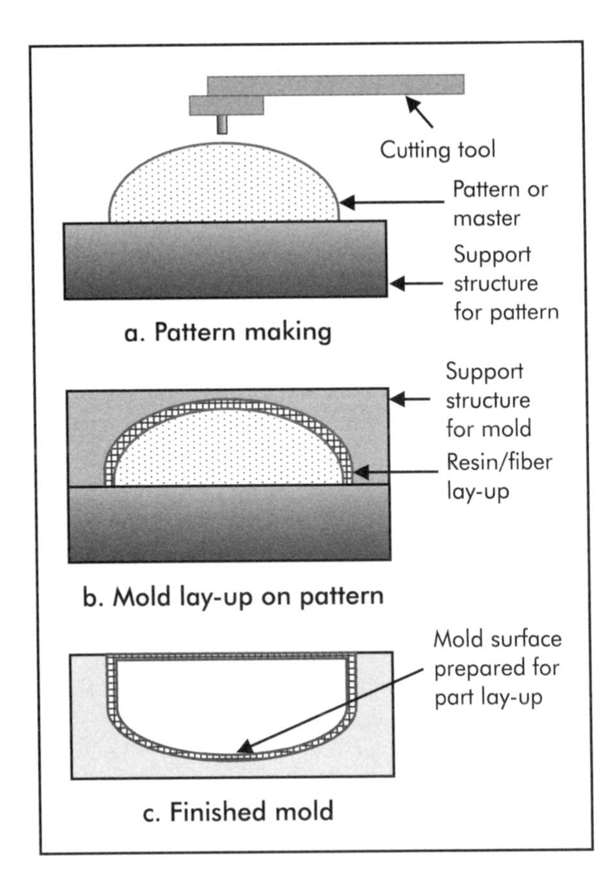

Figure 13-3. Mold-making.

a shape that is exactly like the finished part. The shape that incorporates all of the geometry for one of the part surfaces is called the **master model, pattern** or, occasionally, the **plug**. The model can be made from a variety of materials but the material must be chosen so that the model can be shaped exactly as desired. This shaping is usually done by conventional cutting processes. Today, the use of a multi-axis computer numerical control (CNC) milling machine is the most common cutting and shaping method. However, some handwork is done for finishing.

The surface of the model is shaped and finished to exactly match the surface of the part. Resin shrinkage during molding will result in a part that is smaller than the master model. For some parts, the master model may be made slightly larger to offset shrinkage. The exact amount of shrinkage is complicated and depends on the type of resin, the fiberglass content, the processing temperatures, and other variables. The master model is also modified to enlarge the opening of the mold slightly so the part can be removed. This is done by creating draft angles of about 2° on the sides. A special hard surface coating is sometimes applied to the master model and finished to the final dimensions so it is more durable. When the master model is exactly as desired, the surface of the model is prepared with mold release.

The production mold can be made directly from the master model as shown in Figure 13-3. However, this is a simplification since, in many cases, the mold made from the master model is actually, itself, merely an intermediate from which other models and molds are made. (This is a topic that will not be discussed further here. Books on mold making can explain these intricacies and their advantages. Nevertheless, the methods of making a mold from a master model are the same in all systems.)

The master model serves as a tool on which the mold is laid up. A gel coat is applied just as was described for the lay-up molding process. In this case, however, a special tooling gel coat is used to increase the wear capability of the mold. After curing the gel coat, reinforcements are laid onto the master model and resin is applied. This layer is cured and other layers are applied to obtain the desired thickness. Again, special laminating resins are used to improve the durability of the mold. It is not unusual to use epoxy resins for the tooling because of their higher performance versus polyesters. It is also common to cure the mold at elevated temperatures and/or to post-cure the mold so that a high level of crosslinking will be achieved. This also adds to durability.

The mold is then removed from the master model and, after proper surface preparation, the mold is ready for production use.

If a higher-temperature surface is desired, epoxy molding compound can be coated on the surface, cured, and then machined. This epoxy material is called **seamless molding paste**.

Some mold-makers have used a metal facing to further improve the surface durability of the mold. The metal is usually nickel applied by a sputtering process. While this technique is useful, the increased cost is justified only in those cases where production volumes are high.

As is seen in Figure 13-3, the mold has a support structure or bracing. This structure is needed to provide easy handling for large molds and gives dimensional stability to the mold. Sometimes this support structure is created just by making the mold thick. However, this often results in heavy molds. Therefore, an alternate method of supporting the mold is to apply a network of braces that support the mold surface so it is dimensionally stable while providing convenient grips to use for lifting and moving the part. Because these braces often form a criss-cross pattern that resembles the shape of an egg crate, the system is called an **egg-crate mold**.

Another system for creating molds involves the use of a fluidized bed into which the part or master model is pressed. The bed is solidified so the shape of the part is retained in the bed. Then, after the part is removed, the surface of the bed can be finished, and the bed becomes the mold. The use of patented materials for the bed has made these systems using two-phase technology practical.

Removal Techniques for Parts

Mold release needs to be added to master models and to production molds so that the parts made on these forms can be removed. Proper preparation of the mold surface is especially important at its first use and whenever the mold's surface has been repaired or sanded. In these difficult release occasions, a film-forming material, such as polyvinyl alcohol (PVA), is often applied to the mold surface to guarantee release. The PVA sticks to the part but can be easily removed. (PVA dissolves in water.)

For most occasions, the production mold is conditioned for release by adding a mold release material. The original **release agents** were waxes and these are still frequently used. Other types of release agents have been developed. Some are based on silicone technology and still others use various polymers as the basis for release. These materials are added several times over several heating cycles to condition a new mold. Then, after conditioning, the mold is ready for the first application of resin. As indicated previously, PVA film may be used for this first run. Afterwards, the mold release agents are added after each part is removed. Gradually, the mold becomes conditioned and the application of mold release need not be as frequent. If the part shows a tendency to stick, the mold release should be reapplied.

When a part is to be removed from the mold, as little force as possible should be used. Because of good mold design and/or simplicity of the part, some parts are easily removed. However, complex and large parts are often difficult to remove. One method used to remove the part is to attach the part to a crane and the mold to an anchored base and pull. To do this means the part must have special locations designed for gripping and pulling. Another method is to insert wooden wedges between the mold and the part so that the adhesion is broken. Care must be exercised that neither the mold nor the part is damaged. Air is often blown into the openings created by the wedges. Some molders provide for air holes in the mold so that air can assist the part. Water also can be used to float the part. Some have cooled the inner surface of the part so that it will shrink away from the mold. Another method is to tap on the outside of the mold using rubber mallets. This method has great danger of damaging the mold and part, so it only should be used as a last resort.

QUALITY CONTROL AND SAFETY

All parts made by lay-up and spray-up should be inspected after removal from the mold. There also should be intermediate inspections by operators and quality control inspectors so that problems can be caught as early in the process as possible.

One of the problems with open-molding systems is that the weight of the composite cannot be accurately monitored during the lay-up and spray-up processes. In spite of this, a method of feedback to the operator in charge of spraying or laying up the material is useful. This can be done by weighing the part after demolding and reporting the weight to the operator. If a record of the weight is kept and compared to the "ideal" weight, the operator will be able to see whether the total weight needs to be increased or reduced and adjust the spray-up or lay-up accordingly.

The safety of lay-up and spray-up is concerned first of all with the resin. A material safety data sheet (MSDS) for each material should be filed and easily available. Protective equipment also should be worn when resins are used. Federal standards mandate a maximum concentration for exposure to styrene. This should be monitored.

CASE STUDY 13-1

High-volume Production of Tub and Shower Units

Most molders consider lay-up and spray-up molding to be for low-to-medium production volumes. However, with some creative modifications to the traditional methods, the rate of production of parts can reach hundreds a day in a single factory.

The market demands for high volume and low price of integral tub and shower units force the manufacturers to devise methods that permit high volume with low production cost. These can be met using spray-up molding and accelerating the curing of each layer of the multi-laminate part. The methods for spraying up the gel coat into a composite mold are standard. However, after the gel coat is applied, the mold is hooked onto an overhead chain system that serpentines through the plant and carries the molds from one workstation to the next. The mold passes through an oven and then arrives at a station where the laminating resin and reinforcement are applied. The mold is taken off the chain and sprayed up in the normal fashion. The mold is then reattached to the chain and moves through another oven. The process is repeated at other spray-up stations until all the desired laminate layers have been applied. The part is then fully cured in yet another oven. It then moves to a demolding station where the part is removed and transferred to finishing, inspection, and shipping. The mold moves to an area where it is refurbished and then put back onto the production line.

There are hundreds of molds moving through the plant at the same time. This system achieves high productivity by combining the advantages of the assembly line with the methods of spray-up and then adds elevated-temperature curing to make it all work quickly.

SUMMARY

When a simple method for making a part is desired, lay-up is the method most often used. It uses manual labor to overcome some of the problems of more automated methods. However, because of the ability of manual labor to adjust to the varying conditions and materials, it is the process that has the most opportunity for variation in resin, reinforcement, and curing methods.

The process begins with the application of a gel coat. This is the outer layer of the part and is made from a specialized polyester resin to which various pigments, dyes, fillers, and other additives have been added. The gel coat is applied by spraying in both the lay-up and spray-up methods.

Some improvement in productivity of lay-up can be achieved by using precut broad goods. These can be cut using a variety of methods. After cutting, the broad goods can be kitted together, thus further improving the speed with which the reinforcements can be placed into the mold.

Spray-up is similar to lay-up in many ways but adds the advantage of faster application of the resin and reinforcement. However, it requires more expensive equipment and it is harder to achieve uniform coverage. In both lay-up and spray-up, wet-out of the fibers is achieved by manually rolling the fibers onto a mold to which resin has been applied.

Curing of lay-up and spray-up materials is usually done at room temperature. However, both have the ability to be modified for elevated-temperature curing but the exotherm that occurs with curing must be controlled. This is done by limiting the thickness of each layer of composite material.

The tooling for both lay-up and spray-up is generally made of composites. This tooling is made from a master model shaped like the final part. The mold is made by laying up composite on the master model. This mold is then conditioned with mold release and then can be used to make many parts.

LABORATORY EXPERIMENT 13-1

Making a Laminate

Objective: Practice the methods of lay-up.

Procedure:

1. Formulate a batch of resin by adding initiator and mixing.
2. Lay a sheet of polyethylene terephthalate (PET) film (Mylar®) onto the picture-frame mold.
3. Pour a small amount of initiated resin onto the film and spread it evenly.
4. Place a surfacing veil onto the top of the resin.
5. Apply more resin and roll to wet-out the veil.
6. Apply a layer of chopped strand mat on top of the wetted veil.
7. Add more resin and roll to fully wet the mat.
8. Apply a layer of woven roving on top of the wetted mat.
9. Roll to wet the roving.
10. Add a sheet of chopped strand mat and roll to wet-out the fibers.
11. Finish by brushing on a layer of wax topcoat.
12. Cure by allowing to stand overnight.

QUESTIONS

1. What is "fiber blooming" and how is it prevented?
2. What is the advantage of spraying on a gel coat versus some other application method such as rolling or brushing?
3. What is a "neat" resin?
4. What is the purpose of rolling the fibers after resin has been added?
5. What are thixotropic agents and why are they used?
6. How is the ratio of resin and initiator controlled in the spray-up method?
7. How are additional reinforcements for selected areas added to a spray-up part?
8. What is one advantage of spray-up over lay-up? What is one advantage of lay-up over spray-up?

9. How are the weights of resins and fibers controlled in lay-up?
10. What is a "gel coat" and why is it used?
11. Why is it important to control resin viscosity in any process that involves adding resin to fiber reinforcements?
12. Why is it important to control resin viscosity in a spraying operation?
13. Why is a master model (pattern) made slightly larger than the part it otherwise duplicates?
14. What is the purpose of the rigid support structure that backs molds?
15. Indicate a problem that might occur if the resin is heated too quickly during an elevated-temperature cure cycle.
16. Why are the performance characteristics (properties) higher for a resin used to make a mold than for the same laminate resin to make a part?

BIBLIOGRAPHY

Cook Composites & Polymers. 2005. *Composites Application Guide*, 10th Ed. Kansas City, MO: Cook Composites & Polymers.

Dow Plastics. *Fabricating Tips*. 10/94. Revised Edition. Midland, MI: Dow Plastics.

Society of Manufacturing Engineers (SME). 2005. "Manual Lay-up and Spray-up" DVD from the Composites Manufacturing video series. Dearborn, MI: Society of Manufacturing Engineers, www.sme.org/cmvs.

Strong, A. Brent. 2006. *Plastics: Materials and Processing*, 3rd Ed. Upper Saddle River, NJ: Prentice-Hall, Inc.

14

Open Molding of Advanced Composites

CHAPTER OVERVIEW

This chapter examines the following concepts:

- Process overview
- Prepreg lay-up (manual and automated)
- Vacuum bagging
- Curing (with and without autoclaves)
- Roll wrapping
- Tooling (molds)

PROCESS OVERVIEW

Although the wet lay-up and spray-up methods described in the last chapter could be used with advanced materials, for example, epoxy resin and carbon fibers, typically other processes are used. These other processes, which will be discussed in this chapter, overcome some of the inherent problems associated with the wet lay-up and spray-up methods.

The major problems that have led to the modified manufacturing methods for advanced composites include the following:

- difficulty in aligning the fibers in exactly the directions desired,
- difficulty in optimizing the amount of fibers to achieve the highest properties possible,
- difficulty in controlling the fiber/resin ratio,
- difficulty in getting full fiber wet-out,
- difficulty in reducing the void content, and
- potential problems in mixing the resin and hardener (including emission and toxicity problems).

In contrast to those methods discussed in Chapter 13, the manufacturing methods developed for making advanced composites in an open mold involve differing materials, manufacturing procedures, and equipment. The materials most often used for advanced composite parts are prepregs. These materials were discussed in some detail in Chapter Nine. As a reminder, these are sheet materials, unidirectional fibers, or fabrics, which have been carefully coated with fully initiated resin.

Molded products are made by laying sheets of prepreg into a mold. The prepregs are carefully oriented so that the fibers are in the direction desired. Since prepregs come in set widths based on the width of the machine used to make them, they are cut to fit the shape of the mold. Layers of prepreg are placed, often in different orientations, until the proper thickness is achieved. Most prepregs have a sticky surface because the resin is not fully cured. This facilitates placement of the prepreg layers in the mold. If prepregs are made using thermoplastic resins as matrices, they are not sticky. Thus a heating iron or some other method to soften the resin is used to anchor the prepregs in place so that they do not move after proper placement. Such laid-up prepreg assemblages are, of course, laminar. Curing is usually done under conditions of vacuum and then pressure. These conditions reduce the number of voids and allow the fiber content to be maximized.

As might be obvious, the procedures using prepregs largely eliminate the problems of wet lay-up and spray-up identified previously. How the use of prepregs and other advanced manufacturing techniques solves the problems will be even more obvious as they are discussed. (Both manual and automated prepreg lay-up methods are demonstrated in the *Composites Manufacturing* DVD series, which correlates with this text [Society of Manufacturing Engineers 2005]).

It is good to keep in mind the overall goals of the open-molding system for advanced composites. They are to:

- control fiber orientation and location,
- control the ply thickness,
- control the fiber-resin ratio,
- minimize the void content, and
- reduce internal stresses.

PREPREG LAY-UP

The location where prepreg lay-ups are done is usually an environmentally controlled room. The temperature of the room is kept constant in the range of 72–77° F (22–25° C) and the air is filtered to eliminate dust particles. Entrance to the room can be through a double-entry system to keep the air clean. The workers in the room may have special coats, hats, and shoe covers to further reduce dust and other contamination.

A special and most difficult problem is contamination from silicone. Experience has shown that if silicone contamination is present on the surface of the prepreg, the ability of that layer of prepreg to bond with another is compromised. Also, the ability of the finished part to bond with other parts can be difficult. The subtlety of this contamination is that it is airborne and cannot be easily removed with filtration. The silicone originates in some mold release products. When those are applied to a mold, especially when sprayed onto the mold's surface, the silicone vapor is picked up by the air conditioning system and recirculated. When the vapor is cooled by the air conditioning, it can settle on the prepreg and cause surface contamination. The contamination is difficult to detect until delamination problems become apparent and an investigation reveals that silicone is present at the delamination point. The best method to prevent this contamination is to avoid the use of silicone-containing mold releases. If they must be used, choose a type that is non-aerosol and less volatile.

Prepreg materials are almost always kept in freezers to extend shelf life. Most often the prepreg roll is kept inside a plastic bag. This helps keep the prepreg clean and also reduces the amount of moisture contamination. After removing the bag from the freezer, the bag should remain closed (sealed) while the prepreg material warms to room temperature. When the prepreg is ready to be used, the bag is opened and the prepreg roll is removed. To avoid contamination from body oils and dirt that might be present on the operator's hands, clean cotton gloves should be used to handle prepregs.

Two systems are available for placing the prepreg in the mold. These are manual lay-up and automated lay-up.

Manual Lay-up

The manual lay-up system involves laying plies of precut prepreg into a mold. To begin the lay-up, sufficient material is cut from the roll of prepreg for all of the layers of the part to be made. If material remains on the roll of prepreg after the desired amount is cut off, the roll should be returned to the protective plastic bag and sealed. Then, the bag should be returned to the freezer to maintain the maximum amount of shelf life.

The sheet of prepreg is then placed on a special cutting table where it is cut to the shapes required in the lay-up. Uncured prepreg can be cut manually or by a machine. Manual cutting of uncured prepregs has been the major method used for many years. Standard knives with replaceable blades (Stanley knives) are probably used more widely than any other type to cut prepregs. A template is normally used and the cutting table often has a urethane or other soft plastic cutting surface that does not get permanently marked (self-healing surface). Cutting and trimming also have been done with carbide-edge cutting wheels (like pizza cutters), reciprocating knives, and shears (manual and powered). New designs and materials in these cutting tools (such as ceramic shears) have improved their efficiency and accuracy, but they still remain difficult, inaccurate, and time-consuming to use.

The reciprocating knife is an improvement over the above methods in accuracy and labor costs. It has been used extensively in the textile industry. This method uses a knife that either cuts by vertically impacting the prepreg material like a guillotine into a plastic bristle or soft metal cutting surface below the material, or by penetrating and then remaining buried in the material and slicing through with rapid vertical strokes. Both types of reciprocating knives can cut many layers of material simultaneously. Curves and sharp corners can be easily cut, and the blade can be retraced to skip over areas that are not to be cut.

Ultrasonic vibratory cutting, which is similar to the reciprocating knife process, is a major development for cutting prepreg. In this method, mechanical vibrations are used to drive a knife at ultrasonic speeds (several thousand vibrations per second). The ultrasonic vibrations enhance the cutting action of the blade by reducing friction between the blade and the prepreg. However, the wavelength of the vibrations causes nodes (non-vibration points) along the surface of the long blades, which hinder cutting at those points. Some stacks of material may be too high and if cutting is attempted at one of these nodes, poor cutting results. Compromises must be made between the efficiency of cutting many layers and the nodal problem with the cutters. A reasonable compromise is to limit the total stack thickness of prepregs to about .75 in. (1.9 cm). The major advantages of ultrasonic cutting are high speed and accurate cutting of stacked prepregs.

Another method for cutting several layers at the same time is **die cutting** (also called **steel-rule blanking**). Again, this method has been used in the textile industry for many years. In die cutting, a die made from hardened steel strap, which has had one edge sharpened for cutting, is placed over the material and then punched through the material with a press. The steel strap has been shaped into the outer profile of the part. Sometimes this is called the cookie-cutter method. Advantages include being able to cut many parts of the same size and shape easily, minimum mechanical breakdown potential, being able to cut a wide variety of materials, potential automated cut-part retrieval and lay-up, and good material yield with careful layout of the cutting patterns. The careful layout of the material to avoid waste is called **nesting** as shown in Figure 14-1.

Laser and waterjet cutting are gaining acceptance for cutting uncured as well as cured composites. The advantage of these methods is their speed and accuracy. The disadvantage is the cost of the machines, especially when the alternate methods, such as manual cutting, are by far more economical. Laser cutting of uncured composites can

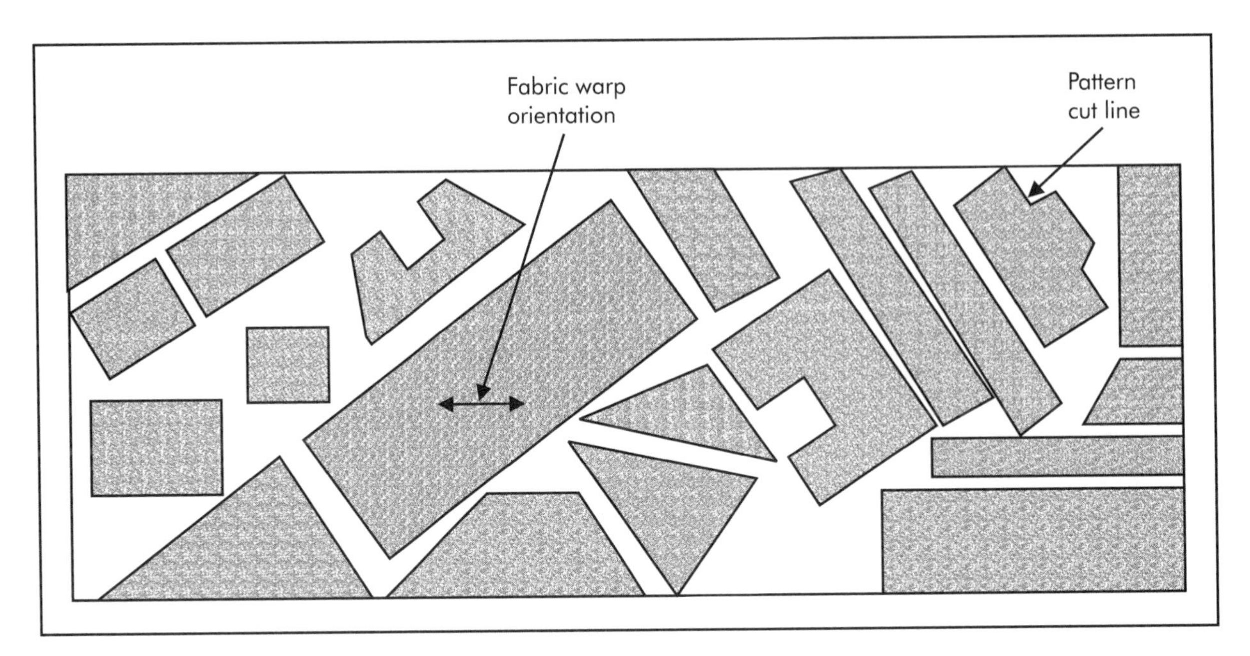

Figure 14-1. Efficient layout of a typical nest pattern.

also scorch or burn the edge of the material. Although this is normally insignificant, it is unacceptable in some applications. Likewise, waterjet cutting may contaminate the material with water. When properly controlled, this is rarely a problem, but many companies are reluctant to use waterjet cutting on uncured materials because of the potential problems water contamination could create.

Several of the cutting methods can be automated using a broad goods cutting machine. The roll of prepreg is fed into the machine and it cuts off the proper amount and slides the material onto a cutting bed. A vacuum is applied to hold the prepreg against the bed and, if desired, several plies of material can be held and cut simultaneously. The material is cut by a gantry-mounted cutting tool (of various types) or by dies that are automatically placed on the material. Nesting is done automatically so the amount of waste material is minimized. The material can even be removed by a robot (using suction tips) and deposited in trays that are ready for the manual lay-up process. A semi-automated cutting table is shown in Figure 14-2.

After the materials have been cut to the proper shape, the pieces are placed into the mold. As before, the prepregs should only be handled with protective cotton gloves. Each prepreg piece is placed carefully making sure the alignment of the sheet is exactly according to the specification of the composite part's design. To facilitate this, a directional reference grid can be marked around the perimeter of the mold on the flange. Another method of assisting in orienting the part is to project the directional grid directly onto the surface of the mold using laser projectors. The ease and accuracy obtained from these projections often justify the cost of the machine.

The prepreg pieces should have sufficient tack that when pressed against the mold surface, they remain in place. The prepreg pieces also should have sufficient drape so they will conform to the shape of the mold. Occasionally, with difficult shapes, the prepreg will have to be cut and shaped. Often,

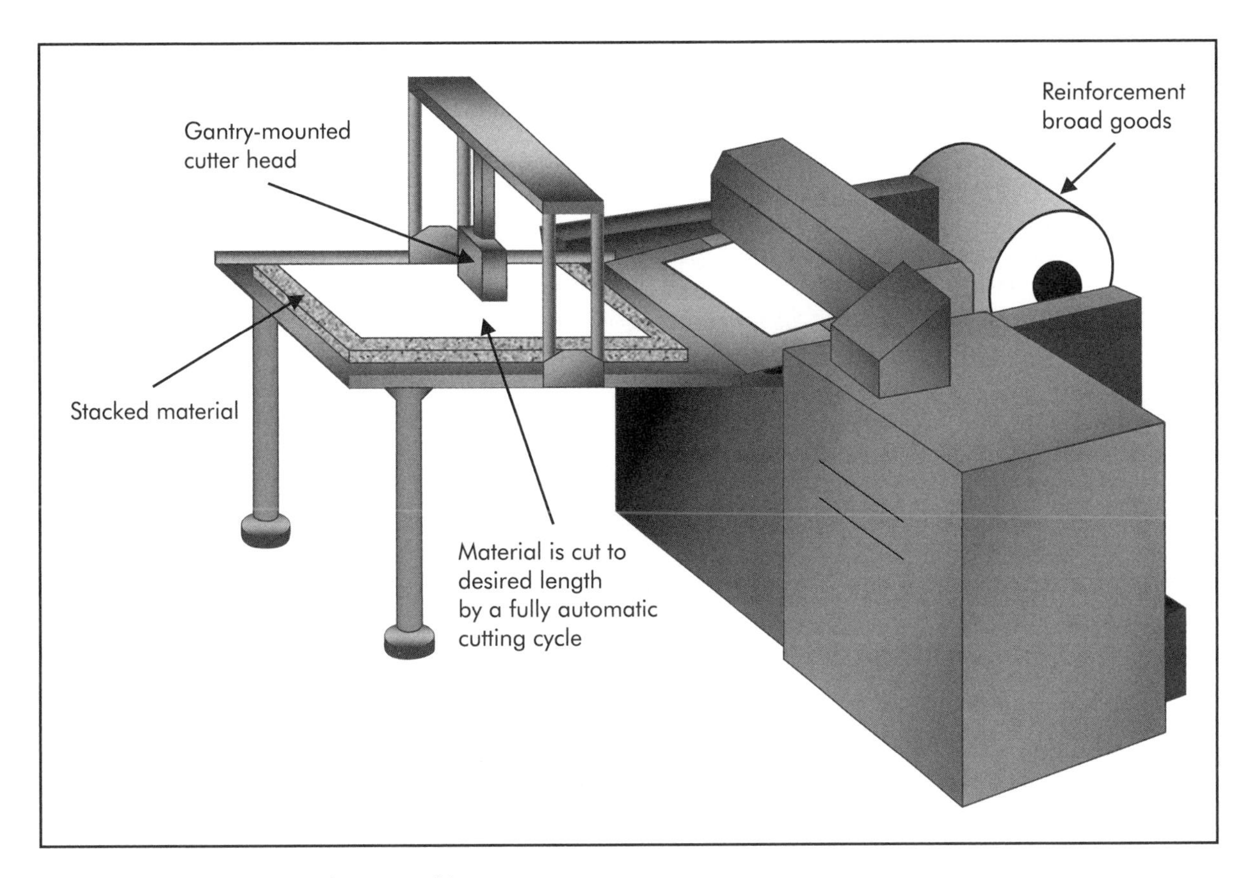

Figure 14-2. Semi-automated cutting table.

some additional flexibility can be obtained by using a prepreg made on a flexible fabric instead of a unidirectional prepreg sheet. There should be no wrinkles, bubbles, or gaps between the prepreg pieces.

Thermoplastic prepregs can be used in open mold lay-up. However, these materials do not have tack or drape. Therefore, they need to be softened so they can be shaped to fit the mold and stay in place. This softening can be done by heating the prepreg with a hot air gun or by tacking it in place with a soldering iron.

Whenever possible, trimming to non-critical dimensions and removal of excess material should be done before the prepreg is cured and while the material is still sufficiently soft to be cut by knives or shears. After curing, special cutting wheels, lasers, waterjets, or diamond cutters are usually required because of the hardness of the composite.

After the first layer of the part has been placed into the mold, subsequent layers are placed on top of the first. Because the design of the composite part often requires that the orientations of the fibers be different from layer to layer, great care must be given to following the design with exact orientation of each prepreg piece.

Automatic Lay-up

The automatic lay-up system begins when a roll of prepreg material is mounted on the payout drive of a tape-laying machine. Typically, the prepreg roll is quite narrow (about 2–4 in. [5–10 cm]). The narrow width allows the machines to place a layer of prepreg over a moderate curvature. If the curvature is too severe, the lay-up machine cannot be used.

Automatic lay-up machines are large and expensive. They can, however, accurately and quickly lay down plies of a prepreg. As indicated, they work best for flat panels and parts with gentle and moderate curvature. A typical example would be a tail or wing section for an airplane.

The machine has a gantry-mounted head that applies and cuts the prepreg. The machine indexes the area within which the prepreg is to be applied. Then, according to the lay-up design specification, the machine presses the leading edge of the strip of prepreg onto the mold surface. The machine then moves along a straight line and feeds and presses the prepreg onto the mold while simultaneously lifting the paper backing. When the edge of the mold is reached, the head cuts the prepreg strip at the proper angle and then lifts and moves to the correct position to begin laying the next trip. This process is repeated until the entire laminate has been laid. An automatic tape-laying machine is illustrated in Figure 14-3.

Tape-laying machines also can be used for thermoplastic prepregs. In this case, the head heats the prepreg as it is applied to the mold surface. The heating softens the thermoplastic matrix and allows the material to stick where it is placed and conform to the minor surface contours.

A recent innovation to tape-laying machines is the separation of the tape-laying and tape-cutting operations. If the head need only lay the tape, it can be reduced in size

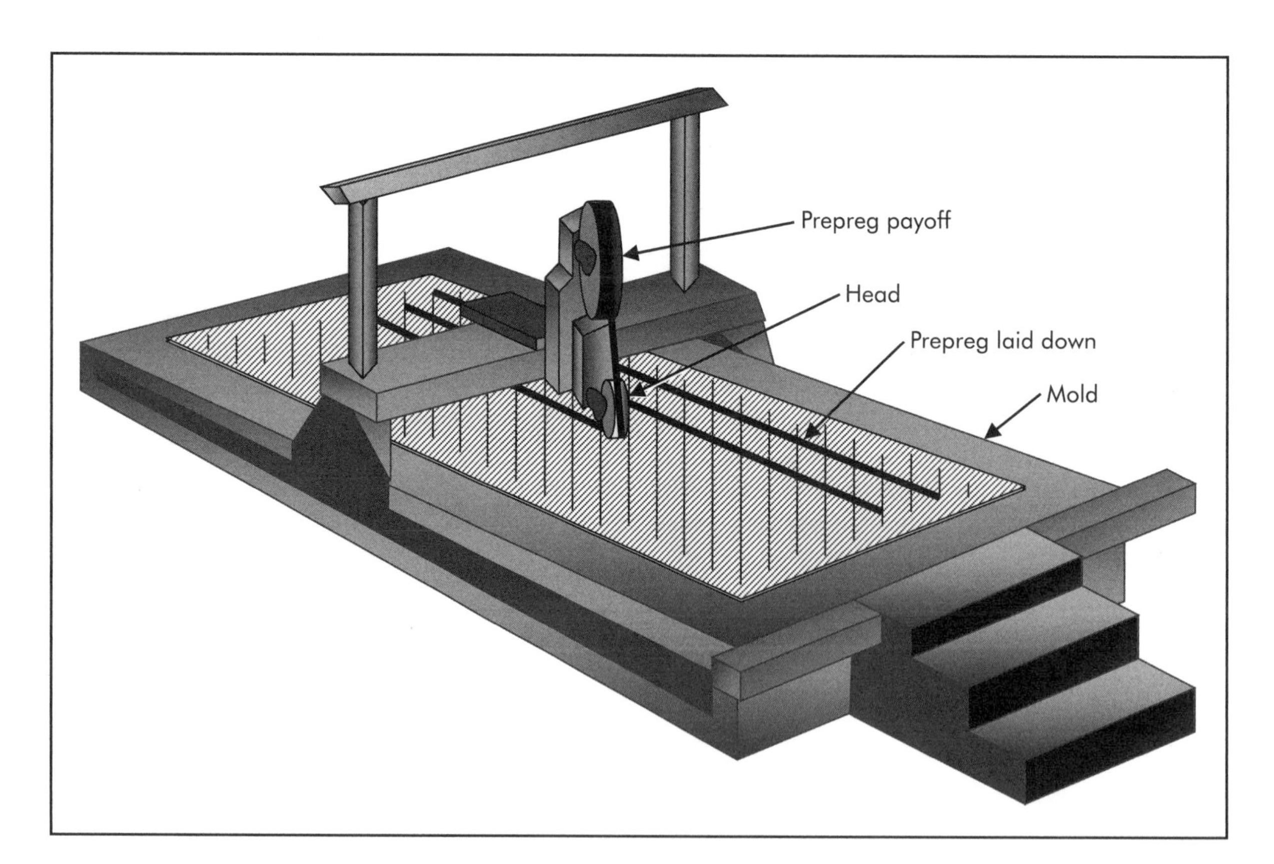

Figure 14-3. Automatic tape-laying machine.

significantly and simplified in its operation. A fly cutter can be placed on the head to simply cut off the tape (but not precisely). The cutting, in this case, precise trimming, can be done by another head, maybe even on a different machine to which the mold is moved after the tape has been laid.

VACUUM BAGGING

Even with the greatest care during the lay-up process, air is inevitably trapped between the prepreg layers. Experience has shown that better parts can be made if some of the air is removed after the lay-up and before the part is cured. This process of air evacuation is called **debulking**. The part also should be debulked during the curing process when the resin is heated and is, therefore, able to move more freely.

If the number of plies in the part is small (up to four), then debulking is usually done by applying the vacuum as the assembly is cured. If the number of plies is large, one or more debulking operations are usually required. This usually occurs after each lay-up of four to six plies. With each debulking, the stack of prepregs is vacuumed. Then the entire bagging system is removed and lay-up is continued. This is repeated until the final plies are placed and the final bagging system is assembled over the stack.

Applying a vacuum, especially during cure when the resin is less viscous, has another benefit in making composite parts. This benefit is the removal of volatiles that may be present in the resin. Such volatiles may include residual solvents used in making the prepreg. Other volatiles may have been absorbed into the resin when it was stirred to formulate it with the hardener (before it was made into the prepreg). The elimination of volatiles helps to lower the void content of the final part.

Still another advantage of vacuum bagging the prepreg system is the movement of resin that occurs during the curing stage as a result of the nature of the bagging system. The molten resin is encouraged to move between the layers and into an absorbent layer in the bagging system, thus eliminating the boundaries that might exist between prepreg layers and, therefore, reducing the chance of delaminations. Also, some resin may move out of the prepreg and be absorbed by materials placed in the bagging system. This absorption causes a moderate loss of resin that increases the fiber/resin ratio and reduces the total weight without significantly reducing the mechanical properties of the composite.

Bagging Methods

The bagging system is illustrated in Figure 14-4. The procedure for assembling the vacuum bag system is as follows.

1. Prepare the mold by coating it with an appropriate mold release.
2. Build up the part by placing layers of prepreg on top of each other in the prescribed pattern and to the prescribed thickness.
3. Place the **release film** or **peel ply** material over the laminate stack. The purpose of the release film is to provide a simple method of pulling all the bagging system off the part after curing. Generally, the release material is porous to permit excess resin to flow through it but it should be strong enough so it can be pulled to release the bagging system and not tear. A common material for the release film is a perforated Teflon® sheet or Teflon-coated glass fabric. The perforations in the release film impart a texture to the part, which is useful for subsequent bonding but can be smoothed by sanding if the outer surface of the part is the finished surface.
4. Apply **bleeder material** on top of the release film. The bleeder material is a mat that absorbs the excess resin.

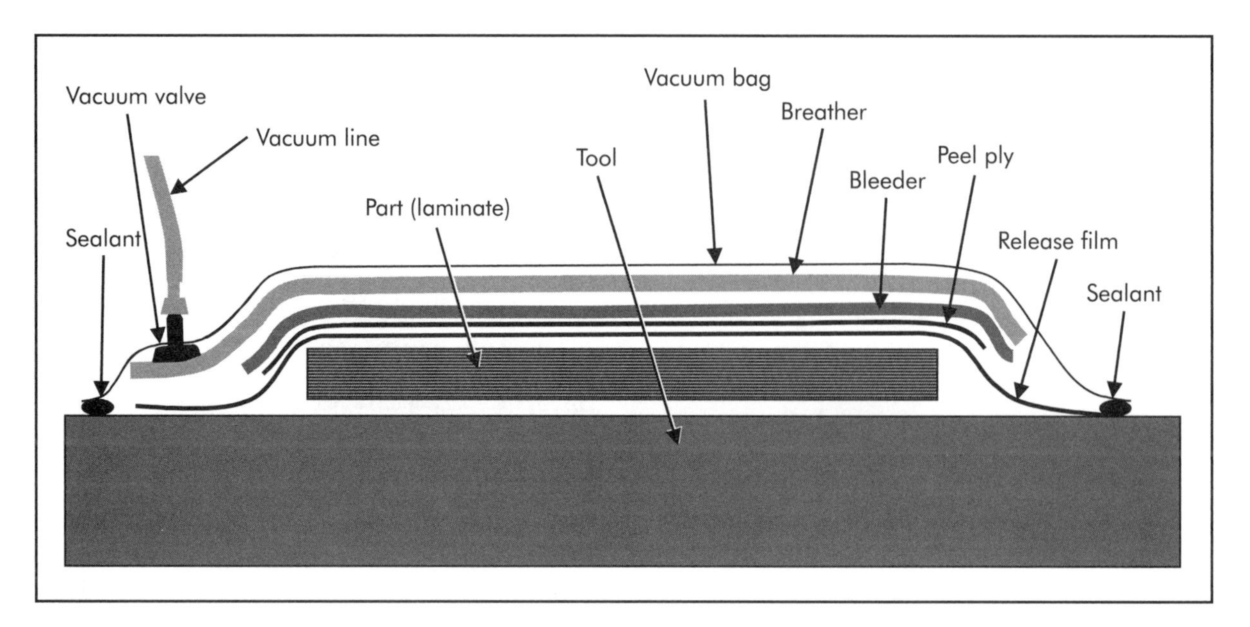

Figure 14-4. Vacuum bag system.

Common bleeder materials are polyester felt or mat, fiberglass mat, and cotton. It is important that the bleeder material has good absorption qualities and does not compact under the pressure of an autoclave. Along with other variables, the resin content of the final part depends on the ability of the bleeder material to absorb a measured amount of resin. Therefore, the type of bleeder and the amount (thickness) must be determined for each laminate to be cured.

5. Apply a **breather material** on top of the bleeder. The breather material acts as a distributor for the air (vacuum) and for escaping volatiles and gasses, as well as a buffer between bag wrinkles and part surfaces. Breather layers should be highly porous and must not collapse under vacuum, heat, or pressure. Typical materials are fiberglass, polyester felt, and cotton. The bleeder materials also can serve as breather materials but the amount of resin absorption should be carefully controlled when this is done.
6. A vacuum valve is placed so that it is on top of the breather material. This will allow the vacuum to be distributed over the entire part. The valve is connected to a vacuum line that leads to a vacuum pump. Normally, there is a resin trap placed in the vacuum line to protect the pump from resin that might get inadvertently sucked into the line.
7. A **sealant** material is placed around the entire assembly. This sealant is usually a sticky polymeric tape, which is to form an airtight seal with the vacuum bag. It must withstand the temperatures experienced during cure.
8. Normally, thermocouple wires and leads for other monitoring devices are inserted above and below the part. These should permit immediate reading of the temperature and pressure within the system and give a record of the curing cycle. To ensure an airtight seal where the thermocouple wires enter the assembly, the outer insulation may be scraped off the wires (especially if it is

cloth insulation). The wires and leads must then be embedded in the sealant material with, usually, additional sealant covering them.

9. Lay the **vacuum bag** over the assembly and push the vacuum valve stem through the bag. A fitting is provided in the valve assembly that tightens on the vacuum bag. The vacuum bag is then pressed onto the sealant to create an airtight seal all around. Vacuum bags can be made out of any polymeric film that is strong enough to hold a vacuum, flexible enough to conform to the shape of the assembly, and able to withstand the temperatures of curing. In practice, the most common material is nylon or a co-extruded nylon that has been heat stabilized. (The vacuum bag is similar to the type of cooking bag used to cook meat in the oven.) If high temperatures are used in the cure (generally above 400° F [204° C]) a more heat-resistant film should be used. The most common high-temperature bagging material is Kapton® polyimide film. Some molders have found that reusable silicon or rubber bags are useful. When the same type of part is made repeatedly, these heavy bags fit over the assembly and can be used multiple times.
10. Activate the vacuum and check the system for leaks.

Several problems are common with bagging systems. They must be addressed early because of the high cost of composite parts that can be ruined. The most common are the following.

- As discussed previously, the system is subjected to a vacuum to debulk the fibers. The vacuum should be carefully monitored during this period to ensure that there are no leaks. Often occurring at the bag/sealant interface, leaks are largely the result of using a sealant that is not thermally stable at curing temperatures. Areas where the bag is folded also are susceptible to leaks.
- Bag breakage is catastrophic because it is often not detected until after the part has been cured and, therefore, it is too late to remedy. Good quality bags that are free of pinholes and internal stresses should be used. Further, they should not be subjected to pinching or nicking. Nylon film is hydroscopic and subject to moisture change in relation to the relative humidity of its environment. Dry, brittle film occurs when the moisture content in the bag falls below 2%. Dry film is especially susceptible to damage and cracking when excessively handled.
- Bridging occurs when the shape of the part, that is, the shape of the mold that makes the part, does not allow the vacuum bagging materials to press against all of the part's surface. These non-pressed areas are not, therefore, properly compressed. For example, a part with a narrow, deep channel may allow bagging materials to form a bridge across the top of the channel. This prevents the transfer of pressure to the bottom of the channel. Similarly, a part may have a corner across which the bagging material might bridge. If the nylon bag itself bridges over a region in the part, the film may be stretched beyond its limits during cure and burst as well as fail to transfer proper pressure to the bridged area. Good bagging techniques, such as pleats intentionally put into areas where bridges may occur allow enough excess film to be present for bag conformity against the part's surface. (See Figure 14-5.) Another technique is the use of a **pressure intensifier** (**caul**) that fits in a potential bridging

area between the bag and the part to ensure that the area receives full pressure. A typical example of the use of a pressure pad is also shown in Figure 14-5. Pressure pads are often made of a rubberized or flexible material that can withstand cure temperatures without degrading.

- The easiest path for resin to take during curing is along the direction of the fibers (**edge bleed**). However, to ensure that the air trapped between the layers is removed and to give good integration of the layers within the composite, the direction in which the molder would like the resin to move is perpendicular to the plane of the fibers (**face bleed**). Therefore, a technique to encourage resin movement in the perpendicular direction and to prevent leakage is to include a **resin dam** around the perimeter of the bagging assembly. Simultaneously, sufficient bleeder is placed perpendicular to the plane of the fibers so that saturation of the bleeder will not occur. (A resin dam is shown in Figure 14-5.) The molder should be cautious, however, not to cause too much resin to flow and thus starve some area of the part. When **resin starving** occurs, there is insufficient resin to fully wet the fibers and therefore loads cannot be transferred properly from the matrix to the reinforcements.

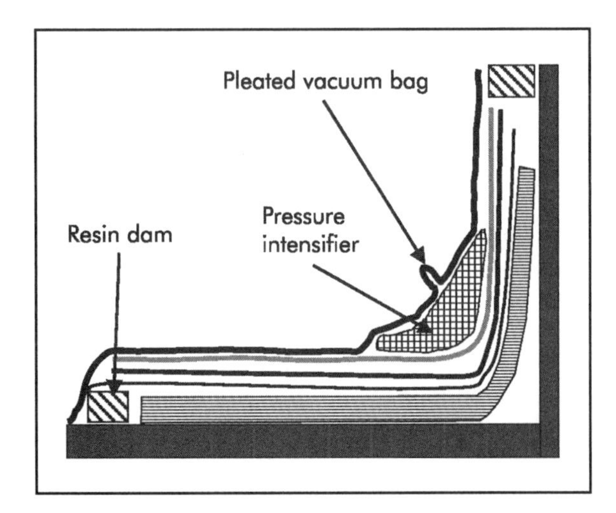

Figure 14-5. Vacuum bag assembly showing use of a pressure intensifier and pleated vacuum bag.

Several of the above problems can be resolved by using a spray-on bag instead of the nylon film. The spray-on bag essentially eliminates the problem of bag breakage and leakage. It always conforms to the shape of the part and gives the proper pressure to all surfaces. Recent developments in spray-on bags have shown that they can have high elongation and, therefore, less susceptibility to tearing under high pressure. They are able to withstand normal curing temperatures and may be reusable if temperatures are not too high. It is even possible that such bags could be used for multiple curings.

Silicone bags also have been used. These are molded to fit the shape of the part and are typically used for many moldings. The silicone bags are more expensive than the nylon or spray-on bags, but are also more durable and faster to apply.

CURING

After the bagging system has been assembled and the part has been debulked, the resin needs to be cured. The temperatures and times for curing are determined by the nature of the resin and the curing agent. Most advanced composite systems are epoxy or some other advanced resin.

Two tasks must be accomplished during cure. The first is curing and that is done by heating the part. The other is to reduce void content. This is done, in part, by debulking during the cure. It is important to ensure that the void content is low because of the serious negative effects of the presence of voids. As a guide, the interlaminar shear strength reduces by about 7% for each 1% of void content present up to a maximum of about 4%. A reasonable goal for void content in the finished laminate is 0.5% or less.

The importance of pressure during curing has been confirmed by studies of thick laminates in which the consolidation was measured directly and in others in which the consolidation was mathematically modeled. These studies clearly show that pressure is a critical factor in consolidation of thick laminates to achieve optimum properties.

Vacuum also helps by consolidating the laminate's layers. This was noted in a study of shear strength for two parts, one cured with a vacuum, and one cured without a vacuum. The shear strength increased by 20% in the part molded with a vacuum.

The molder has two options available with respect to the pressure that may be added during cure to assist in reducing voids. One is to cure in an autoclave. The other is to cure without one and either ignore this pressurization step or use some other method to exert pressure besides an autoclave.

Autoclave Curing

The prepreg method generally uses both a vacuum and an autoclave to assist in consolidating and curing the part. Autoclaves are pressure vessels that allow the simultaneous imposition of pressure by vacuum and heat. The vacuum line is led directly to the part. The major difficulty with autoclaves is their high capitalization cost. This is because they must, by law, pass stringent pressure code regulations. However, because many small parts can be cured simultaneously in an average autoclave, labor and cure costs on a per-part basis are not usually that high.

Most autoclaves are equipped with a temperature and pressure control system that allows for a programmed heating and pressurization cycle (similar to that shown in Figure 4-15 in Chapter Four).

Autoclaves are often filled with an inert gas (generally nitrogen) to suppress any oxidation that may have a tendency to occur during the heated curing cycle. This oxidation may occur in the mold, the part, the bagging system, or at the fittings and connections.

Using an autoclave, assemblies can be bonded together at the same time they are curing. In this application, one purpose of the pressure is to ensure that the layers of the parts remain in intimate contact so that bonding and curing occur properly. This process is called **co-curing**.

Non-autoclave Curing Methods

Autoclaves are not the only method for applying pressure to a composite laminate. In fact, autoclaves are not necessarily the best pressurization method for all laminates. An example is the making of tennis racquets. The preferred manufacturing method for these parts is to wrap the prepreg materials around a bladder. After the correct amount of prepreg material has been wrapped around the bladder, the assembly is placed inside a **clamshell mold**. These molds are often hinged along one edge and are not designed to exert large pressures. They are, rather, designed to define the outer surface of a part when it is pressed against the mold by an internal pressure mechanism. In the case of a tennis racquet, the internal pressure is supplied by pressurizing the bladder. When this is done, the prepregs are pressed outward against the inside of the mold, consolidating and defining the part's outer surface. Normally, the clamshell mold is only slightly larger than the size of the prepreg assembly so that the amount of movement of the prepregs as they are pressed outward is minimal. When the part is cured, the pressure on the bladder is released and it collapses. It can be withdrawn from the part or simply left in place if its presence does not detract from the performance of the part.

A similar system is used to make airplane fuselages. Using filament winding (discussed in Chapter 17), the fuselage is wound around a mandrel over which a bladder has been installed. After winding is completed, the laminate assembly is placed inside a clamshell mold. The bladder is pressurized to push the laminate layers outward against the inside surface of the mold. The mold is heated to cure the resin.

It is possible to use a rubberized material (usually silicone) to pressurize a composite part. This material is similar to a pressure intensifier (discussed previously), however, in the intensifier case, the pressure was applied by a vacuum and/or an autoclave. In this case, called **trapped rubber molding**, the expansion of the rubberized material with heating is the pressurizing force. A typical system uses a molded silicone medium placed inside a metal mold so it fully occupies the cavity except for the molded area that corresponds to the shape of the part. The part is made by laying up prepreg over a plug mold and then placing the plug mold inside the molded cavity in the silicone. After closing, the mold is heated, which then heats the silicone and causes it to expand and compress the prepreg layers.

Still another method uses a rigid mandrel covered with a layer of release material and then with a layer of aluminum. Since the aluminum layer provides the expansion and pressure, it should be thick enough to perform as required. The aluminum layer is covered with another release layer and then overwrapped with the uncured prepreg. The prepreg layer is covered with another release layer and then tightly overwrapped with graphite filament. When the entire assembly is heated, the aluminum expands 22 times more than the graphite filaments, thus consolidating the composite layer as it cures. This method is suitable for cylindrically shaped parts such as tubes and pressure vessels.

It is also possible to exert pressure using matched metal dies rather than autoclaves or any of the other methods discussed here. This method, which is important for high-volume production, is the subject of Chapter 15.

ROLL WRAPPING

The **roll-wrapping** method, also called **tube rolling**, combines elements of filament winding and open molding. The similarity to filament winding is the use of a mandrel as the tool on which the fibers are wrapped. The similarity to open molding is the use of prepreg and the curing of the part with pressure. However, the pressure is not from an autoclave but, rather, from an overwrap of shrink film. Golf shafts and other simple tubular shapes are often made by this method.

First, prepreg material is cut into pieces that are usually about three times as wide as the diameter of the mandrel. If the mandrel is tapered, as it is for a golf shaft, the prepreg pieces are wedge-shaped so that when wrapped around the tapered mandrel, the fibers remain aligned with the length of the shaft. The rolling is done by placing the mandrel on the top of the prepreg and then using the tack of the prepreg with some applied pressure to stick it to the surface of the mandrel. The mandrel is then rolled. Usually, the manual wrapping and rolling step is assisted with a simple rolling table. The prepreg material is manually started around the mandrel and then the mandrel and prepreg are placed between padded plates of a rolling table that moves one plate relative to the other and thus completes the wrapping of the prepreg around the mandrel. If the part is tapered, the wrapping motion will move the thicker side of the mandrel over a greater distance than the thin side as shown in Figure 14-6.

When the thickness of the wrap needs to be greater, additional wraps are placed on the mandrel. The direction of the layers is determined by normal laminate design theory. Care must be taken to ensure that the beginning and ending edges of each wrapped layer

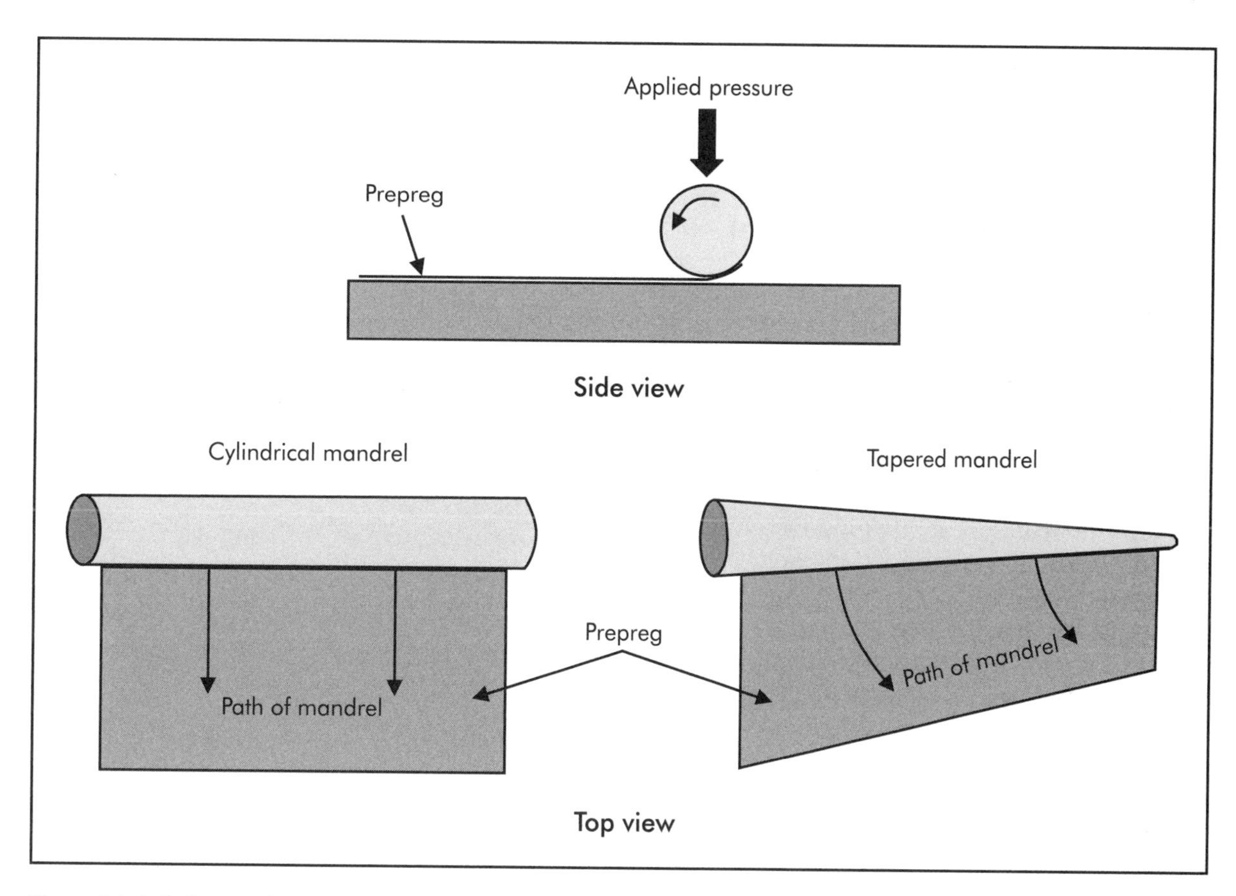

Figure 14-6. Roll wrapping.

are spread uniformly around the circumference of the mandrel. These ends are potential stress-riser points and if they accumulate in one zone of the circumference, that zone would be subject to failure initiation.

After all the designed prepreg has been wrapped onto the mandrel, a layer of shrink tape is wrapped around the assembly. This shrink tape is usually applied by a simple wrapping machine. The wrapped mandrels are then hung vertically in an oven and heated to cure the prepreg. After curing, the shrink tape is removed and the surface is ground smooth.

The simplicity of roll wrapping has been important to the development of some products such as golf club shafts. Such products are price sensitive and the roll wrapping method is highly cost effective. Today, golf club shafts are made by both roll wrapping and multi-spindle filament winding. Those who sell the filament-wound shafts point to the lack of edges in the winding pattern as a performance advantage. Those who sell the roll-wrapped shafts believe that this potential problem is not real and that the lower cost and good reputation of roll wrapping are evidence of its value.

TOOLING (MOLDS)

The molds used in an autoclave are usually made of metal or composite and must, of course, withstand the pressure forces of the autoclave as well as the curing temperature. Therefore, egg crating, which enhances the pressure tolerance of the mold, is common for advanced composite molds. The molds are made using the same general system outlined in Chapter 13. However, for advanced composites, the molds are almost always of high-performance epoxies reinforced with carbon fibers, or of a metal or ceramic consistent with the expansion characteristics of the part. The tooling epoxies usually cure at 350° F (177° C).

The type, composition, and size of the tooling can affect the cure because the heat-up rates of the tooling can vary widely. In some cases, the tooling is heated independently to reduce the problem of variable heating and accelerate the cure by more quickly getting the entire assembly to cure temperature.

The thermal characteristics of the mold are important in advanced composites to avoid building internal stresses. Table 14-1 presents the thermal properties of several materials that are used in both advanced and engineering composite molds.

To reduce the problems associated with thermal expansion and yet give a more durable mold than can be obtained with composites, some molders use metal molds made from Invar because of its low coefficient of thermal expansion. Invar molds are especially common in airplane manufacturing where the value of the parts and their volumes are reasonably high.

The ability to heat a tool outside of an oven or autoclave provides additional manufacturing options and may result in significantly lower costs. Even when used along with an autoclave, the independently heated tooling may reduce cycle times and provide improved thermal control. Independently heated molds are well known but not common. They are generally metal molds that have heating elements embedded in their walls. Recently, however, a new material offers something different that promises simpler molding. That material is carbon foam. It is unique because the material can be the heating element itself. This is done with a simple electrical connection to the mold.

Yet another innovation is the use of chopped carbon fiber prepreg as the mold material. The chopped prepreg is obtained from the manufacturer as a mat. The mat can be molded using lay-up techniques. To obtain good durability, the resin used in the material is bismaleimide (BMI). This method permits the creation of complex shapes while still realizing the low cost of a molding material.

CASE STUDY 14-1

Bladder Molding of a Complex Part

A common method of forming a highly complex part is to lay up prepregs around a bladder, which then is pressurized to consolidate the prepregs. However, forming the bladder using silicone is often a difficult and expensive process. To make the process more feasible for low-cost applications, nylon film sheet can be used for the bladder. The nylon film used is the same film that is common as a vacuum bag material. It is cut into shapes and bonded on its edges, much in the way cloth is made into clothing, forming a net shape that is close to the final shape of the hollow part. A high-performance film adhesive, also common in composite work, is used to bond the edges.

Such a preformed nylon bladder was tested on a complex, three-dimensional part that otherwise would require building in two halves. The part is shown in Figure 14-7. Once the lay-up was complete, the bladder inserted, and the mold closed, the bladder was pressurized and the part was cured. The finished part was found to have excellent consolidation and dimensional accuracy.

SUMMARY

The process of using prepreg materials in an open mold is important for making advanced composite parts. Many small parts

Table 14-1. Tooling materials (typical properties might demonstrate a wide range based upon mix ratios, fiber orientation, lamination sequence, fiber volume, etc.). (Courtesy of *SAMPE Journal*)

Tooling Material	Specific Gravity	Specific Heat, Btu/lb/° F (J/kg/K)	Thermal Mass, Btu/lb/° F (J/kg/K)	Coefficient of Thermal Conductivity, Btu/ft²/hr/° F/in. (J/m²/K/cm × 10⁶)	Coefficient of Thermal Expansion (CTE), μin./in./° F (μcm/cm/° C)
Metals					
Cu-Be (C17510)	8.80	0.10 (417)	0.88 (3,676)	1,680 (13.7)	10.0 (18.0)
Cast aluminum	2.70	0.23 (959)	0.62 (2,589)	1,395 (11.2)	12.9 (23.2)
Steel	7.86	0.11 (460)	0.86 (3,593)	360 (2.9)	6.7 (12.1)
304 Stainless steel	8.02	0.12 (501)	0.96 (4,011)	113 (0.9)	9.6 (17.3)
Nickel	8.90	0.10 (417)	0.89 (3,718)	500 (4.0)	7.4 (13.3)
Zinc	7.14	0.09 (376)	0.64 (2,673)	746 (6.0)	19.0 (34.2)
Invar 36	8.11	0.12 (501)	0.97 (4,052)	73 (0.6)	0.8 (1.4)
Invar 42	8.13	0.12 (501)	0.98 (4,094)	106 (0.9)	2.9 (5.2)
Ceramics					
MgO	2.90–3.58	1.13–1.50 (4,721–6,266)	3.28–5.37 (13,704–22,436)	79 (0.6)	6.1 (11.0)
Al_2O_3	2.90–3.98	0.86–1.03 (3,593–4,303)	2.49–4.10 (10,402–17,129)	22 (0.2)	3.3 (5.9)
State change*	0.45	Not available	Not available	0.83 (0.06)	1.1–3.3 (2.0–5.9)
Plaster					
Gypsum based	1.4–1.6	0.84–1.00 (3,509–4,178)	1.18–1.60 (4,928–6,684)	10 (0.08)	8.3 (14.9)
Composites					
Glass/epoxy	1.8–2.0	0.3 (1,253)	0.54–0.60 (2,256–2,506)	21.8–30.0 (0.17–0.24)	8.0–9.0 (14.4–16.2)
Carbon/epoxy	1.5–1.6	0.3 (1,253)	0.45–0.48 (1,879–2,004)	24.0–42.0 (0.19–0.34)	0.1–3.0 (0.2–5.4)

Table 14-1. *Continued.*

Tooling Material	Specific Gravity	Specific Heat, Btu/lb/° F (J/kg/K)	Thermal Mass, Btu/lb/° F (J/kg/K)	Coefficient of Thermal Conductivity, Btu/ft²/hr/° F/in. (J/m²/K/cm × 10⁶)	Coefficient of Thermal Expansion (CTE), μin./in./° F (μcm/cm/° C)
Graphite					
Monolithic	1.74–2.00	0.27–0.30 (1,128–1,253)	0.47–0.60 (1,965–2,506)	160–220 (1.3–1.7)	0.1–1.0 (0.2–1.8)
Foam					
Polyurethane foam board§	0.24–0.80	Not available	Not available	Not available	27 (48.6)
Carbon foam†	0.1–1.6	Not available	Not available	1.7–173 (0.01–1.3)	2.7–3.2 (4.9–5.7)
Water-soluble materials					
Water soluble‡	0.53–0.59	Not available	Not available	1.1–1.4 (0.008–0.009)	3.6–5.4 (6.5–9.7)

* Two-phase technology Reconfigurable Tooling System (RTS™) material
§ General Plastics, FR-4500 tooling board
† Touchstone Research laboratory, CFOAM®
‡ Advanced Ceramics Research, Inc., Aquapore® and Aquacore® materials

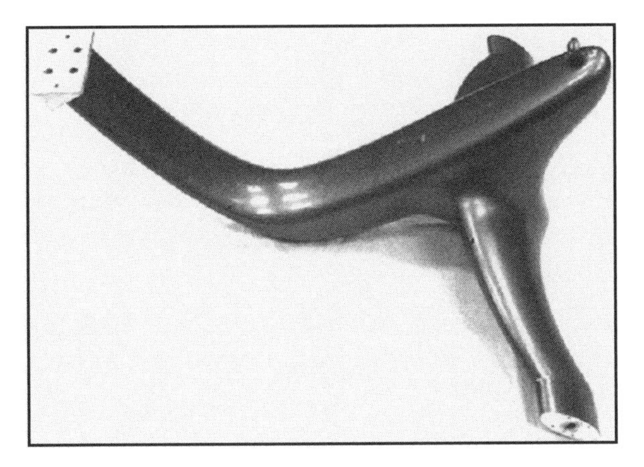

Figure 14-7. Complex part made with a nylon bladder.

and some parts as large as airplane wings and tails are made by this method. The small parts are usually laid up manually whereas the large parts might be done by an automated lay-up machine if the curvature of the part is not great. When done manually, the prepreg pieces are cut prior to laying them into the mold. They might be cut with various fiber orientations so that the design specifications can be met. Usually the prepreg will be unidirectional sheet material, but some parts may use prepreg fabric.

After the prepregs have been laid into the mold, they are bagged and a vacuum is applied so the air between the layers and any other volatiles present will be removed. After debulking, the part is then cured. Most of the time, the cure will be done in an autoclave so that the layers will be packed even tighter, thus reducing void content. The bagging system also allows the resin to move between the layers of the laminate, thus improving the consolidation of the part. After curing, the bagging system is removed and the part is finished.

The assembly of prepreg materials can be pressurized without an autoclave. This is commonly done with pressure methods that include internal bladders or materials that expand within a closed mold to compress the parts.

Roll wrapping is a simple method for making tubular parts. This method uses prepreg that is wrapped around a mandrel. Compaction is done by wrapping the part with shrink tape. This method provides favorable results for many products, especially in the sporting goods industry.

The molds for the processes discussed in this chapter are usually made from epoxy reinforced carbon fibers, but might be made of metals for longer wear. In the aerospace industry, Invar molds are common because they have a low CTE, thus matching the expansion characteristics of the composite part and avoiding stresses that might result from mis-matched materials.

LABORATORY EXPERIMENT 14-1

Making a Laminate as Compared to a Wet Lay-up Composite

Objective: Practice the methods of prepreg lay-up.

Procedure:

1. Obtain a roll of glass fiber prepreg material.
2. Lay a sheet of polyethylene terephthalate (PET) film (Mylar) onto a picture frame mold.
3. Cut sheets of prepreg in 0° and 90° directions. Lay these on top of the film in the mold in the order of 0°, 90°, 90°, and 0°.
4. Place a release material on top of the resin.
5. Place a bleeder material on top of the release.
6. Place a breather material on top of the bleeder.
7. Add a nylon bag and seal it to the mold. Put the vacuum port into the bag before sealing.
8. Apply a vacuum to debulk the assembly.

9. Put the assembly in an autoclave and cure according to the recommended cycle.
10. Remove the bagging materials and then cut the part.
11. Compare with samples made by the wet lay-up method as discussed in Chapter 13.

QUESTIONS

1. Why do thermoplastic prepregs not have tack or drape?
2. Why are epoxies and carbon fibers the most commonly used components of prepregs?
3. What is the purpose of applying a vacuum to a bagging assembly?
4. Indicate the part property that is improved when autoclaves are used for curing the part.
5. Outline a non-autoclave method for curing a prepreg part.
6. What types of parts are made by the roll wrapping method?
7. What are the problems of uniformity with roll wrapping?
8. With advanced composites, why is prepreg molding more common than wet lay-up?

BIBLIOGRAPHY

Beckwith, Scott W. and Dorworth, Louis C. 2006. *SAMPE Journal*, Vol. 42, No. 6, November/December, p. 53.

Boursier, Runo, Callis, Rick, and Porter, John. 2006. "New Composite Tooling: Material and Concept for Aerospace Composite Structures." *SAMPE Journal*, Vol. 42, No. 6, November/December, pp. 60–66.

Christou, Philippe. 2006. "High-temperature-resistant Tools and Master Models: Seamless Molding Paste (SMP) Technology." *SAMPE Journal*, Vol. 42, No. 6, November/December, pp. 7–13.

Cull, Ray A., et al. 1991. "Out of Autoclave Processing of Advanced Composites Utilizing Silicone Elastomers." *SAMPE Symposium*, April 15–18.

Kemp, Daniel N. 1989. "TRM Trims Molding Costs." *Manufacturing Engineering*, March, pp. 53–56.

McCracken, William W. 1993. *Handbook of Lay-up and Bagging*. Whittier, CA: Mar Mac Graphics & Engineering.

Merriman, Douglas J. and Lucas, Rick. 2006. "Carbon Foam Tooling: Self-heating Concept, Evaluation, and Demonstration." *SAMPE Journal*, Vol. 42, No. 6, November/December, pp. 42–49.

Society of Manufacturing Engineers (SME). 2005. "Manual Lay-up and Spray-up" and "Automated Lay-up and Spray-up" DVDs from the Composites Manufacturing video series. Dearborn, MI: Society of Manufacturing Engineers, www.sme.org/cmvs.

Strong, A. Brent. 2006. *Plastics: Materials and Processing*, 3rd Ed. Upper Saddle River, NJ: Prentice-Hall, Inc.

Young, Wen-Bin. 1995. "Compacting Pressure and Cure Cycle for Processing of Thick Composite Laminates." *Composites Science and Technology*, 54, pp. 299–306.

15

Compression Molding

CHAPTER OVERVIEW

This chapter examines the following concepts:

- Process overview
- Equipment
- Bulk molding compound (BMC)/sheet molding compound (SMC) molding
- Preform compression molding
- Prepreg compression molding
- Part complexity, properties, and other performance considerations

PROCESS OVERVIEW

This chapter discusses the most important manufacturing method for composites that employs closed molds. The process is called **compression molding,** or alternately, **matched-die molding**. During this process a specific amount of material—the **charge** of uncured resin and fibers—is placed into the cavity of a matched mold in the open position. The mold is closed by bringing the male and female halves together, and pressure is exerted to squeeze the composite material so it uniformly fills the mold cavity. While under pressure, the material is heated so that it cures.

The molding process can require high pressures and so the molds are mounted in large presses. The presses allow rapid mold cycles and high-volume production. Hence, this process is widely used in the automotive industry and in others where high volumes of small-to-moderate-sized parts are manufactured. Part-to-part reproducibility is good, probably better than any other composite manufacturing process. The squeezing of the material results in low void contents in the finished parts. The amount of scrap is minimal. Some post-molding work is generally required, but that is often not extensive. Compared to other composite molding processes, compression molding has low labor requirements. (Consult the "Compression Molding" video, SME 2005 for a look at the equipment and the molding techniques associated with compression molding. These videos are excellent supplements to lectures and laboratories.)

The compression molding process is not unique to composites. Plastics molders, especially those who process thermosets and elastomers (rubber products), have used compression molding for many years with widespread use dating from the early 1900s. Tires and rubber products were molded using compression molding before 1900. The earliest fully synthetic plastic, phenolic, was invented in 1909 and processed by compression molding to make a variety of products including electrical components and heat-resistant handles. Compression molding continues to be a major manufacturing method for thermoset plastics and elastomers. Therefore, molding techniques are well known and the composites industry benefits from this prior experience with plastics. Composites molders also benefit from the high technology and availability of compression molding equipment. Few, if any, changes have to be made in the equipment used in the plastics industry to accommodate composites molding.

EQUIPMENT

The molding assembly for a typical compression molding system is shown in Figure 15-1. It consists of a heavy metal base onto which slide rods are attached. These rods guide the up and down motion of the molds. A hydraulic motor provides the force to move one of the mold halves. The other mold half is anchored to the base unit. In some machines, the mold opens by the top mold half sliding upward. In other machines the lower mold half slides downward to open. Compression machines are normally rugged and massive, often lasting for decades.

The capacity of the compression press is stated as tons in the English system or as Newtons in the metric system. In both cases, the capacity depends on the area of the hydraulic ram and the pressure that the hydraulic motor can exert. The *press capacity* is also called the *machine rating* and the *machine size*.

The maximum distance between the mold halves when the assembly is opened is called the **daylight opening**. For ease of loading the mold, a large daylight opening is desired. However, for speed in closing the molds, a small daylight opening is faster. Therefore, a compromise in opening is sought. So, to facilitate rapid closing, a two-step hydraulic system is used. In the first part of the closure sequence, the molds close quickly; then, in the second part, closure is much slower, thus allowing for good movement of the material to fill the mold.

The molds are almost always metal because the pressures involved in compression molding are so high. The most common mold material is tool steel and, because of the abrasive nature of composite molding material, the mold faces are often chrome plated. The chrome plating also helps to release the molded part. The molds are usually heated by the platens to which they are anchored. The platens have internal heating units.

The parts are normally ejected from the mold by using **ejector pins** (**knockout pins**). The knockout pins are attached to an ejector plate that slides to push the pins into the cavity and thereby knocks out the part. Care must be taken that the part is fully cured or else the knockout pins might damage a soft part. Similarly, the diameter of the knockout pins should not be too small or the pins may penetrate into the part.

If the mold halves fit together tightly so that no resin can leak between the sides, this is called a **positive-type mold closure**. This type of mold is preferred for composite parts. However, since there is not leakage of excess material and any shortage of material results in a smaller part, it requires that the charge of material be accurate. When the mold halves allow some excess of material to leak between the sides of the mold, this is a **flash-type mold closure**. The abrasive nature and high filler content of many composite materials is not compatible with the flash-type mold because the excess material, called **flash**, may lodge between the platens and prevent the mold from closing.

The specific molding procedures differ slightly according to the nature of the reinforcement used. Three generalized procedures are discussed in this chapter: BMC/SMC molding, preform molding, and prepreg molding.

BULK MOLDING COMPOUND (BMC)/ SHEET MOLDING COMPOUND (SMC) MOLDING

The natures of BMC and SMC and the methods for making these materials were discussed in Chapter Three as part of the

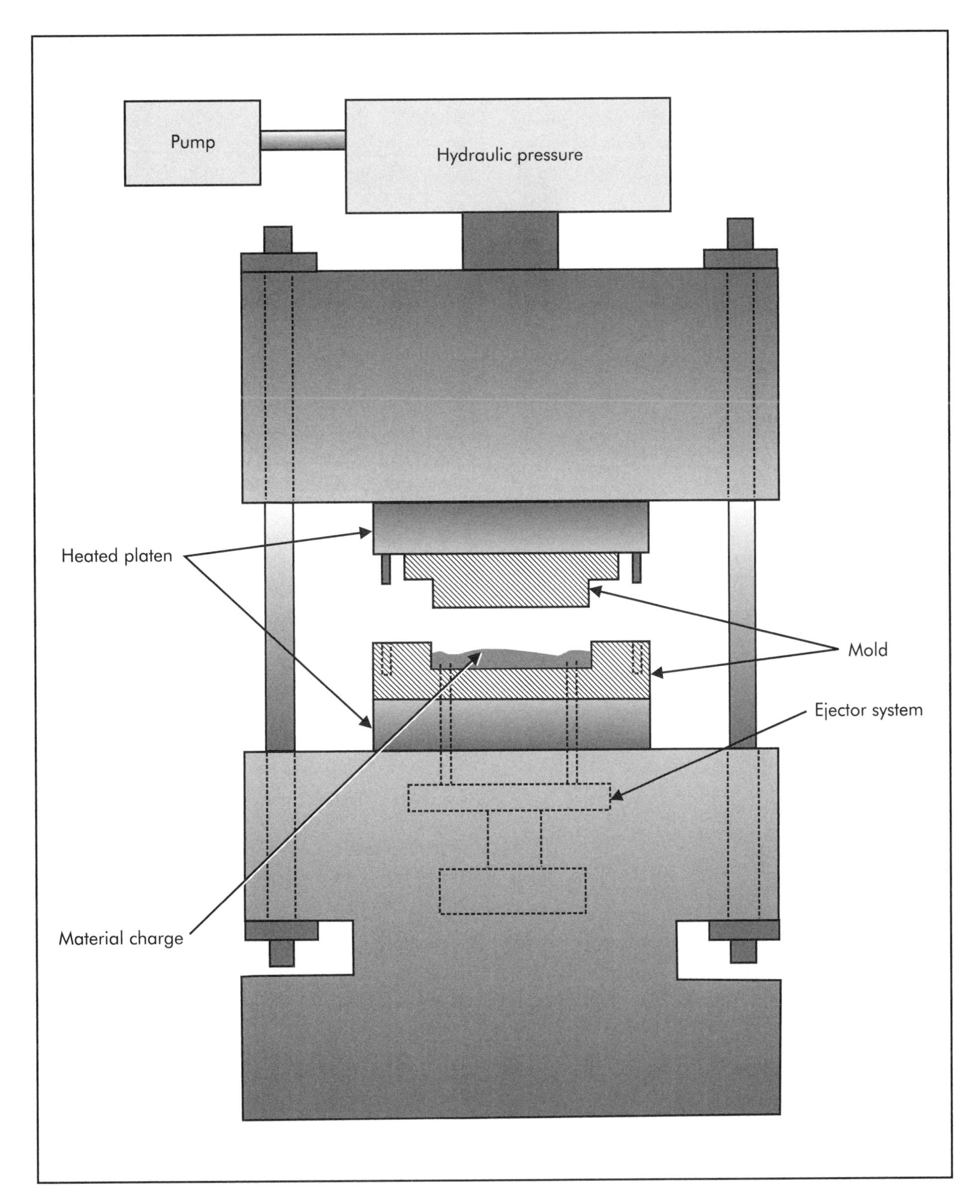

Figure 15-1. The mold assembly for compression molding.

general exploration of polyester resins. By way of review, both of these materials are mixtures of resin (almost always polyester), reinforcement (almost always fiberglass), fillers, and curing agents. BMC and SMC differ principally in the way the fiberglass is added. In BMC, the chopped fiberglass is simply mixed into the initiated resin and filler so that the resultant material is a paste-like or dough-like mass, which is usually formed into logs of about 4 in. (10.2 cm) in diameter. In SMC, the initiated resin and filler paste is spread onto a wide carrier film and then the fiberglass is chopped and sprinkled across the coated film. Another layer of coated film is placed on top of the chopped fiberglass and the sandwich of material is kneaded between rollers to wet-out the fibers.

In both BMC and SMC all the components necessary for complete curing of the part are mixed into the resin when the molding compound is made. Normally, the initiator is heat-activated peroxide so only minimal curing occurs during manufacture of the molding compounds. As a further precaution to prevent premature curing, the molding compounds are refrigerated and a strict maximum shelf life is observed.

BMC is molded by placing a weighted amount of the material into the mold cavity. The molds are then closed and the material is squeezed to fill the mold and heat is applied by the molds to cure the part. Normally, the closing of the mold spreads the material evenly throughout the mold. However, if the viscosity of the mixture is not within an acceptable range, the resin, filler, and fiberglass may not flow to fill the mold properly. If the viscosity of the mixture is too low, the resin may flow but not move the fiberglass with it. On the other hand, if the viscosity is too high, the entire mass may just sit in the center of the mold and not flow at all.

The mixture's viscosity is ever changing until final curing occurs. During the storage time between when the molding compound is made and when it is used, the viscosity increases. This occurs because the resin continues to wet-out the fibers and the filler. There is also some curing that occurs, and some of the additives in the mixture have a natural tendency to increase thickness. The viscosity must rise to a level that is appropriate for molding. This thickening is shown in Figure 15-2.

During molding, the viscosity drops initially because of heating. Then, when curing begins, the viscosity increases. As indicated previously, proper flow occurs within a range of acceptable viscosities and, of course, must occur before the onset of substantial curing. This is also shown in Figure 15-2.

Obtaining acceptable viscosity involves a series of compromises because, as the amount of fiberglass is increased to improve mechanical properties, this raises the viscosity and makes it harder to flow. Using longer fibers to improve properties also makes flow more difficult. Further, the viscosity also increases as a result of fillers added to improve other properties, such as those added to improve flame-retardant properties, or to reduce the overall cost of the mixture. Increasing the molecular weight of the resin to improve properties will also increase viscosity. Adding more styrene will lower viscosity but this leads to greater air pollution and more brittleness. Therefore, great care and some experimentation are necessary in developing BMC materials that flow properly in the mold.

SMC is molded by cutting off the proper amount of sheet from the roll of molding compound, removing the carrier film, and laying the sheet material into the mold. The size of the SMC sheet generally matches the area of the mold (within about 80%). Therefore, little movement of the molding compound in the mold occurs with SMC. If there is the need to add additional reinforcements to some areas of the part, this can be done by placing strips of SMC in those loca-

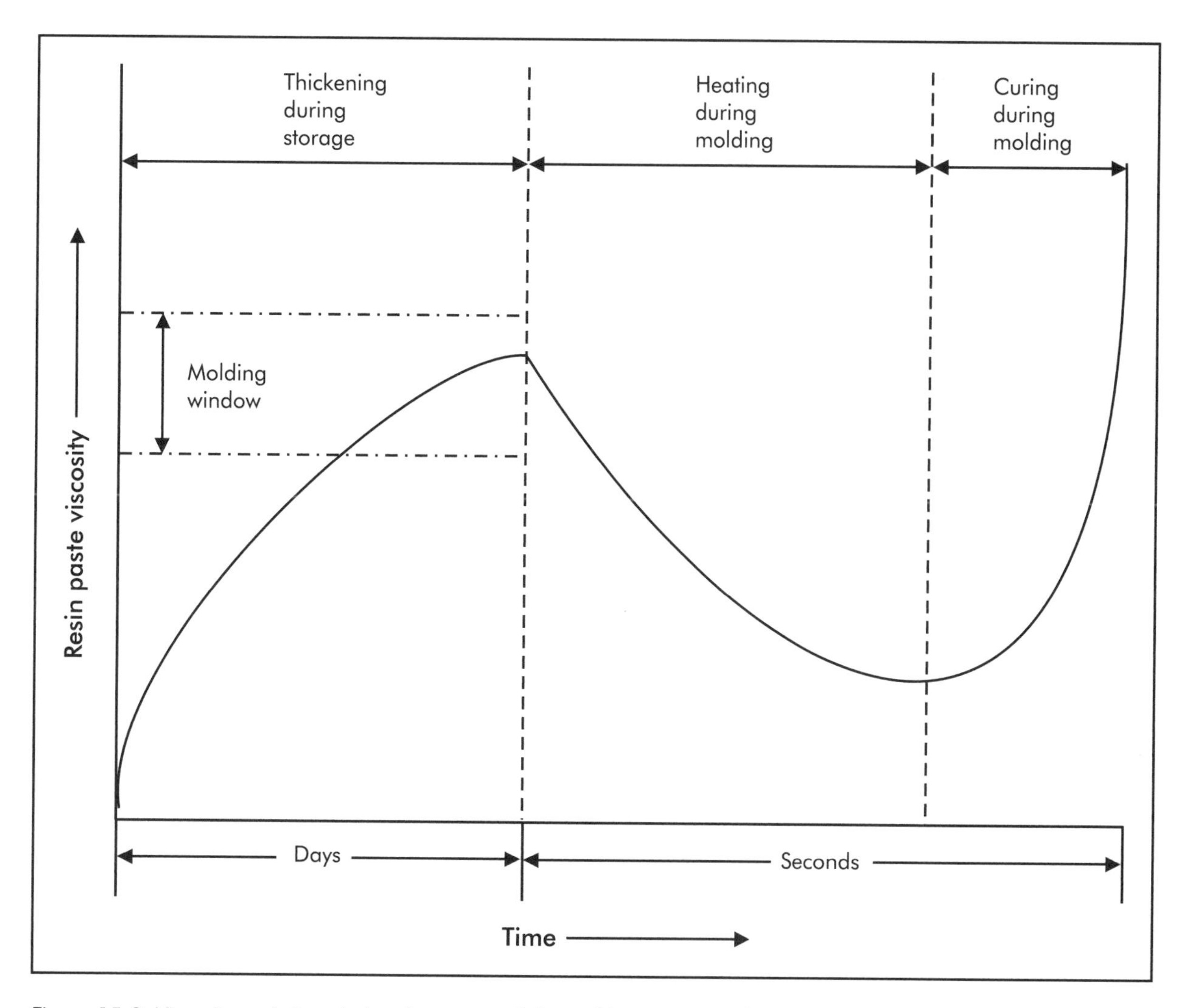

Figure 15-2. Viscosity variation during the stages of the molding compound process.

tions before closing the mold. With SMC, the viscosity problem is solved since the sheet essentially covers the full area.

The molding conditions for both BMC and SMC are essentially the same. Typical settings are: temperature range = 275–335° F (135–168° C); pressure = 500–5,000 psi (3.5–35 MPa); cycle time = 1–10 minutes.

The molding time is long for some applications, like automotive parts, and so some efforts have been made to reduce it. One of the methods used is preheating the molding compound just prior to placing it in the mold. The heating is efficiently done using radio frequency (RF) heating in which the molding compound is subjected to the RF signal in a process not too different from microwave heating for food. Normally, the RF heating raises the temperature to about 130° F (54° C). When this is done, the cure occurs in about 30 seconds. The entire mold cycle can be further reduced when fast closing is used along with a vacuum assist for withdrawing the air and vapors from the mold.

When the part is molded, extraction by using knockout pins is the norm. However, even with these pins some mold release is generally needed. On a well-seasoned mold, the application of mold release is not necessary for each molding cycle.

The most important single parameter affecting the performance of a part made from molding compound is the glass content. The effect of glass content is shown in Table 15-1. Every mechanical property—flexural, tensile, compressive, and impact—improves with increasing fiberglass content. As would be expected, properties that depend chiefly upon the resin, such as water absorption and flammability, do not change with fiberglass content. Somewhat surprisingly, the shrinkage remains constant, probably because the measurement is not fine enough to detect the minor differences due to the fiberglass holding the resin in place. (The major change in shrinkage with fiberglass is between the neat resin and almost any significant fiberglass loading.)

Some epoxy-based molding compounds are commercially available. These materials are made by methods similar to those used for polyester molding compounds. The molding methods are also similar except for adjustments for curing temperature and shrinkage. The properties of the products reflect the improvements in many properties that epoxies bring compared to polyesters. This material has proven to be especially valuable in making parts used in corrosive environments.

PREFORM COMPRESSION MOLDING

In spite of the obvious mechanical property advantages of higher fiberglass content, the viscosity problems in BMC and, to a lesser extent in SMC, along with the need to have filler for economy and improved properties, preclude ever higher fiberglass contents. In those products that need higher fiberglass content, a different molding approach is taken. This approach is called *preform compression molding* and it has some advantages besides just the ability to use more fiberglass.

In preform molding, the dry reinforcing material is preformed to the approximate shape of the part and placed into the open mold. (The fabrication of these preforms was discussed in Chapter Nine.) Resin is added by pouring the liquid onto the top of the preform with some minimal effort to evenly distribute the resin over the entire surface of the preform. The mold is then closed, which causes the resin to flow throughout the preform, wets the fibers, and compacts the preform to the shape of the mold cavity. The part cures

Table 15-1. Properties of molding compound with various percentages of fiberglass content.

Properties	Fiberglass Content		
	15%	22%	30%
Izod impact, ft-lb/in. (J/m^2)	7.0 (374)	9.0 (480)	14.0 (747)
Flexural strength, psi (GPa)	17,000 (117)	21,000 (145)	25,000 (172)
Tensile strength, psi (GPa)	7,000 (48)	10,000 (69)	12,500 (86)
Compressive strength, psi (GPa)	20,000 (138)	24,000 (165)	28,000 (193)
Specific gravity	1.77	1.80	1.85
Water absorption, 24 hours at room temperature	0.25	0.25	0.25
Shrinkage, %	0.02	0.02	0.02
Oxygen index	21.5	21.5	21.5

within the heated mold and, when cured, it is removed with knockout pins.

In general, little filler is used in preform molding. This means that the fiber content of the final part can be much higher than would be possible with molding compounds. The lack of filler also means that viscosity is less of a consideration than with the molding compounds. Also, the mass of the liquid material is less and this results in faster heating of the resin and, therefore, faster cures.

In essence, the fibers in the preform do not move appreciably during molding except for compression of the fibers into the cavity. Therefore, preform molding permits fibers that are continuous (over the entire size of the part) rather than being limited to just chopped fibers as in the molding compounds. The combination of higher fiber content and longer fiber lengths results in much higher mechanical properties for parts than can be achieved with BMC or SMC.

Because the resin is added separately from the reinforcement, the molder has greater choice with respect to the resin used. This means that epoxy resins can be used and, therefore, advanced composites are more common in preform molding than the molding compounds. Other important resins like phenolics can be conveniently used with this method. Note, however, that when a polymer is used that crosslinks by a condensation process, the condensate that forms must be allowed to escape from the mold. This is done by opening the mold slightly during the molding cycle. The condensate is usually a vapor at this stage because of the heating of the mold and simply escapes through the small opening created when the mold is cracked open. This process is called **bumping the mold** or **breathing the mold** or, less often, **burping the mold**.

PREPREG COMPRESSION MOLDING

Placing cut prepreg strips into the mold and then compression molding is another method of molding using matched dies. Fiber-resin contents are, of course, set by the prepreg manufacturers. The fiber lengths can be over the entire length of the part. Orientation is controlled both by the lay-up within the mold and by the closing of the mold since some movement within the mold is common. However, with careful design of the mold cavity and the lay-up stack, in-mold movement can be minimized. Therefore, properties of prepreg-molded parts can be excellent. Engineering and advanced composites are made using this method.

When prepregs are used, the uniformity of the part is usually superior to those made with preforms, BMCs, or SMCs. Prepreg molding is usually reserved for small parts because larger ones are more easily made in open molds with autoclave curing.

Prepreg molding, which is essentially a manual lay-up process, can be used effectively for making parts that contain inserts. The inserts can be placed at the same time as the prepreg material and the prepreg can be shaped around the inserts. During molding, the prepreg flows to create a smooth boundary attachment and gives good bonding to the insert.

When lower properties are acceptable, the prepreg material can be chopped and then placed into the mold. This gives a reasonably random orientation to the fibers; of course, the fibers are shorter than when the prepregs are stacked manually in the mold. However, the chopped prepregs can be placed in the mold by semi-automated methods. Therefore, the labor component of the molding costs can be substantially reduced.

PART COMPLEXITY, PROPERTIES, AND OTHER PERFORMANCE CONSIDERATIONS

Compression-molded parts are generally not able to achieve the high complexity that is possible with the open-molding systems.

For example, compression-molded parts should have relatively few changes in depth. This means that ribs, bosses, and variable cross-sections should be modest in depth and not too narrow. Fins or other thin sections are generally not possible.

Undercuts are difficult with compression-molded parts unless complicated and expensive sliding molds are used. Draft angles should be 1–3°. Sharp corners should be avoided since they are not only stress-riser locations, but they are places where the fibers may not flow during the molding process.

When molding compounds are used, it should be remembered that the fibers are chopped and relatively short and random. Also, the molding compounds contain fillers, which may preclude their use in some applications.

The need to use considerable pressure to mold parts is also a limitation because of the cost of large presses. The pressure requirements increase roughly with the square of the area of the part. Therefore, large parts require enormous presses to mold them.

In spite of the limitations, the advantages of compression molding have been so great that its use continues to increase for both large and small parts. The automotive and truck industries are especially important users of compression molding. Parts as large as entire cabs for trucks are compression molded. (The presses required to do this are several times larger than the truck cabs.) Other automotive and truck parts made using compression molding primarily with BMC and SMC include: grille opening panels, light-housing extensions, rear-end panels, cowls, spoilers, fenders, body extensions, hoods, roofs, underbodies, deck lids, sunroofs, dual-wheel fenders, trim panels, cargo doors, engine covers, window garnish moldings, air deflectors, and rocker panels. Production volumes of 1,000 parts per day are not uncommon.

Another automotive application used matched-die molding because of the tight tolerances that could be achieved. This part was molded using epoxy SMC around a shaft, replacing a machined-steel component. The only machining required on the SMC part was drilling two .008-in. (0.20-mm) diameter holes and finishing the surface to assure ±.0005 in. (±0.013 mm) tolerance.

Non-automotive and truck applications are also prominent including: basketball backboards, snowmobile hoods, outboard motor housings, motorcycle fairings, electrical enclosures, electronic housings, refrigerator cabinets, building panels, furniture, garden-tractor hoods, and lawnmower housings. Aerospace applications include aircraft interior panels, missile exhaust nozzles, nose cones, aircraft bulkheads, and some wing and body components. The good elevated temperature properties of epoxy SMC were utilized in an office copier part that operates at 400° F (204° C) and is required to maintain dimensional stability of ±.005 in. (0.13 mm). It replaces stainless steel and also serves as a thermal barrier to protect more temperature-sensitive components.

CASE STUDY 15-1

Preform Molding of a Major Truck Component

A part that encompasses the entire hood and fender of a large truck was made by preform molding and has become a major commercial success. The technical requirements were met or exceeded and the commercial requirements have proven to be highly favorable.

The preform was made by directing chopped fiberglass and a small amount of binder onto a plenum (rotating screen) and then bonding by heating the preform in an oven. The resin was applied manually. It was a special mix formulated specifically for this application containing calcium carbonate filler and a polyvinyl acetate low-profile additive.

The surface of the finished part was reported to be superior to the surface obtained with SMC. The new truck hood had a Diffracto Sight Index value between 38.1 and 38.8, compared with a value of 130 for a similar Class-A SMC part. (Diffracto is an optical scanning system that produces a highly magnified photo and matching computer-generated grid of a surface. To be considered Class A, a surface must register an Index value of 150 or less.)

The precision matched dies ensured excellent fit of the part to the other components of the truck. Lower requirements for moving materials in the mold (compared to SMC) and the reduced size of the press and strength requirements resulted in savings on equipment and tooling costs. The total volume is over 6,000 parts per year.

SUMMARY

Matched-die or compression molding has an important place among the processes used to make composite parts. The principal advantages of using compression molding are low labor and high volumes, especially where high accuracy and well-defined surfaces are required. The automotive industry is, therefore, especially inviting for this manufacturing method. However, small aerospace parts are also compression molded.

Three different modes of compression molding have been defined in this chapter. These relate to the type of materials used. In the first, BMC and SMC molding compounds are used. Viscosity control is important. Further, because the fibers are chopped, parts requiring high performance are generally not molded using this method. The second method is preform molding. In this method, the dry preform is placed in the mold and resin is then added. The mold is closed to distribute the resin and cure the part. The third method uses prepregs placed in the mold. High-performance parts can be made by this method.

The major *disadvantages* of compression molding include the following

- Regions of high resin flow may orient the fibers in some regions.
- Flow-related problems such as residual stresses, warping, weld (knit) lines, fiber kinks, and breakdown of the resin paste and glass interface are possible.
- Excessive molding pressure can lead to distortion of the part.
- Poor viscosity control may lead to incompletely filled molds.
- Internal defects from poor flow are possible.

The *advantages* of compression molding include the following.

- Either continuous or random chopped fibers can be used.
- Several resin choices are available.
- The matched dies give better dimensional control than open-molding processes.
- Higher pressures give good part definition and eliminate many surface problems such as blistering, cracking, and splitting.
- High production rates are possible.
- Labor is less than in other composite molding processes.
- Air pollution is lower than open-molding processes.
- Automation is possible.
- There is little waste.

LABORATORY EXPERIMENT 15-1

Compression-molded Parts Compared to Open-molded Parts

Objective: Discover the differences in the performance of compression-molded versus open-molded parts.

Procedure:

1. Make a simple compression mold to fabricate flat panels.
2. Obtain a sample amount of BMC (sufficient to make several BMC panels).
3. Mold the BMC according to the directions of the material supplier.
4. After molding, cut the samples for mechanical testing and compare with samples made by the opening-molding process.

QUESTIONS

1. Give three compromises that have to be reached in determining the viscosity of BMC.
2. Give three differences between BMC and SMC.
3. List two differences in molding BMC and SMC.
4. List two advantages of preform molding over molding with BMC/SMC.
5. Distinguish between preform molding and prepreg molding.
6. Which of the molding methods using matched dies is more consistent with making advanced composites? Why?
7. Why are compression molds usually made out of metal?
8. Why does preform molding usually require less pressure than BMC/SMC molding?

BIBLIOGRAPHY

Childs, William I. 1987. "Compression Molding of Structural Composites." SME Technical Paper, EM87-556. Dearborn, MI: Society of Manufacturing Engineers.

Childs, William I. 1989. "SMC Structural Composites: High Strength at Low Cost." *Plastics Engineering*, February, pp. 37–39.

Modern Plastics. 1991. "Liquid Composite Molded Truck Hood has Smooth, Automotive-quality Finish." October, pp. 24–25.

Society of Manufacturing Engineers (SME). 2005. "Compression Molding" DVD from the Composites Manufacturing video series. Dearborn, MI: Society of Manufacturing Engineers, www.sme.org/cmvs.

Strong, A. Brent. 2006. *Plastics: Materials and Processing*, 3rd Ed. Upper Saddle River, NJ: Prentice-Hall, Inc.

16

Resin Infusion Technologies

CHAPTER OVERVIEW

This chapter examines the following concepts:

- Process overview
- Resin infusion technologies
- Equipment and process parameters
- Preform technology for infusion
- Resin characteristics
- Core and flow media materials
- Part design for resin infusion
- Centrifugal casting

PROCESS OVERVIEW

Besides being the best known of the many **resin infusion technologies** for making composite parts, **resin transfer molding (RTM)** is often the general term used for all resin infusion processes. However, in this text, RTM is used in a more narrow sense and "resin infusion" is the more general term. Another phrase occasionally used to describe the resin infusion processes is **liquid molding processes**. (Consult the "Liquid Molding" video, SME 2005 for a look at the equipment and techniques of liquid molding processes. These videos are excellent supplements to lectures and laboratories.)

All of the resin infusion technologies share some common features. In all of them the dry (without resin) fiber preform is placed into a mold and the mold is closed. Resin is then injected into the mold so that the preform is fully wetted with resin. The resin is cured, the mold is opened, and the part is extracted. The basic resin infusion process is illustrated in Figure 16-1.

The advantages of the resin infusion processes, listed in Table 16-1, are immediately apparent. The factors especially appealing in the current composite world are: the better quality possible with resin infusion (good tolerances that are repeatable and excellent surfaces out of the mold); lower emissions; low labor requirements; low requirements for auxiliary equipment (like freezers and autoclaves); and the degree of design flexibility in both part size and complexity. Cycle times depend on the resin system used but typically fall in the range of 5–10 minutes. However, they can be as high as 2–8 hours for large parts like railroad cars and yachts. Inserts can be easily molded in and this means that one-piece, co-cured parts are often easy to make, thus reducing the total number of parts in an assembly.

An important advantage of the process that might not be obvious at first thought is the feature of being able to place continuous fibers (that is, continuous across the part) into the mold and then carrying out the molding operation without moving the fibers. This means that resin infusion can be used to make advanced composite parts where the direction of the fibers is critical to part performance. In addition, the ability to use low-cost materials and the speed

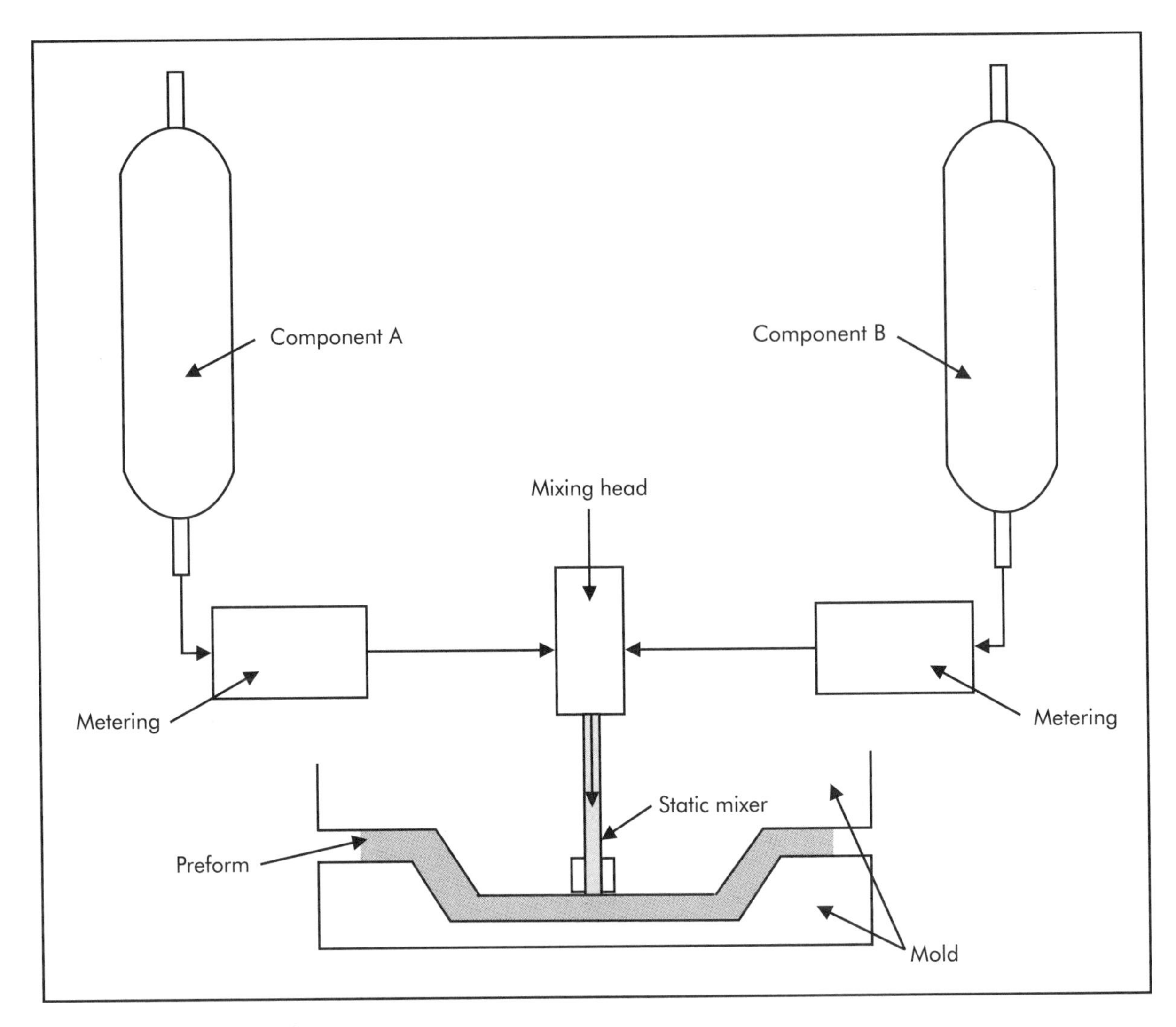

Figure 16-1. Basic resin infusion process.

of manufacture allow resin infusion to be used for engineering composites.

In engineering composites, resin infusion surpasses BMC and at least equals SMC in mechanical performance. This is because the fibers do not need to move in the mold. The operational costs of resin infusion are about equivalent to BMC and SMC but the material is slightly higher because it is difficult to add fillers. (They are sometimes separated from the resin as the resin/filler mixture infiltrates through the fiber preform.) However, other costs such as the reduction in auxiliary equipment needed and the environmental aspects of resin infusion often result in it being the lowest-cost molding process. Hence, the process is likely to be important across the entire spectrum of composites manufacturing. The use of infusion processes is growing rapidly for both new parts and those that were previously made by other composites processes.

The disadvantages of resin infusion processes, also listed in Table 16-1, are not

Table 16-1. Advantages and disadvantages of the resin infusion process (Benjamin and Beckwith 1999).

Advantages	Disadvantages
• Best tolerance control since tooling controls dimensions, high repeatability • Class-A surface finish possible from tool • Surfaces may be gel coated for better surface finish • Smooth finish on both surfaces possible • Cycle times can be short • Molded-in inserts, fittings, ribs, bosses, and reinforcements possible • Low-pressure infusion operation (usually less than 100 psi [700 kPa]) • May increase post-infusion process pressure to higher level (for example, 200–300 psi [1,400–2,000 kPa]) • Prototype tooling costs are relatively low • Volatile emissions (for example, styrene) controlled by closed-mold tooling process • Lower labor intensity and skill levels • Considerable design flexibility: reinforcements, lay-up sequence, core materials, inserts, and mixed materials co-cured in place • Mechanical properties comparable to autoclaved parts with low void contents (<1%) • Wide part size range and ability to handle complexity • Near-net-shape molded parts with little trim required • No bagging required • Heating can be integrated into mold • No refrigeration needed (as required with prepregs and SMC/BMC) • Process can be automated	• Mold and tool design critical to part quality • Tooling costs can be high for large production runs • Mold filling permeability based upon a limited permeability data base • Mold filling software limited or still in development stage • Preform and reinforcement alignment in mold is critical • Production quantities only in the range of 100–5,000 parts • Requires matched, leak-proof molds • Restricted resin choice (due to viscosity) • Air entrapment • Full wet-out of fibers can be difficult (resin flow path)

negligible, but they are manageable, and seem to be under control and diminishing as manufacturers become more familiar with the processes. The most important continuing problem is proper mold design. Mold (tooling) design is critical to achieve complete resin saturation within a reasonable time. Mold filling analysis software is limited to fairly simple shapes (mostly 2D) and by the lack of enough preform permeability data at the present time. However, the situation is getting better and considerable research is aimed at developing more complete models and data for predicting optimal mold designs.

RESIN INFUSION TECHNOLOGIES

The acceleration in the number of parts being made by resin infusion is a result of the number of variations in the process that have been developed. These modifications, each satisfying a particular need or demonstrating a new innovation, have resulted in an "alphabet soup" of names. The most common of the resin infusion processes are listed in Table 16-2. A suggestion has been made by technical leaders in the engineering and advanced composites fields that the resin infusion processes might be divided into two types: pressurized injection (a characteristic of RTM), and vacuum-assisted injection (VAI), which is done at pressures lower than atmospheric. This text prefers to retain the generic "resin infusion" to describe all of the processes, regardless of injection pressure. Nevertheless, other terms such as RTM and VAI will continue to be used.

Resin Transfer Molding (RTM)

Resin transfer molding (RTM) is the process most widely employed and has become the common name for all resin infusion processes, especially in the area of advanced composites. In this discussion, the narrow definition of RTM is used—a pressure-injected, liquid infusion process. RTM grew from a series of related processes for plastics molding. The earliest was transfer molding. This process is similar to compression molding except that the mold is closed when the thermoset plastic material is injected. Injection is accomplished by placing the thermoset resin in a transfer chamber where it is heated to ensure that it is highly fluid but not heated sufficiently to cause significant curing. A plunger pushes the resin through a sprue and runner system into the mold cavity or cavities. The resin is cured in the heated mold and is then ejected by a knockout pin system. This system is similar to the injection molding system for thermoplastics except that the resin is not melted by a screw but, rather, is simply heated in a chamber from which it is transferred to the mold cavities. Transfer molding is, however, often referred to as the "injection molding process for thermosets."

Another precursor process is called reaction injection molding (RIM). In this process, two materials that react rapidly are mixed and injected into a closed mold. The reactive materials usually are the components that make polyurethane. This process is used extensively for automobile bumpers. Occasionally the RIM process will inject into a mold in which some reinforcement is already present.

While transfer molding, RIM, and structural reaction injection molding (SRIM) are similar to RTM, the major difference is that, in RTM, traditional composite resins, such as unsaturated polyesters, epoxies, and vinyl esters, are employed. Also, neither transfer molding nor RIM are reinforced and the amount of reinforcement in SRIM is usually much less than in RTM. The fast cure times of RIM and SRIM allow cycle times of a few seconds to a minute whereas RTM cycle times are usually 5–150 minutes. A further difference is that transfer molding, RIM, and SRIM are high-pressure processes (often to 4,000 psi [28 MPa]), whereas RTM is much

Table 16-2. Comparison of various resin infusion molding processes (Benjamin and Beckwith 1999, Campbell 2004).

Process	Attributes or Features
Resin transfer molding (RTM)	Developed from non-reinforced plastics (transfer molding) and urethane (resin injection molding [RIM] and structural reaction injection molding [SRIM]) technologies Resin injected into matched mold under pressure Excellent surface finishes on both surfaces Can obtain high fiber volumes (55–65%) Process called co-injection resin transfer molding (CIRTM) uses different resin systems during process
Vacuum infusion processing (VIP)	Often termed vacuum-assisted resin transfer molding (VARTM or VRTM) Vacuum pulls liquid resin into preform (no applied pressure) Single-sided tool normally used with vacuum bagging technology Requires lower-viscosity resins Excellent surface finish on tool side only Tooling less expensive than RTM Lower fiber volumes normally obtained (45–55%)
Resin film infusion (RFI)	Resin film is placed in bottom of tool and autoclave heat/pressure cycle is used to melt and force resin into preform Resin film tiles normally based on prepreg B-staged formulations Normally requires matched die tools for complex parts Several industry variations exist: —Resin liquid infusion (RLI) in which liquid resin replaces resin film sheets in tool bottom —SPRINT™ in which resin layers are interspersed in the preform lay-up and flowed during heat up Capable of producing high-quality parts depending on tooling
Expansion RTM	Thermal expansion resin transfer molding (TERTM) uses a matched die with internal core that expands to provide pressure after RTM infusion of preform Rubber-assisted resin transfer molding (RARTM) uses silicone rubber tooling inserts that expand upon heating to provide pressure intensification within the tooling cavity after RTM infusion of the preform
SCRIMP™ or "SCRIMP-like" VIP	Seeman's composite resin infusion molding process (SCRIMP) uses a proprietary infusion tubing and media to speed resin flow and thickness infusion of the resin "SCRIMP-like" VIP processes use various commercially available flow media to achieve results similar to the SCRIMP process

lower. Hence, RTM is truly a composite process whereas the others are more traditional plastics processes.

In comparison to the other resin infusion processes to make composites (as opposed to the largely plastics processes of transfer molding, RIM and SRIM), a key feature of the RTM process is that the resin is injected with moderate pressure. Typical pressures are about 30 psi (207 kPa) but can be as high as 100 psi (690 kPa). The injector must not only be able to deliver the resin under pressure, but the mold must be able to withstand the pressure without distortion. Therefore, metal molds (usually aluminum but not uncommonly steel) are the norm for RTM.

The pressurized injection of RTM has the advantage that resin can be pushed through even difficult and compacted preforms. Fiber volume contents as high as 70% and void contents of about 1% are possible. There is more control over the injection process than when only vacuum is used. However, the increased pressures not only mean that the tooling must be robust, but it must be held firmly shut. Whereas simple clamps can be used in vacuum processes, RTM requires more extensive closure devices. Some manufacturers simply put the RTM mold in a press.

The pressure injection pushes the resin firmly against the mold surfaces resulting in excellent surface quality on all sides, sometimes even a Class-A finish. However, the pressure can also cause fibers to move in the mold, which is called **fiber wash**.

Parts made by RTM can be of complex geometries. A part's complexity is only limited by the ability of the resin to flow into all areas of the mold. The resin uniformity needs to be verified during the development process. This can be done by examining resin/fiber ratios throughout the part. In complex parts, resin movement can trap air in many areas and these should all be vented in the mold.

Vacuum Infusion Processing (VIP)

Vacuum infusion processing (VIP) differs from RTM in the pressure conditions under which the resin is introduced to the mold. In VIP, the pressure is applied by the atmosphere against an evacuated system. Normally the vacuum is applied to strategic locations where the resin is likely to fill last. The vacuum can be applied at multiple locations and each can be sealed (pinched off) when the resin is seen to arrive. Vacuum continues to be applied to other locations, ensuring complete mold filling.

Several names have historically been applied to similar VIP processes. These include vacuum-assisted resin transfer molding (VARTM, VRTM) and vacuum-assisted resin injection (VARI). In some cases, the vacuum is applied along with pressure as a way to further assist in the resin infusion of standard RTM. However, these dual processes (pressure and vacuum) are really more characteristic of standard RTM than of VIP. Therefore, this discussion will focus on the vacuum-only processes.

Because of the lower pressures used in VIP, the molds do not need to be metal. The most common alternate mold type is a vacuum bag on one side and a traditional metal or composite mold on the other. This results in a much lower cost for tooling than in RTM. However, the side of the part defined by the vacuum bag is not the same surface quality as the mold-defined side. Therefore, the surface quality is not as good as with RTM.

Another common mold system uses a thin composite mold. The thin mold is filled with the preform and the resin is infused as in other resin infusion processes. However, the thin composite mold cannot withstand the normal pressures of RTM, but does well under just vacuum pressures. This system is sometimes called **light RTM**. All part surfaces can be finished, but are not as well defined as with traditional RTM. Also, part tolerances are not as finely held as in RTM.

Nevertheless, many parts can be inexpensively made by this process. Since the low pressures allow thin molds to be used, mold support mechanisms are frequently used. These will be discussed in more detail later in this chapter.

Products made by the VIP process include bicycle parts, tennis racquets, racquetball hardware, and component housings for snowmobiles and watercraft.

Resin Film Infusion (RFI)

So far the methods discussed have involved moving essentially low-viscosity, liquid resins into molds and using vacuum and/or external pressure as the driving force for fiber wet-out. However, there are some resins that are so viscous that they will not reasonably infuse through the part in the normal RTM or VIP processes. They are typically so high in viscosity that at room temperature they are films. Therefore, a method was developed to lay the film resin in the mold and then to apply heat and pressure (usually sequentially) so that the resin drops in viscosity and infuses through the preform. Typically, the resin sheet is placed in the mold so that the resin need only infiltrate through a small dimension of the preform (usually the thickness) thus reducing the need for it to travel long distances.

Figure 16-2 shows a photograph of a large (42 ft [12.8 m]) wing part made by RFI. A complex preform was used in the part. In the process used, the fiber preform was laid on top of a resin sheet that was placed in the mold. The mold had internal pressure sections that expanded during the cure to give good definition to the many ribs.

A variation on the standard RFI process is to intersperse the resin sheets with preform layers. This process is called SPRINT™. Clearly, the resin does not need to travel as far as in standard RFI, but the preform must be laminar, thus reducing the complexity of the parts that can be made and eliminating the ability to have totally dry and finished preforms.

Another variation in RFI technology is to place the liquid resin inside the mold before closing. This method is called resin liquid infusion (RLI). In this process, the resin must be lower in viscosity than for traditional RFI but does not require the fibers to be precoated as with prepreg since fiber coating is forced by the pressures involved in the process.

Expansion RTM

The expansion RTM process has two different permutations. In one, called the thermal expansion RTM (TERTM) process, a material that expands when heated is placed as a core within the preform. The resin is infused and the mold is heated. The heating of the mold causes the core material to expand, which forces the resin to wet-out the preform.

The other related process, sometimes called rubber-assisted RTM (RARTM), uses a rubber pad and a metal (usually aluminum) plate. Both are placed inside the mold along with the preform. The pressure on the rubber pad presses against the preform and

Figure 16-2. Lower wing section of a commercial aircraft after RFI processing—all parts co-cured. (Courtesy Scott Beckwith)

causes the resin to completely wet the fibers. This process has the advantage of molding materials with high fiber contents (as much as 75% by volume).

SCRIMP™ and Similar Processes

The full name of the SCRIMP™ process is Seeman's composite resin infusion molding process. It is a patented method of resin distribution coupled with the vacuum bag molding process. By means of various manifolds, the resin is quickly distributed across the part while the vacuum is applied under the bagging material. The resin then saturates the preform. The manifolds are shaped somewhat like the Greek letter omega and are, therefore, called omega channels. Because the resin is distributed widely and quickly, large parts can be made by this process as shown in Figure 16-3.

The process utilizes one-sided tooling and therefore is best suited to applications like boats where only one side is exposed to view. The driving power for resin movement is a vacuum and therefore the parts are usually not too complex. Void contents are somewhat higher than with RTM or other pressure processes. Achievable fiber volumes are usually limited to about 50%. However, the controlled distribution of the resin allows several regions to be simultaneously infused. (In Figure 16-3, there are nine diffusion regions.) Although the knitting together of the resin flows is a potential problem, the choice of low-viscosity resins (unsaturated polyesters and vinyl esters) seems to have eliminated this problem in most parts.

Figure 16-3. Traditional SCRIMP™ processing of 65-ft (19.8-m) boat hull. (Courtesy Scott Beckwith)

The SCRIMP process is licensed widely. Commercial parts manufactured using this process include yachts, railcars, flatbed truck trailers, and piers. Lately, several VIP processes use various commercially available flow media to achieve results similar to the SCRIMP process.

EQUIPMENT AND PROCESS PARAMETERS

Most equipment for resin infusion has been described briefly in characterizing the different resin infusion processes. However, some generalized descriptions are probably valuable to allow comparisons of the various process machines and to understand the range of operating parameters for each of the infusion processes.

Injection Equipment

Commercial resin infusion equipment is available from a variety of sources. These machines usually consist of holding tanks for the resin components, initiators, hardening agents, and sometimes even for the minor additives. The holding tanks are often pressurized so the flows can be accurately metered or they might be equipped with pumps to meter the liquids. The outlets of the tanks are often directed toward an in-line mixer that causes the liquids from the various tanks to flow together and mix prior to entry into the mold. While these in-line mixers are convenient and generally work well, some limitations on the completeness of mixing are present, especially at high flow rates.

Many molders have made their own resin injection systems. While these are often not as sophisticated as the commercial types, many homemade versions seem to work quite well. For simple, small parts, the injection system might be just a plunger in a tube into which the pre-mixed resin system is placed. The plunger can be moved manually or by machine. The problems with this simple system are the need to pre-mix the materials (and the difficulties with mix ratios and problems with handling various liquids), the potential inaccuracy of the injection, the potential lack of sufficient force to properly inject the resin to achieve full wet-out, and the limited pot life when using the pre-mixed resin system.

An interesting variable that has been shown to be effective in a laboratory setting is the sequential injection of resin across the length of the mold. This improves overall injection time and has the potential for improved wet-out, especially in systems where the long resin flow paths may be a problem because of the resin's viscosity or rapid curing.

Molds

Metals are the most common materials for RTM molds. When pressures are lower, as in VIP, the molds can be made of composites or, for improved surfaces, composites coated with metal (usually nickel).

Several mechanical clamps located around the perimeter of the mold are used for closing the halves together. These are usually manually operated but can be machine assisted. These types of clamps are easy to use, inexpensive to acquire, and generally reliable except for under high pressures. Therefore, when the pressures exceed the capability of these clamps, the molds are usually placed in a press or autoclave. This will generally ensure that the mold will stay closed during resin injection, but the convenience of accessibility to the mold during filling is reduced or eliminated. An advantage to the press or autoclave might be, however, that the press can be heated and this would eliminate the need to have separate heaters for the mold.

Two interesting innovations have been used to create backing support for thin molds that might have injection pressures higher than would be normally encountered in the light RTM or VIP process. The first of these methods uses a water bath with pressure to resist any shape changes in the mold. The second uses mechanical plungers placed against strategic locations on the backside of the mold. Since the plungers can be selectively placed, they allow for a variety of mold shapes (as does, of course, the water-backed system).

Venting is a significant issue in all resin infusion molds. Normally the vents are located at the points most distant from the injection port(s). Vent ports should be provided at the top (end) of all long pathways, especially if air could be trapped in such places.

PREFORM TECHNOLOGY FOR INFUSION

Preform technologies were discussed in detail in Chapter Nine where the nature of preforms and the methods of making them were explained. There are, however, some special considerations about preforms that are especially important in the resin infusion processes. In addition to the load-carrying requirements required of all composite reinforcement forms, the preforms used in resin infusion processes must fit inside the mold properly. There should be no gaps or open areas where resin can puddle. Further, the preform must allow the resin to permeate readily and not move extensively when the resin is flowing.

The preform can be a simple 2D structure that is nearly laminar in construction. For example, such a preform has been used in hockey stick blades. The blades could have been made using other processes, such as hand lay-up. However, resin infusion proved to be both economical and reliable for the high-volume part because the preform simplified the retention of the materials in their proper orientation when they were molded. In this case, the preform was lightly stitched to hold its shape.

The preform shown in Figure 16-4a is much more complicated than the simple 2D preform used to make the hockey stick blade. The complex preform is used to make an aircraft hinge, which will be subjected to significant forces in all directions. The completed part, made by RTM is shown in Figure 16-4b. Note the inclusion of metal inserts and the overall excellence of the surfaces.

An important consideration in choosing preforms for resin infusion processes is how the resin will wet-out the preform. In general, the resin will move most easily in the direction of the fibers. (The resin flows along the small gaps in between.) Generally, it is desired to have those pathways interrupted so the resin is forced to flow throughout the preform. This phenomenon can be seen in the differences in the permeation rate of resin through woven material and through knitted/stitched material. The rate is slower in woven material, presumably because it has more crossover points and, therefore, more interruptions in the flow.

Another related consideration is the fiber volume fraction in the preform. Permeation of the resin is strongly related to the openness of the structure as shown in Figure 16-5. Note that at 50% fiber-volume fraction the penetration is extensive and rapid. At 71% fiber-volume fraction that penetration is much less extensive and less rapid. Therefore, in constructing the preform, the openness of the structure must be related to the flow of the resin and to the conditions (pressure, temperature, etc.) that will be used. If the resin flow is too slow, the resin could cure before the preform is fully wetted and the mold filled. This phenomenon is called **resin freeze** or **resin stall out**. It is, of course, to be avoided.

The preform need not be made as a single, stand-alone unit. It is possible to place flat sheets of dry fibers (woven or mat) between flexible (elastomeric) diaphragms that are sealed inside a mold-like frame. The resin can be injected between the sheets and then the entire assembly can be put into a matched die mold where it is shaped

a. Complex aircraft hinge—preform uses stitching technology with textile materials

b. Completed RTM aircraft hinge structure after resin infusion and cure (note metal inserts)

Figure 16-4. RTM aircraft hinge. (Courtesy Scott Beckwith)

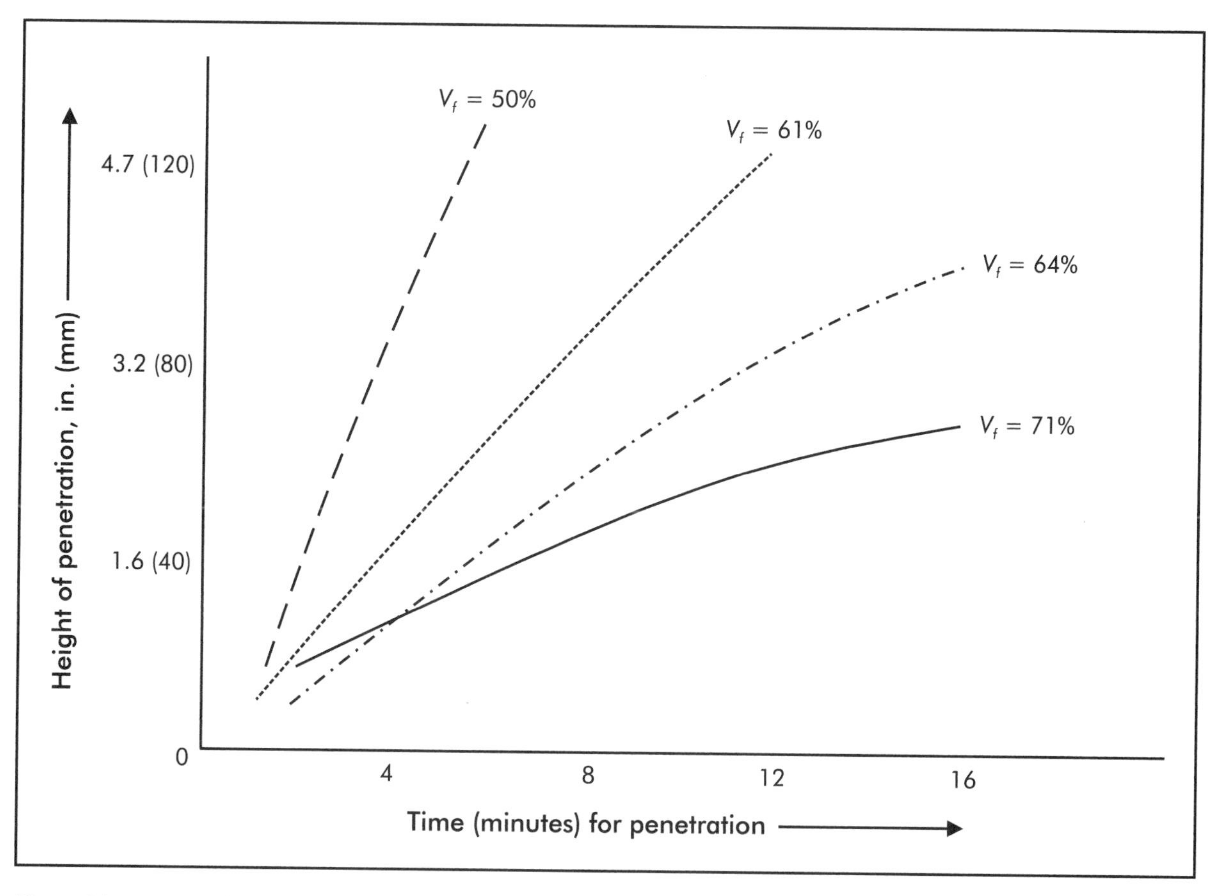

Figure 16-5. Resin infusion as a function of the fiber-volume fraction.

and cured. It is similarly possible to move the reinforcement sheets (woven or mat) continuously between rollers and inject the resin. Downstream rollers and forms are then used to shape the part. This second variation is like a combination of RTM and pultrusion.

Table 16-3 lists some advantages and limitations of various options for preforms. The preform materials considered are chopped or spray-up (mat), cloth laminates, 2D and 3D woven fabrics, 2D and 3D braided materials, warp knits, and stitched preforms.

RESIN CHARACTERISTICS

Most resin types can be used for resin infusion provided that the viscosity and cure characteristics are appropriate. Some are, however, much easier to use because the adjustment of viscosity and cure conditions are simpler. In general, most resin infusion is done with unsaturated polyester, vinyl ester, and epoxy. But even within these three dominant resin types, many grades cannot be used.

Viscosity is the single most important parameter. The normal range of viscosities is 50–1,000 centipoise. That viscosity is a function of the resin type, the molecular weight, and the temperature. To give an idea about viscosities, some common materials and their viscosities at room temperature are listed in Table 16-4. The effect of temperature can be dramatic.

Table 16-3. Preform options and their advantages and limitations with RTM or VIP (Campbell 2004).

Preform or Textile Process	Advantages	Limitations
Chopped spray-up materials	Low-cost preform approach Low-to-moderate in-plane properties Lower fiber volumes Several variations (short chopped-fiber spray-up, P4 glass process and P4A carbon process)	Lower mechanical properties Not used in advanced technology applications
Low crimp uniweave	High in-plane properties Good tailoring of properties Highly automated preform fabrication process	Low transverse and out-of-plane properties Poor fabric stability Labor-intensive ply lay-up
2D woven fabric	Good in-plane properties Good draping Highly automated preform fabrication process Integrally woven shapes possible Suited for large area coverage	Limited tailoring of off-axis properties Low out-of-plane properties
3D woven fabric	Moderate in-plane and out-of-plane properties Automated preform fabrication process Limited woven shapes possible	Limited tailoring of off-axis properties Poor draping
2D braided preforms	Good balance of off-axis properties Automated preform fabrication process Well suited for complex curved parts Good draping	Size limitation for off-axis properties Low out-of-plane properties
3D braided preforms	Good balance of in-plane and out-of-plane properties Automated preform fabrication process Well suited for complex shapes	Slow preform fabrication process Size limitation due to machine availability
Multi-axial warp knit	Good tailoring for balanced in-plane properties Highly automated preform fabrication process Multi-layer, high-throughput material suited for large-area coverage	Low out-of-plane properties
Stitched preform structures	Good in-plane properties Highly automated fabrication process provides excellent damage tolerance and out-of-plane strength Excellent assembly aid	Small reduction in in-plane properties Poor accessibility in complex curved shapes

Table 16-4. Viscosities of common materials at room temperature.

Material	Approximate Viscosity, centipoise
Water	1
Blood	10
Kerosene	10
Antifreeze (ethylene glycol)	15
Motor oil (SAE 10)	50–100
Corn oil	50–100
Motor oil (SAE 30)	150–200
Maple syrup	150–200
Motor oil (SAE 40)	250-500
Glycerine	250–500
Corn syrup or honey	2,000–3,000
Molasses	5,000–10,000
Chocolate syrup	10,000–25,000
Ketchup or mustard	50,000–70,000
Peanut butter	150,000–250,000
Shortening or lard	1,000,000–2,000,000
Caulking compound	5,000,000–10,000,000
Window putty	100,000,000

The pot life of the resin is also important. It must be long enough to ensure that resin freeze does not occur. However, it must be short enough to allow for economical cycle times. The pot life is dependent on temperature, pressure, vacuum, viscosity, and cure additives. An interesting variant of normal processing is to use an ultraviolet (UV) cure material. This allows the temperature to be raised and therefore reduces the material's viscosity without shortening its pot life. When the material has filled the mold, the part is subjected to UV light and cure takes place. This technique is limited in the thickness of the part that can be cured (because of UV light penetration). However, other methods of irradiating the resin have better penetration power and are just as effective, such as microwave, ultrasonic, or radio frequency waves.

The infusion processing parameters for a typical resin are summarized in Table 16-5.

Bismaleimide and high-performance polyimide resins have also been adapted for resin infusion processing. The high-performance polyimide process is proprietary and available for licensing.

CORE AND FLOW MEDIA MATERIALS

Many (but not all) parts made with resin infusion have cores within the composite structure. These cores can be important, as discussed in Chapter 12.

When a core is present, some provision must be made for the resin to pass through

Table 16-5. Resin characteristics for infusion processing.

Resin Parameter	Desired Characteristics
Viscosity	Flow viscosity must be low enough to rapidly wet-out preform Viscosity should be in 50–1,000 centipoise range
Pot life or gel time	Must be sufficient to allow complete wet-out of entire preform Too short—resin stalls out and leaves dry areas Too long—unnecessary cycle time duration
Temperature controls	Higher temperatures can: —Decrease resin viscosity initially —Shorten infusion time too much —Speed up cure reaction —Decrease time to 1,000 centipoise limit Temperature controls (heating) can be placed upon: —Mold cavity —Resin mix bowls —Incoming resin lines —All of the above or only certain regions

or around it so that the composite laminates on the side of the core opposite the injection port can be wetted by the resin. This can be done with channels on the sides of the core or even channels through the core provided the core is not compromised by their presence. What cannot be done, however, is to allow the resin to permeate and stay within the core. If that happens, the amount of resin in the structure will increase dramatically along with the weight and this excess resin serves no useful purpose. The problem of resin infiltration is especially important in honeycomb and open-cell foam materials. To prevent resin infiltration in these materials, the adhesive layer used to bond the honeycomb to the laminate plates should be in place. This will be a barrier to the flow of resin. While the grooves on or through the core can be a problem, it is also possible to tailor the grooves so that resin flow can be managed, even used to improve wet-out of the fibers in the face sheets.

Even when no core is present, resin flow through the entire mold is sometimes a problem. The flow can be facilitated through the use of flow media materials. These materials have a similar function for resin as a breather material does for air in a vacuum bagging assembly. The flow materials function as a pathway for the resin (and for the air) to escape. Some common materials used for flow include a screen layer in which the resin can flow and layers in which the resin is less dense than in the surrounding preform areas. Another method is to use two layers of film and allow the resin to move between the layers to strategic exit points. Some caution must be taken, however, to avoid excessive resin flow through these areas to the exclusion of the preform itself. Another approach is to add wetting agents for the resin that facilitate its movement through the preform.

Several types of flow media materials are listed in Table 16-6. Some are available only

Table 16-6. Flow media material enhancement for infusion processing.

Flow Media Type or Approach	Generally Observed Effects
Omega tube or perforated tubing	Open side of omega tube allows rapid resin flow through tube and lateral flow across the preform surface Perforated holes in tubing act in same manner Spiral cable wrap also a typical substitute
Screen materials	Rigid and semi-rigid screen materials used to increase permeability for resin flow between layers Materials of choice: plastic mesh, window screen, burlap
Double vacuum bagging	Resin infused with secondary bagging layer pulled off preform's surface (creates higher permeability and lower fiber volume initially) Vacuum eventually applied across preform to remove voids and provide compaction after resin infusion is completed
Perforated, slotted cores	Core perforations or cross-cut slotting (saw cuts) in core material Creates z direction resin flow to opposite sides of core for complete wet-out

under license and others are commercially available from standard suppliers.

PART DESIGN FOR RESIN INFUSION

Some considerations for design of parts manufactured by resin infusion include the following.

- Flow is better in the plane than across the plane of the fibers.
- Large parts require long flow times, which may require filling and backfilling of the resin.
- The preform must fit easily within the tool but cannot have large void areas where resin will accumulate.
- Gating needs to minimize flow paths without resulting in excessive meeting of resin fronts (decreasing the potential for poor resin knitting).
- Molds must be carefully vented.
- Provision must be made to determine that the mold is completely filled, possibly by use of flow sensors located strategically in the part (or, at least, in the prototype so the flow model can be accurately made).
- Estimate the amount of fiber volume fraction desired in the final part and make the preform accordingly.
- Design the mold for easy part removal.
- Consider automation in making the preform, placing the preform in the mold, injecting the resin, and removing the part.

- Maintaining the seal on the mold throughout the injection and cure cycles is important.

CENTRIFUGAL CASTING

Centrifugal casting is not usually considered to be a resin infusion process. Centrifugal or spin casting of pipes and tanks is common although not as widely used as filament winding. Some manufacturers use this method effectively for large fiber reinforced plastic (FRP) tanks. In this method, the fibers and resin are applied to the inside of a spinning mandrel that uses the centrifugal force generated to push the resin and fibers against the inside wall. This method builds up the tank in reverse order so the structural layers are created first. The last is the corrosion layer. Fiber angle is usually random because the fibers are often blown into the inside of the structure.

Fillers can be added easily during centrifugal casting. Sand is one commonly used filler that adds weight and lowers cost. Some performance benefits have also been reported with sand fillers. However, others have reported that the sand cuts the reinforcement fibers and, therefore, results in significant loss of strength.

CASE STUDY 16-1

Dodge Viper®

Resin infusion is the principal method of manufacturing the major body panels for the Dodge Viper®. Because this car was the first RTM application for Chrysler, it formed a partnership with the paint supplier and the body-panel manufacturer. Chrysler and its partners believed that RTM would become a major manufacturing method for automobiles, especially for those models having a relatively small production volume. The Viper fit that niche exactly.

The early prototypes were tested for dimensional accuracy and surface smoothness. The mold was adjusted and smoothed until these parameters were within the specifications. Because these adjustments could be made in the original tool set, development time for the first production parts was greatly reduced compared to traditional manufacturing methods. Eventually the tools were of epoxy with a nickel coating.

The production was done using low-pressure (150–200-ton [2,000–2,700 MPa]) presses versus the large 3,000-ton (41,000 MPa) presses needed for metal stamping. This, along with the low-cost mold materials, kept investment low.

The Viper hood was the largest part and required the most complex tooling. To form the sides, sliding molds were required. The outer skin was an especially difficult part because of the high gloss required. Attempts with the low-cost RTM tooling were not successful so a change was made to metal sheet molding compound (SMC) tooling.

The use of RTM allowed the entire production process, from original concept to production automobile, to be done in less time than usual. Saving time may prove to be the most important advantage of all as the needs for small runs of products to meet market segmentation requirements continues to increase. RTM may even prove to be the method by which standard production prototypes are made.

SUMMARY

Table 16-7 provides a convenient summary of the various processing parameters typical for resin infusion processes. These parameters differ somewhat from those of other composite molding processes and so they should be carefully considered when choosing to make a part using resin infusion.

Resin infusion can be compared with other composite molding processes to determine the relative costs and production capabilities. Figure 16-6 is especially helpful in making these comparisons as it illustrates

Table 16-7. RTM and VIP processing variables and their effects (Benjamin and Beckwith 1999).

RTM or VIP Process Parameter	Potential Effects on Processing or Structure
Resin viscosity	50–1,000 centipoise typical processing range for viscosity Processing at 10–100 centipoise, higher temperature also typical Higher viscosity—preforms often hard to wet out Lower viscosity—more rapid infusion may leave dry areas and voids
Resin pot life	Too short—resin fails to completely fill preform Too long—process cycle lengthened unnecessarily
Resin injection pressure	Helps drive resin into mold and preform Applied too fast—may move preform out of position within mold Too high—may cause "fiber" wash of preform, damage mold, tooling, or blow seals Too low—cycle times long, resin could gel during fill period
Resin injection vacuum level	20–28 in. Hg typical process range Helps pull resin into mold and preform Aids in reducing void content through part Assists in holding mold halves closed Aids in removing moisture and volatiles
Multiple injection ports	Commonly used to ensure more complete wet-out of preform structure Sometimes used sequentially to fill long or large-area parts ("pumping" action)
Internal rubber/ elastomeric tooling	Rubber/elastomeric inserts have high expansion (high coefficient of thermal expansion [CTE]) Often called "pressure intensifiers" Used to provide internal compaction pressure Higher fiber volumes (>65%) achievable Low void contents are typical (<1%) Tooling must be more robust to handle induced pressures
Closed mold pressurization	Infusion pressures usually below 100 psi (700 kPa) Pressure often increased to 100–300 psi (700–2,000 kPa) range after resin wet-out Decreases microvoids by collapsing bubble cavities
Fiber sizing or coupling agents	Sizing or coupling-agent chemistry must be compatible with selected resin Sizing level (if too high) can reduce resin flow (lowers effective permeability)
Preform reinforcement integrity	Loose preforms (chopped fiber, lightly tacked layers, low fiber volumes, etc.) tend to "wash" and move around the mold cavity Tight preforms (braids, stitched, fabrics, textiles, etc.) tend to maintain shape and integrity

Table 16-7. (*Continued*)

RTM or VIP Process Parameter	Potential Effects on Processing or Structure
Fiber volume	Resin flow permeability is inversely proportional to fiber volume Higher fiber volumes (>60%) require more work, energy, and time to wet-out preforms Commercial market—usually 25–55% fiber volume Aerospace market—usually 50–70% fiber volume
Core materials	Foam and balsa core materials used extensively Open-cell and honeycomb core materials require method for preventing resin flow into cell structure Closed-cell cores typically have some inherent resin path to opposite sides (flow media, cross-cuts, pore holes, etc.)
Molded-in inserts and fittings	Inclusion and co-curing entirely possible with RTM and VIP processes Resin flow around can leave dry areas and voids

the regions of applicability for several of the processes in terms of the production volume and part-size considerations. In this comparison, RTM is seen to fit nicely in the area of fairly large parts and production volumes. As expected, it is close to compression molding and thermoset filament winding.

Another method of comparing resin infusion processes is by the length of the fibers that can be used. Resin infusion has the advantage that fibers of almost any length are acceptable. So it has wider applicability than injection molding (short fibers only) and even compression molding, which uses shorter fibers because they must move in the mold. Hence, resin infusion has the potential for making parts with higher mechanical performance than either injection molding or compression molding—two processes that are economic competitors to resin infusion.

Still another method of comparing the processes is by the fiber-volume fraction that can be used. Higher fiber volumes usually mean that mechanical performance will be better. Resin infusion has the capability of using fiber volumes of up to 70%. This is higher than any other process.

LABORATORY EXPERIMENT 16-1

Objective: Mold a ping-pong paddle using RTM.

Procedure:

1. Make a mold in which the top and bottom halves are glass plates and the perimeter of the mold is a metal picture frame construction. (A nice shape for the mold is that of a ping-pong paddle.)
2. Make an injection port at the center of the bottom plate. Be sure to put in a path for air to escape and for resin to overflow when the mold is full. Apply mold release to the glass plates or use a film that can serve as a release layer.
3. Make a preform for the paddle and a preform with a core for the handle.
4. Obtain some RTM-grade polyester resin.
5. Use a commercial RTM machine to inject the resin. However, if that is not available, use a metal pipe and a plunger to make the injection. Note and record the filling pattern as a function

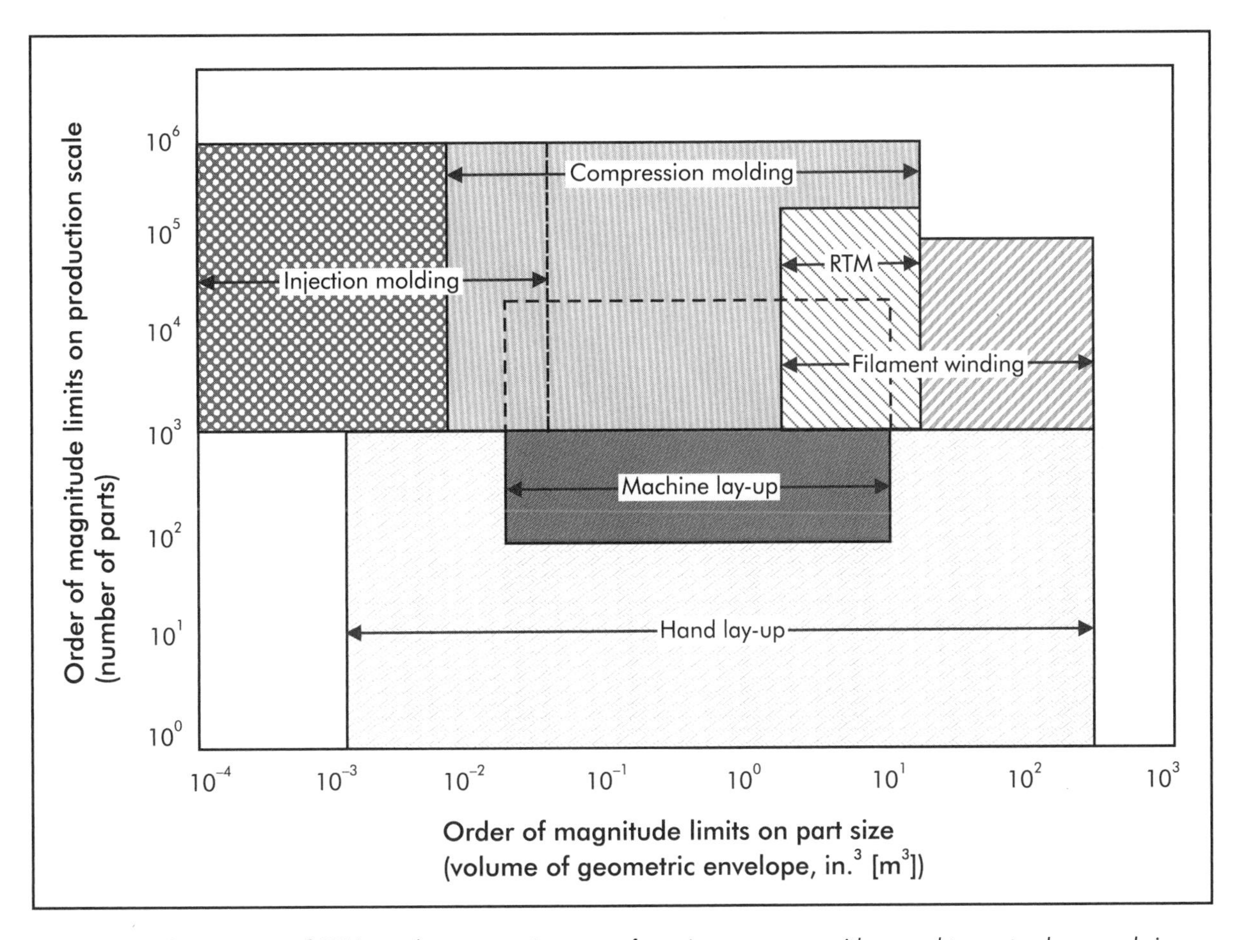

Figure 16-6. Comparison of RTM to other composites manufacturing processes with regard to part volume and size.

of time and pressure. The resin should be fully initiated if the plunger injection system is used. If a commercial system is used, put the initiator in the small tank on the machine. Set the ratio for initiator to resin at about 1.8%.

6. When the injection is complete, allow the part to cure. Remove the part from the mold and trim.

QUESTIONS

1. Discuss the basic differences between RTM and VIP infusion technologies.
2. What is the key similarity between an RFI process and a SCRIMP™ or SCRIMP-like process as far as resin infusion?
3. What are TERTM and RARTM and what key material aspects make these processes achieve higher fiber volumes?
4. Discuss at least six of the key process variables pertaining to the RTM and VIP processes and why they are important.
5. What do preforms do for composites and why are they an important aspect of resin infusion technology?
6. Assume that a 60–65% fiber volume structure with a complex shape for an aerospace application is to be produced.

Discuss the type of preform and process that might be selected and explain why.

7. Visit a home improvement center and select (identify) at least three materials or methods by which flow media might be fabricated for a specific resin infusion process of your choice. Explain how each would be used in the process.
8. Why is resin viscosity important in resin infusion processes?
9. What is the effect of temperature in resin infusion processes?
10. Discuss whether pot life should be short or long in resin infusion and how that might affect cure time.

BIBLIOGRAPHY

Benjamin, William P. and Beckwith, Scott W. (ed.). 1999. *Resin Transfer Molding: SAMPE Monograph, No. 3*. Covina, CA: Society for the Advancement of Plastics Engineering (SAMPE).

Campbell, F. C. 2004. *Manufacturing Processes for Advanced Composites,* Chapter 9, "Liquid Molding." NY: Elsevier Advanced Technology.

Kruckenberg, Teresa M. and Paton, Rowan. 1998. *Resin Transfer Moulding for Aerospace Structures*. Dordrecht, The Netherlands: Kluwer Academic Publishers.

Mazumdar, Sanjay K. 2002. *Composites Manufacturing: Materials, Product, and Process Engineering*. Boca Raton, FL: CRC Press.

Parnas, Richard S. 2000. *Liquid Composite Molding*. Munchen, Germany: Hanser Publishers.

Potter, Kevin. 1997. *Resin Transfer Moulding*. NY: Chapman and Hall.

Quinn, James A. 1999. *Composites Design Manual*. Lancaster, PA: Technomic Publishing Company.

Society of Manufacturing Engineers (SME). 2005. "Liquid Molding" DVD from the Composites Manufacturing video series. Dearborn, MI: Society of Manufacturing Engineers, www.sme.org/cmvs.

17

Filament Winding and Fiber Placement

CHAPTER OVERVIEW

This chapter examines the following concepts:

- Process overview
- Filament winding
- Variations in filament winding
- Fiber placement

PROCESS OVERVIEW

Filament winding is the most important of the composites processes in terms of the number of users and the total number of parts made. **Fiber placement** is a more recent development and is a method that can be used to make highly complex parts, even some that could not reasonably be made with filament winding.

The two filament winding methods that are the focus of this chapter wrap fibers (with resin) around a **mandrel** (a core) as the method of shaping the fibers and resins prior to curing. The methods differ in the way the fibers are placed against the mandrel, which leads to other important variations in the process.

This chapter examines filament winding in detail, including some related processes that expand on its original capabilities. Then fiber placement, the most important process development beyond filament winding, is explored. (Consult the "Filament Winding" video, SME 2005 for a look at the equipment and the techniques of filament winding. These videos are excellent supplements to lectures and laboratories and for general education for those not familiar with this process.)

FILAMENT WINDING

The basic filament winding process can be understood by referring to the illustration of the standard filament winding machine shown in Figure 17-1. The fiber spools are mounted on a rack, called a **creel**, and then strands from many spools are gathered together and fed through a comb or similar alignment device so they make a band of fibers. The number of strands brought together determines the width of the band. The band then enters a resin bath where the fibers are soaked with resin. The resin is fully activated with initiator or hardener so that the only requirements to cure the part are heat and time. The fibers then go through a roller or wiper system to remove the excess resin and then through a ring or some other directing device called a **payoff**. The payoff directs the fibers onto a mandrel.

At initial startup, the process requires that the fibers be manually fed through all the parts of the system. After the fibers have been fed through the system, the turning of the mandrel (to which the fiber strand has been anchored) pulls them from the fiber spools through the system. The payoff is mounted on a carriage or some other transport system that moves laterally along the long axis of the mandrel as the fibers are

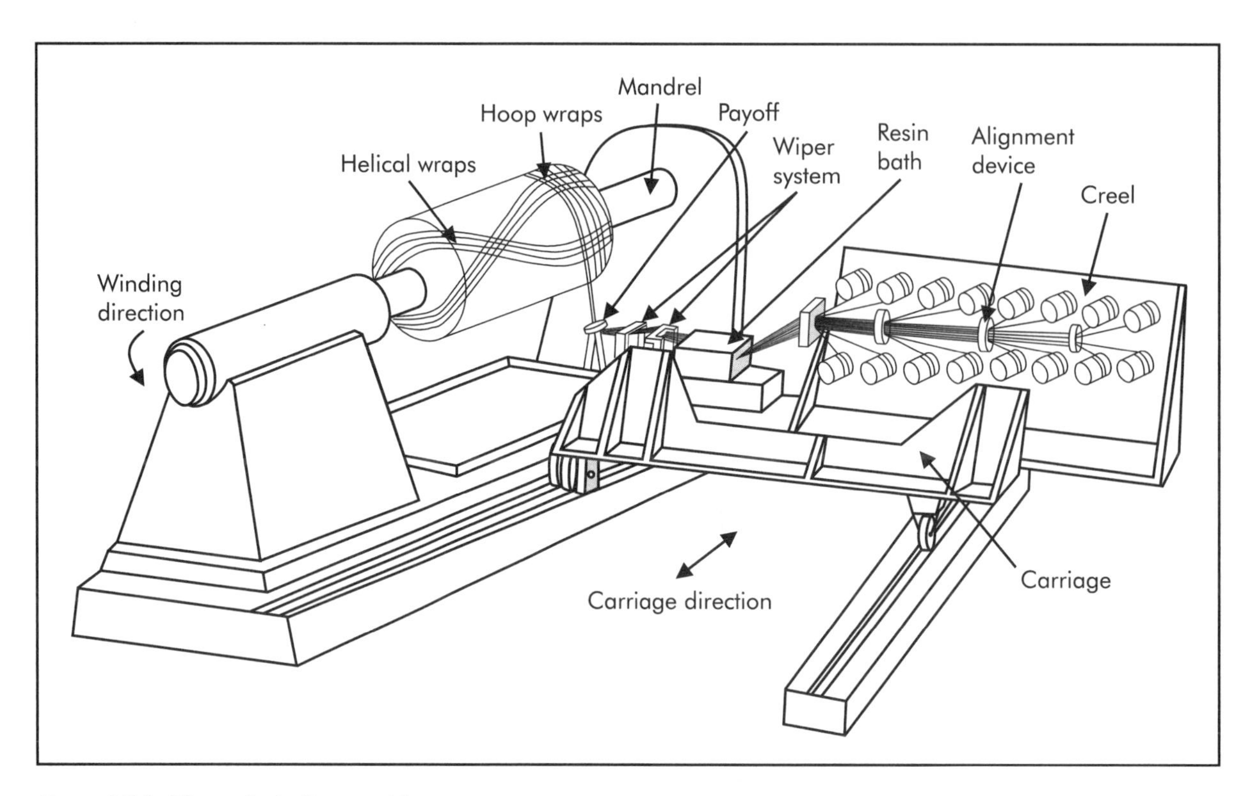

Figure 17-1. Filament winding machine.

being drawn through. The payoff motion is synchronized with the turning of the mandrel to produce patterns of fibers on the mandrel. The angles of the fiber pattern are determined by the relative motion and speeds of the mandrel and the payoff system.

As shown on Figure 17-2, the two general directions of fiber paths are: **hoop windings**, sometimes also called circumferential windings or radial windings, and **helical windings**. The hoops circle the mandrel in paths that are approximately 90° to the long axis of the mandrel. They are like hoops around a barrel. Hoop windings are important to give radial strength to the part being wound. The helical paths are at all angles other than approximately 90°. They move from one end of the mandrel to the other in continuous loops that are slightly offset from the previous loop and, therefore, eventually create a layer of material that covers the entire surface of the part being wound. Helical windings are important to hold the hoops in place and give strength to the ends of the part. They also impart torsional strength and bending strength. Helical windings can be placed at an angle of 0°, that is, from one end of the mandrel to the other. These windings are sometimes called **longitudinal windings** or axial windings. These can all be seen in Figure 17-2.

Layers of helical and hoop paths can be added successively on top of one another. It is possible to have one layer at one helical angle and then the next layer at some other angle. In fact, it is common to have layers that change angle from one to the next. The sequence and angles of the layers are determined in the same type of laminar design system used with other laminar composite shapes made by other processes.

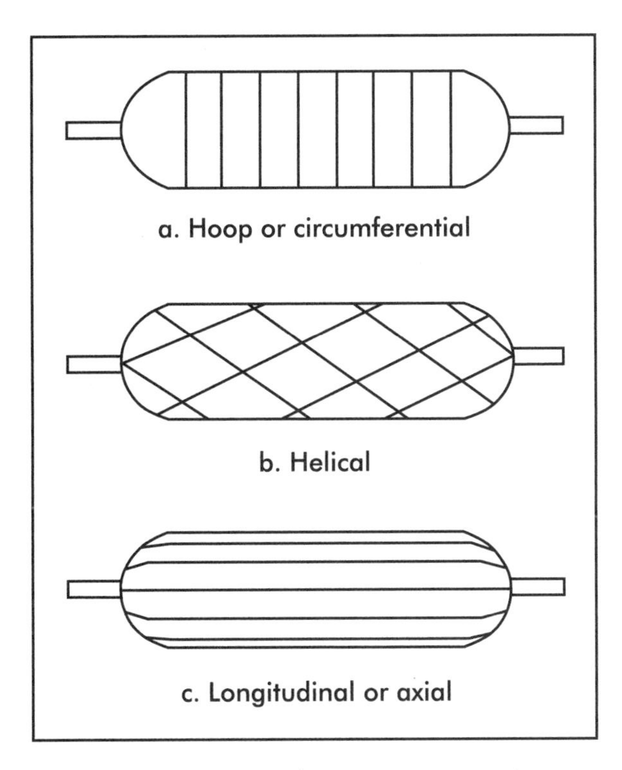

Figure 17-2. Filament winding patterns.

When the winding has been completed according to the laminate's design, the part is cured. This is done with the mandrel in place. Parts can be cured at room temperature or in ovens or autoclaves, depending on the size of the part and the nature of the resin. When parts are cured, they are usually continually rotated so that the resin will not sag to the bottom of the part as the resin viscosity drops during the initial stages of cure. Alternately, if the part cannot be rotated, the parts are hung vertically so that the resin distribution is even around the circumference of the part. Even so, there may be some sagging of resin toward the bottom end of the part.

In normal practice, the mandrel is removed after curing. If the part is open-ended like a section of pipe, the mandrel is usually just pulled out one end. A mechanical extraction device is usually required because the draft angle from one end of the part to the other is usually small or non-existent and thus a strong pulling force is required to extract the mandrel. If the part is close-ended like a pressure vessel, the mandrel is removed by some other technique. Several of those techniques will be discussed later in this chapter. Some parts, such as tanks for firefighters, begin with the metal tank as the mandrel and over-wrap with the fibers and resin to impart additional strength. In cases like those (and others that will be discussed), the mandrel simply remains and is not removed.

The use of a mandrel to shape the fibers and resin means that the general shape of most parts made by filament winding is cylindrical (approximately circular in cross-section). Pipes, pressure vessels, and similar shapes are most commonly made by filament winding. However, by allowing the payoff head to move vertically or twist and rotate, complicated shapes can be made. All, however, must have an axis of rotation. Two statements summarize the capabilities of filament winding with respect to shape.

- If it is round, it can be wound.
- It does *not* have to be round to be wound.

Some typical parts made by the filament winding process include the following: low- and high-pressure pipes, water tanks, chemical storage tanks, reverse osmosis tubes, air tanks for firefighters, scuba tanks, cooling-tower coupling shafts, transmission drive shafts, industrial rollers, compressed natural gas tanks, flywheels, golf shafts, bicycle frames, oars, wind-surfer masts, boat masts, streetlight and utility poles, rocket-motor cases, missile launch tubes, radomes, fuselage and wing sections, aircraft waste tanks, helicopter rotor shafts, and wind-turbine blades. Clearly, the parts made by filament winding are important to a wide variety of markets.

Comparison to Other Composite Manufacturing Processes

The mandrel in filament winding is similar in function to the mold in other composite manufacturing processes such as hand lay-up, spray-up, and resin transfer molding (RTM). Like most of the other composite processes discussed in this book, filament winding is a batch process. That is, one part is made at a time on each mold or mandrel. (Pultrusion, which is discussed in a later chapter, is a continuous process that uses a die to form the composite part and not a mandrel or mold.)

Filament winding is one of the lowest-cost composites manufacturing processes. As will be discussed later in this chapter, it often uses fibers and resin in their lowest-cost forms. Also, the cost of molds or tooling is low as mandrels are almost always lower in cost than a mold. The filament winding machine can be costly, but options allow many parts to be made simultaneously. In addition, the productivity is high so the cost of the machine can be spread over many parts. Labor requirements are low; usually only one operator is required per machine. Therefore, the cost of parts made by filament winding is about one-quarter to one-half the cost of parts made by manual lay-up. The continuous nature of the filaments means that the strength and stiffness of filament-wound parts are excellent. Therefore, the performance of parts is usually better than similar parts made by spray-up or with chopped mat.

Because the filament winding process is automated, manufacturing efficiencies are easily obtained. Fiber lay-down rates, a measure of the processing speed, are dependent upon the part's complexity and the precision with which the fiber angles must be controlled. However, simple parts, such as pipes, are often wound at fiber lay-down rates in the range of 300–3,000 lb/hr (136–1,360 kg/hr). Complex parts and those intended for high-performance markets like aerospace are often fabricated at 10–200 lb/hr (5–90 kg/hr). Non-automated processes, such as hand lay-up, might have one-tenth the production rate of filament winding and, of course, much higher production costs. However, these largely non-automated processes can make parts that are much more varied in shape and complexity than the parts that can be made by filament winding.

Slower speed allows for tighter control over the void content of the parts made by filament winding. Most filament winding designers expect about 3% void content, but this can be as high as 10% in high-production operations and below 1% in high-performance applications. The void content can be reduced by increasing the tension on the fibers during winding and by lowering the resin viscosity so that the fibers are well wetted. Resins can be too low in viscosity, however, and drip excessively from the part as it is being wound. Adding pressure during the cure will also reduce the void content as can changes in the curing cycle to allow for better resin flow during the curing phase.

Filament Winding Materials

Reinforcement Fibers

Almost any of the normal reinforcement fibers used to make composite parts can be used in filament winding. The fibers need to be strong enough that they can be pulled through the system, but that requirement is rarely a limitation. A possible exception could be the case of **staple fibers** that have been made into a continuous filament. If the staple is a natural fiber, it might not have the tenacity to be pulled through the system. More likely, however, is that the system could be adjusted for the weak tenacity of the fibers and so filament winding could still be used. Fiberglass is the most common fiber but carbon/graphite, aramid, and ultra-high-molecular-weight polyethylene (UHMWPE) fibers are also widely used.

In the vast majority of cases, the fibers used in filament winding are continuous. They are pulled from spools through the wet-out system and onto the mandrel in continuous strands. As previously indicated, normally several strands (tows or rovings) are gathered together and fed simultaneously so that a band of fibers is applied to the mandrel. This multiple-strand method permits rapid lay-down of the fibers. The width of the band is a function of the number of tows or rovings that are gathered together. The complexity of the part and the need to have precise fiber alignment can limit the width of the band. For example, small-diameter, closed-end vessels are difficult to wrap smoothly with a wide band.

One variation of the filament winding process uses non-continuous fibers. In this application, fiberglass mat is manually laid against the mandrel as the continuous fibers are being applied. Alternately, chopped fibers can be sprayed onto the mandrel during conventional winding. This technique gives fast lay down of fibers but fiber wet-out is difficult and void contents are high. It is used almost exclusively for inexpensive, low-performance pipe and low-pressure requirement tanks.

No preforms or other special fiber shapes are required for filament winding. The form of fibers used is untwisted, continuous fibers directly from the spools. This is the least expensive form of fibers and, therefore, is a contributor to the overall favorable economics of the filament winding process.

Resins

Almost all types of thermoset resins normally used in composites manufacturing can be used in filament winding. The most common resins are unsaturated polyesters, vinyl esters, and epoxies, although other thermosets and some thermoplastics are also used.

Regardless of the type of resin, grades that are designed specifically for filament winding are normally chosen. These grades are balanced for viscosity, pot life, and reactivity; thus they are suitable to the unique requirements of the filament winding process. The viscosity is generally comparatively low so that wet-out of the fibers is good. The fibers often pass quickly through the resin bath and this means that resin must penetrate the fiber bundles quickly and easily. A typical range of viscosities for filament winding resins is 350–2,000 centipoise. If the viscosity is much lower than 350 centipoise, the resin will flow excessively when on the mandrel, which could result in it dripping off the part and possibly leaving some fibers insufficiently wetted.

The **pot life** of the resin needs to be long because it may take considerable time for a part to be wound. Also, many parts may be wound from the same batch of resin and so the resin bath may be used for a considerable length of time. Not only must the material not appreciably cure, but its viscosity must remain low.

Another problem with the long times involved in winding is the volatility of some resins, such as polyester, where styrene is a major component. This means that volatile gases are being released for a long period of time. Equally troubling is the potential toxicity of some resins. When they are in an open bath for long periods, human exposure to harmful vapors is a potential problem that must be dealt with properly.

One method to extend the pot life is simply to have a heat-activated initiator. This means the crosslinking reaction will not begin until the part is heated. Therefore, the parts are usually moved into ovens or autoclaves for curing. For many applications this is fine. However, for large parts, oven or autoclave heating can be difficult, especially with the requirement that the part continue to be rotated while being cured. Some manufacturers have solved this problem by heating with heat lamps. While this solution is simple, the uniformity of the heat must be carefully adjusted.

When the resin does begin to cure, it is desired that it cure quickly. This will reduce the resin's tendency to drip or sag during the early part of curing. Therefore, the resin should be highly reactive. To achieve this reactivity requires that the initiator react quickly; it may also require that the resin itself react quickly. The latter requirement can make the pot life requirement difficult to meet. Therefore, resins must be balanced for both pot life and high reactivity.

When using a liquid resin and depending on the resin bath to wet-out the fibers, some inherent difficulties arise. The first and most obvious problem is that some fibers might not be wetted in the time that they are in the bath. While it is possible that the resin could continue to soak into the fiber bundles after they leave the bath, most manufacturers prefer to be assured that the wet-out is complete. There are some alternatives that might be used to ensure good fiber wet-out in the resin bath.

- While it is theoretically possible to use smaller bundles of fibers, the fiber manufacturers would be reluctant to make special bundles just for filament winding. If they did, the price of such fibers would likely increase.
- As already discussed, the resin viscosity could be reduced. However, the limitation of having the resin stay in place on the fibers throughout the winding and cure operations limits the ability to reduce the viscosity greatly.
- Mechanical rollers or squeezers can be placed in the resin bath to force the resin into the fiber bundles. This method is widely used, not only as a method to encourage wet-out, but also as a way to control the amount of resin on the fibers. If there is too much, the part will be resin-rich and that wastes money since the excess increases the weight and cost of the part but adds little to its performance.
- The winding speed can be reduced and thus give more time in the resin bath. While this is ultimately the course usually followed, it reduces the throughput of the machine and, therefore, increases total cost.

The use of molten thermoplastic resins in wet-winding operations is difficult and rarely done successfully. The resin must be melted and maintained in a molten state throughout the winding operation. This is difficult because of the heat involved and the long times that are usually required. Thermoplastics might begin to degrade under these conditions. Moreover, thermoplastics are more easily melted with a combination of heat and shear, as is done in an extruder, rather than in a heated resin bath as might be done in an analogous way to thermoset wet winding. However, it is possible to use an extruder to apply the resin to the fibers and thus achieve both the advantages of wet winding and the ease of melting using an extruder. However, just as the wet-out of fibers is likely the factor that limits throughput in traditional wet winding, the wet-out of fibers with thermoplastics applied by an extruder or any other wet method would severely limit throughput. Molten thermoplastic resins have high viscosities so wet-out is extremely difficult, even when the speeds of the fiber through the wet-out zone are slow.

Prepregs

To avoid some of the problems associated with using a liquid resin, filament winding can be done with prepregs. This is called *dry* or *prepreg winding*. With prepregs, most of the problems associated with liquid resins are eliminated. For example, there is no need to mix the resin and the initiator. With no resin bath, the issues of fiber wet-out, resin viscosity, controlling resin/fiber ratios, removing excess resin, and resin dripping off the part during winding are all eliminated. However, pot life, or its equivalent with prepreg, out time, is still a problem. Prepregs contain initiated resin and so they can begin curing unless the initiator is heat activated. Perhaps the biggest advantage is the increase in winding speed possible with prepregs. There is no need to wet-out the fibers and so the time to wind is limited only by the action of the machine.

The biggest disadvantages with prepreg winding are the higher cost of the prepreg versus the liquid resin and the need to compact the prepreg during cure. When the bands of prepreg are laid onto the mandrel, the resin molecules in one layer do not mingle with the resin molecules in other layers. Moreover, there is air trapped between the layers. Therefore, the layers must be compacted during cure. This compaction must occur during the initial phases of cure when the viscosity of the resin drops because of heat and before the viscosity rises because of crosslinking.

When thermoplastic resins have been used in filament winding, they have almost always been prepregs. These thermoplastic prepregs have all of the advantages and the disadvantages of thermoset prepregs and a few additional considerations. Thermoset prepregs are tacky, that is, they are slightly sticky at room temperature. Therefore, they will stick to the mandrel and underlying layers of material when applied. Thermoplastic prepregs are not tacky. Therefore, when they are applied to the mandrel or underlying layers, they tend to slip. This makes winding difficult. To solve this problem, the thermoplastic prepreg is usually heated just prior to its touching the mandrel (or underlying materials). This is done with heat lamps, lasers, or similar high-intensity heating methods. It is also advisable to heat the underlying materials just prior to application of the prepreg to further improve the stickiness of the materials at lay-down.

The problem of compaction that was discussed with thermoset prepregs is also present with thermoplastic prepregs. It is even more critical because the thermoplastic resins are higher viscosity and need the compaction to allow intermingling of the resin molecules across the boundaries of the layers. If the melting of the thermoplastic done just prior to lay-down is complete and accompanied with compaction, as from a roller, then compaction might be adequate and additional heating of the part after winding may not be necessary. This could, of course, be a tremendous cost advantage over thermosets. It may be necessary, however, to anneal the part after winding to get the desired crystallinity in the thermoplastic resin.

With the difficulties involved, the question might arise as to why anyone would use thermoplastics for filament winding. The answer is that it is rarely done but there are some property advantages that cannot be totally ignored. The thermoplastics have higher toughness, can be remelted to correct defects or damage, can be more solvent resistant (depending on the type of resin), and might have a faster curing (consolidation) time.

Additives

Most filament winding is done without additives. However, flow control agents (surfactants) that help prevent the resin from dripping off the part during winding and cure can be useful. Pigments and other minor additives also may be used.

Some applications, especially low-cost pipe, add sand during the winding operation to give bulk and weight to the part. A problem with this practice is the tendency of sharp sand particles to cut the fiberglass, which erodes the performance of the part over time.

Mandrels

The mandrel is the tool around which the impregnated fibers are wrapped. Typically, the mandrel has an axial shaft that allows it to be held and rotated by the winding machine. The mandrel serves the function of a mold and therefore defines the internal geometry (shape and surface) of the part.

Because the mandrel is critical to the filament winding operation, it must meet numerous requirements. Three of the most important considerations regarding the mandrel are the following.

1. The mandrel might be used for hundreds of parts. Each part is removed from the mandrel after cure and this removal often means that the mandrel is mechanically extracted. This extraction often requires that the mandrel be strong and tough.
2. Just as with molds, the tolerances on a mandrel are reflected by the parts to be made on it. When close tolerances are required, the machining of the mandrel might be a major consideration as far as cost and maintenance are concerned.
3. For open-ended parts ,the simplest mandrel is usually the best for part removal and cylinders of steel or aluminum are the most common. These cylindrical mandrels must have smooth, polished surfaces and are often slightly tapered to facilitate easy removal of the part.

For close-ended parts, mandrel removal is more complicated. One method is a collapsible type. The interior support structure collapses, somewhat like an umbrella, and the mandrel is then taken out of the part. Collapsible mandrels also can be used in open-ended parts when the size of the part is such that mechanical removal by pulling is not practical.

Another type of removable mandrel is made of a solid material, such as sand, which is held together by a water-soluble binder, such as polyvinyl alcohol. Soluble salts are also used. After the part is wound and cured, these mandrels are simply dissolved away. A similar concept is to use a meltable mandrel that can be removed after curing by heating it to melting point. Of course, the melting point of the mandrel must be higher than the cure temperature. Further, the materials used to make the part must be able to withstand the higher temperature used to melt the mandrel. Meltable materials used for mandrels include polymers, low-melting metals, and eutectic salts. Dissolvable and meltable mandrels are especially useful for parts that have complex shapes that prohibit easy removal by sliding the mandrel out.

Sometimes the mandrel is made out of a thin, brittle material that can be removed by breaking it out. For example, mandrels made of plaster were used for years to make rocket-motor casings, but this is less common today because of the danger of damage to the parts.

In still other cases, the mandrel is simply left inside the part after curing. Such liners are used for applications such as the high-pressure tanks that contain fuel for natural-gas-powered vehicles. Another example of a mandrel left inside a tank is the aircraft waste tank. In this case, the plastic liner is combined with a thin metal material. The metal gives some strength to the plastic and provides some chemical resistance. Other types of metal liners are overwrapped with composite material to give strength to the liner. This allows the tank to weigh less than an all-metal tank. Such tanks are used to hold air for scuba divers and firefighters.

One of the most important new methods of mandrel removal is to use a **shape memory polymer** for the mandrel. In this technique, the mandrel is molded at a size (diameter) that can be easily removed from the access hole (polar opening) of a closed-end vessel.

First, the shape memory mandrel is heated to a temperature that will allow the molecules of the mandrel to move. While at that temperature, the mandrel is expanded to the size desired for winding. Then it is quickly chilled. The chilling locks the molecules in place in their expanded state. The mandrel is then used for winding as normal. After the winding is complete, the part is cured. Then, after curing, the part and mandrel are heated to above the cure temperature, which allows the mandrel molecules to again soften. The mandrel will then shrink back to the shape it had when it was first molded. This will, of course, allow it to be extracted from the part. The cycle of heating, expanding, chilling, winding, and reheating can be done many times with the same mandrel.

Considerations involving the mandrel include the following.

- There is generally a change in the diameter of the mandrel with the pressure of the wrapping process. As the thickness of the part increases, the accumulated pressure on the mandrel also increases. The constriction pressure could cause the mandrel to collapse. This is especially true with mandrels intended to remain inside the part such as plastic liners. To compensate for the increasing pressure, mandrels for these types of applications have interior pressure that is increased as winding progresses.
- The mandrel can sag because of gravity. Sometimes the part diameter is small and the length is so great that substantial sagging occurs especially in parts such as poles. If the sagging is too much, wrapping of the part could be adversely affected.
- If the part is large and especially when the mandrel is heavy, the part weight becomes a major concern. The machine must be able to rotate such a part easily. Thus the floor on which the entire machine and part assembly rest must be able to support the loads. It should not be assumed that the floor (even if made of concrete) will support the load.
- A calculation of thermal expansion of the part and the mandrel should be made to ensure that the expansions are compatible. The mandrel must not expand so much that it damages the part wound around it. The part will expand slightly, especially if thin, but thick parts are often a problem. One method of dealing with the problem of thermal expansion is to put a layer of elastomeric material over the outer surface of the mandrel before winding. The elastomeric material is squeezed during the curing process by the expansion of the mandrel and thus protects the part from internal expansion pressure. Another method is simply to choose an alternate material for the mandrel.
- The mandrel must be able to withstand curing temperatures.
- Even though one mandrel will be used for many parts, cost is a consideration. A company may have many different part sizes, such as pipe diameters, and each of these requires a different mandrel. Further, the elaborate methods that might be used for part removal might lead to very costly mandrels. Not only is the cost of the mandrel itself important, but the cost of using a particular type of mandrel might be high or low depending on its complexity and the ease of part removal.

Filament Winding Machines

The most common type of filament winding machine is the basic machine shown in Figure 17-1. These basic machines are available in many sizes and are specified by the length of the spindle that can be accommodated, which determines the maximum length of the part, and by the clearance of the mandrel holding fixtures and the machine, which determines the maximum diameter of the part. Winding lengths range up to 100 ft (30 m) or beyond for special orders. The size of the machines can vary over a wide range.

One innovation to the basic machine that has already been mentioned is the addition of other axes of motion. The most common additional axes are for the payoff device to move perpendicular to the plane of the mandrel. This in-and-out motion allows parts of varying surface thicknesses to be wound more easily, but is most helpful in laying down a smooth band when making a curved end as for a closed-end pressure vessel. Another common axis of motion is achieved by rotating the payoff. This also allows improved lay down of the band over the curved ends of tanks. Even more axes of motion are possible, although three- and four-axis machines are the most complicated that are routinely used. More details about the possible axes of motion will be given later in this chapter when fiber placement is discussed.

Several years ago, the makers of filament winding machines realized that adding additional mandrels was relatively inexpensive. This is because the principle change was simply the addition of the turning fittings that hold the additional mandrel. If the parts being made were not too large in diameter, this change could be accommodated on existing machine bases. The major limitations on the number of spindles possible are the space required (diameter of the parts), the possible limitation in turning power from linkages, and the normal loads associated with turning. Machines with 30 spindles are commercially available.

A different type of filament winding machine is the polar winding machine. Typically, the mandrel is mounted vertically in this machine and a rotating arm winds the fibers on the mandrel. The band is moved across the surface of the machine by tipping the mandrel. Polar machines are simple in operation because there is no carriage that must be reversed. Therefore, the problems associated with the ends of the mandrel are avoided. They are also faster than standard horizontal machines. However, polar machines are less common than standard filament winding machines. This is largely because the standard machines are more flexible in the types of parts that can be made. Polar machines cannot make parts that differ greatly from spherical shapes.

The slow down in winding that occurs when the fibers go around a closed-end vessel can cause the fibers to become slack. This problem can be controlled through the use of tensioners. Some type of brake system is used to apply tension to the fibers. Tensioners are usually mounted on the creel or carriage. They sense when the fiber line becomes slack and automatically apply tension to adjust it.

A controller is an important part of the modern filament winding machine, especially when the machine is to be used for more than just a single, simple product. A controller links the carriage speed, the mandrel rotation speed, and other axes of motion so that the desired fiber paths can be laid down.

A valuable feature available on many controllers is the ability to input the fiber path by manually moving a sensing device over the surface of the mandrel in a desired path. After making as many input stops as desired, the controller can smooth the path and display it for verification. Then the controller can calculate the paths that will give precise and complete coverage of the layer to be applied.

Another type of machine makes a continuous pipe that is on an open-ended mandrel. Winding is done from one end using low-angle winding. In this method, the fibers are wound onto the mandrel, which is located inside a ring. The ring rotates and applies the fibers onto the mandrel as the mandrel moves away from the ring. This is somewhat like a braiding operation. The wound pipe is pulled off the mandrel and cured continuously. This type of pipe does not have the burst strength that would come from higher angle (hoop) winding, but it is sufficient for low-pressure systems.

Winding Patterns

The basic winding patterns for the standard machine (hoop and helical) and for the polar machine (polar) have been identified. However, because of the specialty nature of the polar machine, this discussion will focus on the more common standard machine.

Hoop windings are applied to the portions of parts that are essentially cylindrical. This includes areas that are slightly divergent or convergent as well as shapes that are non-circular cross-sections such as square beams or oval tubes. The hoops have a tendency to slide when laid on a highly curved or highly slanted surface.

The hoops are applied by moving the payoff slowly. The lateral movement of the payoff should be just enough to put one edge of the fiber band next to the previous edge. No gaps should occur between the bands as these will become areas of weakness since they do not have reinforcement in those areas. Such areas also may be points of crack initiation.

Helical windings are much more difficult to apply and describe. Because of that difficulty, some conventions have been established for the critical parameters that describe the helical pattern. Those conventions are illustrated in Figure 17-3a. The winding angle is measured from the bandwidth to the axis of the part. Therefore, hoop windings would have a winding angle near 90°. The winding angle is important to resist forces that might be applied to the part. Another convention is that one complete circuit is a full traversing of the part from one end to the other and then back again. The full back-and-forth motion of a complete circuit is shown in Figure 17-3b.

After one complete cycle, the next helical wrap must be applied immediately adjacent to the preceding wrap with no gap between the wraps. To accomplish this feat requires

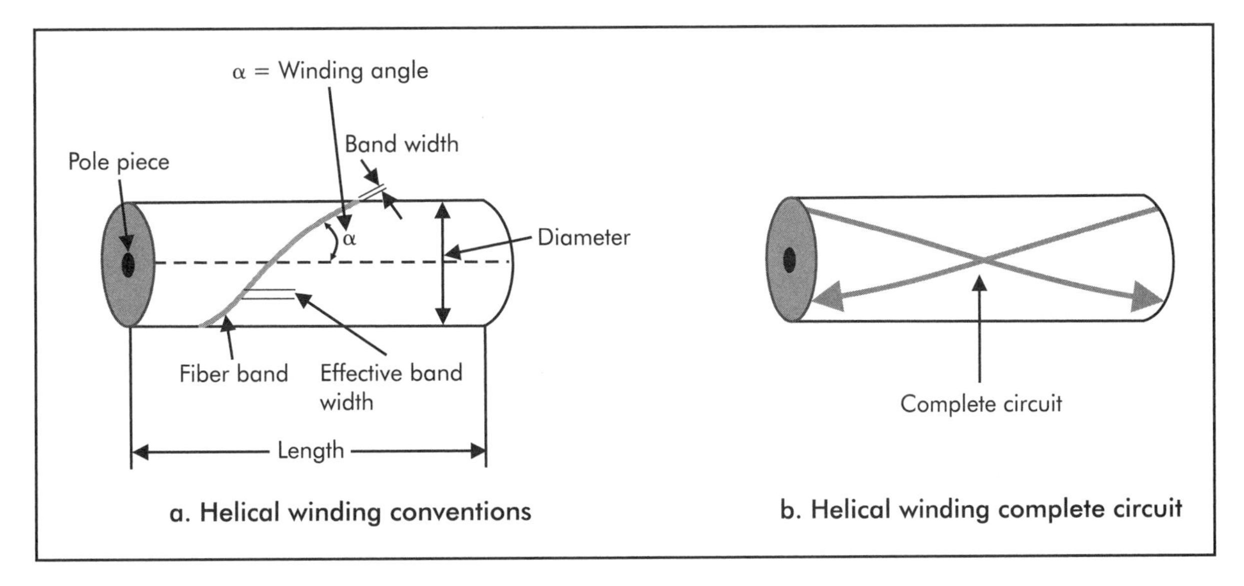

Figure 17-3. Winding pattern conventions.

considerable calculating and is one reason that machine controllers are so valuable. The winding path that makes alignment of the wraps occur depends on several factors, including the following:

- winding angle,
- diameter of the part,
- length of the part,
- band width (actually the critical factor is the effective band width),
- diameter of the shaft that goes through the part and around which the windings are made in closed-end parts (the opening in the end has a fitting in place called the **pole piece**),
- speed of rotation of the mandrel, and
- speed of movement of the payoff.

To achieve complete and even coverage of the mandrel requires that the winding pattern repeat exactly an even number of times. Further, the number of rotations of the mandrel over the entire layer must be a whole number. These requirements are, therefore, also part of the calculation.

When the calculations are done, several combinations of angles and speeds are usually available to give the desired results. Therefore, the angle usually selected, unless some overriding design requirement suggests otherwise, is the angle that gives the **geodesic path**. This path allows the fibers to be placed under tension and remain without slippage. Some examples of geodesic paths are hoop windings around the circumference of a cylinder, a helix on a flat-sided cylinder, and a great circle on a sphere. The great circle is important in close-ended vessels because following that path makes winding much easier. In actual practice, the friction of one layer against those below it will allow deviations of a few degrees of angle off the geodesic path without having slippage problems. The shape of the domes should be designed so that geodesic paths are available.

On an open-ended part, there is, of course, no end or pole piece around which to wrap. This means that when the carriage completes its movement in one direction, there is nothing to anchor the fibers so that the carriage can reverse its path and have the fibers stay in place. One method to allow the machine to reverse and anchor the fibers on the end is to attach many small posts around the ends of the part. These posts are usually about 2 in. (5 cm) high and arranged around the entire diameter of the part on each end. When these are in place, the payoff device takes the fiber band just beyond the end of the part and then turns so the fiber band is hooked by one of the posts. This anchors the fiber band to the post. Then, with the proper rotation of the mandrel, the return path will be exactly right to allow alignment of the bands.

An interesting variation in the winding pattern was done for a spring on the door of a large commercial jetliner. The spring is mounted to the plane so it goes into twisting tension when the door is open and then helps reduce the force necessary to close the door. The spring is made by filament winding but the mandrel is unusual. The mandrel's surface has a series of slots that go around it in a helical fashion. These slots are filled with fiber as the mandrel turns. In this case, the fiber bands are not adjacent to each other but separated by the ridges between the grooves. When cured, the spring is wound off the mandrel.

VARIATIONS IN FILAMENT WINDING

One of the major limitations of filament winding is the shape of the part that can be wound. The part must have an axis of rotation, but also must be generally linear so its surface can be covered by the traversing of the carriage and the payoff. Exceptions to these general shape requirements involve additional equipment that adds other motion capabilities to the standard machine. Several such modifications have been made.

One augmentation to the standard machine is done to make T-fittings. The round portions of the "T" are wound in the normal way except that as the hoop windings are applied to one of the sides, an operator (or automatic robotic arm) pulls the fibers from one side of the "T" to the other. This action is repeated frequently. This creates a crossover that ties the fibers on one side of the "T" to fibers on the other and, therefore, makes the "T" strong.

Another method for creating non-standard shapes is the LOTUS system, which is named for the shapes of the parts that can be made using it. It uses a sophisticated control system and a non-standard winding machine to create curved tubes. In the LOTUS system, the mandrel is held stationary while the fiber source spins around it on a ring and down its length, setting desired angles. Parts as simple as a curved walking cane are difficult to make using standard filament winding but can be made with LOTUS. Other possible parts could include exhaust ducting for engines, bicycle frames, furniture, automobile frames, and medical devices like wheelchairs and walkers.

FIBER PLACEMENT

The desire to use filament winding for a wider variety of shapes led to the development of a highly automated system, called **fiber placement**, which eliminates many of the shape restrictions inherent in standard filament winding. The developers of fiber placement realized that in actual use, filament winding machines were including more axes of motion and using more sophisticated controllers to coordinate all of these motions. Still, many of the limitations inherent in traditional filament winding were present in even these highly sophisticated machines. The most important limitation was that traditional filament winding still used tension when wrapping the mandrel. Further, some axes of motion needed to make some parts were still not available on even the most advanced filament winding machines. Therefore, fiber placement was developed to overcome these limitations.

The concept of fiber placement was to include all of the axes of motion that could be required for essentially any shape and to include in those axes of motion some that pushed the fibers against the mandrel. This latter advancement meant that the fibers did not need to be under tension when applied to the mandrel. Instead, they were placed onto the mandrel. As a result, even concave mandrel surfaces could be made with the fiber placement system.

There are as many as nine axes of rotation on some fiber placement machines in comparison to standard filament winding machines that may have only three or four. While the general physical arrangement of the mandrel and carriage is similar to a standard filament winding machine, the complexity of the carriage and, especially, of the payoff head, are significantly greater. Parts made commercially with fiber placement have undulating surfaces, including areas of concavity. Such parts would not be possible with standard filament winding.

A fiber placement machine requires a highly sophisticated controller and high precision in its head and carriage. These requirements are very costly. A fiber placement machine costs over ten times as much as a standard filament winding machine. Other factors also increase the costs of parts made by fiber placement as compared to filament winding. The fiber lay-down rate for fiber placement is low compared to filament winding, usually only 20 lb/hr (10 kg/hr). Further, although fiber placement is capable of wet winding, prepreg winding is the only method that has been used commercially to any extent. In light of these high capital costs and high operating costs, the number of fiber placement machines in operation is small (less than 20 at this writing). Therefore, the number of commercial parts being made by fiber placement is small in comparison to filament winding.

Most of the parts made by fiber placement are aerospace parts that have unusual shapes. Nevertheless, as the number and size of composite parts increases, especially in aerospace, the demand for highly sophisticated manufacturing will continue to increase. The costs of machines and the throughput for fiber placement are likely to improve. Fiber placement seems destined for high volume and has the possibility of becoming a key manufacturing technology for composite parts in the future.

CASE STUDY 17-1

Starship

A limitation of filament winding has been the lack of definition of the outer surface of parts. After winding and curing, the outer surface of a part is quite uneven because it is not defined by a mold. Therefore, if filament winding were to be used for an application in which the outer surface of the part were required to be smooth, like an aircraft skin, extensive machining (grinding) would have to be done. This problem was solved in the fabrication scheme developed for filament winding of the Beech Starship's fuselage.

The dream of many aerospace engineers has been the fabrication of the fuselage of an airplane in a single winding operation. Engineers saw this as a way to improve the productivity and quality of aircraft manufacturing. So a team was assembled to manufacture a filament-wound fuselage for the Beech Starship. Several of the fuselages were built, but in the end the company elected to manufacture the part by other techniques. Nevertheless, the methods developed for the Starship are the basis for manufacture of a new round of corporate aircraft that promise to revolutionize airplane travel.

The secret to solving the problem of the rough outer layer of a filament-wound part was to use an external clamshell mold to cure the part. To accomplish this, a flexible membrane was installed over the winding mandrel. Then winding was done in the normal way. The first of the Starship fuselages was made continuously over about a 24-hour period.

After winding, the part and mandrel were placed inside the clamshell mold. The mold's inner diameter was just slightly larger than the outside diameter of the part as it was wound. The entire assembly was then placed inside an oven. As heat was applied, the elastomeric membrane placed over the mandrel was inflated from internal pressure. That inflation pressed outward on the overwrapped composite and moved the composite materials slightly outward and against the inner surface of the clamshell mold. The pressure was continued throughout the cure. When the fuselage was removed, the outer surface was smooth because it had been defined by the mold.

The part was removed from the mandrel by cutting it just ahead of the cockpit. (Designs for later aircraft have eliminated the need to cut the fuselage.) The part was then assembled with the wings and engines.

SUMMARY

Making composite parts on a mandrel has proven to be successful. The major manufacturing methods utilizing this concept are filament winding, fiber placement, and roll wrapping. Filament winding is, by far, the most prominent in terms of usage. However, fiber placement is becoming more important because of its ability to make parts of almost any shape.

Filament winding is intended for making parts that have an axis of rotation and, in general, a convex cross-section. (The fibers are applied under tension and if the surface were concave, the fibers would bridge over the concave area.) In spite of this limitation, many important parts are appropriate for filament winding. Some of the most obvious are: pipes, tanks (both open- and close-ended), shafts, industrial rollers, cores for oil drilling risers, well casings, rocket-motor cases, aircraft structures, golf-club shafts, softball bats, bicycle components, and a host of industrial products.

The advantages of filament winding are its high productivity and the capability to automate the process. Filament winding is a conceptually simple process that has received great support from innovations in machine manufacturing and electronics control. No wonder it is such a highly valuable and widespread composites manufacturing system.

LABORATORY EXPERIMENT 17-1

Roll Wrapping a Golf-club Shaft

Objective: Make a golf-club shaft.

Procedure:

1. Secure or make a mandrel shaped like a golf-club shaft. (These are generally machined from a steel rod.)
2. Obtain sheets of prepreg material.
3. Cut the sheets of prepreg to size so that they will wrap around the mandrel. The width of the cuts should be sufficient to obtain four or five thicknesses of prepreg. Because the mandrel is tapered, the prepreg should be cut on the same taper.
4. Wrap the prepreg around the mandrel. If a roll table is not available, the wrapping can be done manually. Cotton gloves should be worn for this task.
5. Wrap shrink tape around the wrapped mandrel. Either cellophane or polyethylene is a good material for this purpose. Usually, the width of the wrapping material is about 2 in. (5 cm). The wrapping is best done by mounting the mandrel in a lathe or other turning device and moving the tape along its surface. Use about .5 in. (1.3 cm) overlap between each of the wrappings.
6. Hang the mandrel vertically in an oven and heat to cure according to the specifications of the prepreg manufacturer.
7. After curing, remove the shrink-wrap from the part.
8. Sand or grind the surface to remove the bumps caused by the shrink-wrap.
9. Wipe the surface clean and then coat with a thin layer of resin or paint and allow to cure.

QUESTIONS

1. Give two reasons why filament winding is so commonly used to make fiberglass-reinforced pipe.
2. Give two reasons why filament winding is used extensively for high-performance pressure vessels.
3. List four factors that affect the helical winding path of a close-ended vessel.
4. Describe how helical wraps are anchored on the ends of an open-ended part.
5. Discuss three major differences between the filament winding and fiber placement processes.
6. Describe roll wrapping and point out three major differences between it and filament winding.

BIBLIOGRAPHY

Peters, S. T., Humphrey, W. D., and Floral, R. F. 1999. *Filament Winding Composite Structure Fabrication*, 2nd Ed. Covina, CA:

Society for the Advancement of Material and Process Engineering (SAMPE).

Society of Manufacturing Engineers (SME). 2005. "Filament Winding" DVD from the Composites Manufacturing video series. Dearborn, MI: Society of Manufacturing Engineers, www.sme.org/cmvs.

Strong, A. Brent. 2006. *Plastics: Materials and Processing*, 3rd Ed. Upper Saddle River, NJ: Prentice-Hall, Inc.

18

Pultrusion

CHAPTER OVERVIEW

This chapter examines the following concepts:

- Process overview
- Reinforcement preforming
- Resin impregnation
- Die forming and curing
- Pulling
- Cutting and trimming
- Shapes and applications

PROCESS OVERVIEW

Pultrusion is a continuous, high-volume manufacturing process used to make parts of constant cross-section. The name of the process—pultrusion—gives an indication of the basic concept used. In pultrusion, the materials are *pulled* through the machine wherein they are formed by a die in a manner similar to the extrusion processes used to shape metal and plastics. (Extrusion processes *push* the material through the die; therein is the fundamental uniqueness of pultrusion.)

As a process, pultrusion dates to the period following World War II when modern composite manufacturing methods were developed. Although similar to extrusion, as already indicated, pultrusion is unique to composites because it depends on the pulling of the reinforcement fibers through the machine. Possibly because of its uniqueness, the relatively high cost of the equipment, and because the shape of products is restricted to constant cross-sections, pultrusion molding is less widely practiced than other composites molding processes.

(The "Pultrusion" video from SME's *Composites Manufacturing* series is valuable as a supplement to lectures and laboratories and for general education, especially since it is an infrequently seen manufacturing process [SME 2005].)

The pultrusion process begins with continuous fibers being drawn from reels and formed into a general shape that allows for orderly movement into the resin bath. The fibers are wetted by the resin and then are further formed as they converge toward the die. If desired, additional reinforcements, often mats or cloth, can be directed into the reinforcement preform to include some fibers in a direction other than just the machine direction. A veil is also commonly added to the reinforcement preform. The wetted reinforcements then enter the heated die where they are cured. Upon exiting the die, the formed part enters the pulling system. The puller provides the force for movement of materials through the entire system. Then, the pultruded part is moved to a cutter and trimming station where finishing processes are performed. A schematic of the process is given in Figure 18-1.

Pultrusion is a continuous process with high material utilization (scrap rates are usually less than 5%). Therefore, the cost

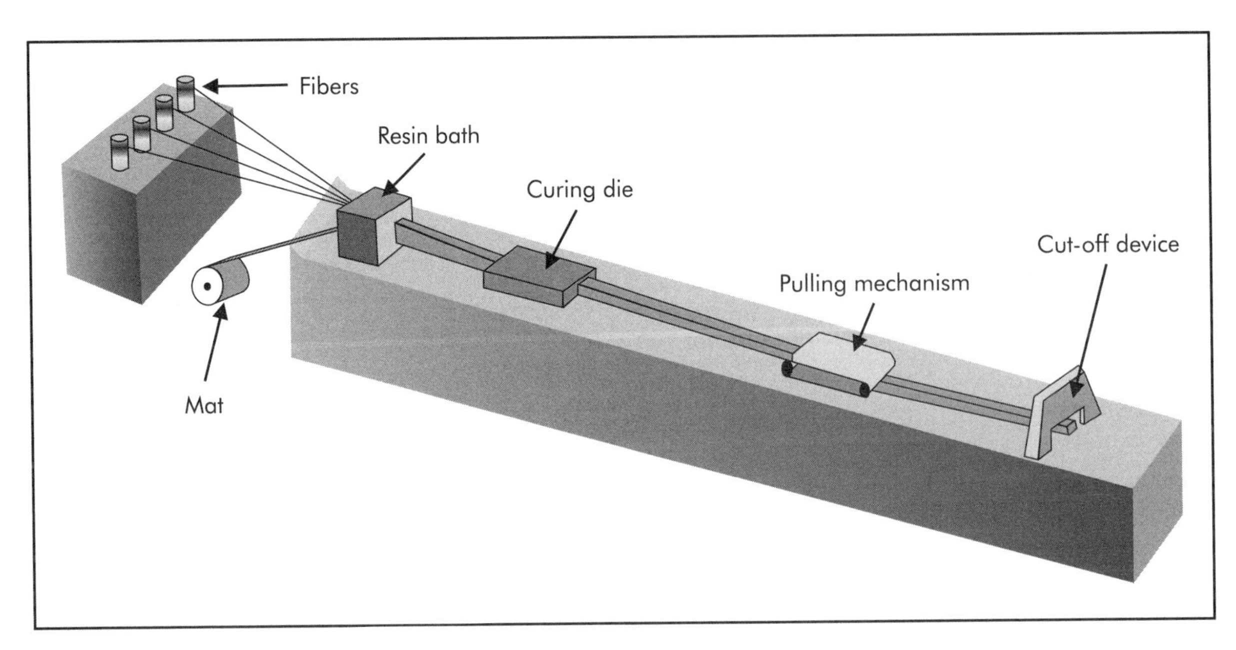

Figure 18-1. Schematic of the pultrusion process.

per part and the productivity in terms of cycle time are the best of any composites manufacturing process as shown in Figure 18-2. Because of these excellent economic and production factors, pultrusion is a rapidly increasing production method. Parts made of metals that have constant cross-sections are now being considered for substitution with composites because of the good performance and low costs of pultruded products. A variety of pultrusion machine designs in many sizes are commercially available from several machine manufacturers. Some of these manufacturers also supply the auxiliary equipment needed for a full pultrusion operation.

REINFORCEMENT PREFORMING

Continuous fiberglass roving in high-count bundles is the principal reinforcement material for most pultruded parts. Roving is convenient to handle because the large fiber bundles build mass in the part without requiring a high number of strands to be used. Another advantage of the pultrusion process is the ability to use a wide variety of different reinforcements. Fiberglass in other forms as well as aramid and carbon fiber are also acceptable. However, pultrusion is less precise in controlling the over-direction of the fibers than prepreg lay-up in open molds. Therefore, advanced performance parts that would typically be made with carbon and aramid fibers are usually not made by pultrusion. However, some success has been achieved with mixtures of reinforcement types to create hybrid composite structures.

The roving is drawn from fiberglass packages that are, generally, mounted on a mobile roving **creel**. The creel is a stand on which the roving packages can be mounted. It is mobile so the strands of reinforcement can be aligned properly with the pultruder. The strands are directed into a collection plate into which holes have been drilled to accommodate them. They are kept separated, yet aligned with the machine.

When the concept of pultrusion is first introduced, people generally assume that all of the fibers are oriented in the machine direction. While most fibers are in the machine

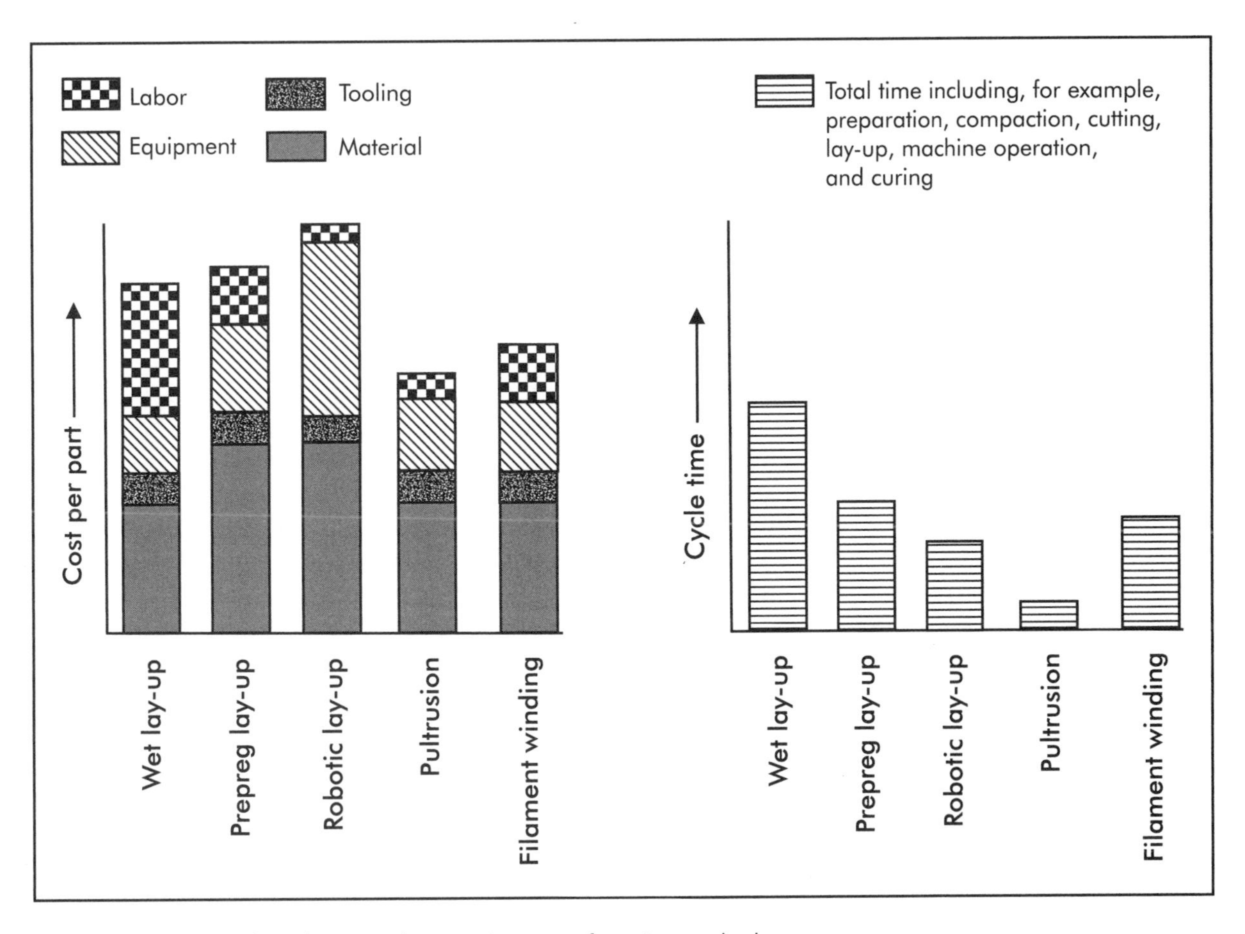

Figure 18-2. Cost and productivity of composites manufacturing methods.

direction in pultruded parts, it is possible to have fibers in other directions by adding broad goods (usually continuous strand mat or cloth) to the fiber stream. This is normally done by introducing relatively narrow widths of the broad goods from the sides (or from the top and bottom) of the main stream of fibers. The broad goods enter the stream just before or just after the resin bath. (When only a small amount of broad goods is used, there is often sufficient resin already on the machine direction fibers. Thus additional resin is not needed to wet out the mat and cloth and so they can enter the fiber stream after the resin bath.)

The broad goods are cut to relatively narrow widths so they can be easily formed to enter the die. These narrow widths are usually made by slitting standard width broad goods. Commercial slitters are available that can handle widths up to 72 in. (183 cm) and 24 in. (61 cm) diameters. The rolls can be cut in widths indexed to .01 in. (0.03 cm) increments.

RESIN IMPREGNATION

Most of the resins used for composites can be used in pultrusion provided they are low enough in viscosity to wet-out the fibers in a bath and that they will cure in a relatively short duration. (Polyimides are not generally suited to pultrusion.) Polyesters have the dominant share of the pultrusion market.

In addition to the obvious cost advantage of polyesters, another characteristic gives them a distinct operational advantage in pultrusion. That characteristic is higher shrinkage upon curing than most other resins. This high shrinkage is from the many crosslinks formed because of the many carbon-carbon double bonds along the chain. In comparison, epoxies and vinyl esters, the other two most common resins in pultrusion, only have crosslink sites at the ends of the chains. The shrinkage is helpful because it enhances the release of the part from the mold, thus reducing the tendency of the part to stick to the mold and the forces needed to pull the part through the mold.

Epoxy resins are occasionally used in pultruded parts. However, they have the disadvantages of lower shrinkage than polyesters and also an inherently greater tendency to stick to the die. (Epoxies are much better adhesives than are polyesters.) To reduce the problems associated with polyesters, pultruders have Teflon®-coated the dies, fittings, and other places where sticking may occur.

Resin Bath

The resin is most often applied in a resin bath. The viscosity of the resin must be low so that wet-out is rapid. This is especially true with relatively thin-walled parts that cure quickly and therefore have a faster pultrusion speed.

The resin bath is often open and so styrene vaporization is quite high. Therefore, lowering the viscosity of the resin by adding styrene is not an optimal solution. Ideally, a better solution is to lower the molecular weight of the resin. If this is done, the number of crosslink sites should be increased so the cured product has sufficient molecular weight to achieve the desired properties. Increasing the number of crosslink sites also increases the shrinkage, which is favorable. The unfavorable consequence of high crosslinking is possible embrittlement.

The resin must have a relatively long **pot life** (time in the bath without curing). Therefore, heat-activated initiators are used. The relatively rapid cure that must take place while the product is in the die is difficult to accomplish with a single initiator. Therefore, multi-initiator systems are often used, especially with thick parts. The mixture of initiators usually allows shorter curing times and gives more complete cures. Typically, three initiators are included in the multi-initiator mixture. One initiator begins the curing at a lower temperature and gels the part into a dimensionally stable B stage. This initial gel formation helps prevent styrene boiling later in the die. The exotherm from this initiator is generally sufficient to begin the initiation of the secondary and finishing initiators. The intermediate initiator provides a smooth continuation of curing. Finishing initiators ensure complete cure and reduce residual monomer content. The selection of the initiator system should be based on resin reactivity, formulation, profile geometry, thickness, die temperature, and pulling speed. The use of a multi-initiator system offers flexibility in the choice of initiators so that these factors can be addressed while the productivity of the pultrusion line is optimized.

After the reinforcements have passed through the resin bath, they move into a forming area where they converge into a shape that is closer to the final shape of the part. This forming guide assists in ensuring full wet-out and squeezes off excess resin, which is recirculated into the resin bath.

Resin Injection

Many of the problems associated with an open resin bath can be eliminated with resin injection. The resin injection system is located just prior to the die. This closed chamber eliminates the vapors from styrene and other resins. (The vapors from the monomers used to make phenolic are especially hazardous.) Resin injection also reduces the concerns of pot life because the resin can be mixed with initiators on-line just prior to injection into the pultrusion system.

A potential problem with resin injection is fiber wet-out. The time for wetting out the fibers is typically less than the time in a resin bath. Also, the reinforcements are generally compressed when the resin is injected because injection occurs just before the die and, therefore, after the forming guide. The best procedure is to inject the resin before full compaction is achieved. Another technique to help wet-out is to heat the resin just prior to injection. This reduces its viscosity and improves fiber wet-out. Further, the injection is done under pressure, which forces the resin into the fiber bundles.

An interesting possibility when using resin injection is to use a resin-like polyurethane that is injected as two components and then polymerizes when the two mix. Reaction time is often fast. Therefore, a pultruded part with a polyurethane or other fast-reacting polymer matrix is possible.

DIE FORMING AND CURING

The wetted material converges and is compacted as it enters the heated die. It then is pulled through the die at a rate appropriate to ensure that the part is cured as it exits the die. The shaped internal cavity of the die gives final shaping to the part as it is cured. Normally, the die will have several heating bands so that the temperature throughout the die can be carefully controlled.

Hollow parts are made by placing a mandrel in the die. The mandrel is attached to the rear of the die, behind the place where the wetted reinforcement enters the die. The reinforcement material converges around the mandrel and is shaped by the mandrel and the inside surface of the die.

The curing process is the rate-determining step for the pultrusion process. Typical speeds for pultrusion are 2–4 ft/minute (0.6–1.2 m/minute) for parts .04–3 in. (0.1–7.6 cm) thick. Sizes of parts can vary from 1 in. to 15 ft (2.5 cm to 4.6 m) in diameter without serious limitations.

Dies are usually made of tool steel and can be costly. Therefore, the size of the die is kept to a minimum. Hence, an alternate method of curing is to allow the part to exit the die before it is fully cured. The part must be rigid enough to pass into a post-curing oven to complete the cure.

Another modification of the standard curing system also allows the part to exit the die before it is fully cured. In this case, the part enters a forming device immediately after the die. The forming device makes minor changes in the shape of the part. In this way, parts of non-uniform cross-section can be manufactured.

Some improvement in the curing time can be obtained by heating the reinforcement material with radio frequency (RF) power. This method can be used with thermal heating or in place of it. The die portion through which the RF power operates is normally made of ceramic so the RF signal can penetrate through the part. This signal imparts energy to the part and causes it to cure. Similar success has been reported when the part was heated with microwave or ultrasonic energy.

PULLING

Two types of pullers are in common use. The first is the **caterpillar pulling system**. It uses two synchronized continuous belts to press against the pultruded part and pull it. The surface of the belts can be shaped to match the shape of the part, thus reducing the potential of crushing or damaging the part when it is pulled. This system has the advantage that the speed of pulling can be easily regulated by adjusting the speed of the caterpillar belts. The continuous nature of the caterpillar system gives a constant and uninterrupted pulling force.

A modification of the caterpillar system uses sets of opposing wheels to pull the system. The wheels have the advantage that they can be easily mounted on pressure pistons so that they are always exerting a uniform pressure against the part as they pull it.

The other commonly used pulling system is called the double-clamp system. There are two pulling devices that alternately clamp and pull the piece. While one is clamping and pulling, the other is retracting to its original position so it is ready to clamp and pull when the other reaches the end of its travel zone. The pull with this system also can be constant and uninterrupted. As with the caterpillar system, the faces of the clamps can be shaped to conform to the profile of the part.

CUTTING AND TRIMMING

Since pultruded parts are continuous, they must be cut for final packaging and shipping. A cutting saw is the most common method of cutting the part. The saw is similar to a woodworking table saw except that it travels in the direction of the part and with the same speed. That way the cut is straight across the part.

The saw is mounted on a sliding table. When the desired length of part has passed a sensor, a signal actuates the saw. A clamp on the saw table is activated to hold onto the part. The saw slides along with the part. When the cut is completed, the clamp loosens and the saw table returns to its initial position.

Since composites are made of both resin and fibers, they are quite abrasive. Therefore, the cutting blade of the saw is usually carbide or diamond tipped so it will wear longer.

SHAPES AND APPLICATIONS

A wide variety of standard pultruded shapes are commercially available. The shapes include rods, tubes, boxes, 90° angles, U-channels, flat strips, and I-beams. The sizes available also vary widely. For example, standard U-channels are available in widths of 2–10 in. (5.1–25.4 cm). Each of the parts has a specification for tensile, compressive, and flexural properties along with indications of hardness, water absorption, and coefficient of thermal expansion. Clearly, the system for marketing these standard parts is similar to that used for metals.

Standard shapes have been used for handrails in Chicago's transit system. The handrails were faster to install than metal systems and will require less maintenance.

One interesting product made by pultrusion is grating for chemical plants and other industrial locations. These gratings have rectangular openings that are about 2 in. (5.1 cm) square and 2 in. (5.1 cm) thick. The square gratings are about 3 ft (1 m) in overall dimension. Two methods are used to make the parts using pultrusion. In the first, the grates are assembled from rods and flat strips in which holes have been drilled to allow the rods to pass through. These have the advantage of using standard pieces but the labor to make the parts is high. Another method is to pultrude the large grating as a continuous piece and then slice off each individual grating. Making a part 3 ft (1 m) square with many internal ribs requires a complex mold and a large pultrusion machine. However, the labor is minimal and the integrity of the grate is superior because all the sections are cured together.

Standard bolts are also pultruded. These are made from rods onto which threads have been cut. Composite bolts also can be made by pultrusion. These studs and bolts have been used in numerous applications including new electrical plants built by General Electric for a European consortium of electrical companies.

While standard shapes are a large business, the majority of pultruded parts are custom shaped. Pultruded parts are used for applications like roll-up door panels for trucks, sidewalls for automobile transport railroad cars, aircraft flooring, automotive leaf springs, ladder rails, door supports for automobiles, and electrical conduits. The variety of parts made by this method is exploding.

Booms for mechanical lifters mounted on trucks and used to access power lines is an important application because of cost, ease of operation, and safety. The composite beam weighs less than its metal counterpart does and therefore it moves more easily with less force. However, the composites are not electrical conductors and so they are much safer than the metal equivalents. Ladders used by utility companies are also made of composites for similar reasons. The non-conductive nature of composites also led to the use of a pultruded buss bar cover for the electrically powered people mover at Disney World.

CASE STUDY 18-1

Optimized Pultruded Beams

Composite structural beams have several inherent advantages over competitive steel beams. Most importantly, composites are not subject to corrosion, which is the major reason that steel beams have to be replaced in bridges. To reduce corrosion, extensive painting and other maintenance are required for steel. Moreover, the composite beams eliminate the electromagnetic and radio-frequency interference common to metal structures. The lighter weight of composite beams may not seem like a big advantage, but bridges could be built with less massive support structures with composites. In remote areas, the lighter weight of the composite structure could vastly simplify transportation and installation. Therefore, a pultruded beam was developed for a bridge in the Daniel Boone National Forest (Kentucky).

The manufacturers, Morrison Molded Fiber Glass Company (now Strongwell), under a development grant from the National Institute of Standards and Technology (NIST), decided to explore another potential benefit over steel. For the bridge application, it was going to explore the possibility of pultruding a beam with a shape and dimensions designed specifically for maximum stiffness and strength. Modeling of bridge beam shapes eventually suggested that a double-flange design with internal ribs would be the best shape. This shape is shown in Figure 18-3.

The new design offers an eight- to tenfold improvement in torsional rotation over standard I-beams. Whereas conventional I-beams provide about 2,500–3,000 ksi (17–20 MPa) flexural modulus, the new beam projected a flexural modulus of 6,000 ksi (41 MPa).

In addition to the innovative structural shape, the composite beam utilized a combination of fiberglass and carbon fiber to achieve its high performance. For cost purposes, the majority of the beam was made of fiberglass. The most efficient use of carbon fiber was to incorporate it in the top and bottom flanges only. The presence of carbon fiber more than doubles the stiffness of the beam. The design also incorporates 0/90° and ±45° stitched fabric primarily in the web.

Because of the desire for low flammability, the resin chosen was phenolic. Working with Georgia-Pacific, a pultrusion-friendly grade of phenolic was developed. The product had a longer pot life and lower viscosity than standard phenolic.

The final beam was 24 in. (61 cm) high and 60 ft (18.3 m) long. The bridge is a footbridge. It was removed from the transportation truck and carried to the location by a team of about 10 installers. (No crane

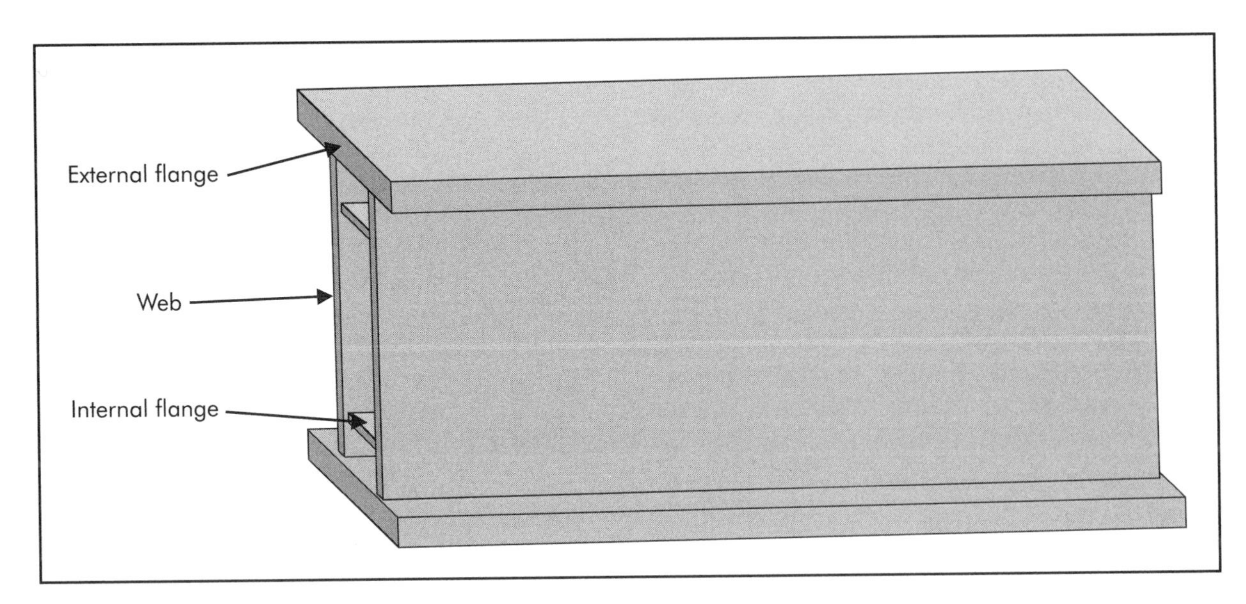

Figure 18-3. Pultruded beam with internal stiffeners.

or other lifting devices were needed.) Higher performance beams that can be used for vehicular traffic are under development.

SUMMARY

Pultrusion has some major advantages as a composites manufacturing method. The material usage is high because of the ability to make the part to net shape. Also, the throughput rate is high. The process can give high reinforcement content. Hence productivity is the best of all composites manufacturing processes.

The disadvantages of pultrusion have limited its use to products in which they are not significant. In pultrusion, the part's cross-sections must be generally uniform. Therefore, shapes such as tubing, angles, I-beams, and rods are commonly done. Custom shapes are also possible but they must be of constant cross-section.

Operational problems like buildup of resin and fiber at the opening of the die and low compaction, which creates voids in the part, need to be addressed in each manufacturing operation. Resin baths can be a major source of air pollution, but that problem can be reduced by covering the bath and nearly eliminated by using resin injection.

In spite of the limitations, pultrusion is a rapidly growing process. As composites become more widely accepted for civil engineering applications, the growth will be even faster. Pultrusion delivers low cost and reliable manufactured parts of relatively simple geometry.

LABORATORY EXPERIMENT 18-1

Analysis of a Pultruded Part

Objective: Discover the nature of fiber placement in a pultruded part.

Procedure:

1. Obtain a pultruded composite part.
2. Weigh the part (or a portion of the part) to obtain a reference weight.
3. Place the sample into a muffle furnace and heat at about 1,000° F (538° C) for 3 hours.
4. After cooling the sample, weigh it to obtain the fiber content.

5. Carefully examine the part to investigate the direction of the fibers. Attempt to see the placement of the fibers to determine how the lateral fibers were added and whether the longitudinal fibers are in the central portion of the part.

QUESTIONS

1. In some pultrusion operations, cloth or mat is pulled into the die along with fibers. What is the purpose of these broad goods?
2. Describe what the cut would be like if the cutting saw was not moving with the pultruded part.
3. Describe how you would introduce a stitched fabric into the pultrusion process.
4. List three considerations for the manufacturing process when using epoxy resin in a pultruded part.
5. Suggest three reasons why composite beams are not immediately certified for use in vehicular bridges.

BIBLIOGRAPHY

Akzo Nobel Polymer Chemicals LLC. 2002. "Pultrusion Guide." Publication PC2002. Chicago, IL: Akzo Nobel Polymer Chemicals LLC.

Brown, Gordon L., Jr. 1987. "Pultrusion—Flexibility for Current and Future Automotive Applications." SME Technical Paper, EM87-346. Dearborn, MI: Society of Manufacturing Engineers.

Fisher, Karen. 1997. "Pultruded Beams Reflect Design-for-manufacture." *High-Performance Composites*, May/June, pp. 23–26.

Society of Manufacturing Engineers (SME). 2005. "Pultrusion" DVD from the Composites Manufacturing video series. Dearborn, MI: Society of Manufacturing Engineers, www.sme.org/cmvs.

Strong, A. Brent. 2006. *Plastics: Materials and Processing*, 3rd Ed. Upper Saddle River, NJ: Prentice-Hall, Inc.

19

Thermoplastic Composites Processing

CHAPTER OVERVIEW

This chapter examines the following concepts:

- Wet-out of thermoplastic composite materials
- Processing short-fiber thermoplastic composites
- Thermoplastic composites molding by traditional thermoset processes
- Thermoplastic composites molding by unique processes

Molding of thermoplastic composites is strongly dependent on the length of the reinforcement. If the fibers are short (whiskers), then the thermoplastic composite can be processed much like a traditional non-reinforced thermoplastic. If the fibers are long, the thermoplastic composite, with some important adjustments, can be processed like a standard thermoset composite. However, the unique properties of thermoplastics, especially high-performance thermoplastics, permit some unique methods of processing that hold great promise for highly efficient and low-cost molding of thermoplastic composites. Some of these have already been realized.

As the various manufacturing methods for thermoplastic composites are examined, some advantages and disadvantages of these materials versus thermoset composites should be kept in mind.

In general, the *advantages* of thermoplastic composites are:

- overall cost savings of up to 25%, generally because of faster mold cycles;
- indefinite shelf life;
- molding conditions are more flexible;
- ability to make either high-performance or engineering (general) materials using the same process;
- ability to remold to repair mistakes;
- many are highly fire resistant with low smoke generation; and
- most high-performance thermoplastics have excellent solvent resistance.

There are some *disadvantages* associated with thermoplastics:

- more difficulty in fiber wet-out;
- higher processing temperatures than thermosets; and
- problems with lay-up because of lack of tackiness.

The inherently high viscosity of thermoplastic materials resulting from the necessarily high molecular weight of the resins causes serious problems when attempting to wet-out the fibers (see Chapter Six, Thermoplastic Composites). Hence, this chapter will first look at wet-out methods. Then the processing of short-fiber thermoplastic resins and long-fiber resins both by traditional thermoset methods and by methods unique to thermoplastic composites will be examined.

WET-OUT OF THERMOPLASTIC COMPOSITE MATERIALS

Since thermoplastics do not crosslink, the molecular weight of the polymer is determined when the polymer is first polymerized. Because the molecular weight must be high to have good properties, the viscosity of the thermoplastics (even when heated) is also high. This means that thermoplastics will not wet the fibers as easily as thermosets. Therefore, in applications where long fibers are needed (over about .25 in. [0.6 cm]), thermosets have a distinct advantage.

Several methods have been explored to solve the problem of wet-out using thermoplastic resins. The first is simply to raise the temperature so the viscosity of the resin falls to a level where wet-out is easy. However, for most thermoplastics, degradation begins to occur before the viscosity is low enough to really make wetting out the fibers easy. Hence, merely raising the temperature generally does not work.

Another obvious method to reduce the viscosity is to dissolve the thermoplastic resin. The resulting polymer/solvent solution can be diluted to obtain a low viscosity and thus facilitate fiber wet-out. However, the problems and costs associated with using a solvent, especially of removing the solvent after the fibers have been impregnated, discourage most manufacturers from using this method. Further, many of the most desirable thermoplastics, especially high-performance thermoplastics, are not readily soluble in any reasonable solvent. Hence, because of costs and practicability, solvent thinning is rarely a good option.

Most polymeric melts will thin when they are sheared. Hence, the combination of heating and shearing can result in lowering the viscosity of the resin and, therefore, allow for good wet-out. This method is used extensively when the fibers are short and the shearing will not seriously degrade their length. Therefore, the most common method for wetting out short fibers is simply to add the fibers to an extruder in which the resin has already been melted. The resin, of course, must be meltable in an extruder. This is possible for most of the common engineering thermoplastics but is not feasible for some of the high-performance thermoplastics that have high melting points.

The mixing of the fibers and the thermoplastic is done by putting the short, whisker-length fibers (about .04 in. [0.1 cm]) into the barrel of the extruder at a point where the resin is melted. Most extruders have a vent port conveniently located downstream of where the fibers are melted and so this is where they are added. By adding the fibers downstream in the extruder, the amount of shearing is minimized, thus protecting the fibers. The fibers and resin are extruded into stands that are chopped to about .25 in. (0.6 cm). This is then used as feed stock in molding operations on conventional thermoplastic processing equipment, as will be discussed later in this chapter.

For parts requiring fibers longer than whiskers, which cannot pass through an extruder, other methods of wet-out are required. One method is to pull the fibers through a bath of molten resin, which is done by the Verton® proprietary process. Then, after cooling, the resin/fiber strands are chopped to the desired length (usually about the same as the extruded pellets). In this case, however, the fibers are the entire length of the pellets. Therefore, they are many times longer than the whisker-length fibers used in the extruded pellets. Somewhat confusingly, these fibers are called "long fibers" by the manufacturer. This term seems confusing in light of the continuous fibers and longer chopped fibers that might be used in materials such as bulk molding compound (BMC) and sheet molding compound (SMC). Therefore, in this text the extruded fibers are called "whiskers" or "short fibers" and the Verton fibers are called "longer" or "pellet-length."

While some care must be taken to enlarge the restrictions in the molding equipment, such as the gates in the mold, longer fibers can be used in many injection molding machines and other traditional thermoplastic molding processes. These fibers fill in a gap between the whisker-length and continuous fibers. The improvement in properties with the longer fibers is shown in Table 19-1 where various properties of samples made from short fibers and the longer fibers are compared.

Another method of wet-out that was attempted in the early days of thermoplastic composites molding was to stack thin sheets of the thermoplastic resin on top of the fibers. The thermoplastic materials can often be obtained as thin films since they can be extruded. This creates a configuration similar to that obtained during the making of normal thermoset prepreg materials. The exception is, of course, that the thermoset is lower in viscosity than the thermoplastic. However, by heating and pressing the thermoplastic, impregnation of the fibers can occur. However, full wet-out is often a problem, especially with high-performance thermoplastics, which are inherently more viscous than most engineering thermoplastics. The heating lowers the viscosity of the thermoplastic resin but is generally not sufficient to obtain penetration of the fiber bundles. Further, the shearing action from pressing the fibers and resin films together is insufficient to give much resin thinning. This method is illustrated in Figure 19-1a, along with other wet-out methods that will be discussed later.

Some resin and prepreg manufacturers have tried dusting the fibers with solid resin and reported excellent wet-out even though some of the powdered resin falls off the fibers. The powdered resin is generally applied after the fibers have been wetted with a small amount of water or water-soluble resin so that there is some adhesion. Wet-out occurs when the fibers are heated in the mold. One manufacturer has also applied a polymeric coating over the dust-covered fibers as a protective skin to help prevent the resin from falling off the fibers. The polymeric coating is very thin and is often a resin like polyvinyl alcohol. The resin coating can be left in place and incorporated into the final product or, if desired, can be easily removed after molding. These two dusting methods are shown in Figure 19-1b and c.

Table 19-1. Comparison of short and longer fibers in injection molded nylon samples.

Property	No Reinforcement	Short Fibers	Pellet-length Fibers
Tensile strength, ksi (MPa)	11.8 (81)	27.0 (186)	28.5 (196)
Tensile elongation, %	60	4	4
Flexural strength, ksi (MPa)	15.0 (103)	38.0 (262)	46.5 (321)
Flexural modulus, Mpi = 10 E^6, (GPa)	0.41 (2.9)	1.30 (9.0)	1.45 (10.0)
Heat distortion temperature (HDT), ° F (° C)	150 (66)	490 (254)	495 (257)

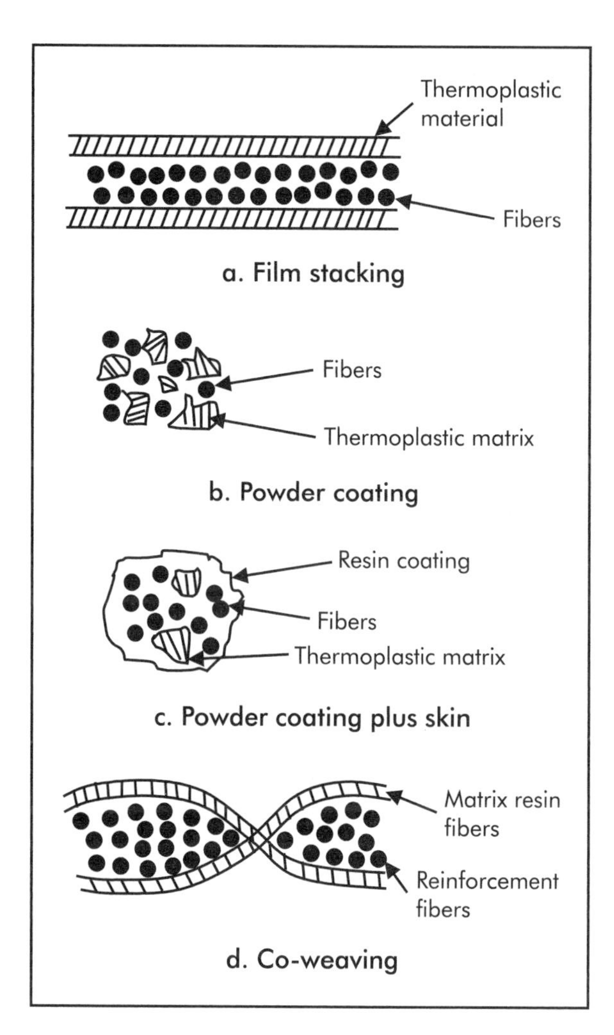

Figure 19-1. Thermoplastic wet-out methods.

An interesting and not fully explored method of wetting the fibers is to co-mingle or co-weave the reinforcement fibers with fibers made of the matrix thermoplastic resin. Many thermoplastic matrix resins can be extruded into fibers, just as they can be extruded into films (as discussed previously). **Co-mingling** is when the reinforcement and thermoplastic fibers are mingled together in the yarns. **Co-weaving** is when the two fiber types are woven or knitted together, usually with one being the warp and the other the weft (see Figure 19-1d). These methods allow cloth, preforms, braids, knits, or some other fibrous form to be made using textile technology. Since textile technology is quite advanced, the creation of such preforms or fabrics could result in more complex parts which, when heated, are simply wetted by the thermoplastic fibers co-mingled with the reinforcements. This technology seems ideal for resin transfer molding (RTM) and related technologies because the problems of resin infusion are largely eliminated.

Another new and potentially dramatic change in the method of wetting out the fibers is the use of cyclic **oligomers** (short-chain polymers). The cyclic oligomers are small enough in molecular weight that their viscosity is acceptably low, thus giving excellent wet-out of the reinforcement fibers. Cyclic oligomers will, however, polymerize during molding, thus increasing in molecular weight. The chemical process involved is a catalyzed breaking of the ring structures of the initial oligomers and then a joining of the opened rings into long, linear polymers.

The process of coating with cyclic oligomers is done by melting the powdered cyclic oligomers and simply mixing with or coating onto the fibers to make a prepreg. Then, when the prepreg is heated during molding, the cyclic oligomers polymerize into chains that are long enough to give the mechanical properties needed. A special, proprietary catalyst is necessary to get the molecular weight high enough to yield good properties. Mechanical properties are largely dependent on molecular weight; thus the oligomers have poor properties and only become good when the molecular weight increases with polymerization. The molecular weight of the cyclic polymers can reach extraordinarily high values. For example, a normal engineering thermoplastic used in traditional plastic applications might have a molecular weight of 50,000 whereas a post-polymerized, cyclic thermoplastic of the same type of resin could reach a molecular weight of 500,000. Thermosets, before crosslinking, also have poor mechanical properties. Therefore, the polymerization reaction of the oligomers is analogous to the crosslinking reaction of thermosets. The major difference between these new cyclic-based thermoplastics and traditional thermosets is that after the thermoplastic polymer is fully formed, it is still processible by melting, whereas the thermoset cannot be remelted.

At this time, only two resins are being created by the cyclic oligomer method. These thermoplastics are polycarbonate (PC) and polybutylene terephthalate (PBT). The polymerization of both types of cyclic oligomers is triggered by increasing the heat after the catalyst has been added. The polymerization activation heat for PC is somewhat higher than the melt point and so the resulting polymer is a liquid and must be cooled after it is polymerized. With PBT the polymerization temperature is lower than the melting point so it simply hardens to a solid in the mold, much as a thermoset hardens in the mold. As with the thermoset, some minor cooling might be necessary to give the part enough stiffness to be removed from the mold. The resins have been demonstrated in pultrusion and RTM.

PROCESSING SHORT-FIBER THERMOPLASTIC COMPOSITES

Extrusion

Since extruders are used to combine many conventional thermoplastic resins and short fibers, it is logical that they also can be used to form these reinforced thermoplastics into shapes other than the pellets used in subsequent molding operations. Such products are made by using a die that is shaped in the cross-section of the desired product rather than the simple round hole used to make pellets. The die is mounted on the end of the extruder. (The die makes a continuous rod that is then chopped to make the pellets.)

Other shapes are made by changing the shape of the opening in the die through which the melt of resin and fiberglass must pass. For example, the die could be square shaped or wide and flat. The die might even have an inner mandrel and thus force the resin-fiber melt into the space between the mandrel and the outer die body to create a hollow tubular shape. Such a die is shown in Figure 19-2.

Note in Figure 19-2a that the fibers are generally aligned in the direction of the flow. This type of alignment is common in extrusion. It even occurs in non-reinforced extrusions where the molecules become aligned because they are forced to flow through a long, narrow section such as the area between the mandrel and the wall of the die. The degree of alignment is increased by making the restricted flow path longer and narrower. (Figure 19-2a is, of course, idealized as actual parts do not have the uniformity of alignment shown, but the alignment is certainly strongly in the direction of flow.) In reinforced thermoplastics, the molecules of the resin and the short fiber reinforcements are aligned.

Figure 19-2b shows the change in orientation that can be achieved with a change in the geometry of the die. By dramatically expanding the flow path, the fiber orientation can be changed laterally to the machine direction. Since the fibers are critical to increasing mechanical properties in the direction in which they are oriented, this change results in a tube that is much stronger in the hoop direction. (However, a machine-direction part would be stronger in tension in the machine direction.)

Figure 19-2c shows that by partially obstructing the flow path, the fibers can be turned to orient perpendicularly outward. This orientation results in a part that is especially strong in compression. While this orientation may not be highly desired in a tube, it is certainly valuable in some sheet materials where the forces on the part are generally perpendicular to the plane of the product.

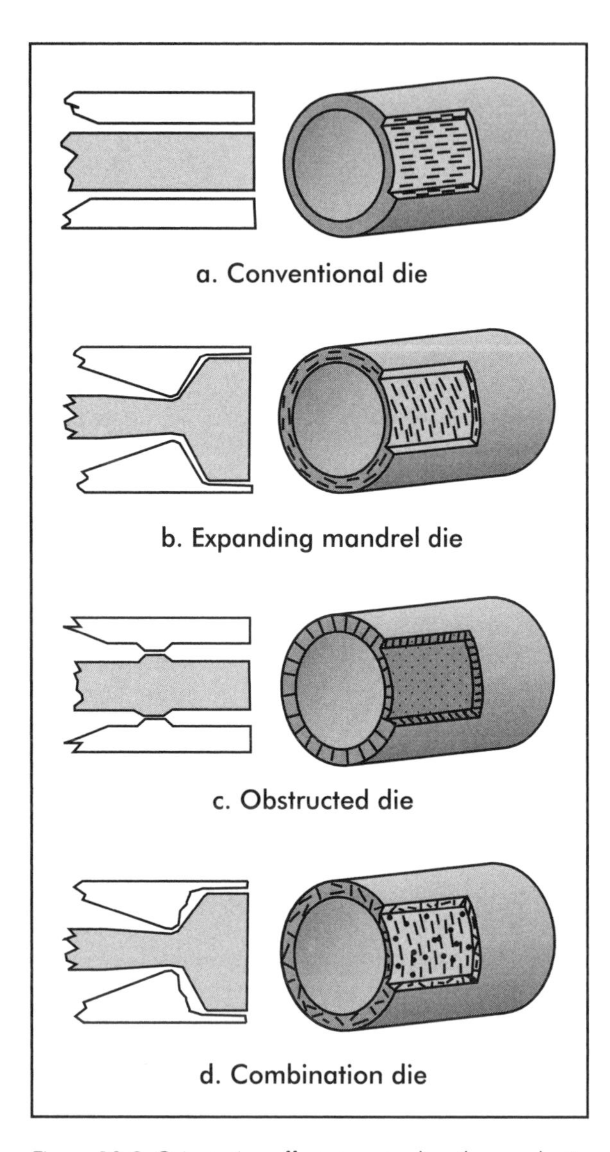

Figure 19-2. Orientation effects in extruding thermoplastic composites.

Figure 19-2d shows a combination die that gives a generally random orientation to the fibers. This is a good compromise when forces on the tube are from all directions.

In all extruded thermoplastic composite parts, the nature of the fiber orientation is important to understand and control as fiber direction has a major effect on the performance of parts.

Injection Molding

For years, injection molders have been using short fibers (whisker length) as reinforcements in thermoplastic resins. Injection molding is widely used because the process is easily automated, it offers high productivity due to short molding cycles, and excellent detail can be obtained. Most engineering thermoplastics are readily processed by injection molding. The properties of these products are excellent for some applications with significant increases possible in tensile strength and flexural stiffness as shown in Table 19-1. Sometimes the increases in a property may be as much as 300% over the non-reinforced thermoplastics. Typical fiber contents range up to 30%, although 40% and even 50% are available for some resins. In every case, the tensile strength and flexural modulus increase with fiber content. This would be expected because the fibers, which are stiff and strong, carry much of the load. Note that elongation dramatically drops when fibers are present.

In a typical injection molding machine, the resin enters the machine from a gravity-fed hopper. A rotating screw advances the material. The screw has deep flights in the zone under the hopper—the **feed zone**. The material then moves into the **melting zone**, in which electrical heaters are placed around the barrel and the root of the screw gets thicker to provide mechanical heating. The plastic is melted through combined thermal and mechanical heating, and is conveyed by the screw into the **metering zone**. The purpose of this zone, in which the root of the screw is thick (the flights are shallow), is twofold: to ensure that all the resin is melted and to build up pressure. Note that an extrusion screw has these same zones and that, when fiberglass is added during extrusion, it is added in the metering zone, after the resin has been melted.

At the proper time in the mold cycle, the screw stops rotating and advances. This motion injects the molten material through the nozzle and a channel, called the **sprue**, into the mold. A check valve at the end of the screw prevents the melt from flowing backward. The melt flows through a network of channels, the **runners**, to the mold **cavities**. The actual entry points into the cavities are called the **gates**. In the simplest mold, the sprue and runners solidify along with the part. They are removed from the part after it is ejected from the mold. Some molds have hot runner systems that do not solidify, thus making trimming of the part unnecessary. The mold is opened after sufficient cooling time is allowed (often only a few seconds). Knockout pins eject the part.

The mold can be either single or multi-cavity. The cavities in multi-cavity molds may be either identical or different, to allow for simultaneous production of various parts. The latter type is called a family mold because several parts that go into one assembly can be molded simultaneously. Molds also may be equipped with unscrewing devices to allow for internally threaded parts, and with sliding inserts to allow production of hollow parts. All such modifications to the simple one-cavity mold add complexity and cost to the mold, but may add productivity to the operation, thus reducing the cost of the final part.

The size of the injection molding machine is determined by the size of the part or parts to be made. Two measurements of machine size are required: shot size and clamping pressure. The shot size is simply the volume of the melt to be injected (parts, runners, and sprue). Typical machine sizes range from 1–50 oz (30–400 g). The clamping pressure is the pressure required to hold the mold closed against the injection pressure. The area of the part and the restrictions in the sprue, runners, and at the gates are major considerations in determining this pressure. Typical values are 10–1,500 tons-force (90–13,500 kN).

The locations in the injection molding machine that orient the fibers are the places where the flow path is restricted. These points are critical because they may be so constricted that the fibers are not able to flow through them at all (or only with great difficulty). Hence, to allow for good flow of the resin-fiber melt, the restriction points must not be too confining. That means the runners, gates, and internal cavity clearances should be much larger than might be used for non-reinforced resins.

Another major consideration in molding reinforced thermoplastics is the width of walls, ribs, or other thin sections in the parts. The fibers must have ample clearance to flow into narrow sections. If the section is too narrow, the fibers bridge over the top and the section either fails to fill or has resin with few, if any, fibers. This same problem can occur in corners that are too sharp.

If it is desired to have the fibers randomly oriented throughout the part, as is usually the case, the flow path should be as wide as possible. This allows the fibers to turn randomly as they flow to fill the area. Randomization is also increased if the temperature of the melt is high and the flow rate is low.

Some problems can occur when two flows of fibers meet as would be the case when multiple injection points (gates) are present in a mold. This situation is shown in Figure 19-3. The flow lines meet in a region between the gates. In this region, the actual meeting of the flows is called a **knit line** or **weld line**. For proper strength of the part, the fibers and resin flowing from one gate must interpenetrate with the fibers and resin from the other gate. If they do not interpenetrate, a boundary will exist between the flows. Because fibers do not cross that boundary, it becomes a place where the mechanical properties are much lower. Interpenetration is improved by raising the temperature of the melt and allowing sufficient time for the flows to mix.

In general, injection molding is a viable method for producing reinforced thermoplastic parts. The advantages in mechanical properties can be outstanding. However, fiber orientation, fiber flow restrictions, and fiber mixing are important considerations, which may require some changes in the equipment. Also, operating conditions may need to be changed (especially raising the temperature of the melt) so that fibers can move freely. Although properties can be much better with reinforced products, raising the temperature usually means that the mold cycles will be longer.

Other Traditional Thermoplastics Molding Processes

While it is possible to mold reinforced thermoplastics by other traditional thermoplastic molding processes such as rotational molding, blow molding, thermoforming, and casting, their use is not common. This is probably because the presence of the fibers reduces the elongation so dramatically. A phenomenon related to low elongation is seen even in some of the molding processes such as blow molding and thermoforming. In both, the semi-molten resin must stretch and flow into the mold. When fibers are present, the stretch is restricted and the fibers actually become points of breakage in the flowing plastic film. Hence, if the reinforced material is stretched too far, breaks occur and the part will have holes or other defects.

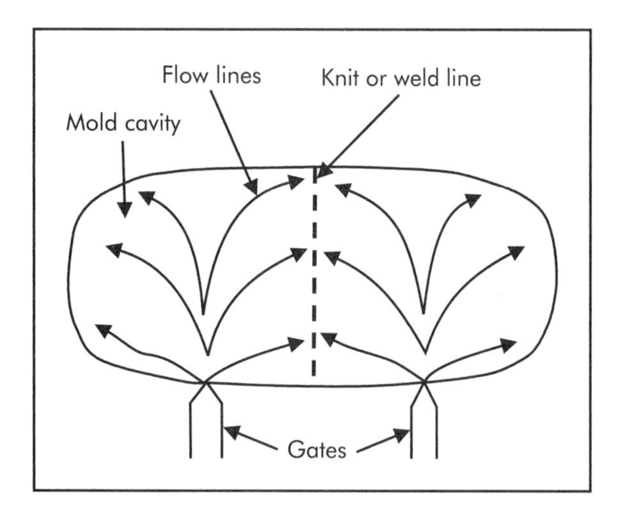

Figure 19-3. Flow lines and knitting in an injection mold with multiple gates.

The reluctance to use reinforced thermoplastics in traditional thermoplastic processes also may be because of the difficulty in getting good flow of the resin-fiber material. For example, in rotational molding, the raw material is a thermoplastic powder that is tumbled inside a heated mold. During the tumbling and heating process, the material partially melts and becomes sticky, thus coating the inside surface of the mold. Eventually the particles fuse into a solid mass that has the shape of the mold. If fibers are present, the fusing of the particles is hindered. In casting, the molten resin is poured into the mold. In this process, the presence of fibers may restrict the entry of the melt into areas of fine detail or into narrow flow paths.

THERMOPLASTIC COMPOSITES MOLDING BY TRADITIONAL THERMOSET PROCESSES

There are critical applications where the properties of the fiber are important and, therefore, the fiber length must be considerably longer than is possible in traditional thermoplastic molding processes. In these applications, the manufacturing method used with thermoplastic composites is likely to mimic one of the thermoset manufacturing methods. The higher performance requirement of the part will also, in general, require that the resin be one of the high-performance resins discussed in Chapter Six rather than one of the traditional engineering thermoplastics.

Even though the molding of thermoplastic composites may mimic the thermoset processes, some critically important differences must be kept in mind. The thermoplastic becomes liquid by heating (melting) and then, after it is shaped in the mold, it is cooled to solidify. Thermosets, in contrast, are heated to cure and turn into a solid. Therefore, curing in the normal sense does not exist with thermoplastics. Further, thermoplastic melts are much more viscous than uncured thermosets.

Open Molding

In general, wet methods are not applicable for thermoplastic resins. The problems associated with melting the resin and then applying the hot, highly viscous material to the fibers effectively prevents wet methods from being used. Therefore, neither wet lay-up nor spray-up can be used effectively with thermoplastic resins. However, thermoplastic prepregs are available and have been used in several commercial products. The prepregs are engineering composites, such as polypropylene resin on fiberglass, and advanced composites, such as polyetherimide on carbon fibers. Unidirectional and fabric prepregs are generally available.

Because thermoplastics are solids at room temperature, thermoplastic prepregs are stiff, without drape, and have no tack. They must be softened (slightly melted) so that the drape can be improved sufficiently to allow them to be applied to the contours of a mold. The heating also improves tack so that one layer will stick to the layer under it. This softening is usually accomplished by heating the prepregs with a hot air gun or direct heating tool, such as a soldering iron.

The manufacture of thermoplastic prepregs must account for the higher temperatures required to melt high-performance thermoplastics and their much greater viscosity compared to thermosets. The most common method of wetting out the fibers for thermoplastic prepreg is to heat the resin and simultaneously shear it by pulling the resin and fibers through a roller system.

Because many thermoplastic resins can be made into fibers, a material that holds great promise is cloth made from both the reinforcement fibers and the fibers of the matrix material. Such a material was presented previously, but it has special relevance in this discussion on molding with prepregs. The hybrid woven material is, essentially, a prepreg since it contains both matrix and reinforcement in a sheet form. Therefore, it can be used in lay-up. One major advantage of this material is that it is highly conformable (good drape). Therefore, the problems of stiffness and lack of drape do not exist. Further, the cloth-like texture of the material means that layers will have less of a tendency to slide relative to each other; therefore, tacking with a heat gun or soldering iron is not needed. The disadvantage of this material is the curing time and the pressure needed to ensure that the resin fibers melt and thoroughly wet the reinforcement fibers.

An advantage of thermoplastic prepregs over thermoset prepregs is that thermoplastics have infinite shelf life. Refrigeration is not necessary and no other precautions are needed to ensure their usefulness over a long period of time.

The bagging step for thermoplastic material should be essentially the same as for thermosets. However, the bagging materials must be able to withstand the higher temperatures and pressures normally required for thermoplastics. For example, typical bags for thermoplastics are polyimides, such as Kapton®, which can withstand 700° F (371° C), or aluminum foil. The other components such as the sealing tape, release, breather, and bleeder, also must be high-temperature tolerant. Higher temperature bagging materials are generally more expensive and more difficult to work with than are their lower-temperature counterparts.

Molding of thermoplastic prepregs must also account for the much higher temperatures required to obtain good flow characteristics and compaction. Whereas the thermosets require the high temperatures to cure the resin, the thermoplastics require the high temperatures to melt it. This is so the resin will form to the shape of the mold and so that the layers of prepreg will join together into a solid mass. Most high-performance thermoplastics melt in the 500–700° F (260–370° C) range, compared with the 250–350° F (121–177° C) curing temperatures for epoxies. Therefore, mold materials and autoclave fittings and couplings must be able to withstand the higher temperatures. Many of the molds used with thermoset composites will not work as they are made of low-temperature materials.

A problem with the lay-down of thermoplastic composite prepregs is that the prepregs tend to be "springy." That is, they seek to return to their original shape. Therefore, if the prepreg has been shaped to fit the contours of the mold, it will not adhere well unless the temperature and pressure have been applied for sufficient time so that these forces of recovery have been allowed to relax.

In addition to the concern about the effect of the high temperatures on the materials used during molding, there is concern about the coefficient of thermal expansion. Thermal expansion becomes more critical in the mold design because of the elevated temperatures. Further complications come from the need for higher forming pressures in thermoplastics because of the higher viscosity of the resin, which must be moved between the layers of prepreg and into the shape of the mold.

Some machines have been developed for the automatic lay-down of thermoplastic prepreg. These machines resemble automatic lay-down machines for thermosets, except that a special heated shoe is attached to the applicator head. As the thermoplastic prepreg tape comes between the head and the mold, the tape is heated by the shoe, which softens it and gives it the tack and drape necessary to stick to the mold or to previously laid layers. A cooling shoe is placed just after the heated shoe to re-solidify the tape and ensure that it remains in place.

Liquid Infusion Molding

Just as wet lay-up is impractical with thermoplastics, using thermoplastics in liquid infusion molding processes like resin transfer molding (RTM) is also very difficult. The resin is hard to liquefy without some device like an extrusion screw. Further, it is hard to maintain as a liquid because of the high heat involved to attempt to lower the resin's viscosity. Even at the high temperatures, the viscosity is usually too high to effectively wet the fibers in the mold. Moreover, the injection pressures must be high to inject such a viscous material as the molten thermoplastic resin. Even if the material can be injected, its high viscosity results in fiber movement and, of course, the problem of wetting out the fibers. To keep the resin liquid while it is being injected, the mold must be heated, which also can be a problem.

In summary, liquid infusion molding with thermoplastics is difficult and rarely done commercially.

Filament Winding

If filament winding is attempted with a resin bath, the problems of maintaining the thermoplastic as a melt will be present. Moreover, the fibers are moving relatively rapidly through the resin bath so the wet-out time is short. Hence, wet filament winding is rarely done in commercial practice.

Fortunately, an alternate method is available that allows filament winding with thermoplastic resins to be done with only minor modifications to the standard thermoset process. Instead of attempting to wet the fibers during the filament winding process, the material used in the winding operation is **prepreg tow**. The pre-impregnated tow material is made in an operation in which the thermoplastic can be heated and/or sheared while the fibers in the tow are spread to facilitate wet-out. Some prepreg tow is even made by coating the tow with powdered thermoplastic resin, although the resin tends to fall off when this is done.

The tow is applied in a standard filament winding machine that has been modified to allow heat and pressure to be applied at the point where the fibers are laid onto the mandrel. This heat softens the prepreg tow and gives it some tack so it will adhere to the mandrel or layers of material that have already been applied. The pressure assures that the bonding will be effective. To facilitate this process, the layers of material already on the mandrel also may be heated so that they, too, will be sticky when the new material is laid down.

Several heating methods have been used in commercial production. In one highly successful operation, the materials already on the mandrel are heated by merely mounting a high-energy heating lamp so it heats the area of the fibers that are about to have more material laid upon them. The prepreg tow about to be laid onto the mandrel and other materials is heated with a laser. It applies high-intensity heat that can be focused directly onto the prepreg tow just before it is laid onto the mandrel. Others have also used high-intensity lamps to heat the prepreg tow, but this has been less successful than the laser.

The tendency of the thermoplastic material to retain a previous shape can be a problem. Therefore, heated consolidation, molding in a clamshell mold, or bagging with an autoclave are often done to ensure that the material accepts the shape of the mandrel and that there is good consolidation of the materials.

Pultrusion

The problems of a wet thermoplastic resin bath are also present in pultrusion. However, they are not as severe as in filament winding because, in pultrusion, the resin-coated fibers pass into the die where they are heated and consolidated. The speed of the pultrusion line is often slow enough that this consolidation results in good fiber wet-out. The shearing action of moving through the die helps thin the resin and improves wet-out. The die temperatures can be staged so the fibers are heated through the die until good wet-out is achieved. Then the die sections beyond that point can be cooled to solidify the part. This cooling is often rapid so the overall time in the mold can be at least as short as with thermosets and possibly shorter.

Still, the continuous heating of the thermoplastic in the resin bath to temperatures high enough to give a good working viscosity can result in degradation and other problems that have already been discussed. One method to alleviate some of these resin bath problems is to use resin injection. In this case, as opposed to liquid infusion methods like RTM, the resin can be injected into a small area in the mold and little resin motion is required. The resin then passes through the die and is further heated and consolidated before being cooled and solidified at the end of the die. The heating of the resin prior to injection can even be done with a small extruder, thus imparting good efficiency to the heating operation. The extruder can also meter the resin into the die.

An interesting capability available for pultruded thermoplastic products is to pultrude standard shapes (rods, channels, I-beams, hat structures, plates, etc.) and then post-form them to create structural members fitted to the specific application. The savings potential would be great in using a high-production method, such as pultrusion. Inventories could be reduced by stocking standard parts and sizes that could be later shaped to fit specific needs. Hence, pultrusion of thermoplastic composites is practical and commercially viable.

Compression Molding

Traditional compression molding where the raw material is placed into a heated mold and then cured under pressure rarely occurs with thermoplastic composites. One conceivable method of using compression molding and a thermoplastic resin would require that the thermoplastic be melted outside the mold and then injected into the mold. The fibers could be in the thermoplastic as part of the melt or they could be placed into the mold and the thermoplastic injected onto them. Either way, external melting of the thermoplastic and subsequent injection into the mold has proven to be difficult.

Alternately, the solid thermoplastic could be placed directly into the mold and then melted by the heated mold. This, too, has proven to be ineffective, largely because melting by this simple contact method is far less efficient than heating by combined shear and thermal heating as is the case with in an injection molding screw.

However, the use of matched die molds for shaping thermoplastics has excellent possibilities for molding thermoplastic composites. The methods used have similarities to compression molding. However, they are different enough that they are considered in the section on unique thermoplastic molding methods, which follows.

THERMOPLASTIC COMPOSITES MOLDING BY UNIQUE PROCESSES

Thermoplastic Composites Molding

The most important process for molding thermoplastic composites is shown in Figure 19-4. The process begins with the formation of thermoplastic composite blanks. These are prepreg sheets of thermoplastic resin with reinforcement fibers that have been laid up according to the design of the final part. That is, the layers of the prepregs are oriented and tacked together as they are to appear in the final part. They might be fully consolidated but need not be if little movement of the sheets is expected. (This is similar to the concept of kitting discussed in Chapter Seven.)

For manufacturing efficiencies, the blanks can be placed onto shelves in a frame so that they can be automatically fed onto a belt that carries them into an oven. The blanks are heated to the point where the resin is softened and will flow easily under moderate pressure but not to the point where the resin melts. This softening point is similar to the **sag point** in traditional thermoplastic vacuum forming, which is the condition of the thermoplastic sheet when it is ready to be thermoformed.

When the blank is ready, it is conveyed into a cold matched die mold. The blank is clamped to hold it in place and to maintain the order and orientation of the sheets within it. The mold closes on the hot thermoplastic composite blank and molds it to the shape desired. Because the blank is clamped, the prepreg must stretch as it is deformed to the molded shape. This type of matched die molding is therefore called **stretch forming**. Because the fibers cannot stretch very far, the complexity of the part that can be molded using stretch forming is limited.

One interesting modification used with stretch forming is the use of prepreg in which the fibers are discontinuous. These prepregs are made from fibers that are chopped but still highly aligned (usually over 99%). Therefore, when the softened prepreg is deformed into the mold, the fibers separate and the prepreg material stretches.

If some movement of the fibers in the blank can be tolerated, the clamping can be light or eliminated. This means that the fibers slip in the mold to conform to its shape. The slippage of the fibers is lubricated by the molten resin. This process is called **flow**

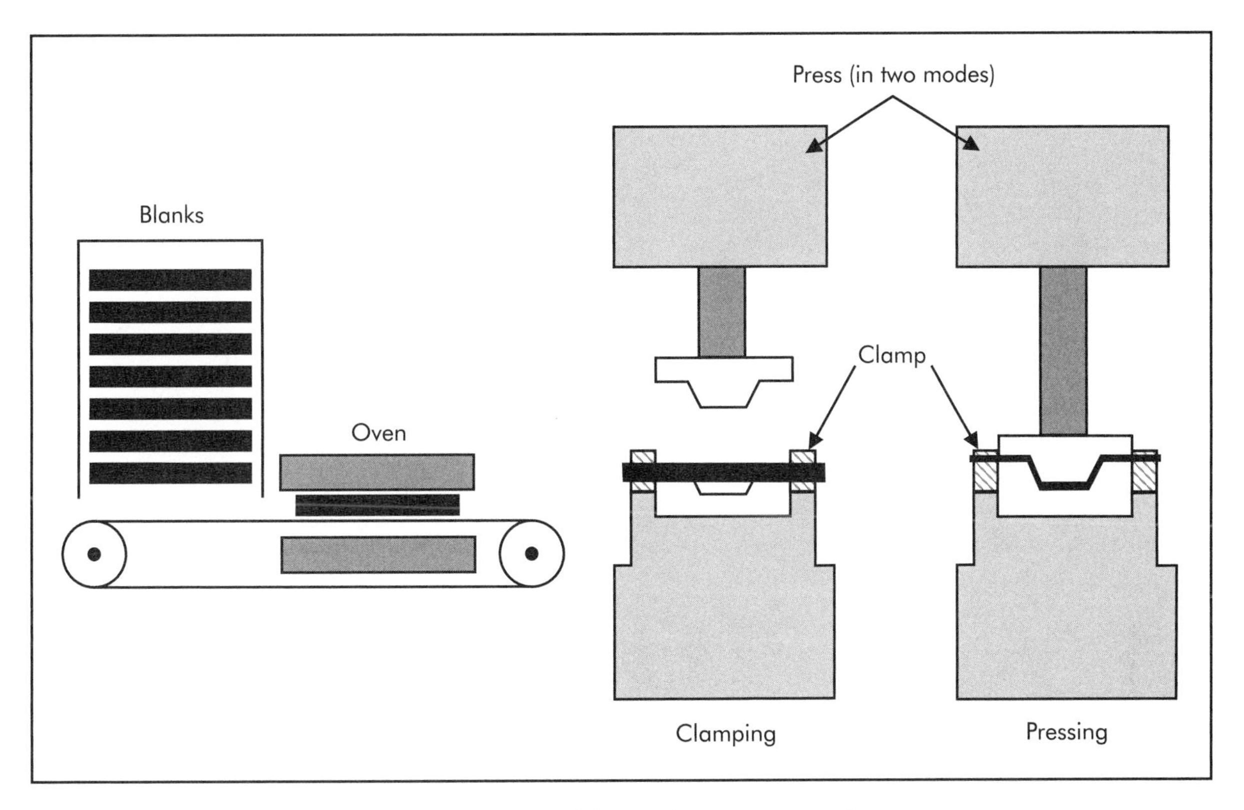

Figure 19-4. Thermoplastic molding with preheating of thermoplastic composite sheets.

forming. Because of the high viscosity of the resin and to control the movement of the fibers as much as possible, flow forming is done slowly (that is, the mold is closed slowly).

Because the mold is cold, the part rapidly cools and solidifies. Often, the time in the mold is only a few seconds. The overall production cycle is typically 2–10 minutes—far less than the typical thermoset curing time in compression molding. To improve repeatability, a mechanical stop is usually installed to limit travel of the press platens.

The pressures involved in matched die molding are generally low. Therefore, a variation of the standard matched die process is to use one flexible mold and one metal mold. Under some circumstances, two flexible molds can be used. This process is called **flexible die forming**. If the flexible die is backed by a hydraulic reservoir, the process is called **hydroforming**. (In practice, the term "hydroforming" is applied to any process in which a flexible die is used.) The flexible die can be shaped to match the metal die. However, for thin panels where the deformations of molding are not great, the flexible die need not be exactly matched to the opposing metal die. The combination of rigid and flexible dies will still provide good consolidation and shaping of the material because of the applied hydrostatic pressure. (Hydrostatic pressure is that pressure resulting from pressing by a flexible material or a fluid—in this case, the flexible die.)

The pressure of the system can be varied to assist in definition of the part. If the flexible die is not completely filling the cavity of the metal die, the pressure often can be increased sufficiently to cause the die to deform and completely fill the cavity (therefore giving good shape definition to the part). This is especially important for highly complex shapes. However, when the process is controlled by pressure only, instead of using a mechanical stop to define the movement of the molds, variability of the parts is often excessive, and can result in squeezing the resin out of the prepreg. Hydroforming can use shaped or unshaped flexible dies.

Even though the die is cold, heat is transferred to it by the heated laminate (blank). This can be a serious problem with the high temperatures required to mold high-performance thermoplastics. Therefore, high-temperature elastomers, such as silicone, can be used for the entire flexible die or as a heat-barrier cap on less expensive rubber dies (but these will quickly become brittle and require replacement). The compressibility (hardness) of the flexible die is typically 60–70 Shore A, which is sufficiently firm for compression while also giving good conformity to the shape of the metal mold. Although pressures vary according to the thickness of the blank and the nature of the resin, they are typically 100–250 psi (0.7–1.72 MPa) for a .25-in. (0.6-cm) thick carbon fiber laminate.

The advantages of using flexible dies in the hydroforming method, rather than matched metal dies, are the inherent lower cost of the flexible dies and, because generally lower forces are required, the decreased likelihood of fiber damage.

Thermofolding

An interesting process can be used with thermoplastic composites because of the ability to reform the material after an initial molding or consolidation. For example, assume that a thermoplastic panel needs to be bent at 45° in its center. The panel could be made by thermoplastic pultrusion, heated in the middle (only), and then bent to the proper angle. That bending operation usually employs a jig or form to ensure that the angle of the finished part is correct. The pressure to bend the material can come from a press and die or might come form a flexible tool that presses the hot material into the forming jig. The process is shown in Figure 19-5.

Thermofolding is done by heating a pre-consolidated thermoplastic laminate (blank) to the softening temperature (sag point). Only the portion of the laminate that will be bent needs to be heated. Therefore, heating can be done with directed heating sources, such as a heating element (like in an oven), a laser, or a high-intensity lamp. This method is, therefore, highly efficient and does not require heating of the entire material. That may be important if fittings or some other material that is thermally sensitive has been attached to the ends.

One of the problems with thermofolding is that the fibers may not stay aligned properly in the bend area. Figure 19-6 illustrates three common problems caused by fiber movement. The movement might create a resin-rich/fiber-deficient zone, it might kink the fibers, or it might cause the fibers to slip and, therefore, the alignment of the design might be compromised. These problems are difficult to overcome, but can be minimized by bending slowly and using good pressure.

Despite the difficulties, thermofolding has been used commercially to make avionics enclosures, kick panels, cargo-floor panels, and x-ray panels.

Incremental Forming

Thermoplastic composites permit a special type of molding for large parts. Instead of forming these parts all in one operation, the parts can be formed incrementally, that is, a section at a time. Because the sections formed are small, substantial tooling cost savings can be realized.

When the amount of deformation of a component is not excessive, incremental forming might be used. The section to be formed

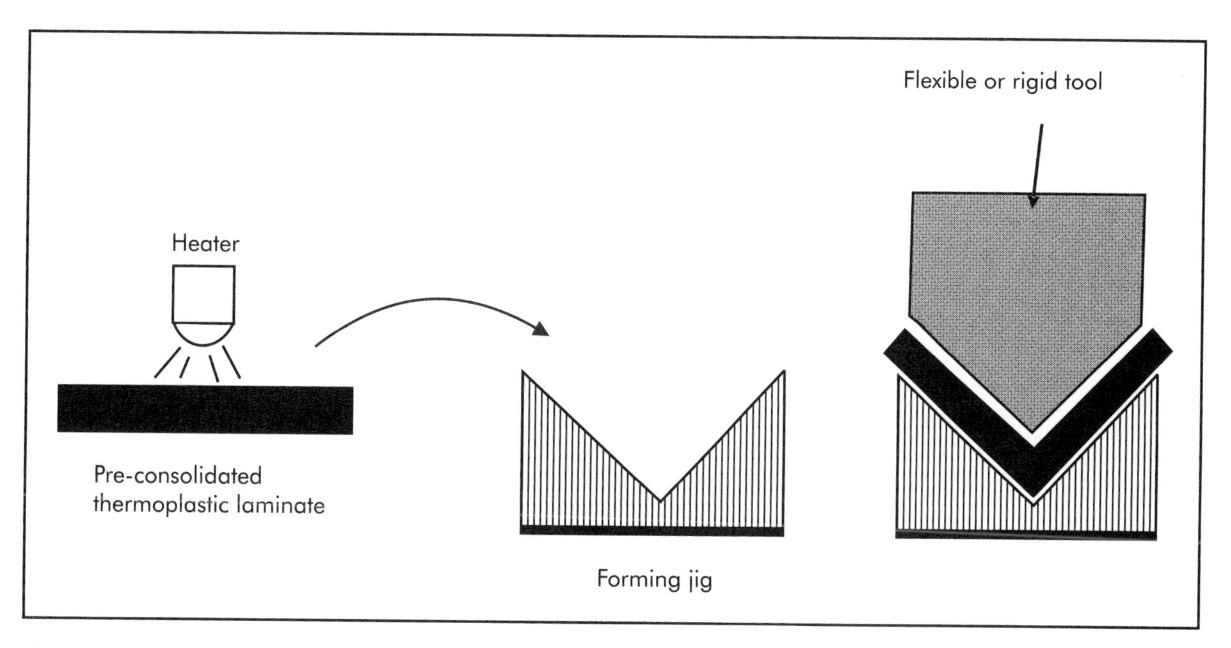

Figure 19-5. Thermofolding.

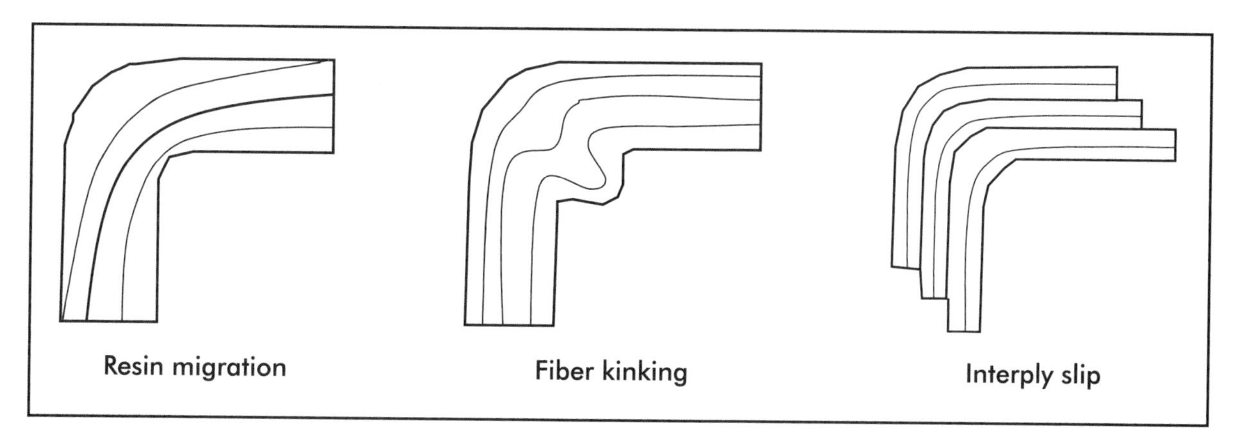

Figure 19-6. Deformation at corners.

is placed in a mold and heated. That part is formed and then the mold is shifted to the adjoining area which is, in turn, heated and formed. In this incremental fashion, the entire part can be formed.

A key to the success of incremental forming is to use transition zones between each of the areas to be formed. These transition zones link the curvature or deformation of one molded area to the next. Hence, the transitions are smooth and the fibers remain in alignment.

Transition Molding

In transition molding, a cold part is forced into the shape desired and held. For this to be possible, the deformation of the thermoplastic composite part must be modest. The deformed part is then heated to a moderate temperature for a long period. This allows the resin molecules to move and relieve the strain induced by the deformation.

For the deformation of air conditioning ducts using thermoplastics and glass fibers, the temperature for stress relief is 350–365° F (177–185° C) and the time for relaxation is 15–20 hours. The molding temperature is unique for each resin and can be critical; that is, it must be within a narrow range. If the temperature is too low, no stress relaxation will occur. If the temperature is too high, the original high surface quality can be lost and delamination may occur.

Roll Forming

A modification of thermoplastic molding applicable for long parts is to move a heated thermoplastic blank through a series of rolls that are set so they impart some shape change to the part. The shape change for each set of rollers can be quite small, but by passing the part through many such roller sets, a major change in the part can be achieved.

CASE STUDY 19-1

Lightweight Ladder

A difficult problem exists in trying to make industrial ladders for utility, cable, and construction companies. The ladders have to be long enough to reach the rooftops of houses, generally about 20 ft (6.1 m). Further, they must be strong enough that, fully extended, they can withstand 200 lb (91 kg) suspended from the center without excessive flexing, and still be light enough (under 55 lb [25 kg]) for one person to carry. Aluminum ladders are too heavy. Even traditional pultruded polyester/fiberglass ladders are too heavy.

A method was developed for producing the ladders that has proven to meet all of the requirements for strength and length with weight under the maximum limit. The method uses thermoplastic composite prepregs. The prepregs are laid in the mold and then, in only specific locations, additional material is added so that the critical failure areas have extra reinforcement. The entire prepreg material is then heated and consolidated with pressure.

The ladder has proven to be a tremendous success. It is light and strong and corrosion resistant, but does not conduct electricity so it can be used by electrical utilities.

SUMMARY

The most common applications for reinforced thermoplastics are relatively small parts where fast molding cycles are critical and the desired properties of the part can be met by short-fiber composites. As the parts become larger, the ability to quickly mold them (usually by injection molding) decreases and the advantages of short-fiber-reinforced thermoplastics diminish.

But what if some other thermoplastic molding process, besides injection molding, were used to make the parts? Where thermoplastic sheets are presently thermoformed and then backed with thermoset material (as, for instance, in luxury spa tubs), a viable alternative could be thicker reinforced thermoplastic sheets. Again, however, as the amount of material increases (in this case, as the required thickness increases), the thermoplastic becomes less competitive, simply because it costs more than unsaturated polyesters.

The thick thermoplastic sheets could be made by the process illustrated in Figure 19-1a. The layered thermoplastic composite sheets are stacked in the proper sequence for the application, dropped onto a conveyor line, heated in an oven, and then molded in a compression mold under moderate pressure. The molding cycle for such a process can be as low as a few seconds since all that is required to happen in the mold is shaping of the softened material and then sufficient cooling to remove it from the mold.

Are there any critical advantages that might tip the balance toward reinforced thermoplastics even though the material costs might be higher? Thermoplastics can be cut, drilled, and machined much more easily than thermosets. The inherent elongation of the thermoplastic resins allows these finishing operations to be accomplished with far less worry about delamination, chipping, or cracking. Thermoplastics also can be welded much more easily than thermosets, although thermosets usually can be adhesively bonded better than thermoplastics. The advantage in welding is the shorter time required for joining.

In some environments and operations, the shorter shelf life of thermosets can be a problem. Thermoplastics have essentially unlimited shelf life.

Another major advantage of high-performance thermoplastics is their general resistance to most solvents. For instance, some of these materials are resistant to airline hydraulic fluids (Skydrol®) and have found applications where contamination from that fluid is likely.

Some aerospace applications have been converted from aluminum or thermoset parts because of the weight savings that can be gained from using thermal welding instead of rivets. There are also lower labor costs resulting from use of the welding technology.

Thermoplastics seem to have special advantage for either the small part that can be injection molded from short fiber reinforced engineering thermoplastics, or for the high-performance part that needs special properties such as flame resistance, solvent resistance, or extra weight savings. However, with the new wet-out technologies now being used, thermoplastics can compete effectively against common polyester fiber reinforced plastics. Several traditional thermoset methods have been modified to accommodate thermoplastics. Other methods are unique to thermoplastics and have been developed to make them competitive with thermosets even for continuous, high-performance composite applications. Considering the advantages in toughness, cycle time, and finishing (drilling, cutting, etc.), the future for thermoplastic composites is bright.

LABORATORY EXPERIMENT 19-1

Thermoforming of Thermoplastic Composites

Objective: Learn about the advantages and limitations of thermoplastic forming.

Procedure:

1. Obtain some thermoplastic composite prepreg material.
2. Lay up the prepreg into a blank that is at least five layers thick. It is best to use different fiber orientations in the blank. A good shape for the blank is a square about 18 in. (0.5 m) on a side.
3. Consolidate the blank in a heated press.
4. Cut the blank into strips that are about 2 in. (5 cm) wide.
5. Heat the center portion of the strip with a focused heater. The temperature for molding will depend upon the nature of the resin. Therefore, it is necessary to obtain the molding specifications from the prepreg manufacturer.
6. When the strip has reached the proper molding temperature, lay it onto a bending jig and press the strip into the bend. The pressing can be done manually with an insulated metal block that has been treated with mold release.
7. Note any fiber deformations that occur in the fold region.
8. Repeat the forming method using a dramatically increased forming speed and note the differences in laminate integrity in the bend area.

QUESTIONS

1. What is a thermoplastic composite?
2. Give three advantages of thermoplastic composites over thermoset composites.
3. Why is fiber wet-out a particular problem with thermoplastic composites?
4. Give three methods commonly used to wet-out thermoplastic composites and explain the advantages and disadvantages of each.
5. Discuss three changes that must be made in filament winding to accommodate thermoplastics.
6. Discuss four advantages of the matched-die-molding system for thermoplastics.

BIBLIOGRAPHY

Dykes, R. J., Mander, S. J., and Bhattacharyya, D. 1989. "Roll Forming Continuous Fiber-reinforced Thermoplastic Sheets: Experimental Analysis. Auckland, New Zealand: Center for Polymer and Composites Research, University of Auckland, private communication.

Offringa, Arnt R. 1996. "Thermoplastic Composites Rapid Processing Applications." *Composites: Part A*, 27A, pp. 329–336.

Society of Manufacturing Engineers (SME). 2005. Composites Manufacturing video series. Dearborn, MI: Society of Manufacturing Engineers, www.sme.org/cmvs.

Strong, A. Brent. 1993. *High Performance and Engineering Thermoplastic Composites*. Lancaster, PA: Technomic Publishing Co.

Strong, A. Brent. 2006. *Plastics: Materials and Processing*, 3rd Ed. Upper Saddle River, NJ: Prentice-Hall, Inc.

20

Damage Prevention and Repair

CHAPTER OVERVIEW

This chapter examines the following concepts:

- Damage and its effects
- Damage prevention
- Damage assessment
- Smart structures
- Repair

DAMAGE AND ITS EFFECTS

The philosophy of damage to composites has been expressed as: *If the damage is visible, it should be repaired or the part should be scrapped. If the damage is not visible, the structure should be capable of withstanding the required residual strength load level.*

It may seem obvious, but there are some significant problems associated with detecting damage in composite parts and their repair to ensure that the required strength or other performance property (per the design) is achieved. Possibly the most important and fundamental problem is that, because composites are specifically structured to maximize directional strength and stiffness, and minimize weight, a deterioration of the strength and stiffness, especially when hidden, may compromise the overall performance of the part. Hence, the matter of damage and its prevention and repair are important topics.

In light of the philosophy and the nature of composite materials, the key issues in dealing with damage and repair are:

- understanding the problem,
- prevention,
- detection,
- damage assessment,
- repair specification,
- repair accomplishment, and
- repair evaluation.

Understanding the Problem

High Impact and Primary Damage

The most straightforward situation is that in which the damage is severe. It is readily obvious and, therefore, using the philosophy of damage to composites, it should be repaired unless the damage is so severe that the part should be scrapped.

Sometimes the assessment is obvious. For example, if a major portion of the wing of an airplane has been shot off by a rocket, it seems reasonable to just replace the entire wing. On the other hand, if a single bullet has penetrated the door of a vehicle, a repair seems logical.

Often, however, the decision to repair or to scrap rests *not* on what is the obvious damage but on the estimate of whether there is significant non-visible damage. The non-visible damage could be so extensive that performance of the part could be seriously compromised. Therefore, even in the case of severe high-impact damage, much of the concern is with the hidden, collateral, or low-impact damage.

Low Impact and Collateral Damage

The sources of damage are, of course, too numerous to enumerate. However, the examples given here are illustrative and will serve as conceptual triggers for understanding the nature of composite damage, and in general prevention and remediation.

Some composite parts are very large. Wind blades over 100 ft (30 m) are common. Rocket-motor cases with diameters of 30 ft (9 m) are routine. Yachts of 60 ft (18 m) are made daily in several factories. Several problems can occur during the manufacturing process itself. Even after manufacturing has been completed, these large parts must be moved within the manufacturing facility, prepared for shipment, shipped, unloaded, and placed into service. In all of these steps, the potential for low-impact damage is high.

An example of the manufacture of a large boat is used to illustrate the potential for damage.

After the gel coat has been sprayed into the mold and cured, the workers will walk on the cured gel coat to apply the veil and then walk on the surface again to apply each of the layers of fiberglass/resin laminate. They may then walk on the surface again to apply wood reinforcements. All of these steps may cause damage to the layers that have already been applied. Then, during demolding, the part may be twisted or bent as it is lifted out of the mold with an overhead crane. This lifting could cause microcracks. If the part sticks in the mold, common practice is to pry the walls of the boat free using wooden wedges which are, occasionally, pounded into the space between the boat and the mold. Microcracks (or worse) are clearly a possibility. Air is often forced into the space to attempt to lift the boat out of the mold. This air could cause flexing that might result in microcracks.

After the boat is taken from the mold, it is moved to an assembly position where other sections of the boat are welded to it. Of course, the boat must be set in place (usually in a cradle). But this can be a problem if the overhead crane does not perfectly line up the boat and the cradle. Then, as each section is assembled, further stresses and bumps are possible. Workers are moving all over the boat and carrying various tools that might be dropped onto the boat.

Finally, the boat is moved out of the plant to a transport vehicle. The boat is loaded (with high potential for bumping) and then secured onto the transport vehicle. During transport, the boat might be hit by road debris such as rocks kicked up by the tires or passing vehicles, or it might be damaged by hail or other falling objects. The boat is unloaded (another high potential for damage) and then placed in service. The boat owner must be constantly careful not to bump against piers, rocks, and other boats. To say the least, the potential for damage is enormous.

In the case of large rocket motors on which careful assessments of damage have been made, the inspections revealed that about 55% of the damage is small in size (less than 1 × .625 × .500 in. [2.5 × 1.6 × 1.3 cm]); about 17% is medium sized and barely detectable with the naked eye, and 28% is large enough to be readily seen.

An assessment of the effect of the medium damage (barely visible) was then made. It was determined that the dry static strength was reduced by 70% and the wet static strength was reduced by 90%. These types of results have led designers to use safety factors of 1.25 to 1.50 and these factors do not account for visible damage, which is always assumed to be repaired. The safety factors allow for the normal property degradation that occurs because of operational and environmental effects. The factors considered often include: temperature and humidity, fatigue and vibration, shock, loading, aging, and damage. However, damage is often the most difficult to assess because it is usually random whereas the other factors are more regular or, at least, predictable.

So, what is the level of kinetic energy from casual impacts? A 2-lb (1-kg) hammer dropped from 3 ft (1 m) results in 6 ft-lb (8 J) of kinetic energy. Energies of that level can cause fiber breakage and matrix microcracks. Even energies four times that level (as, for example, the slight impact from a forklift attempting to raise a composite part and bumping it as it slides the forks in place) can cause damage that is still not detectable with the naked eye.

DAMAGE PREVENTION

Until now this book has focused on the nature of composite materials and the process of composite design and manufacture. The discussions have provided information on the fundamentals of the materials and the optimization of their properties, but not on how the properties of a composite might be changed by damage. The brief discussions about damage largely concerned the impact strength (toughness) of the materials. The recommended inclusion of safety factors to allow for damage and manufacturing defects provided a limited focus.

This section looks at how material choice might be altered for parts where damage is either frequent or catastrophically important. The subtleties in design that can be used to minimize the effects of damage will be considered. Then some specific steps are presented that can be taken (beyond changes in the resin and reinforcements) to minimize the damage to a composite part.

A variety of tests have been used to assess the effect of damage on composite materials and parts. Some of those tests, like the Izod and Charpy impact tests, are direct measurements of toughness. Other tests monitor the changes in properties of laminates after they have been damaged. Sometimes these tests will also subject the laminate to environmental factors, such as water or heat, because the effects of the damage are magnified under those conditions. The conditions may also represent what might be encountered in real-life use.

Figure 20-1 illustrates the way some tests are used to simulate impact. In the open-hole compression test, composite samples using one type of carbon fiber were made with three types of epoxy resin. (The curves for each of the resins were similar so an average curve is shown.) Half of the samples were impacted with sufficient energy to create an internal area of damage. The size of the damaged area was determined using ultrasonic testing. A hole of the same diameter as the damaged area was drilled in the other half of the samples. These samples were then tested in compression to determine the strain to failure. Because it was learned that the effect of the damage was dependent on the ratio of the damage area to the total area of the part, the failure strain was plotted versus the ratio of the hole diameter to the total width of the sample. (Logically, the non-damaged area is carrying most of the compressive load. So, if the damaged area is a major portion of the part, then the effect of the damage will be higher than if the damaged area is small compared to the non-damaged area.)

In Figure 20-1, the curves from the damaged samples and the drilled samples are reasonably similar indicating that drilling a hole is a model for damage. Therefore, samples are routinely tested for the effect of damage by drilling a hole of a standard size (usually .25 in. [0.6 cm]) and then doing a compression test. Open-hole tensile tests are also common when tension is the major force that will be encountered by the composite part in actual use.

Other tests commonly done to test the effect of impact include cyclic compression or tension on open-hole samples in which the forces applied are less than the ultimate stress. These are to simulate use conditions

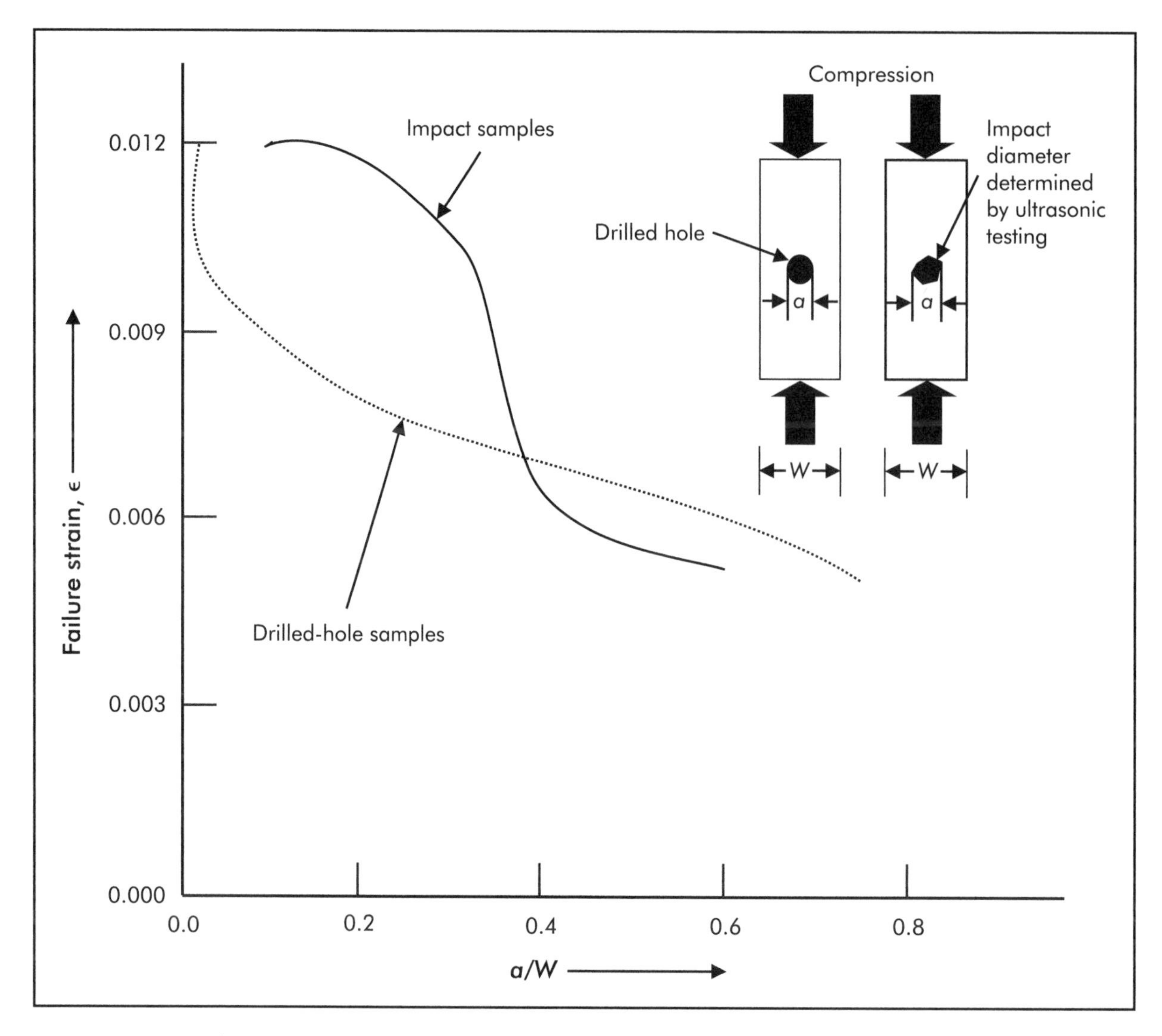

Figure 20-1. Open-hole compression test as a simulation for damage.

in which the forces are repeatedly applied as, for example, in repeated aircraft takeoffs and landings. Sometimes these tests also include thermal cycling (hot and cold).

Material Choice

Using the various impact tests as guides, some resins have been found to be especially damage tolerant. The most obvious choice for improving epoxy toughness is a toughened epoxy. These toughened resins significantly minimize the effects of impact damage. The toughened epoxy's higher strain to failure at higher impact energies is obviously beneficial to composite performance. Thermoplastic resins are even better than toughened epoxies in minimizing the effects of damage. However, as discussed in a previous chapter, the processing of thermoplastics is often significantly different from thermosets; therefore, the option to use a thermoplastic may not be practical.

The depth of the indentation is another way to measure the damage from impacts. In this test, samples of carbon fiber composites made with a standard epoxy and another composite made with thermoplastic are impacted at various energy levels. The criterion for impact resistance is the depth of the indentation. The depth increases with increasing impact energy for both the epoxy and thermoplastic, but the thermoplastic consistently has a smaller indentation for the same impact energy. This measure is especially useful when an ultrasonic tester is not available to determine the actual area of the impact damage.

When the damage area can be determined by an ultrasonic machine, the impact damage for a standard epoxy and for a thermoplastic can be plotted against the impact energy. As is the case with the depth of indentation, the thermoplastic is better in minimizing the damage area for an equivalent impact than is the epoxy.

The choice of reinforcement fiber has a major effect on impact damage. This effect was discussed in Chapter Eight where it was shown that aramid fibers were substantially better in minimizing the damage from impact than were carbon fibers.

The form of the reinforcement also can affect the extent of the impact damage. Composite samples were made with unidirectional tape prepreg and woven fabric prepreg using the same resin and fibers. These samples were then impacted at various levels and the strain to failure was measured. The results revealed that fabrics are better at resisting impact damage than tapes since they have a higher strain to failure. The reason is simply that the crossover points in the fabric are places where energy is transferred; therefore, the impact energy is spread quickly and its effect is minimized in the impact location. However, these crossover points are also places where the fiber is squeezed and this effect reduces fiber strength. Therefore, for strength, tape is chosen; for impact toughness, fabric is chosen.

The effect of 3D stitching is an important factor in impact damage resistance. As might be expected, impact damage is reduced when composites are stitched. This effect arises from the ability of the stitched fibers to withstand the impact force, which is applied in the direction in which those fibers are oriented.

Another method of improving impact damage resistance with fibers is to incorporate fibers of two different types in a laminate. For example, fiberglass strips are sometimes mixed into a carbon fiber laminate. The fiberglass arrests the propagation of cracks and minimizes the extent of damage. A structure of this type is shown in Figure 20-2. The capability of the fiberglass to arrest crack growth arises from the fact that energy is not transferred well across the boundaries of different materials.

A variation combines two fiber types by placing the tougher fiber, such as aramid, on the surfaces most likely to receive an impact and using fiberglass or carbon fiber elsewhere to make a hybrid structure (as discussed in Chapter Nine). This system

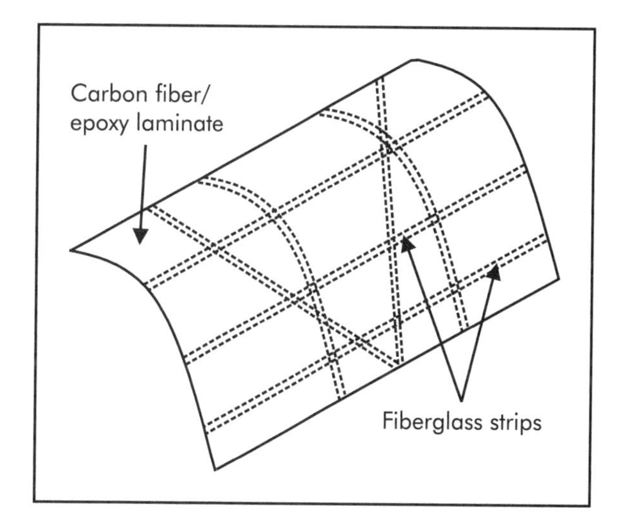

Figure 20-2. Fiberglass used for crack arrestment.

minimizes costs but gives improved impact protection. It also would be possible to combine the aramid and fiberglass in the same fabric by interweaving or mixing the fibers into the same tow.

An issue that needs to be considered is scale-up. Tests have shown that small samples do not have the same impact damage characteristics as do full-scale real parts. The small samples show a lower strain to failure for a particular impact energy; therefore, they predict worse impact damage resistance than is actually the case in full-size samples. The reason for this behavior is likely because the larger size of the real parts allows distribution of the impact force over a wider area, therefore reducing the damage in the immediate vicinity of the impact. This spreading of the energy reduces the local damage. Fortunately, the samples predict worse behavior than the actual parts and so the testing is conservative.

Design Changes

In addition to the material-based concepts for improving damage tolerance already discussed, the design of the composite can be optimized for damage tolerance.

One design factor that impacts damage tolerance is fiber direction. In general, energy is passed most easily from one layer to the next in the laminate if the directions of the fibers in the layers are the same. For example, when two layers of 0° prepreg are adjacent, the energy from an impact is passed directly from one layer to the other. That characteristic means that the energy from an impact is passed deeper into the composite and, as a result, the severity of the damage is worse. Hence, to minimize the effect of an impact, it is better to spread the energy laterally so it will dissipate over a wider area. This is done by making the directions of adjacent layers overtly different. Therefore, from a crack-resistance point of view, it is better to have a [0/90/0/90] rather than [0/0/90/90] or [0/90/90/0].

Actual samples of carbon fiber and epoxy were made to test the effects of layer direction. One sample was a [10/80/10] lay-up and the other was [40/50/10]. The samples were impacted and then the strain to failure was measured. The [10/80/10] had a loss of strain to failure of 27% whereas the [40/50/10] had a loss of 40%. The lower loss of strain to failure indicated that the [10/80/10] was less affected by the impact.

The poor transfer of energy between layers that differ from each other contributes to the improved impact resistance of hybrid composites. Not only can a tough fiber be placed on the surface that is likely to be impacted, but the interface between the tough fiber (usually aramid) and the structural fiber (such as carbon or glass) further dissipates the energy, improving impact toughness. This property has been used effectively in bullet-proof vests where layers of aramid and ultra-high-molecular-weight polyethylene (UHMWPE) fibers have been placed in alternating layers to achieve optimal ballistic protection. The property works when the resins in the layers are different and even when the types of carbon fiber are different, for example, with high-modulus and high-strength fibers in adjacent layers. Of course, the more different the layers are, the greater is the effect.

Fiber direction is also important in pressure vessels. In this case, the problem is associated with the interrelationship of the direction of the fibers and their contribution to the pressure-holding capability of the tank. As discussed in the chapter on filament winding, two types of winding patterns, hoop and helical, are used in composite vessels. The hoops go around the tank at approximately 90° to the long axis of the tank. The helical wraps are at an oblique angle to the axis of the tank. The hoop wraps are much more effective and important in containing the pressure in the tank. Therefore, if the hoop fibers were in the outer layers of the tank and they were impacted, the breakage of these fibers would be much worse than breakage of a similar number of fibers or layers of fibers in helical wraps. Therefore, a rule of thumb for design of pressure tanks is to increase the design tolerance to "bury the hoops"; that is, put the hoops closer to the inner surface of the tank and cover them with helical wraps toward the outer surface so that accidental impact of the hoop wrappings is less likely.

Examination of pressure vessels of several wall thicknesses suggests another design factor that affects impact toughness. As the thickness of the walls of the pressure vessel increase, the damage area, even as a percentage of total thickness, decreases. This is postulated to occur because thick walls have less deflection and therefore less micro-cracking. Also, the damage from an impact on a thick part is more likely to be compressive as opposed to flexural in thin-walled vessels and, because compression strength is usually greater than flexural, the damage is less. The hypothesis was confirmed by wrapping the pressure vessel on a steel core, which has little deflection, and comparing damage in thin-walled vessels also tested without the steel core. In this case, the thin-walled samples on the steel core had less impact damage than the unsupported samples. So, in general, stiff composites having little flexing are more damage tolerant than highly flexed parts.

External Protectors

A straightforward method of reducing the damage from impacts is to cover the composite material with an external material that will absorb some or all of the impact energy. (The external protector can be removed just prior to use.) This method is rarely used except when the composite part is costly or highly susceptible to any decrease in properties. A case where these conditions apply is with the rocket-motor cases for the space shuttle. The high cost of launching materials means that the weight of the rocket motors (and all other space-shuttle components) must be kept to a bare minimum. Therefore, design safety factors are quite low. This means that the integrity of the cases must be guarded carefully.

One external protectant method is simply to put a layer of shock-absorbing material over the composite. Some materials that have worked well are cork with an aramid hard shell on top. The aramid spreads the impact and the cork underneath ensures that the energy is dissipated before it reaches the composite. Sometimes this type of layer remains with the composite as it is bonded to its outer layer to ensure that it remains in place. In addition to the cost of making the protectant layer, another disadvantage is the weight added to the total assembly.

It is possible to make a removable outer protective shell. These are especially useful for space vehicles and missiles since the protective shell can be used while the part is being shipped, stored, and readied for launch. Then, just before the launch, it can be jettisoned. This type of protective shell is usually made of composites, but since there is no particular problem with weight or cost (assuming that the shell is recoverable), any reasonable material is acceptable. For best protection, an energy-absorbing material can be added inside the shell.

On pressure vessels, an external material that has been used successfully is a skirt, which is used to cover critical areas of the tank. These areas are usually the ends and the transitions zones where the vessel changes from hoop to helical wraps. The most common type of skirt is illustrated in Figure 20-3.

DAMAGE ASSESSMENT

The initial method of assessing impact damage is to visually inspect the damaged part. To

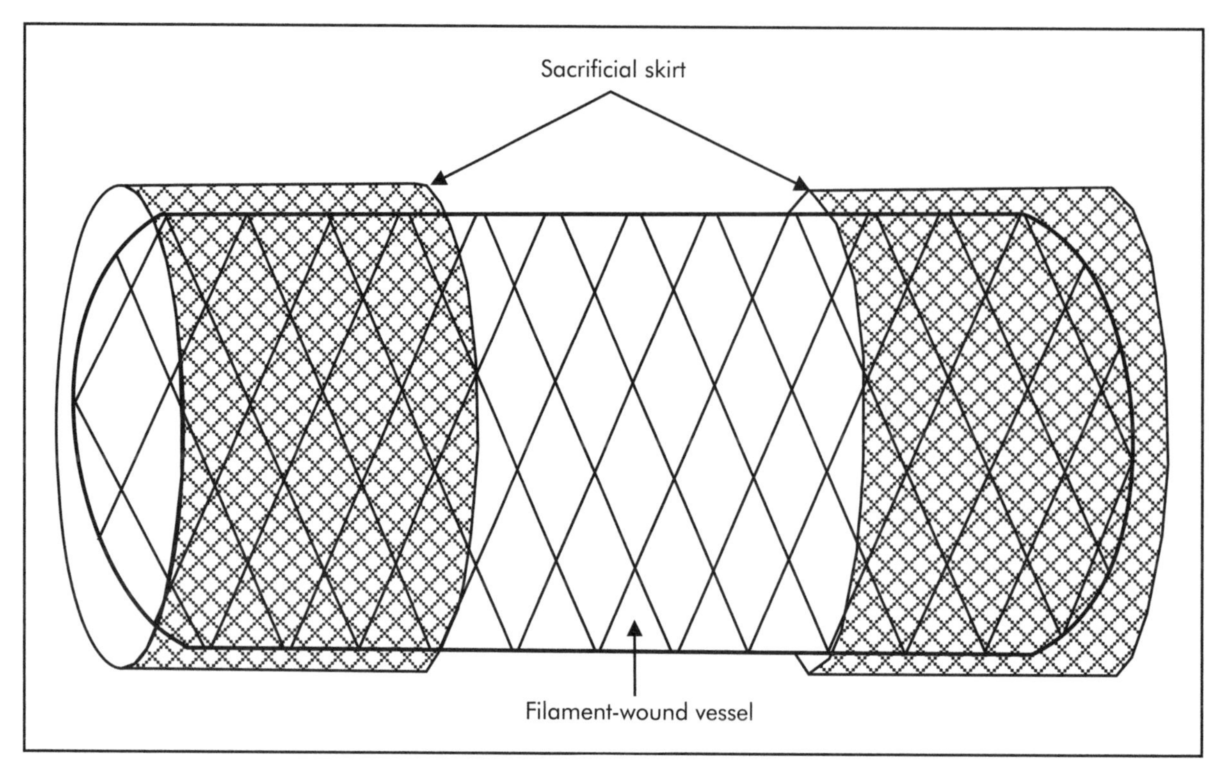

Figure 20-3. A sacrificial skirt as a protectant against impact damage on pressure vessels.

do that the structure must be accessible to inspect for the extent of the damage. One of the well-known and difficult problems with any composite structure is the hidden damage issue. Some impact damage, especially when caused by light impact, may be barely visible on the surface. However, there could be considerable matrix cracking and fiber breakage below the surface. The damage will typically spread in a cone-shaped area away from the impact point and extensive delamination can occur toward the backside of the laminate. If only one side of the laminate can be inspected, it is prudent to assume that the hidden side has received greater damage than is evident from the impact side. If possible, nondestructive testing (NDT) should be used to assist in further assessment. Table 20-1 lists common NDT methods and their applicability. The effectiveness of NDT testing will be discussed later in this chapter.

In an effort to relate the visual assessment to previous damage assessment data in the composites industry, several types of damage have been defined with visual references. These are listed in Figure 20-4. Although not quantitative, these definitions and the accompanying diagrams are useful, especially in field situations, for deciding the amount of repair that must be done to correct the damage. They are also useful in assessing possible causes of the damage.

Experience has shown that visual inspection needs to be done carefully and from several different viewing angles. Of course, one angle is the direct perpendicular view. In this view, the inspector should note (measure) the depth of the damage and attempt to determine the class of damage according to the

Table 20-1. Common non-destructive testing (NDT) methods and their applicability.

Defect	Test Method							
	Visual	Tap Test	Pulse Echo	Through Transmission	X-ray	Thermography	Dye Penetrant	Optical Non-destructive Inspection (NDI) Methods
Abrasion	●	◑	○	○	○	○	○	◑
Dent	●	◑	○	○	○	○	○	◐
Scratch	●	●	◑	◑	○	◑	●	●
Crack	●	●	◑	◐	○	◐	●	●
Cut	●	●	◑	◐	○	●	●	●
Gouge	●	●	◐	◐	○	◐	◑	●
Delamination	○	◐	●	●	◐	●	◑	●
Hole	●	●	●	●	●	●	◑	●
Inclusions	○	○	●	●	●	●	○	●
Porosity	○	◑	◐	●	◑	○	○	◐
Voids	○	◑	◐	●	◑	○	○	●
Debonds	○	●	◐	●	◐	●	○	●

Key: ● = Good ◑ = Some capability
◐ = Fair ○ = Little capability for practical purposes

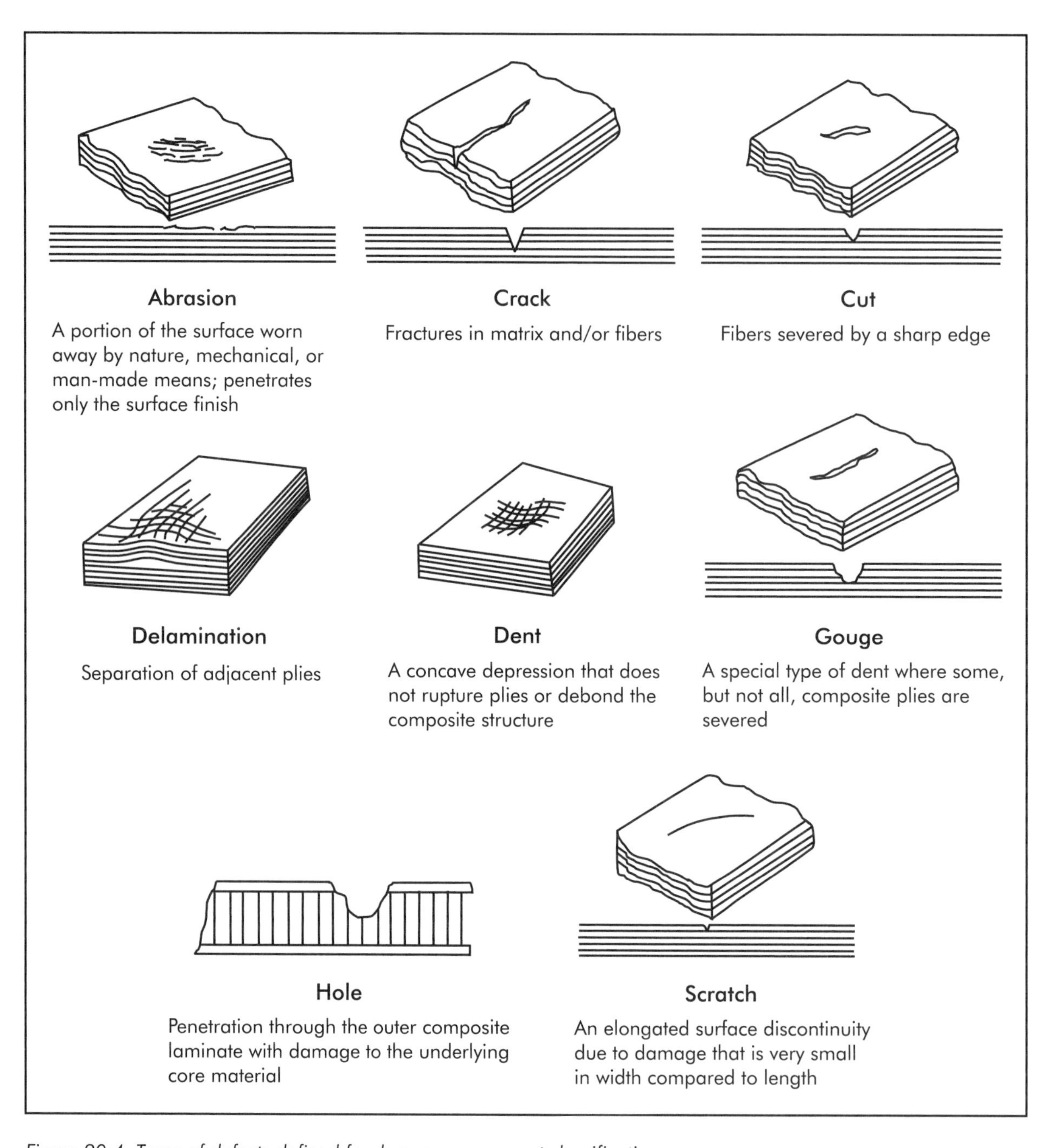

Abrasion
A portion of the surface worn away by nature, mechanical, or man-made means; penetrates only the surface finish
Crack
Fractures in matrix and/or fibers
Cut
Fibers severed by a sharp edge
Delamination
Separation of adjacent plies
Dent
A concave depression that does not rupture plies or debond the composite structure
Gouge
A special type of dent where some, but not all, composite plies are severed
Hole
Penetration through the outer composite laminate with damage to the underlying core material
Scratch
An elongated surface discontinuity due to damage that is very small in width compared to length

Figure 20-4. Types of defects defined for damage assessment classification.

system shown in Figure 20-4. The inspector should also look at the damage with an oblique viewing angle and several different light-source angles so that collateral damage might be detected. Careful comparison of the surface surrounding the damage should be done with specific attention to disruptions in the surface smoothness and inspection for hairline cracks, which would indicate that the damage energy radiated outward.

Visual inspection can be enhanced by using a small tap hammer to note the sound of the composite in the vicinity of the impact point. This is called the **tap test** and it is done by tapping on the surface of the composite part and listening carefully for changes in the sound from one place to another. The standard, against which all sounds are to be compared, is simply the most common of the sounds heard when tapping the part in a known non-damaged area. In other words, the part is assumed to be regular and undamaged in general, with deviations from that sound arising from defects or other non-uniformities. The part is tapped gently back and forth until a subtle change in tone is detected, from a clear sharp ring to a duller thud. Then the inspector backs off a bit, maybe .5 in. (1.3 cm) or so, taps again, and makes a mark indicating the boundary with a felt-tip pen. The steps are performed again from about 1 in. (2.5 cm) away and so on until there is an outline of the irregularly shaped area of damage.

A limitation with the tap test is the thickness of the part. The test works best when the parts are relatively thin and the defect is close to the surface. Also, since the test detects differences in acoustic resonances of a defect compared to surrounding areas, only those defects that have different vibrational characteristics are picked up. Generally, those would be laminar-type defects such as delaminations. A further limitation is on parts with complex geometries where the vibrations might be affected by the shape of the part from one area to another. Attempts have been made to reduce the variability of the tap test by providing a machine-like tapping device and an instrumented detection system. In general, these improvements have not proven to be sufficiently better to offset the cost and convenience of the simple tap method.

The tapping test inspection method actually works quite well, but is clearly limited to small areas where damage is already suspected. If a large area needs to be inspected, tapping gets tedious in a hurry. So for large areas, such as a survey of an entire bridge deck, other inspection methods, such as thermal imaging, may be much more appropriate.

Clearly, experience is a great benefit in using the tap test, both in recognizing the standard sound and in detecting differences from place to place along the part. Some standardization has been attempted by specifying that a standard tap hammer be used in the test (as has been done by the Federal Aviation Administration [FAA]). If this is not readily available, a specific coin (a quarter in the USA), can be used as the tapping instrument. The force of the tap is, of course, still a variable that requires some experience to exercise properly.

Sometimes damage occurs and is not detected, especially if it is slight. The person doing the damage might not realize that damage has occurred or, perhaps, the person seeks to hide the accident. However, even slight damage can result in a significant loss in composite properties. Therefore, a method for detecting even light impacts has been developed. This method, illustrated in Figure 20-5, involves painting the surface of the composite and then over-coating with a clear resin containing easily broken microcapsules. The upper layer is cloudy because of the light diffraction of the microcapsules. If an impact occurs, the microcapsules break, thus revealing the paint layer. The change in surface color is easily detected and would lead to investigations of possible damage.

After initial visual and tap hammer inspection, the methods for NDT discussed in Chapter 10 are the principal means for assessing the extent of the damage to composite parts.

SMART STRUCTURES

Wouldn't it be nice if inanimate objects such as bridges, airplanes, and even houses

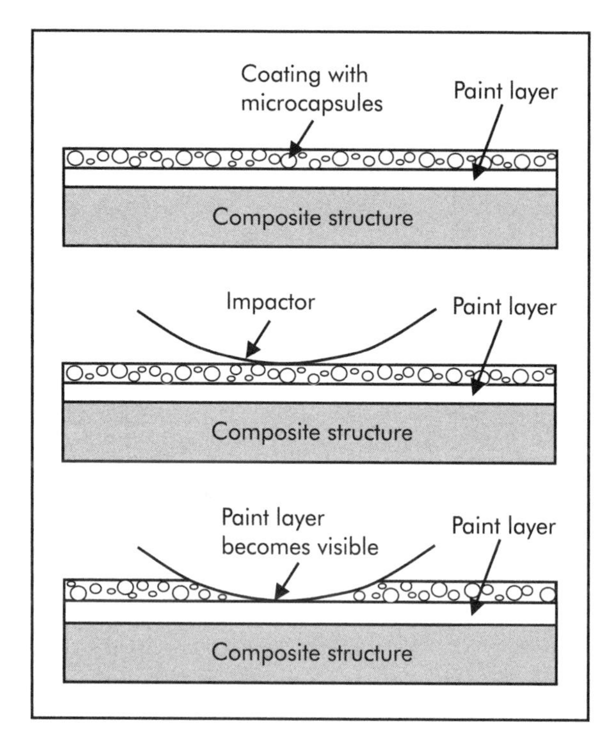

Figure 20-5. Use of paint-filled microcapsules to detect damage.

could sense when they had been damaged and then repair themselves, just as living creatures do? These self-healing structures are already in research and development. They are called **smart structures**.

A smart structure or system incorporates sensors and actuators into the material of the system in such a way that enables it to sense the environment and then respond appropriately in a preprogrammed manner. Typical smart structures or smart material systems are composites, although this is not a requirement. Composite structures especially lend themselves to the inclusion of sensors and actuators because of the way they are made (in laminar structures).

A closely related type of structure is called an **intelligent structure**. Intelligent structures are smart structures that have the added capability of learning and adapting rather than simply responding in a preprogrammed manner. This learning and adapting is usually accomplished by the inclusion of an artificial neural network (ANN) into the smart structure.

Applications

One important application of smart structures would be bridges and other civil engineering structures. Here the structural components (walls, decking, etc.) could sense when repairs are needed, such as from cracks or corrosion, and then report the location and the extent of the problem. The sensors could also activate motors that could dampen the structure during an earthquake to minimize shaking damage.

Alternately, a smart airplane or boat could sense cracks, leaks (either externally or internally as with fuel), or excessive forces. The responses could be to report the problem and its extent, and initiate the application of a sealant, which has been preloaded into the critical space. Or, with more sophistication, a sensor could activate motors that might actually change the shape or stiffness of the outer structure to minimize the effects of the forces and thus prevent damage and/or improve operational performance. A small-scale model of an F/A-18 fighter has already been made and is under test in wind tunnels to verify these performance concepts. NASA has also flown full-scale demonstrators of the F-111 and the Hornet with re-shapeable wings that flex and change their aerofoil shapes.

Most smart structure concepts are still in the research and development phase but some have found their way into actual commercial use. One of the most common applications of smart composites is for the damping of vibrations. Sensors detect excessive vibrations through piezo-electric devices, which then apply an opposing current to stimulate changes in the structure and counteract the vibrations. Noisy home appliances are silenced by smart structures that sense the acoustic waves being generated by the appliance and then cancel those waves through active damping.

- requires a large number of the devices spread throughout the structure to sense and regulate actions.

Components and Techniques

Smart structures can be categorized by the type of device used to do the sensing or actuation. The most common of these devices are: piezo-electric, fiberoptic, shape memory alloys, electro-rheological fluids, and other electro-oriented techniques.

Piezo-electric Devices

Piezo-electric devices are made of either ceramic or polymeric film, both of which are electro-mechanical transducers; that is, they change electrical signals into mechanical motion and vice versa. If an electrical signal is applied across a piezo-electric device, the device shape will change (usually length) because of the change in electrical voltage. Alternately, mechanical deformation of the device will result in a change in the voltage output. Hence, the device not only senses changes in the structure that it might be embedded in, but also gives an electrical signal that can be amplified and used to change the characteristics of the structure in which it resides.

Advantages of piezo-electric devices include:

- ceramic devices can be used as both the sensor and the actuator, and
- rapid response time (although the films are just sensors).

The disadvantages include:

- ceramic devices are brittle and have a small range of mechanical motion, which limits their ability to sense or control large spatial deviations directly;
- some potential applications cannot use electrical devices, thus prohibiting the use of piezo-electric devices; and

Fiberoptic Devices

Fiberoptic devices are similar to the normal fiberoptic cables used to transmit telecommunication signals. When embedded in composite structures, the output light signal of these devices changes in relation to conditions of the surrounding environment. For instance, if the structure moves, the fiberoptic will change in diameter, thus changing some components (frequency, amplitude, phase, polarization, etc.) of the imposed optical signal. These changes in signal can be detected and related to changes in the structure. Other examples of environmental changes that can be detected might be temperature, pressure, rotation, acceleration, acoustics, chemical composition, radiation, fluid flow, and liquid level.

When fiberoptic devices are used for transmitting signals in normal communications, the system operator desires to minimize the very effects that allow them to be used as sensors. That is, the operator wants to minimize changes in the signal, which might be caused by environmental effects on the fiber. Thus, using the fibers as sensors is inherently counter to their use as signal transmitters. This duality obviously complicates the interpretation of the data obtained from using these devices.

The advantages of fiberoptic devices include:

- ability to sense many different effects,
- highly compatible with composites (because both are fibers),
- sense both general and local changes, and
- largely immune to electro-magnetic interference.

The disadvantages of these devices include:

- glass is brittle and easily damaged with relatively small movements, although some polymeric optical fibers may not have this problem, and
- getting the signal out of the device with the proper interpretation (the general nature of the environment must be well characterized—they are really only sensors).

Shape Memory Alloys

Shape memory alloys have a "memory" because of a transition that occurs within them between two separate phases. For example, a metal might switch back and forth between crystal structures, such as between a martensitic phase when cooled and an austenitic phase when heated. This allows the metal to remember the shape it had when it was cool (or hot) even when it is subjected to the other conditions. Therefore, when a material is changed from one phase to the other by changing the temperature, the material wants to change shape. But if constrained, the material will exert tremendous forces on itself. These forces can be used to sense and change the configuration of the assembly in which the shape memory material is placed. These materials are, therefore, used as actuators in smart structure applications. Most of the shape memory alloys are titanium-based, although other materials can also possess these properties.

The advantages of shape memory alloys include:

- ability to exert high forces and generate high strains, and
- they are ductile, thus resisting impact damage.

The disadvantages include:

- limited ability to sense changes,
- sensitive only to temperature changes, which occur slowly, so it is difficult to obtain rapid response,
- susceptible to fatigue, and
- relatively low in efficiency.

Other Techniques

Several other methods have been investigated as components of smart structures. Some of these include:

- electro-rheological fluids, which change their viscosity as a result of an applied electrical field (used primarily for damping);
- electro-strictive materials, which constrict when an electric field is applied;
- magneto-strictive materials, which constrict when a magnetic field is applied; and
- various types of more traditional actuators and sensors in combination.

Barriers to Implementation

Many of the same barriers that inhibit the rapid adoption of composite structures also slow the introduction of smart structures. These barriers include:

- lack of design data (especially the effects of inclusion of sensors and other smart structure components);
- high initial cost (smart structures have the cost of added components to the already high cost of the composite, plus the additional cost of installing the components during manufacture of the smart structure);
- increased durability problems of the composite structure because of the presence of the smart structure component;
- adequate methods for quality control and inspection; and
- repair difficulties.

In all of these, the problems of smart structures are added to the already significant problems of adopting composite materials. This fact is especially true in fields where composites are new, such as in infrastructure applications (for example, civil engineering structures). Some of these challenges will be solved as the promise of smart structures becomes realized. For example, although the initial costs will be high, the costs over the life of the structure might be considerably lower when maintenance, inspection, and structural failure costs are included.

The smart structure devices (components) themselves present additional challenges. For example, some components are notorious for being fragile (especially the ceramic-based devices), which raises the question of long-term durability of the entire system. This also points to the increase in system complexity, which already includes the fibers and matrix. Now sensors, actuators, connectors, processors, communication links, power, and feedback all compound the situation. However, this complexity is not surprising in consideration of another self-sensing and repairing mechanism—the human body. Surely just the sensing components of the body (nervous system) are many times more complex than even the most complex of smart structures.

In addition to the complexity of the system, there are some problems arising with the methods of manufacturing the smart structures. How should the devices be installed? If the device is a fiberoptic line, it can be filament wound along with the reinforcing fibers. However, if the structure is made of chopped strand, that option is not open. Piezo-electric devices are usually installed manually, which raises the question of manufacturing uniformity from structure to structure. Even some seemingly simple manufacturing steps are vastly more difficult because of the presence of the smart structure components. For example, because current technology requires the use of external components, trimming must be done with great care in the presence of leads used to connect the devices to the external portions of the system.

Successful monitoring and modification of physical parameters hinges on the existence and knowledge of a unique relationship between the physical parameters of interest and the properties that can be modified. In other words, what does the data mean and how do you use it? In general, researchers have been able to make test smart structures, put them into service, correctly interpret the data from the sensors, and use that data for structure modification. All of this has been accomplished in real time, at least for some systems. These experiments seem to imply that the promise of smart structures is, in fact, achievable. However, large-scale commercial implementation is still in the future.

Structures that can be made smart while maintaining the majority of their conventional properties offer tremendous advantages over the same structures without the technology. Smart structures are able to respond to changes in the environment, permitting their use in a greater number of applications and/or over broader operating ranges. In addition, structures can be made smart to increase performance while simultaneously saving weight by eliminating over-design. Through in-situ health monitoring of the structure, safety can be increased tremendously and cost cut by extending service life and eliminating the need for frequent in-service structural inspections. Thus the potential for greater safety promises enormous benefits, especially in industries where it is a critical issue, such as the commercial airlines and automotive and construction industries. Liability could be significantly reduced with more accurate failure predictions.

REPAIR

In the early days of composite aircraft structures, it was common to design and build them based on conservative assumptions about material strength and durability. This was because there was significant variation in material properties and many unknowns concerning longevity. Thus the safety or engineering factors were high. In these traditionally "overbuilt" structures, there was a tremendous advantage—inherent damage tolerance.

Today, the design factors are smaller. Laminate schedules and ply orientations are carefully optimized. There is improved control during manufacturing processes such as prepreg or resin injection of fiber preforms, vacuum bagging, and oven or autoclave curing. The resulting products are strong, lightweight laminates with high fiber-to-resin ratios. Laminate thickness is being reduced with these materials and processes; thin-skinned cored structures, using honeycomb, foam, or balsa wood cores are popular. While these can be excellent structures from many points of view, repairs to damaged structures have to be more than just "patches." The repair must actually return the structure to very nearly its original design strength as the rest of the structure cannot just pick-up the load. As the optimal strength limits of the material are approached in well-designed structures, the more critical and difficult repairs become to assure the structure retains its full load-carrying capability throughout its lifetime.

While the initial thrust in making optimized composite structures was initially pursued in aerospace applications, many advanced composite industrial and civil engineering structures are now quite similar in design and fabrication to aircraft structures. These also must be repaired to near original design properties. Therefore, the nature of repair has become increasingly important.

Two main categories of repairs are usually considered.

1. The use of an **external doubler** involves placement of the repair *onto* the damaged area, thus adding to the thickness of the part. This method, at least conceptually, doubles the amount of material. The damaged area need not be removed when using a doubler. A doubler can be either bonded or bolted over the damaged area. Often, especially in thick solid laminates where the composite can tolerate a bolt repair, simply covering the damaged area with another pre-cured composite laminate section or a piece of sheet metal is effective and sufficiently strong to give a long-term structural repair. Close-tolerance bolt holes must be used. Careful attention must be given to the overlap distance, fastener size, the number of rows of fasteners used, and edge distance. Another consideration is the potential for corrosion problems if carbon fiber composites are involved.
2. **Flush-type** bonded repairs are placed *into* the damaged area. In these repairs, a repair patch of uncured laminate is placed into the damaged area and then cured. However, before doing the repair, the damage needs to be removed.

Damage Removal

After the inspection and damage determination is completed, the marked area of damage may or may not need to be removed from the structure, depending on the type of repair to be made.

Damage removal is usually done by either grinding or routing (cutting). For cutting, a wheel or circular saw is recommended. Cutters with reciprocating blades, such as in a jigsaw, should be avoided as the rapid up-and-down motion of the blade causes delaminations along the edges of the cut. The best cutting technique uses a high blade speed and gentle pressure while moving along the cut at a low feed rate. It is best to remove damage in an oval or circular fashion to avoid stress concentrations at corners. If it is necessary to have an odd-shaped damage removal area, ensure that the corners have as large a radius as is practical and avoid sharp corners at all costs.

For carbon fiber structures, a diamond-grit-edge blade is best. These blades are expensive but last a long time, and the cost per cut is the cheapest of any blade. Carbide blades will last a short time, and ordinary high-speed steel blades will be dulled to uselessness almost immediately. For aramid fibers, the best damage removal is by cutting. Minimum fuzz is achieved with a split helix router bit, operated at high speed, typically between 20,000 and 27,000 rpm. A reversed steel-toothed blade in a circular saw also may be used. (DuPont, the maker of Kevlar® aramid fibers, offers a free cutting and machining manual and can provide more information on aramid cutting, drilling, routing, machining, etc., including suppliers of specialty bits.)

Design

After the damaged areas are completely removed, the new repair material can be applied according to the repair design. In an ideal repair, the purpose is to try to match and *not exceed* the original structure's strength, stiffness, and weight. There are obviously other concerns, such as a good long-lasting adhesive bond, the smoothness of the repair, cosmetic appearance, etc. However, structurally matching the original strength, stiffness, and weight are the foremost goals.

In any composite design (including for repairs), it is important to realize that the fibers carry most of the load. The matrix resin is weak, brittle, and serves primarily to transfer the load uniformly through the fibers, keep them in alignment, prevent them from buckling under compressive loads, give the part shape, and protect the part from the environment. These are all important functions, but it is still the fibers that carry the actual structural loads.

In a high-performance advanced composite structure, the fibers are oriented in specific orientations to carry the design loads. The only way to get a repair to carry these same loads is to match the ply orientations of the original structure with the repair plies. Therefore, the original ply orientations and materials must be determined. In the preferred repair design, the original structure will be matched exactly, ply-by-ply.

It is important for the repair to be able to transfer the loads in the remaining original structure into the repair plies, and back out again. The most efficient way to do this is through a large adhesive bonding area, which attaches the repair to the non-damaged structure. The best way to achieve a large bonding area is by creating a gradual, flat taper, called a **scarf**, as the interface between the repair and the non-damaged area. A scarf repair is illustrated in Figure 20-6. This is most often done by sanding away from the edges of the prepared hole.

Stepping is an alternate to a scarf for the interface. A stepped repair is illustrated in Figure 20-7. Although stepping may be occasionally used instead of a smooth scarf, it induces stress concentrations at the corners of the steps and is difficult to achieve without damaging underlying plies.

In conventional marine structures, taper angles of anywhere from 7:1 to 14:1 (length to thickness) are used. In heavily loaded advanced composite structures, flatter angles are often used, anywhere from 20:1 to 60:1 being common. Sometimes aircraft repairs will go out as far as 100:1. Obviously, the flatter the taper angle, the more surface area for the bond and the more gradual the load transfer from the original structure into the repair plies and back out again, and the more undamaged laminate that has to be machined out.

If a relatively flat scarf, for example 40:1, was done that exactly matched the original ply lay-up sequence, somewhere between

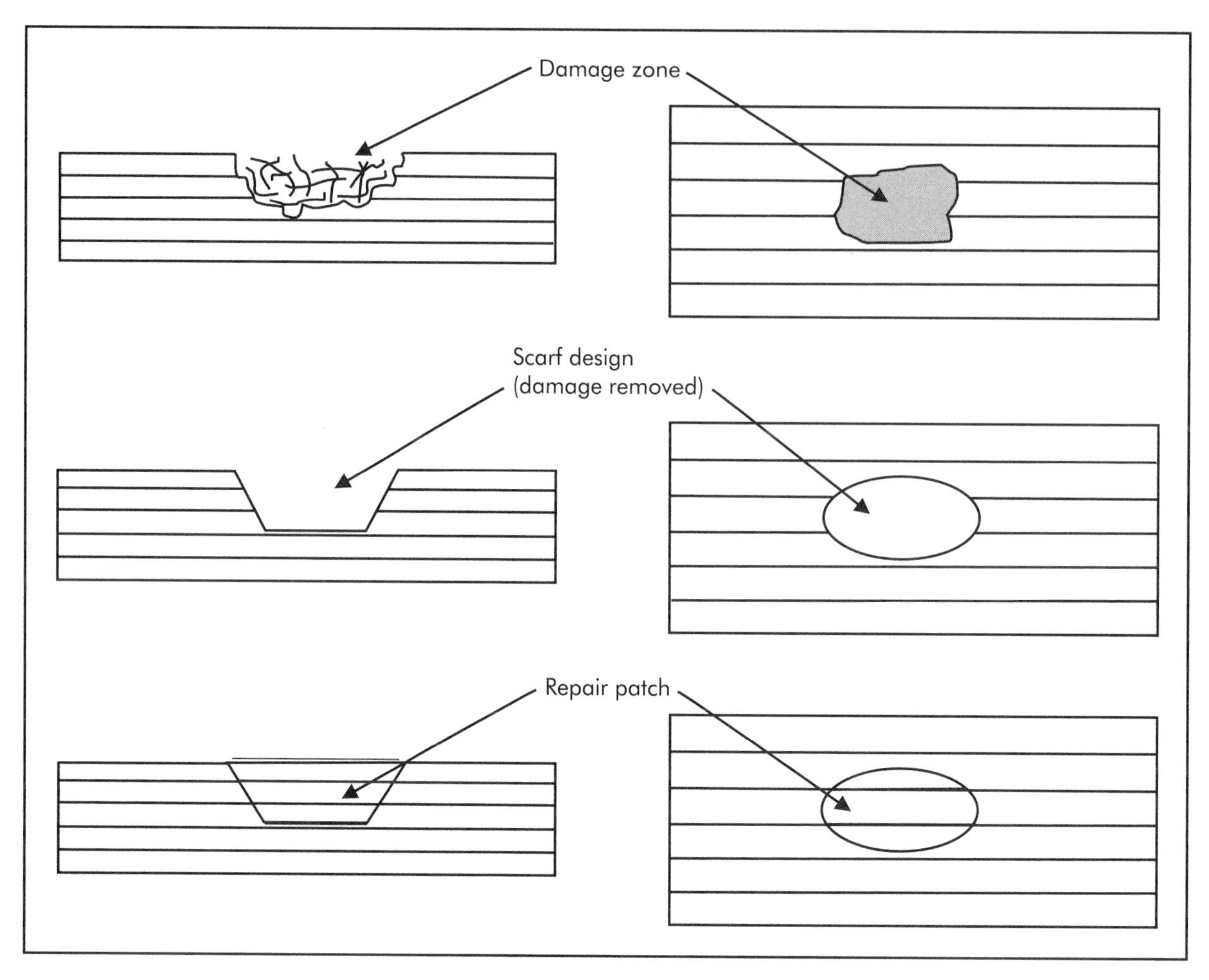

Figure 20-6. Scarf-type repair.

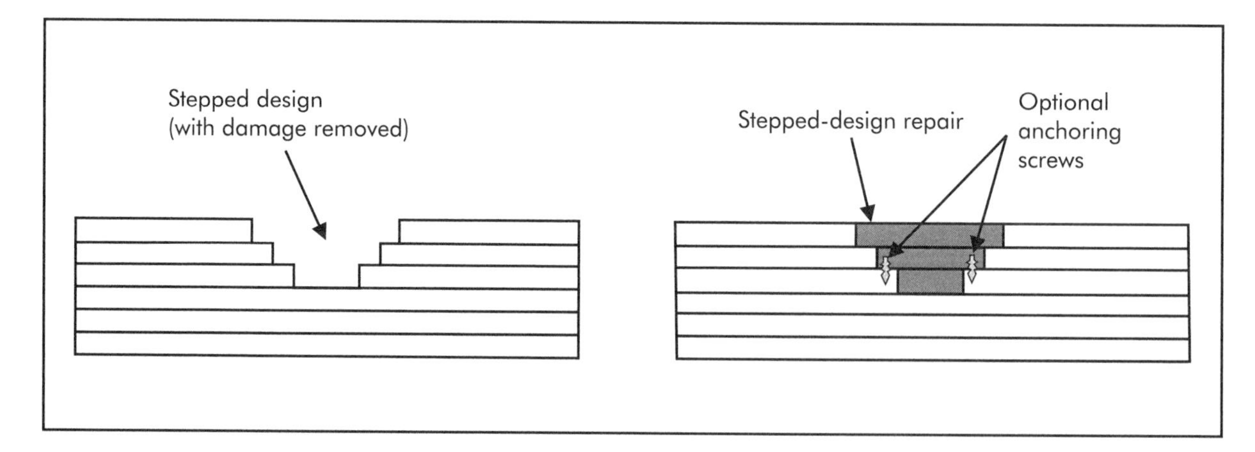

Figure 20-7. Stepped-type repair.

60% and 100% of the original strength (tensile, compressive, and shear) of the part would be maintained. So, for a .25-in. (0.6-cm) thick skin, a 40:1 taper gives a scarf length of .25 × 40 in. (0.6 × 102 cm). The total full strength is not obtained because the fibers in the repair and in the original part are cut. Also, the fibers must transfer their loads through a new structural adhesive bond, which did not exist in the original structure. The fibers themselves do not cross the bond line. Typically, to get back to 100% strength, extra repair plies would need to be added. This increases the weight and thickness of the repair.

Bending stiffness is more a function of the *thickness and ply position* of the structure with some dependence on the type of matrix and fiber (carbon being much stiffer than fiberglass or aramid). All else being equal, bending stiffness goes up approximately with the third power of the increase in thickness. This means that if the repaired area is the same thickness as the original structure, it will have about the same bending stiffness. But if the repair is twice as thick, it will be about *eight times* as stiff! So a full-strength repair, attained by adding extra repair plies, will be *stiffer* than the original. A stiffer repair patch will attract more load and effectively weaken the structure. This may be fine or it may be a disaster!

The consequences of the changes in properties need to be considered carefully. If tensile and/or compressive strength are most important, (for example, in a structural column) then a repair that is stiffer than the original will probably be fine as long as the compliance requirements (in case of an earthquake) are considered. However, a helicopter rotor blade experiences large bending loads. Thus an overly stiff repair halfway out on the blade would be a problem as it would induce large stresses around the periphery of the repair, possibly leading to premature failure.

To give an extreme example of a stiffness-critical structure, consider a pole-vaulter's pole. In use, it is bent through an extreme angle and, of course, also experiences compression and tension loads. If it was damaged and repaired, care would need to be taken to not create a stiff spot, which would cause the pole to unnaturally flex just at the edges of the repair. Such a stiffness discontinuity would cause the pole to fail at the edges of the repair at a much lower bending load than normal. Therefore, to repair a pole-vaulter's pole, a large, flat, scarf area would be prepared with a low taper angle. Such a technique is sometimes called **feathering**. However, since the thickness would be carefully matched to control the bending stiffness, the expected result would be a decrease in tensile and compressive strength.

Bonding

Since heavy structural loads must be transferred across the repaired area through the bond lines, the adhesive bond itself must be structural. The type of adhesive is important, as is the preparation of the surfaces to be bonded. There is no ideal adhesive that will solve any bonding problem. However, epoxies, acrylics, and their derivatives, such as methacrylates, are generally the best for most repair purposes. Polyester resin does not make a good structural adhesive.

Epoxies, while strong, are intolerant of dirty surfaces so meticulous surface preparation is crucial to achieving a long-lasting bond. Bond deterioration due to a poorly prepared surface is seen over time, especially when there is exposure to water. Bond strengths deteriorate much more rapidly when initial surface preparations are poor than with properly prepared surfaces. The difference between a structural bond that lasts one year, and one that lasts 30 years, can literally be the difference in the type of rag used in solvent wiping, or in the cleanliness of the solvent itself.

Surface preparation for bonding is actually a long and involved subject, with many subtle points just now being well understood. But briefly, it is prudent to be *very* clean in preparing the scarfed surface. Do not blow compressed air on it to get rid of dust. Compressed air is full of oil and water. Clean, reagent-grade solvents and clean, one-use only, disposable rags should be used. Surfaces should not be touched with hands after scarfing unless clean gloves are worn. The surface should be thoroughly dry before applying the adhesive. Further, the cleaned surface should be protected with a taped-on piece of paper if the person doing the repair takes a break between steps. The basic sequence for conducting a damage assessment and structural repair is as follows.

1. Gain access to the part from both sides, if possible.
2. Clean and dry the surfaces.
3. Remove any gel coat and/or paint by careful mechanical abrasion, not with paint strippers.
4. Inspect for damage visually and by tapping.
5. Mark out the damaged area.
6. Remove the damaged area of the structure. If it is thick and the damage is not all the way through, then the damaged plies may be ground away until solid plies are reached.
7. Carefully taper (scarf) the surface by sanding away from the edge of the damage. Use at least at a 12:1 taper for lightly loaded structures and up to 40:1 for heavily loaded structures.
8. Ensure that the scarfing is done slowly and carefully by a skilled artisan.
9. After the scarfing is completed, cover the surface with a smooth, taped-down sheet of clean paper or plastic film to protect it from dirt, moisture, grubby fingers, etc.
10. Trace out on additional pieces of clear film the outline of each repair ply so all will neatly fit in the "contour map" created on the surface by the scarfing.
11. Mark on the templates the orientation of each ply.
12. Cut out replacement plies from the same material as the original structure. If this is not available, consult with vendors to find a close substitute. Original materials or very close substitutes are a must for heavily loaded, lightweight, structural repairs with small safety margins.
13. If working with prepreg, apply a layer of film adhesive to the *entirely clean* scarfed area. If wet lay-up is being used, apply a thin layer of the epoxy resin to the scarfed surface.
14. Lay down the innermost, smaller ply first. If there is a hole through the structure, it will need support with a released caul plate on the backside if access is available. If there is not backside access, then one technique is to bond by using an oval hole rather than a circular one with a larger oval on the backside to act as support for the "filler" ply.
15. Continue to lay up one ply at a time, working up toward the larger surface plies. Be careful to match the ply orientations of the original structure.
16. Once the top ply is in place, add any extra repair plies with each being about .5 in. (1.3 cm) larger all the way around than the original top ply. The ply orientation of the extra ply or plies should match the original outer ply.
17. A sacrificial outermost ply of thin fiberglass may be added to act as a sanding ply.
18. Vacuum bag the repair using standard bagging techniques as described in a previous chapter. If using prepregs, if

the repair has more than four or five plies, or if it has a complex contour, then it is strongly recommended to perform intermediate debulking steps using a vacuum bag to pre-compact and consolidate the plies.

19. Cure the repair. Use high temperature, if required, in a controlled oven. If an oven is not available, *carefully* use heat lamps or heat guns and multiple thermocouples to monitor and control temperatures evenly. If a room-temperature wet lay-up is being used to create the repair material, most structural epoxies still require a high-temperature post-cure to develop full strength in a reasonable time. However, some epoxies can be fully cured to reasonable strength numbers at room temperature. These may be fine if high temperatures will not be encountered by the repaired structure in service.
20. After the cure is completed, debag the repair and inspect it for delaminations, de-bonds, or other flaws.
21. Clean it up carefully and sand it smooth (do not cut the fibers in the top ply unless it is a sacrificial sanding ply). Apply paint or gel coat as appropriate.

Doing a repair the right way is not difficult, just time-consuming, requiring a high level of training and attention to detail. However, it is possible to repair critical structures correctly and with good confidence that the repairs will last. It has been done every day in the aircraft business for many years and is now being done in the marine industry commonly.

In addition to curing the adhesive used to make the repair, the layers of material that make up the repair itself must be cured. If the repair is to mimic the properties of the original laminate, it must be made of the same materials. Further, it must be cured using the same cure-cycle parameters as the original laminate. The difficulty with this situation is that the repair is contained within a much larger part, which cannot and probably should not be put into an autoclave. For this purpose, portable curing systems, which cover only the repaired area, have been developed. Figure 20-8 is a diagram of such a heating system. This system uses a vacuum bag (normally reusable) to cover the repair area. The system is sealed with vacuum "tacky" tape and then a vacuum and heat are applied. The cure cycle is set by a controller; thus a precise thermal curve is developed.

When repairs are made to aircraft, the Federal Aviation Administration (FAA) requires that a record of the repair be made. That record must show the cure cycle of the repair material. Thus these portable repair systems usually have a recorder for the temperature and the time.

CASE STUDY 20-1

Self-healing Composites

Living bodies are amazing organisms. When they are damaged they can heal themselves. Inspired by that capability, researchers at the University of Illinois Urbana-Champaign have developed a composite material that can also heal itself. It even bleeds for a short while before completing the healing process.

The self-healing material is made by mixing epoxy resin with a small amount of organometalhalide (Grubb) catalyst and microspheres containing the "healing agent" (dicyclopentadiene [DCPD] monomer). When the epoxy is cured, both the catalyst and the microspheres are encapsulated in and distributed throughout the matrix. If the composite receives a blow of sufficient magnitude to initiate a micro-crack, the microspheres rupture as the crack expands through the matrix. The rupture of the microspheres releases the DCPD monomer, which "bleeds"

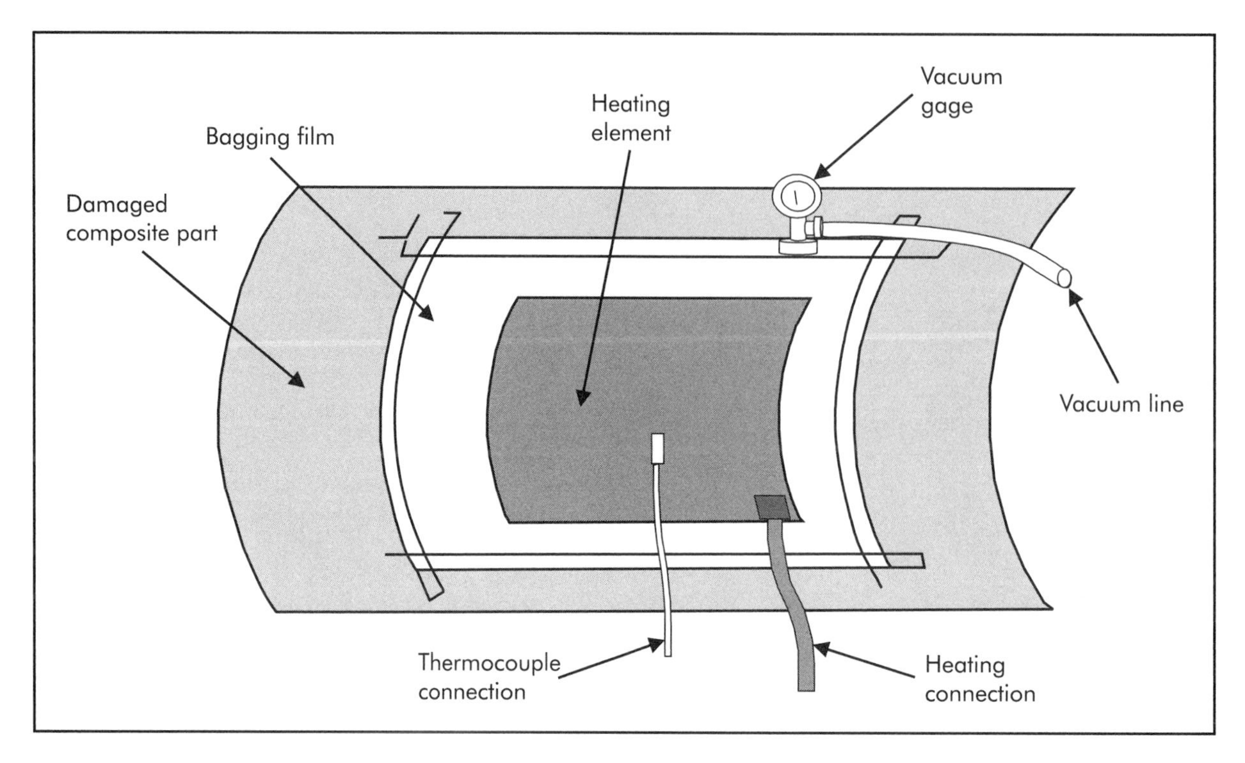

Figure 20-8. Portable heat system for curing the repair area.

in the area of the crack. As the DCPD bleeds, it comes into contact with the catalyst, which is distributed throughout the matrix. The catalyst triggers polymerization. Because of the rapid polymerization capability of DCPD, the resin quickly solidifies and bonds the sides of the crack together. Hence, the crack is "healed." This situation is depicted in Figure 20-9.

The secrets of self-healing technology lie in the choice of material and strength of the walls of the microcapsules. If the walls are too thick, they will not rupture when the crack approaches. If they are too thin, they will be broken in processing. The toughness and stiffness of the walls of the microcapsules and the strength of the bond between the microcapsules and the matrix are relevant control parameters. The amount of catalyst and the concentration of microcapsules are also important considerations.

The healing mechanism has been confirmed by optical and scanning electron microscopy. Mechanical testing has revealed that, with some optimization of the critical parameters, 85% healing efficiency (recapture of original fracture toughness) was obtained.

This autonomic (from the same root as "autonomy") healing technology has obvious applications in products such as tubs and showers, which are susceptible to thermal cracking, electrical applications where cracking might cause electrical failure, in fatigue-prone products, and in any application where hidden cracks might shorten product life.

SUMMARY

Because composites are usually brittle materials, impacts often result in damage to the fibers and the matrix. When composites have been optimized for weight and perfor-

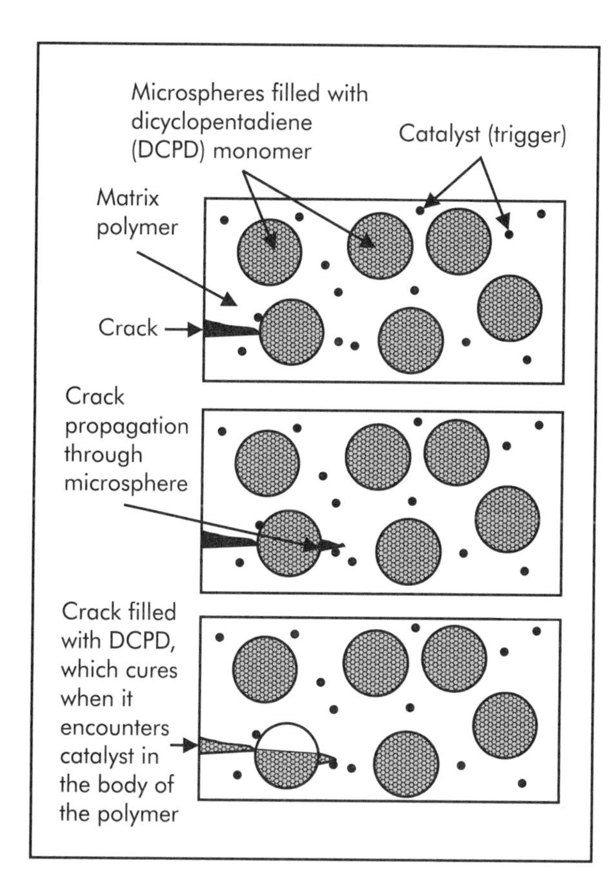

Figure 20-9. Self-healing composite.

mance, and especially in critical applications, this damage can be catastrophic. Therefore, great efforts have gone into preventing damage or into reducing its effects. Sometimes, without much of a performance penalty, material changes can be made to reduce the effect of the damage. It is also sometimes possible to change the design of the part to reduce damage effects.

The reality is that damage is inevitable. So, methods have been developed to inspect and quantify it. These methods also help to classify the damage so that the repair process can be more effective. The repairs attempt to return the properties of the composite part to its original values. This can rarely be done because the fibers in the repair are shorter than in the original. Also, because an adhesive layer holds the repair patch, there is inevitably lower performance than if there were no adhesive present. Nevertheless, repairs can come quite close to their intended targets if they are carefully designed and executed.

LABORATORY EXPERIMENT 20-1

Scarf Repair

Objective: Compare the strength of a repaired composite to the original material.

Procedure:

1. Make two composite laminates that are as nearly identical as possible. It is suggested that the laminates be about 10 × 10 in. (25 × 25 cm) and about .25 in. (0.6 cm) thick and constructed from layers of epoxy prepreg.
2. After curing the laminates, set one aside as a control.
3. The other laminate is supported so that it has space underneath. It is then impacted. (The sample can be supported on a metal ring of a reasonable diameter and then impacted with a weight sufficient to cause visible damage.)
4. Inspect the damage visually and with the tap test to identify the extent of the damage.
5. Remove the damage area using the scarf technique.
6. Place uncured epoxy prepreg into the scarf area using the techniques of fitting and pressing.
7. Cure the repair materials.
8. Cut tensile specimens from the control laminate and the repaired laminate. Be sure the repaired area is in the tensile specimen from the repaired laminate.
9. Perform the tensile test on the specimens and compare the strength, stiffness, and elongation of the original and repaired laminates.

10. Repeat the experiment with different scarfing angles and change the lay-up technique by laying the largest ply down first as opposed to laying the smallest ply down first.

QUESTIONS

1. What is the tap test? List two precautions that should be observed to make the test more accurate.
2. What is a scarf repair and how does it differ from a step repair?
3. Describe feathering. What is the advantage to its use?
4. What are two advantages of using a standard damage classification system?
5. Which lay-up would you expect to have the smaller impact damage, [0/45/0/45] or [0/45/45/0]? Why?
6. Why does a hybrid composite suffer less damage than an otherwise equivalent single fiber composite?
7. Explain how a self-healing polymer works.

BIBLIOGRAPHY

Kessler, M. R. and White, S. R. 2001. "Self-activated Healing of Delamination Damage in Woven Composites." *Composites*, Part A32, pp. 683–699.

Society of Manufacturing Engineers (SME). 2005. Composites Manufacturing video series. Dearborn, MI: Society of Manufacturing Engineers, www.sme.org/cmvs.

Strong, A. Brent. 2006. *Plastics: Materials and Processing*, 3rd Ed. Upper Saddle River, NJ: Prentice-Hall, Inc.

White, S. R., et al. 2001. "Autonomic Healing of Polymer Composites." *Letters to Nature*, Vol. 409, 15 February, pp. 794–797.

21 Factory Issues

CHAPTER OVERVIEW

This chapter examines the following concepts:

- Problem of emissions and how to deal with them
- Governmental regulation
- Material storage
- Contamination in the plant
- Disposal, waste, and recycling
- Factory simulation

PROBLEM OF EMISSIONS AND HOW TO DEAL WITH THEM

Some aspects of the information in this and succeeding sections have already been discussed in earlier chapters. However, because this chapter may be examined by itself and because a stronger focus is needed on some of the points, the material is treated here again in some detail. New material is also presented, which is unique to this chapter.

Purposes of Styrene

Styrene is added to unsaturated polyester resins and vinyl ester resins for two major purposes:

1. to dilute the mixture for effective viscosity control so that the fibers can be wetted; and
2. to serve as a crosslinking agent for the resin.

Materials, like styrene, which simultaneously accomplish both purposes, are called **reactive diluents**. Other reactive diluents exist, such as methyl methacrylate. However, because of the predominance of styrene in the unsaturated polyester and vinyl ester resin market, the focus in this section will be almost exclusively on styrene.

The ability of styrene to accomplish both purposes has resulted in lower costs overall and improved performance and manufacturing efficiencies. Imagine, for instance, a diluent that was non-reactive. This would require that the diluent be removed during molding. It could not be effectively removed after molding because the cured resin would be so stiff that the solvent would be trapped inside. Only long and careful heating would allow the solvent to be extracted from such a solid part and, then, only at the risk of part degradation. Several epoxy and other high-performance resins must be solvated and the solvent removed during curing, and, as a result, are difficult to use and more costly.

Generally, the excellent reactivity of styrene during crosslinking results in it being used up during the reaction. Styrene forms a part of the cured polymer, generally at a lower cost than the backbone resin system itself. When it has been crosslinked, styrene imparts stiffness to most cured systems (and some brittleness) and generally enhances the physical properties of the finished product.

Besides being effective, styrene is also a relatively inexpensive diluent. It is able to easily dissolve most of the important resins and, thus, contributes to easy wet-out of the reinforcement. Hence, styrene is nearly ideal from a functional standpoint. However, styrene is listed as a **hazardous air pollutant (HAP)** because it forms a potentially dangerous vapor. Therefore, the amount of styrene emitted in the manufacture of fiber reinforced plastic (FRP) parts must be monitored and controlled. Recent federal government regulations in the United States and companion regulations in Europe and many other countries are now forcing reductions in the level of styrene vapors permitted in the air, and the exposure time that workers are permitted to accumulate during the course of their work.

Reducing Styrene Content

There are some strategies being pursued by resin manufacturers to reduce the styrene content of their products. However, the explanation surrounding them is complex and beyond the scope of this book. Therefore, the focus here will be on the direction and concepts behind the changes rather than the details.

Modification of Existing Polymer Systems

The initial effort by almost all of the manufacturers of unsaturated polyesters and vinyl esters was to modify their existing resin systems so that less styrene would be needed. This strategy was not only prudent commercially, but it also gave quick results.

The resin manufacturers realized that if the viscosity of the base resin were lower, the amount of styrene needed to achieve good viscosity for fiber wet-out could be reduced. The first method to reduce the viscosity of the base resin was to shorten the molecular chains, that is, to reduce the molecular weight of the resin. Shorter chains have less intermolecular entanglement and, therefore, flow more easily. This means that the viscosity of the resin is lower. The amount of styrene needed in these lower-molecular-weight resin systems has been successfully reduced. However, the reduced level of styrene dictated by viscosity control requirements has also reduced the amount of styrene available for crosslinking. Therefore, the nature of the crosslinking has also changed. The question is, "How has crosslinking changed the resulting polymer's properties?"

Lower overall styrene content and higher crosslink density occur simultaneously with low-molecular-weight resin systems. The lower amount of styrene available for crosslinking results in a "starved" situation during crosslinking. This starving means that the number of styrene molecules in each of the crosslinks is reduced. (Under excess styrene conditions, multiple styrene molecules join together to form long crosslinks, whereas under starved conditions, each crosslink is formed with only one or two styrene molecules.) The reduction in the total amount of styrene contained in the crosslinks, and therefore, in the cured polymer, would normally lead to lower overall brittleness since styrene is inherently brittle. However, the shorter polymer chains mean that the carbon-carbon double bonds occur relatively more frequently, resulting in a higher number of crosslinks being formed overall. This higher crosslink density normally increases brittleness. However, the effects on brittleness from the reduction in styrene content and the increase in crosslink density approximately cancel one another. Thus the brittleness is about the same as it would be in the unmodified resin system.

Generally, the heat distortion temperature and solvent resistance will be increased by higher crosslink density. Other properties, such as blister resistance, ultraviolet light resistance, chalking, cracking, and surface smoothness, also may be affected. The extent of the effects must be evaluated on a case-by-case basis with each resin and application. The effects are much more important in gel coats than in laminating resins. So, before changing gel coat resins, the effects need to be evaluated carefully.

Another modification being made to current resin systems is a change in molecular weight distribution. This is not just a shift to lower molecular weights. Rather, it is a change in the statistical spread of the molecular weights of the polymers. Even when the average molecular weight is still high, the presence of a significant quantity of low-molecular-weight molecules has a tendency to make the molecules slide over each other more easily, thus lowering the viscosity. This lower viscosity allows less styrene to be used. Therefore, the molecular weight of the resin can be reduced and the molecular weight distribution can be skewed to have more low-molecular-weight component. In practice, these changes can be made without significant changes in the properties of the resin. But there is a limit to which these changes are effective so other methods of reducing styrene also may be needed.

Developing New Polymer Systems

Polyesters are made from the reactions of diacids and glycols. Resin producers have found that by varying the types and quantities of diacids and glycols, resin properties can be significantly changed. (For example, the differences between ortho and iso resins lie in the choice of acid that is reacted.) Therefore, many possibilities exist for changing the nature of the base polyester resin.

In addition to different combinations of previously used components, some completely new diacids and glycols are being tried. Some resin manufacturers call this the clean-slate approach. The resin manufacturers are reluctant to disclose the details of these changes because of competitive reasons. However, it is safe to say that all of the major resin manufacturers are working on innovative new resin structures that will enable the styrene content to be reduced. It is anticipated that within the next few years, many new products will be announced.

Without disclosing any proprietary information, a few basic concepts can illustrate some of the directions the resin producers are investigating. Some of the acids and glycols (like neopentyl glycol) are known to be more soluble in styrene than the other common components. This increased solubility comes from their molecular shape and chemical polarity effects. At the same time, these alternate components might give additional benefits such as increased toughness or higher heat distortion temperature. Therefore, the new resins might have not only reduced styrene content but improved physical properties, or at least no major reduction in properties as a result of the decrease in styrene.

Some resin manufacturers have found that new resin systems requiring lower styrene concentrations can be made by varying the sequence of addition of the diacid and the glycol components. These changes occur because mixtures of multiple types of diacids and glycols, which have different reactivities, are often used in making the polyester resins. An example of such a change would occur under these two scenarios. If two diacids are mixed together and then added to the glycols, the distribution of the acids could be random. However, if one of the diacids is added to the glycols and then the second is added later, the result could be preferential positioning of the second diacid at the ends of the polymer. Hence, two very different polymers could be formed.

The sequence of components along the backbone resin also can be altered by changing the addition sequence because of differences in the reactivities of the components. A particular component could be forced to group together or to be dispersed to the ends depending on its relative reactivity with respect to other components and when they are added. Some have called these changes one-stage and two-stage additional polymers.

Using known principles and new technologies, the resin manufacturers are attempting to create shapes and chemical natures that favor lower styrene contents.

Occasionally, a polyester resin is made using a component other than a diacid and glycol, such as dicyclopentadiene (DCPD). This material can be used to react with maleic acid and glycols to form a low-molecular-weight chain of only a few units. Such units are often called **adducts** because a few units are "added together." The net result of mixing DCPD with maleic acid and ethylene glycol is an adduct of DCPD—maleic, glycol, maleic, DCPD. This adduct has low molecular weight and, therefore, a low styrene requirement. The net result is a resin with fast cure cycles, good reinforcement print hiding, good fiberglass wet-out, good moisture resistance, good stiffness, and good air bubble elimination when compared with many conventional resins.

The properties of a vinyl ester also can be changed significantly by modifying the nature of the backbone. Vinyl esters tend to have more options available for doing so than unsaturated polyesters. Of course, the specific natures of the changes are different, but the concepts discussed in this section are similar between vinyl esters and unsaturated polyesters.

Another major area of focus is the blending of unsaturated polyesters and vinyl esters with other polymer types. These blends can be done as monomers (often referred to as *copolymers*) or as blends after full polymerization has been done (referred to as *mixtures*). Some of the interesting blends with unsaturated polyesters include polyurethanes, melamines, phenolics, epoxies, and vinyl esters. In general, blended polymers are higher in cost than the basic polyester, but the properties of the blend can be substantially better, thus allowing for the use of less resin or discovering new applications.

Substituting Other Monomers

Styrene is not the only monomer that acts as a reactive diluent for unsaturated polyesters and vinyl esters. Therefore, some alternate monomers, which have lower vapor pressures, are being investigated as styrene alternatives. One approach is to choose monomers that are similar to styrene in chemical reactivity but that have a higher molecular weight and, therefore, a lower vapor pressure. For example, vinyl toluene is similar to styrene (having just an additional methyl group on the benzene ring), but has a lower vapor pressure and thus lower air pollution potential. Another example is t-butyl styrene. A major limitation to using alternate monomers is their higher cost.

Another approach is to simply combine two or three styrene molecules together and then use these dimers and trimers as the reactive diluents. (Short chains such as these are sometimes called **oligomers**.) Performance of these materials is almost the same in the cured product. Wet-out of fiberglass remains excellent since only enough of the dimers and trimers is added to achieve the same viscosity used with the monomer alone. However, the cost of producing the dimers and trimers is significant.

Some manufacturers have been looking at using completely new monomers that are not like styrene at all. The prime example is methyl methacrylate (MMA), sometimes known as acrylic. This material also has a high vapor pressure and so it, too, is considered a hazardous air pollutant (HAP). However, MMA may be more compatible with some polymer systems and, therefore, allow lower concentrations to be used to achieve acceptable viscosities. MMA also offers some improved properties such as lower smoke generation. Again, cost is a problem.

Suppressants

Yet another strategy is to form a surface layer to suppress the emission of styrene. Layers such as this, often formed by adding wax to the resin mixture, have been used for many years to exclude oxygen so that cures can be improved. Hence, this concept is really using an old technology for a new purpose. The problem with this method has always been the potential for poor bonding wherever the wax is present. This adhesion problem has been alleviated somewhat by creating adhesion-promoting films. These films contain carbon-carbon double bonds, thus allowing crosslinking to occur through the film. Adhesion between materials on both sides of the film is thus promoted.

Film-forming materials work best on horizontal surfaces and when the materials are quiescent (that is, non-turbulent). Hence, if sprayed, they will form best if allowed to sit for some time. Non-horizontal surfaces will form some films, but they are not as effective. To alleviate these problems, some of the film-forming systems contain excess thixotropic agents to ensure that the resin stays on the vertical surfaces and becomes quiescent.

The use of highly thixotropic materials can, in themselves, reduce solvent escape. The paint industry has successfully used this technology for some time. This is why some paints are thick when in the can but smooth and spread easily when brushed, rolled, or sprayed.

GOVERNMENT REGULATION

Business executives laugh (and cringe) at the thought of receiving a call from the government saying "Hello, I'm from the government and I want to help." For most people, the fewer contacts with the government, the better. It seems that all government officials want to do is collect taxes or shut down a company for some alleged violation. And, if a company does receive such a call from the government, who is there to help the company? Seemingly, no one feels qualified enough to handle it. The company is compelled to turn to a lawyer to fight the government, and that means lots of money wasted in what seems like an unnecessary and counter-productive effort.

One of the reasons why businesses are so distrustful of government is the activist role that it (as a bureaucracy) and many of its employees (as individuals) have assumed. While it is good to have some element of society leading in various progressive causes, it seems that these activists assume that their opinion is "right" and that all others are self-serving and reactionary. The idea that they might be wrong just does not occur to them. When officials have this attitude, alienation of those with different opinions is inevitable.

To make matters worse, governmental activists believe that they are actually helping your company and society, even though you may not agree with or appreciate their actions. It is like your mother giving you a spoonful of medicine "for your own good." C. S. Lewis (Lewis 1979) articulated well this problem when he said, "Of all tyrannies, a tyranny sincerely exercised for the good of its victims may be the most oppressive. It may be better to live under robber barons than under omnipotent moral busybodies. The robber baron's cruelty may sometimes sleep, his cupidity may at some point be satiated; but those who torment us for our own good will torment us without end, for they do so with the approval of their own conscience."

Some people seem to have entered governmental service because they see government as a place of power and influence. Not only are they able to enact the changes that they envision, but they also experience the thrill of power when, in other settings, they might not be able to have such power. This power often comes because of their faithfulness to bureaucracy rather than their insight and performance in directing and leading change. But, the picture painted here is not the total view.

Historical Perspective

The role of government has been and can be quite beneficial to industry. A look at past interactions of government and business may help to form some suggestions for improvement of the currently antagonistic situation. Interestingly, the origins of trade associations and professional societies can be seen in the historical view.

In the early Middle Ages (approximately 500 to 1000 A.D.) most people lived on small, meager farms that were grouped around tiny villages. There was little travel because the barbaric tribes that had conquered the Western Roman Empire regularly robbed travelers and traders. Hence, each family was forced to produce its own food, wool, cloth, clothes, leather, shoes, and farm implements. It was a time when the family was the principle production unit.

About the year 1000, the implementation of the feudal system resulted in less thievery and plunder on the highways. People could travel with reasonable safety and goods could be shipped. Cities began to be repopulated and individuals who were especially adept at a trade moved to the city where they could become artisans in that trade.

The artisans generally produced higher-quality products than did the common family producer. They sought to establish a quality standard and raise the price so their special skills could be rewarded in the marketplace. To accomplish this, they formed trade associations, called guilds, which solicited legal protection from the local governments. The governments granted the guilds exclusive rights to sell their products within the confines of the city and allowed them to specify the price of the products; but for those privileges, the guild members had to guarantee a high standard of quality. The guilds were also required to establish a system of training so that the goods could be made long term with high quality. The apprentice system was the result. Individual artisans became the primary makers of goods (since this was before factories were created). The guilds became so powerful that they dictated the nature of government policy toward manufacturing, along with monetary policy (or more) in most medieval cities. The power of the guilds can be appreciated in cities such as Brussels where a visitor to the central square of the city sees that the guild halls share the perimeter of the main plaza with the city hall and the church.

With the coming of the Renaissance, two factors contributed to the weakening and eventual demise of the guild system. The first of these was the rivalry and competition that grew among the members of the guilds. The prime example of this was Michelangelo. He was a member of the sculptor's guild in Florence, but left it when his apprenticeship was completed. He believed that the quality standards established by the guild were restrictive on his creativity and, therefore, he refused to be bound by its rules. Since his work was obviously excellent, the government had little choice but to allow him to sell his goods even though he was not a member of the guild. Hence, the monopoly of the guild was broken. Similar situations occurred throughout Europe over the next hundred years or so and thus the guilds declined in power.

Another factor was equally important in reducing the power of the guilds. This was the rise of power and prestige of the kings and their desire to control the economies of their countries. Kings in this period (roughly 1500–1750) sought to purchase goods at preferred rates and/or with high quality. Therefore, the kings named a particular company as "purveyor to the crown," which gave that company a competitive advantage over all others. Sometimes the kings or their preferred nobles even took an ownership position in these companies. Kings would also give trading companies exclusive territories (such as the Hudson Bay Company and the Dutch East Indies Company) in efforts to increase the total exports of their countries. This system of government manipulation of industry and trade was called mercantilism. Because the crown favored some companies at the expense of others, those companies gained great power and did not need to be nor did they want to be part of a guild. Again, the power of the guilds was weakened.

With the rise of the Industrial Revolution (about 1750–1900), the formation of factories resulted in far too many companies with excellent quality and prices for any king to have only a few favorites. Simultaneously, the merchant class (which included the industrialists) grew in political power so that the kings were forced to retreat from favoring specific companies. The merchant class took over control of the government and forced the kings into limited monarchies. The political and economic policies of this new merchant class were formed by writers, such as Adam Smith, who wrote the classic, *Wealth of Nations,* in 1776. Smith and others advocated that the government stay out of direct involvement in industry and just let the natural forces of the marketplace dictate what would occur. This policy was called *laissez faire*. With the growth of free markets, the ability of guilds to monopolize trade was over.

The economies of the European and American nations during the Industrial Revolution were strong. Clearly industry thrived when given the freedom to do so (subjected only to the natural forces of the marketplace). Eventually, however, the greed and shortsighted practices of industrial leaders of the day resulted in atrocious conditions for workers and high prices and poor quality for consumers. These were the days of the "robber barons," child labor, and the monopolies. The public was outraged and demanded that something be done.

Governments, which were responsive to the publics they served, took two different courses of action to control industry. One, guided by the doctrines of Karl Marx, was to control industry by governmental ownership. In various forms, this system, Socialism, still exists in parts of Europe, Africa, Asia, and Latin America. The other course of action was for government to become an opponent of industry. That was the path taken by some governmental officials in the United States. Today, the activists within government still see industry as a great evil that must be controlled. Hence, government and industry are often confrontational in the United States.

Negotiations between a single company and the government are often not balanced. The government, with nearly unlimited resources, can intimidate almost any company. Hence, companies need some assistance in their negotiations. That is a role that trade associations can fill. For example, the American Composites Manufacturers Association (ACMA) is the entity negotiating with the government on styrene emission standards. There is hope for amicable relationships between governments and industries when interactions are facilitated by associations.

Role of Government

Government officials must first remember that they are ultimately responsible to the public. There should be no disparaging of the "common man" or diminution of the importance of doing what the public feels is correct. This country was founded on the principle that government derives its powers from "the consent of the governed" and so government should be highly responsive to public wants.

Government officials should also remember that, historically and today, the pool of resources on which government depends is based, ultimately, on the industry of the nation. People pay taxes out of the wages they receive from industry. Industry pays taxes out of their profits. Hence, industry is an important (fundamental) constituency of government.

Historical precedent and modern practice suggest that government has a legitimate role in assisting and regulating industry. This should be done fairly, that is, without favoring one company at the expense of its competition. Further, government should help industry improve. One program already working to accomplish this is the Manufacturing Extension Partnership (MEP) sponsored by the Department of Commerce through the National Institute of Standards and Technology (NIST). The MEP receives significant support from the federal government (and usually from state governments) to assist companies. This help is often initiated through assessments conducted by MEP engineers and then establishment of programs designed to help companies where needed.

Government also has the responsibility to create an infrastructure that supports and promotes industry. Roads should be good. Utilities, which are government regulated, must be fair in their pricing and current in their technology. This governmental responsibility can be critical to a company's success as can be seen most clearly in countries where such infrastructure support is not present. Great efforts are underway to build the infrastructure in China to meet the needs of that country's industrial growth.

Although not necessarily a problem in the United States, corruption in government is rampant in some other parts of the world. This should not be tolerated. Where corruption exists, international pressure and internal demands by the populace must urge the government to purge itself.

Role of Industry

Each industrial executive strives to make a profit for the investors in the company. This system, capitalism, has proven to be highly beneficial to the population in general and is, therefore, the dominant industrial system in the world today. However, these profits must be made responsibly. The most responsible method is to have a long-term plan for the company that is responsive to customers and the public in general. Just as with government, the public's needs and desires should be the ultimate focus of the company. Therefore, environmental concerns, good quality, fair prices, and reasonable salaries should be part of management's focus.

Most companies pay from 30 to 50% of their income to the government in the form of various taxes. That means, to a real extent, that the government is a partner in the company. Too often company executives see their profits drained away, not considering what the government has done to "earn" their share of profits. In reality, the government has done much. They have created the climate for success. If you doubt this, go to a third-world country and see the problems that a company must endure just because the infrastructure and economic climate are not friendly to companies. This includes everything from roads to utilities to banks to education. All are affected by the government and its policies toward business.

Industry also has the role of improving the environment in which the company exists. This does not just mean eliminating air and water pollution. It means supporting the little league team, promoting education, and helping the local charities. Why should a company do this? Because these acts will improve the culture in which the company operates. After all, a company is responsible to the ultimate customer—the public.

Role of Associations and Societies

It has already been indicated that associations should represent industry in some governmental negotiations. Conversely, the association should also represent government to its member companies. That two-way communication, often facilitated by the association, is key to improving government/industry relationships.

Other roles for associations (and professional societies) include being a forum where technology can be exhibited. Usually associated with these exhibitions are training and educational functions—another important role for associations. The social aspects of such meetings are also important and should be fostered. Magazines and journals are important for keeping the members informed and provoking new ideas. In the world of composites, the Society for the Advancement of Materials and Process Engineering (SAMPE) has been especially active. In general manufacturing and also composites, the Society of Manufacturing Engineers has been a valuable source for information on manufacturing concepts of all types.

The New Relationship

The cooperation between government, industry, and associations is not without precedent. After all, the three branches of government in the United States have a similar separation and interrelationship. They check each other and balance the power but they do not, necessarily, have mutual hostility. In fact, at most levels, and assuming partisan politics can be overlooked, the three branches interact with extraordinary cooperation. Each branch understands its role and respects the roles of the other branches.

Just as with the branches of government, industry and government are both ultimately responsible to the same people—the public—and so they should seek ways to benefit the public. This common ground will help them see their different roles as important.

A helpful model is to view things on both a macro and micro scale. Government is responsible for the macro scale and does its best work at that level. At this level the climate for doing business is established. Industry operates at the micro level. They have direct control over their products, the interactions with their customers, and their local environment. Only rarely and reluctantly should government enter the microsphere. If it does so, it must remember that it enters only because the public is insisting that it do so. Associations facilitate the interactions between government and industry. They also help industry with functions that are too costly or too difficult for the company to do alone.

How does the individual help? First, an effort should be made to understand the relationships as they should ideally exist. Work within your company or within government to fulfill the role that is appropriate. Second, elect individuals who see the proper role of government. It is sometimes too easy for a politician to criticize and suggest radical solutions that are outside the legitimate realm of governmental action. Turn these politicians away. Third, join an association (or, for individuals, a professional society) and stay active in it. Support it with time and personal contributions.

There is hope for a better world through respect and cooperation!

MATERIAL STORAGE

Another aspect of composites manufacturing that has some governmental control is the storage of hazardous or potentially hazardous materials. The government has suggested certain amounts of material that can be safely stored within the active work zones of a factory. The purpose of these regulations is to prevent excessive exposure of workers to these materials. The reasoning is that when the material is inside the work areas, the potential for leaks and other accidental exposure is high. Therefore, to minimize that danger, the amount of material that can be stored in the areas where people routinely work is regulated.

Many of the governmental restrictions regarding material storage have been reinforced by insurance agencies. They see not only the dangers from exposure but, because many of these materials are highly flammable, the dangers from fire. Companies would be wise to investigate the maximum recommended storage amounts of all the materials they use.

Besides the mandated or regulated storage amounts for materials, some storage practices are just wise. Some examples follow.

- Resins should be stored in temperature-controlled facilities. If the temperature rises excessively, even resins protected with inhibitors can cure prematurely. When the curing reaction begins, the exothermic effect can cause rapid heat buildup and result in a dangerous, possibly flammable situation. Cautious storage practice suggests that means be provided so the resin tank can be emptied into a container large enough to spread its bulk over a large area and thus dissipate the exotherm. A method of doing this is to have the tank in which the resin is stored surrounded by a concrete retaining wall. The wall also functions to retain any leaks or spills that might occur.
- Resins should be stored out of direct sunlight. The resin container should be opaque (they are generally metal) so that ultraviolet (UV) light does not strike directly onto the resin. Even in the absence of peroxides, UV light can initiate the polymerization reaction.
- Most resins, but especially gel coats, should be gently stirred during storage. When fillers are part of the resin mixture, as is often the case with gel coats, these fillers have a tendency to settle out of the mixture. When this happens, they are sometimes difficult to get back into suspension. If the solution is not uniform throughout, the properties of the cured resin and the cure itself can be compromised.
- Normally, peroxide initiators are shipped in non-returnable 1-gal (4-L) polyethylene containers (four containers to a case) and 5-gal (20-L) polyethylene containers. Excessive quantities of peroxides should not be stored because of the inherent instability of these materials. They should be stored below 86° F (30° C). Therefore, a refrigerated storage facility is often required. Two opinions exist relative to the location of the initiator storage facility. Some people think that it should be located outside the main production building so that if an explosion were to occur, the building itself would not be damaged. The other opinion is that the initiators should be stored inside the building because it is air-conditioned and overheating through a failure in the refrigeration unit or a mistake in leaving the unit open will be less likely to cause an explosion. Most experts support the outside storage concept. Therefore, safety alarms should be installed to signal if the temperature of the refrigeration unit exceeds a safe level.
- Great caution should be taken to ensure that peroxide initiators and chemical accelerators, such as cobalt naphthenate, do not mix directly. Further, they should not be stored in the same location.

CONTAMINATION IN THE PLANT

Many contaminants, such as dust, water, mixed materials, and foreign objects, are often obvious. Therefore, they can be eliminated by good upkeep of the facility, putting lids on containers, vacuuming rather than sweeping in the vicinity of easily contaminated materials (like resins), and generally being aware of the problems that can occur if materials are contaminated.

However, some contaminations are much more subtle. The most important of these are contaminants that are easily transferred by contact or those that are airborne and travel through the air conditioning of the plant. The most troublesome of these is silicone-containing mold releases. Silicone materials have a reputation for migrating through the plant, thus cross-contaminating other parts. Silicone contamination makes subsequent bonding of molded materials difficult or impossible because it interferes with the ability of the adhesives to stick to the molded part.

The contact transfer occurs most often because a person touches a part that has been molded using a silicone mold release. Then, this person touches another part or an area where bonding is to be done. This problem is made even worse when cotton gloves are used in handling the parts, as is the general rule in the industry. The gloves retain the silicone for long periods of time and so contamination is routine.

Silicone molecules are carried into the air through spraying or, simply, evaporation of the solvent or water carrier in the mix with the entrainment of some silicone molecules. These silicone molecules then migrate through the air in the plant (sometimes even into the air conditioning system), contaminating parts that are not directly involved with the mold release. Migration through the air is especially great if the mold release is sprayed onto the surface. The problem with cross-contamination is further worsened when low-molecular-weight components are present as these are more easily volatilized.

Some mold release companies have suggested that their mold releases can be "fixed," thus decreasing the amount of cross-contamination. This process is said to create a "semi-permanent" mold release. If that fixing occurs by heating the mold or if they are applied to a heated mold, the problem of contamination may be actually made worse during the initial stages of fixing since heat causes volatilization of the silicone. Even if the fixing of the mold release occurs through chemical reaction, with or without mold heating, the contamination problem could be worse in the initial stages because of the higher reactivity and bonding of the silicones. This is especially true of amine-modified silicones, which are known to be difficult to remove, thus making any contamination more problematic. A manufacturer of these amine-modified silicones, which are used for automobile polishes, notes specifically that the amine group causes the material to deposit and adhere to various automobile finishes and metal surfaces. Hence, adherence to other surfaces would be worse too.

The general agreement that silicones prevent good adhesive bonding is further reflected in the procedures for intentionally causing a delamination to occur. This procedure stipulates that silicones be placed at the site where adhesive bonding is to be prevented. The result will be an induced delamination.

While some successes may be achieved in using silicone-containing mold releases, they are often in circumstances where the surface of the part is specially cleaned before bonding or painting occurs (such as with a special solvent or by grinding off the molded surface to smooth it prior to painting). In theory, if subsequent bonding is not to be done, silicone-containing mold releases can be used. However, the molder must take special care to avoid cross-contamination and clean carefully. Because of these added concerns, most consider the disadvantages too great and simply avoid using silicone-containing mold releases.

Another transfer contamination material is the oil on human hands. Therefore, operators who touch raw materials, such as prepregs, should wear clean cotton gloves to prevent contamination.

DISPOSAL, WASTE, AND RECYCLING

Great caution should be exercised in disposing of any liquid materials. In general, the manufacturer of the material should be consulted for safe disposal methods. Local regulations on disposal also should be consulted.

Rather than attempt to dispose of the resins and peroxides separately, most manufacturers simply mix these materials and cure them. Concentrations do not need to be exact since the quality of the material to be crosslinked is not important. Again, care should be taken so as not to mix the peroxides and the accelerators directly since that mix could explode.

Other liquid materials, such as solvents, which are not readily polymerized, should be disposed according to local regulations. Under no circumstances should solvents be placed into the sewer.

Solid materials, including the cured resin, can usually be disposed to a landfill.

Some regulations in Europe have mandated that all materials used in commerce should be recyclable. Further, the original manufacturer has the obligation to accept the materials back for proper recycling. However, thermoset materials cannot be remelted and reused in a product that was similar to the first use (as can most thermoplastics). Thermoset materials, like most composites, can be ground to small particles and used as fillers. While the value of these fillers is not as high as the original materials, they can often be used in place of the normal fillers in such materials as bulk molding compound (BMC) and sheet molding compound (SMC). Some users have found that the fillers made from recycled composites are actually better than the traditional fillers, such as calcium carbonate. This is because the recycled materials are polymeric and can bond to the resin in the new composite, resulting in slightly improved mechanical properties.

FACTORY SIMULATION

Planning and managing today's manufacturing systems pose nearly insurmountable challenges for manufacturers. Often, competing objectives (cost, throughput, cycle time, etc.) must be met while satisfying multiple constraints (budgets, schedules, technology, etc.). Further, the manufacturer must deal with numerous interdependent variables that all seem critical for proper operation: the number of resources, routings, product mix, schedules, etc. For example, assuming that there are five jobs to be performed on five different machines, it is possible that there are 120 different ways in which the jobs can be produced—and this is a simple situation. Something clearly needs to be done to help the manufacturer deal with this complexity. Simulation has proven to be an effective tool in helping to sort through the complex issues surrounding manufacturing decisions.

Simulation is typically conducted using a computer program. In composites manufacturing, simulation is typically performed on one of two levels: process simulation and system simulation. Each is summarized in this section.

Case Example

World-class manufacturers not only have clearly defined goals and objectives, but they know what it takes to achieve these goals. This requires a thorough understanding of and control over the cause-and-effect relationships of different production decisions. In even the smallest manufacturing systems, success requires understanding how the system operates, knowing what is to be achieved with the system, and being able to identify key leverage points for best achieving the desired objectives. To illustrate the nature of the manufacturing challenge, consider the following example.

ABS Pipe Company manufactures a highly specialized type of filament-wound pipe. It is able to sell its finished products for nearly ten times the cost of the raw materials. The company enjoys virtually unlimited demand for its product so the sky is the limit. In the manufacturing process, the raw materials are combined using a proprietary process, wound onto a mandrel, and then heated and cured into pipe. The pipe is then stenciled with labels and stacked and bundled for shipping. The secondary operations are performed off-line. The process sequence is shown in Figure 21-1.

Due to various inefficiencies, the system is able to run at only about 60% of its theoretical capacity, costing the company millions of dollars a year in lost revenue.

In an effort to improve manufacturing productivity, management studied each step in the process. It was fairly easy to find the slowest step in the line; however, additional study showed that only a small percentage of lost production was due to problems at this "bottleneck" operation. Sometimes a step upstream from the bottleneck would experience a problem, causing the bottleneck

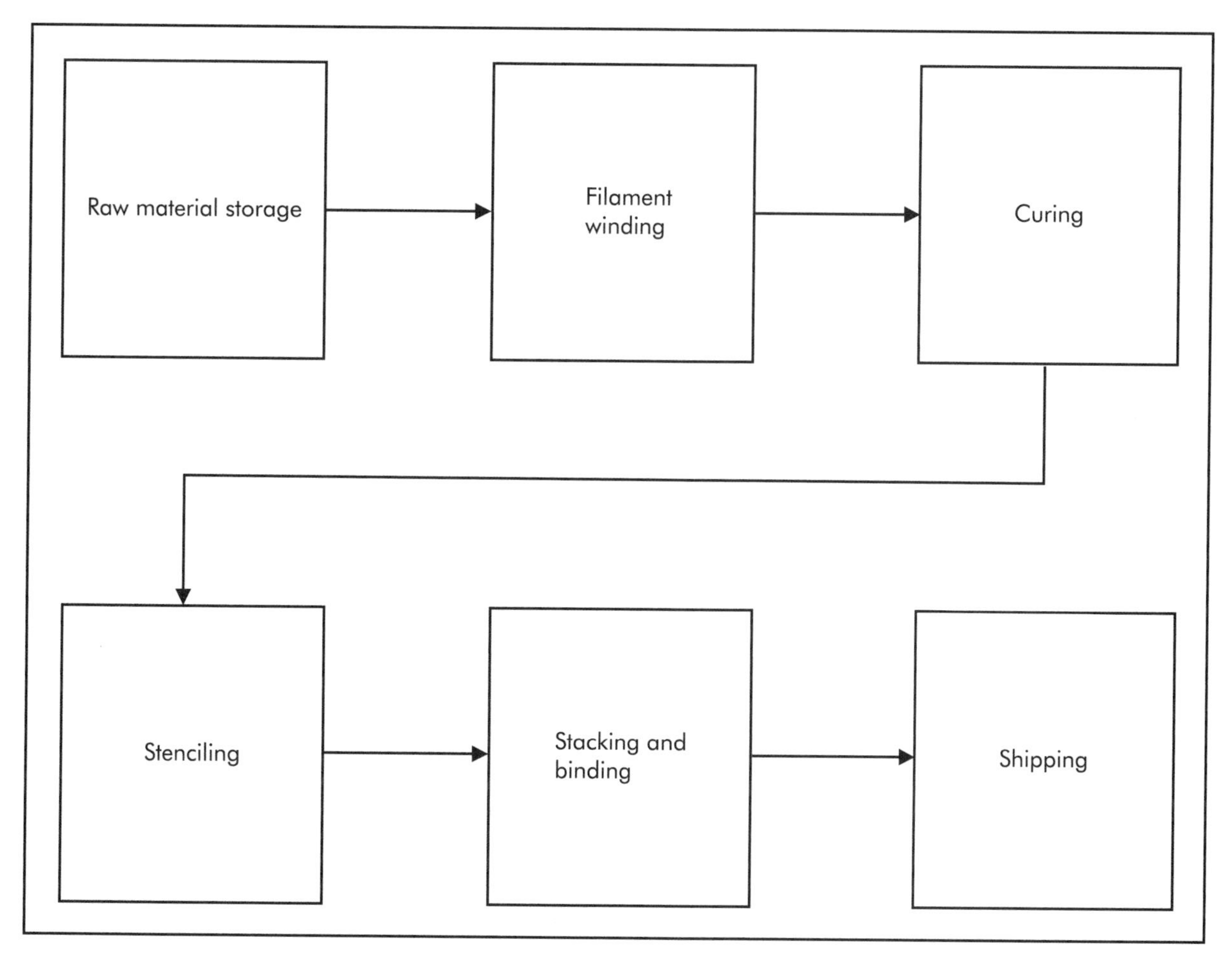

Figure 21-1. Process flow for ABS Pipe Company.

to run out of work. Or, a downstream operation, such as the strapping machine, would go down temporarily, causing work to back up and stop the bottleneck operation from producing. Other times the bottleneck would get so far behind that there was no place to put incoming, newly made pipe. In this case, the workers would stop the entire pipe-making process until the bottleneck was able to catch up. Often the bottleneck would then be idle waiting until the curing oven heated up and was functioning properly again, and until the new pipe had a chance to reach it. Sometimes problems at the bottleneck were actually caused by improper work at a previous location.

In short, there is no single cause for the poor productivity seen at this plant. Rather several separate causes contribute to the problem in complex ways. Management is at a loss to know which of several possible improvements would have the most impact for the least cost. Under consideration are: additional or faster capacity at the bottleneck operation, additional storage space between stations, better rules for when to shut down and start up the filament winder, better quality control, and better training at certain critical operations. The poor performance of the system is costing enormous amounts of money. Management is under pressure to do something, but what should it be?

This example illustrates the nature and difficulty of the decisions that a manufacturer must make. Managers must make decisions that are the "best" in some sense. To do so, however, requires that they have clearly defined goals and understand the system well enough to identify cause-and-effect relationships. The goals are created by strategic planning. Because of its complexity, gaining an understanding of the system is difficult, even with years of experience. Problems of this complexity are best solved using computer simulation.

A natural reaction to the thought of using simulation is "Uh-oh, it sounds too difficult." No one wants to use a tool that is more difficult to use than living with the problem it is intended to solve. Today, however, simulation tools are easier than ever to use. In a matter of minutes, a system map, such as the one at ABS Pipe Company, can be modeled and analyzed. Not surprisingly, most of the effort in simulation involves gathering data and preparing presentations. However, the amount needed is not too great, and the effort required to find the data and prepare for the simulation is valuable in understanding the process.

Process Simulation

Process simulation is used to analyze what happens to material and tooling under a defined set of conditions for a *particular manufacturing procedure*. A filament winding process, for example, might be modeled to simulate pressure forces on the mandrel or the pattern of fiber laydown. Simulation of processes, such as plastic injection molding, has been conducted for years. Mold filling, cooling, and warpage analyses are powerful tools for troubleshooting quality problems and understanding the injection molding process and its effect on part properties.

Software is available to model process control and performance characteristics for a number of different processes. It can be useful for determining machine settings, including some of those involved in the filament winding process. Users can enter information on part design and quality requirements, and the software interprets the data by simulating results, which suggest the ideal machine setup conditions. For injection molding, these include cooling times and profiles for temperatures, ram velocities, and injection and holding pressures. For filament winding, the speeds of the payoff device, the turning of the mandrel (thus giving the angle of laydown), the number of fibers in the bundle (to give the width of the laydown), and the number of wraps at each angle are given.

The benefits of process simulation have been significant, enabling process engineers and molders to quickly optimize machine setup, reduce cycle times, and monitor and correct the various machine processes during production.

System Simulation

When it comes to analyzing and improving the *entire production process*, system simulation is used. Simulation of production systems is becoming increasingly recognized as a quick and effective way to design and improve operational performance. (For a comprehensive treatment of the use of system simulation technology, see Harrell et al. 2004.)

To analyze a manufacturing system, one of the following methods might be chosen:

- Construct a simple flow chart.
- Develop a spreadsheet model.
- Build a computer simulation model.

The choice depends on the complexity of the system and desired precision in the answer. Flow charts and spreadsheet models are fine for modeling simple processes with little or no interdependency or variability. However, what is seemingly a simple process might be quite complex because of interactions often overlooked. Hence, computer simulation is often used even for seemingly simple systems and certainly for any complex one. Further, it is necessary for understanding the interactions of the system over time.

In practice, manufacturing simulation is performed using commercial simulation software, such as ProModel® or ProcessSimulator®, which have modeling building blocks called *modeling constructs*. These packages are specifically designed for easy entry of data describing the dynamic behavior of systems. Using the modeling constructs available in the package, the user builds a model that captures the processing logic (that is, why each step follows after another) and the constraints of the system being studied. As the model is "run," performance statistics are gathered and automatically summarized for analysis. Modern simulation software provides a realistic, graphical animation of the system being modeled to better visualize how it behaves under different conditions.

During the simulation, the user can interactively adjust the animation speed and even make changes to model parameter values to do "what if" analysis on the fly. State-of-the-art simulation technology even provides optimization capability. Not that the simulation itself optimizes, but scenarios that satisfy defined feasibility constraints can be automatically run and analyzed using special goal-seeking algorithms.

Because simulation accounts for interdependencies and variability, it provides insights into the complex dynamics of a system, which are unobtainable using other analysis techniques. Simulation gives manufacturers unlimited freedom to try out different ideas for improvement risk-free. There is virtually no cost, no waste of time, and no disruption to the current system. Furthermore, the results are visual and quantitative with performance statistics reported on all measures of interest.

Application

Simulation helps evaluate the performance of alternative designs and the effectiveness of alternative operating policies. Typical design and operational questions, which simulation might be used to address, are listed as follows.

Design questions:

1. What type and quantity of machines, equipment, and tooling should be used?

2. How many operating personnel are needed?
3. What is the production capability (throughput rate) for a given configuration?
4. What type and size of material handling system should be used?
5. How large should the buffer and storage areas be?
6. What is the best layout for the factory?
7. What automation controls work the best?
8. What is the optimum unit load size?
9. What methods and levels of automation work best?
10. How balanced is the production line?
11. Where are the bottlenecks (bottleneck analysis) in the system?
12. What is the impact of machine downtime on production (reliability analysis)?
13. What is the effect of setup time on production?
14. Should storage be centralized or localized?
15. What is the effect of vehicle or conveyor speed on part flow?

Operational decisions:

1. What is the best way to schedule preventive maintenance?
2. How many shifts are required to meet production requirements?
3. What is the optimum production batch size?
4. What is the optimum sequence for producing a set of jobs?
5. What is the best way to allocate resources for a particular set of tasks?
6. What is the effect of a preventive as opposed to a corrective maintenance policy?
7. How much time does a particular job spend in a system (throughput)?
8. What is the best production control method (kanban, manufacturing resource planning [MRP], etc.) to use?

Economics

Savings from simulation are realized by identifying and eliminating problems and inefficiencies that would have gone unnoticed until system implementation. Cost is also reduced by eliminating overdesign and removing excessive safety factors, which are added when performance projections are uncertain. By identifying and eliminating unnecessary capital investments, and discovering and correcting operating inefficiencies, it is not uncommon for companies to report hundreds of thousands of dollars in savings on a single project through the use of simulation. The return on investment (ROI) for simulation often exceeds 1,000%, with payback periods frequently being only a few months or the time it takes to complete a simulation project.

The real savings from simulation come from allowing designers to make mistakes and work out design errors on the model rather than on the actual system. The concept of reducing costs through working out problems in the design phase rather than after a system has been implemented is best illustrated by the rule of tens. This principle states that the cost to correct a problem increases by a factor of ten for every design stage through which it passes without being detected.

Because simulation accounts for interdependencies and variability, it provides insights that cannot be obtained any other way. For important system decisions of an operational nature, simulation is an invaluable tool. Its usefulness increases as variability and interdependency increase and the importance of the decision becomes greater.

Lastly, simulation actually makes designing systems exciting. A designer cannot only try out new design concepts to see what works best, but the visualization brings realism to the whole process. Through simulation, decision makers can play what-if games with a new system or modified process before it actually gets implemented. This engaging process stimulates creative thinking and results in good design decisions.

CASE STUDY 21-1

Quality in Foreign Manufacturing

Several companies that do offshore manufacturing were interviewed to discover what problems they had with the quality in their offshore operations. Not only is high product and service quality an absolute necessity in domestic or international markets, but it serves as a convenient probe to discovering many of the critical issues in manufacturing. Because of the inherent questions surrounding foreign product quality, good quality may be the deciding factor on whether a foreign company is able to sell its product. Therefore, most companies that have opened international operations expend significant effort to maintain the levels of quality achieved domestically.

Example one is an electronics firm in southern China that produced portable stereos and electronic components for stereo products. Two production lines were operating within this plant. One produced a fairly popular brand sold in the United States, which was acknowledged to be generally of only moderate-to-low quality. The second line was producing similar products for a Japanese name brand, one acknowledged as having good-to-excellent quality. The facility was the same and so were the component feeder lines. However, line one had a measured defect rate of 10–11%. The second line, the Japanese line, had a defect rate of less than 2%.

The difference was in the expectations, techniques, and commitment demanded by the Japanese marketing company who was the customer for the Japanese line. The Japanese firm placed a full-time manufacturing/quality engineer in the Chinese plant to ensure that the quality methods of the Japanese firm were followed. Thus, Japanese quality standards were enforced and, even more, the attitude of quality was present in the commitment of the Japanese company.

Something surprising was discovered after a day of tours and discussions in the plant. The plant manager asked what the company could do to improve the quality of the American line. On being told the best that could be done was to learn from the Japanese engineer, he responded with "But it is not how our company [the Chinese company] does things." Even with daily interactions with the Japanese engineer, workers could not see how to adopt the methods to other situations even though many of the changes would have been simple to do. It was not their way, or more accurately, the way of their management or company culture and values. Similarly, the Japanese firm had significantly higher quality for the same reason—commitment to good quality was imperative and that commitment was in force even from 2,000 miles away.

In example two, a major manufacturing company in Japan was placed under import restrictions to the United States. This company then licensed some of its products to other companies, in other countries, which then made the products and sold them in the U.S. After several years, the Japanese company was again allowed to sell directly to the U.S. However, the licenses were not revoked, thus allowing the Japanese company and the licensee to sell against each other.

In an effort to see if Japanese manufacturing practices were still being used by the licensees, visits were made to the Japanese company and then to a Korean licensee.

The Korean company had almost none of the cultural characteristics of the Japanese company (such as lifetime employment, quality circles, and close attention to quality details). Instead, the Korean company viewed itself as a low-priced alternate to the admittedly superior Japanese product. Moreover, the Korean plant was a hodgepodge of motivational signs and slogans, urging the employees to work harder to become more successful and compete better. The Japanese plant was, in contrast, a study of quiet efficiency, with high automation and technically advanced equipment and procedures. The cultures of the countries seemed to be readily apparent in their manufacturing practices.

The above examples are focused on quality. Usually other issues are even more important in doing business in a lesser-developed country. The World Bank has surveyed companies and found that the majority of difficult issues are outside the control of the company. Table 21-1 shows the types of obstacles commonly encountered in doing business in lesser-developed countries.

SUMMARY

The efficiency of factory operations can mean the difference between profit and loss for a company. If the manufacturing process is poorly laid out or the plant is messy and dirty, the result is not just contamination and poor quality. It is a reflection of the culture, which tolerates the lackadaisical attitude of employees.

The improper storage of liquid materials presents a serious potential for disaster. The resins can harden or, even worse, they can set off an exotherm so high that they burst into flames. Materials, such as peroxides, can explode. Therefore, great care should be taken to ensure proper storage of these materials.

Table 21-1. Obstacles to business in lesser-developed countries.

Obstacle	Survey Score
Corruption	4.8
Inadequate infrastructure	4.7
Financing	4.5
Tax regulations	4.4
Policy instability	4.0
Crime and theft	3.8
Regulations on foreign trade	3.7
Labor regulations	3.6
General uncertainty on cost of regulations	3.5
Safety or environmental regulations	3.3
Inflation	3.2
Startup regulations	2.7
Foreign currency regulations	2.5
Price controls	2.4
Terrorism	2.3

The advantages of simulation are chiefly associated with modeling the process to improve efficiencies. However, the process of assembling the data and putting it into the simulation package assists the system designer in thinking deeply about the process. This has tremendous benefits beyond just simulation.

LABORATORY EXPERIMENT 21-1

Understanding the Dangers of Improper Storage

Objective: Discover the dangers of improper storage of a resin.

Procedure:

1. Obtain the storage suggestions from a resin manufacturer.
2. Using the storage suggestions, develop a checklist that can be used to verify that the procedures of the resin manufacturer are being followed.
3. Visit a composite manufacturing facility and do an inspection of their storage practices.

QUESTIONS

1. What is the danger of storing resin in large quantities?
2. What is the procedure that should be followed if the resin overheats (exotherm)?
3. Discuss whether it is better to store resin inside or outside the plant.
4. What are the problems associated with using a silicone mold release?
5. What is the result of mixing peroxide and a cobalt naphthenate accelerator?
6. Give three reasons for using simulation in a production plant.

BIBLIOGRAPHY

Bateman, Robert E., Bowden, Royce O., Gogg, Thomas J., Harrell, Charles R., and Mott, Jack R. A. 1997. "System Improvement Using Simulation." Orem, UT: PROMODEL Corporation.

Berdine, Robert A. 1993. "FMS: Fumbled Manufacturing Startups?" *Manufacturing Engineering*, July, p. 104.

Cave, Alex, Nahavandi, Saeid, and Kouzani, Abbas. 2002. "Simulation Optimization for Process Scheduling Through Simulated Annealing." *Proceedings of the 2002 Winter Simulation Conference*. Yücesan, E., Chen, C.-H., Snowdon, J. L., and Charnes, J. M., eds. pp. 1,910–1,913.

Glenney, Neil E. and Mackulak, Gerald T. 1985. "Modeling & Simulation Provide Key to CIM Implementation Philosophy." *Industrial Engineering*, May.

Hancock, Walton, Dissen, R., and Merten, A. 1977. "An Example of Simulation to Improve Plant Productivity." AIIE Transactions, March, pp. 2–10.

Harrell, Charles, Ghosh, Biman, and Bowden, Royce. 2004. *Simulation Using ProModel*, 2nd ed. New York: McGraw Hill.

Hopp, Wallace J. and Spearman, M. 2001. *Factory Physics*. New York: Irwin/McGraw-Hill, p. 180.

Kochan, D. 1986. *CAM Developments in Computer Integrated Manufacturing*. Berlin, Germany: Springer-Verlag.

Law, A. M. and McComas, M. G. 1988. "How Simulation Pays Off." *Manufacturing Engineering*, February, pp. 37–39.

Lewis, C. S., Hooper, W., ed. 1979. *God in the Dock: Essays in Theology*. London: Collins.

Nembhard, Harriet Black, Kao, Ming-Shu and Lim, Gino. 1999. "Integrating Discrete Event Simulation with Statistical Process Control Charts for Transitions in a Manufacturing Environment." *Proceedings of the 1999 Winter Simulation Conference*, Phoenix,

AZ. P. A. Farrington, H. B. Nembhard, D. T. Sturrock, and G. W. Evans, eds.

Schriber, T. J. 1987. "The Nature and Role of Simulation in the Design of Manufacturing Systems." *Simulation in CIM and Artificial Intelligence Techniques*, Retti, J. and Wichmann, K. E., eds. San Diego, CA: Society for Computer Simulation, pp. 5–8.

Shannon, Robert E. 1998. "Introduction to the Art and Science of Simulation." In *Proceedings of the 1998 Winter Simulation Conference*. Medeiros, D. J., Watson, E. F., Carson, J.S., and Manivannan, M. S., eds. Piscataway, NJ: Institute of Electrical and Electronics Engineers, pp. 7–14.

Shingo, Shigeo. 1992. *The Shingo Production Management System—Improving Process Functions*. Trans. Dillon, Andrew P. Cambridge, MA: Productivity Press.

Strong, A. Brent. 2006. *Plastics: Materials and Processing*, 3rd Ed. Upper Saddle River, NJ: Prentice-Hall, Inc.

22

The Business of Composites*

CHAPTER OVERVIEW

This chapter examines the following concepts:

- Economics of composites manufacturing processes
- Planning
- Corporate creativity
- The ethical process
- Managing technology

ECONOMICS OF COMPOSITES MANUFACTURING PROCESSES

Income Statement

Most companies use the income statement as a simple method of determining the success of a business. An example income statement is shown in Figure 22-1. (Slightly different versions are also used, but the one shown is a common format. The amounts shown are typical for a mature manufacturing company.) Some of the basics of business economics can be understood from a brief examination of the income statement.

The revenue amount represents the total amount of money that comes to a company in the course of its normal business. For a manufacturing company, which is the case examined here, the revenue is from sales of the company's products. There may occasionally be other sources of revenue for the company, such as sale of an asset, but these are usually termed "extraordinary income" and are not part of the discussion here.

The cost of goods sold is the next line on the income statement. This line is critically important for a manufacturing operation. It is a total of all the costs associated with making the product and getting it ready to be sold. Because of its importance, the cost of goods sold will be examined in more detail later. The value of the cost of goods sold is subtracted from the revenue to determine the gross margin. This is the number that indicates whether the manufacturing costs are in line with the sales of the company. Normally, a gross margin of 40% of revenues is considered to be good, although some products with high values can have gross margins substantially higher. Commodities on which the margin is low may be in the 15–20% range.

Sales expenses are shown below gross margin because they do not relate directly to the costs of making the product. These costs include the salaries and expenses of the sales force, and usually shipping costs.

The next category is general and accounting. This includes the costs of corporate management and the accounting function.

* Some of the material in this chapter was written by A. Brent Strong (sometimes with assistance from other authors) and published in various issues of *Composites Manufacturing* (American Composites Manufacturers Association).

Income Statement	$ ('000)
Revenue (sales)	1,000
Cost of goods sold	600
Gross margin	400
Sales expenses	150
General and accounting	50
Depreciation	20
Interest	10
Taxable income	170
Taxes (40%)	68
Net income	102

Figure 22-1. Example of an income statement.

Depreciation is, at least in theory, money set aside to repurchase the manufacturing equipment when it has passed its useful life. There are formulas for calculating depreciation. Because it is considered an expense of the company and is shown above taxes in the income statement, most companies that are making a profit will choose a formula that maximizes the depreciation (thus reducing taxable income).

The taxes for most areas in the United States are about 40%. This includes federal, state, municipal, and other miscellaneous taxes that might be assessed.

The bottom line is net income. It reflects the profit of the company. These profits can be divided among the stockholders of the company (dividends) or used for other purposes outlined in the company's articles of incorporation and approved by its directors.

Cost of Goods Sold

The fundamental costs that comprise the cost of goods sold are material, labor, and overhead. Each of these categories can be subdivided so that specific costs become more obvious and, therefore, more easily controlled. For example, the material costs for a company that makes composite bath and shower units might be divided into resin, fiberglass, filler, initiator, and minor additives. The costs of these materials can be determined as the total amounts purchased from the various material vendors. Alternately, the material costs might be figured as the total amount of material in the products sold plus the amount of material wasted in the plant. In theory, these two accounting methods should give the same material costs. However, the second method gives a better measure of the wasted material because it notes the waste explicitly. Therefore, in a plant in which the amount of waste is high, the second method of accounting for material costs would be better because it gives more visibility to this problem.

Another variation of the cost model subdivides the labor component of the cost of goods sold. For example, the labor could be classified according to functions, such as: fabrication, assembly, inspection, and design. This model might be used when labor is an especially high component of the cost of goods sold. An example of when this situation might be especially valuable is when two different manufacturing processes are being compared and the labor component is the major difference between the two. Figure 22-2 illustrates a comparison of the labor hours required for filament winding versus hand lay-up. Filament winding has fewer labor hours (1,500 versus 3,000) with the major differences being the much higher time required for hand lay-up. Offsetting this difference is the greater amount of support needed in filament winding. Other operations—quality assurance, kitting, routing, and manufacturing and assembly were comparable in the percentage of time dedicated for each process. However, because the hand

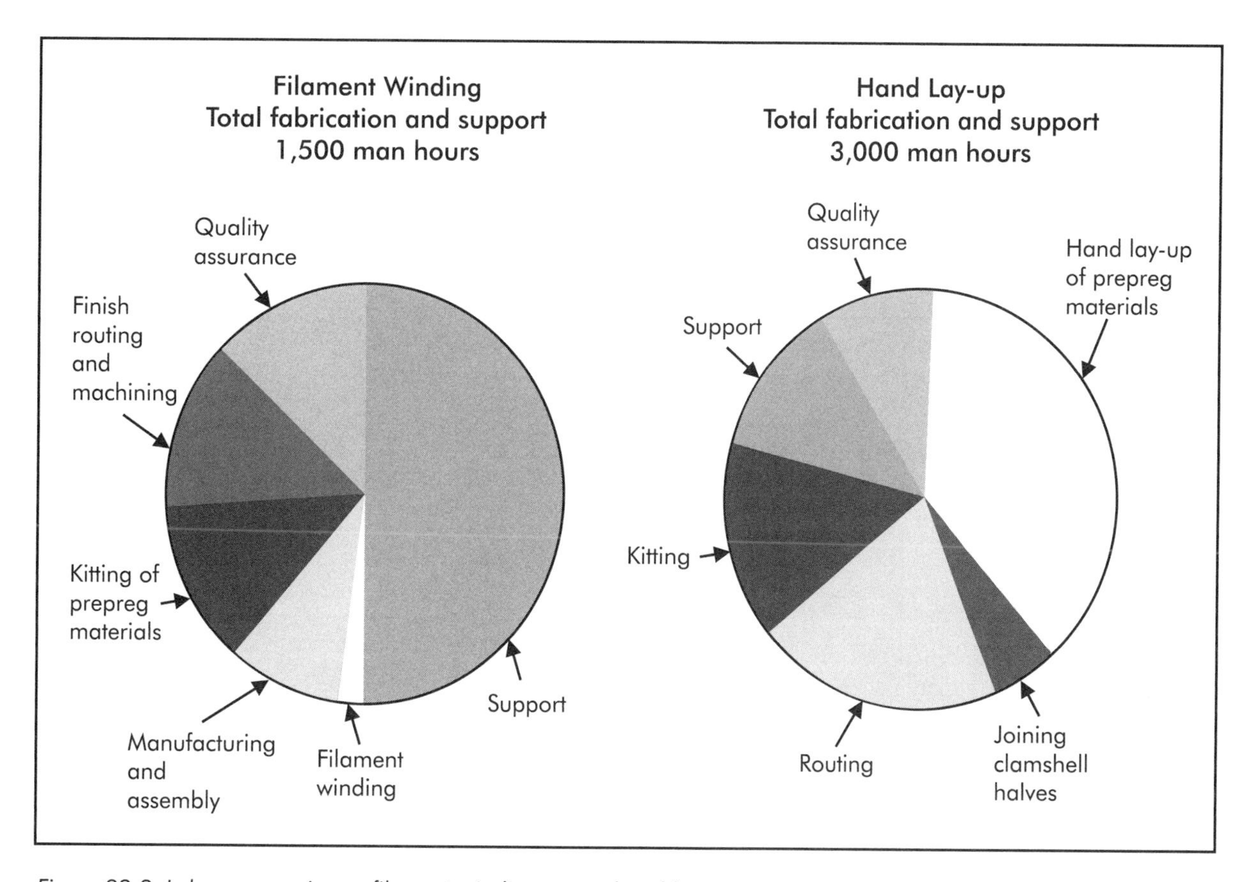

Figure 22-2. Labor comparison—filament winding versus hand lay-up.

lay-up took twice as much overall time, each of these actually consumed more time.

The third component of cost of goods sold, overhead, also can be important, especially when the costs of a machine or tooling for a particular process are considerably higher than for an alternate process. Consider, for example, the case of making a front cross-member for a minivan. Based upon a production volume of 40,000 units, the costs of a fiberglass reinforced plastic cross-member are lower than for a steel equivalent. This is largely because of the high costs of the machines used to make the steel part. However, at high production rates, the steel part becomes less expensive because the machine costs are fixed and can, therefore, be spread over many more parts.

Comparing Composites Manufacturing Methods

Comparisons between various manufacturing processes are difficult because the costs vary according to the nature of the part, the volume of parts made, and the efficiencies of each manufacturer's operation. However, some comparisons are possible based upon generalities over many parts made by a particular process. Figure 22-3 gives a comparison of several composites manufacturing processes, assuming that the same part could be made by all processes. An aluminum process to make an equivalent part is used for comparison.

As might be expected, the more highly automated processes, pultrusion and filament winding, have the lowest overall costs

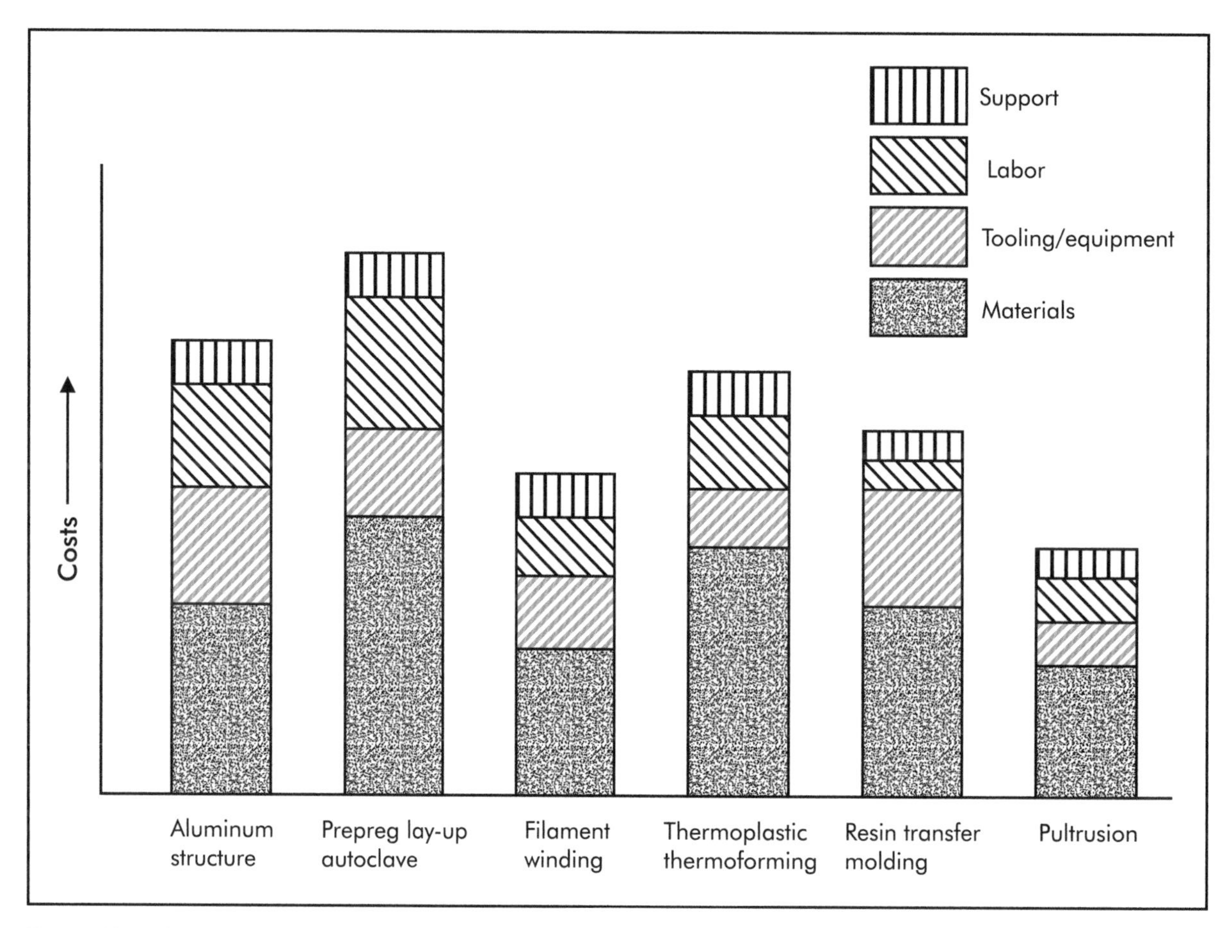

Figure 22-3. Cost comparison of manufacturing methods.

with pultrusion being lower than filament winding largely on the basis of lower-cost tooling. These highly automated processes not only have low labor costs, but they also have low materials costs because they use tow and wet resin, which are the lowest-cost forms of the principal raw materials.

Resin transfer molding is next lowest, largely because the costs of tooling and materials are less than thermoplastic thermoforming, the next higher cost. The highest cost of any of the methods, even higher than aluminum, is prepreg lay-up with an autoclave. This result is not unexpected in light of the high material costs, high tooling/equipment costs (with an autoclave), and high labor costs.

Life-cycle Costs

The costs of any real product are more than just the price of the product as it is originally sold. Over the life of the product there are other costs. And eventually, when the product is to be discarded, there are still more costs. Therefore, the real cost analysis should look at the costs over the entire life cycle of the product. In general, the life cycle costs consist of three components:

1. cost to manufacture and sell,
2. cost to operate, and
3. cost to dispose.

To illustrate these costs, the automobile is used as an example because the three cost components are widely understood.

The cost to manufacture is the cost that has been considered thus far in this chapter. It is the costs associated with materials, labor, and overhead. For composites, the costs of manufacture are generally higher than the costs for steel products. This is simply because steel is a lower-priced raw material and at the high volumes typical of manufacturing, the expensive equipment used to form and join it can be easily justified. However, if the part is highly curved or has some other unusual characteristic, then composites can be justified. Hence, flat panels of automobiles are generally made of steel but highly curved panels are made of composites (usually sheet molding compound [SMC]).

The costs of operation are, of course, paid by the purchaser of the automobile. The manufacturer has little incentive to improve these costs unless the customer demands it or the government mandates it. However, the high cost of fuel is now forcing both customers and government to make these demands, at least on some models. In such an environment, the value of composites rises sharply because of their significantly lower weight compared to steel. Since fuel savings is directly proportional to weight savings, the value of composites can be estimated from a weight savings calculation. On average, a car made of composites would weigh about 75% as much as a steel car. This would translate to a savings about $6.00/lb ($13/kg) of weight reduction over the life of the car. This is at 2007 fuel prices and accounting for the time value of money. The savings could be as much as $1,000 for a moderately sized vehicle.

The cost of disposal is relatively minor for an automobile. After the materials are salvaged, the remainder of the car is disposed to a landfill or chopped into filler. However, if this cost becomes significant because of governmental action, the disposal factor will have to be considered in evaluating the type of materials used in the automobiles.

PLANNING

Predicting the Future

As Yogi Berra said, "Predictions are hard, especially when they are about the future." Yogi's statement is especially true in a time, like now, when change is no longer slow or evolutionary. Change today is *episodic* and *random*. These times defy predictions.

How can you plan for the future at a time when change occurs so quickly? Planning seems futile because our current view of the future is likely to be different from the real future when it arrives. In a strange, almost surrealistic way, the leader who can plan properly in these times will achieve a dramatic competitive advantage. But, the old methods of planning will probably not work today. Therefore, they must be adapted to meet the new reality of rapid and dramatic change.

Operational, Structural, and Strategic Decision-making

There are three types of decisions required in a business. The first type is **operational decisions**—deciding what to do *right now* to run the business. These are simply decisions required to get the job done. They include making the sales call, getting the order, making the product, shipping the product, purchasing, and scheduling. These are the day-to-day (maybe hour-by-hour) decisions of normal operations. Operational decisions are easily made, of low consequence (individually), straightforward, frequent, and self-forcing (that is, they must be made to continue the business). They are often delegated to middle managers or operators and controlled by standard operating procedures.

The second type is **structural decisions**. These are *periodic* decisions that provide *limited guidelines* for daily activities. In other words, these are decisions made on an occasional basis that have to do with the way the business is organized. Typical structural decisions include those affecting company organizational charts and reporting relationships, pricing policy, workflows, operating procedures, etc. Structural decisions are intermittent, somewhat complex, of moderate consequence, relatively easy to make, and self-forcing.

Structural decisions are usually made by middle and upper management. In fact, the essence of the role of managers is to formulate and then implement structural decisions. The usual purposes of structural decisions are related to developing *efficient* and *productive* systems.

The third type is **strategic decisions**, which are related to the company's fundamental purposes. Another way of saying this is that strategic decisions are related to the *values* of the company and they become the *underlying principles* on which all other decisions are made. Strategic decisions are made infrequently, are extremely complex, have severe consequences, and are not self-forcing.

Strategic decisions do not typically come up in the normal operation of a company. The founders of the company probably did strategic planning when they established the original purposes and nature of the company. After that, strategic planning sessions were likely done only as major changes occurred, which forced the senior managers to re-examine the company and its overall direction. Therefore, to be most effective in strategic decision-making, senior management should schedule the times when strategic planning should be done. During these planning times, a formal procedure for analyzing the basic positioning of the company can be useful. Those basics might include the following:

- What does the company want to be?
- How should the company get to this desired position?

Notice that these questions really focus on how the *company* will change. Thinking about the nature of change, you will realize that *management is not about change*. In fact, it is about maintaining the status quo or, at best, moving forward along predictable and logical paths of improvement.

The process of *change is about leadership*. Change occurs because a leader sees the new position and then influences others to adopt the new way of thinking (paradigm) so that the entire organization can move to the envisioned position. This is only accomplished by the leader's repeated reinforcement of the desirability of moving to the new position and reiteration of the benefits that can accrue from that change. Further, the leader must not only instill a vision of the new position, but he/she must also instill a deep understanding of (or adoption of) the basic principles of the company. This is so that when new situations arise during the transition, the employees of the company will know the guidelines that must be used in handling them.

The leader must make sure that all facets of the company are *aligned* with the basic principles. Alignment means that procedures and practices reflect the basic values of the company. If some aspect of the company is not aligned, then confusion occurs in the minds of the employees.

For instance, suppose that the leadership has articulated a certain basic positioning of the company. Suppose the position is that the company will be the foremost composite manufacturer in the area of corrosion-resistant pipes. Further, this position will be achieved by giving full design, installation, and maintenance assistance to the user companies who buy the pipes. However, alignment with this position will not occur if the chief financial officer of the company refuses to allow emergency maintenance trips to customers' plants because of a budget crunch that particular month. Similarly, misalignment might occur if the company drastically reduces its design department because it is having trouble getting a sales price high enough to pay for this service. If these non-aligned moves are required, the company should revise and communicate its new basic strategic position so that consistency can be preserved. To do otherwise is to create confusion in the company and in the marketplace. Knee-jerk decisions destroy the trust of the employees in the leaders of the company.

Strategy is often about winning or losing. It must, therefore, consider the competition. As a result, the best strategy planning should be done in an environment in which data on the competition and the marketplace are readily available. This will, of course, require some effort to assemble this data. Strategic thinking can be stimulated by comparing a company's strengths and weaknesses against the competition (benchmarking).

Often, analogies to sports, games, and even war are useful in communicating strategy to employees. (Some of history's best leaders have come from the military.)

Strategic decision-making can be summarized as follows.

- Strategic decisions concern the *future*.
- Strategic decisions concern *change*.
- Strategic decisions are the realm of *leadership*.
- Strategic decisions are based on *basic values*.

There is a word of caution in regard to strategic planning. Too often companies get into a mode of planning and forget to move forward and actually change the company. Although extremely important for long-term viability of a company, the strategic decision-making process does *not* create immediate business. Therefore, the intelligent leader will remember that strategic decision-making is done infrequently and that the company needs to *move ahead* to realize the vision.

The business makes money because of *sales and production*. These should be guided by the decisions made in planning sessions; however, no amount of planning can take the place of *action* in these two areas.

CORPORATE CREATIVITY

"[A] traditional bureaucratic structure, with its need for predictability, linear logic, conformance to accepted norms, and the dictates of the most recent 'long range' vision statement, is a nearly perfect idea-killing machine" (Dresselhaus 2000). Therefore, to avoid having the corporate climate kill ideas and creativity, leaders need to understand the creative process and how that can be influenced by the corporate climate.

Improving Corporate Creativity

Though creativity is the subject of many books, most of them only focus on improving *personal creativity*. Few discuss changing the environment to promote creativity, especially in companies. The following sections provide some steps that have been demonstrated to improve *corporate creativity*.

1. Make Creativity a Centerpiece of Corporate Strategy

Making creativity center to the corporate strategy may seem obvious, but this certainly is not the case in most companies. In fact, usually the focus is on many other things besides creativity or its practical manifestation, innovation. Leadership worries about cash flow, lagging sales, poor production, or employee morale. While these worries are certainly legitimate, many can be reduced or eliminated by encouraging creativity in problem solving and creating new products. When creativity is a corporate priority and focus, everyone is involved: "The biggest single trend we've observed is the growing acknowledgment of innovation as a centerpiece of corporate strategies and initiatives. What's more, we've noticed that the more senior the executives, the more likely they are to frame their companies' needs in the context of innovation" (Kelley 2001).

2. Reward and Support Creativity

When creativity is the centerpiece of corporate strategy, the company organization changes to keep the stars of creativity in jobs where their creativity is not swallowed up in bureaucratic duties. These creative people can and should be promoted, as reward for their efforts and as role models for others, to positions of leadership (visionaries). However, many people can be promoted to a level of incompetence by moving them out of their area of strength (Peter and Hull 1969). In new leadership roles, creative people require the back up of others (usually assistants) to help them in their managerial duties. The additional costs in payroll are usually worth it.

3. Give Time and Opportunity to be Creative

Some creative companies, like 3M, ask (insist) that all employees devote a meaningful part of their time (such as 10%) to non-assigned projects. Thus, employees have time to think deeply, read broadly, and otherwise practice the skills of creativity. For companies like 3M, which thrive off new ideas, these personal projects are strategically critical. For others, allowing time for creativity can result in major breakthroughs that could give a significant competitive advantage. In all companies, large and small, increased employee involvement in the company and the resulting improvement in employee attitude is worth the small amount of time off of normal projects, even if no breakthrough ideas are ever produced.

Sometimes managers are reluctant to allow employees to be creative because they believe it to be a waste of time. There is so much that must be done in a typical company, that the manager keeps pressure on everyone to keep busy. This situation is described by author, Gordon MacKenzie (MacKenzie 1998). He describes the city fellow who went into the country and, with some disgust, noted the lazy cows that were just standing around in the field, occasionally munching the grass. He said to the dairy farmer, "How can you allow your producers to be so lazy? They must do more." What the city man did not understand is that the cows are performing the miracle of turning grass into milk even though there is not apparent action. Stressing the cows to produce more will not result in more milk but, probably, do just the opposite.

4. Establish a Culture of Paradigm Re-examination

In one of the best of the creativity books, Charles Thompson reports a story that illustrates well how a creative culture can be fostered by senior management through thoughtful actions that challenge creativity in others (Thompson 1992). "One day, Corning's president said to the head of research, 'Glass breaks. Why don't you do something about that?' The directive to the lab then became: 'We're going to prevent glass from breaking.' The lab came up with 25 different ways of preventing glass from breaking; 18 of them worked, and five made money . . . The end result was the now-famous Corelle® line of dinnerware." Corporate leaders should actively try to spark creativity and ask challenging questions.

5. Use Teams and Brainstorm Often

Many companies seem to believe that creativity is something done by isolated geniuses. In truth, teams are usually much more creative than any single individual. Even Thomas Edison, the holder of the most U.S. patents, relied strongly on a creative and hard-working team. Quoting a Honda manager, Charles Thompson wrote in his book (Thompson 1992), "I am always telling the [new product development] team members that our work is not a relay race . . . Every one of us should run all the way from the start to the finish. Like in a rugby game, all of us should run together, passing the ball left and right, and reaching the goal as one united body."

6. Prototype Early and Often

Prototyping is not only a practice that helps identify problems early in the development process; it is also important in conveying the concept that "It is OK to fail, but try to fail when it costs the least." This failure-acceptance climate encourages creativity. The use of prototypes to identify mistakes is the best practice for achieving a realistic view of new products.

7. Create an Atmosphere of Urgency

When there is a strong motivation, most people are more creative. Therefore, corporate leadership should instill a climate that motivates all employees to be more creative. The most effective tactic to create motivation is urgency—urgency from competition, urgency from possible loss of market opportunity, urgency because the greatest profits are gained at the beginning of a product's life, urgency of goals, etc. When the entire climate of a company has this sense of urgency about new product development, it is amazing how effective people can make the development process. Cooperation between departments is improved and jobs just get done better because people are trying to help each other out.

Pyramids and Trees

The task of managing people is difficult. This chapter has, hopefully, suggested that proper management should include some methods to encourage creativity. However, the effective manager also knows that productive companies, especially those doing manufacturing, require significant uniformity and cooperative actions. Therefore, managers must find ways to solicit cooperation from employees, but at the same time encourage individual creativity.

There are two different models for organizations—one demanding obedience and conformity and the other giving more leeway for individual creativity (MacKenzie 1998).

The first organizational analogy is a pyramid with management at the top and the workers down below. Management sees the future and proclaims that, "We must grow or die!" However, growing is difficult because of the shear weight of the structure. Besides, when you step back and think about pyramids, how do you make an actual pyramid larger? This is not an easy task!

To grow a pyramid you would have to add another row at the base all the way around and then add blocks one row at a time on top of the new base. However, to fit with the existing pyramid, the new rows would have to be tapered. Thus, not just any block would fit; you would have to do considerable shaping before fitting it in place. When finished, the pyramid would be higher, but the outer layer would not be attached to the rest. It would be nothing but an outer encasement. That type of growth would be tenuous.

What is the alternate structure that a company might take to facilitate growth? It has been suggested that it be a tree (MacKenzie 1998). Top management is the trunk of the tree, giving support and direction to the entity. This trunk-management is also the conduit for nourishment (cash flow), which comes from the roots. The managers and supervisors are the branches that give support to the actual producers—the leaves and the buds, which will ultimately result in the cash crop. Growth is simple because the structure just extends the branches so that more producers can be supported. Creativity is allowed because the actual producers see a wide view, receive energy from their surroundings (the sun and air), and have no structural limitations except to be attached to the tree itself. All the leaves and buds seem to sense their collective responsibility, but there is certainly room to try something different (should cross-fertilization be possible, for instance).

THE ETHICAL PROCESS

What are Ethics?

Ethics, as a study, is defined as, "The discipline dealing with what is good and bad with moral duty and obligation" and "a set of moral principle or values" (*Webster's Ninth New Collegiate Dictionary* 1983) by which persons conduct their lives. Several attempts have been made to establish such a code for companies, professions, and government. But these attempts have proven to be mostly fruitless because there are so many differing situations and ways to do things wrong that attempts to cover them all is simply impossible. It is virtually impossible to define the many ways unethical or improper acts could be committed. So, if people will be governed only by what they *cannot do*, versus by a set of basic standards as to what they *should do*, then there is really no way to govern at all. Of greater practical value is an understanding of the basis on which a code is built. Even more precisely, there should be an *understanding of the basis* on which decisions are made. If the basis is ethical, then the system or code used to direct every day life will be ethical.

The Process of Ethics

Here is a critical observation. By defining ethics in terms of the basis on which decisions are made, it has been suggested that *ethics are the result of a process based on principle*. It is the process by which decisions are made. When the basis of the process is understood, then moment-to-moment decisions can be made in keeping with these principles.

The process of making ethical decisions is much like the process of making quality decisions in a manufacturing process. There are rules for administering quality, but ultimately they must be based upon some underlying principle, such as "Keep the established rules that create good quality," or "The product must be good enough to meet its intended life," or "Ship only product that delights the customer." These three examples of rules are somewhat different and will cause different actions to be taken in deciding on the quality. However, because they are based on certain principles, each will provide for its own consistency in the decision-making process. To restate the principles for each example more generally, the basis of example one is "Always keep the rules;" in example two it is "Always maintain product integrity;" and in example three, "Always serve the customer."

Another point is that quality is best controlled by controlling the process. The best way to get quality is not just by inspection (100% testing) or by just monitoring the process (statistical process control [SPC]). It is best to have a "target-centered manufacturing process" where the principle is to "reduce variation." This is enhanced by seeing the entire enterprise as a whole and then working hard to see the future and control all aspects of the manufacturing process.

By analogy, ethics are best implemented by seeing them in light of the entire philosophy and climate of the company and according to the underlying basis on which the company does business. The philosophy and, therefore, the principles by which it operates may be good or bad. For instance, assume that a company has making money at any cost as its underlying principle. Following this principle, it can be seen that quality decisions will be made so that any product that can "get by" will be shipped. Similarly, ethical decisions that just "get by," at least in the short term, will be acceptable within the company.

Ethics is a process, much like quality. The basis should be established with careful thought and in light of the vision and values of the company. Then, some reasonable methods should be established to implement the decisions that are made most often. But the most important part of the process, the part that will really make a difference, is the articulation of the basis for the decisions. This basis will help everyone make more consistent and satisfying ethical decisions.

MANAGING TECHNOLOGY

Technology is one of the defining aspects of our time. It is ubiquitous and is woven seamlessly into society. As such, technology dramatically impacts the way we work, play, learn, and live. Given the society-permeating nature of technology, everyone has to come to terms with the effects of technology in their life. Do we like it? Do we hate it? How do we reduce our frustrations with technology and not let ourselves become resentful or even apprehensive about it? How do we gain the knowledge and capability to take full advantage of technology in our lives? This section will share some perspectives on technology, its impacts in the global economy, and why people should exploit it to the benefit of our world.

What is Technology?

The word "technology" seems to have many meanings. To some it means "applied science" or "gadgets, tools, and machines," and to others "a complex social enterprise or process." The origin of the word "technology" is the Greek word *techné*. When originally used, *techné* meant the making of something and included the disciplines of craftsmanship and art. The artistic side of craftsmanship is retained in the word "artisan." The close relationship between craftsmanship and art is further reinforced in the emergence of two English words directly from *techné*—technology and technique. Therefore, it is concluded that technology should enhance and be guided by artistic values. However, today many of the concepts of technology or, at least, our perceptions of technology are opposed to art and beauty. However, the best of technology does contain elements of beauty and art, such as simplicity, symmetry, balance, depth, and elegance. "Elegant solutions" are often found for problems, thus indicating that this connection between art and technology is desired and appreciated.

When John H. Gibbons was the Director of the U. S. Congress' Office of Technology Assessment (1988), he described technology as "applied human knowledge." Another perspective of technology is that it is "human ingenuity in action." Given these perspectives, technology is not just tools or know-how, or thinking, acting, or doing—technology requires thoughtful action(s). Technology has purpose; that purpose is generally to extend human capabilities. For example, a calculator extends human capabilities to quickly do math functions like the square roots of numbers. Likewise, an automobile, using a system of roads, traffic signals, etc., extends human capabilities for transportation.

Technological developments are driven by human needs and wants. People, of course, often confuse wants with needs—we need a car for transportation but we want to drive a Corvette®. A market-driven economic system responds to human wants and needs with products and services. However, all new products and services are not successful or profitable. The deployment and adoption of new technologies are driven by human values—improving the quality of life, profits, recognition, etc. These values differentiate the technologies that people consume and those that are forgotten.

Anti-technology Values

While most people gratefully consume or use technology, some have negative attitudes toward all technology. Luddites (a term from the Industrial Revolution [1800s]) are people who are anti-technology. They paint a negative picture of technology and point to the negative impacts it has on people and their jobs. Luddites generally want us to go back to the "good old days." Of course, how good were the "good old days," really? For example, some Luddites might suggest that going back to the horse and buggy days would improve air quality because it would eliminate the tail pipe emissions from millions of cars and trucks. However, do they realize that in 1900 in New York City alone, the horses for horse-drawn taxis produced 4 million pounds of horse manure daily? Also, the horses had to be stabled away from the cabs at night because the ammonia fumes from the horse urine blistered the paint on the cabs. In addition, air-borne bacteria in the horse manure contributed significantly to a tuberculosis epidemic in New York City. Do we really want to go back?

After having partaken of technology and its benefits, it is difficult to go back and give it up. Who wants to own and drive a car in the summer that does not have air conditioning? Who wants to give up computers and photocopiers and go back to the days of typewriters and carbon paper? Who wants to go back to the days of cross-country travel without airlines, trains, or automobiles? It is even difficult (really, impossible) to assess the point in time that would be considered to be "free" of technology. Technology has been with us from the beginning of civilization and, in fact, is generally seen as the trigger for the creation of civilization itself.

A further appreciation for the attitudes of people toward technology can be gained by understanding the three phases of technology.

The first phase is technological adoption. This phase is successful because the technology to be adopted has increased some human capability and, generally, solved some problem.

The second phase of technology then occurs. In this phase, the problems with the adoption of the technology are seen. Some people lament that the technology was ever adopted. An obvious example of these two phases is seen in the development of nuclear power for making electricity. The technology was initially welcomed. However, after problems at Three-mile Island and concerns about disposal of nuclear waste, no new nuclear plants have been built in the United States for over two decades. It should be noted, however, that in many other countries, especially Belgium and France, many new nuclear plants have been built. Both of these countries use nuclear power as the principle source for electricity, often finding that they have sufficient electrical power to export it to other countries. What is the difference between these European countries and the U.S.? They have entered the third phase of technology adoption. That is the phase where the original technology has been modified to overcome the problems identified in the second phase. However, success in the third phase also requires that the public be positively responsive to the changes. Therefore, the third phase requires not only technological improvement but also attitude change. It is again seen that technology viability is more than just engineering—it must deal with the entire culture of society.

Cultural Lag

People embrace and adopt new technologies at varying rates. There are early adopters—those who acquire and use the new gadgets, devices, and systems as soon as they become available. These earlier adopters are sometimes referred to as "techno-geeks."

Why don't we all embrace and adopt new technologies immediately? While slow adoption of some new technologies can be associated with cost, cultural values and social change often are the reasons why most people lag behind the early adopters. This is referred to as "cultural lag." Some people wait until a new technology is a little more mature before acquiring it. These people embrace new technologies, but they are a little more conservative. They want to make sure of the benefits and see others using it before adopting it. These are the people who waited and purchased VHS VCRs and DVD players as opposed to purchasing beta VCRs and laser disk players. Still others wait until the technology becomes so ubiquitous that they are forced to use it just to continue to function in the modern world. These late adopters are often those who resent the technology and have the highest frustration level with it.

As companies develop new technological devices, they need to be aware of the phenomenon of cultural lag as they develop plans for distribution and sales. Companies should excite the early adopters, reassure the mainstream, and comfort the late adopters. Clearly, cultural lag can impact market penetration.

Technology and Globalization

In society, progress is generally associated with economic well-being—of individuals, families, companies, and nations. Technologies, such as computers and the Internet, have spurred globalization. Technology deployment in many developing nations is impacting the U.S. balance of trade and our economic well-being. Technology is making the world a level playing field. This is allowing people in many countries to acquire the capabilities that we (the U.S.) once thought were particular to our advanced economy (Friedman 2006). Now, concerns are being expressed about the number of manufacturing jobs being sent offshore. Technology is the enabler for many of these offshore jobs because it allows overseas manufacturing to be competitive with manufacturing in developed countries.

Should a country fight technology to save manufacturing jobs? Should we become Luddites? Should we capitulate and give up on domestic manufacturing jobs? To fight for jobs, the U.S. must adopt a progressive attitude toward technology. We must recognize that technology is powerful and dynamic. It alters our perceptions of time and space. It is pervasive and irreversible. It requires specialized knowledge and ways of thinking. It is cumulative and thrives on innovation. We cannot stop the growth of technology any more than a person can stop the waves of the seas; nor can we stop the growth of jobs overseas. However, this change in society should be viewed as an opportunity and not as an evil. Technology is not really the reason why jobs are moving overseas, but it may be the answer to the problem of job drain. It is clear that those who have technological knowledge have power over those that lack knowledge of technology and how it works. Likewise, creative innovators who develop new technologies that are needed and/or wanted are in a position of power. The answer to the problem of job drain is, therefore, more technology, not less.

Competing in the Global Market

Manufacturers in highly developed countries will have difficulty competing head-to-head in labor-intensive manufacturing environments like that of China, which has very low wages, when the only difference is the labor. The U.S. can, however, compete in the areas of design, technology development, and technology management. These areas are connected to ingenuity, critical thinking, and creativity.

William Wulf, President of the National Academy of Engineering (NAE), has expressed alarm that the U.S. school systems are not preparing students to understand technology, how it is developed, and how it impacts life and work (Wulf 2000). He has called for schools to prepare "technologically literate" students. This issue was confirmed by the Committee on Technological Literacy (CTL), a group of experts convened by the NAE and the National Research Council's (NRC's) Center for Education (Pearson and Young 2002). The National Science Board's (NSB's) Task Force on National Workforce Policies for Science and Engineering has also raised concerns about the declining numbers of students pursuing engineering and scientific careers in the U.S. (National Science Board 2003).

It is clear that a nation must have "excellence in discovery and innovation" to continue to lead in the global economy.

CASE STUDY 22-1

Composite Fabricating Economics—Helicopter Fuselage

The decision of whether to make a helicopter fuselage out of aluminum or composites is a good example of composites economics and the division of the cost of goods sold into subdivisions so that the costs can be studied more accurately. Figure 22-4 compares the critical elements of the cost of goods sold for an aluminum versus a composite fuselage. The costs of a composite fuselage are higher for fabrication, materials, and quality, but the much higher costs for assembly of aluminum more than offset the lower costs in the other elements. For example, a composite fuselage requiring 7,000 mechanical fasteners would require 85,000 mechanical fasteners in a corresponding aluminum fuselage. Further, whereas a composite fuselage would have 1,500 parts, the aluminum fuselage would have 10,000 parts. Bonding composite parts as an alternative to mechanical fastening further reduces assembly costs.

Because the fabrication costs for composites are high, this is the logical place to attempt to save costs. One of the methods employed for fuselage structures is fiber placement. This method has high capital costs but is automatic and, therefore, less costly. The savings have been so good that the fuselage of the Osprey, a vertical take-off and landing aircraft, is being made using fiber placement.

SUMMARY

Even though composites are often highly favored for their performance, most real-life situations require that the composite costs also be favorable or, at least, not so high that they are outside the realm of feasibility. Therefore, costs should be analyzed in detail to understand them fully and discover methods that might be used to reduce them.

Other aspects of business are also important to make good composite parts and to continue to be competitive in the ever-shrinking manufacturing world. Planning in a world of change is critical to success. However, that planning needs to be more than just an outline of the path the company will take in the future. Because of the uncertainties of today's world, a system of dynamic planning in which alternate paths are explored is now the preferred way to plan strategically.

Another of the most important aspects of business is creativity. Increasingly, the company that is creative is the company that will develop the products first and, therefore, receive the highest profits. Several practices were identified to help the company become and stay more creative.

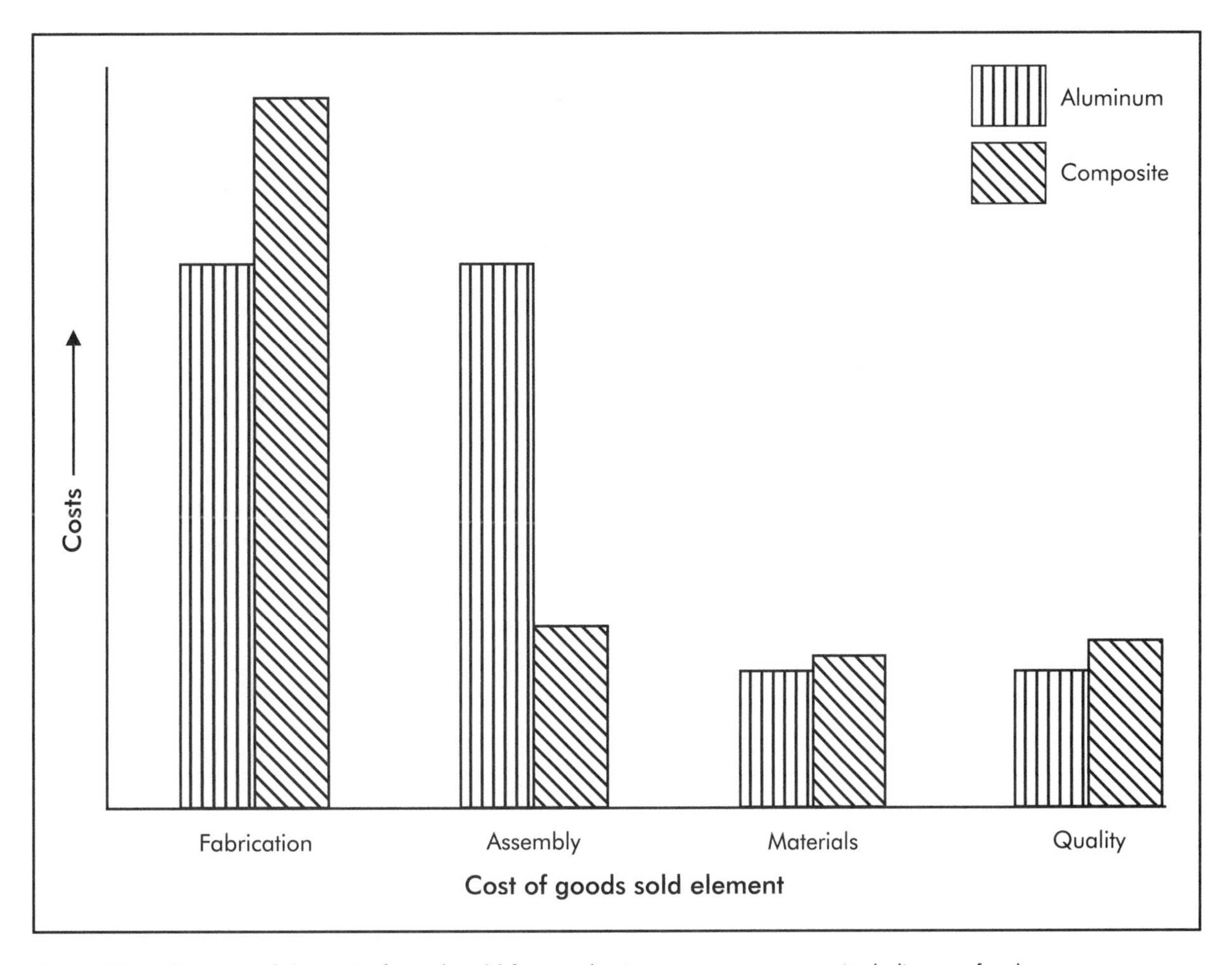

Figure 22-4. Elements of the cost of goods sold for an aluminum versus a composite helicopter fuselage.

Ethics is also a critical factor in long-term profitability. Customers, competitors, and others in the industry expect that a company should be ethical. As identified herein, being ethical is a process and, much like quality, needs to be inherent to the company and its culture. Along with ethical behavior, the company needs to work within an environment of healthy dissent. When done with the intent of helping, this dissent can make teams more effective and the company more responsive to its customers and suppliers.

Technology is a critical element of today's world. No industry is more sensitive to it than the composites industry because it is dependent on recently developed technology. Moreover, the composites industry exists, in part, because it has demonstrated superior technology to other materials and manufacturing methods.

Within this highly technological world, the Internet is a critical tool. Some perspectives on the use of the Internet for improved company performance and marketing were given.

LABORATORY EXPERIMENT 22-1

Composite versus Metal Cost Comparison

Objective: Discover the differences in costs for a simple part made of either a composite or a metal.

Procedure:

1. Using the Internet or the library, search for a case study that compares the costs for the same part made using either a composite or a metal.
2. Summarize the report, indicating the areas in which the composite's manufacturing method is lower in cost and those areas where it is more expensive.
3. Give the reasons for your findings.
4. Suggest methods that might be used to lower the costs of the composite's manufacturing method.

QUESTIONS

1. What is the cost of goods sold?
2. What is meant by "the bottom line?"
3. What are the three types of decisions made in a company and describe when each is appropriate.
4. What is meant by the process of ethics?
5. Describe two similarities between ethics and quality.
6. Give three situations when dissent would be valuable.
7. Give two situations when dissent would be harmful.

BIBLIOGRAPHY

Arbinger Institute. 2000. *Leadership and Self-deception*. San Francisco: Berrett-Koehler Publishers, ISBN 1-57675-094-9.

Business Week. 2002. "Business' Killer App: The Web." Online version at http://www.businessweek.com/technology/content/apr2002/tc20020415_3981.htm. April 15.

Covey, Stephen R. 1989. *Seven Habits of Highly Effective People*. New York: Simon and Schuster, ISBN 0-67166-3984.

Crandall, Rick, ed. 1998. *Break-Out Creativity*. Corte Madera, CA: Select Press, ISBN 0-9644294-7-0.

Dresselhaus, William F. 2000. *ROI: Return on Innovation*. Portland, OR: Dresselhaus Design Group, Inc., ISBN 0-9679209-0-6.

Easum, William M. 1998. *Dancing with Dinosaurs*. Nashville, TN: Abingdon.

Friedman, Thomas L. 2006. *The World is Flat: A Brief History of the Twenty-first Century*. New York: Farrar, Straus, and Giroux.

Gibbons, J. H. 1988. "Technology Education: The New Basic," video. Albany, NY: Delmar Publishers.

Kelley, Thomas. 2001. *The Art of Innovation*. New York: Doubleday, ISBN 0-385-49984-1.

MacKenzie, Gordon. 1998. *Orbiting the Giant Hairball: A Corporate Fool's Guide to Surviving with Grace*. New York: Penguin Putnam, Inc., ISBN 0-670-87983-5.

Nair, Keshavan. 1994. *A Higher Standard of Leadership*. San Francisco: Berrett-Koehler Publishers, ISBN 1-881052-58-3.

National Science Board. 2003. "The Science and Engineering Workforce: Realizing America's Potential." http://www.nsf.gov/nsb/. August 14.

Pearson, G. and Young, A. T., eds. 2002. *Technically Speaking: Why All Americans Need to Know More About Technology*. Washington, DC: National Academy Press.

Peter, Laurence J. and Hull, Raymond. 1969. *The Peter Principle*. New York: W. Morrow.

Roberts, Royston M. 1989. *Serendipity: Accidental Discoveries in Science*. New York: John Wiley and Sons, Inc., ISBN 0-471-60203-5.

Senge, Peter M. 1990. *The Fifth Discipline*. New York: Doubleday.

Strong, A. Brent. 2005. *History of Creativity, Pre-1500*. Dubuque, IA: Kendall-Hunt Publishing Company, ISBN 0-7575-2277-7.

Strong, A. Brent. 2006. *History of Creativity, 1500-Present*. Dubuque, IA: Kendall-Hunt Publishing Company, ISBN 0-7575-2692-6.

Strong, A. Brent. 2006. *Plastics: Materials and Processing*, 3rd Ed. Upper Saddle River, NJ: Prentice-Hall, Inc.

Thompson, Charles. 1992. *What a Great Idea*. New York: Harper Perennial, ISBN 0-06-096901-6.

Webster's Ninth New Collegiate Dictionary. 1983. Springfield, MA: Merriam-Webster.

Wheatley, Margaret J. 1992. *Leadership and the New Science*. San Francisco: Berrett-Koehler Publishers.

Wulf, W. A. 2000. "The Standards for Technological Literacy: A National Academies Perspective. *The Technology Teacher*, 59 (6), pp. 10–12.

23

Composites Applications

CHAPTER OVERVIEW

This chapter examines the following concepts:

- Introduction
- Traditional composites markets
- Learning lessons—space structures
- The ultimate composite structure—IsoTruss®
- Critical market—armor
- Breakthrough markets—commercial and corporate airplanes
- Future-today markets—unmanned vehicles

INTRODUCTION

Not all composites uses can be discussed in any single book. The objective of this chapter is not to attempt such a task, but rather to illustrate some traditional and specific applications, which are instructive for a variety of reasons. The traditional composite uses will be presented by markets in the order of the size of the market. This order will, for some, be surprising even though the market sizes were given in Figure 1-2 (in Chapter One of this text). This surprise is because the largest markets are discussed less often in the popular press and even in classrooms than the smaller but higher-technology markets. However, to really appreciate the full breadth of the composites marketplace, the traditional markets should be considered as well as those that are more leading edge.

Some specific composite products will be discussed. The lessons learned about composite materials, manufacturing processes, and the entire process of making, transporting, and using composite materials are shared. Other discussions will look at products that might be reasonably made internationally, including comparisons of the types of manufacturing that might be done. Some breakthroughs in composites that have the potential to increase their use far beyond the current levels and even beyond what can be reasonably imagined will be discussed. These breakthroughs are the result of years of accumulated familiarity with composites. Some advantages of composites go beyond the normal weight savings and molding capabilities traditionally cited. Further, innovations in composites manufacturing add to the potential of great success with the breakthrough technologies.

TRADITIONAL COMPOSITES MARKETS

Transportation

Land transportation is a major market for fiberglass reinforced composites. The 100% composite body of the 1953 (and subsequent years) Corvette® demonstrated the capabilities of this material to the automobile industry. Weight savings has always been an advantage that composite manufacturers have tried to use as a reason for the adoption of composites in automobiles. While composite's weight savings is certainly an advantage versus steel, the makers of steel have been successful in creating new grades of high-strength steel that allow parts to be made thinner and, therefore, lighter. Hence, large portions of most automobile bodies are still made of steel, especially relatively flat areas that are easy to form. Even though the promise of weight savings is always part of the investigation into using composites as a replacement for steel, the reality is that other properties of composites also must be important for them to be used in automobiles.

The most common body panels made of composites are highly shaped sections, such as at the front and rear, where the molding capability of composites gives them an advantage over metals. Some composite side panels are proven to have good "non-dent" capabilities, especially if the material contains a high amount of thermoplastic. The most difficult body panels for composites to capture have been the large horizontal surfaces like the hood and the trunk (bonnet and boot). This is because composites have had problems with sagging in hot weather. Recently, some cars (such as the 2004 Corvette® Z06 and later models) have used carbon fiber as the reinforcement in hood panels. The greater stiffness of carbon helps prevent sagging. Like the fiberglass sheet molding compounding (SMC) hood used on the regular production Corvette, the carbon fiber hood is made with an inner and an outer panel. The carbon fiber hood weighs 20.5 lb (9.3 kg) and that weight is 10.6 lb (4.8 kg) lighter than the SMC hood. The outer panel is only .048 in. (1.2 mm) thick. Other automotive makers are examining carbon fiber for use in high-performance automobiles. In the 2006 model year, BMW and Volkswagen have both introduced carbon fiber parts into their regular production vehicles.

With increasing frequency and volume, automobile makers seeking improved performance are switching to composites for applications other than body panels. In excess of 150 different automotive applications for fiberglass composites have reached regular automotive production. Desirable properties in addition to weight savings and the ability to be molded include freedom from rust and corrosion, parts consolidation, and lower tooling costs.

Most automotive composite parts are made with SMC or bulk molding compound (BMC). The economics of these materials has been improved by the continuous lowering of cycle times. Early problems with paintability and smoothness have largely been solved for exterior parts. Interior parts have been a strong market for many years. For example, composites heating and air conditioning ducts have the advantages of easy shaping (molding) and low weight while meeting the rather modest strength and stiffness requirements.

Truck tractor bodies and sleeper units, as well as entire trailer bodies are routinely made of composite materials. Many of these truck cabs are made by spray-up with resin transfer molding (RTM) growing rapidly as an alternate process. Truck makers view the weight savings and low maintenance requirements of composites as strong economic incentives for their adoption. Thus composites have made greater penetration into truck than into automobile manufacturing. The use of composites has allowed consolidation of many parts in trucks, further improving their economy.

In light trucks, a recent innovation is the truck pickup box. The composite version reduces the total weight by 50 lb (23 kg). The truck has been field tested for more than 1,000,000 miles (1,609,000 km) over extreme environmental conditions of cold, heat, and off-road operation. The outer panels were made of reinforced nylon, which bolt/snap on for easy removal, allowing repair or replacement with minimum downtime. The structural inner panels are corrosion resistant and tough. The total box load-carrying capacity increased to 1,000 lb (450 kg) versus the 600-lb (270-kg) capacity of the competitive steel box.

Many transportation applications continue to be explored. For example, a drive shaft made from epoxy/carbon fibers with bonded U-joints was found to have an ultimate torsional load capability equivalent to the metal shaft it would replace. However, the composite shaft would result in a 60% weight reduction with only somewhat higher cost than the metal shaft.

Several metal engine parts have been targeted for replacement with composites. Some of the components that seem likely to be replaced include: valve covers, water pumps, pulleys, oil pan, and even some internal parts like pistons and connecting rods. Some engineers have even suggested a nearly total replacement for the engine by targeting the engine block, cylinder head, valve tappets, valve-spring retainers, and intake-valve stems.

The use of composites for automobile frames has been examined in a few experimental automobiles with favorable results. In addition to the weight savings, a major advantage was crashworthiness. Composite structures collapse progressively (as the laminate layers separate) and this gives an increase in the energy-absorbing potential in comparison to metals. A significant proof of the crashworthiness of composite body frames was shown in a MacLaren Formula One racing car. The driver compartment and surrounding structure in the car was made of epoxy/carbon composite material. Initially, the weight savings was the driving force for composites use. After a serious crash that would probably have destroyed a metal compartment, but that did little damage to the composite compartment (the driver was unhurt), composite material is now the standard in Formula One cars.

Items as large as boxcars and light rail cars are now commonly made from fiberglass composites. Fiberglass reinforced plastic (FRP) is also used extensively in the molds for making many of the parts already discussed. The ability to make the composite flame retardant is important in some of these applications.

As more aggressive governmental weight reduction (fuel efficiency) goals are enacted, composites will become more competitive and carbon fibers will begin to compete for the transportation market share. At present, the use of carbon fiber is restricted to a few highly specialized applications and some concept cars. However, aggressive research programs are defining the value and performance of carbon fiber composites. Some of the research being pursued includes: carbon fiber composite durability, joining technologies, processing methods and economics, energy management, and crash energy management.

Construction

The construction market for composites is large and diverse. Some of the products have been used for many years, such as corrugated fiberglass patio/car park covers. This product is made by pultrusion or lay-up of fiberglass mat and unfilled resin. The product competes against sheet metal (often galvanized steel). The composite material's advantages are low cost, light weight, and lack of corrosion.

Another traditional product is the tub or tub/shower unit. These products are made by spray-up, often in a highly integrated factory in which the molds move through the various spray-up booths and curing ovens on an overhead conveyor system in a process reminiscent of an automobile assembly plant. Large whirlpool spas are made by spray-up. However, they are so large that they are rarely moved through a factory the same way as the smaller tub and shower units.

Fiberglass reinforced wall panels are especially important for the institutional market because these can be cleaned and, especially for hospitals, disinfected easily. These panels are often highly decorative. When flame and smoke retardant resins are used (especially phenolics), the panels will meet even the most stringent fire codes. Therefore, they can be used for mass transportation applications.

To date, composite fibers in bridges are limited to those for pedestrian and light-duty automotive traffic and a few components of heavier automotive traffic bridges. The lack of performance data for composites is undoubtedly limiting their application to higher-tolerance automobile bridges. With additional experimental installations and time to collect data, many more bridge installations are likely to occur. A related application is the wrapping of concrete highway overpass pillars with carbon fiber composite. The mechanical properties of the composite are highly complementary to the concrete pillars. Concrete has good compression properties, but is not good in tension. Composites are just the opposite. Hence, the weight of the overpass can be supported by the concrete (in compression) but the concrete itself will be supported by wrappings of composite (in tension). Although actual field tests have not been installed for sufficient time to gather long-term (10–20 years) data, laboratory tests on full-size bridge structures and the limited number of installations currently in testing are encouraging. There are two major purposes for wrapping these columns. The first is to remediate old columns by reinforcing the concrete and preventing it from flaking off. For new construction, wrapping with composite simply extends the life of the columns. The second purpose for wrapping the columns is seismic protection. It has been found that the likelihood of the column collapsing during an earthquake is significantly reduced if the column is wrapped with composite. This application could, of course, be enormous in the volume of fiber used.

A portable, self-supporting bridge to cross ravines and small rivers has been developed by the North Atlantic Treaty Organization (NATO). It uses epoxy and carbon fibers and can cross much wider gaps than comparable metal bridges. The collapsed bridge structure is carried in a truck or attached to a military vehicle (such as a tank). When a ravine or small river is encountered, the carrying vehicle is positioned on the edge of the gap to deploy the bridge. The first step in the deployment is to extend a support beam across the gap. This is a box-shaped composite beam that is roughly 39 in. (100 cm) high and 20 in. (50 cm) wide and comes in 20 ft (6 m) sections. The sections are connected so that as one is extended, the next can be unfolded from the end of the previous section and then lowered to the horizontal. The beam is, therefore, deployed across the gap in this unfolding manner and is cantilevered from the support vehicle. Note that the weight of the beam must be as little as possible. If the beam were too heavy, the weight of it could tilt the support vehicle and the bridge would fall. The weight and length can be calculated using simple cantilever beam theory. After the support beam is deployed across the gap, the bridge tracks are rolled across the beam. These tracks are about 10 ft (3 m) wide and 4 ft (1 m) high.

The mass production of standardized tubes, shafts, I-beams, angle beams, ribs, rails, and other simple parts is a growing market for composites. These parts are used within structures to add strength or to build upon. All of these composite parts are made in highly reproducible processes and, therefore, will have predictable performances.

Marine

In marine applications, boat manufacturers have adopted the name "fiberglass" and made it synonymous with any reinforced plastic composite. Approximately 70% of all outboard pleasure boats 15 ft (5 m) and longer are now constructed of glass fiber reinforced plastic materials. The major benefits of using glass fiber reinforced composites in marine applications include:

- almost any boat design or size is moldable,
- seamless construction,
- high strength and durability,
- minimum maintenance, and
- freedom from corrosion, rust, dry rot, and water logging.

Other marine applications include submersibles, hovercraft, mine-sweepers, sealed pontoons, outboard motor shrouds, canoes, and kayaks.

The most prominent use for composites in the marine industry is for hulls, decks, and other structures. These are generally made of fiberglass and polyester, although some use vinyl ester for improved resistance to water blistering.

In high-performance (racing) boats, composites also have taken a major position. The masts and spars of sailboats are usually made of composites—often carbon fiber and/or aramid. In some high-performance boats, the entire hull and even the deck are made from carbon fiber and epoxy.

Military boats also use composites extensively, although usually not for hulls, except in the case of specialized boats such as mine-sweepers. However, composites are increasingly used for housings for many components on ships because of the reduced maintenance of these structures over metals.

The excellent radar transparency of some composites has led to their use as radomes on ships and for other components in which signal transmission is to be maximized. The major problem encountered in using composites in these structures seems to be ensuring that they are fabricated without bubbles or voids, which may interfere with signal transmission.

Composites have found wide use in pressure hull and buoyancy structures in underwater applications (submersibles). Submersibles have been developed and successfully used (remote control) at depths as great as 5,000 ft (1,524 m) and at hydrostatic pressures of 2,000 psi (13,789 kPa). However, commercial and manned structures have been generally limited to depths of 1,200 ft (366 m). The most common material for these applications is fiberglass epoxy and polyesters, although aramid wound structures also have been successfully used. The structures are normally filament wound with hoop windings as the predominant pattern to optimize the strength at given weights.

Corrosion Products

Pipes, tanks, well casings and the various components that complement these structures represent the majority of the corrosion product business for composites. Corrosion products are usually made with fiberglass. However, the choice of resin is strongly dependent on the nature of the fluids that will be transported or stored. For water transport/storage, polyester resins dominate. Water transport and storage is the highest-volume portion of the corrosion product business. Composite pipes are rapidly replacing other materials such as concrete, fiber-cement, steel, and ductile iron. The combined issues of cost, durability, weight (to transport and install), and corrosion resistance favor composites for the transport and storage of other liquids too. If the liquid is not water and sometimes even when it is, the resin of choice is vinyl ester. Difficult liquids may require epoxy.

Another corrosion product often used in chemical plants is grating. It is used as walkways in many industrial plants, but especially where the chemical environment might corrode metal grating. The grating is usually made by extrusion but some lay-up and even RTM gratings are commercially available.

A market of immense potential but currently limited in actual sales is components for oil drilling and related oil-field activities. Some opportunities for composites include the pipes, tanks, and well casings. Sometimes the high temperatures (up to 400° F [204° C]) and highly corrosive environments require that high-performance resins be used, which are now routinely processed in many applications. Other oil-drilling applications include offshore platforms, tethers to hold the platforms, and the underwater housings used to encase some of the drilling operations. Some significant problems encountered with metal components in these applications are solved by the use of composites. But the composites also have some problems. In particular, abrasion resistance and the composite-metal interface (joints, etc.) have been challenges that need to be worked on. Liners and exterior coatings have been suggested as ways to solve these problems and some promising results have been obtained. The composite-metal interface must be able to withstand shock forces. The use of tough resins in the couplings may yield sufficient performance. In spite of the difficulties, the use of composites for oil drilling, especially in offshore wells, is expected to grow tremendously in the near future.

Electrical

The most common uses for composites in electrical applications utilize the nonconductive nature of these materials. Typical examples include circuit boards and insulators. Under certain conditions, composites can be fabricated to absorb nearly all the radar energy that impinges upon them. This is usually done by fabricating the part with a tapered impedance in which the conductivity varies continuously, or in discrete steps, from essentially zero on the incident surface to the greatest electrical conductivity on the rearward surface.

The use of composites as a substrate for printed circuit boards is one of the most important electrical applications. These boards are made by bonding a substrate to a conductive material, such as copper foil, and then etching the circuit pattern onto the copper. As substrate material for printed circuit boards, composites resist the warp and twist common in printed circuit boards because of the presence of dissimilar materials (substrate and copper foil), which are bonded together to make the board. Therefore, the design of the composite should resist shear forces. Most boards are made of fabric reinforcement material, which gives some shear strength. Typical materials would be:

- paper/phenolic,
- cotton cloth/phenolic,
- nylon cloth/phenolic,
- glass cloth/phenolic,
- glass cloth/melamine,
- glass cloth/silicone,
- glass cloth/epoxy, and
- paper/epoxy.

The choice of the substrate material is dictated by price and its mechanical, electrical, thermal, and physical properties. The most common materials for high-quality substrates are glass cloth and epoxy.

One of the problems encountered in the use of composites for printed circuit boards is the drilling of the holes. Mechanical drills quickly become dull, but are still used extensively. Laser drilling is also used widely. Although more costly for the equipment, the lasers last much longer and generally allow increases in throughput speed.

The future of electronics is toward miniaturization and that trend is especially true in printed circuit boards. Since the substrate must provide the proper electrical, thermal, and mechanical environment for the components, it is part of the miniaturization process. A difficulty in further reducing size, however, is the removal of heat generated in the circuitry. The distance between components poses additional problems. Solving these problems requires greater density in the substrate material. Materials now being investigated for boards include polyimides, aramid, quartz, and ceramics.

Standoff devices are used to ensure that the wires are insulated from the pole. These insulators have been made of ceramics for many years. However, the brittle nature of ceramics has led to the widespread adoption of fiberglass reinforced polyester or epoxy insulators. A related application is the tools used by electric company service people to work on the power lines. These devices need to be non-conductive, tough, and lightweight. Polyester fiberglass poles have proven to be excellent materials for tool handles.

Because carbon fibers conduct electricity, at least to some extent, some electrical applications requiring conductivity have been identified for them. For example, carbon fibers are used as brushes for motors. Carbon fiber laminates are also used for high-performance battery plates. These applications are not, however, as prevalent as the applications in which the composite is nonconductive.

Consumer/Sports

The number and variety of sports equipment uses for composites is large and continues to grow at a tremendous pace. Some of the uses for composites in the sports equipment industry include the following: golf clubs, racquets (tennis, racquetball, and squash), pole-vaulting poles, archery bows and arrows, skis and ski poles, water skis, snowboards, tent poles, fishing poles, bobsleds and bobsled tracks, bicycles, baseball and softball bats, and rifle barrels and stocks.

The use of composites in consumer/sports applications is generally dependent on the particular material's stiffness, strength, and weight savings. In some cases, the ability to tailor the placement of the fibers to match the anticipated loads has been a major advantage. For example, a typical fishing pole design utilizes a hollow tubular construction to minimize weight and optimize the strength and sensitivity of the rod. The rods are wound on a removable, tapered mandrel. The inner spiral (hoop) windings give the strength needed to eliminate the pinging or collapsing of the hollow tube when it is flexed. The second layer of longitudinal fibers gives the rod sensitivity because the vibrations of the fish are readily transmitted along the fibers. The longitudinal fibers also increase the bending strength of the rod. Additional layers are sometimes added to adjust the stiffness and strength to meet specific requirements.

An ambitious program to develop six new bobsleds for the 1988 Winter Olympics was undertaken with composites as the likely construction material. The design parameters included the following considerations: safety, interface with the athletes, structural design, handling and control, mechanical components, runners, and aerodynamics. All current sled designs were reviewed and a new design was proposed with a chassis made of .375 in. (9.53 mm) honeycomb core covered by six layers on each side of graphite/aramid cloth with epoxy resin. The shape of the chassis was refined in wind-tunnel experiments after the other design factors had been implemented. The composite bobsled chassis showed a 40% reduction in aerodynamic drag over sleds used previously and gave good handling and speed performance in actual runs. Now, composite sleds are the standard for bobsledding and other sled competitions.

The use of composites for rifle stocks seems obvious from a stiffness, weight, and maintenance point of view. The barrel is actually constructed of a lightweight metal, which is overwrapped with carbon fiber epoxy composite. The overwrapped barrel is about four times stronger than steel, which means better safety. The composite barrel casing is four to five times stiffer than steel, leading to improved accuracy. The barrel is five to six times lighter than steel and that gives better comfort and mobility. Moreover, since steel barrels expand and contract with temperature change due to a non-zero coefficient of thermal expansion, the flight of the bullet changes as the rifle is used. But, with the proper wrapping of the fibers, the composite casing can have a zero coefficient of thermal expansion. This means the performance and accuracy of the composite rifle does not change with heating from repeated firings.

A large-volume application of composites in the recreational market is for mini-transport carts. This market includes golf carts, transporters for elderly people, in-factory transporters, utility carriers, dune buggies, and even some street mini-cars. The popularity of these vehicles signals tremendous growth in this market.

Appliances

Fiberglass has a natural advantage in water applications because it does not corrode (rust) as many metals do. This means that maintenance is substantially reduced over metal components. The low-maintenance requirement coupled with easy molding and low costs have made the use of composites advantageous to the appliance market.

Most appliance composites are reinforced with fiberglass. Typical applications include housings for motors, ducts, fan blades, and interior panels and compartments. Many of these applications are made with thermoplastic composites because the cost demands are tight and the structural requirements are low. The concept of thermoforming thermoplastic sheets is especially inviting.

Medical

For people who have had a leg or foot amputated, composite prostheses offer a chance for greater mobility and capability over any other artificial limb material. Composite limbs are usually of carbon fibers in epoxy. However, for short limbs, fiberglass/epoxy is also used. These devices are roughly shaped like a foot and leg but are flat. The roundness needed to simulate a normal leg is achieved by covering the composite structure with flexible foam. The composite can, therefore, be thought of as analogous to the bone and tendon structure of the leg. A composite leg is shown in Figure 23-1.

The composite structure is typically tapered in thickness from about 20 layers in the toe to 80 layers at the top. These thicknesses vary depending on the weight and activity of the person. A light person may require as few as 15 layers at the toe and

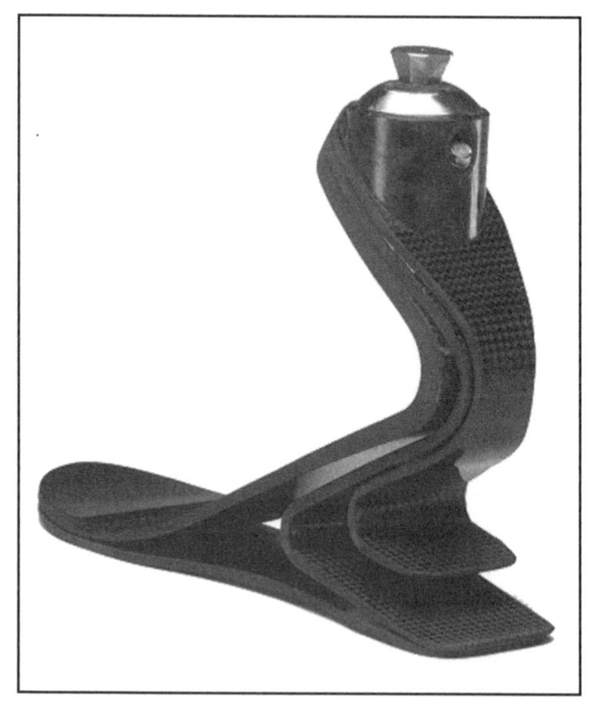

Figure 23-1. Composite leg/foot prosthesis. (Courtesy Applied Composite Technology [ACT])

50 layers at the top. Several standard sizes are offered to meet most weight and activity combinations and legs can be custom built.

The layering of the composite helps give the feel that is desired. The structure is flexible enough to give good springiness, but strong enough to give good flexural strength and fatigue resistance. The near-perfect return of energy (no stress-strain hysteresis) of composites means that the movement and jumping capability with the device are far greater than with a device made of metal. The metal device is much heavier and stiffer.

The problems with the composite material include interlaminar shear, especially at the top where the prosthetic device is bolted to the holder, which is the device that couples the artificial leg to the person. Fatigue failure is also a concern, along with the constant difficulty of any prosthetic remaining attached to the body. However, great progress has been made in these problem areas and artificial limbs allow good athletic capabilities as well as near-normal walking and running. Useful lives of the devices have been several years.

Composites can be tailored to have almost the same mechanical properties as human bones. This capability has proven to be especially valuable in making surrogate human bodies for research into the effects of various traumas. For example, a recent research effort developed a torso and the underlying skeletal structure, which is being used to investigate the effectiveness of bulletproof vests. The surrogate body is instrumented to detect the degree of trauma from a shot fired into the bulletproof vest it is wearing. To verify the extent of the damage, the body can be taken apart and inspected for damage such as broken bones. The ability to tailor composites to the same tensile stiffness and strength as human bones allows these comparisons to be meaningfully made.

The near transparency of composite materials to x-rays has led to their use for tables on which x-rays are taken. Also, the arms and other moving parts of portable x-rays are often made of composites. However, the impetus for their use in this instance is probably for their lighter weight rather than their transparency to x-rays.

Aerospace

Even through the aerospace market is only about 1% of the total composites marketplace, it remains highly important for the entire composites industry because of its high-tech image and the technological developments that continue to emerge from it. The aerospace industry is the largest consumer of carbon fiber as shown in Figure 23-2. Therefore, even though fiberglass is used in some aerospace applications, this

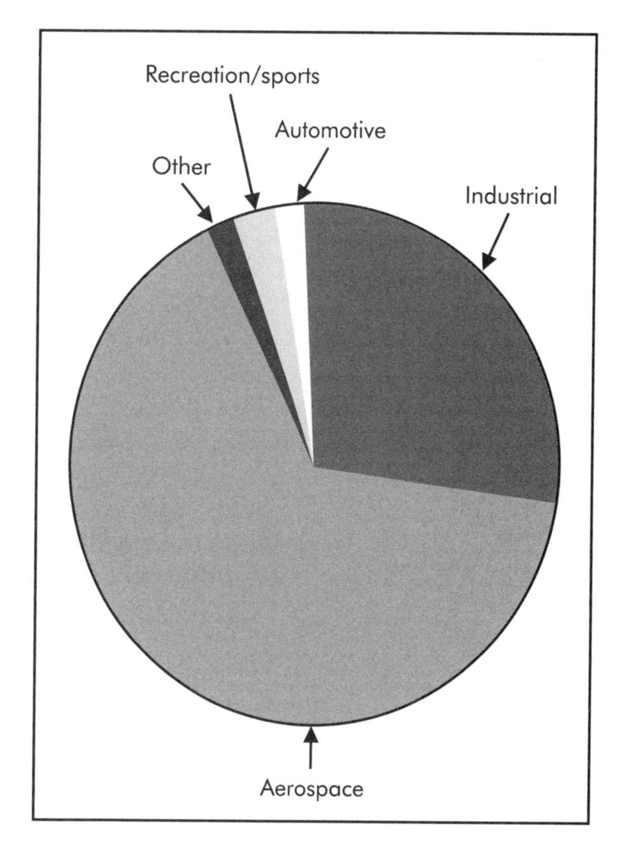

Figure 23-2. Carbon fiber markets.

discussion of the aerospace industry will focus principally on carbon fibers.

For many years the most compelling economic incentives for carbon fiber usage were in space vehicles because of the extremely high cost of launching weight. Estimates suggest that the cost to launch 1 lb of weight into space is about $10,000 ($4,536/kg). Therefore, carbon fibers were used to make the casings of rocket motors, which were the greatest single weight component of many space vehicles. Other components of space vehicles commonly made of carbon fiber include struts and other support members. Carbon fiber is especially useful in support structures because the coefficient of thermal expansion (CTE) of the composite parts can be tailored to be near zero. This is possible because of the negative CTE of carbon fibers. Hence, when high-precision orientation is required, such as for the Hubble telescope, the preferred support material is carbon fiber composite. Some space vehicles, such as the space shuttle, use carbon-carbon composites in parts where its superior heat resistance is critical.

A problem in the use of composites in space is their susceptibility to radiation. Composites crack when exposed to high levels of radiation. Some coatings have been developed to improve the crack resistance of composites, but still the problem persists. However, metals and even ceramics are also affected by the aggressive space environment. Thus the designer must choose which material is least likely to be affected in a way that is detrimental to the particular part being designed.

An application that is similar to space vehicles is missiles. Composites are used extensively in missile motor housings and in the launch tubes from which the missiles are fired, especially when the missiles are mounted on aircraft. While not all of these missile applications use carbon fiber (some are fiberglass), its use is increasing as weight savings and stiffness become ever more critical to performance.

Aircraft parts are also a significant market for carbon fibers. Initially, carbon fiber composites were used in some non-critical components of aircraft such as access doors and engine cowlings. One aircraft application for composites that has been important for many years is waste tanks. Originally, the waste system on aircraft operated with a flush system that required the tanks to withstand positive internal pressure. Later waste system versions operate with a vacuum and therefore require the tanks to withstand an internal negative pressure.

With continued good performance, the use of carbon fiber composites has continued to increase. Military aircraft adopted carbon fiber components more quickly than commercial aircraft. Some of the military aircraft applications included the empennage and speed brake on the F-15; the empennage on the F-16; the wing skins and empennage on the F-18; the wings, empennage, and fuselage of the AV8B; and the wings of the A-6. The advent of stealth technology has pushed carbon fiber composites even more strongly into military aircraft. This is because stealth technology depends, to some extent, on the properties of carbon fiber composites. Therefore, entire planes like the B-2 and the F-22 are made of composites.

Helicopters are also important applications for composites. Most helicopter blades are made from fiberglass or combinations of fiberglass and carbon fiber. The bodies of many helicopters are also made of composites. Especially in the case of military helicopters, carbon fiber is becoming the standard reinforcement. In tilt-rotor aircraft, carbon fiber composites are often the key to acceptable performance.

Industrial

Industrial products are important as the second largest market for carbon fibers.

These products often rely on carbon fiber composites because of their ability to save energy costs. The industrial market has long had many fiberglass applications but, increasingly, carbon fiber products are being used.

The use of fiberglass and, increasingly, carbon fibers in wind-turbine blades is a rapidly growing market, especially in Europe. Operating efficiencies favor large blades. As the blades get larger, the design constraints strongly favor carbon fiber over fiberglass as the reinforcement. This is due to the higher specific strength and specific stiffness of carbon fiber. Composites are not only used for the blades, but wind-tower motor housings and other components are increasingly made of composite materials (usually fiberglass).

A similar application is the use of carbon fiber composites in robot arms or similar large moving machine members. The light weight and stiffness are especially important to ensure efficient movement and accurate location of the arm.

Specific stiffness is the principal factor in the use of carbon fiber composites in industrial rollers. These rollers must not deflect when they apply pressure. However, they must be easily stoppable so they cannot be massively large. Hence, the specific stiffness of carbon is highly desirable.

The application of carbon fiber (pitch-based) for heat removal is poised to become a major market. Carbon fibers are one of the best thermal transport materials, especially if optimized for this application during the fiber manufacturing process. In general, pitch-based carbon fibers are superior to others in heat transport properties. The concept of thermal management with carbon fibers has already been proven in spacecraft. Now, the industrial market is starting to notice. As the size of products decreases, such as in the case of laptop computers, the need to remove heat becomes more critical. Other applications are for items, such as activator motor housings, which are often small and need to be placed in highly confined areas where heat loss by other methods is difficult. The high thermal conductivity of carbon fibers seems to be ideal for these applications. Carbon fiber composite housings will be common for many items.

LEARNING LESSONS—SPACE STRUCTURES

The most important aerospace application is the space shuttle. It has brought great technological innovations. Every launch has been watched with great interest by a large portion of the population of the United States and people in other countries. However, there have been several disasters related to the space shuttle. These disasters are examined here so learning can come from them.

The Columbia Disaster

The board investigating the Columbia Space Shuttle accident announced their preliminary findings and recommendations in 2003. Just as most people suspected, the breakup of the Columbia was related to the loss of or damage to a few carbon-carbon areas, which are an integral part of the thermal protection system (TPS). Even though other missions have flown and landed successfully with some damage to the TPS, the damage sustained was in another part of the TPS system other than the carbon-carbon area. Clearly, something tragically different occurred with the Columbia's last flight. Careful consideration of this disaster leads to important lessons for the National Aeronautics and Space Administration (NASA) and even for the manufacturers of fiber reinforced plastic (FRP).

Lesson 1: Learning from History

It is helpful to consider a story from World War II. During the middle part of the war (1942–1943), the Royal Air Force (RAF) was flying daily bombing runs over

German-occupied territory and airplane losses were devastatingly high. The German anti-aircraft fire was so intense that on almost every run most airplanes were receiving damage. About 20% of the planes were shot down. The RAF could not continue to sustain that level of losses. However, the suggestions for reducing the losses all seemed to have some problems. For instance, the planes could not fly higher because they were not pressurized and, moreover, they had some inherent limitations on altitude. Further, the RAF could not cover the bottoms of the planes with armor plate because the planes would be too heavy to take off.

Finally, a mathematician known for his ability to carefully analyze problems was asked to help. The mathematician began his analysis by examining the planes that were returning. After several days of these examinations, the mathematician made a brilliant discovery, which led to a suggestion that resulted in a significant reduction in the number of planes that were lost.

What was the brilliant discovery? The mathematician reasoned that, taking the entire squadron as a whole, the anti-aircraft damage would be random over the entire under-surface of the airplanes. But, that is not what he saw. After examining the planes that had returned, he found that there were some areas on the planes that were not damaged. How could that be? He then realized that he was looking at only the planes that had survived the flights. The planes that did not survive must have been those planes that received damage in the areas where he had not seen damage. Those areas, he reasoned, must have been the critical areas of the plane and when damage was received there, the plane would be lost.

The mathematician's suggestion was, therefore, quite simple. He recommended that the RAF put armor only in the areas of the planes where no damage had been observed on the surviving planes. These were the critical areas for survival of the aircraft. The areas to be armored were not so many that the weight was excessive and the number of airplane losses was cut in half.

What does history have to do with the Columbia accident? It is logical that the areas on the Columbia where damage occurred during the flight, which resulted in the accident, were critical areas. Previous TPS damage on other flights was obviously not critical and so those flights survived. Why didn't NASA realize how critical damage in specific areas could be? The probable answer is that NASA had not really tested for criticality of various areas (although some critical areas were obvious).

What does the lesson have to do with FRP manufacturing? Some combinations of manufacturing defects and end-use environments produce unanticipated results. Manufacturers should make the effort to clearly identify the areas or the properties of their products that are critical to success. These critical product characteristics should be protected with special care and robustness. That protection should include testing and determination to never compromise the integrity of critical characteristics.

Lesson 2: The Broken Window Theory

The broken window theory grew out of an observation made in New York City in the 1990s. It has direct application to the Columbia accident and the FRP business. The theory is this (Gladwell 2002): if a window is broken and left unrepaired, people walking by will conclude that no one cares and no one is in charge. Soon, more windows will be broken, and the sense of anarchy will spread from the building to the street on which it faces, sending a signal that anything goes. In a city, relatively minor problems, such as graffiti, public disorder, and aggressive panhandling are all the equivalent of broken windows. They are invitations to more serious crimes.

When the officials of New York City began to immediately repair broken windows, remove graffiti, stop people from jumping the turnstiles in subways, clamp down on muggers and various other relatively minor crimes, the incidence of major crimes also declined dramatically. The criminals realized that someone was in control. Even more importantly, an attitude was developed among the city officials and among the populace that even the smallest problem needs to be fixed and everyone has the responsibility to do their best to fix it.

NASA engineers and officials had become accustomed to seeing missing or damaged TPS on returning space shuttles. Given that the design was qualified (tested to be sure that it was able to meet all requirements), the missing/damaged TPS was considered to be either evidence of a combination of circumstances not tested or manufacturing/installation defects.

The TPS has long been an area of concern. The opportunities for a problem to develop are many: extreme temperatures of operation; the generally brittle nature of the materials; the complexity of the fabrication processes; human-intensive installation; complex assessment, refurbishment, and repair processes; and custom tailored installation to accommodate the curved surfaces of the vehicle.

A combination of declining budgets and the growing body of data indicating that some loss of TPS would not cause catastrophic failures induced some measure of tolerance for little problems. This tolerance eventually led to an accumulation of problems, which led to disaster. The broken window did not get fixed and eventually the building was destroyed.

For the FRP manufacturer, the lesson is simply this: the company will work better if it takes care of the small problems. Further, problems that seem to be inherent in the system should be solved because the ultimate result might be a disaster (such as a massive product recall). In summary, it is important to take care of even the small problems.

Lesson 3: Testing and Quality Control are Critical

The Columbia Accident Investigation Board suggested that NASA implement a strong program of testing of the reinforced carbon-carbon composite TPS components of the shuttle. Non-destructive testing on the actual shuttle and destructive testing on the control panel was suggested. Some of the testing methods currently employed include: physical tap, ultrasonic, radiographic, eddy current, weight gain, and visual tests. These should be upgraded and new methods identified to assess the structural integrity of the composite supporting structure and attaching hardware on an ongoing basis. Tests of the shuttle after each flight should be made and directed toward evaluating the longevity and performance of each component, especially those made of composites, so that routine maintenance and replacement can be initiated.

The board also noted that the U.S. Government has the capability of imaging the shuttle while it is in orbit. Such imaging is now required on each flight.

What was the problem with the testing done before the accident? Maybe nothing . . . but the board's report points out potential improvements that may have prevented the accident. Similar cases where testing has been considered to be merely a method of assuring compliance rather than a method of ensuring performance have been common. Every manufacturer should remember that quality testing is a method to assist in making sure that the part will perform as intended and to monitor with ongoing vigilance the capability of the manufacturing process. Testing and quality control are not just a series of hoops to be blindly jumped through.

Lesson 4: Materials are Critical to Success

An old saying asserts: "Everything is made out of something." Clearly the Columbia accident occurred because of failure of the materials from which it was made. One guess from the scenario of failure was that some of the thermal insulation on the structure, which holds the external tank to the shuttle, hit the shuttle's leading edge (reinforced carbon-carbon composite), causing damage to the brittle carbon-carbon areas. This may have been of sufficient severity that a path was created for the re-entry heat to reach unprotected and critical subsurface elements.

In the case of materials for space vehicles, including Columbia, selecting materials and understanding their behavior in the many environments to which they are exposed provides special challenges. Much of the work that underlies the exploration and use of space is directed toward the development of materials that can meet the stringent requirements of survivability in space. At the same time, the materials must be light enough that they can be launched economically. A list of the environments to which space vehicles are subjected follows.

- *Vacuum*. Even though the vacuum of space increases with altitude, its effects are evident at all orbits over about 18 miles (30 km). One major problem with a vacuum environment is the inability to transfer heat through convection, which is the most efficient of the heat transfer mechanisms. Therefore, space structures often have very large surface areas to facilitate heat dissipation through radiation. Vacuum is also an environment that encourages **outgassing** of the plastic. This is the loss of small molecules, such as unreacted monomer or co-reactant, which are often present in the resins. The scarcity of all molecules in the vacuum means the out-gassed molecules are able to travel for significant distances without diversion. These molecules can then be captured by and deposited on another surface, sometimes one that is sensitive to the contamination such as optics or solar arrays.
- *Neutral atoms and atomic oxygen*. The neutral environment consists of a variety of gases, some of which exist as single-atom species rather than the more common double-atom molecules found on earth. The most important of these atomic species is oxygen, which is highly reactive with many organic materials. As can be seen in Table 23-1, the reaction efficiency (that is, the tendency to react with atomic oxygen) is nearly no reaction (designated as 0 in the table) for metals. However, for polymeric materials, the reaction efficiency can range from 0.03 for Teflon® to 3.9 for Mylar® (a polyester).
- *Plasma*. Plasma is an environment of electrons and protons that causes uncontrolled electrical charging and discharging as well as the dielectric breakdown of the materials. These effects are diminished by using stable conductive coatings on the materials such as indium tin oxide along with special efforts to ensure grounding of the devices.
- *Radiation*. Although radiation has a degradation effect on polymers similar to the aging effects observed on earth, the most serious problems from radiation (at least short term) are with electronic components. The solutions are to build the components to be radiation hardened, design in fault tolerance, and/or provide shielding with a metal such as aluminum.
- *Micrometeoroids/debris*. Natural space materials and man-made debris can impact the space structure as it flies through space. The possibility of encountering debris is expanding rapidly

Table 23-1. Atomic oxygen reaction efficiencies for common materials.

Material	Reaction Efficiency (Tendency to React)
Aluminum	0
Gold	0
Polytetrafluorethylene (PTFE) or Teflon7®	0.03–0.50
Carbon	0.9–1.7
Silicone (room-temperature vulcanizing [RTV])	0.443
Polyimide (Kapton7®)	1.4–2.5
Epoxy	1.7–2.5
Polyester (Mylar7®)	1.5–3.9

as the number of launches continues and as the satellites now in orbit fall apart. The shuttle windows have sustained multiple impacts and are periodically replaced.

- *Launch environment*. The highest structural loads a spacecraft normally encounters are during the launch (or in the case of the shuttle, launch and landing). Designing for reduced weight to improve performance (payload to orbit) puts special emphasis on making the structure only just strong enough (with margin) to withstand the expected environment.

Even though these special environments are encountered by space vehicles for which much of the testing cannot be duplicated on the ground, the importance of understanding the material properties and behaviors is important. Studying what happens to materials when exposed to a wide variety of environments, even environments not anticipated, provides significant insight into the initiation of potentially catastrophic events. This insight should lead to an early evaluation of consequences and contingencies, which is something every FRP manufacturer should explicitly include in planning.

The importance of maintaining the properties of the materials cannot be overemphasized at NASA or any FRP manufacturing site. Composites are the materials of choice for many applications, including space applications, because of their desirable properties; and so those properties should be guarded with great care.

Conclusion

Though space vehicles are specialty items, the space program is a rich source of information that can be applied in many other areas. Knowledge is not only obtained from actual excursions into space (generally with great difficulty, cost, and time), but from the space hardware that must first be tested as components, subsystems, and then as part of a total system.

The development cycle for complex vehicles averages about 10 years and the operational life is 5–15 years. Hence, the investment for a space vehicle can be very high. A typical satellite investment is $400 million and a launch vehicle adds to this price tag. Such investments are risky when you consider the things that can go wrong to wipe out the investment with a single detail gone awry. Space vehicles might self-destroy (usually due to a propulsion problem). They might veer off course during launch with potential to hit land and require destruction in flight to preserve public safety. They might be launched into an incorrect orbit. Even if they achieve the right orbit, they might not work as planned. Each of these failure modes is real and, sadly, possible.

Nevertheless, space vehicles continue to be made in ever-increasing numbers. They are low-volume products, but the potential for profit has proven to be very high. For NASA, the contributions to the nation via the products developed initially to meet the unique challenges of space, which were subsequently passed into the general economy (commercialized), have been many. These collateral applications are important to NASA, as reflected in NASA's mission and goals (see Table 23-2). Space vehicles continue to be important for the military (evident in the war in Iraq).

Though there are many applications whose designs, materials, and manufacturing will never produce the highly dramatic and catastrophic failure of the Columbia, there are nevertheless some lessons that can be derived from this tragedy.

1. Anticipate and plan for all of the conceivable potential consequences of combinations of environments and the results of less-than-perfect manufacturing processes.
2. Take care of the small problems as they come along.
3. Stay vigilant that the testing and quality control processes do not become rote. They can be harbingers of bigger things to come.
4. Materials are critical; their selection, design, and manufacturing are keys to a product's ability to perform in not only the anticipated environment, but in ways never intended.

Table 23-2. The mission of the National Aeronautics and Space Administration (NASA) (Harris et al. 2002).

- Advance and communicate scientific knowledge and understanding of Earth, the solar system, and the universe;
- Advance human exploration, use, and development of space; and
- Research, develop, verify, and transfer advanced aeronautics, space, and related technologies.

To fulfill this bold mission, NASA has adopted the following long-term goals:

1. Create a virtual presence throughout our solar system and probe deeper into the mysteries of the universe and life on Earth and beyond.
2. Use our understanding of nature's processes in space to support research endeavors in space and on Earth.
3. Conduct human and robotic missions to planets and other bodies in our solar system to enable human expansion.
4. Provide safe and affordable space access, orbital transfer, and interplanetary transportation capabilities to enable research, human exploration, and the commercial development of space.
5. Develop cutting-edge aeronautics and space systems technologies to support highways in the sky, smart aircraft, and revolutionary space vehicles.

Finally, as composites are developed and modified for space applications, manufacturers should remain alert for the potential value these materials can bring to business.

New materials, processes, designs, quality control, and testing methods could all be keys to remaining competitive.

THE ULTIMATE COMPOSITE STRUCTURE—ISOTRUSS®

Why are Composite Materials Used?

The textbook answer for why composites are used is lighter weight, greater stiffness, and higher strength. In the original applications of composites, and still in most aerospace applications today, the reasons for using composites remain the same. But over the past 50 years, composites have gained market share against other materials for a variety of other reasons too (usually in combination with the advantages in weight, stiffness, and strength). For example, chemical resistance is a key requirement for pipes and tanks, ease of molding is critical for tub/showers and autos, and weatherability and ease of repair are important for boats.

In some non-aerospace applications, weight, stiffness, and strength have been compromised for manufacturing and cost reasons. For example, adding a filler will usually increase weight and decrease strength, yet fillers are common throughout the composites industry. Similarly, the use of chopped fiber increases the ease of manufacturing of many composite products, but also decreases stiffness and strength.

In light of the wide array of important composite properties and the compromises that have been made for manufacturing and cost, it might be asked whether anyone is still trying to optimize composite performance for weight, stiffness, and strength. The answer is "yes." Some unique concepts are allowing this optimization process to continue. Those concepts are generally based on the concepts of "iso" structures and composite trusses. Structures based on these concepts hold a wealth of information on the fundamentals of composite materials even within today's requirements for broad-based properties and manufacturing capabilities. These concepts can help product designers so that they can be more efficient in saving weight and adding stiffness and strength to products.

What Does "Iso" Mean?

In the context of structural materials, "iso" refers to a pattern of interconnected isosceles triangles. Several years ago, NASA developed a flat structure, called isogrid, which contained these iso structures. Structural materials based on this isogrid pattern were some of the most efficient possible (in terms of stiffness and weight).

Several researchers (especially those at Stanford University and Virginia Polytechnic Institute) confirmed NASA's assessment of the structural efficiency of isogrid structures. However, they found that manufacturing the isogrid materials was difficult. The structures could be made by hand, but were difficult to make using automated or semi-automated equipment because of the complexity involved in making the overlaps. Eventually, however, automated manufacturing methods were developed.

What are "Trusses?"

An architectural truss, which is the type discussed in this section, is a rigid framework of beams. Such a framework can be used to support another structure, as is done by the truss network that supports the roof of a house. The truss also can be self-supporting as, for instance, in a bridge. Typically, roof and bridge trusses are made of wood or metal, although some composite bridges have been made.

It is probably obvious that architectural trusses are based on simple geometric forms, such as triangles, which are joined together to form the support network. In this way, the trusses are closely related to isogrids. The difference, of course, is that the isogrids are essentially flat panels that have been reinforced by the grid pattern, whereas the trusses are open structures.

A careful analysis of trusses will show that they, like isogrids, are also essentially two-dimensional structures. The thickness is less important in these structures than the length and height. It might be asked whether a structure could be built that is fully three-dimensional. If so, it might optimize the structural aspects of triangular components. Ideally, the structure would be made of composites so that the weight, stiffness, and strength could be optimized. Such a structure now has been developed. This unique new structure is called IsoTruss®.

Understanding IsoTruss Structures

The IsoTruss structure is built on triangular patterns except that the triangles are connected in a unique three-dimensional manner. This makes IsoTruss an extremely efficient structural form.

The IsoTruss concept is simply this: use composite reinforcement for the structure of the truss and orient the fibers in the directions where they will carry direct loads. Everything else is empty space. This concept uses composites in their most efficient manner—for carrying tensile and compression loads. The fibers are directed, as much as possible, in the directions of the various flexural, torsional, tensile, and compressive loads anticipated. Therefore, the original concepts of optimizing composite materials are achieved. The optimization is directed toward saving weight, gaining stiffness, and maximizing strength. IsoTruss structures have approximately half the weight of metal trusses with the same bending or torsional stiffness at equal diameters. An additional key enabler is the ability to increase the diameter beyond an equivalent optimal diameter in steel tubes. For example, at twice the optimal diameter of a steel structure, an IsoTruss is less than one-twelfth the weight. An important additional consideration is the much lower material cost of IsoTruss structures versus conventional composite structures. This is because much less material is used in an IsoTruss to achieve the same (or better) performance versus a conventional composite structure.

The light weight and strength of the Isotruss structures can be easily demonstrated using full-size samples (hand-made structures) as shown in Figure 23-3. Their tremendous advantages lead to some interesting applications. For example, utility poles and freeway sign supports might well be made from Isotruss structures. A natural market is poles for weather monitoring, especially in areas difficult to access because a fully assembled IsoTruss tower can be easily carried by a helicopter. Ready markets would include construction support beams and even deep-sea oil-drilling platforms. Other structures could include aerospace structures like the space station or even airplane and missile components. Even highly commercial products, such as bicycles and pack frames, are potential applications. In short, the concept is good for most products that require lighter weight, high stiffness, and high strength in a tubular configuration.

An early challenge was manufacturing the IsoTruss structures. They were originally made mostly by hand using mandrels that had points at which the fibers could be tied together. Recent advances have suggested semi-automated and even fully automated manufacturing methods based on filament winding and braiding. These advances have not, however, been implemented outside the laboratory. And that is the lesson to be learned. The costs of manufacturing IsoTruss have proven to be extremely high. The semi-automated methods have not yet been proven and the automated methods are still just dreams. Manufacturing was moved to a country where labor costs are low, but

Figure 23-3. Full-size IsoTruss structure. (Courtesy Brigham Young University)

that has resulted in high transportation costs and difficulties with supervision.

Nevertheless, the future for structures of this type is good. Products as lightweight and strong as these just seem too good to not succeed. However, even though the design is highly optimized, the products still have to be manufactured. That process has proven to be most difficult.

CRITICAL MARKET—ARMOR

Applications

Lightweight and effective armor is essential to saving lives. But that armor needs to be designed to meet the perceived threat. Too little armor and little benefit is gained. Too much and other important factors, such as mobility, are sacrificed.

In WWII, armor was almost always made from traditional materials such as metal plates, but that concept has evolved. A good example to illustrate the move from traditional materials to polymer composites is the evolution of the military helmet. Thin steel was the principal helmet material of WWII and remained in use until the 1980s. The advantages of using steel included the simple technology for manufacture of the helmets, low cost, and reasonably good impact protection. However, these helmets were found to provide only marginal protection against fragments.

During the Vietnam War, the U.S. Army experimented with fiber-reinforced inserts in the helmets. These inserts improved ballistic protection but, at the same time, added weight and reduced the air space between the head and the helmet.

By the early part of the 1980s, the U.S. military had developed the first all-composite, 100% fiber-reinforced helmet. This development resulted in even better ballistic protection without altering the weight of the helmet. Other improvements included better head coverage and more space for ventilation.

Other countries such as the United Kingdom and South Korea decided to use a moderate cost, moderate performance nylon composite helmet. These helmets were manufactured by molding with either 100% nylon fabric or a mixture of nylon and aramid and then coating with a phenolic resin and curing. Nylon had a number of qualities that encouraged its use in helmets. At the time, nylon was locally available. It also had excellent insulation capabilities in both hot and cold climates. However, nylon was found to provide only the bare minimum in ballistic protection and, as a result, was discontinued everywhere except in Great Britain.

Another helmet then began to be used in several parts of the world. These helmets were reinforced only with aramid. The basic material was typically a woven prepreg with phenolic as the matrix resin. In the development of these helmets, engineers and designers learned that, to a certain point, lower resin content performed better in ballistic resistance. Hence, the resin concentration was lowered to approximately 12%.

In the latter part of the 1980s, another helmet began to be used. It was manufactured by laying up ultra-high-molecular-weight polyethylene (UHMWPE) fibers in a 0° and 90° cross-ply, which was then bonded with a thermoset resin. The first to use this technology was the French army. Reduced in weight by about one-half, the helmet now weighed 2 lb (1 kg).

Composite materials have taken a leading position in modern armor such as helmets and bullet-proof vests. Most of the modern composite material applications rely on tough and strong fiber reinforcements, usually aramid (like Kevlar®) or UHMWPE (like Spectra®), which are the two most common types in the United States. But these are not the only materials nor are helmets and bullet-proof vests the only applications.

Some other applications for modern armor are listed in Table 23-3. No attempt is made to quantify these markets because they are, largely, untapped and still in development. Moreover, some of the applications will require a change in people's minds to be fully developed because the need for armor is not now recognized. However, a major change in that perception and a more ready acceptance can be achieved in some of the cases by a simple word change—from the word "armor" to "protection." The protection could be against bullets, shrapnel, knives, explosions, crashes, impacts from falling or flying objects, or even from fires. Current growth in many of these markets suggests that the change in perception is underway. However, to exploit these markets, some of the basic concepts and details of materials and manufacturing used in composite armor should be understood.

Table 23-3. Actual and potential markets for armor/protective materials.

Major Market Category	Specific Product Examples
Military personnel protection	Helmets, combat vests, land-mine demolition gear, special gear for various duties (pilots, tank operators, medics, special forces, artillery gunners)
Military armor systems	Vehicle exteriors, vehicle interiors, aircraft-vulnerable sites, radomes, naval armor, personnel carriers, body bunkers, land-mine containment tubes
Law enforcement	Police and security-guard vests, riot gear, riot shields, armored boats, police helicopters, bomb blanket
Other products	Safety helmets, explosive-containment boxes (aircraft cargo), engine-fragment containment, fuel cells, pressure bottles, armored bank trucks, armored personal cars, ambulances, butcher aprons and arm guards, bus and taxi-driver shields, hunter vests, body armor for at-risk people (public figures, bodyguards, store operators in high-risk areas), fire protection, firearm barrels

Performance

Almost all armor applications begin with an assessment of the protection level afforded by a particular armor material. For body armor, this assessment considers the ability of the material to protect against two major dangers:

1. penetration from various missiles, and
2. blunt trauma.

The leading authority that has established most of the standards regarding armor protection is the National Institute of Justice (NIJ). Its literature sets out the procedures for testing and analyzing for penetration and blunt trauma. Most applications other than body armor focus entirely on penetration unless they are also considering some additional danger such as fire or explosion shock. In general, no agency has taken overall control of the tests; thus manufacturers are largely on their own to specify the protection level in these other areas.

By far, the most common method of reporting armor performance is to use the classifications and testing methods of the NIJ. The standard (NIJ Standard 0101.03) establishes six formal armor classifications plus a seventh special one; all are related to penetration. Each classification lists the type, size, and velocity of the bullet against which protection is provided. Higher classifications give protection against larger bullets at higher velocities. Therefore, a purchaser of the body armor must evaluate the likely threat and choose the type that will meet that threat. For instance, Type II provides multiple-hit protection against .357-Magnum, JSP 158-grain bullets to a maximum velocity of 1,395 ft/sec (425 m/sec), and 9-mm, FMJ 124-grain bullets to a maximum velocity of 1,175 ft/sec (358 m/sec).

To establish the classifications and determine the rating of various body-armor materials and configurations, a special test was run in which rounds of each type were fired into the armor. The result of the test is the velocity at which the probability of penetration is 50%. This value is obtained by firing many shots into the vest and noting the velocity of each round. The number of rounds that penetrate the vest is then plotted for each velocity. The velocity at which 50% of the rounds penetrate, called V50, is found from the graph that is made.

Body-armor manufacturers frequently report the V50 value for a particular product and, therefore, the tendency among consumers is to compare the materials on this basis alone. Great caution should be exercised in making these comparisons since the test procedures can be legitimately varied somewhat. Moreover, armor penetration analysis is complicated and simple correlations can be difficult. V50 tends to increase with fiber or yarn strength (tenacity), fabric weight, and the number of layers, but will decrease as the number of filaments in the yarn increases. The complication is, of course, that all of these and other factors such as lay-up sequence, type of matrix, whether the fabric is woven or cross-plied unidirectional, and the strength of the bond between the fiber and the matrix (that is, the presence of a sizing or coupling agent) can all make major differences in the values.

Energy Dissipation

Underlying all of the factors discussed previously is the concept of energy dissipation. Armor materials work because they dissipate the energy of the bullet or other missile that impacts against the surface of the armor material. The dissipation phenomenon is complicated but some of the mechanisms are reasonably well understood. The most important of these is the ability of the armor material to quickly (nearly immediately) spread the impact energy sideways into a wide area. This capability requires that the fibers must be strong enough so they do not immediately break when impacted.

If the fiber can internally absorb energy by some mechanism, such as internal splitting or heating, then additional energy can be absorbed. Therefore, in addition to high strength, the fibers must have sufficient elongation so they move or give slightly to transfer some of the load to other neighboring fibers connected through the fabric weave or through the matrix. However, if the bond between the fibers and the matrix is too strong, the total energy dissipated is actually reduced. This is because the strong bond eliminates another source of energy dissipation—the sliding of the fiber against the matrix. So, the manufacturers of armor must balance the need to have a strong matrix-fiber bond to transfer the energy throughout the structure with the need for some slippage to give additional energy dissipation. A recent finding is that energy dissipation can be altered by changing the directions of the fibers in adjacent layers. Some sequences seem to pass energy more efficiently from one layer to another, moving the energy laterally in the structure. Another recent finding is that more energy is dissipated when adjacent layers are of a different material (such as aramid next to UHMWPE).

Yet another factor in energy dissipation is breaking up of the bullet itself. This might be caused by shattering against an outside layer of the armor. Ceramics and ceramic composites have proven to be especially useful for this purpose, especially in combination with aramid, UHMWPE, or fiberglass composite as backing materials.

It is probably obvious that many factors must be simultaneously optimized and several compromises must be made to achieve the best performance. Moreover, it seems certain that optimizing for ballistic protection will force some decrease in other properties, such as mechanical strength or stiffness.

Looking into the Future

At this moment, many aspects of the total armor market are changing. New and traditional fibers are being investigated for armor capability alone and in combination. These might be treated in special ways or used in unique lay-up patterns to optimize energy dissipation and cost. Several matrix materials, including thermosets and thermoplastics, are being investigated as well. Some applications eliminate the matrix altogether. The addition of secondary materials, such as ceramics and metals, is also being investigated.

In addition to the many investigations of materials, the methods of manufacturing are also being re-examined. What has traditionally been done by hand lay-up may now be done by resin transfer molding (RTM). This change may require completely new concepts in weaves and preforms, thus requiring another round of material optimization.

On top of all the material and process investigations, the basic shapes of existing products are changing and new products are being created. The armor market is certainly in flux. This is, for some, a time of great opportunity even though the risks are high.

BREAKTHROUGH MARKETS—COMMERCIAL AND CORPORATE AIRPLANES

Aerospace Needs a Breakthrough

The history of aerospace has traditionally been full of exciting innovations in materials, designs, and manufacturing techniques. The Wright brothers' plane, itself, had innovations in the lightweight engine, propeller design, steering and flight control mechanism, and overall design. Their work was soon enhanced by the development of flight control surfaces (like ailerons), which were separate from the primary flight surfaces, the introduction of aluminum structures (requiring new alloys), radar, mass-production techniques (remember Rosey the Riveter), jet engines, computer control (fly by wire), composites (including the introduction of carbon fibers and new resins), new weapons (from bombsights to missiles), and stealth technology. The overall design and efficiencies kept pace with the developments in materials and manufacturing techniques so that the entire aerospace industry was infused with excitement.

A strong conservatism in design and manufacturing has taken away from that innovative spirit. This conservatism has been caused by liability fears and by the nature of the marketplace. The number of aerospace companies has been reduced dramatically by consolidation, thus removing some of the small companies where innovation was most likely to occur. The market is now dominated by a few large companies with tremendous capital investments. Their bureaucracies seem reluctant to change much beyond just making airplanes with seating capacities and ranges to compete in niches not adequately serviced by the current fleet selection. Price and politics are dominant marketing forces, not necessarily innovation and performance. Market share is a major focus and, perhaps because of the erosion in market share Boeing saw throughout the 1990s, the need for a new direction and new innovative spirit has become critical to its recovery of position.

The Boeing 787 Dreamliner

More than any commercial plane ever made, the 787 is dedicated to the use of advanced composites. Up to now the use of composites has been mostly "black aluminum"; that is, they have been used to replace aluminum as the preferred materials for parts. There was little effort to use the advantages of molding and part unification, which are inherent with composites but lacking with aluminum. Therefore, some weight savings and other minor advantages were achieved but the breakthrough advances were not.

The 787 promises impressive improvements in range (up to 8,500 miles [13,600 km]) for a medium-sized airplane, fuel efficiency (20% improvement), speed (up to Mach 0.85, not currently available for medium-sized airplanes), and cargo-carrying capacity (up to 60% more). These performance breakthroughs come directly from the new attitude of Boeing's management toward using composites. As now conceived, the plane will not be "black aluminum." Designers will consider composites as unique materials with properties that can be of advantage to achieving the plane's performance goals. The majority of the primary structure, fuselage and wings, will be largely made of composites. The company is also looking at incorporating smart structures into these components to monitor their health.

Because the composite structures are stronger and stiffer than the previous metal ones used on other airplanes, the 787 will be certified to allow higher pressurization. This means that customers will feel better during flight as confirmed in an actual pressurization study conducted for Boeing at the Farnborough Air Show in the United Kingdom.

As has been the case with previous Boeing aircraft, an international consortium of partners will actually build the various systems for the final assembly. Inevitably, this wide focus will improve composites manufacturing throughout the world. It will also help develop new composite manufacturing techniques that will work their way into other composite products, including fiberglass reinforced plastics (FRP).

The Spectrum Civilian Aircraft

Developing the ability to manufacture lightweight, low-cost aircraft represents serious potential in aircraft manufacturing. Indeed, it would constitute the accomplishment of the dream of composites manufacturing. In the 1980s, the concept of a low-cost, single-unit aircraft body was realized but not fully commercialized. Seven prototypes of the Beech Starship were made by filament winding. These prototypes were lighter, stronger, and had better handling (flying characteristics) than traditionally made versions. However, largely political and bureaucratic problems forced the company to use more traditional manufacturing and design methods.

Even though the Starship was a great technical success, some important new innovations have been made since its time, principally involving integral stiffening. The inclusion of integral stiffeners is an important step in a new manufacturing process that has been used to create the Spectrum aircraft. This aircraft, shown in Figure 23-4, is made of a single shell for the fuselage and a single piece for the wings. These two basic sections are bolted together and then the other components such as the engines, electronics, tubing for fuel, and interior materials are added. The weight is about 70% of a comparable corporate jet made by conventional methods.

What it All Means for Aerospace

There are three areas of opportunity emerging for innovative composite structure development. The most glaring break in the stagnant commercial jet-liner market is the Boeing 787, or Dreamliner. Boeing is on the lookout for new and innovative ways of manufacturing composites. The concept of large, integrally stiffened, and low-cost, co-cured structures has to have great appeal to

Figure 23-4. Spectrum® all carbon fiber composite aircraft. (Courtesy Rocky Mountain Composites)

any airframe company that wants to take the bold steps to develop, qualify, and certify new materials and processes. This opens the door to breakthrough experiments and product development. In addition, materials and processing companies cannot ignore the needs of AirBus to optimize its composites structures for cost savings and increased performance.

The landscape of composites manufacturing is dotted with a myriad of new-generation aircraft emerging to bring private jet convenience to corporate and personal travel. Affordable business jets ease the frustration at airports because of delays and inconvenience experienced since the world changed on 9-11. General Aviation's predicted growth numbers are impressive. The marketplace will not be able to resist the temptation to adopt the new high-performance airplanes currently in various stages of development. It will take innovative companies to directly address the cost of composite structures and advance the technology to the next step of acceptance.

Another area of opportunity lies in the small, 2-4 place, piston-driven aircraft. The adoption of composite materials and processes will truly bring this generation of workhorse aircraft into the age of modern technology.

The small aircraft market is still dominated by an aging fleet that begs for replacement with stronger, lighter, and more efficient airplanes. Computer- and space-age navigational systems will make it possible for these airplanes to be flown with a minimum of training and instruction. The convenience of air taxi services using small out-of-the-way airports seems to merge well with small aircraft. Lightweight, cost-effective structures have every right to be included in basic airplanes as they do in the so-called high-performance genre. In fact, the advancements being made for the new-generation airliners will enhance the viability of the private, piston-driven aircraft. It is reported, for example, that major automotive companies are spending heavily to document materials and processes that can be easily adapted to a number of large, co-cured, integrally stiffened aircraft structures, fuselages, and wings.

The next 100 years of aircraft manufacturing must be simpler and smarter. Simpler means fewer parts and streamlined manufacturing processes. Smarter means the elimination of costs and using materials suited to the application without being exotic. If Boeing and other aerospace/aircraft companies take advantage of the new small-aircraft technology, they could experience a total rejuvenation because of much lower costs and improved performance. It may breathe new life into the industry reminiscent of the 1930s and 1940s.

FUTURE-TODAY MARKETS—UNMANNED VEHICLES

The Scenario

Imagine you are a soldier in the midst of a battle. Your squad has advanced into an enemy town with the object of capturing a large cache of weapons believed to be in a building two blocks ahead. Right now, however, you are under heavy rifle fire and you suspect that the enemy soldiers are hiding in the houses between you and the arms cache. They are probably firing from the windows but you cannot put your head up to see because of the heavy fire. You also hear machine-gun fire but that seems to be a block or two away. You have, in the past, heard the sounds of tanks, but you do not know their location. Right now you are lying on the ground behind a wall. You seem to be safe for the moment but obviously will not be successful in capturing the arms cache unless you move forward, but that seems foolhardy. You know that during previous wars, like World War II, soldiers in your situation would probably ask a companion to cover them and they would jump up and run forward. (At least that is what the movies said they did. They seemed to either get killed or win a Medal of Honor.) Right now, however, you are not sure if that is the wisest course of action.

Just at this moment of indecision and peril, you remember that your company has been given a new tool. It seemed a bit "James Bond-ish" at the time, but now it seems worth a try. You reach in your pocket and take out a small box containing a miniature, hand-sized airplane. It looks like something that you might have flown as a science project in high school. You flip the on switch, the propeller starts and you let it go up into the air above your head. You now look at the carrying box and see that it has a screen on which the street in front of you is pictured from the viewpoint of the little airplane. The box also has controls that allow you to move the plane around; just like playing a computer game. You send the plane forward and see where the enemy soldiers are hiding. You fly the plane to the next block and see that a machine gun guards the street over there, thus telling you that trying to move up that street is foolhardy. You then try the street on the other side and see a tank waiting to block any advance there. Hence, you must move forward, but with the knowledge of exactly where the enemy lies, that seems less formidable.

You use a radio to call an unmanned golf-cart-sized robot vehicle that is carrying a few small tactical missiles. Upon your command, the robot moves, fully guiding itself around obstacles, into position about 100 ft (30 m) behind you. Then, you maneuver the unmanned plane directly over the position of the nearest enemy soldiers hiding in front of you. Taking the global positioning system (GPS) coordinates, you then relay them to the control unit of the robot and instruct it to fire a missile. The enemy position is destroyed, thus allowing you to advance safely. With the use of the plane to scan around corners and over the buildings ahead, and the robot to fire beyond your immediate horizon of vision, you successfully capture the arms cache.

Is the preceding scenario taken from some science fiction movie? It could be! But it is much closer to reality than you might imagine. The world of unmanned aerial vehicles (UAVs) and unmanned ground vehicles (UGVs) is not just futuristic; it is already here in surprisingly sophisticated applications. And, of course, the potential is truly astounding!

Unmanned Air Vehicles

UAVs are any aircraft that operate without an on-board pilot and typically fly with some level of autonomy. Historically, UAVs have their roots in military applications and current applications are still mostly military. Civil activities are limited but have great potential. The U.S. military's involvement with UAVs started in the early 1900s, and they have had active roles in all major conflicts since the Vietnam War. Recent combat in Kosovo, Afghanistan, and Iraq has proven that modern UAVs are vitally important, especially in urban environments.

Current UAVs range in size from full-scale aircraft down to systems that fit in the palm of your hand (see Figure 23-5). These aircraft are loosely classified into three categories based on size: full scale (wingspans in excess of 50 ft [15 m]), tactical (wingspans around 25 ft [8 m]), and mini/micro (wingspans below 10 ft [3 m]).

Full-scale "endurance" UAVs, such as the propeller-driven Predator and the jet-powered Global Hawk, have nearly become household names in recent years. The Predator has been used extensively in the U.S. War Against Terror (Afghanistan and Iraq), primarily for reconnaissance but also for limited attack missions. One crucial advantage of these aircraft is their ability to

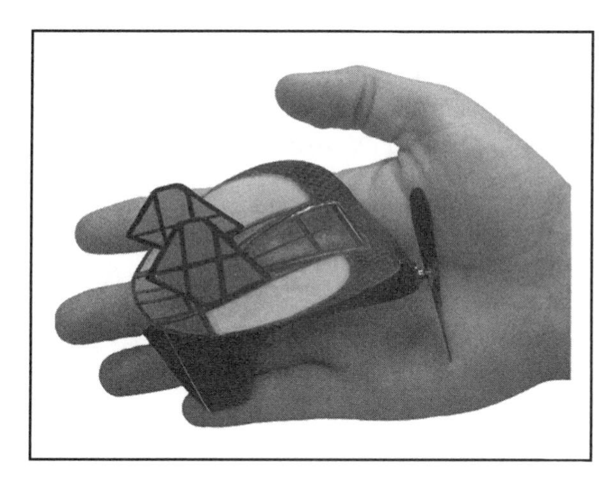

Figure 23-5. Unmanned aircraft made, in part, with carbon fiber composites. (Courtesy American Composites Manufacturers Association [ACMA])

persistently collect data. Both the Predator and the Global Hawk can stay airborne for a day or more. The importance of persistence was seen during the initial phases of the current Iraq conflict. The single Global Hawk that was deployed flew only 3% of the aircraft imaging-collection sorties, yet collected data on 55% of all air-defense-related, time-sensitive targets.

Predator is a lightweight, high-endurance craft with a wingspan of just under 50 ft (15 m) and a payload of 450 lb (204 kg) for up to 40 hours. It provides surveillance imagery from synthetic aperture radar, video cameras, and infrared cameras, which can be sent in real-time not only to the operational commander but also directly to the front-line soldier. These UAVs have been operational since 1995, and in total have flown over 100,000 hours. The UAV ground-control station is built into a single 30-ft (9 m) trailer and houses a pilot and three sensor operators. Although the ground-control station must be located at the place of take-off and landing, the UAV can be monitored and controlled from anywhere in the world via satellite links. An armed version of Predator exists, which carries AGM-114 Hellfire missiles. It has been used for several strikes initiated by the military and the CIA. In December 2002, for the first time in history, a manned and an unmanned aircraft engaged in combat when an Iraqi MiG-25 shot down a Predator after the Predator fired at it.

Global Hawk has a wingspan of over 130 ft (40 m) and is capable of carrying a 3,000-lb (1,361 kg) payload. The sensor suite contains synthetic aperture radar, an electro-optical camera, and an infrared camera. Continuous flight time is 36 hours. The system is designed to perform most routine flight operations autonomously, thus allowing the operators to concentrate on mission execution while simply monitoring the aircraft's performance. Communication with the ground crew can occur through satellite links, meaning the system can be remotely operated from anywhere in the world. The first developmental Global Hawk flew in 1998. Due to the terrorist attacks of 9-11, several of these developmental systems were rushed into service in Afghanistan and Iraq. In total, Global Hawk flew only 15 missions during Operation Iraqi Freedom, yet provided over 4,800 images in near real time. The system located at least 13 surface-to-air missile batteries, 50 SAM launchers, 300 canisters, and 70 missile transporters; it also imaged 300 tanks. During these missions, the UAV and its sensors were operated remotely from Beale Air Force Base in California, thus reducing the logistical requirements in the field by more than 50%.

The next phase for this class of UAV is an armed system that can perform surveillance, suppression of enemy air defenses, and precision strike missions, all without a human in the loop.

Tactical UAVs include Pioneer and Shadow, with wingspans of approximately 15 ft (5 m). Pioneer gained fame during the first Iraqi conflict. After enemy troops realized that the presence of this UAV meant imminent 2,000-lb (907-kg) naval gunfire rounds would soon land on their position, they surrendered to the unmanned aircraft. Flight times are on the order of 5–6 hours, and payload capacities are around 70 lb (32 kg). Payloads typically consist of a gimbaled electro-optical and infrared camera. Line-of-sight data links transmitted video to the ground user in real time. The all-composites Shadow UAV has been extensively used by the U.S. Army in the Afghanistan and Iraq conflicts. The fleet logged over 75,000 flight hours as of early 2006. Brigade-level UAVs, they provide support to the ground-maneuver commanders. Tactical UAVs have the advantage of providing battlefield awareness at a cost approximately 1/100th that of a manned platform, and with significantly reduced risk to human life.

The mini/micro air vehicle (MAV) class is an area of intense growth and interest. Loosely, "mini" refers to crafts with wingspans from approximately 10 ft (3 m) down to 6 in. (15 cm), while "micro" refers to wingspans below 6 in. (15 cm). These aircraft are designed to fill the need of short-range, limited-duration, "over the hill" reconnaissance, providing situational awareness. Likely the most widely employed aircraft of this class is the Pointer, with a wingspan of approximately 8 ft (2 m) and a flight time of 90 minutes. This UAV carries an electro-optical or infrared camera and a short-range (approximately 6 miles [10 km]) line-of-site communications package. The Pointer system, including ground station, fits in several hand-carried cases and can be transported and operated by two users. More recent systems, such as Raven and DragonEye, are similar in operation and capability, but with wingspans on the order of 4 ft (1 m). The requirements for this class of aircraft include a wide range of operational environments, simple user operation, all-weather operation, automatic collision avoidance, and low cost.

An eventual military goal for MAVs is to provide personal situational awareness for every foot soldier. One or more air vehicles would be part of a soldier's normal gear. A wrist-mounted control station and video screen, not much larger than a common wristwatch, would be worn. These MAVs would perform under supervised autonomy, where the user would provide only high-level commands such as takeoff/land, waypoint path following, loiter, etc. For combat in urban or forested environments, the soldier could have his/her own personal eye-in-the-sky automatically loitering over the current position and providing real-time video of the tops of nearby buildings or the next clearing. For this to occur, the airframe and ground station must obviously become even smaller. BlackWidow is probably the most well-known, micro-sized vehicle, having a wingspan of approximately 6 in. (15 cm). Other aircraft, such as the one developed at the University of Florida, are even smaller. Aircraft of this size have demonstrated the ability to locate and image in real time a 3 ft (1 m) target from 1,640 ft (500 m) away.

UAV materials and construction are as varied as the aerodynamic designs. It is apparent, however, that composites will continue to be the material of choice for UAVs of all sizes. In addition to being lightweight, composites offer quick build time, fewer fasteners, and resistance to corrosion (from salt water, for instance). UAVs of the mini class are commonly constructed of metal and/or plastic skeletons with Kevlar skins. Larger UAVs, on the other hand, employ primarily fiberglass and carbon fiber composite materials.

The future of the U.S. military market appears solid. UAV capabilities have expanded from a single "eye-in-the-sky" in the early 1990s to persistent intelligence surveillance and reconnaissance today. The excitement in the military market is not unique to the U.S. According to the American Society of Aeronautics and Astronautics. As of 2005, over 500 UAV platforms were on record and over half of the nations in the world had some type of UAV in their arsenals. Funding by the Department of Defense for UAV development and procurement has quite literally exploded in the past 5 years, from an annual budget of $363 million in 2001 to an expected $3.5 billion in fiscal year 2009. The bulk of this funding has been earmarked for Predator, Global Hawk, and Joint Unmanned Combat Air Systems (J-UCAS), formerly known as Unmanned Combat Air Vehicles (UCAVs). Smaller systems stand to benefit as well from this significant increase in spending.

Potential commercial applications are widely varying, including border patrol, police surveillance, search and rescue, forest fire and wildlife monitoring, detection of fish schools, large-facility security, mail and package delivery, and even recreation. Aerial surveillance currently performed by manned helicopters and aircraft represents an application where unmanned aerial vehicles could have a large impact. For example, law enforcement in U.S. cities with over 250,000 people cumulatively uses nearly 500,000 flight hours each year of piloted aerial surveillance. At a cost of $500/hr, this represents $244 million in spending each year. Statistics are similar for traffic/news applications.

Endurance UAVs equipped with electro-optical and infrared cameras could autonomously fly along the border and detect illegal crossings at remote locations. Large industrial facilities such as oil refineries might use UAVs to perform aerial inspections of the sites, including visual, audio, or chemical monitoring (when mounted with a suitable chemical sensor system). Such chemical systems open the possibilities of using UAVs to enter highly hazardous areas such as damaged nuclear power plants, chemical leak zones, war areas, and other places where the environment must be monitored before actual human entry can be made. It is even possible that UAVs with infrared or other sensors could look for survivors of disasters like avalanches, floods, and earthquakes in areas too remote or hazardous for manned surveillance. Super-sensitive chemical sensors could detect the existence and nature of terrorist bombs and possibly even effect the removal of such devices.

Current technology allows for an autopiloted toy airplane with a wingspan of 12 in. (30 cm) and an on-board video camera to enter the market below the $100 price point. (So, an unskilled 10-year-old boy could send his toy plane down the street to "survey the wildlife" at the neighbor's backyard pool!)

The primary hold-up for release to the civilian market is unresolved issues of operating unmanned vehicles in commercial airspace. Several concerns have delayed the process, including autonomous collision avoidance and the possibility of malicious overriding of the autopilot system. The urgency of appropriate regulations is well recognized, however. In 2004, the Defense Science Board Task Force on UAVs issued a report calling on all U.S. government agencies to work toward making UAVs an official part of military and civil aviation. "The DOD has an urgent need to allow UAVs unencumbered access to the National Airspace System (NAS) outside of restricted areas . . . here in the Unites States and around the world." As soon as these regulations are established, the commercial UAV market is destined to explode.

The "brain" of the UAV is the autopilot, which is becoming ever-more sophisticated and capable, varying as widely as the applications for computers themselves (especially artificial intelligence applications). As UAV systems push to smaller platforms, the autopilots themselves are also becoming smaller.

Unmanned Ground Vehicles

Imagine the following situation: You are driving on the highway and pass a moving vehicle that has no human driver. You are shocked and move a lane away just so that you are safe from this strange vehicle. You are, however, captivated by the concept of a driverless car and you want to observe it. You therefore slow the speed of your car to match the other vehicle. After cruising together for some time, you discover that the unmanned vehicle performs precisely as would a vehicle driven by a human driver. Moreover, you discover that the unmanned vehicle is not being guided by a remote operator (there are no radio transmissions to or from the vehicle). The vehicle is, truly, self-driven.

Is the preceding scenario frightening to you? Does it seem impossible? Hardly . . . a similar situation actually occurred in May 2006 as engineers tested—in remotely piloted mode—unmanned vehicles in the western Utah desert.

The Grand Challenge

Spurred on in part by the success of unmanned air and underwater systems, the United States Congress, in conjunction with the U.S. Department of Defense (DoD), has mandated that one-third of all military land vehicles be autonomous (unmanned) by 2015, and two-thirds by 2025. In response to this mandate, the Defense Advanced Research Projects Agency (DARPA)—the central research and development organization for the DoD—created the Grand Challenge as a field test intended to accelerate research and development in autonomous ground vehicles. DARPA's purpose in holding the Grand Challenge was to bring together individuals and organizations. Representatives from industry, the research and development community, government, the armed services, academia, students, backyard inventors, and automotive enthusiasts gathered in the pursuit of a technological challenge. The end goal would be to help save American lives on the battlefield.

The first Grand Challenge was held in 2004. The goal was to conduct a field test of 15 autonomous ground vehicles that would run from Barstow, California to Primm, Nevada. The winner would receive $1 million. However, all the vehicles foundered in the desert. No vehicle traveled more than 8 miles (13 km).

Undaunted, DARPA sponsored another race—the Grand Challenge 2005. More than 190 teams entered the competition, with 23 competing in the final event. Five teams finished the 132-mile (212-km) course over desert terrain. All of the vehicles were totally autonomous; that is, no radio guidance was allowed. The vehicles had to follow an unknown course without human assistance.

Due to this incredible advancement in technology and vehicle capability, DARPA announced in May 2006 that it would hold the Urban Challenge 2007. This event will feature autonomous ground vehicles executing simulated military supply missions safely and effectively in a mock urban area, such as the scenario given at the beginning of this section.

Even as teams from across the United States prepare for Urban Challenge 2007, numerous military organizations are asking for and awarding contracts to companies for their unmanned vehicle solutions. Although the military has implemented for tactical use a number of unmanned airplanes, relatively few land vehicles are currently in use, and none of them are completely autonomous.

The lack of autonomous ground vehicles is due in part to the complications of traveling across the Earth's surface, which is different from traveling through the air. Planes fly through a three-dimensional (3D) space where all obstacles (other aircraft) are identified. But land vehicles travel on a two-dimensional (2D) plane where obstacles (think of them as disturbances in the 2D plane) may be unidentified and collision avoidance is essential. As noted previously, however, safety in the air will become an increasing problem as the number of UAVs increases.

Humans rely primarily on vision to drive a vehicle, and other senses allow them to travel a designated path and respond to obstacles by maneuvering around them. Unmanned vehicles must also have this ability to sense surroundings, travel a specified course, and maneuver around obstacles.

The state of autonomy in unmanned vehicles is in its infancy. Currently, most unmanned vehicles are known as remotely piloted vehicles (RPVs). All functions of an RPV are controlled by a human operator including propulsion and collision avoidance. Sometimes this is also referred to as telepresence. RGVs are semi-autonomous. They have some ability to automatically react to observations of their surroundings. RGVs are still controlled by a remote operator, but do not require 100% attention. The Mars rovers are good examples of RGVs. The final class discussed here are autonomously guided vehicles (AGVs). They are able to observe their surroundings and react to them based upon an "attitude." AGVs are provided a series of objectives and perform them with little to no human input. The "attitude" gives them the level of aggressiveness that they will use to complete the objectives (wait at the stop light forever or run it).

The levels of autonomy can be thought of as a linear progression from today with RPVs as the norm to tomorrow where AGVs are commonplace.

Conclusion

The technology of air and land unmanned vehicles is moving forward at a dynamic, some might say frantic, pace. The military applications are currently spurring this amazing growth. Eventually, civilian applications will become even larger in volume and complexity. In this field, creative energies have just begun to be applied. The future for unmanned vehicles is exciting but unknown. The one certain thing with unmanned vehicle technology is that innovators within the industry can imagine uses beyond the current horizon.

CASE STUDY 23-1

The Wonder of Flight

Airplanes are special. Few products in the history of the world have been identified so completely with technological progress as has the airplane. That may arise from the innate desire of human beings to fly, perhaps as a demonstration of our ability to conquer all obstacles that nature might impose. Or, it might come from our marvel at the apparent impossibility of a gigantic airplane lifting off the ground so effortlessly and carrying us quickly to a distant destination. Most people are also amazed by the latest innovation in airplane technology and desire to understand how it works, at least a little bit. Those in the composites field identify with airplanes, even though they work outside the industry.

In many ways, the development of the airplane has led to some of the greatest human achievements as well as some moments of great sadness. The airplane has affected wars and peace as has no other technology. The developments of the airplane and its child, space travel, were dominant technologies in the 20th century. However, something else is special about airplanes, which is not quite so readily apparent. It will become clear as the technologies that have enabled human flight are examined.

Wilbur and Orville Wright

In at least one critical way, the Wright brothers were typical of those who dreamed of flying ever since ancient days. They were visionaries and scientists and added yet another critical element to the development of sustained human flight. That new element was entrepreneurship. Wilbur told his father, as the two began their quest, "I believe that flight is possible, and while I am taking up the investigation for pleasure rather than profit, I think there is a slight possibility of achieving fame and fortune from it."

The Wrights were highly innovative, self-taught engineers whose experimental skills were honed in their bicycle shop. They followed in the footsteps of the visionary scientists of the 1800s who learned about flight from using gliders. Wilbur wrote the Smithsonian Institution in 1899 (then headed by another pioneer flight scientist, Samuel Langley) asking for a summary of the information that had already been learned. The Wrights made several interesting gliders and kites to test their theories of wing design and control methods. Almost none of these worked well. Then a breakthrough occurred.

The Wrights had been traveling regularly from their home in Dayton, Ohio, to Kitty Hawk, North Carolina, because of the strong and almost constant wind along the North Carolina shore. On the way back to Ohio, they realized that many more experiments needed to be done but that the prove-out trips to North Carolina to test each new development would be prohibitively expensive. They conceived of a way to test the models in their shop in Ohio. They invented the wind tunnel.

The next breakthrough occurred when they realized that they should break the problem of flight down into several smaller problems. They worked on control, leaving thrust and lift alone. Then, when they felt that they had a method for control, they would work on one of the other elements. For instance, when investigating lift, they tied off the control mechanism so that they could see the effects of lift more clearly. They finally had the details of glider flight worked out and successfully built and flew (many times) full-size gliders in 1902. They seemed to have worked out many of the lift and control problems.

The thrust component was then tackled. The Wright brothers invented the modern design for propellers using wind-tunnel experiments and built them in their bicycle shop. With assistance from their machinist, they designed and built a four-cylinder internal combustion engine, which they mounted on their newest version plane. They seemed to have all the pieces separately worked out, but would they work together? The brothers were confident and somewhat in awe of their good success. Orville wrote, "Isn't it astonishing that all these secrets have been preserved for so many years just so that we could discover them!"

The Wright brothers took their new model to Kitty Hawk where they had become minor celebrities to the locals. A few local witnesses gathered on the sand dunes on December 17, 1903 to see whether the Wrights would be successful. That morning, Orville climbed onto the lower wing of the bi-wing plane and guided it for 12 seconds over 120 ft (37 m) along the shore. A local caught the flight in a photograph that seems to suggest a hesitating feeling of success. That same day, more successes came (best: 852 ft [260 m] for 59 seconds) and with them the confidence of knowing that powered human flight had been achieved. They dubbed the airplane "Flyer I." Then their entrepreneurial spirit kicked in.

Newspaper reports of the Wright brothers' success followed quickly. However, most of these reports were inaccurate and not very credible. The problem was that the Wrights did not follow up with public flights. They wanted to make sure that the technology and patents were carefully worked out before they gave the world a look. They moved their work closer to their home in Dayton and soon (1904) built the Flyer II, which became the first plane to fly in a circle. In 1905, they built the Flyer III, which could stay in the air for over half an hour and perform most of the maneuvers that would be required of a reliable and controllable commercial airplane.

Finally, in 1908, the Wrights began a series of public flights to advertise their Model A bi-plane. They began taking orders and started producing planes at the rate of four a month. However, just looking at their planes was enough to give competitors ideas on how to improve them. The Wrights became so involved in various patent lawsuits that they failed to keep ahead of the technology. Soon, new technologies were developed and being flown by others. The Wrights tried to maintain control over the market, but the advances were happening too fast and were too different from their patented technologies. The death of Wilbur in 1912 from typhoid fever seemed to end the era of the Wrights. Orville continued on, but his enthusiasm had waned.

Military Impetus and Civilian Circus

Airplanes came into warfare use during WWI. Their role was first reconnaissance and then active engagement in battles. European airplane manufacturers seemed to have a lead in providing military planes for WWI. The Fokker was the plane of Germany's Red Baron while the Sopwith Camel was the plane favored by the British. By the end of WWI in 1918, airplanes had demonstrated (crudely, to be sure) all of the roles that military airplanes would play in the future, except troop transport.

Civilian uses for the airplane, except for flying the mail, seemed to focus around air shows, air races, and contests. Barnstormers would fly around the country entertaining people with acrobatic stunts and rides for those who were brave. This was a golden age for airplanes, in large part because they were so unusual and fun. Adding to their allure was the accomplishment of Charles Lindbergh in 1927. He flew an American plane from New York to Paris in 33 hours, non-stop, and unaccompanied. In the days of dead-reckoning navigation, this was truly a major feat. He rightly was given a hero's welcome in France and the United States.

During these golden years, the visionary engineer-entrepreneurs of the United States were busy building airplanes and learning how to improve the technology. Companies established during these years included: Martin Company, Douglas Company, Lockheed Company, Curtiss Aeroplane and Motor Company, TWA, North American Aviation, Northrop Aircraft, and Boeing. Most were started by one or two people using garages and other available space (such as behind a barbershop, a cleaners and dye shop, a small hotel, a church, and an apricot cannery) as their initial manufacturing locations. Each company seemed to be headed by tremendously dedicated individuals who, like the Wrights, had vision, engineering skill, and entrepreneurial instincts. These American companies and similar companies in Europe would prove to be vitally important in the development of the airplane, especially during WWII.

WWII and the Modern Era

While the Germans clearly understood the importance of air power at the beginning of WWII, the allies had strangely lagged behind in military aircraft development. Consequently, the German Luftwaffe was a powerful force left largely unchecked during 1939 and early 1940. The biggest problems encountered by the Luftwaffe seemed to be the weather and its own mistakes. For instance, during the evacuation from Dunkirk, when both the English and French armies were being hurriedly transported from Europe just ahead of the German advance, the Luftwaffe was grounded by the weather. This allowed for a relatively uneventful evacuation across the English Channel. War experts believe that if the Luftwaffe would have attacked the transport ships during those days of evacuation, the British and French armies would have been so decimated that they would not have been able to repel a German invasion of England. But, the weather saved the troops and the golden oratory of Winston Churchill dissuaded Hitler from the invasion, which might have been successful and would have changed the war.

Not too long after those disastrous days of the early war, the British aviation industry developed the Spitfire—a fighter that wreaked havoc on the German bombers that were trying to win the Battle of Britain without an invasion. Eventually, the American aviation industry joined with the British and subjected the Germans to day and night bombing. Similarly, air power also proved to be pivotal in the Pacific part of the war, eventually leading to devastating bombing runs on the Japanese mainland, undoubtedly shortening the war.

When WWII was over, many engineer-entrepreneurs within the major airplane manufacturing companies were given permission to explore the possibilities of aircraft for civilian applications. Many of these engineers laid the foundation of the vast and extremely successful modern aviation industry. Others felt restricted by the size of the major airplane companies. So, they began to explore applications on their own, eventually developing the products and technology that would be the basis of the fiberglass reinforced plastics industry. Still others recognized the need for new materials and manufacturing techniques, many of which were developed by engineer-entrepreneurs in big and small companies.

With the growth of the civilian aircraft business and the many people who flew as passengers, safety and government regulation became major factors in building airplanes. These restrictive factors were best handled by ever larger and more complex companies building ever larger and more complex planes. These companies began to acquire the small entrepreneurs and then to merge together among themselves. The emergence of space as an important part of the aerospace industry further accelerated this consolidation of aerospace companies.

The Current State

Aerospace is now dominated by a few very large companies. Innovations are made incrementally and over long periods of time. For safety, this is probably good. But for dynamic market development and progress, this is likely bad. Entrepreneurs within the large aerospace companies find it hard to get support for something really different and innovative.

Even those aerospace engineer-entrepreneurs outside the large companies who want to develop a new aerospace technology find that major developments require tremendous infusions of capital. There are, of course, small aerospace contracts for these people. However, these contracts are highly restricted and usually suppress rather than encourage innovation because of the need to meet rigid specifications.

The large aerospace firms and similar firms in other industries need to learn a few lessons. First, they must find a way to develop internal innovations from engineer-entrepreneurs. This will probably require that the big companies give profit possibilities and company time to those employees who want to be innovative. Also, the big companies need to be willing to take risks. That means they must be less focused on their quarterly reports and more on their earnings in 10 years. It may also mean that lawyers need to find ways to protect the companies from liability suits. The investors of Wall Street must recognize this vision and be willing to not only allow it to happen but encourage it. When that occurs, the large companies will promote management that understands how to make the companies grow through innovation. The recent development of the Boeing 787 Dreamliner suggests that some of the entrepreneurial spirit has returned to the big companies. Some of the fighter aircraft also show this innovative spirit. But, these seem to be exceptions rather than the rule.

Where then are the visionary engineer-entrepreneurs of today? The true visionary engineer-entrepreneurs are in small businesses. There they can experiment and innovate. They can test new markets, fail, and then recover to try again. Like the Wright brothers, these modern engineer-entrepreneurs learn from the past but also recognize that they need to develop new science and technology. They are surprisingly agile in the way they approach products and markets. They are the future of industrial growth. (In fact, over the last 10 years, the net production of new jobs in the United States has been exclusively in the small business sector. Big business has produced no new growth even though it represents two-thirds of the gross national product.)

Many of the visionary engineer-entrepreneurs are in FRP and advanced composites, but probably not in the big companies. They are owners of small shops where products are manufactured for new and exciting markets. They may even be making old products but using new materials to make them better.

SUMMARY

The breadth of composites part usage is immense. These products are used in markets as diverse as: automobiles, buses, and trucks; various construction products such as panels and wrappings for concrete bridge pillars; pipes and tanks that defy corrosion; electrical circuit boards; golf clubs and racquets of many types; housings for appliances; medical devices like artificial limbs; and, of course, a multitude of aerospace products.

The growth of composites use continues to be strong. Fiberglass-based composites continue to gain market share in all of the traditional markets. Carbon fiber-based composites are poised for breakthroughs in aerospace, which might require a complete rethinking of the supply-and-demand picture.

Composite materials are well characterized and are being refined constantly. New developments seem to happen daily. The manufacturing methods are also changing to provide far greater productivity. The potential for composites has never been brighter.

LABORATORY EXPERIMENT 23-1

Objective: Become acquainted with a professional organization dealing with composites technologies.

Procedure:

1. Attend a meeting of a professional organization like the Society for the Advancement of Materials and Process Engineering (SAMPE). The SAMPE website (SAMPE.org) can be consulted for information about local chapter officers who can then be called to get the date, time, and location of the next meeting. It may be possible for the entire class to attend together.
2. If it is not possible to attend a meeting of a professional organization, contact one and request information.
3. Obtain a copy of the proceedings of a conference of a professional society and give a summary of one of the papers presented at the conference.

QUESTIONS

1. Why is the 787 Dreamliner such an innovative airplane?
2. What are two breakthrough technologies of the small aircraft?
3. What are two major motivations for using composites in automobiles?
4. What is the chief reason for composites not being adopted to a greater extent in automobiles?

5. Give three ways to make a composite tank and discuss the relative advantages and disadvantages of each. Which would be favored in a developing nation?
6. Describe the mechanism involved in stopping a bullet in a bullet-proof vest.
7. Give two reasons why a composite rifle might be preferred in an Olympic shooting competition.
8. Recognizing that the Wright brothers were technical and entrepreneurial, why isn't their airplane company still in existence?

BIBLIOGRAPHY

The Associated Press. 2005. "Critics Scrutinize Cost of Shuttle." http://www.usatoday.com/news/nation/2003-02-04-shuttle-critics_x.htm.

Curry, Andrew. 2003. "Taking Flight: How the Dropouts from Dayton Changed the World Forever." *U.S. News & World Report*, July 21.

Epstein, George and Heimerdinger, M. William, eds. 2003. "Celebrating the 100th Anniversary of the Wright Brothers' First Flight." SAMPE Tribute to the Centennial of Flight. Los Angeles, CA: Society for the Advancement of Material and Process Engineering.

Gladwell, Malcolm. 2002. *The Tipping Point*. Boston: Little, Brown and Company, p. 141.

Harris, Charles E., Shuart, Mark J., and Gray, Hugh R. 2002. "Emerging Materials for Revolutionary Aerospace Vehicle Structures and Propulsion Systems." *SAMPE Journal*, Vol. 38, No. 6, November/December, pp. 33–43.

National Aeronautics and Space Administration. 2002. "Celebrating a Century of Flight." NASA Publication SP-2002-09-511-HQ, ISBN 0-16-067541-3. Washington, DC: National Aeronautics and Space Administration.

Strong, A. Brent. 2006. *Plastics: Materials and Processing*, 3rd Ed. Upper Saddle River, NJ: Prentice-Hall, Inc.

Tribble, Alan. 1995. *The Space Environment, Implications for Spacecraft Design*. Princeton, NJ: Princeton University Press.

Warren, C. David. 2001. "Carbon Fiber in Future Vehicles." *SAMPE Journal*, Vol. 37, No. 2, March/April.

Glossary

A

ablation The gradual eroding of a material, which is often the result of slow burning combined with abrasion from rapidly flowing hot gases over the surface of the ablative material.

accelerators Chemicals that react with organic peroxides to create free radicals. They are often used to achieve room-temperature curing. Cobalt compounds are the most common type. (Also called ***promoters***.)

acid number A measure of the amount of residual diacid monomer in a polyester polymerization reaction. Residual diacid is generally detrimental to polymer properties and so maximum values for the acid number are often set in the polymerization process.

acoustic emissions A non-destructive method to monitor the structure of a composite part by recording acoustic emissions from a sample and then comparing those to emissions from a known good sample.

activation energy The tendency or lack thereof, for a bond to resist being broken by a chemical reaction.

addition polymerization The method of linking monomers to form polymer molecules, which employs an activation of the monomer, usually with a peroxide free-radical. It occurs in a series of chain-reaction-like steps in which other monomers are added to the growing chain.

additives Materials combined with the basic composite material (resin system and reinforcements) to improve some specific property.

adduct A material added to a normal mixture to alter the reactions or products of its reactive components. The resultant material may be called the *adduct*, which is short for *adduct product*.

adherend The material that is to be bonded when an adhesive is used as the bonding agent.

adipic acid A non-aromatic diacid that imparts toughness and weatherability to polyester thermosets.

advanced composites A group of composite materials using either thermoset and thermoplastic resins. They are generally characterized by the use of reinforcements having higher mechanical properties than fiberglass and resins with higher thermal, chemical, and mechanical properties than unsaturated polyesters or vinyl esters. Also called ***high-performance composites***.

aliphatic Polymers that have no aromatic content.

alkyd An alternate name for ***unsaturated polyester*** materials. The name derives from a contraction of alcohol and acid.

allylic A type of unsaturated polyester that contains a particular arrangement of atoms, including a carbon-carbon double bond.

amorphous region The regions in a polymer in which the molecules are not packed into a crystalline structure.

anisotropic Describes the majority of composite materials, which are not equally oriented in all directions (also called *non-isotropic*).

annealing Exposing a part to an elevated temperature (but not near its melting or decomposition temperature) for an extended period to advance the amount of cure (crosslinking). This generally improves the mechanical properties of the part. Sometimes called ***post-curing***.

anti-foaming agent An additive that prevents or reduces the tendency of a resin mixture to foam.

antioxidant Materials that prevent or retard the reaction of a polymer with oxygen, thus preventing the degradation of properties from oxidation.

aramid fiber A high-performance reinforcement that is an aromatic polyamide. The most common brand of aramid fiber is Kevlar®.

aromatic A group of carbons arranged in a ring with hydrogen molecules attached, which impart strength, stiffness, flame retardant characteristics, and other properties to polymers.

aromatic heterocycle A molecule in which a heterocyclic group is bonded to an aromatic group.

aspect ratio The ratio of the length to the diameter of a fiber.

attraction A secondary bond between molecules occurring especially when the molecules become solids.

B

basket weave A weaving pattern that has two warp fibers and two weft fibers jointly alternating in a regular over-and-under pattern.

bias The direction that is 45° off the principal or machine direction of a fabric. Positive bias is generally to the right of the machine direction and negative bias is generally to the left.

bisphenol-A A monomer that can be used to improve the water/chemical resistance, toughness, and strength of polyester thermosets.

bleeder material Part of a vacuum bagging system whose function is to absorb excess resin.

blooming A process in which fibers migrate through a gel coat and reach the surface of a composite part.

bond energy The energy required to completely separate two bonded atoms and, therefore, to break the bond.

bond strength The stability of a bond, usually expressed in terms of bond energy.

braids Fabrics made by alternating several strands in an over-and-under pattern.

breather material Part of a vacuum bagging system whose function is to allow air circulation throughout the assembly.

breathing the mold See ***bumping the mold*** or ***burping the mold***.

bridge molecules The ***co-reactants***, like styrene, which serve as the links in crosslinked polymers that use addition reactions or free-radicals as the crosslinking method.

broad goods A general name for fabrics or other planar materials.

B-staged resin A stable intermediate stage of a resin when it is partially crosslinked.

B-staging The process of partially reacting a resin so that it reaches a stable B-stage.

bulk molding compound (BMC) A mixture of polyester resin, filler, and fiberglass made by low-intensity mixing. It is generally used in compression molding for applications such as automotive panels and parts.

bumping the mold A process of opening the mold slightly and quickly during the cure of some resins to allow the condensate to escape. Also called ***breathing the mold*** or ***burping the mold***.

burping the mold See ***breathing the mold*** or ***bumping the mold***.

burst test A method of determining the pressure-holding capability of a pressure vessel by filling it (usually with water) and then adding fluid until it bursts.

C

carbon-carbon composites A composite in which carbon fibers are surrounded by a matrix, which is also carbon. Also called ***carbon matrix composites***.

carbon fiber A reinforcement material consisting mostly of carbon atoms arranged in a hexagonal planar structure which, therefore, has high strength and stiffness. Today the term is interchangeable with ***graphite fiber***.

carbon matrix composites See ***carbon-carbon composites***.

carbonization A step in the manufacture of carbon matrix composites in which the non-carbon atoms are driven out of the material through a high-temperature pyrolysis process. This is also the name for a step in the manufacture of carbon fibers during which the planar carbon rings are formed through elimination of hydrogen molecules and atomic rearrangement.

catalyst A material that, without actually being consumed in the reaction, enhances the reactivity of a chemical reaction. This improves the speed and effectiveness of the reaction; peroxide ***initiators*** are sometimes erroneously called catalysts.

caterpillar pulling system A puller type used in pultrusion, it involves two continuous synchronized belts that press against the part and pull it along.

caul A pad used in a bagging assembly to intensify the pressure in a particular area of a part within a vacuum bagging system. Also called a ***pressure intensifier***.

cavities The areas in a mold into which the melt flows and solidifies to make the part.

centrifugal casting A process in which fibers are placed inside a mold, usually of cylindrical shape, and then rotated as the resin is introduced to forcefully infiltrate the fibers. Also called ***spin casting***.

ceramic A material characterized by ionic bonds formed between metal and non-metal atoms (actually ions), which have high melting points, high stiffness, and low toughness.

C-glass A type of fiberglass formulated to be especially resistant to attack from chemicals. It is only rarely used in composites.

char The material formed when some plastics, such as phenolic, degrade because of pyrolysis (burning in a low oxygen atmosphere); it is similar to the charcoal residue when wood burns.

charge A specific amount of material introduced into the mold.

chemical polarity See ***polarity***.

chemical vapor deposition A method for impregnating the porous carbon matrix composite (made porous by a carbonization reaction) with new resin so that the pores will be filled and the properties improved.

chlorendic acid A diacid monomer added to polyester thermosets to impart flame retardant and chemical-resistance properties.

chopped strand mat A sheet material made by chopping fiberglass into staple lengths and directing the fibers onto a belt where they are sprayed with a binder to form a mat structure.

cis isomer The arrangement of atoms in which two key groups are on the same side of a carbon-carbon double bond. ***Maleic acid*** is a cis isomer.

clamshell mold Molds that are often connected on one edge and open simply to allow the part to be put in or taken out. Normally, the clam-shell mold requires only minimal pressure and is used chiefly to define the outer surface of a part.

co-curing A process in which parts are cured and bonded during a single heating cycle.

coefficient of thermal expansion (CTE) A measure of the amount of dimensional change that occurs in a sample under differing thermal conditions.

colorants Materials that give color to a composite. They can be either a ***dye*** or ***pigment***.

co-mingling Mixing fibers of different types into a yarn; often done with reinforcement fibers and with fibers made of the thermoplastic resin, but also can be done with two types of reinforcement fibers.

co-monomer A name for the bridging molecules in crosslinking reactions. Styrene is the most common co-monomer for unsaturated polyester crosslinking reactions. Also called a ***co-reactant*** or ***crosslinking agent***.

composite materials Solid materials with at least two components, each one being in a separate phase, and generally consisting of fibrous reinforcements being surrounded and held in place by a resinous material.

compounder A resin processor that takes neat resin from the manufacturer and adds materials such as reinforcements, colorants, and other minor constituents. This addition is usually done with extruders and the resulting product is in the form of resin pellets.

comprehensive stitching Stitching across the entire surface of a laminate.

compression-after-impact test A test in which the sample is impacted and then tested in compression to determine the decrease in properties that might occur because of fiber and matrix damage.

compression molding A composites manufacturing method in which the raw material is placed into a mold and then a matching mold compresses the material while it is being heated to cure. Also called ***matched-die molding***.

compression properties Those characteristics determined by pressing on the ends of a columnar sample.

condensation A type of chemical reaction in which a small molecule (such as H_2O) is created in the reaction and separates from the rest of the products and reactants by condensing out of the solution (or, in other words, the small molecule separates as droplets).

condensation polymerization The method of linking monomers to form polymer molecules, which employs two different monomers, each having two active end groups that will react with each other over and over again when mixed. Each reaction results in the elimination or condensation of a small by-product molecule.

continuous fibers Those fibers that traverse the entire dimension of a part.

continuous strand mat A sheet structure made by swirling continuous fiberglass strands onto a belt and then spraying them with a binder to hold them into a mat structure.

co-reactant See ***co-monomer*** and ***crosslinking agent***.

cost of goods sold The cost of manufacturing goods, which includes the costs of materials, labor, and overhead.

coupling agent A material that improves bonding between two materials, such as a fiber and a matrix, because one end of the coupling agent is chemically similar to the fiber and the other end is chemically similar to the matrix.

covalent bond A bond between atoms in which the electrons are shared by the atoms. This type of bond is characteristic of polymeric materials.

co-weaving Mixing of different fibers in a cloth, knit, or other textile material, which can be done with reinforcement fibers or with fibers made of the matrix resin.

crack deflection A method of toughening a composite in which the reinforcement acts as a barrier to the passing of a crack. This method is similar to ***crack pinning***.

crack pinning The ability of a material, like a reinforcement, to arrest the growth of a crack through a matrix. This is highly important in ceramic matrix composites.

cracking A series of surface breaks in a composite part generally present in the gel coat but may also extend into the laminate layer.

creel A stand or platform on which fiber rolls can be mounted so that the fiber can be withdrawn easily.

creep The gradual movement of polymer chains when subjected to modest loads over long periods of time.

crimping The result of the crossover of fibers in a woven, knitted, or braided fabric. Crimps decrease strength in most reinforcement fibers.

crosslink A bond in thermoset resins that joins polymer bonds together.

crosslink density The number of crosslinks, usually expressed as a ratio of the actual number to the theoretical maximum.

crosslinking The series of chemical reactions that create bonds or links between polymer molecules, thus creating thermoset materials.

crosslinking agent The solvent or ***co-reactant*** that forms bridges between polymer molecules; the bridges are the crosslinks.

crowfoot satin weave A weaving pattern in which the warp strands skip over or under several weft strands in a random fashion.

crystalline region A zone in a polymer in which groups of molecules pack closely together. These zones exhibit properties that are somewhat analogous to crystals in metals and ceramics.

cure time The time to complete the cure of a composite part.

cured hard Curing of the matrix completely to the outer surface. The cure of some resins, especially polyesters and vinyl esters, is inhibited by oxygen and the evaporation of styrene; therefore, they do not cure completely to the surface. This can be prevented by adding a coating material, like wax, to the polymer mix.

curing The process of forming crosslinks.

curing agent A ***hardener***; a molecule that will react with an epoxy to form a crosslink.

D

damping The speed at which vibrations decay.

dart drop impact test A test for toughness in which a weight is dropped onto the surface of the sample and the energy absorbed during the impact is measured.

daylight opening The distance between the opened molds in a compression molding machine.

debulking Removal of air from a stack of prepreg materials. Debulking also might be used to consolidate a preform.

decomposition point The temperature at which a material begins to decompose or degrade. This results from breaking the primary bonds of the material.

denier The weight, in grams, of 9,000 meters of fiber.

diacid An organic compound that has two acid groups (COOH). Often used as a monomer in making polyester thermosets, the diacid is often the component that has the unsaturation needed for crosslinking.

diallylphthalate (DAP) and **diallyl isophthalate (DAIP)** Allylic materials often used in molding compounds.

dicyclopentadiene (DCPD) A chemical that can be used as a monomer along with diacids and glycols to make polyester resins; it imparts some improvements in properties and cost.

die cutting A method of cutting textiles and uncured prepregs in which a steel die is used as the cutting tool. The die is made in the shape of the part to be cut. The lower edge of the die is sharp. It is placed on the broad goods and then pressed through to the material, thus making the cut. Also called ***steel-rule blanking***.

diethylene glycol A glycol monomer that adds toughness to thermoset polyesters.

differential scanning calorimetry (DSC) A method to determine the glass transition temperature of a resin or composite and, thus, monitor the completeness of cure.

difunctional Having two reactive groups.

diglycidyl ether A general name given to molecules that contain two glycidyl groups linked to the remainder of the molecule through ether linkages.

dipropylene glycol A glycol monomer that adds toughness to thermoset polyesters.

double-clamp system A pulling system used in pultrusion, which uses two clamps that alternately grasp and pull and then release and retract. The action of the two clamping stations together gives a smooth pull.

double composite structure A sandwich structure consisting of composite laminate face sheets bonded on the top and bottom of a core material.

drape The ability of a fabric to hang without creases. In a prepreg it is its ability to be shaped.

drape test A method of assessing the cure state of a prepreg by measuring the amount that it drapes over a standard mandrel.

dye An organic material that imparts color to a composite. Also see ***colorants***.

E

eddy current testing A non-destructive test method that can be used on electrically conductive composites. The method examines the surface currents created because of the internal structure of the composite.

edge bleed The flow of resin along the plane of the fibers in a bagging system.

egg-crate mold A mold in which the dimensional stability of the mold surface is improved by supporting the back with a series of interlocking braces.

E-glass A type of fiberglass that is especially useful in electrical applications. Because it is the least expensive type of fiberglass with generally acceptable properties, E-glass is the most widely used type of glass fiber for composites.

ejector pins Pins mounted in the compression molding press system, which are used to eject the part from the mold. Also called ***knockout pins***.

elastic deformation A stress-strain deformation in which the material returns to its original shape when the stress is relieved.

electron conjugation Electrons spread over several atoms, as occurs in aromatic molecules where there are alternating single and double bonds, which results in chemical stability of the bonds.

elongation The strain to failure, usually expressed as a percentage.

engineering composites Composite materials that compete with many low-performance metals but do not have the high properties of advanced composites. The fibers are usually short and the resins are typically inexpensive. The group includes all fiber-reinforced plastics (FRP) materials and most short-fiber-reinforced thermosets. Also called ***industrial composites***.

epoxy group A three-member ring containing two carbon molecules and an oxygen molecule, which is characteristic of the epoxy polymer. Also called an ***oxirane group***.

epoxy molar mass The molecular weight of an epoxy molecule divided by the number of active groups. A low epoxy molar mass number indicates that there are few atoms between active groups and, therefore, the number of crosslinks per unit of weight is high.

equilibrium toughness The toughness determined under equilibrium conditions and, generally, during a tensile test. The area under the tensile curve is the equilibrium toughness.

ester A grouping that contains COOC atoms arranged in a particular pattern. This grouping is formed by the reaction of an organic acid (or diacid) and an alcohol (or glycol).

ethylene glycol The most common and least expensive of the glycol monomers used in making polyester thermosets.

exotherm peak The time it takes to reach the maximum temperature during cure.

extended chain polyethylene (ECPE) fibers An alternate name for ***ultra-high-molecular-weight polyethylene (UHMWPE) fibers***. The term emphasizes the essentially linear nature of the polymer chains, which are stretched and then formed into fibers.

external doubler A repair system in which the repair material, usually a cured laminate, is simply bolted or bonded over the damaged area.

F

face bleed The flow of the resin in a direction perpendicular to the plane of the fibers in a bagging system.

fatigue The decrease in composite properties because of an intermittent load applied over a long period of time.

feathering Extending the edges of a scarf repair so that the entry angle is gradual.

feed zone The portion of the injection molding screw (and the extrusion screw), which moves the unmelted resin ahead in the barrel. It is located under the feed hopper.

fiber A material characterized by a single long axis (length) and other axes that are negligibly small in comparison. These materials are excellent reinforcements as they can be strong and stiff in the major axis direction. Also called a ***strand*** or ***filament***.

fiber placement A composites manufacturing method that places fibers onto a mandrel using a multi-axis head. The process allows the automatic forming of many more shapes than could be made with filament winding.

fiber surface treatment A step in the fiber manufacturing process in which the surface of the fiber is modified, generally so that bonding with the matrix is improved.

fiber-volume fraction (V_f) The ratio of the actual volume of fibers to the total volume of the composite. Also called the *fiber-volume ratio*.

fiber wash The inadvertent movement of fibers caused by resin injection in resin transfer molding (RTM) and other resin infusion processes.

fiberglass reinforced plastics (FRP) Generally made from unsaturated polyester and vinyl ester resins with fiberglass reinforcement, the material should be contrasted with advanced or high-performance composites. Also called **thermoset engineering composites**.

filament A single ***strand*** of ***fiber***. Usually filaments are the output of a single hole in the fiber-forming device (spinneret).

filament winding A composites processing method in which continuous fibers are coated with a resin and wrapped around a mandrel and then cured.

fill In a woven fabric, these are the fibers interwoven above and below the warp fibers. Also called the ***weft***, ***woof***, or transverse fibers.

filler Solid materials ground to fine powders and added to the resin mix to reduce overall cost and, occasionally, to impart some other beneficial property.

finish A coating placed on a fiber to improve the bonding between the fiber and the matrix. A finish is, today, essentially the same as a ***sizing***.

first-ply failure The initial failure condition of a composite, which is used in designing the composite.

flash Excess material that leaks out of a mold.

flash-type mold closure The type of mold closure system in compression molding that allows flashing.

flexible die forming A process in which one or more dies in a matched die set is made of a flexible (rubber-like) material for molding of thermoplastic composites. Also called ***hydroforming***.

flexural properties The properties of a material determined when the material is bent. The test sample is supported between two posts and then pressed on the upper side mid-way between the supports.

flow forming A method in matched-die molding of thermoplastic composites in which the fibers are allowed to slide.

flush-type repair A repair method in which the repair materials are placed into the damaged area (after removal of the damaged material). This type of repair requires curing of the repair material.

free-radical A highly reactive chemical species, usually characterized by an unbonded electron. These materials are useful in starting various chemical reactions and often are made by the decomposition of a peroxide molecule.

FRP See ***fiber reinforced plastics***.

fumaric acid A common unsaturated diacid used in making unsaturated polyesters. It is a ***trans isomer***.

functional specification A document containing the requirements a part must meet; the requirements are based on the function of the part in actual use.

G

gate The entry to the cavity in an injection mold.

gel coat The outer layer of a composite material. It serves as a protective coating.

gel permeation chromatography An analytical tool that assists in determining relative molecular weight by observing the migration of small portions of polymers on a solvent-soaked paper.

gel time The time to reach gelation or the gel point.

gelation or **gel point** The onset of solidification; the condition when crosslinking begins to dominate the nature of the thermoset resin.

geodesic path The path on a surface that allows fibers to be placed under tension and remain in place without slippage.

glass A ceramic material in which the atoms (or ions) do not form crystal structures but, rather, exist in a rigid disordered network.

glass transition temperature (T_g) A thermal transition that occurs in solid polymeric materials, which marks the onset, with increased temperature, of coordinated multi-atom movements within the polymer. Above the transition, the material can be quite pliable or leathery whereas below it the material is rigid.

glycidyl group An organic group containing the epoxy ring and another carbon atom; it is attached to other organic groups through the additional carbon atom.

glycol An organic compound that has two alcohol groups (OH), it is often used as a monomer in making polyester thermosets.

graphite fiber Originally a reinforcement made from carbon fibers, but today the term is interchangeable with ***carbon fiber***. It is a reinforcement material consisting mostly of carbon atoms arranged in a hexagonal planar structure which, therefore, has high strength and stiffness.

graphitization A step in the manufacture of carbon fibers in which the carbonized fibers are further heated to drive off the nitrogen and hydrogen, thus converting the fiber to above 99% carbon.

green preform A mixture of resin and, usually, reinforcement that has not been cured.

H

halogens The chemical atoms fluorine (F), chlorine (Cl), bromine (Br), and iodine (I), which are often added to polymers to improve fire-retardant properties.

hardener A molecule that will react with an epoxy to form a crosslink. A ***curing agent*** for epoxy molecules.

harness weave A satin weave in which the skips are long.

hat stiffener A stiffener that has a hat-like cross-section.

hazardous air pollutant (HAP) Materials that emit potentially dangerous vapors.

heat capacity The measure of a material's tendency to absorb heat.

heat distortion temperature (HDT) The maximum temperature that a plastic or composite material can tolerate without distorting in a mechanical application. It is determined by a specific test.

helical windings The pattern of fibers laid onto the mandrel during filament winding at angles other than 90° to the axis of the mandrel.

heterocyclic A ring group having more than one type of atom, such as the imide ring, which contains both carbon and nitrogen atoms.

hexa A curing agent used with novolac phenolics.

high-performance composites See ***advanced composites***.

high-temperature vulcanization (HTV) A type of resin, often an epoxy, which requires application of heat to effect a cure.

homopolymerization The reaction of a polymer with itself or with another molecule of the same type to extend the polymer chain.

hoop windings The pattern of fibers laid onto the mandrel in filament winding at a 90° angle to the axis of the mandrel.

hybrid reinforcements Schemes that combine two or more different fiber types.

hydroforming A process in which one or more dies in a matched-die set is made of a flexible (rubber-like) material for molding of thermoplastic composites. Also called ***flexible-die forming***.

I

impact strength See ***impact toughness***.

impact toughness The toughness (energy absorbed) when a material is impacted. Also called ***impact strength***.

impregnation The step in the making of carbon matrix composites in which new resin is introduced into the porous matrix after the part has been pyrolyzed.

incremental forming A method of forming thermoplastic composites in which a large part is formed in sections. Also called ***transition molding***.

industrial composites See ***engineering composites***.

inhibitor A chemical that absorbs or otherwise deactivates free-radicals and thus extends the storage life of polyester resins.

initiator A chemical that starts a chain-like chemical reaction; peroxides are common materials of this type.

in-situ reaction sintering A process for making ceramic matrix composites in which two components in the ceramic material react during the sintering step and create a second phase, which acts as a reinforcement.

intelligent structure A structure with sensors that can learn and adapt rather than simply respond in a preprogrammed manner.

Interface Theory A view of the interactions of fiber and matrix that involves the actual touching (bonding) of the fiber and matrix or, if there is a coupling agent present, of the fiber and matrix with the coupling agent. The bonding is along a two-dimensional planar surface.

interlaminar cracking Matrix cracks between two plies.

Interphase Theory A view of the interactions of fiber and matrix with a coupling agent in which the coupling agent forms a three-dimensional network into which the resin molecules penetrate and thus form a phase between the coupling agent and the matrix. The same type of network would also apply to the fiber-coupling agent interaction.

intralaminar cracking Matrix cracks within a single ply.

ion A positively or negatively charged atom.

ionic bond A bond between positively and negatively charged ions, it is a bond typical in ceramic materials.

ISO 9000 A certification program that requires certain quality records to be maintained.

isocyanate A combination of atoms in which the elements NCO are double-bonded.

isogrid A pattern of stiffeners wherein they are arranged as equilateral triangles.

isophthalic acid (iso) A saturated, aromatic diacid often included in the mix of monomers to make unsaturated polyester resins. It imparts superior toughness and water/chemical resistance.

isotropic When the structure of a material is the same in all directions.

J

J-stiffener A stiffener in which the cross-section is shaped like the letter J.

K

kitting Cutting and assembling materials for a lay-up before the actual molding operation to simplify molding and improve overall efficiency.

knit line The area in an injection-molded part where flows from two or more cavity entry points meet. It is important to have good interpenetration of material across the knit line. Also called the ***weld line***.

knitted fabric A material in which the traditional warp and weft fibers are looped to create a fabric with high drape and conformability.

knockout pins See ***ejector pins***.

L

latent A property that is hidden under certain conditions. For example, a hardener that will not react or reacts very slowly at room temperature and then reacts quickly at elevated temperatures is said to be a latent hardener.

lay-up A system of manufacturing composites in which the reinforcement is placed into the mold and then the resin is added and worked into the fibers to cause wet-out.

leno weave A fabric in which two warp fibers twist around a fill fiber.

light resin transfer molding (RTM) A resin infusion process in which the mold is made of a lightweight composite.

liquid molding processes A general term occasionally used to refer to all resin infusion processes.

load transfer A method of crack stoppage, especially useful in ceramic matrix composites, in which the reinforcement has a higher strain to failure than the matrix. Thus, when a crack encounters the reinforcement, the energy of the crack's propagation is transferred to the reinforcement and the crack stops growing.

longitudinal direction In woven fabrics, this is the machine direction.

longitudinal windings The pattern of fibers laid down on the mandrel during filament winding, which is 0° to the axis of the mandrel.

low-profile or low-shrink system A method of using certain fillers to compensate for the natural shrinkage of polyesters, therefore creating a smooth, defect-free molded surface.

M

machine direction The principal direction on a machine in which the product runs. In woven fabrics, it is the same as the ***warp*** direction.

maleic acid A common unsaturated diacid used in making unsaturated polyesters. It is a ***cis isomer***.

maleic anhydride A common monomer used in making unsaturated polyesters. It is easily converted into maleic and fumaric acids to form diacids.

mandrel A core onto which the composite material is placed and then cured.

master model The shaped material from which a mold is made.

matched-die molding See ***compression molding***.

matrix The continuous phase of a composite material that surrounds and binds together fiber reinforcements.

melt-processible resin A resin that can be processed by melting; a thermoplastic.

melting point The temperature at which a solid turns into a liquid. This is the same as the freezing point for most materials.

melting zone That portion of the injection molding (and extrusion) screw that has a changing flight depth. It is surrounded with heating coils so that the resin will be melted in this area.

metal A material formed from either a single type of metal atom or from mixtures of metals. The metal atoms (ions) are held together by metallic bonds, resulting in a material that has a relatively high melting point, high ductility, and high strength.

metal(lic) bond The type of bonding present when positive (metal) ions are held in a lattice by a sea of electrons.

metering zone That portion of an injection molding (and extrusion) screw that has shallow flights so that the shear forces are high on the resin. This zone ensures that all the resin is melted and creates pressure to pump the resin out of the machine.

microcracks A method in which the growth of a main crack is arrested by the energy absorbed when microcracks are created at the point of the main crack.

micro-indentation test A method for determining the fiber-matrix bond strength.

mil gage A small device that allows the operator who is applying a gel coat to monitor the thickness of the gel coat.

milled fibers Fibers that have been chopped and then milled (passed between rollers) to reduce their length. The resultant materials are usually ***whiskers***. These are useful in reinforcing pastes and in other resins used for repair and potting.

modulus The slope of the stress-strain curve; the stiffness of a material.

molding compound A combination of fibers, resin, and filler premixed, which can be later molded.

molecular weight A measure of the length of the polymer chain or, in other words, the number of monomer units that have been joined together.

monofilament A single filament. This term emphasizes the solitary nature of filaments.

monomer A single atomic unit that can be made into a polymer.

N

neat resin A resin that contains no reinforcement or filler.

neopentyl glycol A glycol monomer that adds weathering and chemical/water resistance to thermoset polyesters.

nesting Arranging broad goods (textiles and prepregs) so that the various shapes fit closely together, thus minimizing the amount of wasted material.

net shape Making a pre-molding and resin addition reinforcement package that is near the finished product's shape after molding.

nondestructive tests (NDT) Tests that do not destroy the sample in the process of the testing.

novolac A B-staged phenolic made by the reaction of an excess of phenol in an acid environment. A two-step resin, the resulting thermoplastic requires the addition of a curing agent (typically hexa) to form a full thermosetting phenolic.

nucleophile An atom that seeks a positive charge. Usually the nucleophilic atom will have a negative or slightly negative charge.

O

olefinic unsaturation Another name for the carbon-carbon double bond group that is commonly the basis for polyester thermosets.

oligomer A low-molecular-weight polymer with short chains that will later be polymerized into higher-molecular-weight materials. Also called ***pre-polymer***.

one-step resin A resole phenolic that will cure from the B-stage with heating.

open-hole tensile test A tensile test in which a hole is drilled in the middle of the sample so that the effect of fiber shortening can be determined.

openness A relative measure of the space between fibers in a fabric.

operational decisions Those decisions that have to be made immediately for the continued operation of the company.

optical holography A non-destructive test method to examine the structure of a stressed composite by using photography created using lasers.

orthogonal structure A pattern of stiffeners in which the stiffeners intersect at right angles.

orthophthalic acid (ortho) A saturated, aromatic diacid often included in the mix of monomers used to make unsaturated polyester resins. Ortho is a low-cost monomer that has excellent styrene compatibility. It is referred to as the general-purpose polyester thermoset resin.

orthophthalic anhydride A monomer that converts to ortho under polymerization conditions.

orthotropic Describes composite materials having fibers directed in all directions (random planar orientations).

out-gassing The loss of monomer and other gaseous components of a plastic or resin in a vacuum environment.

out time The amount of time that a prepreg is out of cold storage.

oxirane group See ***epoxy group***.

P

pattern The master from which a mold is made.

payoff The guiding device through which the fibers pass just prior to being placed onto the mandrel in a filament winding operation. The payoff is moved relative to the mandrel to create the fiber lay-down patterns.

peel ply Part of the vacuum bagging system that provides release of the entire bagging assembly from off the top of the part. Also called a ***release film***.

pendant When a group is attached by dangling from the polymer backbone.

permanent set The non-recovered change in shape of a part due to creep.

permeability The tendency or rate at which a gaseous material passes through or into a solid.

phenolic A thermoset resin made by the reaction of phenol and formaldehyde. It is often B-staged to form either resoles or novolacs and then cured to achieve flame-retardant, low-smoke, stiff, highly crosslinked parts.

pigment Inorganic **colorants**.

pitch The material left behind when more volatile components have been removed from coal tar or crude oil. It is occasionally used as the precursor material for forming carbon matrix composites and other carbon-rich materials (such as carbon fibers).

plain weave A weaving pattern characterized by the warp and weft fibers alternating in a regular over-and-under pattern.

plastic A non-natural polymeric material that has been formed and shaped. The term sometimes refers to the matrix material, especially for composites made from fiberglass and low-priced resins.

plug The shaped material used as a master pattern from which a mold is copied.

polarity A measure of the chemical nature of a molecule as determined by the unevenness of the distribution of the electrons within it. Sites within the molecule having high concentrations of electrons tend to be attractive to sites in other molecules where the electrons are relatively sparse. This phenomenon affects solvent resistance and other chemical-dependent natures of the polymers. Also called ***chemical polarity***.

pole piece The fitting in the end of a closed-end vessel on which the shaft is mounted. The shaft turns the part in a filament winding operation.

polyacrylonitrile (PAN) A thermoplastic material that is a starting material for carbon fibers.

polyamide-imide (PAI) A high-performance thermoplastic.

polybenzimidazole (PBI) A high-performance thermoplastic.

polyester A polymer that contains ester linkages.

polyester thermoset (PT) A group of polymers, also called ***unsaturated polyesters*** or ***alkyds***, which are the most common of all thermoset materials used in composites. They are normally crosslinked by the free-radical crosslinking mechanism.

polyetherimide (PEI) A high-performance thermoplastic.

polymer A material composed of many atomic units that combine into a long chain-like structure.

polymer molecule A single polymeric chain. Nature rarely allows encountering a single molecule but, rather, numerous polymer molecules in a single mass. Hence, the value of the concept of a polymer molecule is in simplifying the representation of the polymer.

polymerization The process that converts small atomic units (monomers) into long-chain molecules (polymers).

polyol A molecule having multiple OH (alcohol) groups.

positive-type mold closure A mold-closure system in compression molding in which the mold halves fit tightly together and do not allow flashing to occur.

post-curing See ***annealing***.

pot life The working time of the resin while the fiber wet-out and other pre-curing operations are carried out; the time in a resin bath without appreciable crosslinking.

potting The process of encapsulating an assembly or part in resin. It is often used to protect wire ends and other small electrical components.

preform An assemblage of reinforcement plies shaped to be near the final shape of the product. It lacks only the resin and curing for completion.

premix A general term for bulk molding compound (BMC) or sheet molding compound (SMC) materials.

pre-polymer See ***oligomer***.

prepreg A sheet made of fibers or cloth onto which a mixture of liquid resin and curing agent (hardener) has been coated.

prepreg tow A prepreg material made by coating a resin, which is a thermoset or thermoplastic, onto a tow or bundle of fibers. This material is especially useful in filament winding operations where wet resin application from a bath is either difficult or undesirable.

pressure intensifier See ***caul***.

processing viscosity The viscosity of a mixture measured during the course of the polymerization reaction; it is a control parameter to monitor the extent of polymerization.

promoters See ***accelerators***.

propylene glycol A glycol monomer that is low in cost and gives excellent styrene compatibility in polyester thermosets.

pseudo-isotropic Materials that are reinforced along the principal x, y, and z axes.

pyrolysis A step in the manufacture of carbon matrix composites in which the cured composite is heated in the absence of oxygen to drive off the non-carbon atoms.

Q

quartz A pure, crystalline form of silicon dioxide. It is stronger and stiffer than other fiberglass types and is a good transmitter of electromagnetic radiation.

quench A chemical added or a procedure carried out to stop the curing or polymerization process.

R

radiography A non-destructive test method that uses x-rays to examine the inner structure of a composite.

rayon A type of fiber, made from regenerated cellulose, which can be used as a starting material for making carbon fibers.

reactive diluent Chemicals that are solvents and co-reactants in the curing process of polyester thermosets. They are consumed in the curing reaction.

reinforcement The separated phase of a composite material that provides strength, stiffness, and other properties. It is usually a fibrous material.

release agent Material that is applied to a mold surface to assist in release of the molded part.

release film Part of a vacuum bag system that provides release of the entire bagging assembly from off the top of the part. Also called a ***peel ply***.

resin A polymeric material that has not yet been formed into its final shape. Most resins are liquids or low-melting solids at room temperature. The term sometimes refers to the matrix material even after shaping, especially if it is a high-performance composite.

resin film infusion (RFI) A process in which a resin film is placed inside the mold along with the preform. Then heat and pressure are applied to cause the film to melt and infuse the preform.

resin freeze When a resin cures prematurely and, therefore, does not completely fill the mold during resin infusion. Also called ***resin stall-out***.

resin infusion technology A general term describing all processes in which a dry preform is placed into a mold, the mold is closed, and then resin is introduced into the mold to wet the fibers and form the part.

resin stall-out See ***resin freeze***.

resin starving Transferring (bleeding) too much resin from an area of the part so that the fibers are not properly wetted.

resin transfer molding (RTM) A method of manufacturing composites in which a dry preform is placed into the mold, the mold is closed, and then resin is injected into the mold (under pressure) to wet the fibers. The resin is then cured to form the part. RTM is also a general term that describes all resin infusion processes.

resole A B-staged phenolic resin created in an excess of formaldehyde in an alkaline environment. A one-step resin, it is a thermoset and will fully cure upon heating, without the addition of a curing agent.

ripening The gradual thickening of a molding compound, especially when thickening is desired, so that a certain viscosity can be achieved prior to molding.

roll wrapping A composite molding method in which prepreg materials are wrapped around a mandrel and then cured. Also called ***tube rolling***.

room temperature vulcanization (RTV) A resin type, often an epoxy, that will cure at room temperature.

roving A group of fiberglass filaments.

runners The network of channels in an injection mold through which the melt flows from the sprue to the cavities.

S

sag point The temperature at which a thermoplastic resin is soft enough to flow under a moderate pressure but is not fully melted.

sandwich structure A structure consisting of two face sheets of composite laminate, a core material, and adhesive to bond the face sheets to the core.

satin weave A weaving pattern in which the warp or weft fibers skip over or under several of the other types of fibers. This fabric has high drape.

scarf A type of repair patch in which the sides of the repair are sloped (tapered), thus increasing the amount of surface area on which the bonding of the repair can occur.

sea of electrons The highly mobile group of electrons that move about and bind together the positively charged ions in a metal material.

sealant Part of a vacuum bagging system whose function is to adhere to the vacuum bag and create an airtight assembly.

seamless molding paste An epoxy-based material used to coat the surface of a mold and which, after curing, can be machined to give a high-temperature-resistant mold.

secondary bonding Adding more layers of laminate material onto a previously cured laminate. Attractive forces between the molecules cause adherence, especially when the molecules are solids.

selective stitching Stitching in only selected areas of a laminate (as, for example, in a joint).

selvage Edges of a roll of material, often treated to prevent unraveling.

semi-crystalline A polymer with some regions that are crystalline and some that are not.

S-glass This type of glass and its variant, S2-glass, are formulated to give higher strength than other types of fiberglass. It is, therefore, used in some aerospace and other high-performance applications.

shape memory polymer A polymer that can be molded at one shape, raised to a softening point, modified in shape, and then chilled to lock the molecules in the new shape. Then, at some later time, the material can be reheated to the softening temperature, which causes the molecules to relax and assume their original molded shape.

shearography A non-destructive test method that uses laser photography and shearing software to analyze the structure of a composite.

sheet molding compound (SMC) A mixture of polyester resin, filler, and fiberglass in which the fiberglass is mixed with the resin/filler paste by doctoring the paste onto a film and then chopping the fiberglass so that it covers the entire surface. Then a second film with paste is laid on top and the material is worked by wheels to wet the fibers.

shelf life The time that the uncured resin can be stored. This time is limited by the natural tendency of many resins to spontaneously (although usually slowly) cure.

sintering A process in which a powdered material (resin, ceramic, or metal) is heated to a temperature just below its melting point, which causes the particles to join together (consolidate) and form a solid mass.

sizing A material coated onto the surface of a fiber to prevent damage to the fiber during processing. Such a coating is sometimes called a *size* and is generally not distinguishable today from a ***finish***.

smart structure A structure or system that incorporates sensors and/or actuators so that the environment and state of the structure can be monitored.

specific modulus The tensile modulus of a material divided by the density, thereby achieving a measure of the combined effects of modulus and weight.

specific stiffness The modulus of a material divided by its density. The result is an indication of the stiffness as a function of the weight.

specific strength The tensile strength of a material divided by the density, thereby achieving a measure of the combined effects of modulus and weight.

spin casting See ***centrifugal casting***.

spinneret The metal plate in which holes are drilled and used as a forming device to make fibers. The liquid raw material passes through the holes to form the fiber filaments.

spinning The process of forming a fiber. While synthetic fibers are not actually spun (turned), the terminology comes from the ancient method of forming natural fibers, like wool and cotton, in which the individual strands of fiber were actually spun to entangle them.

spray-up A system of composite manufacturing in which the reinforcements are chopped and then entrained in a resin spray that is directed into the mold. The fibers are then rolled to ensure good fiber wet-out.

sprue The channel that connects the mold to the nozzle of an injection molding machine.

stabilization A step in the manufacture of carbon fibers in which the rings are formed through crosslinking and atomic rearrangement.

staple fibers Continuous fibers that have been chopped into lengths of approximately .5–4 in. (1.2–10 cm). Staple also may be natural fibers that are inherently in the length range of the chopped continuous fibers.

steel-rule blanking See ***die cutting***.

stepped repair A repair design in which the edges of the flush repair are stepped, thus increasing the amount of bonding area when compared to a straight-sided repair.

steric interaction The influence of the sizes and shapes of atoms and molecules, the electrical charge distribution, and the geometry of bond angles on the courses of chemical reactions.

stiffness The resistance of a material to deformation; the modulus.

stitched fabric A laminate in which the fabrics are stitched with a fiber, thus imparting three-dimensional character to the laminate. The stitches are oriented in the direction perpendicular to the plane of the fabrics.

stoichiometric ratio The relative amount of resin and hardener that will give complete reaction of all active sites.

strain The deformation of a sample in a mechanical test, which is usually expressed as the change in length divided by the original length. The strain to failure also can be called the ***elongation***.

strand A general and somewhat imprecise term that usually refers to a bundle or group of untwisted continuous filaments. The term has also been used interchangeably with ***fiber*** and ***filament***.

strategic decisions Determinations that are fundamental to the nature of the company and affect company values and underlying principles.

strategic mapping A method of thinking that allows for the essential unpredictability of the world.

stress The force per unit area applied to a sample in a mechanical test.

stretch forming A type of matched-die molding of thermoplastic composites in which the blanks are clamped and the fibers stretch to fill the mold's shape.

structural decisions Determinations that are made periodically to provide limited guidelines for daily activities.

surface free energy The energy present due to the imbalance of forces on the molecules at the surface of a material (in contrast to the balanced forces that exist on molecules in the bulk of the material).

surface tension The surface free energy for a liquid-gas interface.

surfactant A material that changes the energy of a surface and facilitates the coating of the surface by the resin mixture.

T

tack test A method of assessing the cure state by noting the force required to lift a prepreg sheet from a standard surface.

tap test A test used to assess impact damage. In the test, a small metal hammer (or even a coin) is used to lightly tap the surface of the composite to determine changes in sound between known good areas and possibly damaged areas.

tensile properties Those properties determined by pulling on the ends of a sample.

terephthalic acid (tere) A saturated, aromatic diacid occasionally included in the mix of monomers to make unsaturated polyesters. Tere imparts high heat distortion temperature tolerance to the polymer.

tetrabromophthalic anhydride A halogen-containing diacid that imparts flame-retardant properties to polyester thermosets.

textile fibers The set of organic fibers used extensively in textiles (such as clothing, carpets, etc.) and as reinforcements in some applications such as conveyor belts and tires. Composites that use textile-fiber reinforcements offer modest improvements over non-reinforced materials.

thermofolding A thermoplastic molding method in which only a portion of the part is softened and then bent at the softened region.

thermography A non-destructive test in which sample flaws are detected by comparing thermal photographs (usually taken with infrared cameras).

thermoplastic Resins that are solids at room temperature, which can be melted, shaped, and then cooled to solidify. Additional heating of the shaped part will result in remelting.

thermoset Resins that are liquids at room temperature, which can be reacted to form bonds between the molecules, thus causing the material to solidify. Additional heating of the shaped part will not result in remelting.

thermoset engineering composites See ***fiberglass reinforced plastics (FRP)***.

thermoset polyester A polyester resin that is unsaturated; it contains a carbon-carbon double bond and, therefore, can be crosslinked and become a thermoset.

thinner A material added to a resin mixture to reduce its viscosity.

thixotrope An additive that imparts high viscosity to the resin mixture.

time-temperature transposition The principle on which accelerated thermal aging is based. It assumes that raising the temperature of a material is equivalent in aging as extending the time.

toughened epoxies An epoxy resin that has rubber or thermoplastic resins incorporated into it or that has increased flexibility in its structure.

toughener A material added to a resin mixture to increase the toughness of the resultant composite. These additives usually have high elongation (like rubber). Also called ***toughening agent***.

tow A group of filaments gathered together. The term usually applies to all advanced fibers but not, typically, to fiberglass.

toughening agent See ***toughener***.

toughness The ability of a material to absorb energy, especially when the energy results from an impact. Alternately, it is the resistance of a material to the initiation and propagation of cracks.

trans-isomer The arrangement of atoms in which two key groups are on opposite sides of a carbon-carbon double bond. ***Fumaric acid*** is a trans isomer.

trans-laminar reinforced composite A laminate in which at least 5% of the fibers are oriented perpendicular to the plane of the fabrics. This is often done by stitching the laminate.

transformation toughening A method of toughening a ceramic matrix composite in which the reinforcement experiences a phase or crystalline structure change induced by the formation of a crack. When the phase change increases the volume of the material in the vicinity of the crack end, the crack's growth is arrested.

transition molding See ***incremental forming***.

transverse direction Especially in woven fabrics, the direction that is perpendicular to the principal or machine direction.

trapped rubber molding A process in which a rubberized medium, contained within a closed mold, is used to exert pressure on a laminate.

triaxial weave A fabric in which the warp fibers are oriented along the bias directions of the fabric to give high strength in the bias directions.

tube rolling See ***roll wrapping***.

twill weave A weaving pattern in which the over-and-under pattern of the warp and weft create a diagonal pattern in the fabric.

two-step resin novolac phenolic that will cure from the B-stage with the addition of a curing agent.

U

ultra-high-molecular-weight polyethylene (UHMWPE) fiber A type of high-performance fiber made from high-molecular-weight polyethylene. The polymer chains are stretched and then formed into fibers. Also called ***extended chain polyethylene (ECPE) fibers***.

ultrasonic testing A non-destructive method used to detect flaws in a composite sample by monitoring ultrasonic signals sent through it.

ultrasonic vibratory cutting A method of cutting textiles and uncured prepregs in which a knife is driven by ultrasonic vibrations.

ultraviolet (UV) The component of light that is in the ultraviolet range of frequencies. These are the most damaging light components for polymers.

unidirectional tapes Prepregs made of unidirectional fibers and a suitable resin.

unsaturated polyester (UP) See ***polyester thermoset (PT)*** and ***alkyd***.

unsaturation The presence of a carbon-carbon double bond.

UV inhibitor A material that enhances the resistance of a composite to the adverse results of ultraviolet (UV) light exposure.

V

vacuum bag Part of a vacuum bagging system whose function is to form an airtight seal over an entire assembly.

veil A lightweight fiberglass fabric used to hide fiber patterns and give a smooth surface to the composite lay-up.

vibration The oscillation of a material struck with an impact.

vinyl ester A group of polymers that have properties somewhat like the epoxies but that can be cured like unsaturated polyesters.

viscosity The thickness of a liquid. Resin viscosity is increased with molecular weight and aromatic content.

W

warp In a woven fabric, these are the fibers arrayed in the machine direction.

weft In a woven fabric, these are the fibers interwoven above and below the warp fibers. Also called the ***fill***, ***woof***, or transverse fibers.

weld line See ***knit line***.

wet-out The complete and uniform coating of fibers by the resin.

whiskers Short reinforcement fibers typically about .04 in. (0.1 cm) long.

woof See ***fill*** or ***weft***.

work of adhesion The work required to separate two particles, which is equivalent to the energy of the bond between the two particles.

woven roving A heavy fiberglass fabric used to rapidly build up reinforcement weight. It is made by loosely weaving large roving strands.

Y

yarn A bundle of filaments that has been twisted to assist in keeping the filaments in a group.

Young's modulus The modulus in a tensile stress-strain curve.

Z

***z*-fiber construction** A laminate system that is similar to stitching but uses rods or pins oriented perpendicular to the plane of the fabrics to hold the laminates together.

Index

B

C

D

E

F

G

H

I

J

K

L

M

N

O

P

Q

R

S

T

U

V

W

X

Y

Z